Lecture Notes in Physics

Edited by H. Araki, Kyoto, J. Ehlers, München, K. Hepp, Zürich
R. Kippenhahn, München, H. A. Weidenmüller, Heidelberg
J. Wess, Karlsruhe and J. Zittartz, Köln
Managing Editor: W. Beiglböck

264

Tenth International Conference on Numerical Methods in Fluid Dynamics

Proceedings of the Conference
Held at the Beijing Science Hall, Beijing, China
June 23–27, 1986

Edited by F. G. Zhuang and Y. L. Zhu

Springer-Verlag
Berlin Heidelberg New York London Paris Tokyo

Editors

F.G. Zhuang
Chinese Aerodynamics Research Society
P.O. Box 2425, Beijing, China

Y.L. Zhu
The Computing Center of Academia Sinica
Beijing, China

ISBN 3-540-17172-X Springer-Verlag Berlin Heidelberg New York
ISBN 0-387-17172-X Springer-Verlag New York Berlin Heidelberg

Library of Congress Cataloging-in-Publication Data. International Conference on Numerical Methods in Fluid Dynamics (10th: 1986: Beijing Science Hall) Tenth International Conference on Numerical Methods in Fluid Dynamics. (Lecture notes in physics; 264) 1. Fluid dynamics—Congresses. 2. Numerical analysis—Congresses. I. Chuang, Feng-kan, 1925-. II. Chu, Yu-lan. III. Title. IV. Title: 10th International Conference on Numerical Methods in Fluid Dynamics. V. Series.
QA911.I54 1986 620.1'064 86-27937
ISBN 0-387-17172-X (U.S.)

This work is subject to copyright. All rights are reserved, whether the whole or part of the material is concerned, specifically those of translation, reprinting, re-use of illustrations, broadcasting, reproduction by photocopying machine or similar means, and storage in data banks. Under § 54 of the German Copyright Law where copies are made for other than private use, a fee is payable to "Verwertungsgesellschaft Wort", Munich.

© Springer-Verlag Berlin Heidelberg 1986
Printed in Germany

Printing: Druckhaus Beltz, Hemsbach/Bergstr.; Bookbinding: J. Schäffer OHG, Grünstadt
2153/3140-543210

PREFACE

This issue of Lecture Notes in Physics contains the Proceedings of the Tenth International Conference on Numerical Methods in Fluid Dynamics, held at the Beijing Science Hall in China, June 23-27, 1986. The Proceedings include all the papers presented at the Conference, namely, the inaugural lecture by K. Feng, the invited lectures by V.P. Dymnikov, M.Y. Hussaini, P. Kutler, M. Napolitano, N. Satofuka, F.G. Zhuang, and H.X. Zhang, as well as 108 contributed papers arranged in alphabetical order of the first author's name. The contributed papers were selected from abstracts submitted from all over the world by four Committees on Paper Selection based in China, Europe, the U.S.A., and the U.S.S.R. and headed by the editors (China), Temam (Europe), Holt (U.S.A.), Chernyi and Rusanov (U.S.S.R.).

The Conference was attended by over 200 scientists. In addition to the strong representation from China, a large number of scientists from the U.S.A., Japan, France, Italy, West Germany, the U.S.S.R., the Netherlands, Ireland, Canada, the United Kingdom, Belgium, Sweden, Australia, Brazil, and Norway participated at the Conference. A list of the participants is given at the end of the Proceedings.

We served as general conference cochairmen and are indebted to the many colleagues who helped with the details of the meeting. In particular, our thanks go to all the members of the International Organizing Committee and the Local Committee for the Conference, who were in charge of all academic activities, as well as to Mr. C.S. He of the Chinese Aerodynamics Research Society and Mr. Y. Cao of the China International Conference Center for Science and Technology, who supervised all of the local arrangements.

Financial support for the Conference was provided by the China Aerodynamics Research and Development Center. Peking University, the Computer Center of Academia Sinica and the Institute of Computer Technology of Academia Sinica helped the Conference in many ways. We greatly appreciate their supports.

We are also indebted to Prof. W. Beiglböck and Ms. C. Pendl for valuable assistance in preparing these Proceedings.

August 1986   F.G. Zhuang and Y.L. Zhu
                                                         (Editors)

编者前言

　　第十届国际流体力学数值方法会议於一九八六年六月廿三～廿七日在北京科学会堂召开．本书是该会议的论文集．其中我们收集了会议中的全部文章．它们是冯康教授的开幕学术演讲，Dymnikov 博士，Hussaini 博士，Kutler 博士，Napolitano 教授，Satofuka 教授，庄逢甘和张涵信教授的特邀报告，以及１０８篇"入选"文章．在此文集中，每一类文章是以第一作者的姓名按字母顺序编排的．"入选"文章是由四个选文委员会根据提交来的文章选定的．这四个选文委员会分别设在中国，欧洲，美国和苏联，主席是本卷的编者〔中国〕，Temam 教授〔欧洲〕，Holt 教授〔美国〕，Chernyi 和 Rusanov 教授〔苏联〕．

　　参加此届会议，除了大量的中国科学家以外，还有来自美国，日本，法国，意大利，联邦德国，苏联，荷兰，爱尔兰，加拿大，英国，比利时，瑞典，澳大利亚，巴西，挪威的许多科学家，共计二百多位．本会议录的末尾给出了出席者的名单．

　　作为此届大会主席，我们在此衷心感谢在会议的组织工作中给了我们各种帮助的所有同事们，特别是在学术活动方面做了许多工作的国际和国内组织委员会的委员们，和负责会务工作的中国空气动力学研究会的贺长胜等同事和中国科协国际会议中心的曹跃等同事．

　　中国空气动力研究和发展中心给了这次会议以财政上的支持．北京大学，中国科学院计算中心和计算所也对会议给予了很多支持．我们在此一并表示感谢．

　　最后，我们还要对 Beiglbock 教授和 Pendl 女士在准备此会议录方面所给予的热情帮助表示诚挚的谢意．

庄逢甘　　朱幼兰

一九八六年八月

ACKNOWLEDGEMENTS

At the end of the 10th International Conference on Numerical Methods in Fluid Dynamics, Professor Henri Cabannes, of Mécanique Théorique, Université Pierre et Marie Curie, Paris, stepped down as Secretary of the Organizing Committee, to be replaced by Dr. Soubbaramayer. Professor Cabannes has served in this capacity since the beginning of the 3rd International Conference on Numerical Methods in Fluid Dynamics, which he organized in Paris in 1972, and has worked without respite to ensure the success of the conference series. He established a permanent office for the Organizing Committee in Paris, attended to all correspondence connected with reports of past conferences and preparation of forthcoming conferences, gave invaluable guidance to the committee on such matters as the choice of sites for the conference, selection of speakers and financial support. Because of the unusual international character of the conference, Professor Cabannes has had to exercise considerable diplomatic skill and use a great deal of his valuable time in resolving organizational and personnel problems which arose during the 14 years of his tenure. Past participants in the conference will surely wish to join the Organizing Committee in giving warm thanks to Professor Cabannes for his long and consistent service to the conference. We hope that his advice will continue to be available to the committee far into the future.

October 1986   The Organizing Committee

INTERNATIONAL CONFERENCE ON
NUMERICAL METHODS IN FLUID DYNAMICS

First Conference:	Novosibirsk, USSR, 1969
Second Conference:	Berkeley, California, USA, 1970
Third Conference:	Paris, France, 1972
Fourth Conference:	Boulder, Colorado, USA, 1974
Fifth Conference:	Enschede, the Netherlands, 1976
Sixth Conference:	Tbilisi, USSR, 1978
Seventh Conference:	Stanford University and NASA/Ames, USA, 1980
Eighth Conference:	Aachen, West Germany, 1982
Ninth Conference:	Saclay, France, 1984
Tenth Conference:	Beijing, China, 1986

Inaugural Talk

    Feng, K.: Symplectic Geometry and Numerical Methods in Fluid Dynamics... 1

Invited Lectures

    Dymnikov, V.P.: On Some Problems of Dynamic Meteorology 8

    Hussaini, M. Y.: Some Recent Developments in Spectral Methods 18

    Kutler, P.: A Perspective of Computational Fluid Dynamics 30

    Napolitano, M.: Simulation of Compressible Inviscid Flows: The Italian Contribution 47

    Satofuka, N.: Method of Lines Approach to the Numerical Solution of Fluid Dynamic Equations 57

    Zhuang, F.G. and Zhang, H.X.: On a Marching Iteration Method in Solving Gas Dynamic Equations 70

Contributed Papers

    Aki, T.: Computation of Unsteady Shock Wave Motion by the Modified Flux TVD Scheme 86

    Armfield, S.W. and Fletcher, C.A.J.: Swirling Diffuser Flow Using a Reduced Navier-Stokes Formulation 91

    Azmy, Y.Y. and Dorning, J.J.: Numerical Studies of Bifurcations in the Confined Benard Problem... 96

    Bardos, C.: Diffusion and Rosseland Approximation Property of the Boundary Layer...105

    Barton, J.M. and Yoon, S.K.: Finite Difference Solution of the 3-D Euler Equations Using a Multistage Runge-Kutta Method...................108

    Bassi, F., Grasso, F. and Savini, M.: Solution of the Compressible Navier-Stokes Equations by Using Embedded Adaptive Meshes...............113

    Bercovier, M. and Engelman, M.: Simulation of Large Incompressible Flows by the Finite Element Method......................................120

    Bramley, J.S. and Sloan, D.M.: A Downstream Boundary Condition for the Numerical Solution of Viscous Flow....................................123

    Browning, G.L. and Kreiss, H. O.: Scaling and Computation of Smooth Atmospheric Motions..128

    Bruneau, C.H., Chattot, J.J., Laminie, J. and Temam, R.: Computation of Vortex Flows past a Flat Plate at High Angle of Attack................134

    Bullister, E.T., Cartage, T., Deville, M. and Patera, A.T.: Spectral Simulation of Thermal Convection in Complex Geometries...................141

Carter, J.E., Davis, R.L., Edwards, D.E. and Hafez, M.M.:
  Three-Dimensional Separated Viscous Flow Analyses 147

Chang, J.L.C., Yang, R.-J. and Kwak, D.: A Full Navier-Stokes
  Simulation of Complex Internal Flows...................................154

Chen, Y.W., Zhang, Y.K., Shen, M.Y. and Huang, D.T.: A Strong
  Inviscid-Viscous Interaction Solution of a Plane Transonic
  Cascade Flow..161

Cheng, S.-I.: Computation of Turbulent Spot Evolution.................... 165

Choi, Y.-H. and Merkle, C.L.: Computation of Low Mach Number
  Flows with Buoyancy...169

Choudhury, S. and Nicolaides, R.A.: Vortex Multipole Methods
  for Viscous Incompressible Flows......................................174

Clark, R.A.: Free-Lagrangian Hydrodynamics Using Massless
  Tracer Points...181

Coakley, T.J.: Impact of Turbulence Modeling on Numerical
  Accuracy and Efficiency of Compressible Flow Simulations..............186

Couët, B., Strumolo, G.S. and Dukler, A.E.: Numerical Modelling
  of a Bubble Rising Through Viscous Fluid..............................192

Cuvelier, C. and Driessen, J.M.: Thermocapillary Free Boundaries
  in Crystal Growth...197

Dadone, A.: A Quasi-Conservative COIN Lambda Formulation..................200

Daiguji, H., Motohashi, Y. and Yamamoto, S.: An Implicit
  Time-Marching Method for Solving the 3-D Compressible Euler Equations...205

Dang, K. and Morchoisne, Y.F.: Large Eddy Simulation of a
  Narrow Source of Passive Scalar in Homogeneous Strained Turbulence......211

Deconinck, H., Hirsch, Ch. and Peuteman, J.: Characteristic
  Decomposition Methods for the Multi-Dimensional Euler Equations.........216

Dennis, S.C.R. and Wing, Q.: Generalized Finite Differences
  for Operators of Navier-Stokes Type...................................222

Dinh, Q.V., Periaux, J., Terrasson, G. and Glowinski, R.: On the
  Coupling of Incompressible Viscous Flows and Incompressible
  Potential Flows via Domain Decomposition..............................229

Dong, S.S., Wang, Z.X. and Lee, H.: Free Mass-Lump Method for
  Two-Dimensional Compressible Flow.....................................235

Drummond, J.P.: Spectral Methods for Modeling Chemically
  Reacting Flow Fields..242

Dwyer, H.S., Soliman, M. and Hafez, M.: Time Accurate
  Solutions of the Navier-Stokes Equations for Reacting Flows...........247

Eguchi, Y. and Fuchs, L.: A Finite Element Method for
  Simulation of Unsteady Flows..252

Eiseman, P.R.: Alternating Direction Adaptive Grid
  Generation for Three-Dimensional Regions..............................258

Erlebacher, G.: Transition Phenomena over a Flat Plate for
  Compressible Flows... 264

Favini, B. and Zannetti, L.: On Conservative Properties
  and Non-Conservative Forms of Euler Solvers........................ 270

Förster, K. and Li, F.W.: A Numerical Scheme for the Unsteady
  Transonic Flow Around an Oscillating Airfoil....................... 276

Fromm, J.E.: Free Surface Calculation of Capillary Spreading......... 283

Fuchs, L.: A Combined Numerical Scheme for Transonic Flows........... 290

Hamakiotes, C.C. and Berger, S.A.: Fully Developed Pulsatile
  Flow in a Curved Pipe.. 297

Hartwich, P.-M., Hsu, C.-H. and Liu, C.-H.: Implicit Hybrid
  Schemes for the Flux-Difference Split, Three-Dimensional
  Navier-Stokes Equations.. 303

Hemker, P.W., Koren, B. and Spekreijse, S.P.: A Nonlinear
  Multigrid Method for the Efficient Solution of the Steady
  Euler Equations.. 308

Holt, M. and Pace, C.: Calculation of Flow in a Supersonic
  Compression Corner by the Dorodnitsyn Finite Element Method........ 314

Hou, T.X.: The Solution of System of Nonlinear Algebraic
  Equations Generated in Boundary Points Calculation................. 320

Huang, D.: A Test Problem for Unsteady Shock Wave Calculation........ 324

Huang, M.K.: Applications of Numerical Conformal Mapping Technique... 329

Jameson, A. and Baker, T.J.: Euler Calculations for a
  Complete Aircraft.. 334

Jami, A. and Kermarec, M.: On the Convergence of Particle
  Methods Applied to the Euler and Free Surface Equations............ 345

Johnson, G.M., Swisshelm, J.M., Pryor, D.V. and Ziebarth, J.P.:
  Multitasked Embedded Multigrid for Three-Dimensional
  Flow Simulation.. 350

Kamenetsky, V.F. and Turchak, L.I.: Numerical Simulation of
  Some Separated Flows... 357

Kanda, H. and Oshima, K.: Numerical Study of the Entrance
  Flow of a Circular Pipe.. 363

Kaul, U.K.: A Numerical Method to Assess the Feedback
  in a Free Shear Layer.. 369

Khosla, P.K. and Rubin, S.G.: Consistent Strongly Implicit
  Iterative Procedures... 375

Korczak, K.Z.: An Isoparametric Spectral Element Method in
  Simulation of Incompressible Complex Flows......................... 381

Krause, E., Menne, S. and Liu, C.H.: Initiation of Breakdown
  in Slender Compressible Vortices 386

Ku, H.C., Hirsh, R.S. and Taylor, T.D.: A Pseudospectral Method for Solution of the Three-Dimensional Incompressible Navier-Stokes Equations.. 391

Kwak, D., Rogers, S.E., Kaul, U.K. and Chang, J.L.C.: A Numerical Study of Incompressible Juncture Flows.............................. 398

Lee, W.H. and Kwak, D.: On the PIC Method for Elastic-Plastic Flow...... 403

Li, C.P.: Implicit Methods for Computing Chemically Reacting Flow....... 409

Li, Y.F. and Qian, E.P.: A "Large-Particle" Difference Method with Second Order Accuracy for Computation of Two-Dimensional Unsteady Flows.. 416

Ling, B.Y. and Cole, J.D.: Airfoil Design at Sonic Velocity............ 422

Löhner, R., Patnaik, G., Boris, J.P., Oran, E.S. and Book, D.L.: Applications of the Method of Flux-Corrected Transport to Generalized Meshes.. 428

Lombard, C.K., Bardina, J., Venkatapathy, E., Yang, J.Y., Luh, R.C.C., Nagaraj, N. and Raiszadeh, F.: Accurate, Efficient and Productive Methodology for Solving Turbulent Viscous Flows in Complex Geometry.. 435

Ma, Y.W. and Fu, D.X.: A Simple and Efficient Implicit Scheme for the Compressible Navier-Stokes Equations........................ 442

Madhavan, N.S. and Swaminathan, V.: On an Implicit Numerical Scheme for Two-Dimensional Steady Navier-Stokes Equations............ 448

Malik, M.R.: Numerical Simulation of Transition in a Three-Dimensional Boundary Layer.. 455

Mansutti, D., Bulgarelli, U., Piva, R. and Graziani, G.: A Discrete Vector Potential Method for Unsteady 3-D Navier-Stokes Equations.. 462

Mathieu, J., Ravier, P., Boujot, J., Gendre, P. and Hittinger, M.: Interaction Between Structure and Free Surface Fluid with Large Displacements by Finite Elements................................ 467

Melnik, R.E., Brook, J.W. and DelGuidice, P.: Computation of Turbulent Separated Flow with an Integral Boundary Layer Method...... 473

Mitra, N.K., Kiehm, P. and Fiebig, M.: Numerical Investigations of the Structure of Three-Dimensional Confined Wakes Behind a Circular Cylinder.. 481

Morton, K.W. and Paisley, M.F.: On the Cell-Centre and Cell-Vertex Approaches to the Steady Euler Equations and the Use of Shock Fitting.. 488

Nakahashi, K.: FDM-FEM Zonal Approach for Computations of Compressible Viscous Flows.. 494

Nishikawa, N., Suzuki, T. and Suzuki, A.: Numerical Simulation of Splash of Droplet.. 499

Nordström, J.: Energy Absorbing Boundary Conditions for the Navier-Stokes Equation.. 505

Oshima, K., Oshima, Y., Izutsu, N., Ishii, Y. and Noguchi, T.:
Interaction of Vortical Flow Regions.................................. 511

Osswald, G.A., Ghia, K.N. and Ghia, U.: Simulation of Buffetting
Stall for a Cambered Joukowski Airfoil Using a Fully Implicit
Method.. 516

Perez, E., Periaux, J., Rosenblum, J.P., Stoufflet, B., Dervieux, A.
and Lallemand, M.H.: Adaptive Full-Multigrid Finite Element
Methods for Solving the Two-Dimensional Euler Equations............ 523

Qin, N. and Richards, B.E.: Simulation of Hypersonic Viscous Flows
Around a Cone-Delta-Wing Combination by an Implicit Method with
Multigrid Acceleration.. 528

Reister, H., and Schwamborn, D.: Viscous Pressure Wave Boundary
Layer Interaction... 533

Ruas, V.: Some Nonstandard Finite Element Methods for the
Numerical Solution of Viscous Flow Problems........................... 538

Rusanov, V.V.: Exact Solution of Nonlinear Difference Equations
for Discrete Shock Waves.. 545

Salmond, D.J.: A Cell-Vertex Multigrid Scheme for Solution of the
Euler Equations for Transonic Flow past a Wing........................ 549

Savu, G. and Trifu, O.: Numerical Prediction of the Aerodynamic
Behaviour of Porous Airfoils.. 554

Selmin, V. and Quartapelle, L.: Finite Element Solution to the
Euler Equations... 559

Shaw, G. and Wesseling, P.: Multigrid Solution of the Compressible
Navier-Stokes Equations on a Vector Computer.......................... 566

Sheveley, Yu.D.: Using of an Arbitrary Coordinate for Three-
Dimensional Fluid Dynamic Problems.................................... 572

Shokin, Yu. I.: On Conservatism of Difference Schemes of Gas
Dynamics.. 578

Strani, M. and Sabetta, F.: A Numerical Analysis of a Nonlinear
Eigenvalue Problem Occurring in Viscous Oscillations of a
Supported Drop.. 584

Su, M.D.: Algebraic Model of Large Eddy Simulation.................... 589

Takemoto, Y. and Nakamura, Y.: A Three-Dimensional Incompressible
Flow Solver... 594

Teng, Z.-H.: Variable-Elliptic-Vortex Method for Incompressible
Flow Simulation... 600

Thomas, J.W., Schweitzer, R., Heroux, M., McCormick, S. and Thomas, A.M.:
Application of the Fast Adaptive Composite Grid Method to Computational
Fluid Dynamics 606

Ting, L. and Liu, G.C.: Merging of Vortices with Decaying Cores
and Numerical Solutions of the Navier-Stokes Equations................ 612

Tokunaga, H., Satofuka, N. and Miyagawa, H.: Direct Simulation of Shear Flow Turbulence in a Plane Channel by Sixth Order Accurate Method of Lines with New Sixth Order Accurate Multi-Grid Poisson Solver.. 617

Verstappen, R., ten Thije, J., de Vries, R.W. and Zandbergen, P.J.: Solutions of the Navier-Stokes Equations Using an Efficient Spectral Method.. 622

Walters, R.W., Thomas, J.L. and Van Leer, B.: An Implicit Flux-Split Algorithm for the Compressible Euler and Navier-Stokes Equations.. 628

Wang, L.X. and Luo, S.J.: Numerical Solution of Transonic Small Disturbance Pressure Equation and Its Applications.................... 636

Wang, R.Q., Han, Y.G., Zhou, B.M. and Sun, J.A.: A New Switch-Scheme for Convection-Diffusion Equations........................... 642

Warming, R.F. and Beam, R.M.: Stability of Semidiscrete Approximations for Hyperbolic Initial-Boundary-Value Problems: An Eigenvalue Analysis... 647

Weiland, C. and Pfitzner, M.: 3-D and 2-D Solutions of the Quasi-Conservative Euler Equations................................... 654

Wu, J.H.: An Unconditionally L_∞ - Stable Method of Fractional Steps for Numerical Solution of Convective Diffusion Problems........ 660

Yang, J.Y.: A Hybrid Upwind Scheme for the Computation of Shock-on-Shock Interaction Around Blunt Bodies.................... 666

Yang, Z.H. and Keller, H.B.: Multiple Laminar Flows Through Curved Pipes... 672

Yee, H.C.: Numerical Experiments with a Symmetric High-Resolution Shock-Capturing Scheme............................... 677

Zeng, Q.C., Zhang, X.H., Yuan, C.G. and Liang, X.Z.: A Design and Test of a Numerical Coupled Land-Atmosphere-Ocean Model.......... 684

Zhang, H.X. and Zheng, M.: A Mixed Antidissipative Method Solving Three-Dimensional Separated Flow............................ 689

Zhang, J.: Pointwise Finite Element Method and Its Applications to Compressible Flows..................................... 694

Zhang, J.B.: Unsteady Transonic Flows Around Oscillating Wings.......... 700

Zhou, L.X. and Zhang J.: A Lagrangian-Eulerian Particle Model for Turbulent Two-Phase Flows with Reacting Particles........... 705

Zhu, Z.Q. and Sobieczky, H.: Analysis of Transonic Wings Including Viscous Interaction....................................... 710

List of Participants... 715

SYMPLECTIC GEOMETRY AND NUMERICAL METHODS
IN FLUID DYNAMICS

K. Feng

Academia Sinica Computing Center, Beijing, China

1. INTRODUCTION

It is an honor and a pleasure for me to present the inaugural talk at the Tenth International Conference on Numerical Methods in Fluid Dynamics in Beijing. I want to thank the Organizing Committee, its Secratory, Prof. H. Cabannes, the Conference Chairman, Prof. F.G.Zhuang, and the Co-chairman, Prof. Y.L.Zhu for the kind invitation.

We present a brief survey of considerations and results of a study [1,2,3,4,6], undertaken by the author and his group, on the links between the <u>Hamiltonian formalism</u> and the <u>numerical methods</u> for solving dynamical problems expressed in the form of the <u>canonical system</u> of differential equations

$$\frac{dp_i}{dt} = -\frac{\partial H}{\partial q_i}, \quad \frac{dq_i}{dt} = \frac{\partial H}{\partial p_i}, \quad i = 1,\cdots, n \tag{1.1}$$

with given <u>Hamiltonian function</u> $H(p_1,\cdots, p_n, q_1,\cdots, q_n)$.

The canonical system (1.1) with remarkable elegance and symmetry was introduced by Hamilton as a general mathematical scheme, first for problems of geometrical optics in 1824, then for conservative dynamical problems in 1834. The approach was followed and developed further by Jacobi into a well-established mathematical formalism for analytical dynamics, which is an alternative of, and equivalent to, the Newtonian and Lagrangian formalisms. The geometrization of the Hamiltonian formalism was undertaken by Poincare in 1890's and by Cartan, Birkhoff, Weyl, Siegel, etc., in the 20th century; this gave rise a new dicipline, called <u>symplectic geometry</u>, which serves as the mathematical foundation of the Hamiltonian formalism.

It is known that, Hamiltonian formalism, apart from its classical links with analytical mechanics, geometrical optics, calculus of variations and non-linear PDE of first order, has inherent connections also with unitary representations of Lie groups, geometric quantization, pseudo-differential and Fourier integral operators, classification of singularities, integrability of non-linear evolution equations, optimal control theory, etc.. It is also under extension to infinite dimensions for various field theories, including fluid dynamics, elasticity, electrodynamics, plasma physics, relativity, etc.. Now it is almost certain that all real physical processes with negligible dissipation can be described, in some way or other, by Hamiltonian formalism, so the latter is becoming one of the most useful tools in the mathematical arsenal of physical and engineering sciences. In this way, a systematic study of numerical methods of Hamiltonian systems is motivated and would eventually lead to more general applicability and more direct accessibility of the Hamiltonian formalism. We try to conceive, design, analyse and evaluate difference schemes and algorithms specifically within the framework of symplectic geometry. The approach proves to be quite successful as one might expect, we actually derive in this way numerous "unconventional" difference schemes. Due to historical reasons, classical symplectic geometry, however, lacks the "computational" component in the modern sense. Our present study might be considered as an attempt to fill the blank.

In the following, vectors are always represented by column matrices, matrix transpose is denoted by prime '. Let $z=(z_1,\cdots,z_n, z_{n+1},\cdots,z_{2n})'=(p_1,\cdots,p_n,q_1,\cdots,q_n)'$,

$$H_z = [\frac{\partial H}{\partial p_1}, \cdots, \frac{\partial H}{\partial p_n}, \frac{\partial H}{\partial q_1}, \cdots, \frac{\partial H}{\partial q_n}]',$$

$$J_{2n} = J = \begin{bmatrix} 0 & I_n \\ -I_n & 0 \end{bmatrix}, \qquad J' = J^{-1} = -J.$$

(1.1) can be written as

$$\frac{dz}{dt} = J^{-1} H_z, \qquad (1.2)$$

defined in phase space R^{2n} with a standard <u>symplectic structure</u> given by the non-singular anti-symmetric closed differential 2-form

$$\omega = \Sigma\, dz_i \wedge dz_{n+i} = \Sigma\, dp_i \wedge dq_i.$$

According to <u>Darboux</u> Theorem, the symplectic structure given by any non-singular anti-symmetric closed differential 2-form can be brought to the above standard form, at least locally, by suitable change of co-ordinates.

The <u>Fundamental Theorem on Hamiltonian Formalism</u> says that the solution $z(t)$ of the canonical system (1.2) can be generated by a <u>one-parameter group</u> $G(t)$, depending on the given Hamiltonian H, of <u>canonical transformations</u> of R^{2n} (locally in t and z) such that

$$z(t) = G(t)\, z(0).$$

A transformation $z \to \hat{z}$ of R^{2n} is called <u>canonical</u>, or <u>symplectic</u>, if it is a local diffeomorphism whose Jacobian $\frac{\partial \hat{z}}{\partial z} = M$ is everywhere symplectic, i.e.

$$M'JM = J, \quad \text{i.e.} \quad M \in Sp(2n).$$

The canonicity of $G(t)$ implies the preservation of 2-form ω, 4-form $\omega \wedge \omega$, $\cdots$, 2n-form $\omega \wedge \omega \wedge \cdots \wedge \omega$. They constitute the class of <u>conservation laws of phase area</u> of even dimensions for the Hamiltonian system (1.2).

Moreover, the Hamiltonian system possesses another class of conservation laws related to the <u>energy</u> $H(z)$. A function $\varphi(z)$ is said to be an <u>invariant integral</u> of (1.2) if it is invariant under (1.2)

$$\varphi(z(t)) \equiv \varphi(z(0))$$

which is equivalent to
$$\{ \varphi, H \} = 0,$$
where the <u>Poisson Bracket</u> for two functions $\varphi(z)$, $\psi(z)$ are defined as
$$\{ \varphi, \psi \} = \varphi'_z J^{-1} \psi_z .$$
H itself is always an invariant integral, see, e.g., [5].

The above digressions on Hamiltonian systems suggest the following <u>guidelines</u> for the numerical study of dynamical problems: The problem should be expressed in some suitable <u>Hamiltonian formalism</u>. The numerical schemes should preserve as much as possible the characteristic properties and inner symmetries of the original system. The transition from the k-th time step z^k to the next (k+1)-th time step z^{k+1} should be <u>canonical</u> for all k and, moreover, the invariant integrals of the original system should <u>remain invariant</u> under these transitions.

2. CANONICAL DIFFERENCE SCHEMES FOR LINEAR CANONICAL SYSTEMS

Consider the case for which the Hamiltonian is a quadratic form
$$H(z) = \tfrac{1}{2} z'Sz, \quad S' = S, \quad H_z = Sz, \qquad (2.1)$$
then the canonical system
$$\frac{dz}{dt} = Lz, \qquad L = J^{-1}S \qquad (2.2)$$
is <u>linear</u>, where L is <u>infinitesimally symplectic</u>, i.e. L satisfies $L'J + JL = 0$. The solution of (2.2) is
$$z(t) = G(t) z(0),$$
where $G(t) = \exp tL$, as the <u>exponential transform</u> of infinitesimally symplectic tL, is symplectic.

It is easily seen that the weighted Euler scheme
$$\frac{1}{\tau} (z^{k+1} - z^k) = L(\alpha z^{k+1} + (1 - \alpha) z^k)$$
for the linear system (2.2) is <u>symplectic</u> if and only if $\alpha = \frac{1}{2}$, i.e. it is the case of <u>time-centered Euler Scheme</u> with the transition matrix F_τ,
$$z^{k+1} = F_\tau z^k, \qquad F_\tau = \varphi(\tau L), \qquad \varphi(\lambda) = \frac{1 + \frac{\lambda}{2}}{1 - \frac{\lambda}{2}}, \qquad (2.3)$$

F_τ, as the <u>Cayley transform</u> of infinitesimally symplectic τL, is symplectic. The 2nd order canonical Euler scheme (2.3) can be generalized to canonical schemes of arbitrary high order [2,3]. For example, by taking the matrix transform function $\varphi(\lambda)$ in (2.3) to be the diagonal <u>Padé approximants</u> $P_m(\lambda)/P_m(-\lambda)$ to the exponential function $\exp \lambda$, where
$$P_0(\lambda)=1, \; P_1(\lambda)=2+\lambda, \; P_2(\lambda)=12+6\lambda+\lambda^2, \ldots, \; P_m(\lambda)=2(2m-1)P_{m-1}(\lambda)+\lambda^2 P_{m-2}(\lambda),$$
we can prove that the difference schemes
$$z^{k+1} = \frac{P_m(\tau L)}{P_m(-\tau L)} z^k \qquad m = 1,2,\cdots \qquad (2.4)$$
for the system (2.2) are symplectic, A-stable, of 2m-th order of accuracy, and having

the same set of quadratic invariant integrals including $H(z)$ as that of system (2.2). The case $m=1$ is the time-centered Euler scheme (2.3).

For the general non-linear canonical system (1.2), the time-centered Euler scheme
$$\frac{1}{\tau}(z^{k+1} - z^k) = J^{-1} H_z(\frac{1}{2}(z^{k+1} + z^k)) \tag{2.5}$$
is <u>canonical</u>. However, unlike the linear case, the invariant integrals $\varphi(z)$ of system (1.2), including $H(z)$, are conserved only approximately
$$\varphi(z^{k+1}) - \varphi(z^k) = O(\tau^3).$$
The time-centered Euler schemes (2.3), (2.5) and their canonical generalizations (2.4) are all <u>implicit</u>. For the case of <u>separable</u> Hamiltonian
$$H(p, q) = U(p) + V(q),$$
one can construct <u>time-staggered</u> schemes which are <u>canonical</u>, of 2nd order accuracy and <u>practically</u> <u>explicit</u> [1,2], e.g.,
$$\begin{aligned}\frac{1}{\tau}(p^{k+1} - p^k) &= -V_q(q^{k+\frac{1}{2}}), \\ \frac{1}{\tau}(q^{k+1+\frac{1}{2}} - q^{k+\frac{1}{2}}) &= U_p(p^{k+1}).\end{aligned} \tag{2.6}$$
The p's are set at integer times $t = k\tau$, q's at half-integer times $t=(k + \frac{1}{2})\tau$. We need averaging, e.g., using
$$q^k = \frac{1}{2}(q^{k-\frac{1}{2}} + q^{k+\frac{1}{2}})$$
to compute the invariant integrals $\varphi(p, q)$ and get
$$\varphi(p^{k+1}, q^{k+1}) - \varphi(p^k, q^k) = O(\tau^3).$$

For the comparison of stability for the linear system (2.2) and the canonical schemes (2.4), (2.6) and the application of (2.6) to the wave equation, see [1].

3. CONSTRUCTION OF CANONICAL DIFFERENCE SCHEMES VIA GENERATING FUNCTIONS

A major component of the transformation theory in symplectic geometry is the method of <u>generating functions</u>, see, e.g., [5], which also play a central role for the construction of canonical difference schemes. In [2,4] a constructive general theory of generating functions is given, roughly as follows: Let
$$T = \begin{bmatrix} A & B \\ C & D \end{bmatrix}, \quad T^{-1} = \begin{bmatrix} A_1 & B_1 \\ C_1 & D_1 \end{bmatrix},$$
T be a non-singular real matrix of order $4n$ satisfying
$$T' \begin{bmatrix} 0 & I_{2n} \\ -I_{2n} & 0 \end{bmatrix} T = \mu \begin{bmatrix} -J_{2n} & 0 \\ 0 & J_{2n} \end{bmatrix}, \quad \text{for some } \mu \neq 0. \tag{3.1}$$
T defines a linear transformation in product space $R^{2n} \times R^{2n}$ by
$$\begin{aligned}\hat{w} &= A\hat{z} + Bz \\ w &= C\hat{z} + Dz,\end{aligned} \quad \begin{bmatrix}\hat{z} \\ z\end{bmatrix}, \begin{bmatrix}\hat{w} \\ w\end{bmatrix} \in R^{2n} \times R^{2n}. \tag{3.2}$$
Let $z \to \hat{z} = g(z,t)$ be a <u>time-dependent canonical transformation</u> defined by
$$g(z, t) = M_0 G(z, -t) \tag{3.3}$$

where $G(z,t)$ is the one-parameter group of canonical transformations for the canonical system (1.2) with given Hamiltonian $H(z)$; M_0 is a constant symplectic matrix$\in Sp(2n)$. The Jacobian
$$M(z,t) = \frac{\partial g(z,t)}{\partial z} \in Sp(2n), \quad M(z,0) = M_0.$$
If the transversality condition
$$|CM_0 + D| \neq 0 \tag{3.4}$$
holds, then there exist, for sufficiently small $|t|$ and in (some neighborhood of) R^{2n}, a <u>time-dependent gradient transformation</u> $w \to \hat{w} = f(w,t)$ with Jacobian $\frac{\partial f(w,t)}{\partial w} = N(w,t) \in S_m(2n)$, i.e., everywhere <u>symmetric</u>, and a <u>time-dependent generating function</u> $\phi(w,t)$ with gradient $\phi_w(w,t) = f(w,t)$ such that
$$[\hat{w} - f(w,t)]_{\hat{w}=A\hat{z}+Bz,\, w=C\hat{z}+Dz} = 0$$
is an implicit representation of the canonical transformation $\hat{z} = g(z,t)$. The generating function $\phi(w,t)$ satisfies the <u>Hamilton-Jacobi equation</u>
$$\phi_t(w,t) = -\mu H(C_1 \phi_w(w,t) + D_1 w), \quad w = C\hat{z} + Dz. \tag{3.5}$$

By recursions we can determine explicitly all possible time-dependent generating functions for Hamiltonians $H(z)$ analytic in z:
$$\phi(w,t) = \sum_{k=0}^{\infty} \phi^{(k)}(w)\, t^k, \tag{3.6}$$
$$\phi^{(0)}(w) = \tfrac{1}{2} w' N_0 w, \quad N_0 = (AM_0 + B)(CM_0 + D)^{-1},$$
$$\phi^{(1)}(w) = -\mu H(E_0 w), \quad E_0 = (CM_0 + D)^{-1},$$
$$k \geq 1: \quad \phi^{(k+1)}(w) = -\frac{1}{k+1} \sum_{m=1}^{k} \frac{\mu}{m!} \sum_{i_1,\ldots,i_m=1}^{2n} H_{z_{i_1},\ldots,z_{i_m}}(E_0 w) \cdot \sum_{\substack{k_1+\cdots+k_m=k \\ k_j \geq 1}} (C_1 \phi_w^{(k_1)}(w))_{i_1} \cdots (C_1 \phi_w^{(k_m)}(w))_{i_m}.$$

Choose
$$T = \begin{bmatrix} -I_n & 0 & 0 & 0 \\ 0 & 0 & I_n & 0 \\ 0 & I_n & 0 & 0 \\ 0 & 0 & 0 & I_n \end{bmatrix}, \quad \mu = 1, \quad M_0 = J_{2n},$$
we get generating function of the 1st type for the case $\left|\frac{\partial \hat{q}}{\partial p}\right| \neq 0$:
$$\phi = \phi(\hat{q}, q, t), \quad -\hat{p} = \phi_{\hat{q}}, \quad p = \phi_q, \quad \phi_t + H(\phi_q, q) = 0.$$

Choose
$$T = \begin{bmatrix} -I_n & 0 & 0 & 0 \\ 0 & 0 & 0 & -I_n \\ 0 & I_n & 0 & 0 \\ 0 & 0 & I_n & 0 \end{bmatrix}, \quad \mu = 1, \quad M_0 = I_{2n},$$
we get generating function of the 2nd type for the case $\left|\frac{\partial \hat{q}}{\partial q}\right| \neq 0$:

$$\phi = \phi(\hat{q},p,t), \quad -\hat{p} = \phi_{\hat{q}}, \quad -q = \phi_p, \quad \phi_t + H(p, -\phi_p) = 0.$$

Choose

$$T = \begin{bmatrix} -J_{2n} & J_{2n} \\ \frac{1}{2} I_{2n} & \frac{1}{2} I_{2n} \end{bmatrix}, \quad \mu = -1, \quad M_0 = I_{2n}$$

we get the generating function of a new type—the <u>Euler type</u>—

$$\phi = \phi(w,t), \quad w = \frac{1}{2}(\hat{z}+z), \quad \hat{w} = J(z-\hat{z}) = \phi_w, \quad \phi_t - H(w - \frac{1}{2} J\phi_w) = 0.$$

The <u>general methodology</u> for the construction of canonical difference schemes is as follows: Choose some suitable type of generating function with its explicit expression (3.6) truncate or approximate it in some way and take gradient of this approximation, then we get automatically the implicit representation of some canonical transformation for the transition of the difference scheme. In this way one can get an abundance of canonical difference schemes. This methodology is <u>unconventional</u> in the ordinary sense, but <u>natural</u> from the point of view of symplectic geometry [4]. As an illustration we choose the <u>Euler type</u> generating function $\phi(w,t)$, which is odd in t:

Take sufficiently small $\tau > 0$ as the time-step, define

$$\psi^{(2m)}(w,\tau) = \sum_{k=1}^{m} \phi^{(2k-1)}(w) \tau^{2k-1}, \quad m = 1,2,\cdots. \qquad (3.7)$$

Then the gradient transformation

$$w \to \hat{w} = \psi_w^{(2m)}(w,\tau) \qquad (3.8)$$

represents implicitly a canonical scheme $\hat{z} = z^k \to z^{k+1} = z$ of 2m-th order accuracy upon substitution (3.2). The case m = 1 is the centered Euler scheme (2.5). For linear canonical system (2.2), the generating function is the quadratic form

$$\phi(w,t) = \frac{1}{2} w'(2J \tanh(\frac{\tau}{2} L))w, \quad L = J^{-1}S, \quad S' = S,$$

$$\tanh \lambda = \lambda - \frac{1}{3} \lambda^3 + \frac{2}{15} \lambda^5 - \frac{17}{312} \lambda^7 + \cdots = \sum_{k=1}^{\infty} a_{2k-1} \lambda^{2k-1},$$

(3.8) becomes symplectic difference schemes

$$z^{k+1} - z^k = (\sum_{k=1}^{m} a_{2k-1} (\frac{\tau}{2} L)^{2k-1})(z^{k+1} + z^k).$$

The case m = 1 is the centered Euler scheme (2.3).

4. HAMILTONIAN FORMALISM IN INFINITE DIMENSIONS FOR FLUIDS

Physical processes in continuous media are dynamical systems of infinite dimensions, the corresponding symplectic geometry has not yet been fully developed theoretically. The constructive theory of generating functions [2,4] of the 2nd type and of the Euler type and the corresponding construction of canonical difference schemes (involving time discretization only) has been generalized by Qing and Li to the case of phase space of infinite dimensions of the form $B^* \times B$, where B is a reflexive Banach space, B^* its dual [6], the Hamiltonian canonical systems are of the form

$$\frac{\partial p}{\partial t} = - \frac{\delta H}{\delta q}, \quad \frac{\partial q}{\partial t} = \frac{\delta H}{\delta p}$$

$H = H(p, q)$ is a functional, the right hand sides are variational derivatives. In case B is self-dual, the generalization is valid also for the generating functions of the 1st kind.

The problem of the Hamiltonian structure for the equations of ideal fluids has a long history, dated back to 1850's. There are several different approaches to its solution. We mention here only the oldest one-the representation of velocity by Clebsch variables. Take, e.g., the case of compressible ideal fluid,

$$\vec{v} = \frac{\lambda}{\rho} \text{ grad } \mu + \text{grad } \phi$$

where ρ is the density, λ, μ, ϕ are Clebsch potentials. Then the flow equations can be put in the canonical form

$$\frac{\partial \phi}{\partial t} = - \frac{\delta H}{\delta \rho}, \quad \frac{\partial \mu}{\partial t} = - \frac{\delta H}{\delta \lambda},$$

$$\frac{\partial \rho}{\partial t} = \frac{\delta H}{\delta \phi}, \quad \frac{\partial \lambda}{\partial t} = \frac{\delta H}{\delta \mu},$$

$$H = \int \{ \frac{1}{2} \rho \vec{v}^2 + e(\rho) \} \, dr, \quad e(\rho) = \text{internal energy}.$$

Buneman used this formalism (modified) in computer simulation with a scheme staggered both in space and time and pointed out the inherent computational advantages of the canonical formalism [7].

References

[1] Feng Kang, On difference schemes and symplectic geometry, Proceedings of 1984 Beijing International Symposium on Differential Geometry and Differential Equations-COMPUTATION OF PARTIAL DIFFERENTIAL EQUATIONS, Ed. Feng Kang, Science Press, Beijing, 1985, pp42-58.
[2] Feng Kang, Difference schemes for Hamiltonian formalism and symplectic geometry, Jour. of Comput. Math., 4:3 (1986).
[3] Feng Kang, Wu Hua-mo, Qing Meng-zhou, Symplectic difference schemes for the linear Hamiltonian canonical systems, to appear.
[4] Feng Kang, Wu Hua-mo, Qing Meng-zhou, Wang Dao-liu, Construction of canonical difference schemes for Hamiltonian formalism via generating functions, to appear.
[5] V.I. Arnold, Mathematical Methods of Classical Mechanics, New York, 1978.
[6] Qing Meng-zhou, Li Chun-wang, Symplectic difference schemes of Hamiltonian systems in infinite dimensions, to appear.
[7] O. Buneman, Advantages of Hamiltonian Formulations in Computer Simulations, in "Mathematical Methods in Hydrodynamics and Integrability of Dynamical Systems", Tabor, et., Amer. Inst. Phys., U.S.A., 1981.

ON SOME PROBLEMS OF DYNAMIC METEOROLOGY

V.P. Dymnikov
Department of Numerical Mathematics
USSR Academy of Sciences
Moscow, USSR

In the present paper we will consider some problems of modelling large-scale atmospheric processes (with characteristic spacial scales $\sim 10^6 - 10^7$ m). What are the peculiarities of such processes? First of all, these processes go on on the rotating Earth and the Coriolis force for them is practically balanced by the force of the pressure gradient (quasigeostroficity), further they are quasitwo-dimensional (the characteristic scale by vertical is two orders less than the horizontal one), these processes are quasistatic (the gravitational force is balanced by the vertical component force of the pressure gradient). If we consider the characteristic time scales $t > t_x$ (the Earth rotation period, equal to a day), then the main energetic wave processes in the atmosphere in midlatitudes will be the Rossby-Blinova waves, generated due to the change of the Coriolis parameter with latitude and the absolute vorticity conservation. Inspite of the fact that the large-scale processes in atmosphere are quasitwo-dimensional, it has been believed for a long time, that the Rossby waves get the main energy through the processes of the baroclinic instability, dectermined by the spacial temperature stratification of the atmosphere. It has not been until recently that the great role of the barotropic instability in the formation of the law frequency variability of the atmospheric circulation has become clear.

The rapid development of numerical methods in meteorology problems was determined, of course, by its most practical problems-the problem of weather forecast and the problem of modelling climate and its changes. The study of the physical mechanisms forming the atmosphere circulation variations at various time scales, has led to the formation of <u>diagnostic</u> investigations, connected with quite definite numerical algorythms. It is important to note that the problem

of short-range and middle-range weather forecast and the problem of
modelling climate and its changes have jointed recently - in both problems global models of the general atmosphere circulation play the
leading role. Generally speaking it is suprising, as for the short-
range forecast the central problem is the prediction of amplitude
and the phase of the Rossby waves and the development of synoptic
eddies (let us note, that in this problem it is necessary to describe
each eddy separately), and in climate problems it is necessary to
describe the eddies statistes. However, as synoptic eddies are one
of the basic energy generators in large scales, their description
is made explicitly also in models of climate variations. Let us also
note, that in the middle-range weather forecast one of the main
shortcomings is the so-called climate drift (reaching the model's
statistically steady state). All this leads to the fact that when constructing the forecast models we should take care of the description
of local wave processes as well as o their statistics. While constructing the sehemes one should make to them certain requirements of the
fulfilment of the analogues of numerous consevation laws. The complex of such requiremnts to the models of general atmosphere circulation is formulated, for instance, in /1/.

The most important problem of modern meteorology is the problem
of the pedictability of the atmospheric processes. If we consider
differential-difference approximation of the system of atmosphere
hydrothermodynamics equations (difference by special variables), then it
will be an open dissipative dynamic system (with the number of degrees
of freedom $10^6 - 10^5$ im modern models). The weather forecast is
the calculation of the phase trajectory of this system. The atmosphere circulation can be rouhly divided in two regimes - the regime
with strongly developed zonal flow and the regime with the developed
quasistationary ridge (blocking regime). The second regime is stable
for a long period of time, the ridge itself is quasibarotropic and
the prediction of its existance must be higher. However, though
the energy flux from baroclinic eddies to this pattern is small, this
flow practically completely compensate the dissipation. That is why
a very high resolution /2/ is necessary for the description of the large-
scale structure in this case too. With such a high resolution it is
quite natural that questions of the convergency of the difference
schemes solution to the solution of initial nonlinear differential
equations arise.

Further we will dwell on two important examples of numerical
algorythms, used in meteorology problems. The first algorithm has

been constructed to solve the atmosphere circulation problem, and probably convergency problems can be solved for it. The second example is connected with the investigation of, as it has been already mentioned above, an extremely important problem in modern meteorology, namely, the problem of circulation regimes stability.

1. Taking as a vertical coordinate the pressure, normalized to its value on the Earth's surface and using the quasistatic approximation we will write down the equations of atmosphere hydrothermodynamics in the spherical coordinate system

$$\frac{du}{dt} - (\ell + \frac{u}{a} tg\varphi)v + \frac{1}{a\cos\varphi}\left(\frac{\partial \phi}{\partial \lambda} + \frac{RT}{\pi}\frac{\partial \pi}{\partial \lambda}\right) = F_n ,$$

$$\frac{dv}{dt} + (\ell + \frac{u}{a} tg\varphi)u + \frac{1}{a}\left(\frac{\partial \phi}{\partial \varphi} + \frac{RT}{\pi}\frac{\partial \pi}{\partial \varphi}\right) = F_v ,$$

$$\frac{\partial \pi}{\partial t} + \frac{1}{a\cos\varphi}\left(\frac{\partial \pi u}{\partial \lambda} + \frac{\partial \pi v \cos\varphi}{\partial \varphi}\right) + \frac{\partial \pi \dot\sigma}{\partial \sigma} = 0 , \qquad (1)$$

$$\frac{dT}{dt} - \frac{RT}{C_p\pi}\left[\pi\dot\sigma + \sigma\left(\frac{\partial \pi}{\partial t} + \frac{u}{a\cos\varphi}\frac{\partial \pi}{\partial \lambda} + \frac{v}{a}\frac{\partial \pi}{\partial \varphi}\right)\right] = F_z + \mathcal{E} ,$$

$$\frac{dq}{dt} = F_q - \mathcal{E}_1 , \qquad \frac{\partial \phi}{\partial \sigma} = -\frac{RT}{\sigma} ,$$

$$\frac{d}{dt} = \frac{\partial}{\partial t} + \frac{u}{a\cos\varphi}\frac{\partial}{\partial \lambda} + \frac{v}{a}\frac{\partial}{\partial \varphi} + \dot\sigma\frac{\partial}{\partial \sigma} . \quad \text{where}$$

In system (1) F_n, F_v are the velocities of the change of momentum due to the Reynolds stress, F_r, F_g are small-scale diffusion of temperature and humidity, $\mathcal{E}$ is a diabatic heating, $\mathcal{E}_1$ - is a quantity of condensated water vapour.

As boundary conditions for (1) by longitude the periodicity condition is taken. The Earth's surface is the coordinate surface (σ.=1). The corresponding kinematic condition can be written in the form

$$\dot\sigma = 0 \quad \text{at} \quad \sigma = 1 \qquad (2)$$
$$\dot\sigma = 0 \quad \text{at} \quad \sigma = 0$$

At $\sigma = 1$, besides condition (2) the distribution of geopotential

$$\phi = gz_s = \phi_s \quad \text{at} \quad \sigma = 1 \qquad (3)$$

is also specified. The above-used σ - coordinate system was suggested by Phillips. This coordinate system is most popular nowadays when solving problems of general atmosphere circulation. The main advantage of the system is the correct description of orographical inhomogeneities of the Earth's surface, but as all drawbacks continue advantages, then in the domains of orographical inhomogeneities of the Earth's surface, where the spacial gradients Φ and $\overline{\pi}$ are great, arises the serious calculating problem of determining the terms

$$\left(\frac{\partial \Phi}{\partial \lambda} + \frac{RT}{\pi}\frac{\partial \pi}{\partial \lambda}\right), \quad \left(\frac{\partial \Phi}{\partial \varphi} + \frac{RT}{\pi}\frac{\partial \pi}{\partial \varphi}\right)$$

which are small differences of large values /3/.

The equations system (1-3) possesses the number of integral properties, the account of which is necessary while constructing finite-difference schemes.

a) the law of mass conservation

$$\frac{\partial}{\partial t}\int_G \pi \, dG = 0, \quad dG = a^2 \cos\varphi \, d\varphi \, d\lambda .$$

b) the law of the angular momentum conservation

$$\frac{\partial}{\partial t}\int_0^1\int_G \pi M \, dG \, d\sigma = \int_G \left(\Phi_s \frac{\partial \pi}{\partial \lambda} + a\cos\varphi \, \tau_\lambda|_{\sigma=1}\right) dG \quad (*)$$

where $M = a\cos\varphi(u + a\Omega\cos\varphi)$ is the total angular momentum, $\tau_\lambda|_{\sigma=1}$ - is the λ - component of wind stress at the lower boundary. Averaging (*) by a sufficienty large period of time we can obtain the approximate correlation

$$\int_G \overline{\tau_\lambda|_{\sigma=1}}^{\,t} \cos\varphi \, dG \simeq 0$$

c) the balance of atmosphere moisture

$$\frac{\partial}{\partial t}\int_0^1\int_G \pi q \, dG \, d\sigma = \int_G \left[E_s - \int_0^1 \pi \varepsilon_1 \, d\sigma\right] dG$$

d) the law of the total energy conservation in the adiabatic approximation

$$\frac{\partial}{\partial t}\int_G \pi\left[\Phi_s + \int_0^1 (K + c_p T) \, d\sigma\right] dG = 0,$$

where $K = (u^2 + v^2)/2$.

At $\Phi_s = 0$ this correlation has a quadratic form if we substitute variables

$$u_1 = \sqrt{\pi}\cdot u, \quad v_1 = \sqrt{\pi}\, v, \quad T_1 = \sqrt{\pi}\, T .$$

There is some "asymptotical" conservation laws, which, apparantly, influence the statistic characteristics of the solution of system /4,5/, for example, the law of potential enstrophy conservation in the approximation of the barotropic incompressible fluid or shallow water. The construction of the differential-difference schemes, possessing such conservation laws, is given in /4,5/.

Further we will show the method of constructing the schemes, possessing the accurate quadratic conservation law in the adiabatic approximation. The method is based on reducing system (1-3) to a so-called symmetrizied form and using symmetric difference approximations by space and the Crank-Nickolson scheme by time (perhaps in the combination with the splitting-up method).

Let us illustrate this method on a simple example. We will consider two equations

$$\frac{\partial T}{\partial t} + u\frac{\partial T}{\partial x} = 0, \quad \frac{\partial \pi}{\partial t} + \frac{\partial \pi u}{\partial x} = 0,$$

$$\pi > 0, \quad t \in [0, \bar{t}], \quad x \in [0, 2\pi].$$

as boundary conditions we will take the periodicity conditions. Both these equations can be reduced to the form

$$\frac{\partial \sqrt{\pi} T}{\partial t} + \frac{1}{2}\left(u\frac{\partial \sqrt{\pi} T}{\partial x} + \frac{\partial \sqrt{\pi} u T}{\partial x}\right) = 0,$$

$$\frac{\partial \sqrt{\pi}}{\partial t} + \frac{1}{2}\left(u\frac{\partial \sqrt{\pi}}{\partial x} + \frac{\partial \sqrt{\pi} u}{\partial x}\right) = 0.$$

Substituting variables $\varphi_1 = \sqrt{\pi} \cdot T$, $\varphi_2 = \sqrt{\pi}$ we will obtain

$$\frac{\partial \varphi_i}{\partial t} + \frac{1}{2}\left(u\frac{\partial \varphi_i}{\partial x} + \frac{\partial \varphi_i u}{\partial x}\right) = 0, \quad i = 1,2.$$

It is not difficult to see that $\frac{\partial}{\partial t}\int \varphi_i^2 dx = 0$ (the linear conservation law is reduced to the quadratic one). It is easy to verify /6/, that symmetric difference apprixamtion on a uniform mesh of the operator $L\varphi \equiv u\frac{\partial \varphi}{\partial x} + \frac{\partial \varphi u}{\partial x}$ will give a skew symmetric matrix. By constructing such approximations we will have

$$\frac{\partial \varphi^h}{\partial t} + K_h \varphi^h = 0, \quad (K_h \varphi^h, \varphi^h)_h = 0,$$

$$(\varphi^h, \psi^h)_h = \sum_i' \varphi_i^h \psi_i^h$$

It is important to note that the skew symmetry of operator will be conserved regardless of what we take for u^h i.e. it does not depentd on the type of the projector, converting u in to u^h

and consequently, we can construct schemes satisfying, for example, linear conservation laws /6/.

Using the Crank-Nickolson scheme $\frac{\varphi^{hj+1} - \varphi^{hj}}{\tau} + K_h \frac{\varphi^{hj+1} + \varphi^{hj}}{2} = 0,$ we will get the correlation

$$\left(\varphi^{hj+1}, \varphi^{hj+1}\right)_h = \left(\varphi^{hj}, \varphi^{hj}\right)$$

This method was realized when constructing approximation (1-3) /1/ and allowed to solve the number of new interesting problems, for example, to investigate the connection between the spacial and time spectra of the solution of system (1-3), as it allows to solve the problem with arbitrary coefficients of viscosity and diffusion /1/. It is also important to note, that the implicity and symmetrization of the spacial operator of problem (1-3) can turn out to be a sufficient condition also for the proof of the convergency of the difference problem solution to the differential problem solution (on the class of smooth functions). At least, it is sufficient to prove the convergency theorem for the equations of the Burgers type and two-dimensional barotropic viscous fluid. Let us remind, that convergency theorems for the two-dimensional fluid with using the Arakawa spacial approximations and implicit approimations by time were proved in / 7 /.

2. Numerical experiments with the models of general atmosphere circulation have shown that the remote response of atmosphere circulation on heating anomalies have the quasibarotropic wave structure. It is very essential that the response structure has a geographically localized structure, i.e. the mechanism of the formation of such a response can be connected with the perturbation of the operator, linearized relatively to the basic state (it is clear that first of all one should take into account unstable modes). It is known that the barotropic zonally averaged flow in the atmosphere in midlatitudes is stable (stability according to Lyapunov). However, this statement is not valid as far as the zonal-nonsymmetric flow is concerned. It is also clear that the longitude geographical localization of normal modes can be connected only with the zonaly no nonsymmetric flow.

The equation for the barotropic component of the absolute vorticity will have the form:

$$\frac{\partial \Omega}{\partial t} + J(\psi, \Omega) = -\eta \Delta \psi + K \Delta^2 \psi + F,$$

Here

$\Omega = \Delta \psi + \ell$, $\ell = 2\omega \sin\varphi$, ψ — the stream function,

$u = -\frac{\partial \psi}{\partial \varphi}$, $v = \frac{1}{\cos\varphi} \frac{\partial \psi}{\partial \lambda}$,

$J(\psi, \theta) = \frac{1}{\cos\varphi} \left(\frac{\partial \psi}{\partial \lambda} \cdot \frac{\partial \theta}{\partial \varphi} - \frac{\partial \psi}{\partial \varphi} \cdot \frac{\partial \theta}{\partial \lambda} \right),$

$\Delta \psi = \frac{1}{\cos\varphi} \frac{\partial}{\partial \varphi} \cos\varphi \frac{\partial \psi}{\partial \varphi} + \frac{1}{\cos^2\varphi} \frac{\partial^2 \psi}{\partial \lambda^2},$

η is the dissipation coefficient in the boundary layer; K is the coefficient of the horizontal dissipation. Let $\psi = \bar{\psi} + \psi'$, where $\bar{\psi}$ is the stationary solution, the stability of which is being investigated

$$J(\bar{\psi}, \bar{\Omega}) = -\eta \Delta \bar{\psi} + K \Delta^2 \bar{\psi} + F .$$

We can show / 8 /, that solution $\bar{\psi}$ is globally stable using norm $\alpha_1^2 \|\nabla \psi\|_{L_2} + \alpha_2^2 \|\Delta \psi\|_{L_2}$ if $\eta \geq 2\sqrt{mn}$ where m, $n = \|f_{m,n}(|\nabla \bar{\psi}|, |\Delta \bar{\psi}|)\|$ are the functions of the main flow. But these evaluations allow only separate roughly stable and unstable regimes. For more accurate estimations it is necessary to solve the spectral problem for the operator, linearized relatively to
Let $S = \{x \in R^3 ; |x| = 1\}$ - be a unit radius sphere and $L_2(S)$ be a unitary Gilbert space of single-valued functions on S with the inner product

$$(f, g) = \frac{1}{S} \int_S f \bar{g} \, ds = \frac{1}{4\pi} \int_0^{2\pi} \int_{-1}^{1} f \bar{g} \, d\lambda \, d\mu$$

and norm $\|f\| = (f, f)^{1/2}$

$\mu = \sin\varphi$, φ is a latitude. The Laplacian symmetric operator on the sphere generates the complete orthonormal functions system in

$$\Delta Y_{m,n} = \chi_n Y_{mn} ,$$
$$Y_{mn}(\lambda, \mu) = P_{mn}(\mu) e^{im\lambda} , \quad n = 1, ..., N ; \quad |m| \leq n \quad (5)$$

Let us write the linearized operator of problem (4) in the form:

$$\frac{\partial \zeta}{\partial t} + J(\Delta^{-1}\zeta, \bar{\zeta} + 2\mu) + J(\bar{\psi}, \zeta) + \eta \zeta - K \Delta \zeta = 0,$$

where

$$\zeta = \Delta \psi , \quad \psi = \omega a^2 \psi' , \quad t = \omega^{-1} t' , \quad \eta = \omega \eta' ,$$

ω is an angular velocity of the Earth's rotation, a is the Earth's radius. Let

$$\gamma \equiv (m, n) = (m_\gamma, n_\gamma) , \quad \bar{\gamma} \equiv (-m, n) = (-m_\gamma, n_\gamma) ,$$

$$X_N = \bigcup_{n \leq N} H_{2n+1}$$

where H_{2n+1} is a subspace of the dimensionality $2n+1$ which is the linear manifold of its eigen functions $\{Y_{m,n}\}$,

$m = -n, -n+1, ..., n-1, n$.

Let us approximate $\vec{\varphi}, \vec{\zeta}, \vec{\varsigma}$ by the finite spherical sums:

$$\vec{\varphi} = \vec{\varphi}_L = \sum_{\ell=0}^{L}\sideset{}{'}\sum_{|K|\leq\ell} \vec{\varphi}_{K\ell} y_{K\ell} = \sideset{}{'}\sum_{\beta}^{L} \vec{\varphi}_\beta y_\beta ,$$

$$\vec{\zeta} = \Delta \vec{\varphi}_L = \vec{\zeta}_L = \sideset{}{'}\sum_{\beta}^{L} \vec{\zeta}_\beta y_\beta = -\sideset{}{'}\sum_{\beta}^{L} \chi_{n\beta} \vec{\varphi}_\beta y_\beta , \qquad (6)$$

$$\vec{\varsigma} = \vec{\varsigma}_N = \sideset{}{'}\sum_{\delta}^{N} \vec{\varsigma}_\delta y_\delta$$

where $\vec{\varphi}, \vec{\zeta} \in X_L , \quad \vec{\varsigma} \in X_N , \qquad L \leq N.$

If we multiply (6) scalarly by $y_\delta \in X_N$ then we shall get the system of linear ordinary differential equations for $(N+1)^2$ dimensional vector-column, consisting of the Fourier complex-valued coefficients $\vec{\varsigma}_\delta$ ($n_\delta = 0,\ldots, N$; $|m_\delta| \leq n_\delta$)

$$\frac{d}{dt}\vec{\varsigma} + A_N(\vec{\varphi})\vec{\varsigma} = 0 ,$$

where the elements of matrix /8/:

$$A_{\delta\nu} = \int_S J(G_\nu, y_\nu) \bar{y}_\delta \, dS + D_{\delta\nu},$$

$$G_\nu = \chi_{n_\nu}^{-1} \vec{\varsigma}(\lambda, \mu) + \vec{\varphi}(\lambda,\mu) , \qquad (7)$$

$$D_{\delta\nu} = [\kappa \chi_{n_\delta} + \eta] - i\, 2m_\delta / \chi_{n\delta}] \delta_{\delta\nu} .$$

And now the central problem is the study of the convergency of the eigenvalues and eigen vectors of matrix A_N to the eigenvalues and vectors of the operator

$$A\vec{\varsigma} \equiv J(\Delta^{-1}\vec{\varsigma}, \vec{\varsigma}+2\mu) + J(\vec{\varphi},\vec{\varsigma}) + \eta\cdot\vec{\varsigma} - \kappa\Delta\vec{\varsigma} \qquad (8)$$

It was shown in paper /8/ that if $\eta \neq 0$ and $\kappa \neq 0$ then the point spectrum of operator (8) consists of not more than the countable set of isolated eigenvalues of the finite multiplicity which don't have finite limit points. Let $E(\omega)$ be a finite dimensional rooted subspace corresponding to the isolated eigenvalue ω multiplicity m of the closed operator A, $E_N(X_N)$ be a direct sum of the rooted subspaces of the operator A_N, corresponding to the eigenvalues ω_i ($i=1,\ldots,m$), $F(\theta)$ be an arbitary holomorphic function, determined in the vicinity of ω. Then the following convergency theorem /8/ is valid: if $\vec{\varphi}(\lambda,\mu)$ is sufficiently smooth function such that $\Delta\bar{u}, \Delta\bar{v}, (\nabla\vec{\varsigma})_i \in Lip(\alpha)$, $\alpha \in (0,2)$, $D(A) \subset H_2^{4+\alpha}(S)$ then at sufficiently large N operator A_N has exactly m of eigenvalues $\omega_1(N),\ldots,\omega_m(N)$,

repeated with account for their multiplicity and at $N > N_0$
the estimations are valid

$$\left| F(\omega) - \frac{1}{m} \sum_{i=1}^{m} F(\omega_i(N)) \right| \leq c N^{-\alpha},$$

$$\max_{1 \leq i \leq m} \left| \omega - \omega_i(N) \right|^{\rho} \leq c N^{-\alpha},$$

(ρ is the height of the eigenvalue ω).

The analogous estimations can be write out also for eigenvectors /8/.

In conclusion we will note, that many problems of dynamic meteorology are only formaly described by the equations of the parabolic type, being in reality hyperbolic (equations of moisture and cloud amonut, ozone, adjont equations in the diagnosis problem etc). These substances, possessing large spacial gradients, require the construction of special monotonic schemes-here we deal with the classical problem of constructing monotonic schemes for hyperbolic equations. Now dwelling on this problem, we will note that the sufficiently detailed review of such schemes, applied in meteorological problems is given in /5/.

References

1. MarchukG.I., Dymnikov V.P. atc. Mathematical modelling of atmosphere and ocean general circulation. L., Hydrometeoizdat, 1984, 320p.
2. Migakoda K., Gordon T. et al Simulation of a Blocking Event in January 1977. Mon.Wea.Rev., 1983, 111, N 4, p. 846-869.
3. Gorby G.A., Gilchrist A, Newson R.C. A general circulation model of the atmosphere snitable for long period integration. Quart. J.Roy. Met.Soc., 1972, v. 98, p. 809-833.
4. Arakawa A. Computational design for long-term numerical integration of the equations of fluid motion: two-dimensional incompressible flow. Part 1. J.Comp.Phys., 1966, 1, p. 119-143.
5. Arakawa A. Lamb V.R. A potential enstrophy and energy conserving scheme for the shallow water equations. Mon.Wea.Rev., 1981, v.109, p. 18-36.

6. Dymnikov V.P. Numerical methods in geophysical hydrodynamics, 1984, m., the Department of numerical mathematics of the USSR Academy of Sciences.
7. Backlanovskaya V.F. The investigation of the grids method for the two-dimensional equations of the havier-stoks type with woh-negetive viseosity 11, 1984, The journal of numerical mathematics and mathematical physics, v. 24, N 12, pp. 1827-1841.
8. Dymnikov V.P., Skiba Yu.N. Barotropic instability of zonally nonsymmetric atmospheric flows. Preprint of the Dep. of numerical math., M., 1985, 58p.

SOME RECENT DEVELOPMENTS IN SPECTRAL METHODS

M. Y. Hussaini
ICASE
NASA Langley Research Center
Hampton, VA 23665

1. Introduction

Spectral methods consist of expanding the solution to a problem in terms of basis functions which are global, infinitely differentiable and preferably orthogonal [1,2]. This choice of basis functions is what distinguishes them from the finite difference and finite element methods. In the case of finite element methods, the domain is divided into small elements, and a basis function is specified in each element. They are thus local in character. The case with finite difference methods is similar.

In addition to the basis functions, a key element of spectral methods is the set of test functions or weight functions. The test functions are used to enforce minimization of the residual resulting from the substitution of the series expansion of the solution into the differential equation. The choice of test functions distinguishes between essentially two types of spectral methods -- spectral Galerkin, and spectral collocation. In the Galerkin approach, the test functions are usually the same as the basis functions, whereas in the collocation approach the test functions are translated Dirac delta functions. In other words, the Galerkin approach satisfies the differential equations in the least square sense. In the spectral collocation approach the equations are satisfied exactly at the selected, so-called collocation points. It should be noted that the basis functions are employed solely for the purpose of approximating derivatives. This approach, which became feasible with the advent of computers, is the easiest and the most efficient for nonlinear problems, and is the focus of the present discussion.

2. Basic Aspects

The principal advantage of spectral methods lies in their potential for rapidly convergent approximations. In practical terms, it means that they achieve accurate results with substantially fewer points than are required by typical finite difference methods. Suppose u_N is a numerical approximation to a function $u(x)$. With a given set of basis functions Φ_n, it takes the form

$$u_N(x) = \sum_{n=0}^{N} a_n \Phi_n(x). \tag{1}$$

The expansion coefficients a_n are obtained by enforcing the condition

$$u_N(x_j) = u(x_j). \tag{2}$$

where x_j are the selected, so-called collocation points, which are usually the extrema of Φ_N. Figure 1 provides a graphic distinction between a second-order accurate central difference approximation and a Legendre spectral approximation to the first derivative of the function

$$u(x) = 1 + \sin(2\pi x + \frac{\pi}{4}) \qquad \text{on } [-1, 1]$$

whose values are given at a finite number of grid points. The finite difference approximation for the derivative at the origin, for instance, is estimated by interpolating a parabola through the origin and the two adjacent points, and is thus local in character. The spectral approximation estimates the derivative of the original function by the derivative of the polynomial which interpolates all the available points. Note that the error of the finite difference discretization decreases as $1/N^2$, whereas the error of the spectral discretization decreases exponentially. In the case of a differential equation, a further step is involved, that of finding an approximation for the differential operator in terms of the grid point values $u_N(x_j)$.

Another advantage of spectral methods is their minimal phase error. Consider the periodic solution to the problem $u_t + u_x = 0$ with $u(x,0) = \sin(\pi \cos(x))$. Figure 2 shows the lagging phase of the finite difference solution, while the Fourier spectral solution has zero phase error. A fourth-order Runge-Kutta method is used for temporal discretization in both the cases. For realistic problems with variable coefficients or nonlinear terms, the phase error for spectral methods is, of course, nonzero, but still relatively small.

These are some of the essential aspects of spectral methods which make them the prevailing tool in the study of stability, transition, and turbulence. Some of the drawbacks which have inhibited their wider use are: 1) time-step restriction imposed by the standard spectral grid, 2) sensitivity to singularities and 3) restriction to simple geometry. Progress has been made on all counts [3]. The present work will be confined to the recent developments in overcoming the first obstacle.

3. Iterative Spectral Methods

For evolution problems, explicit time-stepping can be extremely inefficient. This is so because the typical time-step limitation for spectral methods is proportional to $1/N^2$ for the advection equation and $1/N^4$ for the diffusion equation (where N is the number of modes) [1]. Hence, implicit time-stepping becomes a necessity. This results in a set of algebraic equations, which are in general amenable to iterative solution techniques only. Also, elliptic equations governing practical problems virtually require implicit iterative techniques. As the condition number of the relevant matrices are large, preconditioned iterative schemes including multigrid procedures are the attractive choices. In this section, we discuss the fundamentals of iterative spectral methods with reference to an elementary example, and then their application to the three dimensional incompressible Navier Stokes equations for the study of stability and transition in a channel and a boundary layer.

For the purpose of illustration let us consider the equation,

$$u_x = f, \qquad (3)$$

periodic on $[0, 2\pi]$. For the Fourier method, the standard choice of collocation points is

$$x_j = \frac{2\pi j}{N}, \qquad j = 0, 1, 2, ..., N-1, \qquad (4)$$

Setting $u_j = u(x_j)$, the discrete Fourier series for u may be represented by the discrete transform pair

$$u_j = \sum_{p=-\frac{N}{2}}^{\frac{N}{2}-1} \hat{u}_p\, e^{ipx_j} \qquad\qquad j = 0, 1, ..., N-1$$

$$\hat{u}_p = \frac{1}{N}\sum_{j=0}^{N-1} u_j\, e^{-ipx_j}, \qquad p = -\frac{N}{2}, ..., \frac{N}{2}-1$$

The expression for the derivative u_x at the collocation points is

$$u_x(x_j) = \sum_{p=-\frac{N}{2}+1}^{\frac{N}{2}+1} ip\, \hat{u}_p e^{ipx_j}, \qquad j = 0, 1, ..., N-1$$

Thus the Fourier collocation discretization of the equation may be written

$$LU = F, \qquad (5)$$

where $U = (u_0, u_1, ..., u_{N-1})$, $F = (f_0, f_1, ..., f_{N-1})$, and $L = C^{-1}DC$ with C being the discrete Fourier transform operator, C^{-1} the inverse transform, and D the diagonal matrix denoting the first derivative operator in the Fourier space. Specifically,

$$C_{jk} = e^{-2\pi i k \frac{(j-N/2)}{N}}, \qquad j, k = 0, 1, ..., N-1 \qquad (6)$$

and

$$D_{jj} = i\,(j - N/2) \qquad\qquad \text{for } j = 1, 2, ..., N-1 \qquad (7)$$

$$= 0 \qquad\qquad \text{for } j = 0$$

The eigenvalues of L are $\lambda(p) = ip$, $p = -N/2 + 1, ..., N/2 - 1$, and the largest one grows as $N/2$. A preconditioned Richardson iterative procedure for solving Eq. (5) is

$$V \leftarrow V + \omega H^{-1}\,(F - LV) \qquad (8)$$

where the preconditioning matrix H is an approximation to L, is sparse, and is readily invertible. An obvious choice for H is a finite difference approximation L_{FD} to the first derivative. With the various possibilities for L_{FD}, the eigenvalue spectrum of $L_{FD}^{-1} L$ is given in Table 1. Apparently, the staggered grid leads to the most effective treatment of the first derivative. This kind of preconditioning was successfully used in the semi-implicit time-stepping algorithm for the Navier Stokes equations discussed in the section on Navier Stokes Algorithms. The eigenvalue trends of that complicated set of vector equations are surprisingly well predicted by this extremely simple scalar periodic problem.

Next, let us consider the second order equation

$$-u_{xx} = f \qquad\qquad \text{on } [0, 2\pi] \qquad (9)$$

with periodic boundary conditions. A Fourier collocation discretization of this equation is the same as Eq. (5) except for the diagonal matrix D which represents now the second derivative operator in the Fourier space.

$$D_{jj} = -(j - \frac{N}{2})^2, \quad j = 1, 2, ..., N-1$$

$$= 0, \quad j = 0$$

The eigenvalues of L are $\lambda(p) = p^2$, $p = -N/2 +1, ..., N/2-1$. To make the case for the multigrid procedure (consisting of a fine-grid operator and a coarse-grid correction) as a preconditioner, we assume H to be the identity matrix I in the iterative scheme (8). The iterative scheme is convergent if the eigenvalues, $(1 - \omega\lambda)$, of the iteration matrix $[I-\omega L]$ satisfy

$$|1 - \omega\lambda| < 1.$$

Each iteration damps the error component corresponding to λ by a factor $v(\lambda) = |1-\omega\lambda|$. The optimal choice of λ is that which balances damping of the lowest-frequency and the highest-frequency errors, i.e.,

$$(1 - \omega\lambda_{max}) = - (1 - \omega\lambda_{min})$$

This yields

$$\omega_{SG} = \frac{2}{(\lambda_{max} + \lambda_{min})},$$

and the spectral radius

$$\mu_{SG} = \frac{(\lambda_{max} - \lambda_{min})}{(\lambda_{max} + \lambda_{min})}.$$

In the present instance, $\lambda_{max} = N^2/4$, $\lambda_{min} = 1$, and thus $\mu_{SG} \approx 1 - 8/N^2$. This implies order N^2 iterations are needed for convergence. This poor performance is due to balancing the damping of the lowest frequency eigenfunction with the highest-frequency one. The multigrid procedure exploits the fact that the lowest-frequency modes ($|p| < N/4$) can be damped efficiently on coarser grids, and settles for a relaxation parameter value which balances the damping of the mid-frequency mode ($|p| = N/4$) with the highest-frequency one ($|p| = N/2$). Table 2 provides the comparison of single-grid and multigrid damping factors for $N=64$. The high frequencies from 16 to 32 are damped effectively in the multigrid procedure, whereas the frequencies lower than 16 are hardly damped at all. But then some of these low frequencies (from 8 to 16) can be efficiently damped on the coarser grid with $N=32$. Further coarser grids can be employed till relaxation becomes so cheap that all the remaining modes can be damped. In concrete terms, the ingredients of a multigrid technique are a fine-grid operator, a relaxation scheme, a restriction operator which interpolates a function from the fine grid to the coarse grid, a coarse-grid operator, and a prolongation operator interpolating a function from the coarse grid to the fine grid. The fine grid problem for the present example may be written

$$L^f U^f = F^f \tag{10}$$

Let V^f denote the fine-grid approximation. After the high-frequency content of the error $V^f - U^f$ has been sufficiently damped, attention shifts to the coarse grid. The coarse-grid problem is

$$L^c U^c = F^c \qquad (11)$$

where

$$F^c = R \, [F^f - L^f V^f],$$

R being the restriction operator. After a satisfactory approximation V^c is obtained, the coarse-grid correction $(V^c - RV^f)$ is interpolated onto the fine grid by the prolongation operator P, yielding the corrected fine-grid solution

$$V^f \leftarrow V^f + P \, (V^c - RV^f) \qquad (12)$$

The details of spectral multigrid techniques are furnished in [4]. Spectral multigrid techniques have been used to solve a variety of problems including the transonic full potential equation [5,6]. Additional applications of spectral methods to compressible flows are described in [7]. In the next section, we describe a multigrid algorithm for the incompressible Navier Stokes equations.

4. Navier Stokes Algorithms

This section is devoted to a description of recently developed algorithms for the simulation of instability and transition to turbulence in a flat-plate boundary layer. These algorithms deal with the primitive variable formulation of the Navier Stokes equations, and are based on the iterative methods discussed above in the simplest context. They are capable of handling geometric terms and variable viscosity.

The Navier Stokes equations in the so-called rotation form are

$$q_t = q \times \omega + \nabla \cdot (\mu \nabla q) - \nabla P \qquad \text{in } \Omega$$

$$\nabla \cdot q = 0 \qquad \text{in } \Omega$$

$$q(x,0) = q_0(x) \qquad \text{in } \Omega \qquad (13)$$

and

$$q = g \qquad \text{on } \partial \Omega$$

where $q = (u,v,w)$ is the velocity vector, $\omega = \nabla \times q$ the vorticity, $P = p + 1/2 \, |q|^2$ the total pressure, μ the variable viscosity, Ω the interior of the domain, and $\partial \Omega$ its boundary. In the stability and transition problems under study, the domain Ω is cartesian and semi-infinite: periodic in the two horizontal directions (x,z), and bounded by a wall at $y=0$. Fourier collocation can be used in the periodic directions (x,z), and Chebyshev collocation is used in the vertical (y) direction. The collocation points in the periodic directions are given by a relation similar to Eq. (4). The vertical extent of the domain $0 < y < \infty$ is mapped onto $-1 < \xi < +1$. The velocities are defined and the momentum equations enforced at the points

$$\xi_j = \cos(\frac{\pi j}{N_y}), \qquad j = 0, 1, ..., N_y$$

The pressure is defined at the half points

$$\xi_{j+\frac{1}{2}} = \cos[\frac{\pi \, (j+1/2)}{N_y}], \qquad j = 0, 1, ..., N_y - 1$$

and the continuity equation is enforced at these points. The staggered grid avoids artificial pressure boundary conditions, and precludes spurious pressure modes.

After a Fourier transform in x and z, the temporal discretization (backward Euler for pressure, Crank-Nicolson for normal diffusion, and third or fourth-order Runge-Kutta for the remaining terms) of Eqs. (13) leads to

$$[I - MDM] Q + \Delta t A_0 \nabla \Pi = Q_e \tag{14}$$

$$- A_+ \nabla \cdot Q = 0$$

where

$$Q = \{\hat{q}_0^{n+1}, \hat{q}_1^{n+1}, ..., \hat{q}_{N_y}^{n+1}\}$$

$$\Pi = \{\hat{p}_{1/2}^{n+1}, \hat{p}_{3/2}^{n+1}, ..., \hat{p}_{N-1/2}^{n+1}\} \tag{15}$$

$$\nabla = \{ik_x, \frac{\partial}{\partial y}, ik_z\}$$

M is the Chebyshev derivative operator, D the diagonal matrix with $1/2\mu\Delta t$ as its elements, and A_0 is the interpolation operator from the half points to cell faces, A_+ vice versa. Obviously, the equations for each pair of horizontal wave number (k_x, k_z) are independent, and they can be written as the system

$$LX = F$$

where $X = [Q, \Pi]$. The iterative solution of this equation is carried out by preconditioning the system with a finite difference approximation on the Chebyshev grid, and applying a standard iterative technique such as Richardson, minimum residual or multigrid [8].

The method described above solves the implicit equations together as a set. The extension of this method to the more general cases of interest such as those involving two or more inhomogeneous directions is not straightforward. An alternative is the operator-splitting technique or the fractional step scheme [9]. This method yields implicit matrices which are positive definite and are easily amenable to iterative methods. In the first step, one solves the advection-diffusion equation

$$q_t^* = q^* \times \omega^* + \nabla \cdot (\mu \nabla q^*) \tag{16}$$

subject to the initial and boundary conditions

$$q^*(x, t_n) = q(x, t_n).$$

$$q^* = g^* \qquad \text{on } \partial\Omega$$

Note that g* has yet to be defined. In the second step, one solves for the pressure correction

$$q_t^{**} = \nabla P^{**} \tag{17}$$

$$\nabla \cdot q^{**} = 0$$

subject to the conditions

$$q^{**}(x,t^*) = q^*(x,t^*) \qquad \text{in } \Omega$$

$$q^{**} \cdot \hat{n} = g \cdot \hat{n} \qquad \text{in } \partial\Omega$$

where $\hat{n}$ is the unit normal to the boundary. Further, the tangential component of the Eq. (17) holds on the boundary, i.e.,

$$q_t^{**} \cdot \hat{t} = -\nabla P^{**} \cdot \hat{t} \qquad \text{in } \partial\Omega$$

where $\hat{t}$ is a unit tangent vector to the boundary. Now g^* is defined [9] as (using Taylor expansion in t)

$$g^* \cdot \hat{n} = (g^n + \Delta t \, g_t^n) \cdot \hat{n}$$

$$g^* \cdot \hat{t} = [g^n + \Delta t \, (g_t^n + \nabla P^n)] \cdot \hat{t}.$$

Eq. (16) is discretized in the usual spectral collocation manner. After a temporal and spatial discretization of Eq. (17), the boundary conditions are built into the relevant matrix operators, and then a discrete divergence is taken. This results in a discrete Poisson equation (with as many algebraic equations as unknowns) for pressure, which can be solved by standard iterative techniques including the multigrid method.

5. Applications

These algorithms have been used to study the incipient stages of the transition process in channel flows [8] and parallel boundary layer flows [10]. Some representative results are provided here. The channel flow results pertain to the secondary instability associated with the so-called center modes. Unlike the Tollmien-Schlichting modes (sometimes alluded to as wall modes), the center modes always decay with a rather high decay rate. Their phase velocity is near unity and their maxima occurs near the center of the channel. The simulation had a Reynolds number of 5000 based on half-channel width, and the initial conditions consisted of a 20% amplitude two-dimensional center mode with two 1.5% amplitude skewed modes. The harmonic contents of the solution were monitored, and the grid was refined as deemed necessary. The finest grid was 144×96×108. Plotted in Figure 3 are streamwise vorticity (left side) and spanwise vorticity (right side) contours on the streamwise planes at $x = 3/8, 7/16, 1/2$ and $9/16$ of the fundamental wavelength. The so-called peak plane would intersect these streamwise planes along a vertical line in the center of the frames. The structure of the vortex loop can be deduced from these plots. It differs in detail from that of the wall modes. The vortex structures are significant over only a small portion of the wavelength in the streamwise direction, whereas in the case of wall modes they cover almost the whole wavelength. Furthermore, the pinching of the vortex loop in the peak plane appears to be less acute in the case of the center modes. The harmonic history is displayed in Figure 4. The evolution of the secondary instability is apparent and it appears similar to that of the Tollmien-Schlichting modes. What is more interesting is the steep growth of the (0,2) and (2,2) modes which may lead to a strong tertiary instability. To resolve it in detail would require an even finer grid.

The parallel heated water boundary layer cases had a Reynold number of 1100 based on the displacement thickness, and the initial amplitudes of the two-dimensional and three-dimensional Tollmien-Schlichting waves were 2.7% and 0.4% respectively.

Three different situations were studied: 1) uncontrolled case, 2) heated fixed temperature case, and 3) heated active temperature case. In the heated fixed temperature case, the temperature was kept fixed at the initial value pertinent to the mean flow conditions, and the temperature evolution was totally neglected. In the heated active temperature case, the temperature evolution was taken into account by solving the temperature equation along with the momentum equations. In both the cases the wall temperature was 2.75% above the free stream temperature. Figures 5,6 and 7 display the harmonic histories. The fixed temperature case overpredicts the weakening effect of heating on the secondary instability. Figure 8 shows the spanwise vorticity contours on the peak plane. In the uncontrolled case (Figure 8 top left) a kink develops in the high-shear layer at time t equal to three Tollmien-Schlichting periods. It is generally accepted that a irrevocably quick succession of events follows thereafter leading to a turbulent spot formation. Heating the wall to 2.75% above the free stream temperature diffuses the high-shear layer as is obvious from Figure 8 (bottom left). However, within the subsequent one and one fourth period, turbulent spot formation appears to become imminent (Figure 8 top right). In the fixed temperature case, it is clear from Figure 8 (bottom right) that the high-shear layer formation is mellowed down even up to four and one fourth periods. This shows that the effect of temperature evolution is significant and deleterious in the nonlinear regime whereas it is quite negligible in the linear regime.

Acknowledgments

The assistance of D. L. Dwoyer, G. Erlebacher and T. A. Zang is appreciated.

References

1. Gottlieb, D., Orszag, S. A. 1977. "Numerical Analysis of Spectral Methods: Theory and Applications," CBMS-NSF Regional Conference Series in Applied Mathematics, SIAM.

2. Canuto, C., Hussaini, M. Y., Quarteroni, A., Zang, T. A. 1987. *Spectral Methods in Fluid Dynamics*, Springer Verlag.

3. Hussaini, M. Y., Zang, T. A. 1987. "Spectral methods in fluid dynamics," *Annual Review of Fluid Mechanics*, volume 19.

4. Zang, T. A., Wong, Y-S, Hussaini, M. Y. 1982. "Spectral multigrid methods for elliptic equations," *J. Comp. Phys.* 48:485-501 and 54:489-507.

5. Streett, C. L., Zang, T. A., Hussaini, M. Y. 1985. "Spectral multigrid methods with applications to transonic potential flow." *J. Comput. Phys.* 57:43,76.

6. Hussaini, M. Y., Salas, M. D., Zang, T. A. 1985. "Spectral methods for inviscid, compressible flows," *Advances in Computational Transonics,* pp.875-912, Habashi, W. G. (ed.), Pineridge Press, Swansea.

7. Hussaini, M. Y., Kopriva, D. A., Salas, M. D., Zang, T. A. 1985. "Spectral methods for Euler equations: Part 1. Fourier methods and shock-capturing," *AIAA J.* 23:64-70. "Part 2. Chebyshev methods and shock-fitting," *AIAA J.* 23:234-40.

8. Zang, T. A., Hussaini, M. Y., 1985. "Numerical Experiments on subcritical transition mechanisms," AIAA Paper No. 85-0296.

9. Zang, T. A., Hussaini, M. Y. 1986. "On spectral multigrid methods for time-dependent Navier-Stokes equations, *Appl. Math. Comp.*, to appear.

10. Zang, T. A., Hussaini, M. Y. 1985. "Numerical experiments on the stability of controlled shear flows," AIAA Paper No. 85-1698.

Table 1. Preconditioned Eigenvalues for One-dimensional First Derivative Model Problem

Preconditioning	Eigenvalues	
Central Differences	$\dfrac{k\Delta x}{\sin(k\Delta x)}$	
High Mode Cutoff	$\dfrac{k\Delta x}{\sin(k\Delta x)}$	$\|k\Delta x\| \leq (2\pi/3)$
	0	$(2\pi/3) < \|k\Delta x\| \leq \pi$
One-sided Differences	$e^{-i(k\Delta x/2)} \cdot \dfrac{k\Delta x/2}{\sin((k\Delta x)/2)}$	
Staggered Grid	$\dfrac{(k\Delta x)/2}{\sin((k\Delta x)/2}$	

Table 2. Damping Factors for N = 64

p	Single-Grid	Multigrid
1	.9980	.9984
2	.9922	.9938
4	.9688	.9750
8	.8751	.9000
12	.7190	.7750
16	.5005	.6000
20	.2195	.3750
24	.1239	.1000
28	.5298	.2250
32	.9980	.6000

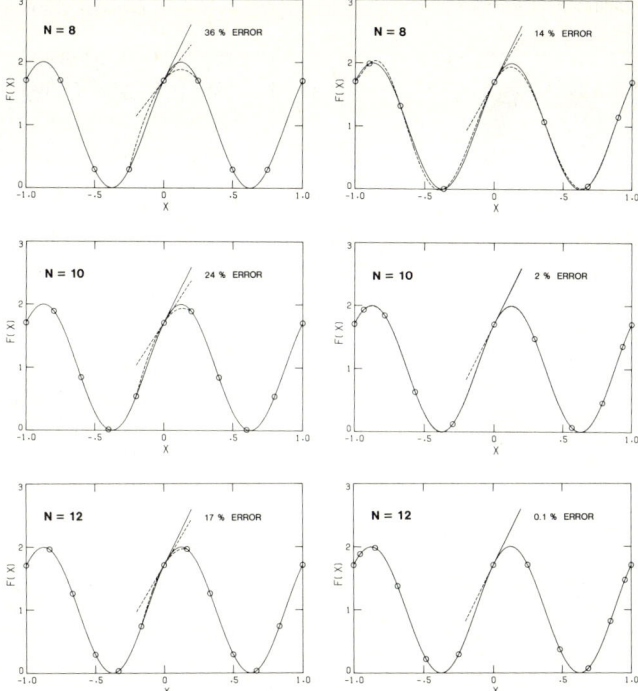

Figure 1: Comparison of finite difference (left) and Chebyshev spectral (right) differentiation. The solid curves represent the exact function and the dashed curves their numerical approximations. The solid lines are the exact tangents at x = 0 and the dashed lines the approximate tangents. The error in slope is noted as is the number of intervals N.

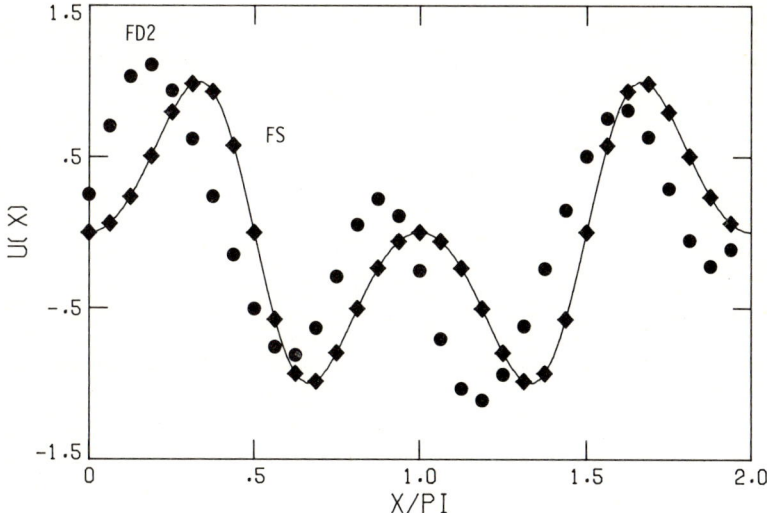

Figure 2: Finite difference (FD2) and Fourier spectral (FS) approximations after one period to a simple wave equation whose exact solution is represented by the curve.

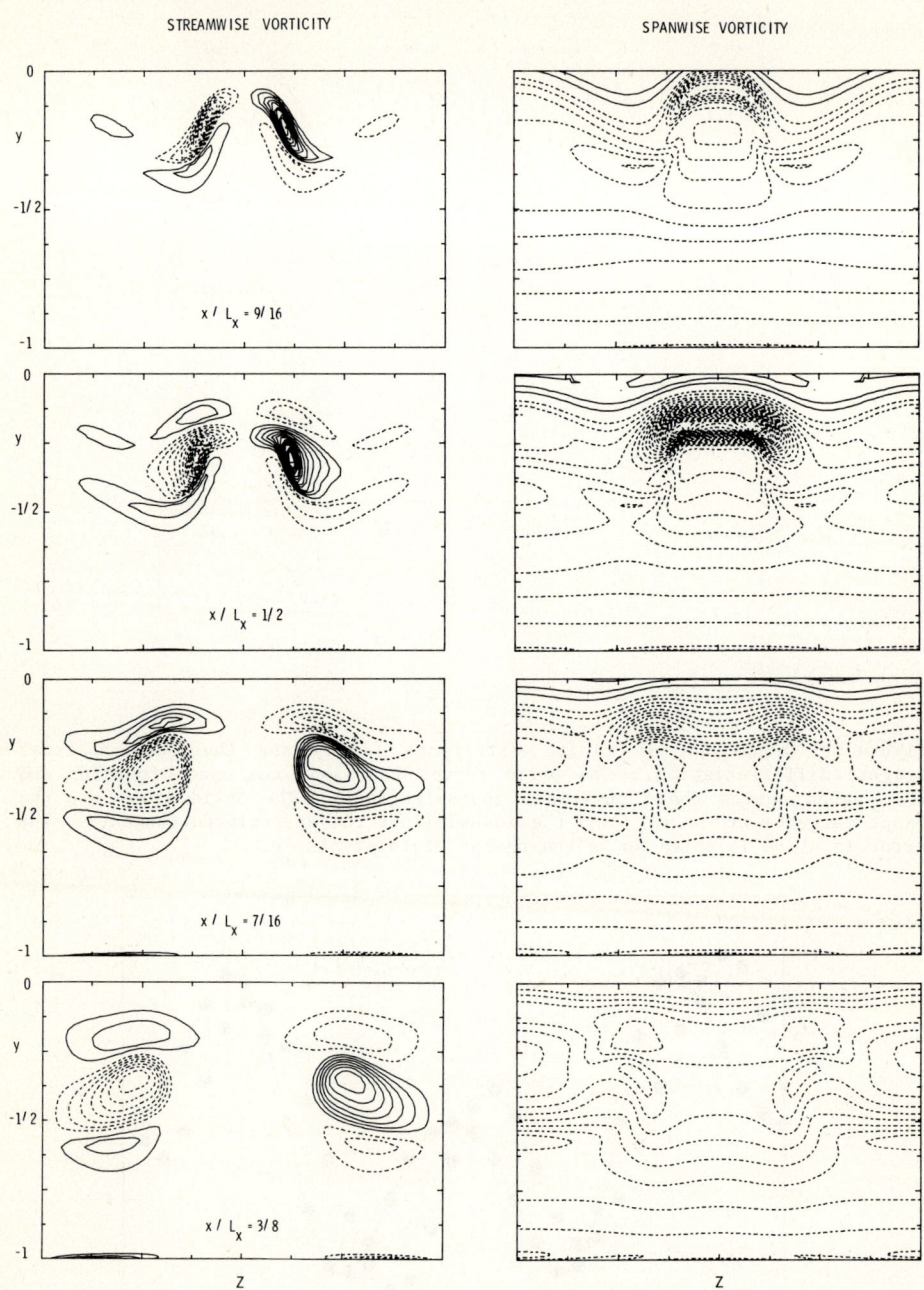

Figure 3: Streamwise (left) and spanwise (right) vorticity at four streamwise locations for a channel flow center mode transition. Only the lower half of the channel is shown.

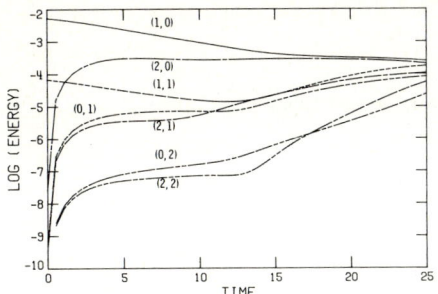

Figure 4: Harmonic history for a Re = 5000 center mode simulation.

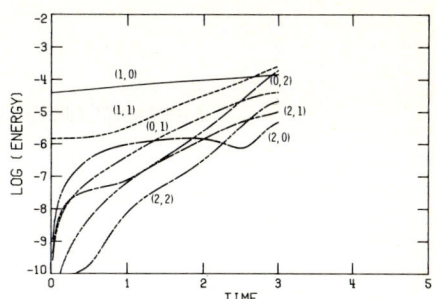

Figure 5: Harmonic history for a Re = 1100 boundary layer.

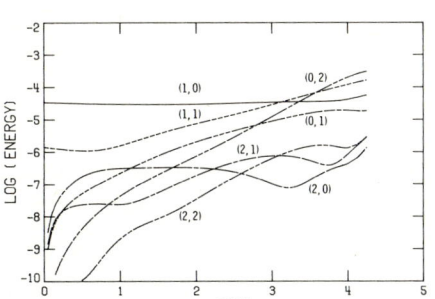

Figure 6: Harmonic history for a Re = 1100 heated boundary layer.

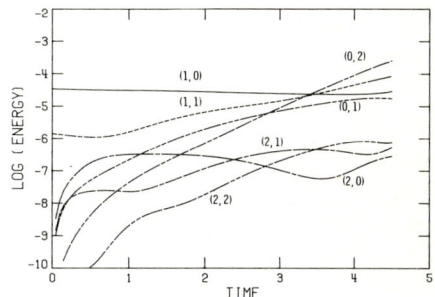

Figure 7: Same as Fig. 6 but for fixed temperature.

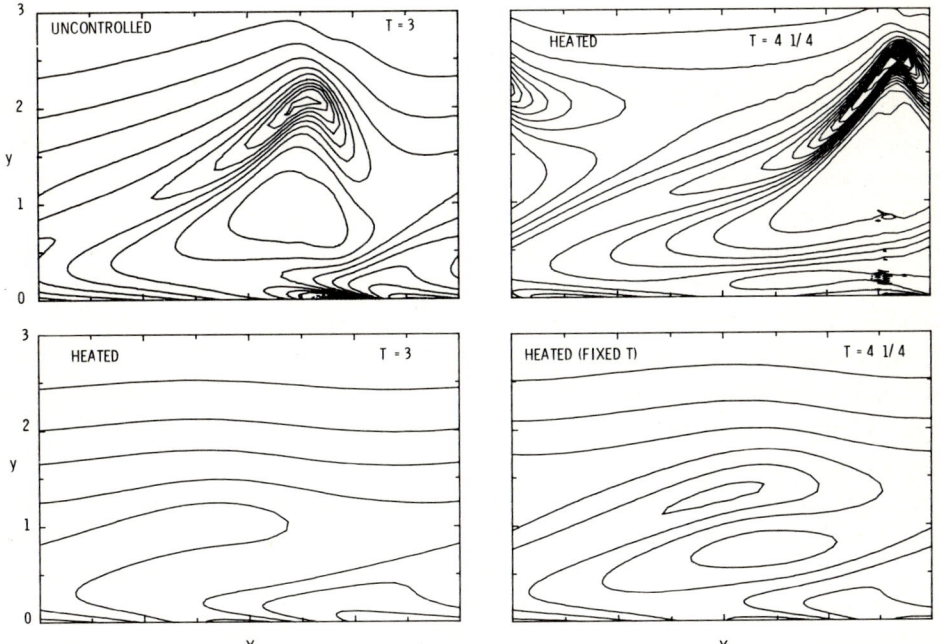

Figure 8: Vertical shear in the peak plane for Re = 1100 boundary layer simulation.

A PERSPECTIVE OF COMPUTATIONAL FLUID DYNAMICS

Paul Kutler
NASA Ames Research Center
Moffett Field, California 94035, U.S.A.

SUMMARY

Computational fluid dynamics (CFD) is maturing, and is at a stage in its technological life cycle in which it is now routinely applied to some rather complicated problems; it is starting to create an impact on the design cycle of aerospace flight vehicles and their components. CFD is also being used to better understand the fluid physics of flows heretofore not understood, such as three-dimensional separation. CFD is also being used to complement and is being complemented by experiments. In this paper, the primary and secondary pacing items that governed CFD in the past are reviewed and updated. The future prospects of CFD are explored which will offer people working in the discipline challenges that should extend the technological life cycle to further increase the capabilities of a proven and demonstrated technology.

INTRODUCTION

Fifteen years ago computational fluid dynamics (CFD) was in its infancy, and on the first part of its technological life cycle curve (fig. 1). The facets that comprise the discipline of CFD, such as algorithm development, grid generation, geometry definition, boundary and initial conditions, turbulence modeling, pre- and post-data processing, computer technology, etc., were all ripe for technological advances. Many simple problems, easily amenable for solution using CFD, were unsolved. At the same time, there were very few researchers working in the discipline and on its various facets (fig. 2). Also, very little computer power was devoted to, or available for, solving CFD problems. Because of the demonstrated potential of CFD, and the understood limitations of experimental testing (see Chapman, Mark, and Pirtle (ref. 1)), the discipline of CFD developed at a rapid pace. Today, because of the manpower and computer resources devoted to CFD, technological advances in the discipline have matured to the point at which CFD is becoming routine, and thus near the growth peak of its technological life-cycle curve. Most of the simple problems have been solved, and only the very complex and difficult ones remain. To extend the life-cycle curve of CFD, it will be necessary, for example, to seek new disciplines that can be coupled with CFD, to utilize CFD for

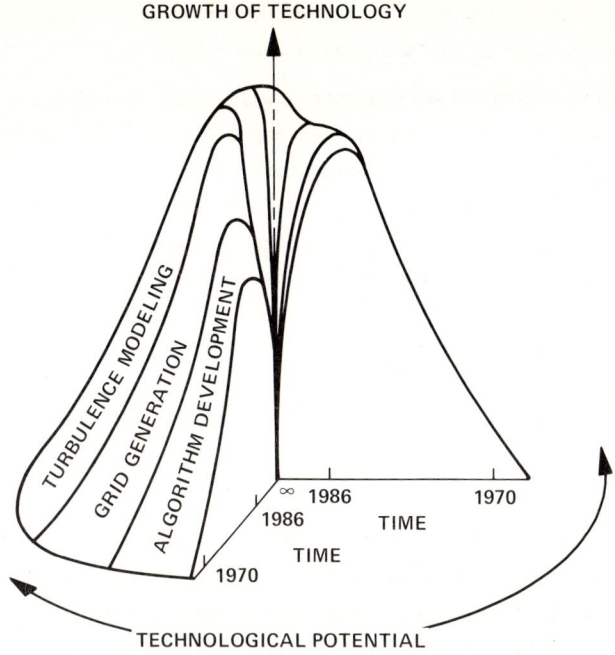

Figure 1. Rate of growth of CFD technology.

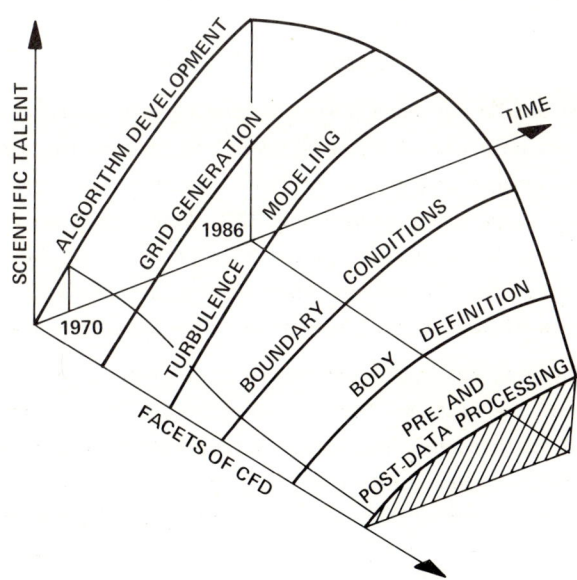

Figure 2. Scientific talent devoted to facets of CFD as a function of time.

understanding or discovering new fluid-flow phenomena, or to apply CFD to new aeronautical challenges offered by future aerospace vehicles.

In the future, aerospace manufacturers will rely extensively on numerical simulations because of 1) the demonstrated capability of CFD, 2) the increasing number of supercomputers available for performing CFD simulations, 3) the lack of ground-based experimental facilities in the flow regimes of interest, and 4) the saturation of existing experimental facilities. Instead of simply utilizing national laboratories for their unique experimental facilities, vehicle designers will be able to perform corresponding numerical simulations on national computational facilities, thereby complementing their experimental test programs.

STATUS OF PRIMARY AND SECONDARY PACING ITEMS

Chapman (ref. 2), in 1981, outlined pacing items for CFD that included three-dimensional (3D) grid generation, turbulence modeling, algorithm development, and computer-mainframe design advances. In a 1983 AIAA Paper (later published in the AIAA Journal), Kutler (ref. 3) updated the list and classified the pacing items according to primary and secondary items. The primary items included grid generation, turbulence modeling, computer power, and solution methodologies; the secondary items included algorithm development, complex-geometry definition, and pre- and post-data processing. In this section, the pacing items of the past are reviewed and updated, and the major accomplishments are summarized.

Computer Technology

The continued demand for more powerful computers by computational fluid dynamicists to perform not only direct Navier-Stokes simulations for simple geometries, but also Reynolds-averaged Navier-Stokes (RANS) calculations for complex 3D configurations, continues to pace the progress of CFD. Also, the problems of the future (e.g., in hypersonics) that require the solution not only of the Navier-Stokes equations, but also of the finite-rate chemistry equations, will require a computer that operates orders of magnitude faster than those currently available (fig. 3).

The balance between the speed and memory of a given computer is essential for the effective use of the machine. Too much memory without adequate speed, or too little memory with an abundance of speed, is inefficient. A summary of the existing and planned scientific supercomputing systems is shown in table 1. It is, however, safe to say that, regardless of the power offered by the latest state-of-the-art computer, the computational fluid dynamicists will easily saturate it and demand more power.

A simple advantage that computers/software have over wind tunnels in performing simulations is that they can be continually and inexpensively upgraded to enhance their capabilities over time whereas the wind tunnels cannot. Computing centers such

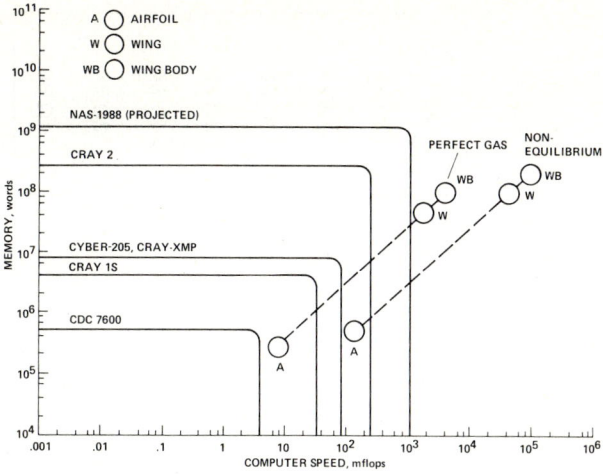

Figure 3. Computer speed and memory requirements (15-min runs, 1985 algorithms).

as the Numerical Aerodynamic Simulation (NAS) Facility, located at NASA Ames, and the National Science Foundation (NSF) centers, distributed around the United States, are cheaper to build and to upgrade than are new wind tunnel facilities. Computational centers, and the simulations performed with them, can help eliminate the saturation of existing wind tunnels, and thus make the use of these experimental facilities more effective and efficient.

Turbulence Physics and Modeling

The simulation of viscous flows is being attacked from both ends of the computational spectrum. At one end are the scientists who solve the RANS equations with a suitable turbulence model derived from theoretical analysis and data obtained from building-block experiments. At the other end of the spectrum are the scientists who solve the complete Navier-Stokes equations with either a sub-grid model for the small scales of turbulence (large eddy simulation), or no model at all (direct simulation). As time progresses and CFD and computer technology mature, the two approaches will merge.

The development of suitable turbulence models for the RANS equations to date remains highly problem-dependent. According to Marvin (ref. 4), to improve the turbulence-model development process, it should be considered at the early stage of code development. The first step in such a process is to identify those flows pacing the development of the aerodynamic computations. The second step is to develop models through a phased approach of building-block studies that combine theory,

TABLE 1. Characteristics of existing and planned computer systems.

SYSTEM	CLOCK CYCLE, nsec	MAIN MEMORY SIZE, megabytes/64-bit megawords	SECONDARY MEMORY SIZE, megabytes/64-bit megawords	MAXIMUM/ SUSTAINED VECTOR RATE, mflops[a]	NO. OF PROCESSORS	NO. OF PIPES/PROC.	DATE AVAILABLE
CRAY XMP	9.5	128/16	1024/128	760[b]/400[c]	4	1	85
CRAY 2 (C2)	4.1	2048/256	—	1620[b]/400[c]	4	1	2Q 85
CRAY YMP	NA[d]	NA	NA	3XC2[e]	NA	NA	2Q 87
CRAY 3	NA	NA	NA	10XC2[e]	NA	NA	1Q 88
ETA 10	7.0	256/32	2048/256	5000/2000[c]	NA	2	3Q 86
AMDAHL 1400	7.5	256/32	—	1133/NA	1	4	4Q 84
HITACHI S820/20	14	256/32	1024/128	630/NA	1	1	3Q 84
NEC SX-2	6	256/32	2048/256	1300/NA	1	4	1Q 86

[a] DATA FOR FULL 64-BIT PRECISION
[b] DEMONSTRATED PERFORMANCE
[c] ESTIMATE FOR OPTIMIZED CFD CODE
[d] INFORMATION NOT AVAILABLE
[e] ESTIMATED PERFORMANCE IMPROVEMENT OVER CRAY 2

experiment (requiring new instrumentation for extracting refined data), and computations. The final step is to provide verification and/or limits of the modeling through benchmark experiments over a practical range of Reynolds and Mach numbers.

The application of different turbulence models in different regions of the flow (e.g., in a zonal approach), might be required to obtain better accuracy of the simulation process. Also, as the Mach number of the free-stream flow increases, the incompressible turbulence models developed to date will probably not work. Thus it will be necessary to develop new models, including previously neglected Reynolds stress terms, for treating these flows.

As the direct numerical simulations progress, it is becoming possible to utilize the numerical data generated from those solutions to create turbulence models. Such data provides much more information than can be gleaned from an experiment, thus permitting the construction of better turbulence models. The limitation is with the simplified and low Reynold's number flows attainable by direct simulations.

Solution Methodology Development

Two noteworthy solution methods that have resulted in advancements of the state of the art in CFD include a zonal procedure developed by Rai (ref. 5) for treating a rotor-stator combination (i.e., a multiple moving-body problem), and a tetrahedron procedure developed by Jameson (ref. 6) for treating commercial aircraft configurations. Rai's procedure is applicable for moving bodies and required the development of boundary condition procedures for transmitting information from one moving grid to the other.

Jameson's procedure permits the easy treatment of complicated configurations because the flow region of interest is discretized using tetrahedrons. He developed a boundary-condition procedure for conserving fluxes across the volume surfaces that did not degrade the accuracy of the solution. He also developed a data-management system for handling the randomly ordered control volumes.

The results from both of these procedures are formidable. A typical viscous-flow result for the rotor stator is shown in figure 4, for blunt-nose blades. The instantaneous velocity vectors, with the free stream subtracted out, are displayed, clearly showing the vortex pattern behind the blades.

To facilitate the use of one's computer program by aerospace vehicle designers, code developers should strive to make the code robust. In essence, this means that, with very few limitations, the program should be capable of yielding results of which the accuracy is predictable, and with little intervention from the user. It should have few flow restrictions, except for the limits of applicability of the governing equation set, and should be capable of treating complex and varied geometries.

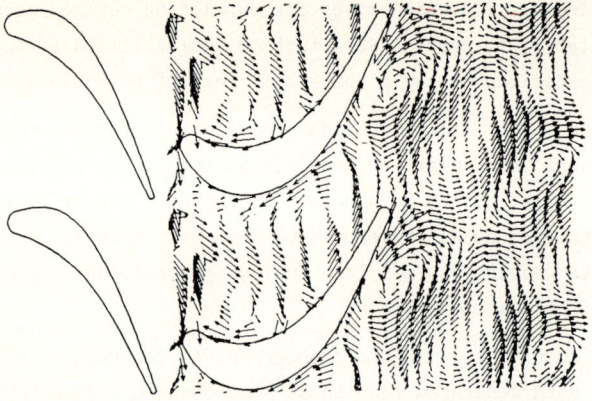

Figure 4. Instantaneous-velocity vector for rotor-stator configuration.

Algorithm Development

Considerable human talent has been devoted to the development of more accurate and faster algorithms for solving the gas-dynamic equations, and has resulted in considerable progress. There is, however, still more progress to be made in this facet of CFD. The current set of equations being attacked by algorithm developers is the Navier-Stokes (including both complete and Reynold's-averaged). For unsteady solutions of these equations, existing algorithms are capable of efficiently obtaining solutions; however, more improvement is still possible in developing algorithms for obtaining convergence to the steady state.

Also, to devise schemes for faster convergence, developers are striving for more accuarate and robust algorithms, particularly in the area of shock-capturing properties. They are trying to minimize the number of free parameters that users must select to utilize an algorithm optimally. Total variational diminishing (TVD) and upwinding schemes satisfy this criterion. On a per-point basis, however, such schemes require more computer time, but the results they yield are worth this time. Figure 5 shows the TVD results of Yee (ref. 7) for a planar blast wave passing over an airfoil. These results clearly depict the intricate wave pattern of the flow, including the slip surface generated from the triple point near the trailing edge of the airfoil. This solution was obtained without the use of a solution-adaptive grid, but with such a grid, the results could be enhanced even further.

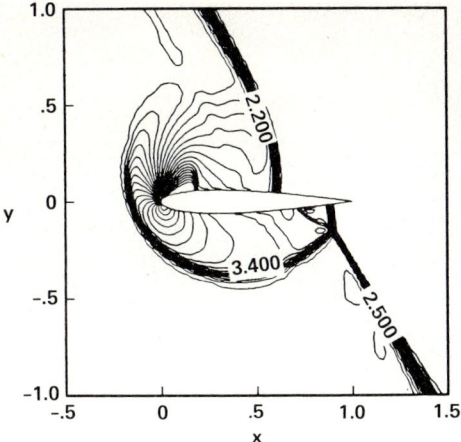

Figure 5. Density contours for blast wave striking an airfoil.

The supercomputers now available are causing some researchers to explore old procedures for solving the Navier-Stokes equations that some time ago seemed impractical because of computer limitations. Beam and Bailey (ref. 8) are looking at Newton's method, which does not involve any approximation, as do most conventional methods. Such a procedure involves the inversion of a rather large matrix, but could be used to evaluate the effects of approximations made in other procedures.

Geometry Definition

Considerable progress has been made in the geometry-definition facet of CFD (i.e., translating a complicated configuration into data understandable by the flow solver), but it is still not a routine process. As the sophistication of computer programs increases for treating complicated configurations, and their routine use by designers also increases, the demand for easily applied geometry-definition procedures will also increase. To date, flows about configurations such as commercial aircraft (ref. 6), fighter planes (ref. 9), spacecraft, and complicated components of those vehicles (ref. 10) are being computed and hence require such a geometry tool. Figure 6 is typical of the complex configurations being studied today. The amount of time required to input the geometry describing configurations like these into the computer, however, is on the order of several months.

It is important that this process becomes routine (i.e, require only a few hours, and possibly only minutes), and thus not slow the progress of CFD. This will require the development of sophisticated software. Such software should be designer-friendly for versatile use and employ high-level computer graphics.

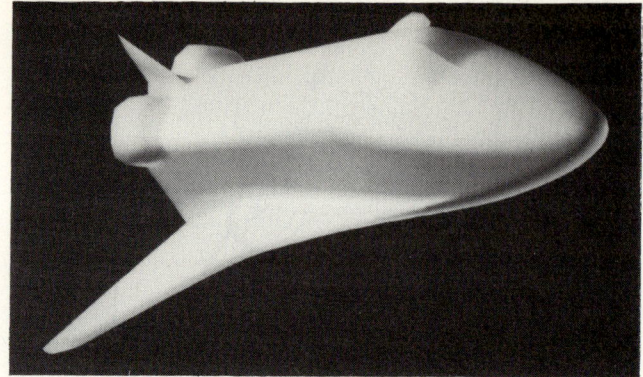

Figure 6. Computer-generated geometry for Space Shuttle.

Grid Generation

The process of grid generation has received considerable attention by scientists because of the need to efficiently and effectively distribute grid points to generate the most accurate solution possible. To date, complicated configurations have been treated computationally. Examples are shown in figures 7 and 8, in which different grid topologies have been used to discretize the flow about the Space Shuttle (a single module grid) and a generic aircraft (a multiple module or block grid). The grid for the Shuttle used a hyperbolic solver, whereas the grid for the aircraft used an elliptic solver.

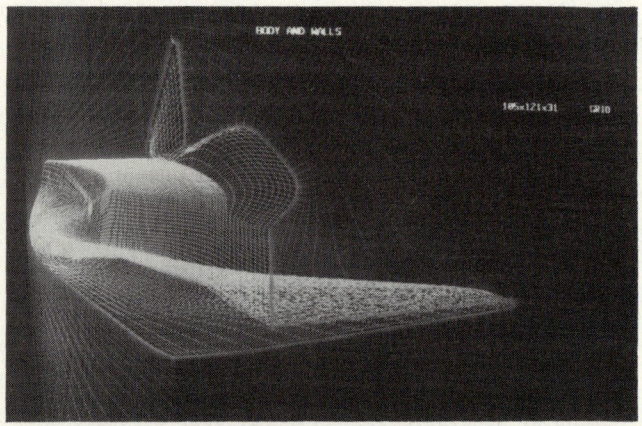

Figure 7. Single module grid for Space Shuttle.

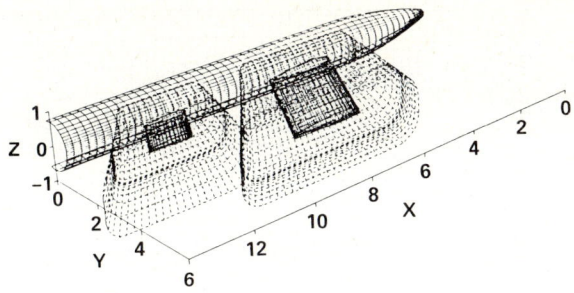

Figure 8. 3D wing/body/tail overset grid configuration.

Because of the CFD scientist's desire for accuracy and efficiency, solution-adapative grid procedures have gained popularity. Typical of the grid enhancements possible using such procedures are the results of Nakahashi and Deiwert (ref. 11), as shown in figure 9 for a two-dimensional airfoil.

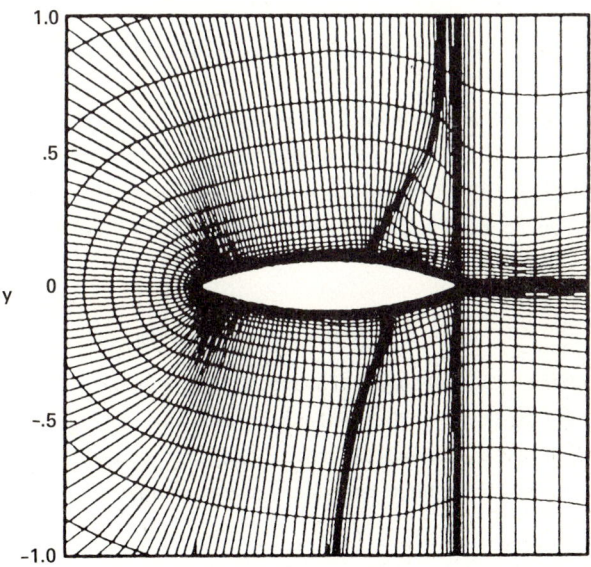

Figure 9. Solution-adaptive grid for airfoil with buffet.

Unsteady problems involving the motion of one body relative to another have created another set of grid-generation problems. An example of such a problem is the store drop from an aircraft. To treat this problem, a component-adaptive, overlapping grid system is used. A typical grid generated by Dougherty (ref. 12) for this problem is shown in figure 10.

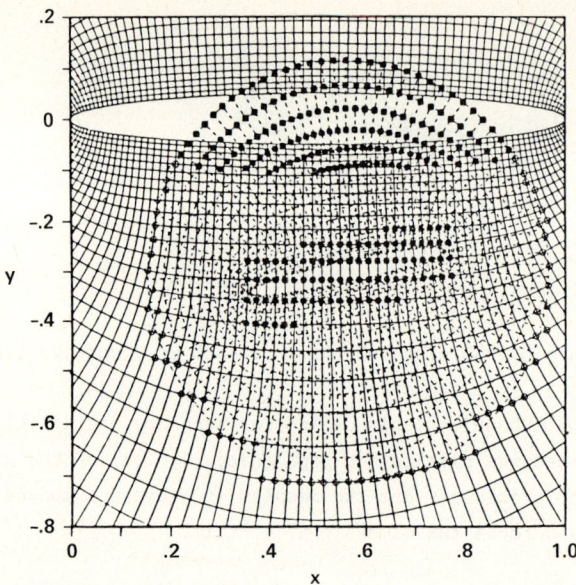

Figure 10. Overset grid for store-separation simulation.

The automation of the grid-generation procedures to facilitate discretizing the flows about complicated configurations should be of paramount importance to those working in this facet of CFD. This is an area in which expert systems might help to eliminate the need for routine application of conventional grid-generation procedures.

Pre- and Post-Data Processing

With the placement of supercomputers around the world, and the use of these machines for solving complicated 3D problems, comes the necessity for managing enormous amounts of data. To accomplish this, the user community has relied heavily on high-resolution, high-throughput computer-graphics devices.

Sophisticated software packages (e.g., that by Buning (ref. 13)), have been designed which permit the viewing of the results in either static or dynamic motion for the analysis and understanding of the flow-field data. Such software packages, however, are passive in the sense that they display only what they are told to display. In such a process, it is possible that some of the interesting or undiscovered fluid physics might be masked. Therefore, what is needed is an active or "smart" software display package that searches the data base for interesting flow phenomena and displays them. This would require, for example, the program to look for different combinations of the flow variables and their gradients, or derivatives of their gradients, to uncover interesting regions of the flow that might be lurking in the small-scale portions of the mesh, such as secondary, separated flow regions, and call the viewers attention to it and display it.

FUTURE PROSPECTS FOR CFD

Numerical simulations can now be performed on many complex configurations. The capability of CFD is such that many complicated problems can be solved if the resources are channeled into the effort. Scientists with the freedom to select their computational research tasks are now faced with the decision as to which problems to solve. To aid in their decision, certain criteria might be suggested, such as 1) is the problem of national importance, 2) will its solution lead to a new design tool, 3) will it aid in the understanding of complex fluid physics or the discovery of new flow phenomena, 4) will it push the state of the art in computational fluid dynamics, and 5) is the problem tractable in a finite amount of time.

The design and construction of future CFD-applications software is becoming continually more complicated because of the complex problems being addressed, requiring teams of researchers. Because of the complexity of the codes, it is critical that the eventual user be involved in the software development stage. This requires that the code builders get "close to the customers." The customers can make constructive suggestions in the program's design, can familiarize themselves with the program, and will thus be more willing to use it when it is completed. Because of the involvement of the customer, the resulting code will be "designer-friendly." Subsequent use of the code in the vehicle-design process, however, will depend on the confidence level designers have of the code and the predictable accuracy of the results generated by the code. That confidence level is enhanced by involving the vehicle designer in the program's development.

Because of the complexity of CFD software, it is vital that the proper program documentation exists. It should not be the duty of the research scientist who conceived the program to provide the documentation (although it should be his or her responsibility), but rather a programmer well versed in such duties who works with the scientist in the development process. The understanding and use of complex codes can also be taught by the computer. Expert systems can easily be constructed to train potential users how to efficiently utilize complex CFD software.

A considerable number of challenging technical areas exist for which CFD will be beneficial, and sometimes mandatory. Two areas of particular importance that will be addressed here are unsteady flows and interdisciplinary physics.

Unsteady-flow problems result from the instability of separated flow regions, or from the relative motion of one body with respect to another. Two problems in which relative body motion causes unsteady flow are the helicopter and rotating turbomachinery. Both problems satisfy most of the criteria outlined at the beginning of this section. To date, neither problem has been simulated computationally (only components of the problem), but formidable efforts are under way that will lead to their eventual numerical simulation.

Researchers such as Davis and Chang (ref. 14), and McCroskey and Bader (ref. 15), are developing Euler or RANS programs for computing the unsteady flow about helicopter rotors. Unsteady problems such as the blade-vortex interaction have been simulated using these codes (fig. 11). Inclusion of the helicopter fuselage and tail rotor in these calculations poses considerable challenge for the scientists, but the problem is not technology-limited, and will eventually be simulated using CFD. It will thus lead to a better understanding of helicopter aerodynamics and improved designs.

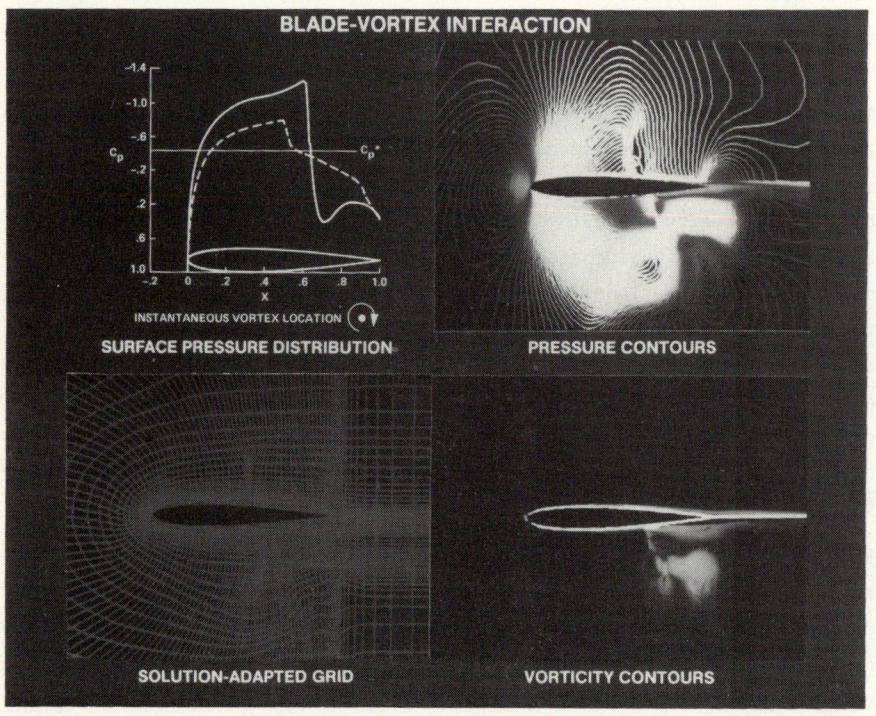

Figure 11. Two-dimensional blade-vortex interaction.

The rotor-stator problem in two dimensions has been successfully simulated by Rai (ref. 5). The solution of this problem required advancements in the state of the art in CFD regarding boundary conditions and grid generation. Extension of this technology to three dimensions is formidable but tractable. Simulation of the 3D problem could yield solutions for flows through propellers, pumps, compressors, and turbines, and eventually lead to the development of more efficient propulsors and jet engines.

The simulation of unsteady viscous flows about realistic aircraft configurations is now possible using computational tools. Several computer codes for simulating these flows have been developed by various researchers throughout the United States. With these codes, it will now be possible to begin studying unsteady flow problems that result when high-performance aircraft/spacecraft fly at large angles of attack. These unsteady flows include assymetric vortex shedding and vortex breakdown or bursting. It will also be possible to use these codes for predicting aircraft/spacecraft performance near their performance boundaries. Numerical results by Rizk (ref. 16) for flow over the Space Shuttle at Mach 1.4 and 0° angle of attack, are shown in figure 12, and demonstrate the capability of today's technology.

Figure 12. Pressure contours from viscous solution for Space Shuttle.

Interdisciplinary physics will not only offer significant challenges to the research scientist, but also challenge the power of existing or planned computational facilities. The mating of different technical disciplines into one computer program for more relevant design applications should tax both the scientist and the machine. In the past, simple couplings have occurred, such as linking a structural response code to a fluid dynamics code to study aeroelastic problems (see Goorjian, et al. (ref. 17)), or the coupling of a flow code with an optimization routine for wing design (see Cosentino (ref. 18)). In the future, other disciplines will be linked with flow codes such as propulsion and controls.

As the need grows for high-speed flight, the gas-dynamic equations routinely solved today governing those flows will increase in complexity because of strong shocks and thermal and chemical nonequilibrium phenomena. Fluid dynamicists and chemists will begin working together to couple their disciplines and study problems involving dissociation and ionization, reaction rates, radiation physics, and equilibrium constants. Efficient algorithms for computing such flows will have to be invented to treat "stiff" equations. These inderdisciplinary equation sets will challenge algorithm developers in the future.

Turbulence models, based on compressible flow theory, will have to be developed because the existing models, based on incompressible flow, break down for high-speed flight. In addition, it will be possible to begin using numerically generated data (i.e., from direct simulations) to extract data required for the development of turbulence models. This is happening on a limited basis for incompressible flows. CFD will play an important role in the high-speed flight regime because of the lack of ground-based experimental facilities.

The introduction of high-speed flight and its associated problems to the CFD community will serve to extend the rate of growth of the technological life cycle of the discipline. This will offer scientists working in the various facets of CFD opportunities and challenges similar to those that existed over the last decade for transonics.

It is important that in the future fluid dynamicists, whether they use computational or experimental tools for their trade, work closely together. The synergy to be garnered is too valuable not to take advantage of such a cooperative arrangement. Computer codes require validation experiments, and experiments require supplemental computations. Research laboratories, whose basic product is information, and which possess the facilities for experimental, computational, and flight testing, will offer the greatest possiblity for synergy, and will produce the most valuable technical product.

CONCLUSIONS

Computers are assuming an increasingly important role in aerospace research and development. Considerable resources (manpower and computer) have been channeled into CFD. As a result, CFD is rapidly becoming an extremely powerful tool in the design process, as well as in the understanding of complex fluid physics. Substantial payoffs in both areas have been demonstrated. On the other hand, the rate of technological growth in CFD is naturally decreasing because of the substantial resources devoted to its development. It is therefore mandatory that new challenges be offered to the CFD research scientist to extend the life cycle rate of growth of CFD technology. Unsteady flow problems and interdisciplinary physics can serve this purpose. Supercomputing centers such as NAS, located at NASA Ames Research Center, and the NSF Centers, located throughout the United States, will play a key role in advancing the state of the art in CFD, and will be a critical element in the national base of aeronautical facilities.

Lack of uniqueness, or the narrowing gap, in computing facilities between foreign countries and the United States, plus lack of uniqueness in CFD technology and talent, make CFD competitive on an international level. This intensity of global competition in "high tech" areas such as CFD could pose a serious challenge to the aeronautical preeminence of the United States.

REFERENCES

1. Chapman, D.R.; Mark, H.; and Pirtle, M.W.: Computers vs. Wind Tunnels for Aerodynamic Flow Simulations. Astronaut. & Aeronaut., 1975, 13, 22-35.

2. Chapman, D.R.: Trends and Pacing Items in Computational Aerodynamics. Lecture Notes in Physics, 1981, 141, Ed. Reynolds, W.C. and MacCormack, R.W., Springer-Verlag, Berlin Heidelberg, Germany.

3. Kutler, P.: A Perspective of Theoretical and Applied Computational Fluid Dynamics. AIAA Journal, 1985, 23.

4. Marvin, J.G.: Future Requirements of Wind Tunnels for Computational Fluid Dynamics Code Verifications. AIAA Paper 86-0752-CP, 1986.

5. Rai, M.M.: Navier-Stokes Simulations of Rotor-Stator Interaction Using Patched and Overlaid Grids. AIAA Paper 85-1519, 1985.

6. Jameson, A.; Baker, T.J.; and Weatherill, N.P.: Calculation of Inviscid Transonic Flow Over a Complete Aircraft. AIAA Paper 86-0103, 1986.

7. Yee, H.: Numerical Experiments with a Symmetric High-Resolution Shock-Capturing Scheme. Proc. 10th Intl. Conf. on Num. Meth. in Fluid Dynamics, June 1986, Beijing, China.

8. Beam, R.M.; and Bailey, H.E.: pvt. communications, January 1986.

9. Edwards, T.A.: Definition and Verification of a Complex Aircraft for Aerodynamic Calculations. AIAA Paper 86-0431, 1986.

10. Kwak, D.; Chang, J.L.C.; and Shanks, S.P.: A Solution Procedure for Three-Dimensional Incompressible Navier-Stokes Equation and Its Application. Proc. 9th Int. Conf. on Num. Meth. in FLuid Dynamics, 1984.

11. Nakahashi, K.; and Deiwert, G.S.: A Self-Adaptive-Grid Method with Application to Airfoil Flow. AIAA Paper 85-1525, 1986.

12. Dougherty, F.C.; Benek, J.A.; and Steger, J.L.: On Applications of Chimera Grid Schemes to Store Separation. NASA TM 88193, 1985.

13. Buning, P.G.; and Steger, J.L.: Graphics and Flow Visualization in Computational Fluid Dynamics. AIAA Paper 85-1507, 1985.

14. Davis, S.S.; and Chang, I.-C.: The Critical Role of Computational Fluid Dynamics in Rotory-Wing Aerodynamics. AIAA Paper 86-0336, 1986.

15. McCroskey, W.J.; and Baeder, J.D.: Some Recent Advances in Computational Aerodynamics for Helicopter Applications. NASA TM 86777, 1985.

16. Rizk, Y.M.; and Ben-Shmuel, S.: Computation of the Viscous Flow Around the Shuttle Orbiter at Low Supersonic Speeds. AIAA Paper 85-0168, 1985.

17. Guruswamy, G.P.; Goorjian, P.M.; and Tu, E.L.: Unsteady Transonics of a Wing with Tip Store. AIAA Paper 86-0010, 1986.

18. Cosentino, G.; and Holst, T.L.: Numerical Optimization Design of Advanced Transonic Wing Configurations. AIAA Paper 85-0424, 1985.

SIMULATION OF COMPRESSIBLE INVISCID FLOWS: THE ITALIAN CONTRIBUTION

M. Napolitano

University of Bari, Bari, Italy

INTRODUCTION

The aim of this lecture is to provide a unitary description of the work recently performed in the area of the numerical simulation of compressible inviscid flows by a few researchers of Italian origin and/or nationality. More precisely, the attention will be focused on the so-called lambda formulation as developed and improved by Moretti, Pandolfi, Zannetti, Colasurdo and Gabutti /1-5/, in New York and Torino, and by Dadone and the author, in Bari /6-9/.
Before addressing the topics that will be covered, it seems appropriate to outline briefly the state of the art in the field of the numerical solution of the Euler equations. At present, two basic methodologies are most widely employed to compute transonic flows of aerodynamic interest. They both use the conservation-law form of the Euler equations in order to properly capture and propagate shocks and differ mainly in the spatial discretization employed. The first approach, developed by Hall, Jameson, Murman and other authors, uses central differences throughout and relies on artificial dissipation to resolve the supersonic regions of the flow field and to reduce the Gibbs phenomenon around shocks to an acceptable level. The second approach, pioneered by Godunov and later perfected by Harten, Osher, Roe, Van Leer, Steger and Warming, Woodward and Colella, and others, uses some kind of upwind differences in order to properly account for the direction of propagation of waves. This second methodology is certainly more complicated, because it requires, at every gridpoint and time step, to determine the eigenvalues of the Jacobian matrix (see, e.g. /10/) or to solve one or more Riemann problems (see, e.g., /11,12/); on the other hand, it is more robust and requires a minimum amount of tuning, if second or higher order accuracy is sought (see, e.g., /12/), or no tuning at all, if first order accuracy is accepted around shocks.
The lambda formulation, of interest here, also uses upwind differences -- and can be shown to coincide with both the flux vector splitting and the flux difference splitting methods in the linear case -- but employs the nonconservative form of the Euler equations. These are recast in terms of generalized Riemann variables and take the form of nonlinear advection equations (see, e.g., /7/), so that the appropriate upwind differences to be used in the discrete problem are immediately apparent and very straightforward to implement. Therefore, the lambda formulation: i) is easier to understand and much simpler to code than any conservative upwind method based on flux vector or flux difference splitting concepts, and is expected to produce more efficient computer programs when using similar time integration schemes; ii) appears to provide more accurate results in the smooth regions of the flow, using comparable spatial discretizations; iii) combines in

the most natural and consistent way with boundary treatments based on the theory of characteristics, which have been shown to improve the stability and accuracy also of schemes using central differences. Finally, the lambda formulation, even under homentropic flow assumption, has been found to be endowed with a kind of "shock-capturing capability" (see, e.g., /1,2,6,7/), insofar as flows with shock-type discontinuities have been obtained without resorting to any "ad hoc" treatment around them, or additional dissipation. Needless to say, due to the nonconservative form of the equations, these discontinuities do not satisfy the Rankine Hugoniot conditions and, more importantly, do not propagate correctly inside the flow field. Admittedly, such a problem has been underestimated at first by some users of the lambda formulation /3,6/, because in several (steady) transonic flow calculations, shocks having very good strength and positions were "captured" using the lambda method /3,6,7/. However, numerical experiments have shown that these "shocks" do not move upstream in the flow field even when the supersonic bubble is embedded in a subsonic region, so that the lambda formulation is definitely unreliable in the computation of transonic flows, no matter how weak the shocks present in the flow. Fortunately, such a very serious drawback has been overcome and various approaches have already been proposed and applied successfully to compute shocks satisfying the correct jump conditions and capable of moving upstream, when warranted. Therefore, the lambda formulation, already extremely valuable to compute smooth flows, has now become an efficient, reliable and very competitive tool for simulating general transonic and supersonic flows.
In the remaining of this lecture, the basic lambda formulation will be provided in some detail, together with an outline of the improvements obtained in both the accuracy /5,8,13,14/ and the efficiency /6,7,9,15/ of the original approach. Some attention will be then devoted to the treatment of boundary conditions, a major strength of the lambda approach. Finally, the most important and controversial issue, namely that of computing flows with shocks or other discontinuities, will be addressed and typical results obtained by various authors will be presented.

LAMBDA FORMULATION EQUATIONS

The nondimensional continuity, momentum, and entropy equations governing the inviscid motion of a perfect gas are given in vector form as:

$$\delta (a_t + \underline{V} \cdot \underline{\nabla}a) + a \underline{\nabla} \cdot \underline{V} - a (S_t + \underline{V} \cdot \underline{\nabla}S) = 0 \qquad (1)$$

$$\underline{V}_t + (\underline{V} \cdot \underline{\nabla})\underline{V} + \delta a \underline{\nabla}a - a^2 \underline{\nabla}S = 0 \qquad (2)$$

$$S_t + \underline{V} \cdot \underline{\nabla}S = 0 \qquad (3)$$

where a is the speed of sound, $\underline{V}$ is the velocity vector, S is the entropy, $\underline{\nabla}$ is the gradient operator, subscript t indicates partial derivatives with respect to time, t, and $\delta = 2/(\gamma - 1)$, γ being the specific heats ratio. It is noteworthy that eqn (3), multiplied by a, has been added to the continuity equation in order to obtain appropriate lambda equations in a simple way. In fact, by summing and subtracting eqn (1) to and from each component of eqn (2), and combining the resulting scalar equations, as done in Ref. 7, the two-dimensional lambda-formulation equations for a general orthogonal curvilinear coordinate system are obtained as:

$$C_t + D_t + (u+a)(C_\xi - aS_\xi) + (u-a)(D_\xi + aS_\xi) + v(C_\eta + D_\eta) = -2\alpha_1 + 2\beta_1 \quad (4)$$

$$E_t + F_t + u(E_\xi + F_\xi) + (v+a)(E_\eta - aS_\eta) + (v-a)(E_\eta + aS_\eta) = -2\alpha_2 + 2\beta_2 \quad (5)$$

$$(C_t - D_t + E_t - F_t)/2 - 2aS_t + (u+a)(C_\xi - aS_\xi) -$$
$$(u-a)(D_\xi + aS_\xi) + (v+a)(E_\eta - aS_\eta) - (v-a)(F_\eta + aS_\eta) = 2\zeta \quad (6)$$

$$S_t + uS_\xi + vS_\eta = 0 \quad (7)$$

$$C - D - E + F = 0 \quad (8)$$

In eqns (4-8), ξ, η are the arclengths measured along the two orthogonal curvilinear coordinates, and u, v are the components of the velocity vector $\underline{V}$; furthermore, C, D, E, and F are the four bicharacteristic (generalized Riemann) variables, namely,

$$C = u + \delta a, \quad D = u - \delta a, \quad E = v + \delta a, \quad F = v - \delta a \quad (9a,b,c,d)$$

and $\alpha_1, \ldots, \beta_2$ and ζ are coefficients containing u, v and a, as undifferentiated source terms, and spatial derivatives of the scale factors of the coordinate system (see Ref. 7 for their expressions). A few comments are needed. First of all, the spatial derivatives in eqns (4-7) are clearly associated with the advection of the four bicharacteristic variables and the entropy along bicharacteristic lines; in other words, eqns (4-7) have the form of nonlinear advection equations, whose terms can be discretized using appropriate upwind differences, simply according to the signs of their advection speeds. Therefore, an appropriate upwind scheme for integrating eqns (4-8) numerically will be "consistent with the physical phenomena" /3/ and will enjoy the intrinsic accuracy and robustness of the classical method of characteristics, while retaining the simplicity of finite difference methods. Secondly, the four bicharacteristic variables are linearly dependent, being defined in terms of only three physical variables, so that the algebraic eqn (8) is needed when using an implicit numerical integration scheme (see, e.g., Refs, 6-9,16). On the other hand, if an explicit integration scheme is to be used, it is more convenient to rewrite eqns (4,5,6 and 8) as follows /14/:

$$u_t = -(u+a)(C_\xi - aS_\xi)/2 - \ldots\ldots\ldots - \alpha_1 + \beta_1 \quad (10)$$

$$v_t = -u(E_\xi + F_\xi)/2 - \ldots\ldots\ldots\ldots - \alpha_2 + \beta_2 \quad (11)$$

$$\delta a_t = aS_t - (u+a)(C_\xi - aS_\xi)/2 - \ldots\ldots + \zeta \quad (12)$$

$$S_t = -uS_\xi - vS_\eta \quad (13)$$

where all the advection terms have been brought to the right hand side and the time variations of the four physical variables u, v, a and S can be computed by evaluating the advection terms explicitly, using appropriate upwind differences. Notice that there are some differences between the notation of Ref. 14 and that used in this paper. Finally, it is important to point out that several forms of the lambda-formulation equations are available in the literature. In particular, it is possible to obtain a more general formulation by using a nonorthogonal coordinate system and by taking the dot product of eqn (2) times two orthogonal unit vectors $\underline{n}$ and $\underline{\tau}$, which are independent of those defining the curvilinear coordinates /5,14/. The main advantage of these lambda equations is in that the boundary conditions are particularly easy to implement if one chooses $\underline{\tau}$ and $\underline{n}$ respectively tangent and normal to the boundary. Of course, the numerical accuracy invariably deteriorates when the coordinate system becomes more and more skewed; moreover, the use of a nonorthogonal coordinate system produces additional spatial derivatives -- with respect to those contained in eqns (4-8) -- which are not clearly associated with the advection of physical disturbances along bicharacteristic lines. Therefore, in the author's opinion, orthogonal coordinates are to be preferred in conjunction with any numerical method based on the lambda formulation.

THE PERTURBATIVE APPROACH

Dadone and Napolitano have found that using the incompressible potential flow net as the computational grid for solving compressible flows by means of the lambda scheme produced a significant accuracy improvement with respect to using a standard (e.g., polar) orthogonal grid /7/. This result is most likely due to the fact that such a grid is somewhat "optimal" to compute the incompressible flow and thus reduces the negative influence on the accuracy of the solution of the large gradients induced by the geometry, e.g., those around the stagnation points of the flow. Therefore, it seemed logical and appropriate to derive a lambda formulation solving only for the differences between the sought compressible flow and the incompressible potential flow past the same body. Such an approach is particularly convenient when combined with a numerical grid generation technique, such as that of Ref. 17, which provides the incompressible potential flow solution at no extra cost, as done in Refs. 8 and 18, which provide the details of the approach. Here, it is worth mentioning that the governing equations have the same structure as eqns (4-8), with additional linear terms in the right hand side, containing the unknowns multiplied by coefficients which depend only on the geometry and the potential flow solution and therefore need to be computed once and for all at the beginning of the calculation process. Therefore any numerical scheme suitable for eqns (4-8) can be used at a negligible extra cost to solve the corresponding perturbative equations. Most notably, in contrast with such a minimal increase in effort, the numerical results have been shown to provide a reduction in the critical total pressure (entropy) errors of up to an order of magnitude, or more, see, e.g., Refs. 8 and 18. It is noteworthy that the perturbative approach can be implemented very easily in any code using the lambda formulation, by simply redefining the bicharacteristic variables and adding the additional terms to the right hand side of the equations. Also, a suitable rescaling of the "incompressible" velocity field, so as to satisfy mass conservation around stagnation points, has been shown to further improve the accuracy of the results, see Ref. 19, for details.

SOME NUMERICAL INTEGRATION SCHEMES

Several numerical schemes have been proposed to solve eqns (4-8) or (10-13), of both explicit /1-5,13,14/ and implicit /6-9,15,16,18/ type. Here, due to the limited space available, only the main features of the schemes developed for two- and three-dimensional flows will be briefly described. Moreover, nothing at all will be said about the special quasi-one-dimensional flow case, for which the interested reader is referred in particular to Refs. 5, 13-15, and 18.
As far as the implicit schemes are concerned, the governing equations (4-8) are discretized and linearized in time using a two level first-order-accurate Euler scheme and the delta approach of Beam and Warming /20/. The resulting semi-discrete equations are then discretized in space using two-point, first-order-accurate, upwind differences for the incremental variables (i.e., the unknowns) and three-point, second-order-accurate, upwind differences for the terms in the right hand side of the equations, which are evaluated at the old time level. In this way, the large linear system to be solved at every time step is more compact and diagonally dominant and, at convergence, the sought steady state solution is still second-order-accurate in space. For the case of one-dimensional flows, the linear system is block-tridiagonal and can be solved directly by standard block-tridiagonal Gaussian elimination. For the case of two- and three-dimensional flows, the linear system has 5 and 7 nonzero diagonals, respectively, so that an approximate factorization of ADI (Alternating Direction Implicit) type /6-8/ or a directional mutilation of the (line) Gauss-Seidel type /9,16/ are used to solve it approximately.
As far as the explicit schemes are concerned, after the early methods /1-3/, the goal has been that of achieving second order accuracy in both space and time, and of extending the unitary CFL stability condition, as done at first by Gabutti /4/. Todate, the most widely employed technique is that originally proposed by Moretti and Zannetti /5/ and later further improved by Moretti /14/. The method, which is very similar to that previously proposed by Zhu and Chen /21/, is a two-step (predictor-corrector) scheme of the leap-frog type, which has a CFL condition greater than one and achieves second order accuracy in both space and time, while using upwind differences based on only two gridpoints at each step.

THE TREATMENT OF BOUNDARY CONDITIONS

In the last decade, a lot of theoretical and computational effort has been devoted to the study of the so-called numerical boundary conditions, namely, those boundary conditions not dictated by the physics of the problem (i.e., the differential equations) but required only by the use of central differences in the discrete problem. Of course, theoretical analyses and numerical experiments can help to obtain numerical boundary conditions having a minimal negative effect on the accuracy and stability of the solution procedure. However, it is undoubtedly a strong point in favor of upwind schemes, in general, and of the lambda formulation, in particular, to require all and only the boundary conditions of the differential problem. In fact, by considering the lambda-formulation equations (4-8) or (9-12), it can be seen that, at any given boundary gridpoint, the number of boundary conditions is exactly equal to the number of spatial derivatives associated with waves propagating from outside the computational domain. Of course, different ways of implementing the boundary conditions are possible. For example, in the implicit schemes of Refs. 7 and 8, at the downstream gridpoints, where only one boundary condition, namely, the pressure, is prescribed, the governing equations are combined linearly so as to eliminate the only derivative requiring unknowns from outside the computational domain and the pressure boundary condition is enforced strictly to close the problem. For the explicit schemes, see, e.g., /14/, it is more convenient to use the boundary conditions directly in the governing equations (e.g., by setting the time derivative of the pressure equal to zero) to obtain appropriate values for the derivatives requiring external gridpoints and then to solve for the boundary gridpoints exactly like for the internal ones. This procedure, however, can produce a slow shift in the boundary value after a large number of time steps, due to round-off accumulation, insofar as the boundary condition is not enforced exactly. In this respect, a very effective and easy-to-implement remedy is provided by the so-called post-correction procedure of de Neef and Moretti /22/. For more details on boundary conditions, see Ref. 7, for the implicit schemes, and Ref. 14, for the explicit ones.

THE TREATMENT OF SHOCKS AND OTHER DISCONTINUITIES

As anticipated in the introduction, four approaches have been proposed to make the classical lambda formulation capable of computing flows with shocks.
Moretti, who has always been a strong advocate of shock-fitting /22/, has recently developed a very simple and reliable discontinuity detecting and tracking procedure /13-15/. For one-dimensional flows, such a procedure has been combined with a box-type lambda scheme to provide an extremely accurate and efficient technique for computing steady flows with shocks /15/ and with the method of Ref. 5 to compute unsteady flows with any sort of discontinuities (shocks, contact surfaces, etc.) /13/. More recently, this second methodology (numerical scheme plus discontinuity-tracking procedure) has been successfully extended to two dimensions /14/. Figure 1, shows the formation and propagation of a lambda shock. The results are quite clean, with the lambda shock being positioned extremely well, and the overall solution is in good agreement with results obtained using a conservative shock-capturing scheme. Further results demonstrating the validity of this approach are given in Ref. 14. The rationale behind Moretti's procedure is simple and quite clear: within the framework of inviscid flow theory, shocks and contact surfaces are discontinuities, which occupy a very small part of the computational domain and can be resolved "exactly" only if fitted. Therefore, if a simple and reliable discontinuity-fitting procedure is available, there should be no doubt about its superiority with respect to any shock-capturing method.

Furthermore, the lambda scheme being extremely well suited to treat boundaries in hyperbolic flows, it combines very favourably with a shock-tracking procedure, which uses the Rankine-Hugoniot equations to compute shocks as internal boundaries.
Pandolfi, having experienced flux difference splitting methods for both unsteady transonic /23/ and steady supersonic /24/ flows, has developed a different approach, based on a hybrid formulation /25/. The simpler, more efficient and accurate lambda method is employed in all smooth regions of the flow, whereas the more cumbersome, but conservative, flux difference splitting method is used only where needed, i.e., in the regions containing and immediately surrounding shocks. It is noteworthy that a first-order-accurate flux difference splitting method is employed by Pandolfi in order to preserve monotonicity around shocks without any tuning. The lower accuracy being limited to a fixed, small number of gridpoints, the second order accuracy of the lambda scheme used in the smooth regions of the flow is formally maintained. A result of Pandolfi's hybrid method is given in figure 2, where the pressure coefficients on the upper and lower sides of a NACA 0012 airfoil are given for Mach = 0.85 flow at 1 degree angle of attack. An advocate of the lambda formulation will look at such a methodology as a smarter way of implementing a flux difference splitting method, by improving its simplicity, efficiency, and accuracy in the major part of the computational domain.
Dadone and Magi /26/ and Dadone /27/ have provided the so-called quasi-conservative lambda formulation. From a careful analysis of the flux difference splitting method, as compared to the classical lambda formulation, it is shown that this second approach will allow for upstream propagation through the shocks, if reformulated as a "cell" method instead of a "gridpoint" method (see Ref. 26, for details). Therefore, the wave pattern in the cell containing a shock is analyzed in detail and used to provide a local correction to the standard lambda formulation. The modified approach has been shown to allow the "captured shoks" to move upstream, when warranted, and has proven to be a reliable tool for computing steady transonic flows. A result of the quasi-conservative lambda method is given in figure 3, for the case of transonic flow past a hump in a channel. The pressure coefficient along the wall obtained using the quasi-conservative lambda formulation is seen to coincide within plotting accuracy with that obtained using a flux difference splitting method and a finer mesh (see Ref. 27, for details).
Finally, Favini and Zannetti /28/ have provided a rigorous analysis to show that the flux vector splitting method of Steger and Warming /10/ can be formulated as a nonconservative method plus an additional term containing the flux of the Jacobian matrix. Such a term is associated with entropy production and thus is essential in the shock regions, to properly capture and propagate shocks, but can actually be an undesirable source of numerical errors in the smooth regions of the flow. The implication of the analysis of Favini and Zannetti is that a hybrid formulation using the lambda scheme, locally corrected to become a conservative method in the regions of shocks, is not only consistent, but is likely to be more accurate overall than a conservative scheme, while requiring considerably less computational effort. An example of the method of Favini and Zannetti is given in figure 4, where the

pressure coefficients (multiplied by ten) on a Joukowski airfoil are given for Mach = 0.7 flow at 3 degrees angle of attack. The shock is rather smeared, since it is computed using the first-order-accurate flux vector splitting method of Ref. 10.
It should be emphasized that all four methods described above only require minor modifications to a running code using the lambda formulation, but are essential to handle transonic and supersonic flows.

CONCLUSIONS

The aim of this lecture was to describe some recent work on the simulation of compressible inviscid flows using the nonconservative lambda formulation and to show that such an approach, after years of improvements and hard work, has become a very simple, accurate, efficient, and reliable tool for computing flows with both stationary and moving shocks. It is hoped that the lambda formulation will receive in the future an increasing attention by the CFD community and that many more researchers and engineers will employ such an approach to perform their research and design activities.

DEDICATION

This work is dedicated to the memory of R.T. Davis, who has shown to all of us how to combine effectively a deep knowledge of fluid dynamics with a rare insight into numerical analysis and has been for me a guide, a model, and a friend.

ACKNOWLEDGEMENTS

The author is indebted to profs Moretti, Pandolfi, Zannetti, Dadone, and Quartapelle, for their precious suggestions and comments, and to all the authors of the figures presented in the paper.

REFERENCES

1. Pandolfi, M. and Zannetti, L., "Some Tests on Finite Difference Algorithms for Computing Boundaries in Hyperbolic Flows", Notes on Numerical Fluid Dynamics, Vol. 1, Vieweg, 1978, pp. 68-88.
2. Moretti, G., "The λ-Scheme", Computers and Fluids, Vol. 7, 1979, pp. 191-205.
3. Zannetti, L. and Colasurdo, G., "Unsteady Compressible Flow: A Computational Method Consistent with the Physical Phenomena", AIAA Journal, Vol. 19, July 1981, pp. 851-856.
4. Gabutti, B., "On Two Upwind Finite-Difference Schemes for Hyperbolic Equations in Non-Conservative Form", Computers and Fluids, Vol. 11, 1983, pp. 207-230.
5. Moretti, G. and Zannetti, L., "A New and Improved Computational Technique for Two-Dimensional Unsteady Compressible Flows", AIAA Journal, Vol. 22, June 1984, pp. 758-765.
6. Dadone, A. and Napolitano, M., "An Implicit Lambda Scheme", AIAA Journal, Vol. 21, October 1983, pp. 1391-1399.
7. Dadone, A. and Napolitano, M., "An Efficient ADI Lambda Formulation", Computers and Fluids, Vol. 13, 1985, pp. 383-395.

8. Dadone, A. and Napolitano, M., "A Perturbative Lambda Formulation", Lectures Notes in Physics, Vol. 218, Springer Verlag, 1985, pp. 175-179; also, AIAA Journal, Vol. 24, March 1986, pp. 411-417.

9. Napolitano, M. and Dadone, A., "Implicit Lambda Methods for Three-Dimensional Compressible Flows", AIAA Journal, Vol. 23, September 1985, pp. 1343-1347.

10. Steger, J. L. and Warming, R. F., "Flux Vector Splitting of the Inviscid Gasdynamic Equations with Application to Finite-Difference Methods", J. Comput. Phys., Vol. 40, 1981, pp. 263-293.

11. Roe, P. L., "Approximate Riemann Solvers, Parameter Vectors and Difference Schemes", J. Comput. Phys., Vol. 43, 1981, pp. 357-372.

12. Woodward, P. and Colella, P, "The Piecewise Parabolic Method", J. Comput. Phys., Vol. 54, 1984, pp. 115-173.

13. Moretti, G. and Di Piano, M. T., "An Improved Lambda-Scheme for One-Dimensional Flows", NASA CR 3712, 1983.

14. Moretti, G., "Numerical Studies of Two-Dimensional Flows", NASA CR 3930, 1985. Also, "A Technique for Integrating Two-Dimensional Euler Equations", Computers and Fluids, to appear.

15. Moretti, G., "Fast Euler Solver for Steady, One-Dimensional Flows", Computers and Fluids, Vol. 13, 1985, pp. 61-81.

16. Fortunato, B. and Napolitano, M., "Numerical Solution of 3-D Compressible Internal Flows Using a Nonisentropic Implicit Lambda Method", Notes on Numerical Fluid Mechanics, Vol. 13, Vieweg, 1986, pp. 105-112.

17. Davis, R. T., "Notes on Numerical Methods for Coordinate Generation Based on a Mapping Technique", VKI Lecture Series n. 5, 1981.

18. Dadone, A. and Napolitano, M. "Accurate and Efficient Solutions of Compressible Internal Flows", Journal of Propulsion and Power, Vol. 1, n. 6, 1986, pp. 456-463.

19. Dadone, A., "A Quasi-Conservative COIN Lambda Formulation", 10th ICNMFD, Beijing, June 1986.

20. Beam, R. M. and Warming, R. F., "An Implicit Factored Scheme for the Compressible Navier-Stokes Equations", AIAA Journal, Vol. 16, April 1978, pp. 393-402.

21. Zhu, Y. and Chen, B., "Difference Methods for Initial-Boundary-Value Problems and Computation of Flow around Bodies", Computers and Fluids, Vol. 9, 1981, pp. 339-363.

22. de Neef, T. and Moretti, G., "Shock Fitting for Everybody", Computers and Fluids, Vol. 8, 1980, pp. 327-334.

23. Pandolfi, M., "A Contribution to the Numerical Prediction of Unsteady Flows", AIAA Journal, Vol. 22, May 1984, pp. 602-610.

24. Pandolfi, M., "Computation of Steady Supersonic Flows by a Flux-Difference/Splitting Method", Computers and Fluids, Vol. 13, 1985, pp. 37-46.

25. Pandolfi, M., "The Merging of Two Different Ideas: A Shock Fitting Performed by a Shock Capturing", International Symposium on Computational Fluid Dynamics-Tokyo, September 1985.

26. Dadone A. and Magi, V., "A Quasi-Conservative Lambda Formulation", AIAA Paper 85-0088, 1985; also, AIAA Journal, to appear.

27. Dadone, A., "Accurate and Efficient Solutions of Transonic Internal Flows", AIAA Paper 85-1334, 1985.

28. Favini, B. and Zannetti, L., "On Conservation Properties and Non-Conservative Forms of Euler Solvers", 10th ICNMFD, Beijing, June 1986.

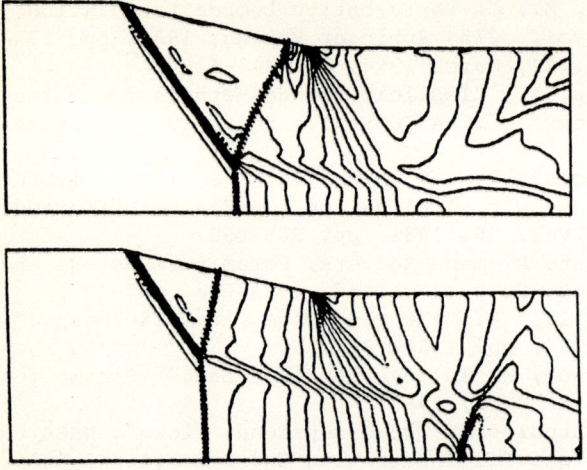

Figure 1. Formation and propagation of a lambda shock computed by Moretti using a lambda formulation and shock fitting.

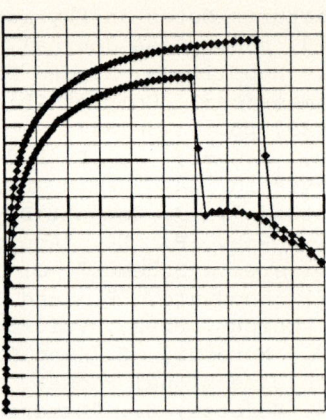

Figure 2. NACA 0012 airfoil c_p distributions as computed by Pandolfi using his hybrid approach.

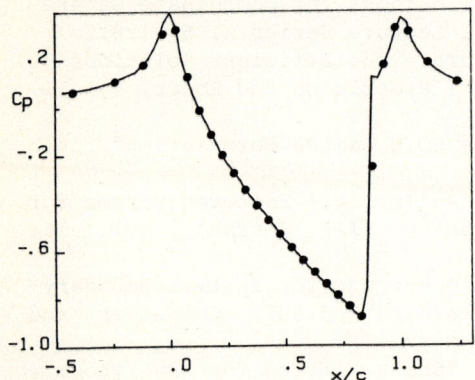

Figure 3. c_p distribution at the wall for transonic flow in a channel with a hump, using the quasi-conservative lambda formulation.

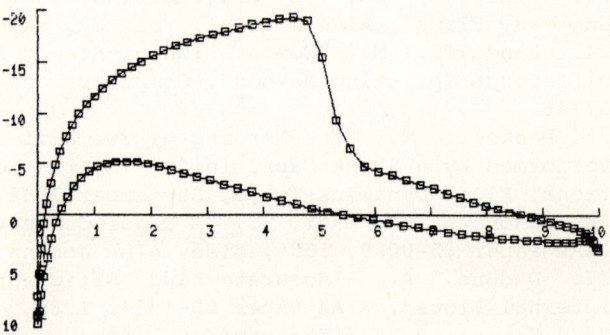

Figure 4. Joukowsky airfoil c_p distributions as computed by Favini and Zannetti.

METHOD OF LINES APPROACH TO THE NUMERICAL SOLUTION OF FLUID DYNAMIC EQUATIONS

Nobuyuki SATOFUKA

Department of Mechanical Engineering
Kyoto Institute of Technology
Matsugasaki, Sakyo-ku
Kyoto 606, Japan

ABSTRACT

A method of lines approach is proposed for solving the fluid dynamic equations. The method is based on a combination of the central finite difference approximation to the space variables with a rational Runge-Kutta or a classical Runge-Kutta time integration scheme. Numerical results for both compressible and incompressible flow problems are presented to demonstrate the utility of the present approach.

1. INTRODUCTION

As is well known, motions of fluids under standard pressure and temperature are described by a system of partial differential equations (PDEs), viz. the Navier-Stokes equations. Since analytic solutions for these nonlinear PDEs are seldom known for practical problems, the numerical solution is an obvious approach. The numerical solution of nonlinear PDEs is a complicated and highly problem dependent process as well as a very diverse and disjointed area of methematics. A number of different numerical techniques and methods are developed and many more are being developed. Typically, these techniques and methods are developed for a single flow problem or at best a narrow class of problems. Development of codes for general classes of flow problems is in an infant state and almost nonexistent.

In this paper, we discuss our attempt to develop a numerical method for general classes of flow problems using the method of lines approach. Since the Navier-Stikes equations are systems of parabolic PDEs in general, the method of lines is ideally suited[1]. The utility of this approach, however, is not limited to the parabolic PDEs [2,3]. In the method of lines, we first discretize the space variables to obtain a system of ordinary differential equations (ODEs) in time. The resulting ODEs are then integrated by an appropriate ODE solver. Jameson and his co-workers have proposed to use an explicit four-stage Runge-Kutta time stepping procedure[4] while Beam and Warming have proposed to use implicit backward Euler or trapezoidal scheme[5]. We propose, in this paper, to use a new explicit scheme for the integration of the fluid dynamic equations. The present method consists of the central finite difference approximation to space variables combined with a rational Runge-Kutta (RRK)[6] or a classical RK4 time integration scheme.

The present paper is organized in the following way. The outline of the method of lines when applied to a simple quasi-linear parabolic model PDE is described briefly in Section 2, application to compressible flows and the results for both inviscid and viscous flow problems are discussed in Section 3; and some numerical results for incompressible Navier-Stokes equations are presented in Section 4 to demonstrate the utility of the present method.

2. THE METHOD OF LINES

In this section, we will describe the basic elements of the method of lines when it is applied to the following single quasi-linear parabolic PDE,

$$u_t = f(x, t, u, u_x, u_{xx}), \quad 0 < x < 1, \quad t > 0, \quad (1)$$

under boundary conditions,

$$A u + B u_x = C \quad \text{at} \quad x = 0, 1, \quad (2)$$

and the initial condition,

$$u(x, 0) = h(x). \quad (3)$$

<u>Spatial Discretization</u>

In the method of lines approach, we first discretize the spatial derivatives in Eq. (1). If we discretized the problem on the mesh

$$x_i = i \Delta x, \quad i = 0, 1, \cdots, N+1, \quad \Delta x = 1/(N+1),$$

and replace the spatial derivatives by second order central finite difference approximations, $u_i(t)$, an approximation to $u(x_i,t)$ satisfies the system of ODEs,

$$\frac{du_i}{dt} = f\left[x_i, t, u_i, \frac{u_{i+1} - u_{i-1}}{2\Delta x}, \frac{u_{i+1} - 2u_i + u_{i-1}}{\Delta x^2}\right]$$

$$= F(t, u_{i+1}, u_i, u_{i-1}). \quad (4)$$

ODEs for $u_0(t)$ and $u_{N+1}(t)$ may be obtained boundary conditions, Eq. (2). There are, of course, many alternatives in the way of spatial discretization. For example, higher order[7,8] or upwind differencing[9] could be used. Also, the spatial variation could be described by writing,

$$\frac{\partial u_i(t)}{\partial x} = \sum_{j=0}^{N+1} a_{ij} u_j(t) \quad . \tag{5}$$

The weighting coefficients a_{ij} can be determined in a way similar to the Lagrangian interpolation. The coefficients for the second derivatives b_{ij} can be computed as $b_{ij} = \sum_{k=0}^{N+1} a_{ik} a_{kj}$. This approach is called as the modified differential quadrature (MDQ) and has been tried in some applications[10] requiring higher order accuracy.

Time Integration

Numerical integration of such system of ODEs, Eq. (4), subject to the initial condition,

$$u_i(0) = h(x_i) \quad , \tag{6}$$

is a simple task using an appropriate ODE integrator. In our approach, we have chosen two explicit multi-stage time integration schemes, namely, a rational Runge-Kutta (RRK) scheme proposed by Wambecq[6] and classical 4-stage Runge-Kutta schemes. Other schemes such as implicit backward Euler scheme may be usefull for some class of problems. However, we prefer to the explicit schemes since they are easier to vectorize and simpler to program.

As applied to Eq. (4), RRK scheme can be written in the following two-stage form, dropping the subscript, i,

$$g_1 = \Delta t \, W(u^n) \quad , \quad g_2 = \Delta t \, W(u^n + c_2 g_1) \quad ,$$
$$u^{n+1} = u^n + [\, 2g_1(g_1, g_3) - g_3(g_1, g_1) \,]/(g_3, g_3) \quad , \tag{7}$$

where $g_3 = b_1 g_1 + b_2 g_2$, $b_1 + b_2 = 1$ and (c,d) denotes the scalar product of two vectors c and d. In Eq. (7) u^n and u^{n+1} indicates the values of u at the n and n+1 time step, respectively. The RRK scheme is generally of order 1 but is of order 2 if in addition, $b_2 c_2 = -1/2$. The scheme is fully explicit and $A(\alpha)$-stable if $b_2 c_2 \leq -1/[2\cos\alpha(2 - \cos\alpha)]$. Therefore, the scheme is free from severe stability restriction to which most explicit schemes are subject.

The general 4-stage Runge-Kutta type scheme can be written as,

$$u^{(0)} = u^n \quad , \quad u^{(1)} = u^{(0)} + \alpha_1 \Delta t \, W^{(0)} \quad ,$$
$$u^{(2)} = u^{(0)} + \alpha_2 \Delta t \, W^{(1)} \quad , \quad u^{(3)} = u^{(0)} + \alpha_3 \Delta t \, W^{(2)} \quad , \tag{8}$$
$$u^{(4)} = u^{(0)} + \Delta t \, W^{(3)} \quad , \quad u^{n+1} = u^{(4)} \quad .$$

The 4-stage scheme is at most fourth-order accurate in time. If the objective is simply to obtain a steady state as rapidly as possible, the order of accuracy is not important. For time dependent flow problems, however, a high order of accuracy is important. Therefore, values of coefficients α_m should be chosen depending on the objective.

3. COMPRESSIBLE FLOW PROBLEM

<u>Governing Equations</u>

The compressible Navier-Stokes equations subject to general transformation can be written in dimensionless, conservation-law form as:

$$\frac{\partial \vec{q}}{\partial t} + \frac{\partial \vec{E}}{\partial \xi} + \frac{\partial \vec{F}}{\partial \eta} = \frac{1}{Re} \left(\frac{\partial \vec{R}}{\partial \xi} + \frac{\partial \vec{S}}{\partial \eta} \right) , \qquad (9)$$

where $\vec{q}, \vec{E}, \vec{F}, \vec{R},$ and $\vec{S}$ are

$$\vec{q} = \frac{1}{J} \begin{Bmatrix} \rho \\ \rho u \\ \rho v \\ e \end{Bmatrix} , \quad \vec{E} = \frac{1}{J} \begin{Bmatrix} \rho u \\ \rho u U + \xi_x p \\ \rho v U + \xi_y p \\ (e+p)U \end{Bmatrix} , \quad \vec{F} = \frac{1}{J} \begin{Bmatrix} \rho v \\ \rho u V + \eta_x p \\ \rho v V + \eta_y p \\ (e+p)V \end{Bmatrix} ,$$

$$\vec{R} = \frac{1}{J} \begin{Bmatrix} 0 \\ \xi_x \sigma_x + \xi_y \tau_{xy} \\ \xi_x \tau_{xy} + \xi_y \sigma_y \\ \xi_x R_4 + \xi_y S_4 \end{Bmatrix} , \quad \vec{S} = \frac{1}{J} \begin{Bmatrix} 0 \\ \eta_x \sigma_x + \eta_y \tau_{xy} \\ \eta_x \tau_{xy} + \eta_y \sigma_y \\ \eta_x R_4 + \eta_y S_4 \end{Bmatrix} ,$$

(10)

with

$$\sigma_x = \lambda \left(\frac{\partial u}{\partial x} + \frac{\partial v}{\partial y} \right) + 2\mu \frac{\partial u}{\partial x} , \quad \sigma_y = \lambda \left(\frac{\partial u}{\partial x} + \frac{\partial v}{\partial y} \right) + 2\mu \frac{\partial v}{\partial y} ,$$

$$\tau_{xy} = \mu \left(\frac{\partial u}{\partial y} + \frac{\partial v}{\partial x} \right) , \qquad (11)$$

$$R_4 = u\sigma_x + v\tau_{xy} + \frac{\gamma}{\gamma-1} \frac{\kappa}{Pr} \frac{\partial T}{\partial x} , \quad S_4 = u\tau_{xy} + v\sigma_y + \frac{\gamma}{\gamma-1} \frac{\kappa}{Pr} \frac{\partial T}{\partial y} .$$

In the conservative variables of Eq. (10), the pressure p is nondimensionalized by p_0, the density ρ by ρ_0, and the velocity components u and v in the x and y directions by $a_0/\sqrt{\gamma}$. Re represents the reference Reynolds number defined as $Re = a_0 \rho_0 L/(\sqrt{\gamma} \mu_0)$, Pr the Prandtl number, κ the coefficient of thermal conductivity, μ the viscosity coefficient, and γ the ratio of specific heats. The coefficient of thermal conductivity and the viscosity coefficient are nondimensionalized with respect to their reference values. The pressure, density and velocity components are related to the total energy per unit volume e by the following equation for an ideal gas

$$e = p/(\gamma - 1) + \rho(u^2 + v^2)/2 . \qquad (12)$$

The so-called contravariant velocity components U and V in the ξ and η directions, respectively, are given by

$$U = \xi_x u + \xi_y v \quad , \quad V = \eta_x u + \eta_y v \quad . \tag{13}$$

In the general coordinate transformation, ξ varies around the body surface, and η varies away from it. The transformation Jacobian in Eq. (10) is defined by

$$J = \xi_x \eta_y - \xi_y \eta_x \quad . \tag{14}$$

The geometrical factors (metrics) in Eqs. (10)-(14) resulting from the coordinate transformation are defined as follows in terms of the derivatives of Cartesian coordinates of the grid points;

$$\xi_x = J y_\eta \quad , \quad \xi_y = -J x_\eta \quad , \quad \eta_x = -J y_\xi \quad , \quad \eta_y = J x_\xi \quad . \tag{15}$$

In general, the metrics of Eq. (15) are not known analytically and must be determined numerically at the beginning of the calculation and stored.

For inviscid flow calculations, the viscous terms $\vec{R}$ and $\vec{S}$ are eliminated. Equation (9) then reduces to the Euler equations. In order to eliminate spurious oscillations, which will be triggered by discontinuities in the solutions such as shock waves, we added the dissipative terms D_ξ and D_η to the right hand side of Eq. (9). Following Jameson and Baker[4], the dissipation is introduced by a combination of second and fourth order differences as

$$D_\xi = d_{i+1/2,j} - d_{i-1/2,j} \quad , \tag{16}$$

where the dissipative flux $d_{i+1/2,j}$ is defined by

$$d_{i+1/2,j} = J_{i+1/2,j} [\, \varepsilon^{(2)}_{i+1/2,j} \Delta_\xi (\vec{q}/J)_{i,j} - \varepsilon^{(4)}_{i+1/2,j} \Delta_\xi^3 (\vec{q}/J)_{i-1,j} \,]/\Delta t \quad . \tag{17}$$

Here Δ_ξ is the forward difference operator and $\varepsilon^{(2)}$ and $\varepsilon^{(4)}$ are defined as

$$\varepsilon^{(2)}_{i+1/2,j} = \omega^{(2)} \max[\nu_{i+1,j}, \nu_{i,j}] \quad , \quad \varepsilon^{(4)}_{i+1/2,j} = \max[0, \omega^{(4)} - \varepsilon^{(2)}_{i+1/2,j}] \quad , \tag{18}$$

where $\omega^{(2)}$ and $\omega^{(4)}$ are adaptive coefficients. As an effective sensor of the presence of a shock wave, we take the second difference of the pressure. Therefore, $\nu_{i,j}$ is defined as

$$\nu_{i,j} = |p_{i+1,j} - 2p_{i,j} + p_{i-1,j}|/(p_{i+1,j} + 2p_{i,j} + p_{i-1,j}) \quad . \tag{19}$$

Numerical Results

This section presents some typical results for two dimensional compressible flows. The first example is the inviscid flow over the NACA0012 airfoil at Mach number

$M_\infty = 0.8$ for an angle of attack $\alpha = 1.25°$. The calculation is carried out on a C-type grid with 129x33 points (97 points around the airfoil). The outer boundaries are located at approximately eight chord length from the airfoil. The treatment of the far field boundary condition is based on the introduction of Riemann invariants for a one dimensional flow normal to the boundary. At a solid boundary, the normal pressure gradient $\partial p/\partial n$ at the wall can be estimated from the momentum equation using tangency condition. The pressure at the wall is extrapolated from the pressure at the adjacent grid point, using the known value $\partial p/\partial n$. In order to accelerate the convergence to steady state solution, the local time stepping and the residual averaging techniques are incorporated into the basic scheme. With the implicit residual averaging, the time steps are chosen at each grid point so that the Courant number is constant everywhere, typically 5.0.

Figure 1 shows the grid and the iso-Mach lines computed by using RRK scheme with residual averaging. For comparison, we also recalculated the flow field by the implicit Beam-Warming code[11]. The calculation was started from an initial condition of uniform flow. The flow field contains a fairly strong shock on the upper surface and a weak shock wave on the lower surface. The pressure distribution obtained by the present method is compared in Fig. 2 with that by the Beam-Warming scheme. The results agree quite well with each other except slight discrepancy in the upper surface shock region. With the present grid (129x33), the total CPU time needed to obtain fully converged solution (1600 time steps) is 19.6s on a FUJITSU VP-200.

The first example for the viscous flow calculations deals with supersonic flow over the NACA0012 airfoil at an angle of attack $\alpha = 10°$. Figure 3 displays the Mach number contours for the case with $M_\infty = 2.0$, $Re = 10^3$. In this range of values of Mach and Reynolds numbers, the shock wave thickness is not negligible and should not be assumed as a discontinuity any longer. Therefore the structure of the shock wave should be resolved by using enough grid points in the transitional region of finite thickness. To resolve very thin shock structure, a fine grid with 257x257 points is used. A definite bow shock is clearly captured without any form of artificial viscosity.

The second example is a cascade of NACA65(12)10 compressor blade. Numerical solutions were obtained for a transonic Mach number $M_\infty = 0.76$, flow angle $\beta_1 = 45°$, stagger angle $\gamma_s = 28.5°$, solidity $\sigma = 1.0$, and Reynolds number $Re = 3 \times 10^5$. The algebraic two-layer eddy-viscosity model proposed by Baldwin and Lomax[12] is used to simulate turbulent flows. The grid had 129x33 points (97 points along the blade). Computed Mach number contours are shown in Fig. 4. A small separated flow region is apparent on the suction surface. Figure 5 shows the computed Cp distribution and the blade profile. The computed Cp distribution is compared with two sets of experimental data with and without sidewall boundary-layer removal and porous endwall suction[13]. The numerical solution agree with the experimental data taken with endwall suction. The computed solution indicated a pressure ratio $p_2/p_1 = 1.229$ and turning angle of 23°. With endwall suction the experimental pressure ratio was reported as $p_2/p_1 = 1.128$. The experimental turning angle, which is independent of the suction, was reported as 22°.

4. INCOMPRESSIBLE FLOW PROBLEM

Governing Equations

For incompressible flows, the vectors $\vec{q}$, $\vec{E}$, $\vec{F}$, $\vec{R}$, and $\vec{S}$ in the Navier-Stokes equations, Eq. (9) can be simplified as

$$\vec{q} = \frac{1}{J}\begin{pmatrix} 0 \\ u \\ v \end{pmatrix}, \quad \vec{E} = \frac{1}{J}\begin{pmatrix} u \\ uU + \xi_x p \\ vU + \xi_y p \end{pmatrix}, \quad \vec{F} = \frac{1}{J}\begin{pmatrix} v \\ uV + \eta_x p \\ vV + \eta_y p \end{pmatrix},$$

$$\vec{R} = \frac{1}{J}\begin{pmatrix} 0 \\ U_\xi \\ V_\xi \end{pmatrix}, \quad \vec{S} = \frac{1}{J}\begin{pmatrix} 0 \\ U_\eta \\ V_\eta \end{pmatrix}. \tag{20}$$

In two dimensions, it is sometimes convenient to solve the Navier-Stokes equations in terms of the vorticity and stream function. With the stream function Ψ related to the velocity by $u = \partial\Psi/\partial y$ and $v = -\partial\Psi/\partial x$ and to the z component of vorticity by $\omega = \partial v/\partial x - \partial u/\partial y$, the Navier-Stokes equations are

$$\omega_t + \frac{1}{J}(\Psi_\eta \omega_\xi - \Psi_\xi \omega_\eta) = \frac{1}{Re \cdot J^2}(\alpha\omega_{\xi\xi} - 2\beta\omega_{\xi\eta} + \gamma\omega_{\eta\eta} + \sigma\omega_\xi + \tau\omega_\eta), \tag{21}$$

$$\alpha\Psi_{\xi\xi} - 2\beta\Psi_{\xi\eta} + \gamma\Psi_{\eta\eta} + \sigma\Psi_\xi + \tau\Psi_\eta = -J^2\omega, \tag{22}$$

where coefficients α, β, γ, σ, and τ are the functions of metric coefficients x_ξ, x_η, y_ξ, and y_η.

Steady and Unsteady Solutions

The pseudocompressibility technique consists of adding an artificial term to the continuity equation in vector $\vec{q}$,

$$\vec{q} = (p/\delta, u, v)^T / J. \tag{23}$$

The parameter p/δ becomes an artificial compressibility term. If the iteration is taken to a steady state ($\partial/\partial t \to 0$), the influence of this artificial term becomes negligible. The Navier-Stokes equations with the artificial compressibility term can be solved similarly to the compressible form of equations.

As an example of result obtained by the pseudocompressibility approach, computed isobars and velocity vectors of a steady incompressible flow past NACA0012 airfoil for Re = 5000 at zero angle of attack are shown in Fig. 6. The results are computed by using 385x65 grid points.

In case of time dependent flow problems, we have to solve either ω-Ψ equations (21) and (22) or u, v, and p equations. In this case the Poisson equation should be solved by using either direct or iterative method. Figure 7 shows the computed stream lines for the unsteady flow past the NACA0012 airfoil at an angle of attack α = 30° and Re = 1000. Comparison with the results of Ta Phouc[14] shows quite excellent agreement.

Higher Order Solution

For problem that requires higher order accuracy in both space and time such as numerical simulation of turbulence, higher order method of lines is an obvious approach. Higher order method of lines using a combination of MDQ method and RK4 time integration scheme has been applied to wide variety of flow problems[15-17].

As an example of simulation for high Reynolds number flow, Figure 8 shows the vorticity contours at t = 3 for ν = 0.0001 obtained by using the 10th order MDQ method on a 512X512 grid. The integral-scale Reynolds number Re is about 25500 at t = 0. The higher order method of lines approach is about 5 times as fast as the pseudo-spectral method.

5. CONCLUSIONS

A method of lines approach has been presented for the numerical solution of fluid dynamic equations. For wide variety of flow problems, the present approach gives efficient and reliable solutions. In addition the method easily adapt to vector computers.

REFERENCES

[1] Machura, M. and Sweet, R.A., A survey of software for partial differential equations, ACM Trans. Math. Software 6, 1980, pp. 461-488.
[2] Jones, D.J., The Numerical Solution of Elliptic Equations by the Method of Lines, Computer Physics Communication 4, 1972, pp. 165-172.
[3] Jones, D.J., South, J.C., and Klunker, E.B., On the Numerical Solution of Elliptic Partial Differential Equations by the Method of Lines, J. Comp. Phys. 9, 1972, pp. 496-527.
[4] Jameson, A. and Baker, T.J., Solution of the Euler Equations for Complex Configuration, AIAA paper 83-1929, 1983.
[5] Beam, R. and Warming, R.F., An Implicit Factored Scheme for the Compressible Navier-Stokes Equations, AIAA paper 77-645, 1977.
[6] Wambecq, A., Rational Runge-Kutta Methods for Solving Systems of Ordinary Differential Equations, Computing 20, 1978, pp. 333-342.
[7] Hicks, J.S. and Wei, J., Numerical Solution of Parabolic Partial Differential Equations With Two-Point Boundary Conditions by Use of the Method of Lines, J. Association for Computing Machinery 14, 3, 1967, pp. 549-562.
[8] Hyman, J.M., A Method of Lines Approach to the Numerical Solution of Conservation Laws, 3rd IMACS Int. Symp. Computer Methods for Partial Differential Equation, 1979.
[9] Heydweiller, J.C. and Sincovec, R.F., A Stable Scheme for the Solution of Hyperbolic Equations Using the Method of LInes, J. Comp. Phys. 22, 1976, pp. 377-388.
[10] Satofuka, N., Modified Differential Quadrature Method for Numerical Solution of Multi-Dimensional Flow Problems, Int. Symp. Appl. Math. Inf. Sci., 1982.
[11] Pulliam, T.H., Jespersen, D.C. and Childs, R.E., An Enhanced Version of an Implicit Code for the Euler Equations, AIAA Paper 83-0314, 1983.
[12] Baldwin, B.S. and Lomax, H., Thin Layer Approximation and Algebraic Model for

Separated Turbulent Flows, AIAA Paper 78-257, 1978.
[13] Briggs, W.B., Effect of Mach Number on the Flow and Application of Compressibility Corrections in a Two-Dimensional Subsonic-Transonic Compression Cascade Having Varied Porous-Wall Suction at the Blade Tips, NASA TN-2649, 1952.
[14] Ta Phuoc, L., Daube, O., Monnet, P. and Coutanceau, M., A Comparison of Numerical Simulation and Experimental Visualization of the Early Stage of the Flow Generated by an Impulsively Started Elliptic Cylinder, Numerical Methods in Laminar and Turbulent Flow, 1983, pp. 269-279.
[15] Satofuka, N. and Nishida, H., A New Method for the Numerical Simulation of Turbulence, BAIL III Conf., 1984.
[16] Satofuka, N., Nakamura, H. and Nishida, H., Higher Order Method of Lines for the Numerical Simulation of Turbulence, IC9NMFD, 1984.
[17] Tokunaga, H., Satofuka, N. and Tanimura, Y., Higher Order Accurate Difference Method with New Direct Poisson Solver and its Application to Direct Simulation of Boundary Layer Instability, Int. Symp. Comp. Fluid Dynamics-Tokyo, 1985.

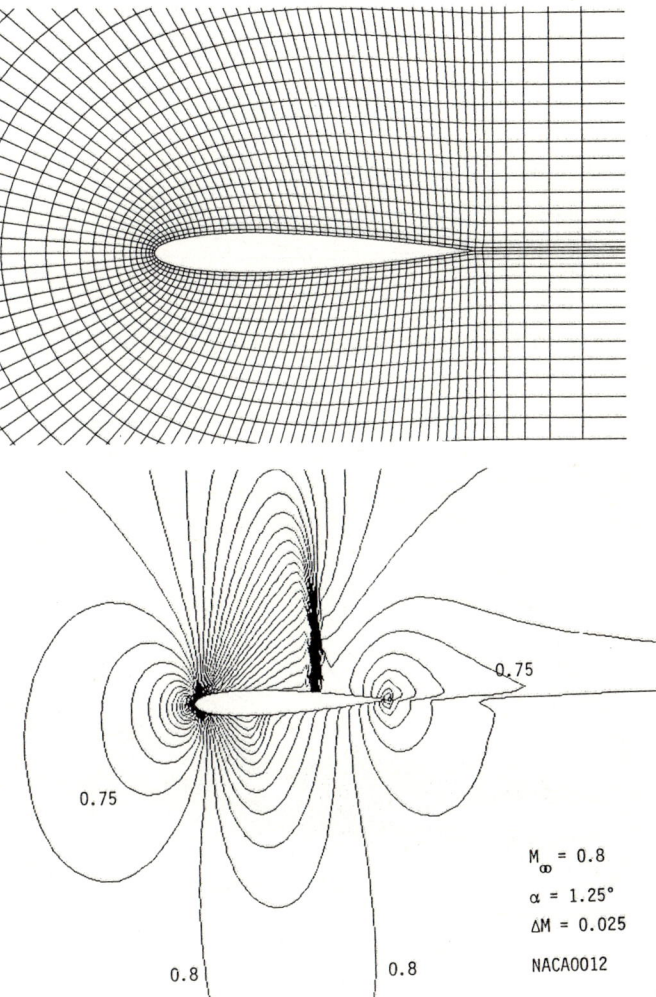

Figure 1   Computed grid and iso-Mach lines for NACA0012 at free stream Mach number $M_\infty = 0.8$ and angle of attack $\alpha = 1.25°$

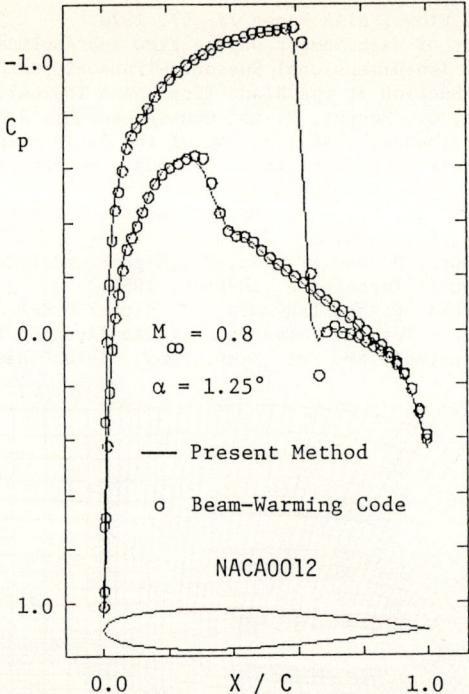

Figure 2   Pressure distribution for NACA0012.

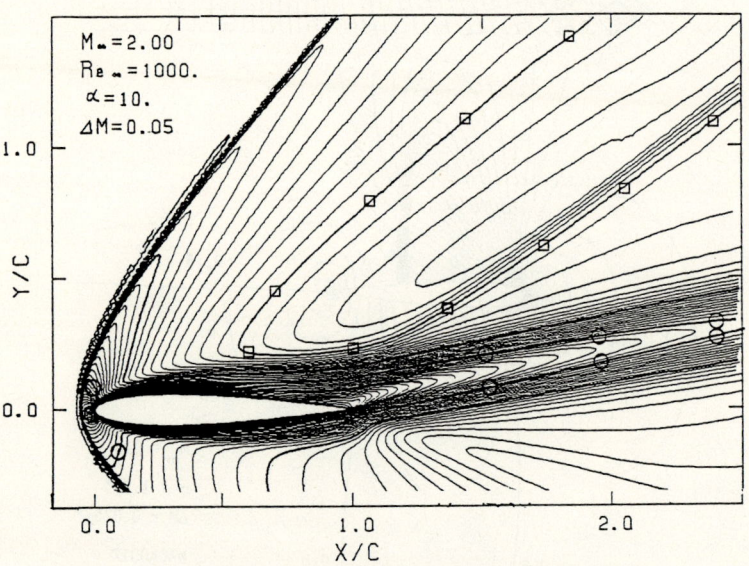

Figure 3   Iso-Mach lines computed for NACA0012 at $M_\infty = 2.0$, Re = 1000, and $\alpha = 10°$.

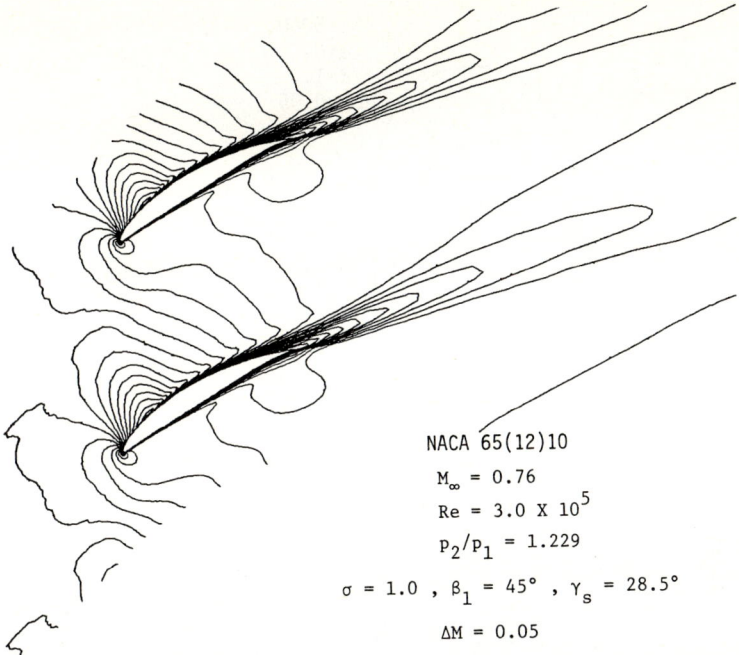

Figure 4   Iso-Mach lines computed for NACA 65(12)65 cascade.

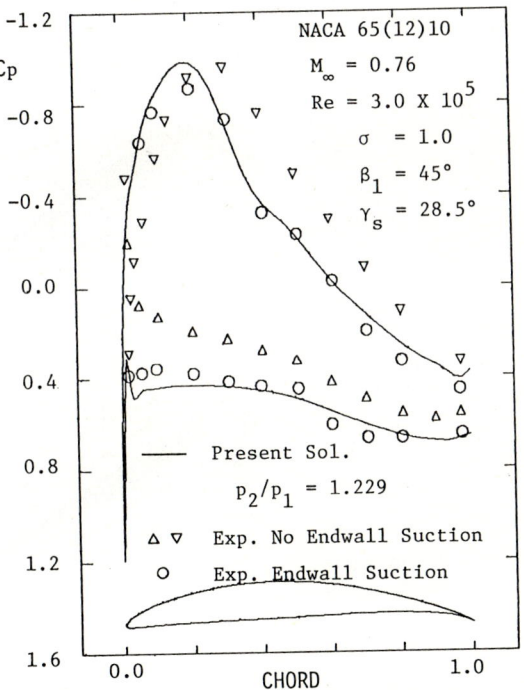

Figure 5   Pressure distribution for NACA 65(12)10 cascade.

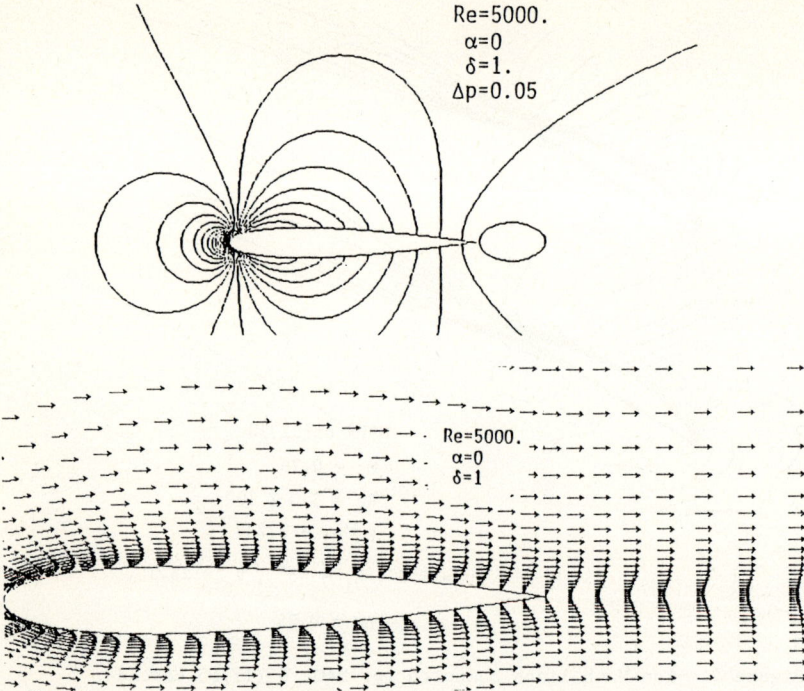

Figure 6 Isobars and velocity vectors computed for a steady incompressible flow past NACA0012 at Re = 5000 and α = 0°.

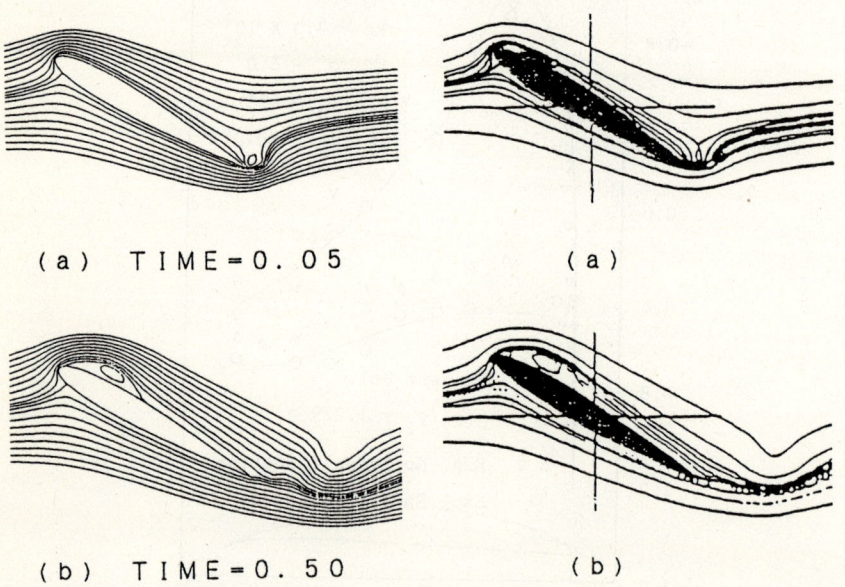

Figure 7 Stream lines computed for an unsteady flow past NACA0012 at Re = 1000 and α = 30° and compared with results of Ta Phouc.[14]

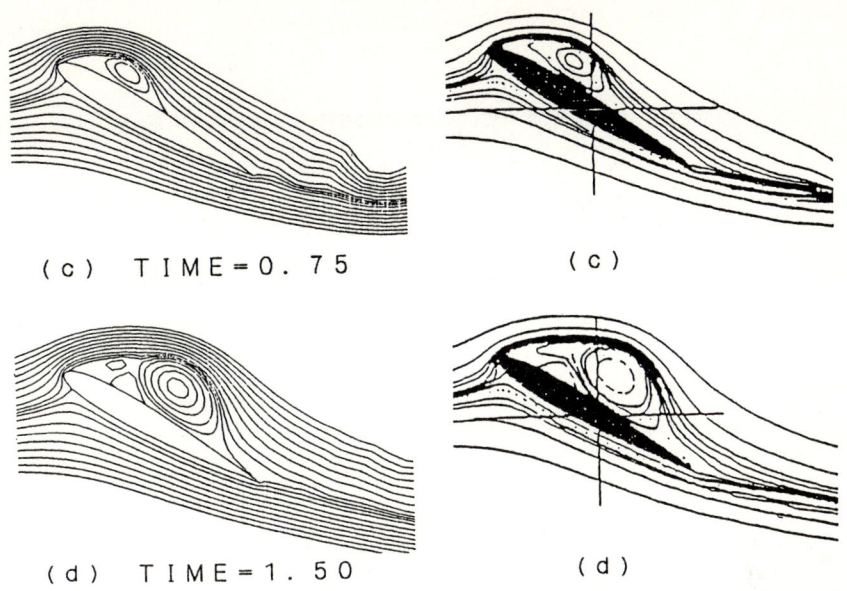

Figure 7 Concluded.

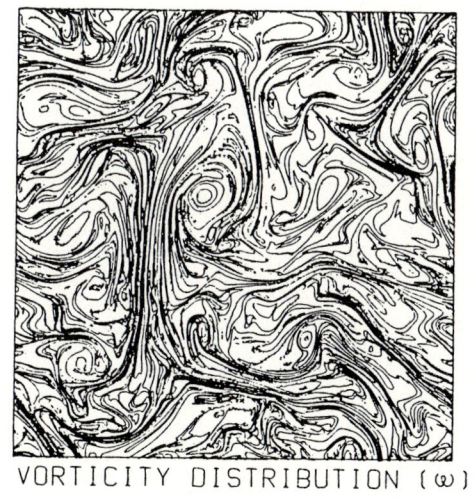

Figure 8 Vorticity contours at t = 3 for ν = 0.0001.

ON A MARCHING ITERATION METHOD IN SOLVING GAS DYNAMIC EQUATIONS

F.G. Zhuang and H.X. Zhang

China Aerodynamics Research and Development Center
Mianyang, Sichuan, China

1. Introduction

The rapid development of super-computers has promoted the research in computational aerodynamics and the requirement of complete numerical flow simulation demands even super-super computers to be developed. It must be pointed out that the development of numerical methods in computational aerodynamics is nevertheless an important aspect, and there still exists a great potential in numerical algorithm improvements. A brief review of the recent methods developed suggested it is essential to develop a new practical finite difference scheme to meet the following objectives, namely, it must be physically well founded, possess good shock capturing capability, high convergence rate, i.e., high numerical efficiency, high accuracy, great robustness and adaptivity to wide class of flows. In general we have explicit, implicit or hybrid difference schemes. The explicit schemes usually need the time step Δt to be less than a certain value determined by the CFL condition to meet the stability requirement. In a flow region where the parameters vary rapidly, it is mandatory to have a very fine mesh. Thus the allowable value for Δt is further reduced and it takes a large CPU time to reach the desired steady state. This is in certain cases not practical besides non-economical. Multiple grid method is a possible outlet. But to relax the restriction on the choice of Δt is of basic importance, and in this respect it seems reasonable to use implicit scheme. In addition one often tries to use non-iterative method together with this scheme, thus the difference operators are further modified by approximate factorization, ADI or time splitting technique. If central differences are used for derivatives, we obtain a system of equations with block triangular matrix coefficients which can be easily inverted[1]. Here we shall encounter two problems, the first is that the difference scheme for the convective terms is not dissipative in the sense of Kreiss, and we have to overcome the nonlinear instability with the introduction of artificial viscosity terms, the forms of which and the coefficients contained there in being quite empirical depend on the flow problems to be solved. It is clear that this is a defect fundamental in the theory. The second is with the adoption of approximate factorization technique etc, the size of Δt is limited at least in the final period of convergence in order to ensure the good accuracy of the numerical solution, or in other words, the solution depends on Δt. The first problem can be resolved with the upwind difference schemes, especially with a flux splitting method[2], then there is no need to introduce the artificial viscosity and at the same time, the solution can still be obtained with a simple non-iterative method. As to the second one, we may abandon the use of approximate factorization etc. and employ Gauss-Seidel iteration, yet the artificial viscosity terms will have to be retained. It seems then natural that we would like to have a combination of these two remedies, i.e., using flux-splitting and Gauss - Seidel iteration method together to eliminate the restriction on Δt and remove the artificial viscosity terms and in (3) we propose a new marching-iteration method thus the rate of convergence is greatly increased, besides a pressure relaxation factor is introduced to further increase the convergence rate. In the second section following we shall present basic description of the method, including the difference schemes and treatment of boundary conditions which also plays a key role in the effectiveness of the numerical algorithm. The third section deals with the pressure relaxation factor and in the fourth section some numerical examples are given and compared with the results obtained with different methods. Finally we give a synthetic discussion.

II. Basic Aspects of Implicit Marching-Iteration Method

For simplicity in illustrating the basic aspect of the present method, we consider a laminar flow over a two-dimensional flat plate, with simplified N.S. equations as governing equations which, when written in a non-dimensional conservation form, are as follows

$$\frac{\partial U}{\partial t} + \frac{\partial F}{\partial x} + \frac{\partial G}{\partial y} = \frac{\partial}{\partial y}\left(\frac{\mu}{Re_\infty}\frac{\partial \Theta}{\partial y}\right) \qquad (2.1)$$

where

$$U = (\rho,\ \rho u,\ \rho v,\ \rho e)^T \qquad (2.2)$$

$$F = (\rho u,\ \rho u^2 + p,\ \rho uv,\ (\rho e + p)u)^T \qquad (2.3)$$

$$G = (\rho v,\ \rho vu,\ \rho v^2 + p,\ (\rho e + p)v)^T \qquad (2.4)$$

$$\Theta = (0,\ u,\ 0,\ H + (\frac{1}{Pr} - 1)h\ )^T \qquad (2.5)$$

and the usual notations are used. x and y are non-dimensional Cartesian coordinates with x axis along the body surface and y axis normal to the surface. The Characteristic length is the length of the flat plate L, and the other characteristic quantities are the free stream density ρ_∞ and the free stream velocity V_∞. Only for the evaluation of non-dimensional viscosity coefficient μ, its free stream value μ_∞ is used as characteristic reference quantity, and the Sutherland formula is used. We may consider equation (2.1) as an equation for the vector U, and F, G and Θ as functions of U.

The extension to full N. S. equations is straightforward.

Assume the equation (2.1) is satisfied at n+1-th level (which may be understood as n+1-th time level or n+1-th iteration), i.e.,

$$\left(\frac{\partial U}{\partial t}\right)^{n+1} + \frac{\partial F^{n+1}}{\partial x} + \frac{\partial G^{n+1}}{\partial y} = \frac{\partial}{\partial y}\left(\frac{\mu^{n+1}}{Re_\infty}\frac{\partial \Theta^{n+1}}{\partial y}\right) \qquad (2.6)$$

and we define the increment of U from n-th to n+1-th iteration δU^{n+1} by $\delta U^{n+1} = U^{n+1} - U^n$, then using Taylor's expansion, we have obviously

$$\begin{cases} F^{n+1} = F^n + A^n\ \delta U^{n+1} + \cdots \\ G^{n+1} = G^n + B^n\ \delta U^{n+1} + \cdots \\ \Theta^{n+1} = \Theta^n + C^n\ \delta U^{n+1} + \cdots \\ \mu^{n+1} = \mu^n + \mu'(T)E^n \delta U^{n+1} + \cdots \end{cases} \qquad (2.7)$$

where

$$A = \frac{\partial F}{\partial U},\quad B = \frac{\partial G}{\partial U},\quad C = \frac{\partial \Theta}{\partial U},\quad E = \frac{\partial T}{\partial U} \qquad (2.8)$$

are Jacobian matrices for F, G, Θ, T respectively, and

$$F = AU,\qquad G = BU. \qquad (2.9)$$

If we use $(U^{n+1} - U^n)/\Delta t_n$ as an approximation to $(\partial U/\partial t)^{n+1}$, and Δt_n the time step, substituting the expressions (2.7) into (2.6), rearranging and neglecting higher order terms, we have

$$\frac{\delta U^{n+1}}{\Delta t} + \frac{\partial}{\partial x}(A^n \delta U^{n+1}) + \frac{\partial}{\partial y}(B^n \delta U^{n+1}) - \frac{\partial}{\partial y}\left[\frac{\mu^n}{Re_\infty}\frac{\partial}{\partial y}(C^n \delta U^{n+1})\right]$$
$$- \frac{\partial}{\partial y}\left[\frac{1}{Re_\infty}(\frac{\partial \beta}{\partial y}\cdot \alpha\cdot)\right]^n \delta U^{n+1} = -\frac{\partial}{\partial x}(A^n U^n) + \frac{\partial}{\partial y}(B^n U^n) - \frac{\partial}{\partial y}(\frac{\mu\ C^n}{Re_\infty}\frac{\partial U^n}{\partial y})] \qquad (2.10)$$

where

$$\alpha = \frac{\gamma-1}{\rho}\mu'(T)\begin{pmatrix} \frac{u^2+v^2}{2} - \frac{a^2}{\gamma(\gamma-1)} & 0 & 0 & 0 \\ 0 & -u & 0 & 0 \\ 0 & 0 & -v & 0 \\ 0 & 0 & 0 & 1 \end{pmatrix}$$

$$\beta = \begin{pmatrix} 0 & 0 & 0 & 0 \\ u & u & u & u \\ 0 & 0 & 0 & 0 \\ \tilde{H} & \tilde{H} & \tilde{H} & \tilde{H} \end{pmatrix}, \qquad \tilde{H} = H + (\frac{1}{P_r} - 1)h .$$

Through similar transformation, matrices A and B can be expressed as

$$A = S^{-1} \Lambda_A S$$
$$B = T^{-1} \Lambda_B T .$$

Λ_A and Λ_B are the characteristic diagonal matrices for A and B whose expressions are

$$\Lambda_A = \text{DIAG}(u, u+a, u, u-a) \tag{2.11}$$
$$\Lambda_B = \text{DIAG}(v, v+a, v, v-a) \tag{2.12}$$

and they may be splitted as following

$$\Lambda_A = \Lambda_{A+} + \Lambda_{A-} \tag{2.13}$$
$$\Lambda_B = \Lambda_{B+} + \Lambda_{B-} . \tag{2.14}$$

Λ_{A+} is a diagonal matrix composed of non-negative eigenvalues of A and the negative eigenvalues replaced by zeros; Λ_{A-} is a diagonal matrix composed of non-positive eigenvalues of A and the positive eigenvalues replaced by zeros. Similar definitions are applied to Λ_{B+} and Λ_{B-}.

As a method of flux splitting, we put

$$A = A_+ + A_- \tag{2.15}$$
$$B = B_+ + B_- \tag{2.16}$$

where

$$A_+ = S^{-1} \Lambda_{A+} S, \qquad A_- = S^{-1} \Lambda_{A-} S \tag{2.17}$$
$$B_+ = T^{-1} \Lambda_{B+} T, \qquad B_- = T^{-1} \Lambda_{B-} T . \tag{2.18}$$

Substituting (2.15) and (2.16) in (2.10) and performing derivative calculations with the rules of signal propagation, we obtain the following equation

$$[\frac{1}{\Delta t_n} + \frac{\partial}{\partial x_-} A_+^n + \frac{\partial}{\partial x_+} A_-^n + \frac{\partial}{\partial y_-} B_+^n + \frac{\partial}{\partial y_+} B_-^n - \frac{\partial}{\partial y_c} \frac{\mu^n}{Re_\infty} \frac{\partial}{\partial y_c} C^n - \frac{\partial}{\partial y_c} \frac{1}{Re_\infty} (\frac{\partial \beta}{\partial y} \alpha)^n] \delta U^{n+1}$$

$$= -[\frac{\partial}{\partial x_-} A_+^n + \frac{\partial}{\partial x_+} A_-^n + \frac{\partial}{\partial y_-} B_+^n + \frac{\partial}{\partial y_+} B_-^n - \frac{\partial}{\partial y_c} \frac{\mu^n}{Re_\infty} C^n \frac{\partial}{\partial y}] U^n . \tag{2.19}$$

Here $\frac{\partial}{\partial x_+}$, $\frac{\partial}{\partial y_+}$ will be replaced by forward differences and $\frac{\partial}{\partial x_-}$, $\frac{\partial}{\partial y_-}$ will be replaced by backward differences, subscript c denotes that central difference is to be used, i.e., in computing viscous terms we use central difference scheme.

We immediately recognize that as $\Delta t_n \to 0$ or sufficiently small, equation (2.19) reduces to an explicit form; as $\Delta t_n \to \infty$ equation (2.19) represents an iteration scheme for steady equations, in fact, it is in the form of Newton's iteration. The flexibility of the method is evident. In general, if all the physical quantities are known at the n-th level, then the equation (2.19) yields an implicit scheme for the solution of δU^{n+1}. It seems difficult to solve equation (2.19), therefore many authors employ approximate factorization technique etc., thus reducing the solution procedure to non-iterative method in several steps. In so doing, the value of Δt_n cannot be too large to effect the accuracy of the solution despite the fact that from the stability point of view, the value of Δt_n is not limited by the CFL condition, the related CFL number could be larger than 1 by several orders of magnitude, and at the same time the rate of convergence is greatly enhanced. We notice that equation (2.19) possesses the form of implicit scheme as suggested by MacCormack, which is

$$\{ \text{Numerical method} \} \delta U^{n+1} = \{ \text{Physics} \},$$

the right hand side actually representing the residue of steady equations. Here we must emphasize two points. For non-steady flow, it is obvious, the left hand side cannot be merely numerical but it must possess the intrinsic physics contained in equation (2.19) as propagation of disturbances; for the limit steady flow, it may seem that we only need the stability criterion to be satisfied and the rapid rate of convergence and do not require the scheme to possess actual physical evolution property. In order to simplify the numerical calculations, the latter can be done in most cases, for example, we may omit the last term in the left hand side of the equation, However this may lead to different steady solutions in the limit, if we start with different initial conditions; and it is clear that if we let $\Delta t_n \to \infty$, we should arrive at an iteration form for steady equation, and the two sides of the equation must be compatible, otherwise the converged solution may not be the solution required in a strict sense.

If we are only interested in the steady solution, we may let $\Delta t_n \to \infty$ (Δt_n could be retained, and the following procedure does apply to this case without causing additional difficulties), and the equation (2.19) is now written as

$$[\frac{\partial}{\partial x_+} A_-^n + \frac{\partial}{\partial x_-} A_+^n + \frac{\partial}{\partial y_+} B_-^n + \frac{\partial}{\partial y_-} B_+^n - \frac{\partial}{\partial y_c} \frac{\mu^n}{Re_\infty} \frac{\partial}{\partial y_c} C^n - \frac{\partial}{\partial y_c} \frac{1}{Re_\infty} (\frac{\partial \beta}{\partial y} \cdot \alpha)^n] \delta U^{n+1} = K^n \quad (2.20)$$

where

$$K^n = -(\frac{\partial}{\partial x_+} A_-^n + \frac{\partial}{\partial x_-} A_+^n + \frac{\partial}{\partial y_+} B_-^n + \frac{\partial}{\partial y_-} B_+^n - \frac{\partial}{\partial y_c} \frac{\mu^n}{Re_\infty} C^n \frac{\partial}{\partial y}) U^n.$$

Suppose we consider marching iteration in the x direction. We may move the x derivatives on the left hand side to the right hand side, such that

$$[\frac{\partial}{\partial y_+} B_-^n + \frac{\partial}{\partial y_-} B_+^n - \frac{\partial}{\partial y_c} \frac{\mu^n}{Re_\infty} \frac{\partial}{\partial y_c} C^n - \frac{\partial}{\partial y_c} \frac{1}{Re_\infty} (\frac{\partial \beta}{\partial y} \cdot \alpha)^n] \delta U^{n+1}$$

$$= K^n - \frac{\partial}{\partial x_+} A_-^n \delta U^{n+1} - \frac{\partial}{\partial x_-} A_+^n \delta U^{n+1}. \quad (2.21)$$

If we introduce in the x-y plane a uniform mesh system with grid lines parallel to the coordinate axes (see Fig. 1) and notice that at grid points i,j

$$\frac{\partial}{\partial x_+} A_-^n \delta U^{n+1} \approx A_{-\ i+1,j}^n \delta U_{i+1,j}^{n+1} - A_{-i,j}^n \delta U_{i,j}^{n+1} \quad (2.22a)$$

$$\frac{\partial}{\partial x_-} A_+^n \delta U^{n+1} \approx A_{+\ i,j}^n \delta U_{i,j}^{n+1} - A_{+i-1,j}^n \delta U_{i-1,j}^{n+1} \quad (2.22b)$$

and if we march in the positive x direction, the expression in (2.22a)

$$A_{-i+1,j}^n \delta U_{i+1,j}^{n+1}$$

is replaced by

$$A_{-i+1,j}^n \delta U_{i+1,j}^n$$

i.e., along the line (i+1), the values $\delta U_{i+1,j}^{n+1}$ at n+1-th level are replaced by that of n-th level; while the expressions in (2.22b) are either known from previous calculations or are to be found in the ongoing calculations. With this artifice the marching calculations in the positive x direction can be made. A similar device can be adopted for marching in the negative x direction. In short we arrive in both cases at a system of equations of $U_{i,j}$ for fixed i. Using boundary conditions, or in other words, imbedding these conditions into the system, we can simultaneously solve for the values of $U_{i,j}$ at different j. It may be seen that as the iteration process converges, i.e., $\delta U^n = \delta U^{n+1} = 0$, we get numerical solutions of exact steady NS equations. The process starts with a given initial flow fields, then we proceed marching in the positive x direction and then in the negative x direction; the process is repeated until certain convergence criterion for δU^n is satisfied. In order to get a solution with second order accuracy, we employ Beam-Warming's two step second order difference scheme for the evaluation of K^n. First order accuracy difference schemes are used for terms containing δU^{n+1}. Thus it makes the iteration simple and reliable and at the same time we have a second order accuracy in the steady case.

From equation (2.21), we have the following difference equations for marching in the positive x direction:

Predictor step:
$$\widetilde{A}^n_{i,j}\, \overline{\delta U^{n+1}_{i,j+1}} + \widetilde{B}^n_{i,j}\, \overline{\delta U^{n+1}_{i,j}} + \widetilde{C}^n_{i,j}\, \overline{\delta U^{n+1}_{i,j-1}} = \widetilde{D}^n_{i,j} \qquad (2.23)$$

where

$$\widetilde{A}^n_{i,j} = \frac{\Delta x}{\Delta y} B^n_{-i,j+1} - \frac{\Delta x}{\Delta y^2} V^n_{i,j+\frac{1}{2}} - (D)^n_{i,j+\frac{1}{2}} \frac{\Delta x}{\Delta y^2}$$

$$\widetilde{B}^n_{i,j} = A^n_{+i,j} - A^n_{-i,j} - B^n_{-i,j}\frac{\Delta x}{\Delta y} + B^n_{+i,j}\frac{\Delta x}{\Delta y} - (V^n_{i,j+\frac{1}{2}} - V^n_{i,j-\frac{1}{2}})\frac{\Delta x}{\Delta y^2}$$
$$+ (D^n_{i,j+\frac{1}{2}} + D^n_{i,j-\frac{1}{2}})\frac{\Delta x}{\Delta y^2}$$

$$\widetilde{C}^n_{i,j} = -\frac{\Delta x}{\Delta y} B^n_{+i,j-1} + \frac{\Delta x}{\Delta y^2} V^n_{i,j-\frac{1}{2}} - (D)^n_{i,j-\frac{1}{2}} \frac{\Delta x}{\Delta y^2}$$

$$\widetilde{D}^n_{i,j} = \Delta x\, K^n - A^n_{-i+1,j}\, \delta U^n_{i+1,j} + A^n_{+i-1,j}\, \delta U^{n+1}_{i-1,j}$$

$$D^n_{i,j+\frac{1}{2}} = \frac{1}{Re_\infty}\, \frac{\mu_{i,j+1} + \mu_{i,j}}{2}\, C^n_{i,j+1}$$

$$V^n_{i,j+\frac{1}{2}} = \frac{1}{Re_\infty}\, \frac{(\beta_{i,j+1} - \beta^n_{i,j})(\alpha^n_{i,j+1} + \alpha^n_{i,j})}{4} .$$

Corrector step:
$$\widetilde{A}^{\overline{n+1}}_{i,j}\, \delta U^{n+1}_{i,j+1} + \widetilde{B}^{\overline{n+1}}_{i,j}\, \delta U^{n+1}_{i,j} + \widetilde{C}^{\overline{n+1}}_{i,j}\, \delta U^{n+1}_{i,j-1} = \widetilde{D}^{\overline{n+1}}_{i,j} \qquad (2.24)$$

where

$$\widetilde{A}^{\overline{n+1}}_{i,j} = \frac{\Delta x}{\Delta y} \overline{B^{n+1}_{-i,j+1}} - \frac{\Delta x}{\Delta y^2} \overline{V^{n+1}_{i,j+\frac{1}{2}}} - (D)^{\overline{n+1}}_{i,j+\frac{1}{2}} \frac{\Delta x}{\Delta y^2}$$

$$\widetilde{B}^{\overline{n+1}}_{i,j} = \overline{A^{n+1}_{+i,j}} - \overline{A^{n+1}_{-i,j}} - \overline{B^{n+1}_{-i,j}}\frac{\Delta x}{\Delta y} + \overline{B^{n+1}_{+i,j}}\frac{\Delta x}{\Delta y}$$
$$- (\overline{V^{n+1}_{i,j+\frac{1}{2}}} - \overline{V^{n+1}_{i,j-\frac{1}{2}}})\frac{\Delta x}{\Delta y^2} + (\overline{D^{n+1}_{i,j+\frac{1}{2}}} + \overline{D^{n+1}_{i,j-\frac{1}{2}}})\frac{\Delta x}{\Delta y^2}$$

$$\widetilde{C}^{\overline{n+1}}_{i,j} = -\frac{\Delta x}{\Delta y} \overline{B^{n+1}_{+i,j-1}} + \frac{\Delta x}{\Delta y^2} \overline{V^{n+1}_{i,j-\frac{1}{2}}} - (D)^{\overline{n+1}}_{i,j-\frac{1}{2}} \frac{\Delta x}{\Delta y^2}$$

$$\widetilde{D}^{\overline{n+1}}_{i,j} = \frac{\Delta x}{2}(K^n + \overline{K^{n+1}}) - \overline{A^{n+1}_{-i+1,j}}\, \delta U^{n+1}_{i+1,j} + \overline{A^{n+1}_{+i-1,j}}\, \delta U^{n+1}_{i-1,j}.$$

In the above expressions, the difference formulas of the derivatives of f in K^n and $\overline{K^{n+1}}$ are carefully arranged to insure second order accuracy in the steady state.

Similarly we may get from equation (2.21) the difference equations used in marching along negative x direction.

As a complete treatment of a physical problem, the boundary conditions must be discussed and imbedded in the difference equations. Consider the typical flow problem, the viscous flow over a flat plate (Figure 1). Just for illustration the computational domain is taken to be a rectangle with a uniform mesh, index i from 1 to N represents the positions along x axis and index j from 1 to M represents positions along y axis. The flat plate lies on a segment of x axis. Thus the following boundary conditions may be applied:

(1) At the inlet section i=1, the profile for U is known, i.e.,

$$U = F(y) \tag{2.25}$$

is given.

(2) At the upper boundary (j = M) we may consider the variation of physical quantities in the y direction can be neglected, i.e.,

$$\frac{\partial U}{\partial y} = 0 . \tag{2.26}$$

(3) At the exit boundary (i = N), we neglect the variations of flow parameters in the x direction, i.e.,

$$\frac{\partial U}{\partial x} = 0 . \tag{2.27}$$

(4) On the surface we have non-slip conditions and assume that the temperature distribution is known, i.e.,

$$u = v = 0 , \qquad T = T_w . \tag{2.28}$$

As a supplementary condition for computational purpose we consider the normal pressure gradient at the surface is zero, i.e.,

$$p_{i,1} = p_{i,2} . \tag{2.29}$$

This is a good approximation, especially, for the flow at high Re. Using the expressions from (2.25) to (2.29), after some manipulations we get

$$\begin{cases} (\delta U)_{1,j} = 0 \\ (\delta U)_{i,M} = (\delta U)_{i,M-1} \\ (\delta U)_{N,j} = (\delta U)_{N-1,j} \\ (\delta U)_{i,1} = M(\delta U)_{i,2} \end{cases} \tag{2.30}$$

where

$$M = \begin{pmatrix} \frac{(u^2+v^2)_{i,2}}{2e_{i,1}} & -\frac{u_{i,2}}{e_{i,1}} & -\frac{v_{i,2}}{e_{i,1}} & \frac{1}{e_{i,1}} \\ 0 & 0 & 0 & 0 \\ 0 & 0 & 0 & 0 \\ \frac{1}{2}(u^2+v^2)_{i,2} & -u_{i,2} & -v_{i,2} & 1 \end{pmatrix} .$$

Based on the equations (2.30) and (2.34) we now obtain the marching-iteration equation in the positive x direction for the predictor step

$$\begin{pmatrix} I & -M & 0 & & & & \\ \tilde{C}_{i,2} & \tilde{B}_{i,2} & \tilde{A}_{i,2} & & & & \\ & \tilde{C}_{i,3} & \tilde{B}_{i,3} & \tilde{A}_{i,3} & & & \\ & & \ddots & \ddots & \ddots & & \\ & & & \tilde{C}_{i,M-2} & \tilde{B}_{i,M-2} & \tilde{A}_{i,M-2} & \\ & & & & \tilde{C}_{i,M-1} & \tilde{B}_{i,M-1} & \tilde{A}_{i,M-1} \\ & & & & 0 & I & -I \end{pmatrix}^n \begin{pmatrix} \delta U_{i,1} \\ \delta U_{i,2} \\ \delta U_{i,3} \\ \vdots \\ \delta U_{i,M-2} \\ \delta U_{i,M-1} \\ \delta U_{i,M} \end{pmatrix}^{n+1} = \begin{pmatrix} 0 \\ \tilde{D}_{i,2} \\ \tilde{D}_{i,3} \\ \vdots \\ \tilde{D}_{i,M-2} \\ \tilde{D}_{i,M-1} \\ 0 \end{pmatrix}^n . \tag{2.31}$$

In the above equation if n is replaced by $\overline{n+1}$ and $\overline{n+1}$ by n+1, then the system of equations of corrector step is obtained. Similarly, equations for marching iteration in the negative x direction can be obtained. Obviously the coefficient matrix of the equation is of block triangular type, of which the solution method is well known.

III. The Marching Iteration Method with Pressure Relaxation Factor

The pressure relaxation factor introduced here is in the same form as pressure splitting factor introduced by many previous authors. In the marching calculation of steady simplified N.S. equations, it often leads to the so called departure solution if the pressure terms are not properly treated. Therefore in order to extend the region of application of the present method and to improve the rate of convergence, it is beneficial to study the problem of pressure splitting. In this section only the steady flow is to be studied. Due to the pressure splitting, we do not have conservation equations in the strict sense. For simplicity, we use nonconservative form of the equations of motion to illustrate the basic concept of the present technique. To use the conservative form will make some of the expressions more complicated but the essence of the content remains the same. Now introduce the non-conservative variable,

$$V = (\rho, u, v, p)^T \tag{3.1}$$

the governing equation (2.1) becomes

$$\rho A \frac{\partial V}{\partial x} + \rho B \frac{\partial V}{\partial y} - C \frac{\partial}{\partial y} \left(\frac{\mu}{Re_\infty} E \frac{\partial V}{\partial y} \right) = 0 \tag{3.2}$$

where

$$A = \begin{pmatrix} u & \rho & 0 & 0 \\ 0 & u & 0 & \frac{1}{\rho} \\ 0 & 0 & u & 0 \\ 0 & \gamma p & 0 & u \end{pmatrix}, \quad B = \begin{pmatrix} v & 0 & \rho & 0 \\ 0 & v & 0 & 0 \\ 0 & 0 & v & \frac{1}{\rho} \\ 0 & 0 & \gamma p & v \end{pmatrix}$$

$$C = \begin{pmatrix} 1 & 0 & 0 & 0 \\ -u & 1 & 0 & 0 \\ -v & 0 & 1 & 0 \\ \frac{\gamma-1}{2}\rho(u^2+v^2) & -(\gamma-1)\rho u & -(\gamma-1)\rho v & (\gamma-1)\rho \end{pmatrix}$$

$$E = \begin{pmatrix} 0 & 0 & 0 & 0 \\ 0 & 1 & 0 & 0 \\ 0 & 0 & 0 & 0 \\ -\frac{\gamma}{p_r}\frac{1}{\gamma-1}\frac{p}{\rho^2} & u & v & \frac{\gamma}{p_r}\frac{1}{\gamma-1}\frac{1}{\rho} \end{pmatrix}.$$

(3.2) gives the X-momentum equation

$$\rho u \frac{\partial u}{\partial x} + \rho v \frac{\partial u}{\partial y} + \frac{\partial p}{\partial x} = \frac{\partial}{\partial y}\left(\frac{\mu}{Re_\infty} \frac{\partial u}{\partial y} \right)$$

Suppose we are concerning with flows with flow directions nearly parallel to x axis and consider the marching iteration in that direction, then we need only to split the pressure p into $\omega p + (1-\omega)p$ in the x momentum equation, thus the equation becomes

$$\rho u \frac{\partial u}{\partial x} + \rho \frac{\partial u}{\partial y} + \omega \frac{\partial p}{\partial x} = (\omega-1)\frac{\partial p}{\partial x} + \frac{\partial}{\partial y}\left(\frac{\mu}{Re_\infty} \frac{\partial u}{\partial y} \right).$$

The pressure term in the energy equation needs not to be splitted either. The equation (3.2) may be rewritten as

$$\rho \hat{A} \frac{\partial V}{\partial x} + \rho \hat{B} \frac{\partial V}{\partial y} - C \frac{\partial}{\partial y}\left(\frac{\mu}{Re_\infty} E \frac{\partial V}{\partial y} \right) = (\omega-1)L \tag{3.3}$$

where

$$\hat{A} = \begin{pmatrix} u & \rho & 0 & 0 \\ 0 & u & 0 & \omega/\rho \\ 0 & 0 & u & 0 \\ 0 & \gamma p & 0 & u \end{pmatrix}$$

$$\hat{B} = B$$

$$L = \begin{pmatrix} 0 & 0 & 0 & 0 \\ 0 & 0 & 0 & 1 \\ 0 & 0 & 0 & 0 \\ 0 & 0 & 0 & 0 \end{pmatrix} \quad \frac{\partial V}{\partial x} = N \frac{\partial p}{\partial x} \tag{3.4}$$

and

$$N = (0,1,0,0)^T .$$

Same as in the preceding section, we may further express $\hat{A}$ and $\hat{B}$ as

$$\hat{A} = Q \Lambda_{\hat{A}} Q^{-1}$$
$$\hat{B} = R \Lambda_{\hat{B}} R^{-1}$$

where

$$\Lambda_{\hat{A}} = \begin{vmatrix} u & & & 0 \\ & u+\sqrt{\omega}\, u & & \\ & & u & \\ 0 & & & u-\sqrt{\omega}\, u \end{vmatrix}$$

$$\Lambda_{\hat{B}} = \begin{vmatrix} v & & & 0 \\ & v+a & & \\ & & v & \\ 0 & & & v-a \end{vmatrix} ,$$

and then split these coefficient matrices

$$\hat{A} = \hat{A}_+ + \hat{A}_-$$
$$\hat{B} = \hat{B}_+ + \hat{B}_- . \tag{3.5}$$

Finally based on the rules of signal propagation, equation (3.3) becomes

$$\rho \hat{A}_+ \frac{\partial V}{\partial x_-} + \rho \hat{B}_+ \frac{\partial V}{\partial y_-} + \rho \hat{B}_- \frac{\partial V}{\partial y_+} - [C \frac{\partial}{\partial y}(\frac{\mu}{Re_\infty} E \frac{\partial V}{\partial y})]_c$$
$$= -\rho \hat{A}_- \frac{\partial V}{\partial x_+} + (\omega-1)N \frac{\partial p}{\partial x_c} . \tag{3.6}$$

Here for the viscous terms we again use central difference schemes. The eigenvalues of the coefficient matrix of $\partial v/\partial x$ in (3.4) are zero, so in the above expression, for the term ($\partial p/\partial x$) the central difference is also employed. If in the right hand side of equation (3.6), we use their values at previous iteration and also Taylor expansions, then we obtain the following equation used for marching iteration in the positive x direction

$$\{(\rho \hat{A}_+)^n \frac{\partial}{\partial x_-} + (\rho \hat{B}_+)^n \frac{\partial}{\partial y_-} + (\rho \hat{B}_-)^n \frac{\partial}{\partial y_+} - C[\frac{\partial}{\partial y}(\frac{\mu}{Re_\infty} E \frac{\partial}{\partial y})]^n\} V^{n+1}_{i,j}$$
$$= -(\rho \hat{A}_-)^n (\frac{\partial V}{\partial x_+})^n_{i,j} - (\rho \hat{A}_-)^n [\frac{\partial(\delta V)^n}{\partial x_+}]_{i,j} + (\omega-1)N(\frac{\partial p}{\partial x_c})^n_{i,j} \tag{3.7}$$

where

$$\delta V^{n+1} = V^{n+1} - V^n . \tag{3.8}$$

Equation (3.7) may also be written as

$$[(\rho \hat{A}_+)^n \frac{\partial}{\partial x_-} + (\rho \hat{B}_+)^n \frac{\partial}{\partial y_-} + (\rho \hat{B}_-)^n \frac{\partial}{\partial y_-} - (C^n \frac{\partial}{\partial y} \frac{\mu}{Re_\infty} E^n \frac{\partial}{\partial y})_c] \delta V_{i,j}^{n+1}$$

$$= -[(\rho A_+)^n \frac{\partial}{\partial x_-} + (\rho A_-)^n \frac{\partial}{\partial x_+} + (\rho B_+)^n \frac{\partial}{\partial y_-} + (\rho B_-)^n \frac{\partial}{\partial y_+} - (C \frac{\partial}{\partial y} \frac{\mu}{Re_\infty} E \frac{\partial}{\partial y_c})^n] V_{i,j}^n$$

$$- (\rho A_-)^n \frac{\partial}{\partial x_+} \delta V_{i,j}^n. \tag{3.9}$$

Similarly, we may give the corresponding equations for marching iteration in the negative x direction.

We notice immediately that after the pressure splitting, the eigenvalues of $\hat{A}$ become u, $u+\sqrt{\omega}\,a$, u and $u-\sqrt{\omega}\,a$. This corresponds to that the speed of propagation of disturbance in the x direction, i.e., the sound speed changes from a to $\sqrt{\omega}\,a$. Thus we expect that through the control of propagation speed the rate of approaching to the steady state may be increased. Next we observe that in the case of steady small disturbance potential flow, the pressure is related to the x derivative of the potential, and hence the pressure splitting expression corresponds to potential splitting, i.e.,

$$\varphi = (1-\omega)\varphi^n + \omega\varphi^{n+1}.$$

Indeed, we have clearly demonstrated that ω plays a role as a relaxation factor. Now we have an active control of the ω factor, and may differentiate the following cases for different choices of the factor:

1. Define $M_X = u/a$. For $M_X < 1$ we take $\omega \leq M_X^2$ and for $M_X > 1$ we take $\omega = 1$, and $u > 0$ in both cases, i.e., the flow is not separated.

It is easily verified that now $\hat{A}_- = 0$. Thus if the pressure field is given (either it is an assumed initial field or it is obtained from previous iteration), then from equation (3.8), we may carry out marching iteration in the positive x direction, i.e., at first we march in the positive x direction to obtain all the flow parameters including the new pressure field, then we can again march in the positive x-direction with this new pressure field. Remember that we have started with the simplified N.S. equations, then we have the conclusion that for the marching solution of the simplified N.S. equations to be effective, only the initial pressure field is needed (this is, of course, also valid for the Euler equations). Fortunately, in most problems the pressure field is the easiest to be guessed among all the fields of related physical quantities. This property alone will expand the practical use of the simplified N.S. equations. Of course for the complete N.S. equations, a complete description of the flow fields must be given in order to start the marching iteration calculation described in the present paper. It should be pointed out, in the actual flow field where $M_X < 1$, the signal is not only propagated in the positive x direction but also in the negative x direction. The present artifice leads to a decrease in the propagation speed in the positive x direction, and what is more important, to reverse the propagation in the negative x direction. This fact may not be satisfactory and may cause a decrease in the convergence rate. But since the pressure field is not very sensitive, this may still serve as a quick means in the engineering applications. As a special case, if in the region near the surface where $M_X < 1$, we assume $(\partial p/\partial y) = 0$, $p(x,y) = [p(x)]_{M_X=1}$, then we may use equation (3.8) and come to proceeding the marching calculation. This is the so called sublayer approximation and it is a special case treated here.

2. The case $\omega = 1$.

In this case if we rewrite Eqs. (3.8) and (3.9) in the conservative form, the marching iteration calculation in the positive and negative x directions becomes exactly the same as in the preceding section. In fact through the transformation Jacobian $\partial V/\partial U$ and $(\partial V/\partial U)^{-1}$, we may easily find the relations between the coefficient matrices of conservative and nonconservative forms.

3. For flow regions $M_X < 1$, we take $\omega > 1$ and for the flow regions $M_X > 1$ we take $\omega = 1$.

As a means to increase the rate of convergence, a possible avenue is to increase the signal propagation speed. For the region $M_x < 1$, the choice of $\omega > 1$ may lead to increase in the signal speeds in both positive and negative directions. But if ω is too large, this may affect the accuracy of the solution. For the region $M_x > 1$, the choice of either $\omega < 1$ or $\omega > 1$ may lead to the signal speed increases in one direction but decreases in the other, so it is really no benefit to choose ω different from 1, thus $\omega = 1$ seems an optimum.

4. For flow regions $M_x > 1$, we take $\omega > M_x^2$.

It is well known that in the region where $M_x > 1$, the signal propagates only in the positive x-direction. But if we choose a value of ω such that $\omega > M_x^2$, then in the Eq. (3.8) or (3.9) we may find the signal propagates also in the negative x direction, thus the flow field now possesses an elliptical character. In the calculation of transonic flow over a supercritical profile, adoption of this technique renders the whole flow fields becoming elliptical type. Obviously this is a variant of artificial density method mentioned in the literature.

IV. Numerical Calculations and Discussions

In order to test the practicality and generality of the present method, we give some numerical examples in the following.

1. We calculated the viscous flow around a flat plate using simplified N.S. equations[4]. Two cases were computed.
(1) $M_\infty = 3$, the Reynolds number based on the length of the flat plate $Re_\infty = 10000$, $T_\infty' = 216.65°K$, the wall is adiabatic.
(2) $M_\infty = 3$, $Re_\infty = 1.68 \times 10^4$, $T_\infty' = 216.65°K$ and also the wall is adiabatic. In Fig.2 the pressure distribution in the first case is given and compared with the numerical solutions of N.S. equations given by Hodge and Carter. In Fig.3 the corresponding velocity and temperature profiles at x=1.0 are shown and also compared with Carter's results. In Fig.4 the pressure distribution in the second case is given and compared with Carter's. All the numerical results agree closely to each other. For the steady case we only need 20 ~ 30 steps of marching iteration to acheive the numerical accuracy which is obtained with the existing implicit methods in 150-200 steps while the computational works in each step for different methods are about the same. This clearly shows the superiority of the present method in numerical efficiency. This method also demonstrates that the shock can be captured automatically with flux splitting method and without the addition of artificial viscosity. And what is more important is that the results are independent of Δt chosen, the CFL number may vary between 10 and 10^9.

2. A calculation of viscous supersonic jet is made. The Mach number at the exit of the nozzle Mj=3, external flow Mach number Me=2, $\gamma=1.4$, the ratio between the exit pressure Pj and external pressure Pe is 15. Fig.5 shows the calculated shapes of external shock and drum shock, and the location of plume boundary. Fig. 7 gives the axial pressure distribution and pressure distribution along plume boundary. For comparison we also give the calculated results of ref. 5. We see the agreement is satisfactory. Since the calculation is limited to the region in front of the Mach disc where $A_- = 0$, one step of marching calculation is sufficient to get final results. In the open literature there are many papers using time dependent method which is of course requiring more CPU time and more expensive.

3. In order to test the capability of the method in the calculation of separated flow, we calculated a supersonic flow past a two-dimensional compression corner with $M_\infty = 3$, $Re_\infty = 1.68 \times 10^4$ (the characteristic length is from the leading edge to the corner point), $\omega=10°$, $T_\infty' = 216.65°K$ and an adiabatic wall. The converged solution of complete flow field is obtained in only 120 steps. The results are shown in Figs. 7 a, b with frictional drag coefficient and pressure distribution, compared favorably with MacCormack's.

To exhibit the generality of the present method we give further numerical examples for ideal gases.

4. First we give a numerical calculation of hypersonic flow $M_\infty = 8$ around a

sphere[7]. Here we employ the pressure relaxation factor $\omega < M_x$. The calculation was made with gas dynamics equations in spherical coordinates. Convergent results are obtained after 8 steps of marching iteration. Compared to the time dependent method, the computation time required in the present method is one order of magnitude less. Fig.8 gives the pressure distributions along the body surface and the shock. Fig.9 gives the locations of sonic line and detached shock waves. In the figures there are shown results obtained with time dependent method [6]. We also calculated the flow at M = 5 and 10 without using ω, see Figs.10, 11 and 12. In the latter cases, 14 steps are required for convergence. Next is a calculation of flow around a sphere cone $M_\infty = 8$ [7]. The results are shown in Figs. 13 and 14 and compared with those from ref. (6). It is observed that a unified program can be used in subsonic region and supersonic region. With about 40 steps of marching iteration, a complete convergent solution can be obtained.

The usefulness of the unified program is best illustrated with the last numerical example, a calculation of the flow around indented shape nose cone with $M_\infty = 9$. Here we have two subsonic regions, that means we require two iteration regions. After the convergent results in the first iteration region are obtained, the marching method gives the numerical results in the downstream supersonic region, then comes the second subsonic region (again here iterations are needed), and finally comes another supersonic region where marching alone is sufficient. Fig. 15 gives the present results compared with those obtained with explicit second order split coefficient matrix method. The agreement is satisfactory, but the present method requires only a fraction of the computational work of the previous methods.

After reviewing these calculations the following preliminary conclusions can be drawn:

1. The present method can provide a better convergent rate than some of the prevailing methods, a better adaptation to wide class of flow problems. We can acheive second order accuracy, and the accuracy of the solution is not limited by CFL number.

2. The marching iteration method is established on the rules of signal propagation, the difference equations are set up according to the signs of the eigenvalues of corresponding matrices. They are second order accurate, and with this scheme shocks can be automatically captured without adding artificial viscosity. Since the method is implicit, the marching process is unconditionally stable.

3. The method is physically well founded because the signal propagation process is simulated.

4. The present method can be used in non-viscous flows and also in viscous flows. It may be applied when the signals are propagated in one direction and also when the signals are propagated in both forward and backward directions. In the first case, marching along that direction once, the solution is obtained. In the second case, the solution is obtained by sweeping in both directions successively, and this process is termed marching iteration. Thus we now have a unified method and a unified program to calculate flow regions with different characteristics.

5. Compared with the ordinary time dependent implicit method, the present one may save the computing time to a large extent in solving N.S. equations and Euler equations.

6. We introduced a pressure relaxation factor ω. It has been shown that with a given initial pressure field, marching iteration process can be applied to a variety of flow problems.

7. The present paper investigates the possibility of accommodating the propagation speed of signal and its effects. The choice of ω such that $\omega > 1$ in the subsonic flow and $\omega = 1$ for supersonic flow may improve the rate of convergence. Also the introduction of ω in the study of transonic flow past a supercritical airfoil produces the same effect as the incorporation of artificial density does.

8. This method can be easily extended to the three dimensional case, and may be combined with the use of multigrid method to further enhance the rate of convergence of the numerical calculations. Some of the more complex flow problems are being studied with the present method to verify the correctness of the above conclusions.

During the course of preparing the lecture the discussions with Mr. Yu Zechu,

Mr. Guo Zhiquan, Mr. Zheng Min and Mr. He Fangshang were very helpful. Part of the numerical computations were done by Mr. Wang Zhijian and Mr. Shen Qing. To all of them the authors express their hearty thanks.

References

1. Beam, R.M. and Warming R.F., An Implicit Factored Scheme for the Compressible Navier - Stokes Equations, AIAA J., Vol. 16, No. 4, 1978 pp. 393-402.

2. Steger, J.L. and Warming, R.F., Flux Vector Splitting of the Inviscid Gasdynamic Equations with Application to Finite Difference Methods, Journal of computational physics, Vol. 40, No. 2, April 1981, pp. 263-293.

3. Zhang Hanxin: The Marching - Iteration Method in Solving N.S. Equations. CARDC Report, 1985 (in Chinese)

4. Zheng Min : Marching Iteration Method in Solving Two - Dimensional Supersonic Viscous Flow Problems, CARDC Report 1985 (in Chinese)

5. Dash, S.J. and Thorpe, R.D., AIAA paper 80-1254.

6. Lyubimov, A.N. and Rusanov, V.V., Gas Flow around Blunt Bodies, "Nauka", Moscow, 1970 (in Russian).

7. Guo Zhiquan: Marching - Iteration Method Used in Solving Steady Euler Equations. CARDC Report 1985 (in Chinese).

Notes added in proof:

Prof. Zhang exchanged part of the work reported here as development in computational aerodynamics in recent years at CARDC with Prof. MacCormack during his visit to the Center in 1985. We found some similarity between our work and his and also some differences in the concept of development of the method and details of numerical algorithm.

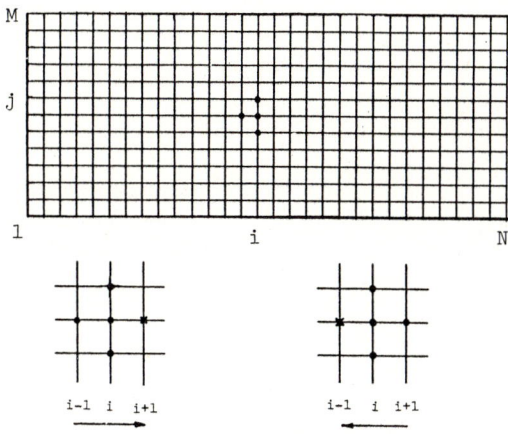

Fig.1 Viscous flow over a flat plate

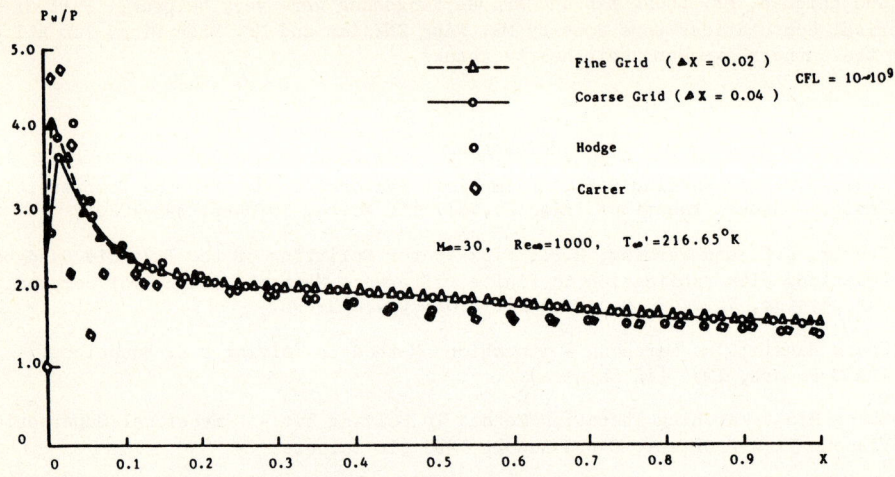

Fig.2 Pressure distribution on the flat plate

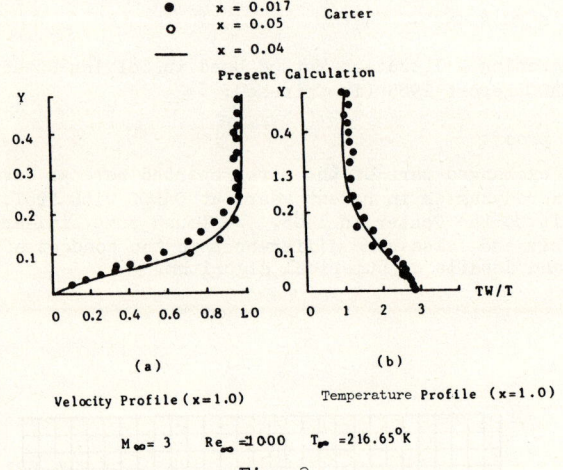

Fig. 3

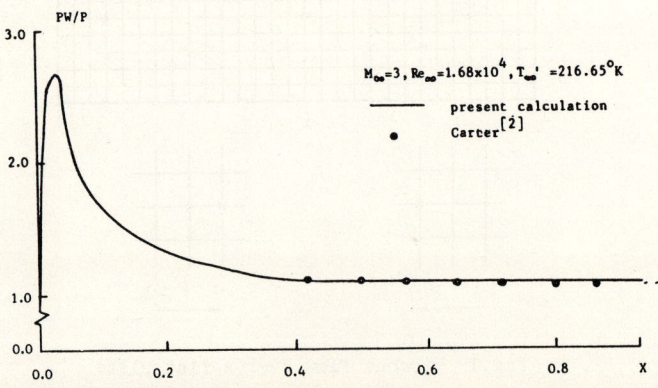

Fig.4 Pressure distribution on the flat plate

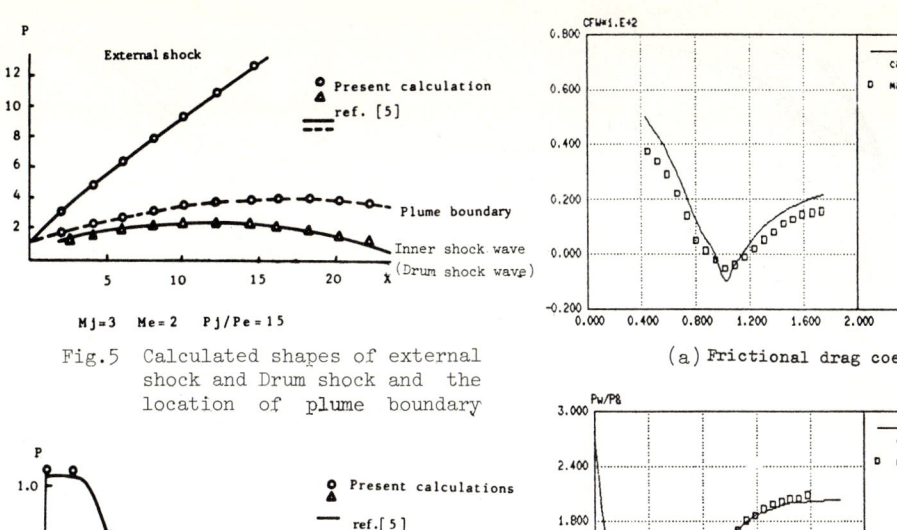

Fig.5 Calculated shapes of external shock and Drum shock and the location of plume boundary

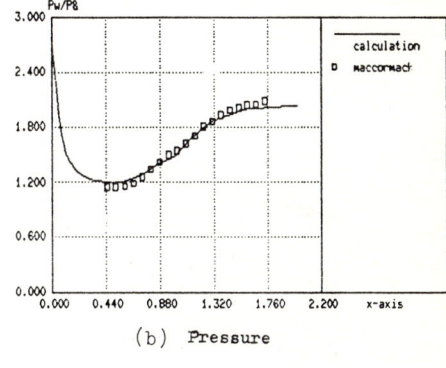

Fig.6 Axial pressure distribution and pressure distribution along the plume boundary

(a) Frictional drag coefficient

(b) Pressure

Fig.7 Distribution of pressure and frictional drag on the wall ($M=3$, $Re_L=1.68\times10^4$, adiabatic wall)

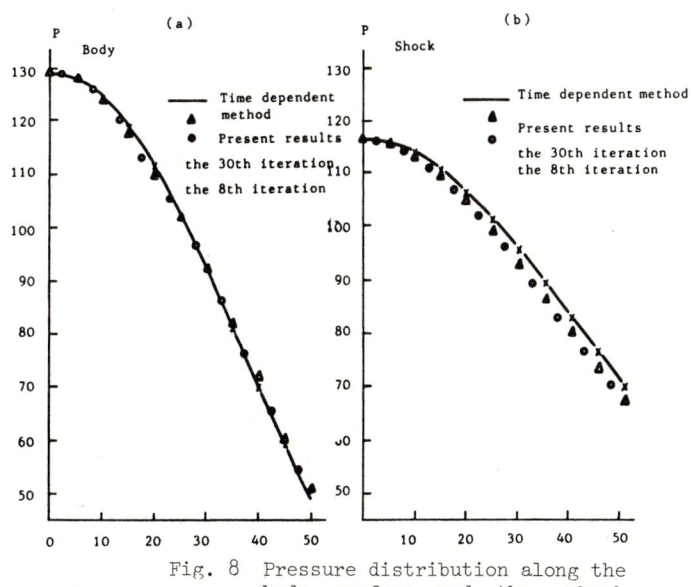

Fig. 8 Pressure distribution along the body surface and the shock

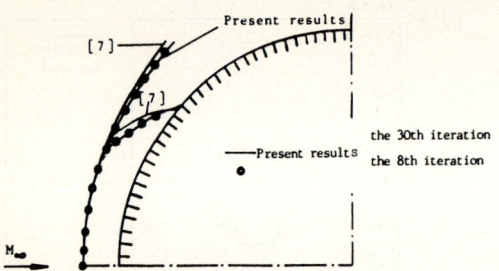

Fig. 9 Locations of sonic line and detached shock wave

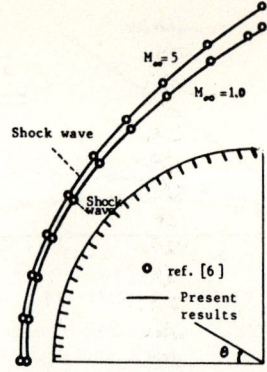

Fig. 10 Location of shock waves for flow around sphere at $M_\infty = 5$ and $M_\infty = 10$

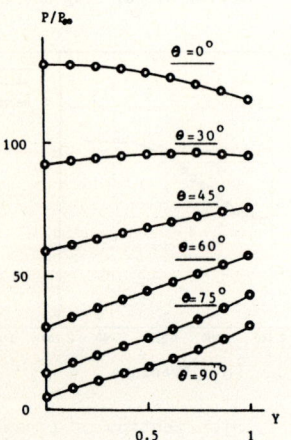

Fig. 11 Pressure distribution curves for flow around a sphere $M_\infty = 10$

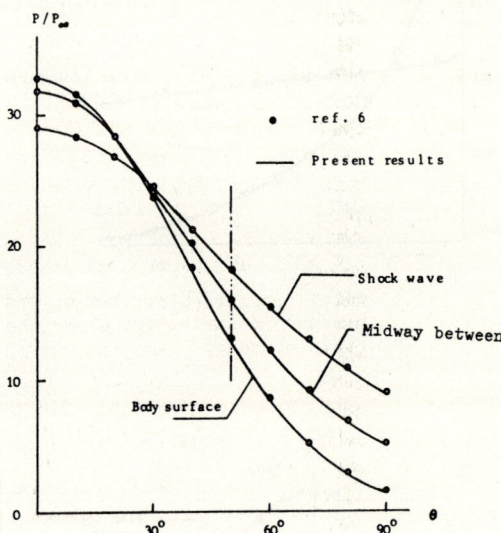

Fig. 12 Pressure distribution curves $p \sim \theta$ ($M_\infty = 5$) for flow around a sphere

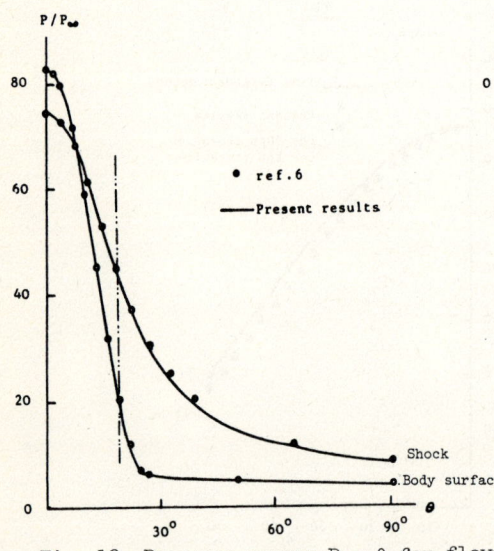

Fig. 13 Pressure curves $P \sim \theta$ for flow around a sphere cone ($M_\infty = 8$)

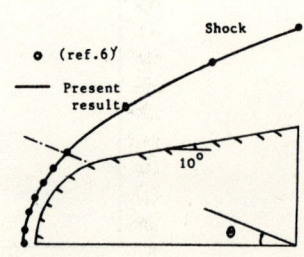

Fig. 14 Shock in flow around a sphere cone ($M_\infty = 8$)

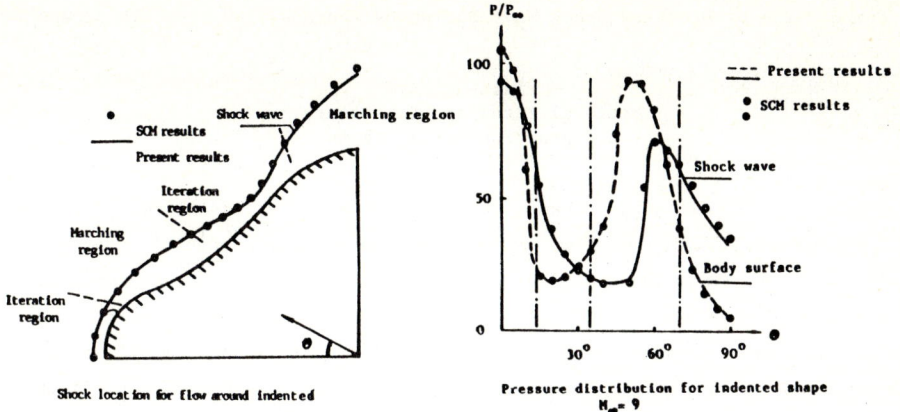

Fig. 15 Shock shape and pressure distribution for a flow around an indented nose cone

Computation of Unsteady Shock Wave Motion by the Modified Flux TVD Scheme

Takayuki Aki
National Aerospace Laboratory
Chofu, Tokyo 182, Japan

I. Introduction   Among the recent developments of difference methods for solving the Riemann initial value problem of hyperbolic equations, several schemes which satisfy the total variation diminishing(TVD) constraints on the field data reveal a superior property as the solver and therefore become attractive tools to attack Eulerian gasdynamic problems. The modified flux TVD scheme proposed by Harten [1-2] has the second order accuracy (almost everywhere) and an algorithmic simplicity. Efforts to develop the scheme towards the computation of practical importance have been dedicated by Yee et al. [3-5]. Their results demonstrate the superiority of the scheme to compute steady state flows. For unsteady and multiple wave interacting flow problems, applicability of the scheme to resolve gas dynamic discontinuities has been left still unclear. The present paper discusses such an applicability of the scheme based on results of numerical experiments.

II. Numerical Method   The Euler equations in two spatial dimensions can be written in the conservation law form as

$$Q_\tau + F_\xi + G_\eta = 0 \tag{1}$$

where a generalized coordinate transformation of the form $\xi=\xi(x,y)$ and $\eta=\eta(x,y)$, and $\tau=t$ has been used. The Q, F, and G are all the column vectors with 4 elements. The elements of Q are the conservative variables and those of F and G are the corresponding fluxes. All of the elements are scaled with the appropriate measures of the coordinate transformation.

By using the fractional step method, the computation can be implemented as follows:

$$Q^*_{i,j} = Q^n_{i,j} - \lambda_\xi (\tilde{F}^n_{i+1/2,j} - \tilde{F}^n_{i-1/2,j})$$
$$Q^{n+1}_{i,j} = Q^*_{i,j} - \lambda_\eta (\tilde{G}^*_{i,j+1/2} - \tilde{G}^*_{i,j-1/2}) \tag{2}$$

where $\lambda_\xi = \Delta\tau/\Delta\xi$ and $\lambda_\eta = \Delta\tau/\Delta\eta$. The numerical fluxes $\tilde{F}^n_{i+1/2,j}$ etc. are of similar form. For simplicity, the index of time n and that of the another spatial grid are suppressed in the followings. Denote R as the matrix whose columns are eigenvectors of Jacobian $A=\partial F/\partial Q$, and denote R^{-1} as the inverse of R. Eigenvalues of A are denoted by a^k (k=1,2,3,4). Typically, $\tilde{F}_{i+1/2}$ can be expressed as

$$\tilde{F}_{i+1/2} = 1/2 (F_i + F_{i+1} + R_{i+1/2} \Phi_{i+1/2}) \tag{3}$$

where the k-th component of the $\Phi_{i+1/2}$ is

$$\phi^k_{i+1/2} = \sigma(a^k_{i+1/2})(g^k_i + g^k_{i+1}) - \Gamma(a^k_{i+1/2} + \beta^k_{i+1/2})\alpha^k_{i+1/2} \tag{4a}$$

with
$$g_i^k = (1 + \omega^k \theta_i^k) \tilde{g}_i^k \quad (4b)$$

$$\tilde{g}_i^k = s\,\text{Max}(0, \text{Min}(|\alpha_{i+1/2}^k|, s\alpha_{i-1/2}^k))$$
$$s = \text{sgn}(\alpha_{i+1/2}^k) \quad (4c)$$

$$\theta_i^k = |\alpha_{i+1/2}^k - \alpha_{i-1/2}^k| / (|\alpha_{i+1/2}^k| + |\alpha_{i-1/2}^k|) \quad (4d)$$

and
$$\Gamma(z) = |z| \qquad |z| \geq \delta$$
$$= (z^2 + \delta^2)/2\delta \qquad |z| < \delta \quad (4e)$$

Here δ is a constant number and

$$\beta_{i+1/2}^k = \sigma(a_{i+1/2}^k)(g_{i+1}^k - g_i^k)/\alpha_{i+1/2}^k \qquad \alpha_{i+1/2}^k \neq 0$$
$$= 0 \qquad \alpha_{i+1/2}^k = 0 \quad (4f)$$

The functional form of σ is selected as

$$\sigma(z) = 1/2(\Gamma(z) - \lambda z^2) \quad (4g)$$

Lastly,
$$\alpha_{i+1/2}^k = R_{i+1/2}^{-1}((JQ)_{i+1}^k - (JQ)_i^k)/(J_{i+1} - J_i)/2 \quad (4h)$$

where J is the metric Jacobian of coordinate transformation. In Eq. (4), the element values at the point with the half integer are evaluated by some average between two adjacent regular points. For obtaining $\tilde{G}_{j+1/2}$, R and a should be replaced with those corresponding to $B = \partial G/\partial Q$. An equation of state for an ideal gas is needed to complete the system of Eq.(1).

III. Results   A particular concern on the scheme is resolvability of discontinuities resulted by multiply interacting waves. Thus the shock propagation problems through a 90° bent in a two dimensional duct are firstly selected instead of usual shock diffraction problems, e.g. those by an obstacle placed in an infinite domain. For the selected problems, the experimental data are available [6].

Assuming an infinite length of the test gas slab, the computation starts with the initial shock position at the upstream juncture of bent to the straight duct. The flow field variables at the inflow and along the wall surfaces are corrected every time advance by solving a set of characteristic compatibility and boundary conditions appropriate to the respective boundaries. The constant numbers appeared in Eqs.(4b) and (4e) whose values must be tuned are taken as $\omega^k = 1$ for all k and $\delta = 0.1$ throughout the present computations. Although the sophisticated form of δ other than a constant can be used[8], the difference in the result was less marked and incommensurable to the elaboration to use. Two methods for the averaging in Eq. (4) were tested, i.e. the arithmetic and Roe's[9] ones. No remarkable differences in the results can be found.

Figs. 1-2 display isobars (left) and isopycnics (right) for shocks with incident $M_s = 2.35$, and 4, respectively. The test gas is taken as to be air in Fig. 1 and CO_2 in Fig. 2. The inner radius of the bent is 80 mm with the width 40 mm. The grid distribution along the walls is shown by the stick marks. Across the duct width direction, 41 grid points are allotted. Note that there are no disturbances in front and behind of the Mach stem and diffracted wave and smooth transitions across them over two mesh widths. Increasing the M_s an enhanced complexity results in the structure of flow field. A computation of the same flow conditions with those in Figs.1-2 has been performed by Sommerfeld et al [7] utilizing a multidimensional upwind method without operator splitting. For M_s=2.35, their result loses a slip line emanating from the triple point and a discontinuity from the inner wall at the colliding point of the Mach reflection, see also Fig.4. The present one success to resolve these discontinuities but suffering to smear over excess grid points. A similar situation appears in results for M_s=4. These comparisons indicate that the present scheme has a superiority over the mentioned one.

IV. Discussions toward further studies   Even the above results, however, are far from satisfactory as compared to the experiment. Sources of the unsatisfaction may be considered those attributable to the grid system, operator splitting, boundary treatment, effects of gaseous imperfection, and so on. That, first of all, much more work is needed to model and formulate the discontinuities possible in multi-dimensional Eulerian gasdynamic flows can be seen as follows.

Shown in Fig.3 is an infinite fringe interferogram obtained by a shock tube experiment[10] at an instance soon after the onset of the Mach reflection during the early stage of propagation. The onset process could be hardly reproduced by the computation on the grid used in Fig.1. And the slip line at the instance corresponding to Fig.3 were very poorly captured, although some success were obtained for the capturing of the slip line at the later grown stage as shown in Fig.1. Capturing of slip lines is a severe task for the computation based on the conservation law form. For example, much of computation has been performed for the problem of shock diffraction by a convex surface such as a circular cylinder. The present author, however, has not been become aware of result with the success to capture definitely the slip line associated with the Mach reflection. It may be a rather surprising consequence that the slip line can be captured as shown in Figs.1-2 by the scheme taking no account of the physics inherent to the two dimensional flows.

Results obtained by using the finest grids in the present computations are shown in Figs.5-6 as those at the instances corresponding to Figs.3-4. The grid system is composed of the points 81 across the width and 400 along the surface of the bent itself, whereas it was 41x120 in the case of Fig.1. Although the slip line resolved in Fig.5 becomes better than that in the result obtained by the coarse grids, it can be scarcely said that a remarkable improvement can be attained. In Fig.6, one can observe the slip line resolved equally with the shocks. On the contrary, the discontinuity following the reflected shock has been retained still in low resolution. Thus the present scheme seems to be of no critical use for studies of the shock dynamics in two or more dimensions.

The above results show that difference schemes with truly two dimensional upwinding may be considered to overcome difficulties encountering in genuinely two dimensional flows in which some informations at a point can propagate homogeneously in directions and strengths. A typical example is the present one and many others may be found in unsteady/transient flows with interacting shocks.

Further studies should be directed toward explorations of scheme with proper modeling to the flow such as in [11] and without loss of the desirable mathematical properties by virtue of flux limiter [12].

References
 [1] Harten,A., J. Comp. Phys. Vol.49,357-393(1983).
 [2] Harten,A., SIAM J. Numer. Anal. Vol.21,1-23(1984).
 [3] Yee,H.C., Warming,R.F., and Harten,A., Lecture Notes in Physics, Vol.170,1982.
 [4] Yee,H.C., Warming,R.F., and Harten,A., J. Comp. Phys. Vol.57,327-360(1985).
 [5] Yee,H.C., and Harten,A., AIAA 85-1513,1985.
 [6] Takayama, K., Proc. 11 Int. Symp. Shock Tubes and Waves(1978).
 [7] Somerfeld,M., Nishida,M., and Takayama,K., Preprint,1985.
 [8] Harten,A. and Hyman,J,M., J. Comp. Phys. Vol.50,235-269(1983).
 [9] Roe,P.L., J. Comp. Phys. Vol.43,357-372(1981).
[10] Takayama,K., Private comunication.
[11] Roe,P.L., J. Comp. Phys. Vol.63,458-476(1986).
[12] Sweby,P.K., Lectures in Applied Mathematics, Vol.22,Part 2,1985.

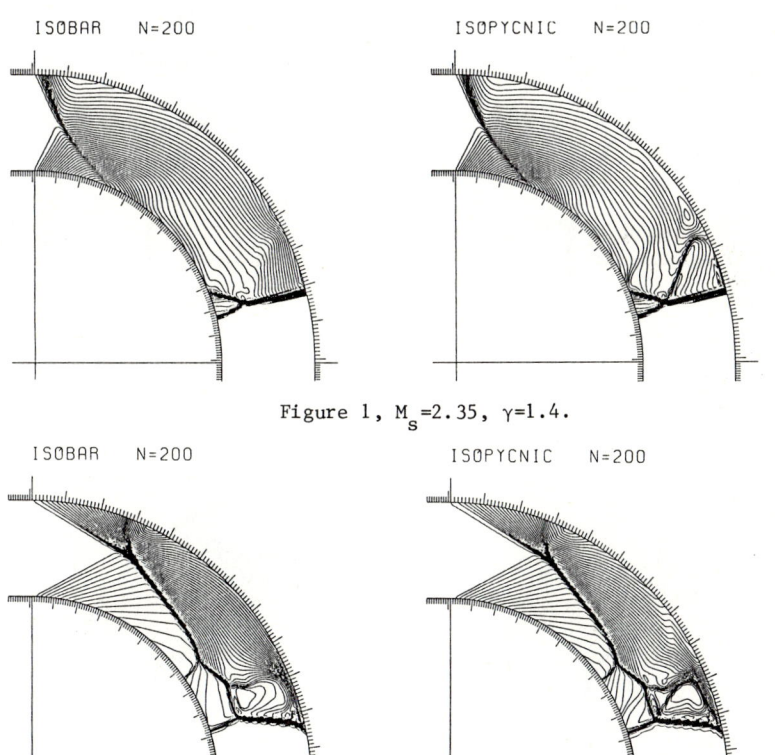

Figure 1, M_s=2.35, γ=1.4.

Figure 2, M_s=4, γ=1.28.

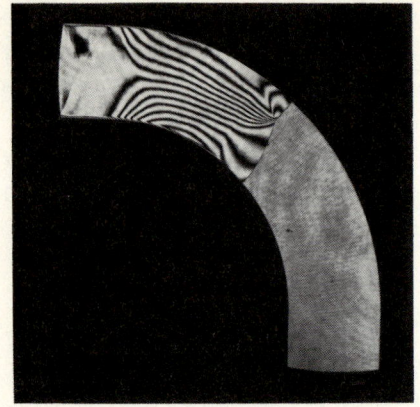
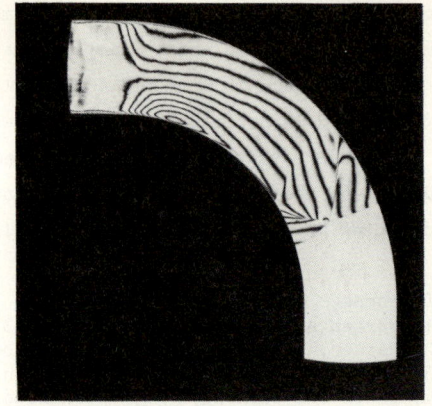

Figure 3, M_s=2.2, Air.

Figure 4, M_s=2.2, Air.

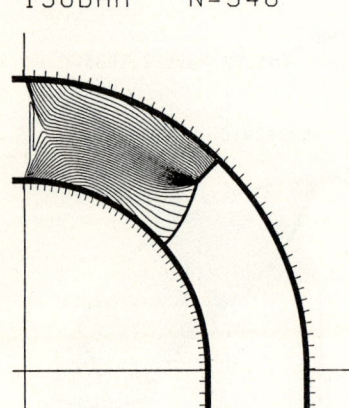
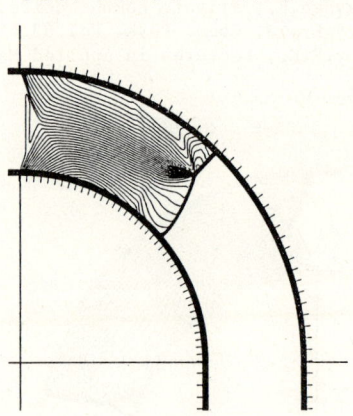

Figure 5, M_s=2.2, γ=1.4.

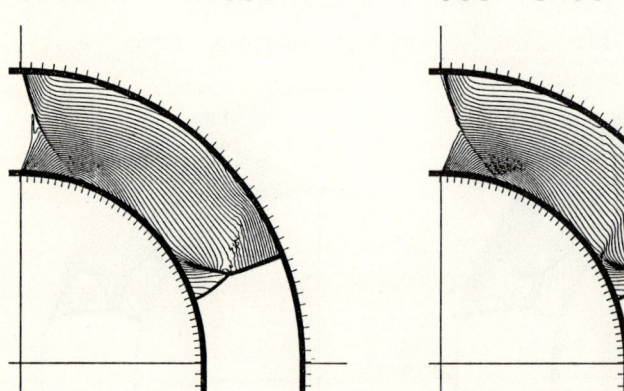

Figure 6, M_s=2.2, γ=1.4.

SWIRLING DIFFUSER FLOW USING A REDUCED NAVIER-STOKES FORMULATION

S.W. Armfield and C.A.J. Fletcher
University of Sydney, Australia

1. INTRODUCTION

Internal swirling flow, for which axial diffusion is much smaller than transverse diffusion, may be modelled accurately by a reduced form of the Navier-stokes (RNS) equations. The RNS equations are formed, via an order-of-magnitude analysis[1], by dropping axial diffusion terms, non-swirl terms from the radial momentum equation and splitting the pressure into a centre-line component and a radial correction. The RNS equations are non-elliptic[2] with respect to the main flow direction (x).

Such an RNS formulation is appropriate[3] for conical diffuser flows up to total diffuser angles of about $15°$. For non-reversing flows with small inlet swirl ratios ($W_{av}/U_{av} \lesssim 0.3$) the RNS solutions from a single downstream march are accurate.[1,4]

With larger inlet swirl ratios the flow through a small total-angle conical diffuser demonstrates a reduced axial velocity component on the axis of symmetry. If the inlet swirl ratio is sufficiently large the flow will reverse at the diffuser axis.

For the high swirl case it is appropriate to embed the basic RNS formulation in a multiple downstream march algorithm. In the multiple march algorithm the pressure gradient term in the axial momentum equation is discretised so as to allow upstream influence to occur reflecting the weak elliptic behaviour of the high swirl RNS equations. The global pressure field is stored and upgraded on each downstream march; the velocity field is not stored since it is computed afresh on each downstream march.

For low values of inlet swirl ratio both algebraic and k-ε eddy viscosity models provide accurate representation of turbulence effects when used in an RNS algorithm[5]. The present paper is concerned with whether algebraic and k-ε turbulence models can provide accurate predictions of higher swirl flows in an RNS context.

2. GOVERNING EQUATIONS

The mesh flow is considered to be axisymmetric, steady, incompressible and turbulent. In spherical coordinates, $(x,\theta,\phi;u,v,w)$ and Fig.1a, an appropriate reduced form of the Navier-Stokes equations is

$$(x^2 u)_x + (x\sin\theta v)_\theta /\sin\theta = 0 \qquad (1)$$

$$uu_x + vv_\theta/x - (v^2+w^2)/x = -P_x + (u_{\theta\theta}+\cot\theta u_\theta)/(Rex^2)$$
$$+ (\nu_t x \sin\theta u_\theta)_\theta / x^3 \sin\theta \qquad (2)$$

$$w^2 \cot\theta = P_\theta \qquad (3)$$

$$uw_x + uw/x + vw_\theta/x + vw\cot\theta/x = (w_{\theta\theta}+\cot\theta w_\theta -w/\sin^2\theta)/(Rex^2)$$
$$+ (\nu_t x^2 \sin^2\theta \{w_\theta - w\cot\theta\})_\theta /x^4 \sin^2\theta \qquad (4)$$

with the Reynolds number, $Re = \rho D\bar{U}^0/\nu$, D is the inlet diameter and $\bar{U}^0$ is the mean inlet axial velocity. In eqs.(2) and (4), ν_t is the eddy viscosity. This is either given by the two-layer Cebeci-Smith model[1] or obtained from a k-ε turbulence model. In the k-ε turbulence model the system, (1) to (4), is supplemented by reduced transport equations[5] for k and ε.

In the near-wall region an anisotropic representation[6] gives separate axial, ν_x, and circumferential, ν_θ, eddy viscosities. Away from the wall $\nu_t = \nu_x = \nu_\theta$. The differing eddy viscosities, ν_x and ν_θ, are generated from a mixing length model[1] which forms the near-wall layer of the Cebeci-Smith turbulence model. In the k-ε turbulence model, the anisotropic representation is embedded in the wall functions, i.e. the boundary conditions for the k and ε transport equations.

Equations (1) to (4), supplemented by k,ε equations as required, are solved in the domain, $x_1 \leq x \leq x_2$, $0 \leq \theta \leq \theta_w$, where x_1 and x_2 are the spherical radii at the diffuser inlet and outlet, respectively, and θ_w is the diffuser half angle (i.e. wall value).

Boundary conditions are

  i) inlet (x_1,θ): $u = u^1$, $v = v^1$, $w = w^1$, $P = P^1$
 ii) diffuser wall (x,θ_w): $u = v = w = P_\theta = 0$ (5)
iii) diffuser centre line $(x,0)$: $v = w = u_\theta = 0$
 iv) (x_2,θ): $u_{xx} = v_{xx} = P_{xx} = 0$.

Equations (1) to (4) are discretised using finite difference expressions on a variable mesh (Fig.1b). Velocity derivatives in θ are centrally differenced, over two points in eq.(1) to facilitate a radial march, and over three points in eqs.(2) to (4). Velocity derivatives in x are upwind differenced to facilitate a downstream march. In eq.(3) P_θ is discretised with a one-sided difference expression to permit a radial march from the axis to the wall. In eq.(2) P_x is discretised by

$(P_i^{n+1,\nu-1} - P_i^{n,\nu})/\Delta x$ where $\nu-1$ indicates the previous downstream march. It is the appearance of this term that provides the mechanism for upstream influence, but necessitates the storage of the global pressure distribution.

3. METHOD OF SOLUTION

A line-by-line iterative procedure is used based on repeated downstream (x) marches. At each downstream location, x_{n+1}, eqs. (1) to (4) and transport equations for k and ε if required, are solved for $\{u,v,w,P,k,\varepsilon\}^{n+1,\nu}$ for all radial locations r_i. Here ν denotes the current downstream march. The iteration is repeated until $||P^{\nu}-P^{\nu-1}||_\infty < 10^{-4}$, typically.

At each downstream station the discrete form of eq. (2) is manipulated into tridiagonal form,

$$A_i u_{i-1}^{n+1} + B_i u_i^{n+1} + C_i u_{i+1}^{n+1} = D_i, \qquad (6)$$

where D_i contain the undifferentiated terms and the pressure gradient terms. The pressure solution is split into $P = P_{C/L} + \Delta P_r$, where $P_{C/L}^{(x)}$ is the pressure distribution on the diffuser centre-line and $\Delta P_r^{(x,r)}$ is the correction accounting for the radial (θ) variation. This is obtained from a one-dimensional integration of eq. (1) in the θ direction from the centre-line to the wall.

The centre-line pressure, $P_{C/L}^{n+1}$, is obtained from a global mass conservation condition. This is obtained by integrating eq. (1) over the interval, $0 \le \theta \le \theta_w$, to give

$$\int_0^{\theta_w} \sin\theta x^2 u\, d\theta = \dot{m}^1, \qquad (7)$$

where $\dot{m}^1$ is the inlet mass flow, and obtained from eq. (5a). Factorisation of eq. (6) into $LU\, u^{n+1} = D_i$ allows the solution of eq. (6) to be written $u^{n+1} = U^{-1}L^{-1}E + U^{-1}L^{-1}FP_{C/L}^{n+1}$. Substitution into eq. (7) gives

$$P_{C/L}^{n+1} = (\dot{m}^1 - \int_0^{\theta_w} \sin\theta x^2 U^{-1}L^{-1}Ed\theta\}/\{\int_0^{\theta_w} \sin\theta x^2 U^{-1}L^{-1}Fd\theta\}. \qquad (8)$$

Evaluations of eq. (8) gives $P^{n+1} = P_{C/L}^{n+1} + \Delta P_r^{n+1}$ and substitution into eq.(6) gives u^{n+1}. The radial velocity components, v^{n+1}, are obtained by a radial march of the discrete from of eq.(1) from the centre-line to the wall. The discrete form of eq.(4) can be manipulated into a tridiagonal system which is solved for w^{n+1}. Thus at each downstream location each equation is treated sequentially, to obtain a different component of the solution.

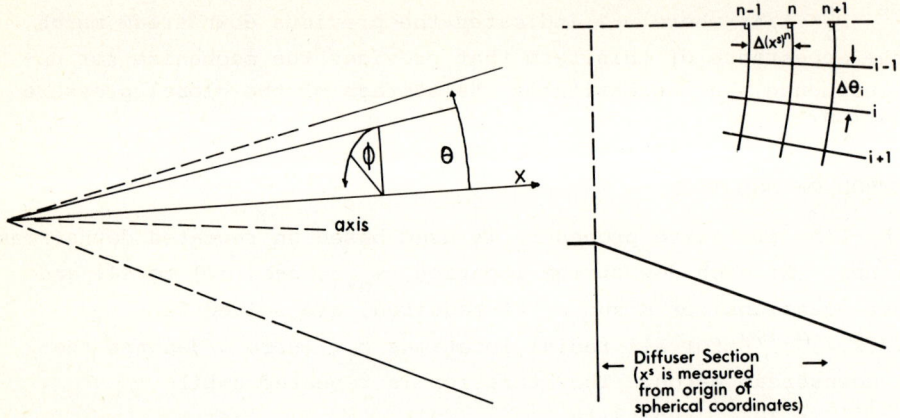

Fig. 1a Spherical Coordinates   Fig. 1b Discretisation

4. RESULTS

Solutions are presented in Figs. 2 to 4 for a conical diffuser with a total angle of 8° and a Reynolds number of 3.88×10^5. The experimental results of So[7] at $x/D = 0.6$, which is just downstream of the diffuser entry, are used to generate the initial data for the computational solution. The computational solutions have been obtained on a non-uniform rectangular grid, $80(x) \times 35(r)$. At the diffuser wall $\Delta r = 0.002$ and at the axis $\Delta r = 0.054$.

The axial velcoity distribution at $x/D = 4.00$ (Fig.2) indicates that the algebraic eddy viscosity turbulence model provides a more accurate prediction than does the k-ε turbulence model. The corresponding swirl comparison is indicated in Fig.3. The poor k-ε results are associated with too large a value of ν_t close to the axis. The pressure solution produced by the algebraic eddy viscosity model (Fig.4) is satisfactory.

Overall, the present RNS formulation gives accurate predictions of mean flow quantities for intermediate inlet swirl ratios, with a two-layer Cebeci-Smith turbulence model. The k-ε model requires either local parameter adjustment or additional curvature corrections to generate accurate solutions.

REFERENCES

1. S.W. Armfield and C.A.J. Fletcher, "Numerical simulation of internal swirling flows in diffusers", *Int.J. Num. Methods Fluids*, to appear, 1986.
2. S.W. Armfield and C.A.J. Fletcher, "Pressure-related instabilities of reduced Navier-Stokes equations for internal flows", *Comm. Appl. Num. Methods*, to appear, 1986.
3. S.W. Armfield, "Internal swirling flows in diffusers", Ph.D. thesis, University of Sydney, in preparation, 1986.
4. S.W. Armfield and C.A.J. Fletcher, *Computational Techniques and Applications: CTAC 85* (eds J. Noye and R. May) North-Holland, 1986, 431-442.
5. S.W. Armfield and C.A.J. Fletcher, *Proc. Int. Symp. Fluid. Dyn., Tokyo* (ed. K. Oshima), North-Holland, 1986, 740-751.
6. M.L. Koonsinlin and F.C. Lockwood, *AIAA J.* 12 (1974) 547-554
7. K.L. So, Rep. No.75, M.I.T. Gas Turbine Lab, Sept. 1964.

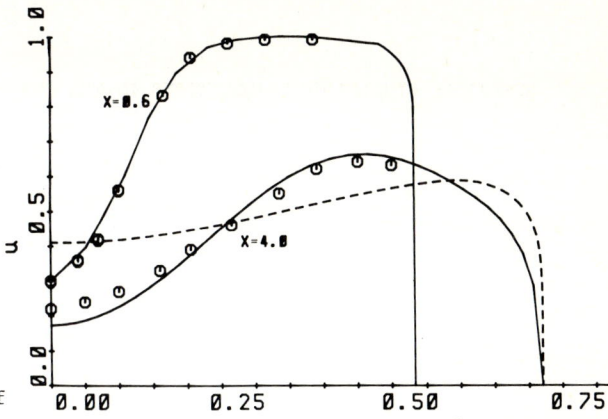

Fig.2 Radial distribution of axial velocity

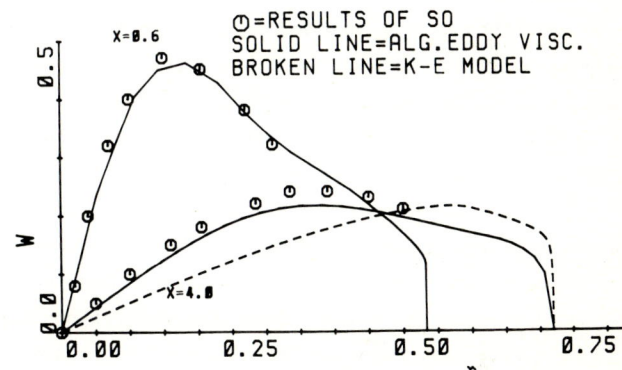

Fig. 3 Radial distribution of swirl velocity

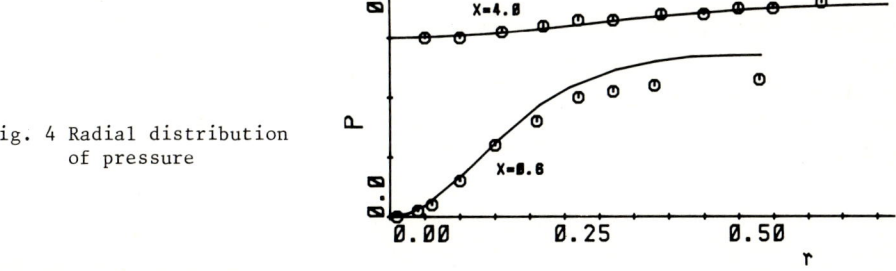

Fig. 4 Radial distribution of pressure

NUMERICAL STUDIES OF BIFURCATIONS IN THE CONFINED BÉNARD PROBLEM[*]

Y. Y. Azmy and J. J. Dorning[**]
Department of Nuclear Engineering and Engineering Physics
University of Virginia
Charlottesville, VA 22901

1. INTRODUCTION

Many physical phenomena which are of great practical interest are governed by nonlinear partial differential equations. Traditionally, linearized versions of these equations were solved and used in the design and operation of devices and machines utilizing such physical phenomena. However, with recent advances in the study of nonlinear equations (differential as well as difference equations) many old, yet unresolved, related questions now are being reconsidered. Numerous papers analyzing the temporal behavior of simple models of complicated nonlinear systems, attempt to predict and/or explain the evolution of the full nonlinear system (e.g. period doublings, onset of chaos,...etc.) on the basis of the analysis of the simple system. Simultaneously, new experiments have been devised in order to validate and guide theoretical efforts into the realm of nonlinear phenomena.

More recently, numerical methods have been devised to investigate the nonlinear phenomena in many systems of practical interest. Unlike conventional numerical studies which were directed simply toward obtaining a solution, the new endeavors were oriented towards calculating singular points, multiple solution branches, linear and nonlinear stability of equilibria, and many other important nonlinear characteristics. The ability of numerical methods to account for the spatial dependence of the solution to a physical nonlinear problem of practical interest in addition to its temporal dependence bring them closer to real systems than analytical investigations where the spatial effects are often averaged out or very crudely approximated. On the other hand, numerical methods have a finite

---

[*] This research was supported by the US Office of Naval Research under Contract No. N00014-85-K-0382 and by the US Department of Energy under ANL Contract No. A85-4032.

[**] Also: Center for Advanced Studies, University of Virginia.

accuracy which is limited by the computer memory size and CPU time available. In order to predict the nonlinear physical phenomena correctly and accurately within the practical limits of current computers it is essential to formulate accurate coarse-mesh methods for nonlinear equations.

Nodal methods have been devised to obtain highly accurate numerical solutions to linear and nonlinear problems on very coarse meshes. The nodal integral approach[1] resulted in a practical formalism for nodal methods which was easily applied[1] to the Navier-Stokes equations. Extensive testing[2,3] of the resulting numerical method and comparison of its accuracy to experiment and very fine mesh calculations has established its superior accuracy and computational efficiency on coarse meshes. Based on these results, we have undertaken a numerical investigation of the nonlinear phenomena in the confined Bénard problem using a nodal integral method for natural thermal convection problems described by the Boussinesq equations.

In Sec. 2 we outline the nodal integral method for the fluids equations in the Boussinesq approximation, and discuss the accuracy of that method. Then in Sec. 3 we present numerical results obtained for the Bénard problem in a tilted square cavity. These results indicate that the second pitchfork bifurcation does not unfold into a cusp catastrophe as the first pitchfork does when the cavity is tilted. Next, in Sec. 4 we report the results of our investigation of the behavior of the singular points for the Bénard problem as the cavity aspect ratio is varied. Finally, the main results and conclusions are summarized briefly in Sec. 5.

2. THE NODAL INTEGRAL METHOD FOR NATURAL THERMAL CONVECTION PROBLEMS

Practical situations often occur where the sole driving force of the flow field is the buoyancy force arising from thermally induced density differences at different locations of the fluid. It is possible to describe many such flow fields by the Boussinesq equations in which the fluid is treated as though it were incompressible in all terms except the buoyancy terms in the momentum equations. In two-dimensional Cartesian geometry, the steady-state, dimensionless Boussinesq equations are

$$\beta \frac{\partial u}{\partial x} + \frac{\partial v}{\partial y} = 0,$$

$$\beta u \frac{\partial u}{\partial x} + v \frac{\partial u}{\partial y} - \beta \frac{\partial \tau_x}{\partial x} - Pr \frac{\partial^2 u}{\partial y^2} - Ra\, Pr\, T \sin\theta = 0,$$

$$\beta u \frac{\partial v}{\partial x} + v \frac{\partial v}{\partial y} - \beta^2 Pr \frac{\partial^2 v}{\partial x^2} - \frac{\partial \tau_y}{\partial y} - Ra\, Pr\, T \cos\theta = 0,$$

$$\beta u \frac{\partial T}{\partial x} + v \frac{\partial T}{\partial y} - [\beta^2 \frac{\partial^2 T}{\partial x^2} + \frac{\partial^2 T}{\partial y^2}] = 0, \quad (x,y) \in [0,1] \times [0,1],$$

where β is the aspect ratio = B/A; Θ is the angle between the gravity vector and the negative y-axis measured in the clockwise direction, u, v, and T are the dimensionless x- and y- components of the velocity, and the temperature field, respectively. The normal stresses $\tau_x \equiv \Pr \beta \, \partial u/\partial x - p$, and $\tau_y \equiv \Pr \partial v/\partial y - p$, where p is the static pressure, have been introduced in order to achieve partial decoupling of the transverse-averaged equations. The dimensionless numbers are, Ra $\equiv$ Rayleigh number = $(\alpha \, \Delta T \, g \, B^3)/\kappa\nu$, and Pr $\equiv$ Prandlt number = ν/κ, where α is the coefficient of thermal expansion of the fluid, κ its thermal diffusion coefficient and ν its kinematic viscosity, ΔT is the temperature difference across the cavity, and g is the acceleration due to gravity.

Since the derivation of the nodal integral method equations from the continuous-variable equations has been described elsewhere,[4] only a brief sketch of the derivation is included here. The nodal integral method formalism consists of five main steps:

(1) Division of the system into M "nodes" or computational cells.
(2) Transverse-averaging the continuous-variable equations within each node to obtain two sets each of four linear ODE's in the transverse-averaged variables.
(3) Exact solution of the linear ODE's in terms of elementary solutions and particular integrals arising from Legendre series expansions of the terms which include distributed sources, transverse-leakage, terms that arise from the transverse-averaging procedure and the nonlinear convection terms.
(4) Derivation of the discrete-variable equations by imposing the continuity conditions on the primitive variables and their first spatial derivatives at node boundaries.
(5) Evaluation of the source-terms expansion coefficients by requiring the uniqueness of the nodal average of each of the primitive variables, and the nodal balance of mass, momentum, and thermal energy sources.

The resulting algebraic set of equations for the node-surface-averaged field variables is closed by consistently deriving boundary conditions on the transverse-averaged variables from the physical boundary conditions. The final set of equations is quadratically nonlinear, and is solved via a Newton-Raphson iterative procedure.

In order to verify the accuracy of this new method for natural thermal convection problems it was used to solve a recently established benchmark problem[5], the double-glazing problem. This is equivalent to the tilted, confined Bénard problem where the tilt angle equals $90°$. Comparison[4] of several quantities (e.g. the average Nusselt number on the hot wall,...,etc.), that were calculated at various values of Ra (between 10^3 and 10^6) on coarse meshes ranging from 4 x 4 to 16 x 16, with their values extrapolated from very fine finite-difference meshes (up to 80 x 80) confirmed[4] the very high accuracy of the nodal integral method previously established in other applications. This made it possible to carry out the numerical study of the bifurcations described below using a relatively coarse, 6 x 6 mesh with reasonable confidence in the accuracy of the results. Clearly, the accuracy of the numerical results would improve if a finer mesh were used; however, the parametric study of solution surfaces requires numerous calculations which are very costly and the mesh used seemed quite adequate based on various comparisons,[4,7] except for cavities with very low aspect ratios.

3. NUMERICAL DETERMINATION OF SOLUTION SURFACES FOR THE TILTED SQUARE CAVITY

The classic Bénard problem of natural thermal convection between infinite parallel horizontal plates with the bottom plate at a higher termperature than the top plate has been analyzed analytically, and many interesting nonlinear phenomena have been understood as a result of these analyses. The confined Bénard problem of natural thermal convection in a rectangular cavity with a temperature gradient parallel to the gravity vector is of greater practical, experimental and numerical interest. First, it is more readily useful as a simple model of natural thermal convection in components or subregions of complicated fluids systems, e.g. blocked meteorological systems, shut down sodium pool nuclear reactors, and other systems which are best modelled with finite horizontal dimensions. Second, it is obviously easier to investigate experimentally in order to measure the singular points and observe the transitions in the nature of the flow field as certain parameters are varied. Finally, it is more amenable to numerical methods most of which require covering the system by a finite mesh, especially since the accuracy of numerical solutions usually deteriorates as the mesh spacing increases. Hence it is not surprising that much attention currently is focused on studying the confined Bénard problem and exploring the very rich nonlinear structures it harbors.

The results presented here have been obtained via the nodal integral method using the same computer code developed for the double glazing problem mentioned in Sec. 2, but with a variable tilt angle, Θ. This was necessary in order to examine the unfolding and preservation of pitchfork bifurcations which occur at $\Theta = 0°$ (i.e., in the confined Bénard problem) and to study the more general tilted-cavity confined Bénard problem. To decrease the computational costs, the high efficiency of the nodal integral method was supplemented by programming a Newton-Euler continuation method in Ra and in Θ, to generate the initial guess at a given point in parameter space (Ra,Θ) using the previously converged solution at a different Ra or Θ. This resulted in a significant reduction in the number of Newton-Raphson iterations required to converge the initial guess to the solution. This, combined with the coarse-mesh (6 x 6) we were able to use due to the high accuracy of the nodal integral method, allowed us to calculate, and quantitatively map the solution surface for the square cavity (β = 1) over the Ra-Θ plane at a relatively low cost.

The first two bifurcation points or critical Rayleigh numbers for $\Theta = 0°$ were located first. This was not done by solving extended systems,[6] which give the conditions for the bifurcation points, because these extended systems require excessive computer storage. The first bifurcation point at which the basic, no-flow, pure-conduction, linear-temperature solution becomes unstable is a pitchfork bifurcation from which two stable single-vortex solutions, one counter-clockwise rotating and one clockwise rotating, emerge[6] was located by obtaining the counter-clockwise (clockwise) solution for $\Theta > 0°$ ($\Theta < 0°$) for low Ra, continuing it to large Ra keeping Θ fixed, and then continuing that solution to $\Theta = 0°$ keeping Ra fixed. These upper and lower branch solutions for $\Theta = 0°$ were then separately continued in decreasing Ra until each reached the no-flow solution at the bifurcation point $Ra^* = 2,524$. The solutions on the upper and lower branches of the second pitchfork for $\Theta = 0°$ which are known to be two-vortex, up-welling and down-welling solutions[6] were obtained for Ra greater than previously calculated estimates[6] of the second bifurcation point Ra^{**} by starting from initial solution guesses which have characteristics different than the single vortex solutions. These up-welling and down-welling solutions were then continued in decreasing Ra until each reached the no flow solution at the second bifurcation point Ra^{**} (0°) = 6,807. The values for Ra^* and Ra^{**} (0) for Pr = 1.7 based on 6 x 6 (36 element) nodal integral method calculations are in good agreement with the values reported recently based on fairly extensive finite element solutions to the extended systems for the bifurcation points.[6]

When the tilt angle was treated as an imperfection parameter and varied from zero, the first pitchfork bifurcation unfolded into a structurally stable, cusp catastrophe as a result of the introduction of this symmetry-breaking parameter. See Fig. 1. This is in agreement with previously reported results.[6] The second bifurcation at Ra^{**} (0°) did not unfold as the tilt angle was varied from zero, however. Rather, the structurally unstable pitchfork bifurcation in Ra was preserved even when $\Theta \neq 0°$ indicating that variation of yet one or more other parameters will be necessary to unfold this pitchfork into a structurally stable configuration. See Fig. 1.

To develope in detail the solution surface topology associated with these two bifurcation points this surface was mapped for the square cavity in the Ra-Θ plane. This was done using Euler-Newton continuation to calculate solutions on each surface from other solutions on the same surface. Figure 1 depicts the solution surface we obtained on a 6 x 6 mesh for Pr = 1.7, represented by the x-component of the velocity, averaged over the lower left quadrant of the cavity as a function of Ra and Θ. Increments used in the Ra-direction were 500, while nonuniform increments were used in the Θ-direction in order to resolve with adequate clarity the complicated structure of the surface in the vicinity of the Ra-axis. The surface is distorted because, although non-uniform spacing was used in θ, the graphics package used normalizes all intervals between parameter values to the same length in each direction. The figure illustrates the unfolding of the first bifurcation point by the tilt angle, and the preservation of the second. The diagrams at the bottom of Fig. 1 show: (a) the structurally stable cusp formed by the projection onto the Ra-θ plane, of the limit points of the solutions

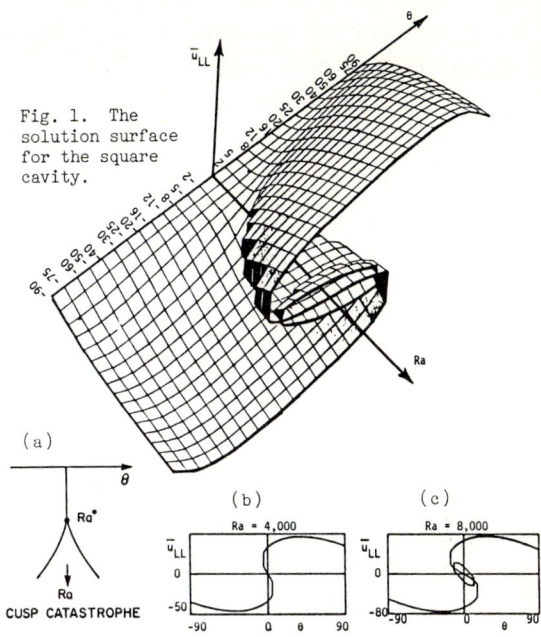

Fig. 1. The solution surface for the square cavity.

unfolded from the first bifurcation point by the imperfection parameter, (b) the hysteresis curve as a function of tilt angle for $Ra^* < Ra < Ra^{**}(0°)$ with the dashed arrows indicating the points at which the stable, clockwise rotating solution "snaps" in a transient problem in which the cavity tilt angle is varied continuously in time, to the stable, counter-clockwise rotating solution and vice versa, and (c) a similar hysteresis curve for $Ra > Ra^{**}(0°)$ with the bifurcation points as a function of θ indicated on the unstable middle branch of the hysteresis curve.

4. TRAJECTORIES OF BIFURCATION POINTS AS A FUNCTION OF THE ASPECT RATIO

In Sec. 3 we reported the topology of the solution surface and the trajectories of singular points as functions of Ra and θ in the case of a square cavity. Clearly, the cavity tilt angle θ has an important effect on the locations and nature of the singular points and the characteristics of the stable solution. Another important parameter in this problem is the aspect ratio $\beta \equiv$ height/width, $\beta \in (0,1]$, because as it is varied the stable flow field (for Ra > the lowest value of Ra_{crit}) changes qualitatively as well as quantitatively. Clearly, one would expect intuitively that the most stable flow in a wide rectangular cavity ($\beta \ll 1$)

would consist of some number of horizontally adjacent vortices rather than one vortex as was the case for the square cavity. Of course the different flow fields possess different properties (such as the total heat transfer, the position and temperature of the hottest spot,...,etc.) which can have a significant impact on the practical implications of the fluid flow model. Hence, the behaviour of the solution surfaces (see Fig. 1) as β is varied also was studied. More specifically, the trajectories of Ra^* and $Ra^{**}(0^\circ)$ as a function of β were determined.

As in the work described in Sec. 3 the calculation of the critical Rayleigh numbers was done interactively. That is, for a particular value of β we used a certain flow field (one vortex or two vortex, for example) as an initial guess, and varied Ra to obtain bounds on Ra_{crit} (the lower (upper) bound corresponding to the largest (smallest) value of Ra for which we were able to converge to a no-flow (non no-flow) solution). We used this approach because of limitations on the storage capacity available. The alternative approach,[6] the analytic calculation of the singular points, requires solving an extended system which is more than twice as large as the algebraic set of equations we solve. The results of our calculations are presented in Fig. 2, for Pr = .5. The first critical Ra which occurs at Ra = 2,524 at β = 1 starts decreasing as β is decreased and then increases slowly until it reaches a maximum around β = .5. This part of this curve separates the region, above it, in which the single vortex solution exists, from the region below it where this solution does not exist. At or close to this maximum a smooth transition from the one-vortex solution to a three-vortex solution occurs as β is further decreased. Hence this curve now separates the region above it, in which the three-vortex solution exists, from the region below it where this solution does not exist.

This smooth transition does not appear to be associated with a singular point. Similarly, after this curve changes its direction of curvature at β ~ .34, its direction of curvature turns downward again at β ~ .24 where the three-vortex solution undergoes a smooth transition to a five-vortex solution beyond which this curve separates the region above it where the five-vortex solution exists from

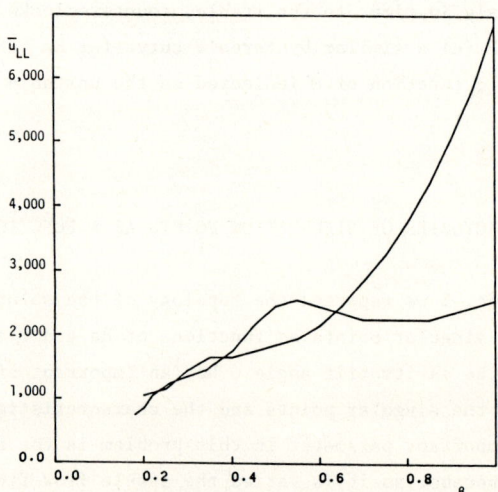

Fig. 2. The dependence of the first two critical Rayleigh numbers on the aspect ratio.

that below it where this solution does not exist. Values of $\beta \lesssim .2$ were not studied since the coarseness of the mesh used here would make questionable any results obtained in which larger numbers of horizontally arrayed vortices occur. In fact, the numerical accuracies of the values of the bifurcation point calculated here as a function of β probably has become quite poor significantly before β is reduced to $\sim .2$. An analogous sequence of transitions occurred when we carried out the same procedure starting from the two vortex solution, for which $Ra_{crit} = 6,807$ at $\beta = 1$. Namely, Ra_{crit} decreases as β is decreased, and then undergoes a change of direction of curvature at $\beta \cong .3$ where the flow field becomes a four-vortex solution, and Ra_{crit} continues to decrease. These results are in good qualitative agreement with the results obtained by Jackson and Winters[7] for the rigid/rigid case which is the only case we have studied.

We, as well as previous authors,[7] expected that the transition from the one vortex flow field to the three-vortex flow field as β is decreased occurs when the bifurcation points corresponding to the two flow fields cross one another on the Ra axis. In fact, curves analogous to those in Fig. 2 were obtained in earlier work where the fact that the three-vortex segment of the present curve was not born in a crossing of a one-vortex curve and a three-vortex curve (see Fig. 16 of Ref. 7) was believed to be an inaccuracy resulting from the discretization error. Our calculated flow fields and map of the solution surface suggests that the three-vortex solution does not correspond to the crossing of bifurcation points, but rather that the transition from one to three vortex fields occurs continuously as we decrease the aspect ratio. Of course, it cannot be guaranteed that this behaviour is not just a common artifact of the very different numerical methods used here and in the earlier work. Figure 3 shows the complicated unfolding of two pitchforks for $\beta = .5$ as Θ is varied from $0°$. The inner pitchfork corresponds to the three-vortex solution, which unfolds, with the upper branch moving up and to the right as Θ is increased from zero. The solution changes continuously into the one-vortex solution for $\Theta \cong .7$. On the other hand the outer pitchfork corresponds to a "peanut" shaped, single-vortex solution. This pitchfork unfolds with Θ as an imperfection parameter, to give the complicated knot-shaped curve shown in Fig. 3, compared to the S-shaped hysteresis curve for the same

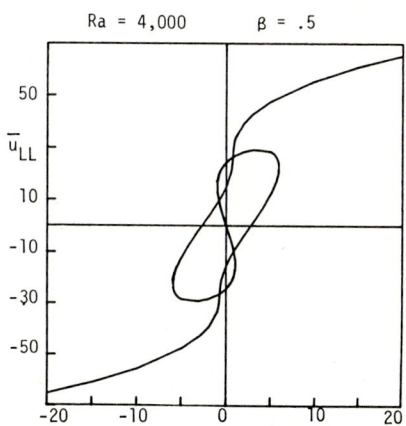

Fig. 3. The unfolding of two pitchfork bifurcations at Ra=4,000 and $\beta=0.5$.

Ra, and $\beta > .55$. That the crossing points in the "knot" are not bifurcation points is clear from the fact that the two solution curves correspond to two different flow fields; i.e., the solutions at the crossing point are not identical. This conclusion was confirmed by plotting the vorticity averaged over the lower left quadrant vs. Θ. Indeed the solution curves represented thusly did not intersect at the values of Θ as in Fig. 3, indicating that the solution curves actually "cross over" one another rather than intersect.

5. CONCLUSIONS

We have used our very efficient, highly accurate nodal integral method to investigate numerically the bifurcation phenomena that occur in the confined Bénard problem. We were able to calculate the multiple solution surface in Ra-Θ plane, and the first two pitchfork bifurcation points for the square cavity. The first of these unfolds into a cusp catastrophe, while the second is preserved as a structurally unstable pitchfork. We next studied the effect of the aspect ratio on the critical Rayleigh numbers and on the flow field. Our course-mesh results for Ra_{crit} as a function of β are in good agreement with previously obtained fairly large finite element results. We have found that the transitions in the flow field from the one-vortex to the three-vortex field, and three to five, etc. occur smoothly both as the aspect ratio β and the tilt angle Θ are varied, and not through bifurcations in these parameters.

REFERENCES

1. Y. Y. Azmy and J. J. Dorning, "A Nodal Integral Approach to the Numerical Solution of Partial Differential Equations," <u>Advances in Reactor Computations</u>, <u>II</u>, Am. Nucl. Soc., LaGrange Park, IL, 1983, p. 893.

2. Y. Y. Azmy and J. J. Dorning, "Nodal Integral Method Solutions and Singular Points for Driven Cavity Problems," Proc. Third Intern. Conf. on Numerical Methods in Laminar and Turbulent Flow, Eds., C. Taylor, J. A. Johnson and W. R. Smith, Pineridge Press, Swansea, U.K., 1983, p. 583.

3. Y. Y. Azmy and J. J. Dorning, "Arc-Length Continuation through Limit Points of Nodal Integral Method Solutions to the Navier-Stokes Equations," Proc. Second Intern. Conf. on Numerical Methods for Nonlinear Problems, Eds., C. Taylor, E. Hinton, D.R.J. Owen and E. Onate, Pineridge Press, Swansea, U.K., 1984, p. 672.

4. Y. Y. Azmy and J. J. Dorning, "Nodal Integral Method Solutions of Cavity Thermal Convection Problems," Proc. Fourth Intern. Conf. on Numerical Methods in Laminar and Turbulent Flow, Eds., C. Taylor, M. D. Olson, P. M. Gresho and W. G. Habashi, Pineridge Press, Swansea, U.K., 1985, p. 753.

5. G. De Vahl Davis, <u>Int. J. Num. Meth. Fluids</u>, <u>3</u>, 249 (1983).

6. K. A. Cliffe and K. H. Winters, <u>J. Comp. Phys.</u>, <u>54</u>, 531 (1984).

7. C. P. Jackson and K. H. Winters, <u>Int. J. Num. Meth. Fluids</u>, <u>4</u>, 127 (1984).

DIFFUSION AND ROSSELAND APPROXIMATION
PROPERTY OF THE BOUNDARY LAYER

Claude Bardos

C.M.A.E.N.S.

45 Rue d'Ulm, Paris, 75005, France

In this paper I intend to describe a series of results obtained jointly with several coworkers of the Paris Group: B. Perthame, F. Golse and R. Sentis; they concern the following equation,

$$\partial_t I + \Omega \nabla_x I + \sigma(\nu,T)(I-B(\nu,T)) = 0 \qquad (1)$$

$$\partial_t T + \ll \sigma(\nu,T)(B(\nu,T)-I) \gg = 0 \qquad (2)$$

The above system describes the evolution of a density of photons in a stellar atmosphere. I is the intensity of photons with position x, angular velocity Ω and frequency ν; T denotes the temperature, $B(\nu,T)$ is a Planckian function given by:

$$B(\nu,T) = 2h\nu^3/c^2(\exp(h\nu/kT) - 1) \qquad (3)$$

where c, h and k are physical constants. $\sigma(\nu,T)$ is a positive, ν and T dependant function which gives the opacity of the system (1), (2). For the approach of the system (1), (2) one should consider the following questions

(i) Existence, uniqueness and stability of the solution

(ii) Approximation of the solution by a function which is independant of the variables ν and Ω.

(iii) Computation of the boundary condition for this equation when the media X ($x \in X$) is limited by a boundary ∂X.

Some of the above ideas have already been used for other type of equations (Boltzmann or neutron transport equation). The neutron transport equation turns out to be an easy starting point because it is linear and in some cases it gives rise to explicit computations. However in the present situation we do have a real nonlinear problem and an important part of this non linearity comes from the opacity function.

The existence and uniqueness of the solution may be easily obtained (cf.[1]) if one assumes (Hypothesis of monotonicity) that the functions $T \to (\sigma,T)$ and $T \to -\sigma(\nu,T)B(\nu,T)$ are non decreasing. Then the problem (1), (2) can be written in an abstract form

$\partial_t U + A(U) = 0$ where U denotes the vector (I,T) and where A is an accretive operator in the space $L^1(S_2 \times x \times R_\nu^+) \times L^1(X)$ when the hypothesis of monotonicity are not satisfied, one may use (cf.[2]) a compactness theorem to obtain the existence of the solution.

To construct the Rosseland approximation one introduce a small parameter ε which is related to the mean free path. The equations (1) and (2) become

$$\partial_t I_\varepsilon + \Omega \nabla_x I_\varepsilon + \sigma(\nu,T_\varepsilon)(I_\varepsilon - B(\nu,T_\varepsilon)) = 0 \qquad (1)_\varepsilon$$

$$\partial_t T_\varepsilon + \ll \sigma(\nu,T_\varepsilon)(B(\nu,T_\varepsilon)-I_\varepsilon)) \gg = 0 \qquad (2)_\varepsilon$$

Then one makes the following:

$$I'_\varepsilon = I_0 + \varepsilon I_1 + \varepsilon^2 I_2 \qquad I'_\varepsilon = B(\nu,T') \qquad (4)$$

and computes T' in such a way that the problem $(1)_\varepsilon, (2)_\varepsilon$ is satisfied up to the order one in ε. This computation gives for T' a "porous media type" equation

$$\partial_t(T' + dT'^4) - \Delta F(T') = 0, \qquad (5)$$

where d denotes a physical constant and where F is given by the equation

$$\partial_T F(T) = \int_0^\infty (\partial_T B(\nu,T)/\sigma(\nu,T) d\nu . \qquad (6)$$

When the monotonicity assumptions are satisfied, the error between I_ε and $B(\nu,T')$ can be estimated, in the more general case, the compacity method can be used for opacity which are independant of ν (this situation is often called the grey problem).

For the problem (5) a boundary condition has to be prescribed, this boundary condition is obtained by the solution of a Milne problem.

$$\mu \partial_x I + \sigma(\nu,T)(I-B(\nu,T)) = 0 \qquad (7)$$

$$\ll \sigma(\nu,T)(B(\nu,T)-I) \gg = 0 \qquad (8)$$

in the semi infinite domain :

$$\nu \geq 0, \qquad x \geq 0, \qquad -1 \leq \mu \leq 1 .$$

The strategy consists in showing that with an incoming flux prescribed:

$$I(0,\mu,\nu) = h(\mu, \nu) \quad \text{given for} \quad \mu \geq 0, \qquad (9)$$

the problem (7), (8) do have a unique solution and that, for x going to infinity, this solutions converges to a Planckian function $B(\nu, T_\infty)$ with T_∞ constant. The rate of convergence provides the size of the boundary layer. If we assume that the functions $T \to \sigma(\nu,T)$ and $T \to -\sigma(\nu,T)B(\sigma,T)$ are not decreasing, one can apply a monotonicity method using the accretiveness of the operator defined by (7), (8) and (9) in the space $L^1(]-1, 1[\times R_x^+ \times R_\nu^+) \times L^1(R_\lambda^+)$, this leaves to an existence-uniqueness proof; however the convergence problem is not completely solved. It turns out that the problem can be completely treated when the opacity is factorised according to the formula:

$$\sigma_\nu(T) = \sigma(\nu) K(T).$$

In this case one has the following.

<u>Theorem</u> (F. Golse to appear). We assume that the functions $\sigma(\nu)$ and $K(T)$ satisfy the following hypothesis: $0 < K_{min} < K(T) < K_{max} < \infty$, $0 < \sigma(\nu) < \sigma_{max}$ and $\sigma(\nu) > c\nu^{-\beta}$ ($\beta > 0$, ν going to infinity), then the problem (7), (8), with the boundary condition

(9), has a unique solution which converges, for x going to infinity, to a Planckian with the rate $\exp(-\lambda x^\rho)$, for some positive (β dependant) ρ. When $\beta = 0$, ρ is equal to one.

This theorem is important because it shows the appearance of a boundary layer which is, for $\rho > 0$, of fractional order.

The numerical approach of (7), (8), (9) uses mainly perturbation methods; starting from linearised version where classical methods, like the invariant imbedding, can be applied.

REFERENCES

Most of the results described are contained in the two following papers

[1] C. Bardos, F.Golse and B. Perthame, Rosseland approximation for radiative transfert equation (submitted for publication in the Comm. on Pure and Appl. Math.)

[2] F.Golse, B.Perthame and R. Sentis, Un resultat de compacite pour les equations de transport et application au calcul de la valeur propre principale de l'operateur de transport C.R. acad Sc. T. 301, 1(1985) 341-344. The basic ideas, from the physical point of view, can be found in the paper of Larsen Pomraning and Badhan: Asymptotic analysis of radiatif transfert problem J.Quant Spect. and Radiatif Trans. Vo 129, n°4, 285-310, 1983.

FINITE DIFFERENCE SOLUTION OF THE 3-D EULER EQUATIONS USING A MULTISTAGE RUNGE-KUTTA METHOD

J. M. Barton   S. K. Yoon
Sverdrup Technology, Inc.
Aeromechanics Department
Middleburg Heights, OH 44130/USA

Abstract

The strengths of the multistage Runge-Kutta methods and finite difference discretization have been combined to produce a new approach for solving the 3-D Euler equations in transonic flow. The integration scheme and adaptive nonlinear dissipation, originally applied to finite volume methods, have been applied to a vertex formulation of the Euler equations written in a general curvilinear coordinate system. The new approach eliminates the mesh sensitivity, boundary condition inaccuracies, and difficulties with extensions to viscous flows commonly encountered with finite volume methods. It likewise provides an alternative to previously used finite difference integration methods whose stability or accuracy limitations render them marginally useful for long term code development. The new method is applied to transonic flow over a hemisphere-cylinder and flow through a high speed propeller. The calculations compare well with measurements and previous calculations. Future developments of the method are outlined.

1.   Introduction

The multistage integration methods combined with nonlinear dissipation have been quite successful in finite volume applications. The finite volume or cell-centered methods, however, have exhibited some inherent difficulties. They are quite sensitive to mesh stretching. In addition, if phantom or fictitious cells are used, boundary conditions are applied only in an average sense; thus, the accuracy of the boundary condition implementation is dependent on the near-boundary mesh spacing. Finally, viscous formulations require coordinates of cell centers for evaluating viscous derivatives. In most finite volume methods the coordinates are stored at cell corners, necessitating a considerable amount of interpolation. Vertex formulations eliminate all three difficulties. As noted earlier, however, most methods commonly employed for integration of the finite difference equations exhibit serious deficiencies which limit their usefulness.

The objective of the present work is to demonstrate a new approach for solving the 3-D Euler equations. The desirable facets of both finite difference and finite volume methods are combined with the goal of gaining the strengths from both. A multistage Runge-Kutta method is applied to solve the semi-discrete governing equations in a general curvilinear coordinate system. The nonlinear dissipation proposed by Jameson, et al.[1,2,3,4] is employed to stabilize the calculations. Results are computed for a hemisphere-cylinder and a high speed propeller.

2. Numerical Method

Governing Equations

The unsteady, compressible Euler equations are written in a cylindrical base coordinate system, which is the absolute (non-rotating) reference frame. The equations are transformed to a general (rotating) curvilinear coordinate system by a time-dependent mapping. The curvilinear system is defined such that the boundaries lie in constant coordinate surfaces, i.e., a body-fitted frame is used. More details concerning the equations and transformation are contained in refs. 5 and 6.

In the relative (rotating) frame the grid is fixed and the flow is steady. The four stage Runge-Kutta scheme is applied. It is fourth-order accurate in time (for linear problems) and is stable for Courant numbers less than or equal to $2\sqrt{2}$. The nonlinear dissipation of ref. 1 was applied in a finite difference method by Pulliam[7]. The same formulation is used here as three one-dimensional operators in the transformed space. For the present results, smoothing was included in every stage of the Runge-Kutta integration. Future work will eliminate smoothing in at least two of the stages, reducing computer time by about 32%.

The boundary conditions represent an isolated body in an infinite ideal fluid. The symmetry of a geometry is expedited to reduce the magnitude of the problem. Specifically, the hemisphere-cylinder is treated axisymmetrically. The propeller is ascribed periodic symmetry so that the flow through only one blade passage is comnputed. Dependent variables on the periodic surface (upstream of the leading edge, downstream of the trailing edge, and above the tip) are obtained by averaging two points on each side of the boundary. Tangency is enforced on all solid boundaries and the surface pressure is obtained by extrapolation from the interior. The inflow and freestream boundaries are placed sufficiently far away to enforce freestream conditions. At the axis of symmetry the radial velocity is zero and radial derivatives of all other variables are zero. At outflow the transformed radial momentum equation is simplified by neglecting the radial velocity and dropping the unsteady term. The remaining equation is solved for the radial pressure distribution by a trapezoidal rule integration, starting from a specified value at the freestream boundary. Freestream conditions were specified everywhere to provide initial conditions.

3. Applications

Two different applications are considered: flow over a hemisphere-cylinder and flow through a modern technology high speed propeller. The smoothing coefficients were kept the same for all cases and a constant time step was used. No enthalpy damping or residual smoothing was employed. All cases were run for a Courant number of 2.5, and the L_2-norm of the residual [L(U) in Eq. (1)] decayed between three and four orders of magnitude.

Hemisphere-Cylinder

Flow over a hemisphere-cylinder was computed for a freestream Mach number of 0.6. The grid had 45 points axially and 35 points radially. The cylinder extended seven diameters downstream of the hemisphere-cylinder junction. The inflow and freestream boundaries were nine diameters from the body. A hyperbolic tangent stretching (radially) produced a smooth non-uniform grid spacing, clustering points near the body. Computed and measured surface pressure is presented in fig. 1. The measurements were taken from ref. 8.

SR-3 Propeller

Three cases were computed at M_∞ = 0.8 and an advance ratio of 3.06. Results are compared with the measurements of ref. 9 in fig. 2. Though the current results over-predict the measurements, they show a marked improvement over the ADI results of ref. 6. It has been shown in refs. 5 and 10 that further refining the grid and solving the field equations on the periodic surface (instead of averaging) bring the computations into closer agreement with the measured data. The Euler analysis should indicate less power required for the propeller when compared to viscous results, i.e., measurements. It is believed that the current nonlinear dissipation with the correct boundary conditions[11] will produce less diffusive results.

Acknowledgment

The work reported herein was supported under contract NAS3-24105 with the NASA Lewis Research Center.

REFERENCES

1. Jameson, A., Schmidt, W. and Turkel, E., "Numerical Solutions of the Euler Equations by Finite Volume Methods Using Runge-Kutta Time-Stepping Schemes," AIAA Paper 81-1259, June 1981.
2. Jameson, A. and Baker, T.J., "Multigrid Solution of the Euler Equations for Aircraft Configurations," AIAA Paper 84-0093, Jan. 1984.
3. Holmes, D.G. and Tong, S.S., "A Three-Dimensional Euler Solver for Turbomachinery Blade Rows," ASME Paper 84-GT-79, June 1984.
4. Celestina, M.L., Mulac, R. and Adamczyk, J.J., "A Numerical Simulation of the Inviscid Flow Through a Counterrotating Propeller," ASME Paper 86-GT-138, June 1986.
5. Barton, J.M., Yamamoto, O. and Bober, L.J., "Inviscid Analysis of Advanced Turboprop Propeller Flow Fields," AIAA Paper 85-1263, July 1985.
6. Bober, L.J., Chaussee, D.S. and Kutler, P., "Prediction of High Speed Propeller Flow Fields Using a Three-Dimensional Euler Analysis," AIAA Paper 83-0188, Jan. 1983.
7. Pulliam, T.H., "Euler and Thin Layer Navier-Stokes Codes: ARC2D, ARC3D," in Computational Fluid Dynamics," UTSI Pub. No. E02-4005-023-84, Mar. 1984.
8. Hsieh, T., "An Investigation of Flow About a Hemisphere-Cylinder at 0- to 90-deg Incidence in the Mach Number Range From 0.6 to 1.5," AEDC-TR-76-112, July 1976.
9. Rohrbach, C., Metzger, F.B., Black, D.M. and Ladden, R.M., "Evaluation of Wind Tunnel Performance Testing of an Advanced 45

Degree Swept Eight-Bladed Propeller at Mach Numbers From 0.45 to 0.85," NASA CR-3505, Mar. 1982.
10. Barton, J.M., Yamamoto, O. and Bober, L.J., "Inviscid Analysis of Transonic Propeller Flows," to be published in J. Prop. and Power, 1986.
11. Barton, J.M. and Yoon, S., "Improvements in a Multistage Runge-Kutta Method for the 3-D Euler Equations," to be presented at the First World Congress on Computational Mechanics, Austin, TX, Sept. 1986.

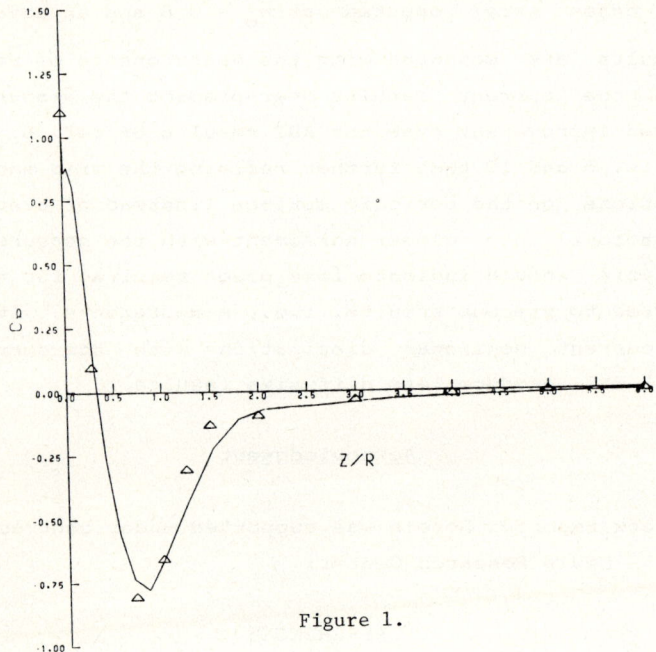

Figure 1.

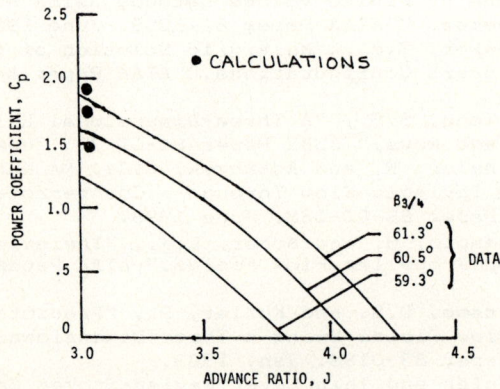

Figure 2.

SOLUTION OF THE COMPRESSIBLE NAVIER STOKES EQUATIONS BY USING EMBEDDED ADAPTIVE MESHES

F. Bassi[^], F. Grasso[^^], M. Savini[^^^]

[^] Dept. of Energetics, Polytechnic of Milan, Milan, Italy, 20133
[^^] Inst. of Gasdynamics, University of Naples, Naples, Italy, 80125
[^^^] Inst. of Research in Propulsion and Energetics, CNR, Peschiera Borromeo (Milan), Italy, 20068

INTRODUCTION

The accuracy of a numerical solution depends on the degree of spatial resolution. In general the finer the spatial discretization the more accurate is the numerical solution. However the computational efficiency decreases when global mesh refinement is employed. Such a loss in efficiency is reduced by use of adaptive techniques, such as zonal, dynamic clustering and embedding.

Several papers dealing with the use of zonal (either patched or overlaid) grid methods [1]-[2] have recently appeared, whereby the computational domain is subdivided into zones and the grid for each zone is generated independently according to the required resolution. However zonal boundaries must be properly treated to ensure conservation.

The clustering of grid lines by moving mesh points toward regions of high gradients, where greater resolution is required, is employed in Refs. [3]-[5]. However this latter technique can produce inaccurate results when excessive grid skewing and stretching occur.

Adaptive embedded grid approaches have been used in Refs. [6]-[9]. When using local mesh enrichment, the accuracy is enhanced in regions of interest, at the expense of moderate increase of computational effort. Berger and Jameson [6] describe an adaptive grid refinement method for the solution of the steady Euler equations. They generate locally refined rectangular grids in regions where the error in the numerical solution is largest, and use Richardson extrapolation for estimating such an error. Dannenhoffer and Baron [9] have presented a grid adaptation technique and have shown the dependency of the computed solution of inviscid transonic flows upon the topological properties of the embedded grid regions when using different adaptation criteria.

In the present work an adaptive embedded grid approach that exploits multigrid strategy is presented for the solution of the compressible Navier Stokes equations. Embedded meshes are generated in regions where an adaptation criterion, based on the cumulative distribution function of pressure gradient, is not satisfied. The computational strategy is as follows.

Initially the solution is obtained on the coarse grid until quasi convergence is reached. Then the adaptation criterion is checked at each cell. Those cells where the criterion is not satisfied are flagged (to indicate where mesh refinement is needed), and finer cells are generated by halving the mesh both in x and y. The grid of different classes are not assembled in a global one, but are kept independently and smoothness properties are maintained on each grid. A solution vector is defined on each grid, and time stepping is performed separately on each one [6]. This has the advantage that different algorithms can be used in different regions depending on the accuracy required . Coupling of the grids at the interfaces is ensured by properly transfering the fluxes computed on the fine grids to the coarser ones, so as to satisfy conservation.

The method has been applied to transonic laminar flows through a double throat nozzle at different Reynolds numbers, and a remarkable improvement in the details of the solution is shown.

In the next sections the governing equations and numerical algorithm are described, then some results and concluding remarks given.

NUMERICAL SOLUTION

A fully conservative formulation of the compressible Navier Stokes equations is employed. In cartesian coordinates the system of dimensionless governing equations is obtained by assuming as reference quantities:

$x_r, y_r = D$ $t_r = D/c_o$ $\rho_r = \rho_o$ $P_r = \gamma P_o$

$u_r = c_o$ $E_r = c_o^2$ $\mu_r = \rho_o c_o D/Re$ $\lambda_r = c_p \mu_r / Pr$

where D, c_o, Re, Pr are respectively the half throat nozzle height, the speed of sound, the Reynolds and Prandtl numbers.

The equations are:

$$\frac{\partial}{\partial t} \int_V w\, dV + \oint_{\partial V} (f n_x + g n_y) dS = \oint_{\partial V} (P n_x + Q n_y) dS$$

where

$w = [\rho, \rho u, \rho v, \rho E]^T$

$f = [\rho u, \rho u u + p, \rho u v, \rho u E + u p]^T$

$g = [\rho v, \rho u v, \rho v v + p, \rho v E + v p]^T$

$P = [0, \sigma_{xx}, \sigma_{xy}, u\sigma_{xx} + v\sigma_{xy} - q_x]^T$

$Q = [0, \sigma_{xy}, \sigma_{yy}, u\sigma_{xy} + v\sigma_{yy} - q_y]^T$

$p = (\gamma-1)[E - 1/2(u^2 + v^2)]$

$\underline{\sigma} = 1/Re[(\nabla \underline{u} + \nabla \underline{u}^T) - 2/3 \nabla \cdot \underline{u} \underline{U}]$

$\underline{q} = -\gamma \nabla e / Re Pr$

Note that constant diffusion properties are assumed.

The computational solution is obtained in two or more phases (with more than one level of embedded grids).

Phase I

First the solution on the base grid is obtained by using an explicit finite volume method whereby the computational domain is subdivided into trapezoidal cells. The system of governing equations is reduced to a system of ordinary differential equations by using the method of lines so as to decouple the temporal and spatial terms. Moreover mean value theorem and mid-point rule are employed to evaluate volume and surface integrals. Time integration is performed by a three-stage Runge-Kutta algorithm that has high damping properties of the high frequency error components and large propagation velocity of the low frequency errors:

$$w^{(k)} = w^{(o)} + \alpha_k \frac{\Delta t}{V} R^{(k-1)} = w^{(o)} + \alpha_k \frac{\Delta t}{V} [-C(w^{k-1}) + D(w^o) + AD(w^o)] \quad (1)$$

where C, D, and AD are respectively the net inviscid, diffusion and adaptive dissipation contributions and they are evaluated as in Ref. [10], and $\alpha_1 = \alpha_2 = .6$

Phase II

Eqn. (1) is solved until quasi convergence is reached. At this point the adaptation criterion is checked and embedded fine grids are constructed as follows. Let f, $\hat{f}$, and F be respectively the modulus of pressure gradient weighted by the cell volume, its average value over the entire computational domain, and the cumulative distribution function (i.e. the fractional number of points having a refinement parameter $f/\hat{f}$ greater than a threshold value T), where

$$f = ([\Sigma_\beta (pn_x \Delta S)_\beta]^2 + [\Sigma_\beta (pn_y \Delta S)_\beta]^2)^{1/2}$$

$$\hat{f} = \Sigma_n f/N$$

Once $F = F(f/\hat{f})$ is constructed, the mesh is adapted in the regions where $f/\hat{f}$ is greater than T. T has been set equal to the value corresponding to the maximum curvature of F with the constraints: i) $T > 1$; ii) $F < .25$ (see Fig. 1).

The cells where the adaptation criterion is not satisfied are flagged, and four embedded finer cells are created for each flagged cell. Further constraints are verified to ensure that each adaptation region be simply connected, and eventually more cells are flagged.

Grids of different levels (M_ℓ) are kept independently in a tree-like structure and interact through interface boundaries. This has the advantage that, once a solution vector is defined on M_ℓ, then time stepping can be performed separately on each grid, and in principle different algorithms (differing in the order of time and/or spatial accuracy) can be used. For computational efficiency data on M_ℓ are organized by introducing a system of pointers governing data communication among grids of different levels as well as on the same grid level.

Once the embedded grid is created, its solution vector is initialized by bilinear interpolation of the coarse mesh solution. Thereafter time integration is performed advancing first the embedded grid solution and then the coarse one. Different strategies may be employed. In Ref. [6] a constant mesh ratio is used and several time steps are taken on the embedded grid for every one coarse grid step.

In the present work a multigrid phylosophy is followed and the solution is advanced in time as follows.

Fine Grid Solution

$$w_F^{(k)} = w_F^{(o)} + \alpha_k \left[\frac{\Delta t}{V} R^{(k-1)} \right]_F \qquad (2)$$

Injection Step

In the regions of adaptation the fine grid solution is injected onto the coarse one by interpolation:

$$w_C^{(o)} = I_F^C \, w_F^{(n+1)}$$

Coarse Grid Solution

$$w_C^{(k)} = w_C^{(o)} + \alpha_k \left[\frac{\Delta t}{V} R^{(k-1)} \right]_C \qquad (3)$$

Conservation is ensured in regions of adaptation by adding a forcing function P to Eqn. (3) given by:

$$P_C = \Sigma_F R_F^{(n+1)} - R_C^{(o)}$$

Correction Step

The fine grid solution is updated by transfering the coarse grid correction in the following way:

$$w_F = w_F^{(n+1)} + I_C^F \left[w_C^{(n+1)} - w_C^{(n)} \right]$$

Interface Treatment

For computational purposes fictitious embedded cells are generated along the boundary of the adaptation regions, i.e. in the proximity of the interfaces between grids of different levels. Interpolation of coarse grid values, to update the fictitious boundary conditions, give rise to high frequency oscillations. Numerical experiments indicate that the oscillations are minimized by second order interpolation. Moreover, at the interfaces, conservation is strictly enforced by using, at each stage of the coarse grid integration step, the fluxes evaluated on the fine embedded meshes at time level (n+1).

Phase III

Once quasi convergence is reached both on coarse and (intermediate) embedded grids, the adaptation criterion is checked to refine the intermediate meshes by generating a second level of embedded grids, and the same computational strategy as that of Phase II is followed.

RESULTS AND DISCUSSION

The effectiveness of the proposed approach has been tested for the solution of laminar flows through a double throat nozzle at different values of the Reynolds number (Re = 400, 1600, 6400), for which several computational results are available.

The computational domain has been discretized in 102x22 (base) grid cells generated as in Ref. [10]. At the wall no slip boundary conditions are imposed on the velocity, and fixed temperature is set; symmetry conditions are enforced at the centerline. Total enthalpy, entropy and flow direction are imposed at inlet. The missing boundary condition is obtained by extrapolation of Riemann invariant from the interior. At the outlet section the flow is supersonic and extrapolation boundary conditions are imposed.

Figs. 3a-3d show the computed results for Re = 400. It is interesting to note the improvement in the accuracy of the results when using embedded grids. However no noticeable differences have been found between the solution (S1) obtained with only one level of embedded grid and the one (S2) obtained with two levels of mesh embedding.

Though not marked some oscillations arise due to interface treatment. However these oscillations are not observed when the coarse grid solution is represented (Fig. 3d). Hence the method can be viewed as a local multigrid where the coarse grid is not a mean for accelerating the fine grid solution, but it is the latter that is used to improve the coarse grid solution accuracy [11].

The same trends are observed for Re = 1600, 6400 (Figs. 4a-5d). However the oscillations are more evident in the immediate vicinity of the oblique shock. Similarly no oscillations appear when plotting only the coarse grid solution. Good convergence of the solution appears from the nearly constant values of the mass flow rate (Fig. 2a-2c).

In conclusion the accuracy of the solution is improved in the regions of high gradients with a relatively small increase in required storage resources by using adaptive (multi)grid embedding. Special care is necessary in treating interface boundaries: strict conservation must be enforced and higher interpolation accuracy must be used when updating fine grid boundary points.

REFERENCES

[1] Rai, M.M., Journal of Computational Physics, 62, 472-503, 1986.
[2] Hessenius, K.A., and Pulliam, T.H., AIAA Paper No. 82-0969, 1982.
[3] Thompson, J.F., AIAA Journal 22, 1505-1523, 1982.
[4] Rai, M.M., and Anderson, D.A., AIAA Computational Fluid Dynamics Conference, Alto, Ca, 1981.
[5] Gnoffo, P.A., Numerical Grid Generation, Edited by Thompson, J.F., North Holland, 1982.
[6] Berger, M.J., and Jameson, A., 9th ICNMFD, Saclay, France, 1984.
[7] Pouletty, C., These de Docteur-Ingenieur, Ecole Centrale de Paris, 1985.
[8] Bristeau, M.O., Glowinski, R., Mantel, B., Periaux, J., and Pouletty, C., Notes in Numerical Fluid Mechanics, to appear.
[9] Dannenhoffer, J.F., and Baron, J.R., AIAA Paper No. 85-0484, 1985.

[10] Bassi, F., Grasso, F., Jameson, A., Martinelli, A., Savini, M., Notes in Numerical Fluid Mechanics, to appear.
[11] Ni, R.H., VKI-LS 86-02, 1986.

ACKNOWLEDGEMENT

Research supported by CNR-PFE2 84.02644.59, and MPI 40% 84.

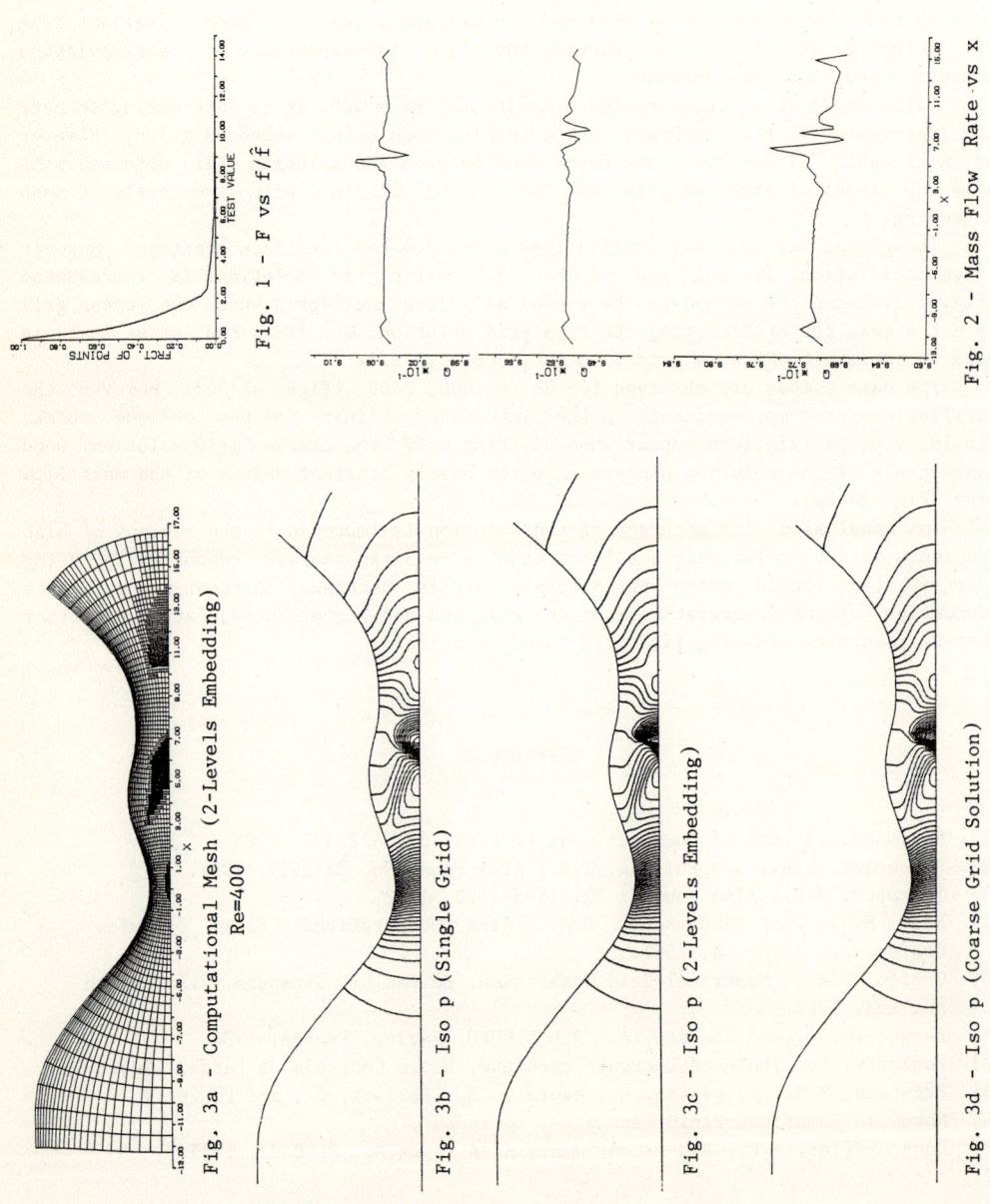

Fig. 1 - F vs f/f̂

Fig. 2 - Mass Flow Rate vs x

Fig. 3a - Computational Mesh (2-Levels Embedding) Re=400

Fig. 3b - Iso p (Single Grid)

Fig. 3c - Iso p (2-Levels Embedding)

Fig. 3d - Iso p (Coarse Grid Solution)

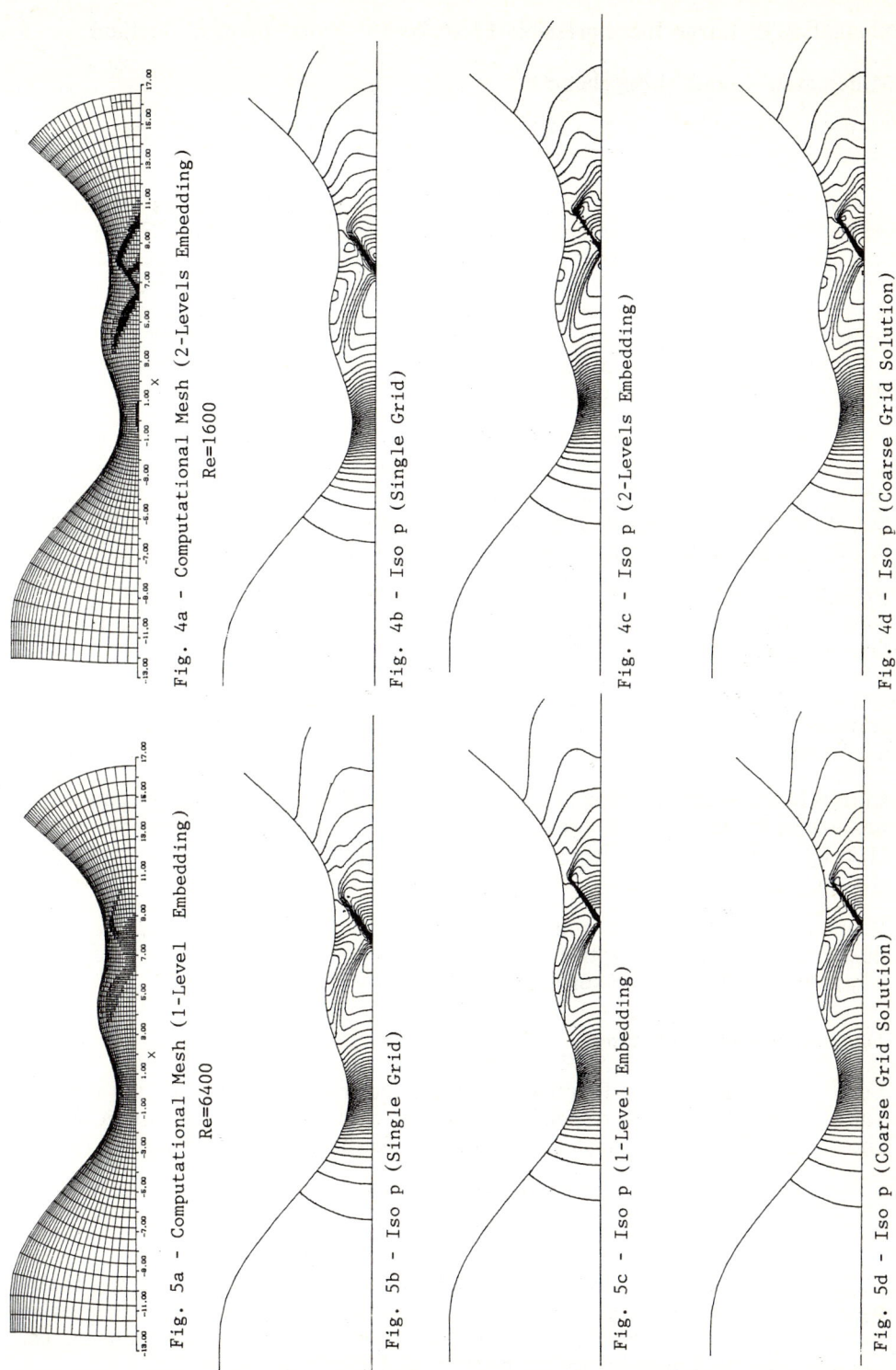

Fig. 4a - Computational Mesh (2-Levels Embedding) Re=1600

Fig. 4b - Iso p (Single Grid)

Fig. 4c - Iso p (2-Levels Embedding)

Fig. 4d - Iso p (Coarse Grid Solution)

Fig. 5a - Computational Mesh (1-Level Embedding) Re=6400

Fig. 5b - Iso p (Single Grid)

Fig. 5c - Iso p (1-Level Embedding)

Fig. 5d - Iso p (Coarse Grid Solution)

Simulation of Large Incompressible Flows by the Finite Element Method

M.Bercovier [x], and M.Engelman [x]

[x] Dep. of Comp. Sciences,The Hebrew University of Jerusalem, 91904 Israel
[x] FDI 1600 Orrington Ave.,Evanston, IL 60201.

One of the blocking factor in the use of computational fluid mechanics is the complexity of the domains to be described. The Finite Element Method (FEM), more and more linked to Computer Aided Design (CAD) Systems, with its systematic modularity and its use of advanced mesh generation techniques can bring an answer to this problem. Moreover the demand of numerical simulations at the design stage keeps growing due to the generalisation of low cost engineering work stations in CAD and this demand is comming from design engineers and not from structural analysis specialists or fluid dynamics specialist.

The FIDAP fluid dynamics analysis package has been designed with this tremendous potential in mind. It uses the FEM to simulate many classes of incompressible flows. Two-dimensional,axisymmetric and three dimensional steady state or transient simulations in complex geometry are possible,with the analysis being limited in size by considerations on computer ressources only.

FIDAP's emphasis on modularity allows for the implementations of new algorithms ,new capacities or new elements easily, an important feature in the still evolving field of computational fluid mechanics.

The models and equations FIDAP is currently solving are: Isothermal Newtonian and non Newtonian flows, creeping flows, forced or mixed convection problems, advection diffusion problems,free surface flows all in two or three dimension

Currently a $k - \varepsilon$ model is being implemented. Such flows will carry two additional unknowns. In mixed formulations there will be up to seven unknowns per node ! One faces here the problem of choosing between modularity and computing effectivness.

Boundary conditions on top of the classical ones include applied heat fluxes,radiative and convective fluxes,normal and tangential boundary velocities (slip) and free surface with surface tension effects.

Finite element approximations

Let V be a space of trial functions for the velocities (and eventually the temperatures), let W be the corresponding space for the hydrostatic pressure, it is well known that those spaces must satisfy the Babuska-Brezzi stability condition (1):

$$\sup_{v \in V} \frac{|\int p \, \mathrm{div} v \, d\Omega|}{||v||_V} \geq k |p|_W$$

If the pressure space is discontinuous then one can use the corresponding penalty method (2). Two dimensional elements include linear and quadratic triangles as well as quadrilaterals. Presently the choice of three dimensional elements is limited to ones based on 8 node bricks and on 27 node Lagrangean interpolation.

The model problem

In a rectangular box including an isolated electronic component, a forced convection problem is considered, the air being forced between two circuit boards which are both at fixed temperature. The component in the center represents a module which is dissipating heat. For the purpose of the simulation it is held at a temperature higher than that of the boards. The aim here is to see how the heat from this module is dissipated. It was treated as a fully coupled flow. Since the Reynolds number is about 150 and the Peclet number is 10 it is a fairly simple flow.

The mesh generated by FIDAP is of 4266 elements (8 node bricks) and 3446 elements, the total number of equations was 11850. It 0 mn of Cray 1S to solve.

Analysis of the results

Three dimensional flows can be very complex, the anlysis of a simulation necessitates the use of good graphic tools. The post-processor of FIDAP allows for the definition of a cut by any plane. On this plane the stream functions,vorticy,temperature,pressure or any user defined function can be represented by contour lines (with colour depending on the output graphic device). The velocities at points of this arbitrary plane in its normal direction can be plotted in isometric views. Comprehension of three dimensional flows requires the representation of many such views.

Transient flows are even more probelmatic to study and there is a definite need for animation capabilities. To illustrate this point a short colour film will be given at the conference.

Conclusions

Super computers together with low cost 32 bit machines with graphics render more actual the challenge of designing a general purpose Fluid Dynamic code. Such an approach was the key to the introduction of computational structural mechanics into the design cycle. There FEM codes have replaced experiments and are evaluation tools at every stage. Except for a very limited domain (low Reynolds numbers for instance) this "black box" concept cannot be aplied in fluid mechanics. Nevertheless the success of the modular approach of FIDAP, the addition of new models and the extension of its pre and post-processing capabilities bring a new bridge between the fluid mechanics engineer and the design floor shop. It is now the task of the scientific community to design acceptable models in domains like turbulent flows,multiphase flows slightly compressible flows and so on. At the same time faster and robust algorithms for highly non linear problems must be devised.

References

1. Brezzi F. 'On the Existence Uniqueness and Approximation of Saddle Point Problems arising from Lagrangian Multipliers.' RAIRO (An. Numerique) 8,129-151,1974.

2. Bercovier M.and Engelman M., 'A Finite Element for the Numerical Solution of Viscous Incompressible Flows .' J. of Comp. Phys.,36, 181-201,1979.

122

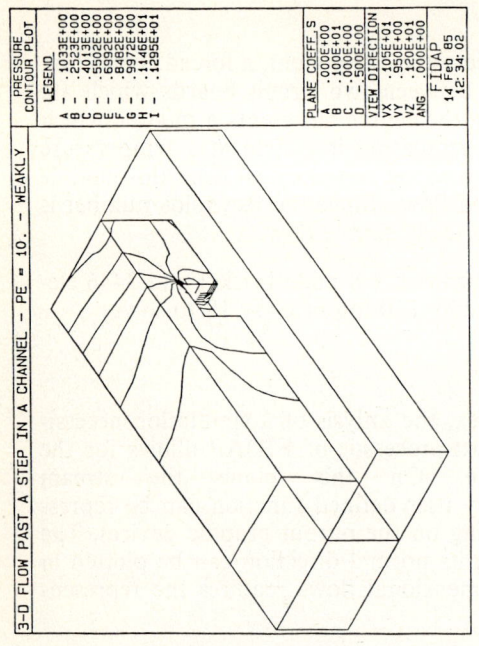
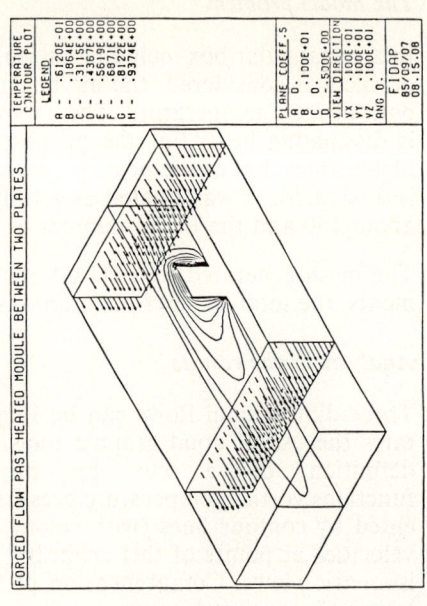
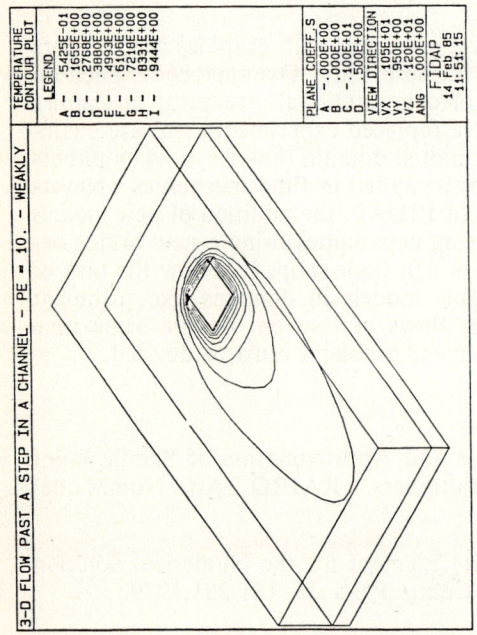
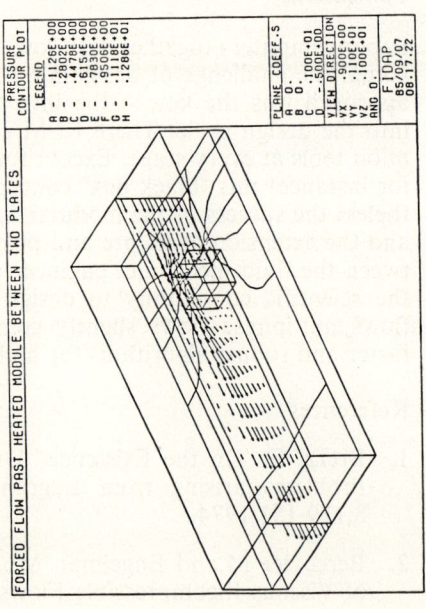

A DOWNSTREAM BOUNDARY CONDITION FOR THE NUMERICAL SOLUTION OF VISCOUS FLOW

J.S. Bramley and D.M. Sloan
University of Strathclyde
Glasgow Scotland

The stationary perturbations of Poiseuille flow in a two dimensional channel take the form

$$\psi = \psi_p + \sum_{i=1}^{\infty} \epsilon f_i(y) e^{-\alpha_i x}, \qquad (1)$$

where ψ is the stream function, ψ_p the stream function for Poiseuille flow, x the coordinate down the channel, y the coordinate across the channel, ϵ a small parameter and α the eigenvalue. Bramley and Dennis [1] calculated the eigenvalues for a selection of Reynolds numbers, R, between 0 and 2000. The dominant eigenvalues for R between 50 and 2000 are all real. Negative eigenvalues are associated with upstream disturbances and positive eigenvalues are associated with downstream disturbances. The upstream flow of the forward facing step problem of Dennis and Smith [5] is symmetric with respect to velocity. The first eigenvalue dominates the second symmetric eigenvalue so only the first one is required:

$$\psi = \psi_p + \epsilon f_1(y) e^{-\alpha_1 x}, \qquad (2)$$

from which we obtain

$$\frac{\partial \psi}{\partial x} = \alpha_1 (\psi_p - \psi). \qquad (3)$$

The boundary condition

$$\frac{\partial \zeta}{\partial x} = \alpha_1 (\zeta_p - \zeta) \qquad (4)$$

for the vorticity ζ is obtained in a similar manner. Equations (3) and (4) have been used as upstream boundary conditions with great success and the region of computation has been reduced as shown in Bramley and Dennis [2,3] and Bramley and Sloan [4].

The upstream eigenvalues vary very little with R while downstream eigenvalues vary like 1/R. This is the reason why downstream disturbances decay over a longer length scale than the upstream disturbances for all but small Reynolds numbers. Bramley and Dennis [2,3] use a logarithmic transformation downstream but still with the larger Reynolds number of 2000 there are problems associated with taking the downstream boundary far enough downstream so that the vorticity and steamfunction can be put equal to the Poiseuille equivalents. Bramley and Sloan [4] use an algebraic stretching within the boundary fitted coordinates and it is to this problem that the present work is an extension. [4] uses Thompson's [6] numerical grid generation algorithm to transform the region shown in figure 1 into a rectangle in the (ξ,η) plane. The generating system is

$$\xi_{xx} + \xi_{yy} = 0. \qquad \eta_{xx} + \eta_{yy} = 0. \qquad (5)$$

Calculations are performed on the rectangular domain, so dependent and independent variables are interchanged in (5) to give

$$\alpha x_{\xi\xi} - 2\beta x_{\xi\eta} + \gamma x_{\eta\eta} = 0, \quad \alpha y_{\xi\xi} - 2\beta y_{\xi\eta} + \gamma y_{\eta\eta} = 0 \quad (6)$$

where $\alpha = x_\eta^2 + y_\eta^2$, $\beta = x_\xi x_\eta + y_\xi y_\eta$,
$\gamma = x_\xi^2 + y_\xi^2$, $J = x_\xi y_\eta - x_\eta y_\xi$.

Equations (6) are solved using second-order central differences on a square grid. The equations of the boundaries of figure 1 are used to determine suitable values of x and y at the boundary nodes in the (ξ,η) rectangle and the values of x and y at internal nodes are then determined by solving a Dirichlet problem. The nonlinear difference equations are solved iteratively using a point SOR method and, at each cycle of the iteration, the coefficients $\alpha,\beta,\ldots$ are evaluated using previous approximations. A typical configuration is shown in figure 2.

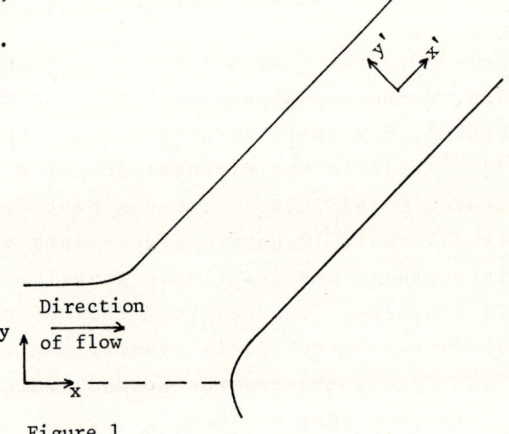

Figure 1

The governing equations are also transformed and solved using a point SOR method as explained in [4]. The

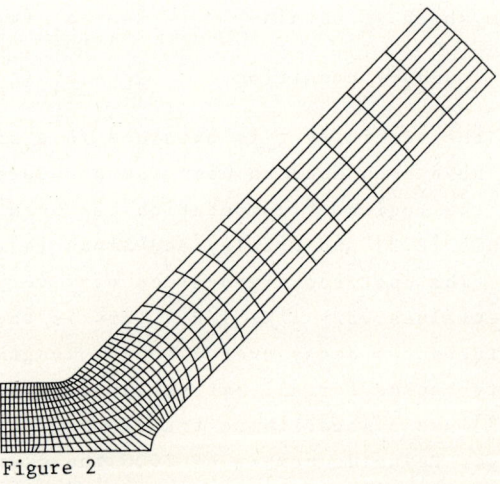

Figure 2

boundary conditions are described in that paper, with the exception that we now describe a new downstream boundary condition.

In the bifurcating channel the downstream perturbation from the Poiseuille flow is not symmetric. The eigenvalues associated with the flow downstream calculated by Bramley and Dennis [1] are mostly in pairs of nearly equal magnitude so we take

$$\psi = \psi_p + \epsilon f_1(y')e^{-\alpha_1 x'} + \epsilon f_2(y')e^{-\alpha_2 x'} \qquad (7)$$

where α_1 and α_2 are the two most dominant eigenvalues, one symmetric and one antisymmetric. The dashed coordinates (x',y') are along and perpendicular to the exit channel respectively as shown in figure 1. Differentiating twice and eliminating the y' dependence we obtain

$$\frac{\partial^2 \psi}{\partial x'^2} + (\alpha_1+\alpha_2)\frac{\partial \psi}{\partial x'} = \alpha_1 \alpha_2 (\psi_p - \psi). \qquad (8)$$

Well away from the bend the transformed coordinate system will be assumed to be orthogonal so we perform a one dimensional transformation from x' to ξ, and from (8) we obtain

$$\frac{1}{(x_\xi)^2}\frac{\partial^2 \psi}{\partial \xi^2} + \left[\frac{\alpha_1+\alpha_2}{x_\xi} - \frac{x_{\xi\xi}}{(x_\xi)^3}\right]\frac{\partial \psi}{\partial \xi} = \alpha_1 \alpha_2 (\psi_p - \psi) \qquad (9)$$

A similar equation can be obtained for ζ:

$$\frac{1}{(x_\xi)^2}\frac{\partial^2 \zeta}{\partial \xi^2} + \left[\frac{\alpha_1+\alpha_2}{x_\xi} - \frac{x_{\xi\xi}}{(x_\xi)^3}\right]\frac{\partial \zeta}{\partial \xi} = \alpha_1 \alpha_2 (\zeta_p - \zeta). \qquad (10)$$

Equations (9) and (10) are now used as boundary conditions at the downstream boundary by using second order finite differences. The results are presented in figures 3-6 in the form of graphs of vorticity along the upper boundary of the channel. In figures 3-5 the solid curves give the vorticity for a solution at which the boundary condition is Poiseuille flow applied at a distance downstream where the curves end, while + gives the solution with the derivative boundary conditions (9) and (10) applied at approximately x=20 and × the same condition applied at x=40. It will be seen that there is reasonable agreement that deteriorates as the Reynolds number increases. In these figures the results denoted by ◊ are calculated using one downstream eigenvalue in equations (3) and (4). Figure 6 gives a solution for Reynolds number 50, the solid curve is for Poiseuille flow applied at x=20 while × gives the solution with derivative boundary conditions (9) and (10) applied at x=8. The downstream boundary condition reduces the computational domain but the computer time is still large due to the fact that the Von Neuman boundary condition downstream is very slow to converge. This could no doubt be speeded up.

References

1. J.S.Bramley & S.C.R.Dennis, (1982) J. Comp. Phys. 47, 179-198.
2. J.S.Bramley & S.C.R.Dennis, Eighth International Conference on Numerical Methods in Fluid Dynamics, Lecture Notes in Physics 170, Springer-Verlag 1982.
3. J.S.Bramley & S.C.R. Dennis, (1984) Comps and Fluids, 12, 339-355.

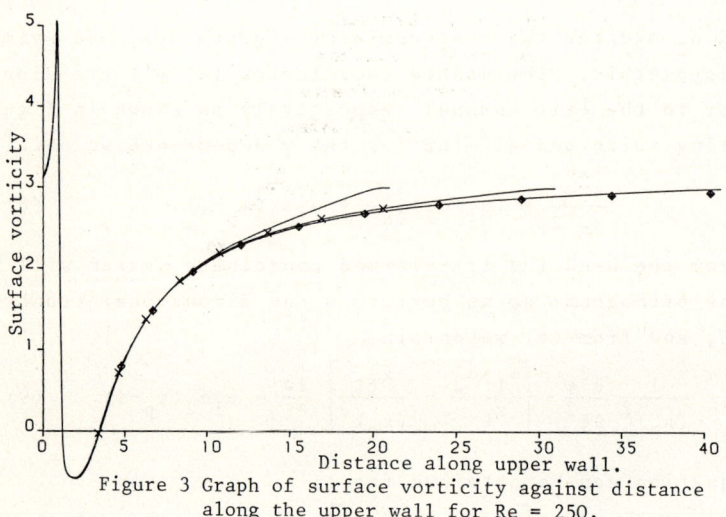

Figure 3 Graph of surface vorticity against distance along the upper wall for Re = 250.

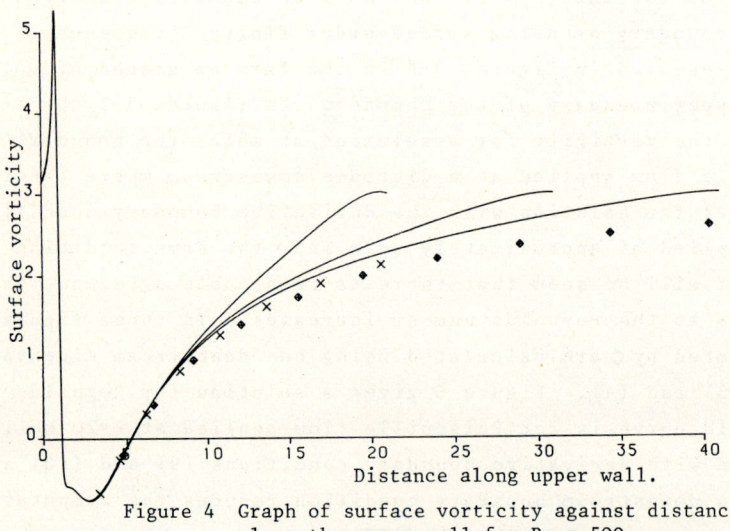

Figure 4 Graph of surface vorticity against distance along the upper wall for Re = 500.

4. J.S.Bramley & D.M.Sloan - submitted for publication.
5. S.C.R.Dennis & F.T.Smith, (1980) Proc. R. Soc. London. Ser. A 372, 393.
6. J.F.Thompson et al. (1977) J. Comp. Phys. 24, 274-302.

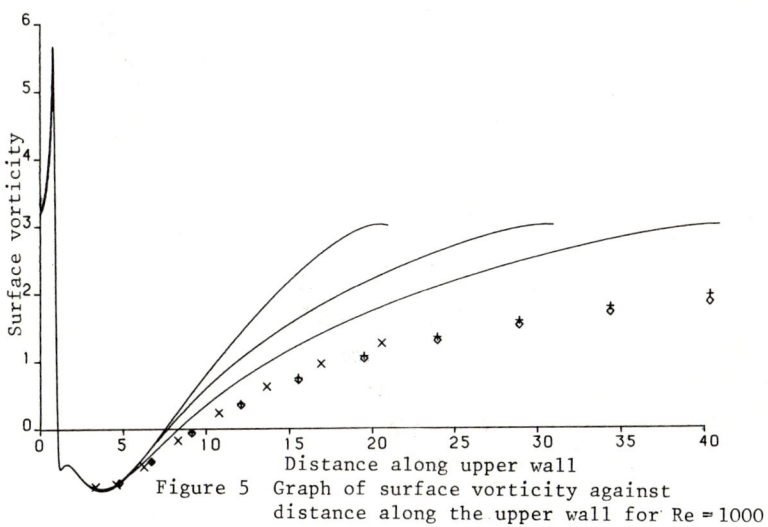

Figure 5 Graph of surface vorticity against distance along the upper wall for Re = 1000

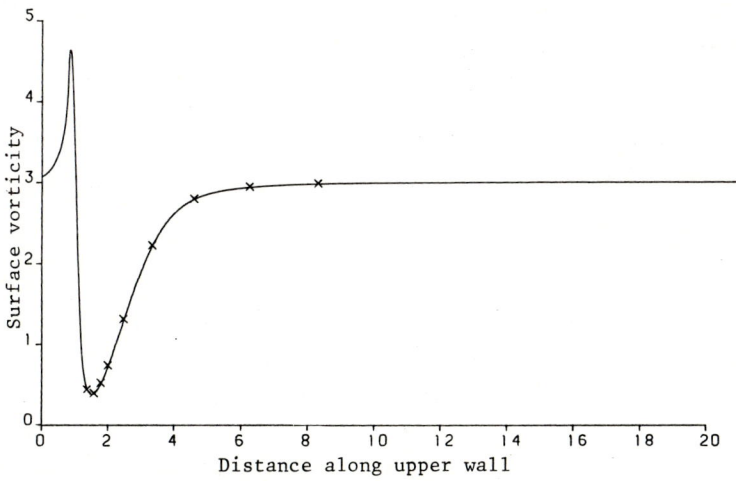

Figure 6 Graph of surface vorticity against distance along the upper wall for Re = 50

SCALING AND COMPUTATION OF SMOOTH ATMOSPHERIC MOTIONS

G. L. Browning
N.C.A.R.
Boulder, CO 80307 U.S.A.

H. O. Kreiss
Cal. Tech.
Pasadena, CA 91125 U.S.A.

1. Introduction

Currently most large scale weather models are based on the primitive equations, i.e. the Eulerian equations of gas dynamics modified solely by the assumption of hydrostatic equilibrium. In the late seventies Oliger and Sundstrom (1978) showed that any boundary conditions used in conjunction with the primitive equations form an ill posed initial-boundary value problem. Thus, the difficulties at the boundaries of limited area forecast models based on the primitive equations are not surprising. We introduce an alternate system of equations which accurately describes the weather systems of interest and when combined with appropriate boundary conditions, forms a well posed initial-boundary value problem.

2. Scaling of the basic equations

The adiabatic inviscid Eulerian equations in Cartesian coordinates x, y, and z directed eastward, northward, and upward, respectively, can be written as

$$\frac{ds}{dt} = 0,$$
$$\frac{dp}{dt} + \gamma p \nabla \cdot V = 0, \qquad (2.1)$$
$$\frac{dV}{dt} + \rho^{-1} \nabla p + f (k \times V) + g k = 0,$$

where t is time, $V = (u,v,w)^T$ is velocity, ρ is density, p is pressure, and $s = \rho p^{-1/\gamma}$ is proportional to the reciprocal of the potential temperature. Also, $f = f(y)$ is the Coriolis parameter, $g = 9.8 \text{m s}^{-2}$ is the constant gravity acceleration, $\gamma = 1.4$ is the adiabatic exponent, and $k = (0,0,1)^T$ is the unit vector in the vertical direction. We assume that f is given by the tangent plane approximation $f = 2\Omega[\sin\theta_0 + (y/r) \cos\theta_0]$ where $2\Omega \approx 10^{-4} \text{ s}^{-1}$ is the earth's angular speed, θ_0 is the latitude of the coordinate origin, and $r \approx 10^7 \text{m}$

is the radius of the earth.

We shall now introduce dimensionless variables to identify the relative magnitude of all terms in the equations. We change independent variables via the relations

$$x = L_1 x' \; , \; y = L_2 y' \; , \; z = D z' \; , \; t = T t' \; , \tag{2.2}$$

where L_1, L_2, D, and T are the representative scales along the x, y, z, and t axes, respectively. We also change dependent variables. For the velocity we introduce the relations

$$u = U u' \; , \; v = V v' \; , \; w = W w' \tag{2.3}$$

Density and pressure can be written in the form

$$p = P_0 [p_0(z) + S_1 p'] \; , \; \rho = R_0 [\rho_0(z) + S_1 \rho'] \; , \; 0 < 10 \, S_1 \leq 1 \; , \tag{2.4}$$

with $P_0 \partial p_0 / \partial z + g R_0 \rho_0 = 0$, $P_0 = 10^5 \, kg \, m^{-1} \, s^{-2}$, $R_0 = 1 \, kg \, m^{-3}$. The restriction on the size of S_1 represents an assumption that the deviation of the pressure from its horizontal mean is small. We also scale the gravity and Coriolis parameter as

$$g = G \, g' \; (G = 10 \, m \, s^{-2}) \; , \tag{2.5}$$

$$f = 2 \, \Omega \, f' = 2 \, \Omega \, [\sin \theta_o + (L_2/r) \cos \theta_o \, y'] = 2 \, \Omega \, [f_o + (L_2/r) \beta \, y'] \; . \tag{2.6}$$

We assume that $T = L_1 U^{-1} = L_2 V^{-1}$, so that for each equation the dimensionless time derivative will be of the same order of magnitude as the horizontal advection terms.

Substituting the relations (2.2)-(2.6) into (2.1) and then applying the bounded derivative theory (Kreiss, 1980), we can show that in the case of midlatitude adiabatic hydrostatic motions with equal horizontal length scales we obtain the dimensionless system

$$\frac{ds}{dt} + \tilde{s}(z) w = 0 \; , \tag{2.7a}$$

$$\epsilon^{3-n} \frac{d\phi}{dt} + p_0 \rho_0^{-1} [\gamma d + \epsilon^{2-n} \tilde{p}(z) w] = 0 \; , \tag{2.7b}$$

$$\epsilon^n \frac{du}{dt} + \phi_x - f v = 0 \; , \tag{2.7c}$$

$$\epsilon^n \frac{dv}{dt} + \phi_y + fu = 0, \qquad (2.7d)$$

$$\eta \frac{dw}{dt} + \phi_z + 0.1\tilde{s}\,\phi + gs = 0, \qquad (2.7e)$$

where $\frac{d}{dt} = \frac{\partial}{\partial t} + u\frac{\partial}{\partial x} + v\frac{\partial}{\partial y} + \epsilon^{2-n} w\frac{\partial}{\partial z}$, $d = u_x + v_y + \epsilon^{2-n} w_z$, $\tilde{s}$, $\tilde{\rho}$, and $\tilde{p}$ are known functions of z, $\epsilon = 10^{-1}$, and $\eta = \epsilon^{4+2n}$. We have dropped the prime notation with the understanding that all variables are dimensionless. When n = 1 we obtain Charney's scaling (Charney, 1948) and when n = 2 we obtain Burger's scaling (Phillips, 1963).

3. The Modified System

It is well known how to choose well posed boundary conditions for the initial-boundary value problem for system (2.7), since it is hyperbolic. The primitive equations are derived from (2.7) by neglecting the term $\eta \frac{dw}{dt}$. Clearly the boundary difficulties with the primitive equations must be caused by this assumption of hydrostatic equilbrium. The problem is that even though η is small, no estimate to prove that $\frac{dw}{dt}$ is bounded independently of ε has been made. The necessary estimate can only be made by considering a more complicated reduced system than the primitive equations. Before deriving the correct reduced system, we will introduce an alternative to system (2.7) which can be used for smooth atmospheric motions.

By setting $\eta = 0$ in order to obtain the primitive equations the speed of the vertical sound waves is increased to infinity. Instead we slow these waves down, i.e. we consider the modified system

$$\frac{ds}{dt} + \tilde{s}(z)w = 0, \qquad (3.1a)$$

$$\epsilon^{3-n} \frac{d\phi}{dt} + p_0\rho_0^{-1}[\gamma d + \epsilon^{2-n}\tilde{p}(z)w] = 0, \qquad (3.1b)$$

$$\epsilon^n \frac{du}{dt} + \phi_x - fv = 0, \qquad (3.1c)$$

$$\epsilon^n \frac{dv}{dt} + \phi_y + fu = 0, \qquad (3.1d)$$

$$\alpha^{-1}\eta \frac{dw}{dt} + \phi_z + 0.1\tilde{s}\,\phi + gs = 0, \qquad (3.1e)$$

where $\alpha \leq 1$. The choice of the new vertical sound speed is determined by the requirement that the solutions of the new system be smooth up to the boundary (Browning and Kreiss, 1985). The easiest way to see what value to choose for α is to note from (3.1b) that

$$\epsilon^n(u_x + v_y) + \epsilon^2(w_z + \gamma^{-1}\tilde{p}w) = O(\epsilon^3).$$

Applying the operator d/dt to this expression we find that

$$\phi_{xx} + \phi_{yy} + \alpha\epsilon^2\eta^{-1}[\phi_{zz} + (\gamma^{-1}\tilde{p} + 0.1\tilde{s})\phi_z + 0.1(\gamma^{-1}\tilde{p}\tilde{s} + \tilde{s}_z)\phi] =$$

$$f_s - \alpha\epsilon^2\eta^{-1}g(s_z + \gamma^{-1}\tilde{p}s) + 2\epsilon^n(u_x v_y - u_y v_x) + O(\epsilon^2).$$

By neglecting the term of order ϵ^2 we obtain an elliptic equation for ϕ. In the smoothness argument it is crucial that the coefficients of the elliptic equation be of order unity in order to estimate ϕ correctly. Thus we must choose

$$\alpha = \epsilon^{-2}\eta = (S_2^{-1}S_3)(S_1S_5)^{-1} = (DL^{-1})^2. \tag{3.2}$$

We first consider the case that n = 1. We want to show that any smooth solution of system (2.7) with this scaling can be computed accurately by using (3.1) with $\alpha = \epsilon^4$. Let a subscript 0 on a dependent variable denote a given smooth solution of (2.7). Then we write the solution of (3.1) in the form

$$s = s_0 + \epsilon^2 F + s_1, \quad F = -g^{-1}(1 - \epsilon^{-2}\eta)\frac{dw_0}{dt},$$

$$\phi = \phi_0 + \phi_1, \quad (u, v, w)^T = (u_0, v_0, w_0)^T + (u_1, v_1, w_1)^T. \tag{3.3}$$

Note that since $\eta \leq 10^{-4}\epsilon^2$, F is of order unity. Substituting the expressions (3.3) into (3.1) we find that the perturbation variables with subscript 1 satisfy the linear system

$$\frac{ds_1}{dt} + \tilde{s}w_1 = -\epsilon^2\frac{dF}{dt}, \tag{3.4a}$$

$$\epsilon^{3-n}\frac{d\phi_1}{dt} + \gamma p_0\rho_0^{-1}[u_{1x} + v_{1y} + \epsilon^{2-n}(w_{1z} + \gamma^{-1}\tilde{p}w_1)] = 0, \tag{3.4b}$$

$$\epsilon^n\frac{du_1}{dt} + \phi_{1x} - fv_1 = 0, \tag{3.4c}$$

$$\epsilon^n\frac{dv_1}{dt} + \phi_{1y} + fu_1 = 0, \tag{3.4d}$$

$$\epsilon^2\frac{dw_1}{dt} + \phi_{1z} + 0.1\tilde{s}\phi_1 + gs_1 = 0, \tag{3.4e}$$

where $\frac{d}{dt} = \frac{\partial}{\partial t} + u_0 \frac{\partial}{\partial x} + v_0 \frac{\partial}{\partial y} + \epsilon^{2-n} w_0 \frac{\partial}{\partial z}$. We have dropped a number of undifferentiated terms which have no influence on the arguments to follow. To further simplify the presentation we freeze the almost constant coefficients and consider the system

$$\frac{ds_1}{dt} - w_1 = -\epsilon^2 \frac{dF}{dt}, \tag{3.5a}$$

$$\epsilon^{3-n} \frac{d\phi_1}{dt} + u_{1x} + v_{1y} + \epsilon^{2-n} w_{1z} = 0, \tag{3.5b}$$

$$\epsilon^n \frac{du_1}{dt} + \phi_{1x} - f_o v_1 = 0, \tag{3.5c}$$

$$\epsilon^n \frac{dv_1}{dt} + \phi_{1y} + f_o u_1 = 0, \tag{3.5d}$$

$$\epsilon^2 \frac{dw_1}{dt} + \phi_{1z} + s_1 = 0. \tag{3.5e}$$

To complete the proof of our first assertion it will suffice to show that the solution of (3.5) with initial data equal to zero is small. To obtain the estimates we need we first symmetrize (3.5) by introducing the new variables

$$s_2 = s_1, \phi_2 = \epsilon^{1/2} \phi_1, u_2 = \epsilon^{n-1} u_1, v_2 = \epsilon^{n-1} v_1, w_2 = \epsilon w_1, \tag{3.6}$$

which gives us the symmetric system

$$\frac{ds_2}{dt} - \epsilon^{-1} w_2 = -\epsilon^2 \frac{dF}{dt}, \tag{3.7a}$$

$$\frac{d\phi_2}{dt} + \epsilon^{-3/2}(u_{2x} + v_{2y} + w_{2z}) = 0, \tag{3.7b}$$

$$\frac{du_2}{dt} + \epsilon^{-3/2} \phi_{2x} - \epsilon^{-n} f_o v_2 = 0, \tag{3.7c}$$

$$\frac{dv_2}{dt} + \epsilon^{-3/2} \phi_{2y} + \epsilon^{-n} f_o u_2 = 0, \tag{3.7d}$$

$$\frac{dw_2}{dt} + \epsilon^{-3/2} \phi_{2z} + \epsilon^{-1} s_2 = 0, \tag{3.7e}$$

with initial data equal to zero. Using standard L_2 energy estimates for the initial value problem we obtain in every finite time interval $0 \leq t \leq \bar{t}$ the estimate

$$||s_2|| + ||\phi_2|| + ||u_2|| + ||v_2|| + ||w_2|| \leq O(\epsilon^{3-n})$$

and a similar estimate for all space and time derivatives. From (3.6) this gives us for the unsymmetrized variables s_1, u_1, and v_1 the estimate

$$||s_1|| + ||u_1|| + ||v_1|| \leq O(\epsilon^{3-n})$$

and a similar estimate for all of their derivatives. Using the differential equation (3.5a) we can then obtain for w_1 the estimate

$$||w_1|| \leq O(\epsilon^{3-n})$$

and a similar estimate for all of its derivatives. Finally we can use the elliptic equation for ϕ that we derived earlier to obtain estimates of the same form for ϕ_1 and all of its derivatives.

If we were only interested in computing motions with $n = 2$, then we could choose $\alpha = \epsilon^6$ and repeat the argument we used for the case $n = 1$ to show that we could also compute those motions accurately. However, in practive both motions can be present and the question arises as to what will happen if we also use the value $\alpha = \epsilon^4$ when $n = 2$. In that case $\alpha^{-1}\eta = \epsilon^4$ and by using essentially the same proof as for the case $n = 1$ we can show that system (3.1) with $\alpha = \epsilon^4$ and $n = 2$ approximates solutions of (2.7) with $n = 2$ up to an error term of order ϵ^3. We can also prove that (3.1) can be used for any smooth stratified atmospheric flow.

References

Browning, G. and Kreiss, H.-O.,1985: Numerical problems connected with weather prediction. Progress in Scientific Computing: Workshop on Supercomputational Fluid Dynamics, Birkhauser Publishers.
Charney, J. G., 1984: On the scale of atmospheric motions. Geofysiske Publikasjoner, 17, 1-17.
Kreiss, H.-O., 1980: Problems with different time scales for partial differential equations. Comm. Pure Appl. Math., 33, 399-400.
Oliger, J. and Sundstrom, A., 1978: Theoretical and practical aspects of some initial-boundary value problems in fluid dynamics. SIAM J. Appl. Math., 35, 419-446.
Phillips, N.A., 1963: Geostrophic motion. Rev. Geophysics, 1, 123-176.

Computation of Vortex Flows past a Flat Plate

at High Angle of Attack. [†]

C.H. Bruneau*, J.J. Chattot**, J. Laminie*, R. Temam*

*Laboratoire d'Analyse Numérique, C.N.R.S. et Université Paris-Sud,
91405 Orsay (France)
**MATRA, 37, Avenue Louis Bréguet, 78140 Vélizy (France)

([†]Work performed with financial support of D.R.E.T. and under grants from the CCVR Ecole Polytechnique, France, and the Minnesota Supercomputer Institute, U.S.A.).

Summary.

The finite element method described in a former paper presented at the 9th ICNMFD ([1]) is applied here to solve the full, steady Euler equations in 3D. The steady flow past a rectangular flat plate of small aspect ratio at high angle of attack is computed for several meshes and angles of attack. These calculations have been made without either adding artificial viscosity or using a Kutta condition. They show how vortices develop spontaneously around the tip of the plate and propagate after the trailing edge.

Introduction.

Vortical phenomena occur in flows around wings in different ways. We are interested in flows around a thin plate of small aspect ratio for high angles of attack. Actually in this application the plate has no thickness and we expect a strong vortex structure to develop at the tip of the plate. So this does not allow the use of the potential equation and needs a more general model like Euler or Navier-Stokes equations.

It is not yet clear what is the best model for the calculation of steady, three-dimensional separated vortex flows. Our aim is to show that the Euler equations are able to represent open separation with large vortices, although they do not contain viscous terms. Numerically an explicit viscosity term is often added or/and a Kutta-Joukowski condition is specified at sharp edges of a lifting surface. In this paper we present a method that does not require either condition for subsonic flows. The vortices are captured, develop spontaneously at the tip of the plate and are convected past the trailing edge. Particular care must be taken in imposing boundary conditions, especially downstream where the pressure can not be fixed as is usually done for two-dimensional subsonic calculations.

Physical problem.

The aim is to compute the flow around a flat plate at incidence for subsonic mach number M_∞. The plate of small aspect ratio $Æ = 0.5$ without thickness is set at an angle α with the incoming flow. The plate is imbedded in an orthogonal domain as shown Figure 1, and the incoming flow is given by : $\vec{q}_\infty = (u_\infty, v_\infty, w_\infty) = (q_\infty \cos \alpha, 0, q_\infty \sin \alpha)$ with $M_\infty^2 = \dfrac{q_\infty^2}{a_\infty^2}$

where $a_\infty^2 = (\gamma - 1)\left(H - \dfrac{q_\infty^2}{2}\right)$. By symmetry we compute only the flow around the right half plate. In addition there is a tangency condition on both sides of the plate.

In this work we take $M_\infty = 0.7$ and a high angle of attack $\alpha = 15°$ or $\alpha = 30°$.

Outline of the method.

We solve the full, steady Euler equations in conservative form :

(1) $\begin{cases} \dfrac{\partial \rho u}{\partial x} + \dfrac{\partial \rho v}{\partial y} + \dfrac{\partial \rho w}{\partial z} = 0 & \text{Conservation of mass} \\[2pt] \dfrac{\partial \rho u^2 + p}{\partial x} + \dfrac{\partial \rho uv}{\partial y} + \dfrac{\partial \rho uw}{\partial z} = 0 & \\[2pt] \dfrac{\partial \rho uv}{\partial x} + \dfrac{\partial \rho v^2 + p}{\partial y} + \dfrac{\partial \rho vw}{\partial z} = 0 & \text{Conservation of momentum} \\[2pt] \dfrac{\partial \rho uw}{\partial x} + \dfrac{\partial \rho vw}{\partial y} + \dfrac{\partial \rho w^2 + p}{\partial z} = 0 & \end{cases}$

(2) $\dfrac{\gamma p}{(\gamma - 1)\rho} + \dfrac{q^2}{2} = H$ Bernoulli's equation

for ρ, $\vec{q} = (u,v,w)$, p the density, the velocity and the pressure respectively where H is the total enthalpy and γ the ratio of specific heats.

- A fixed point algorithm is based on equation (2), computing the density from u,v,w,p solution of system (1) with given ρ.

- The non-linear system (1) for u,v,w,p (ρ given from the previous iteration of the fixed point algorithm) is transformed into a linear system for $\tilde{u}$, $\tilde{v}$, $\tilde{w}$, $\tilde{p}$ by means of Newton's method (cf. [2]).

- The linear system for $\tilde{u}$, $\tilde{v}$, $\tilde{w}$, $\tilde{p}$ is solved through the minimization of a least-square functional. This transforms the first order system into an equivalent symmetric second order system (cf. [3]) for the same variables.

- The unknowns $\tilde{u}$, $\tilde{v}$, $\tilde{w}$, $\tilde{p}$ and the conservative variables ($\rho\vec{q}$, etc.) are approximated through a Q1 linear finite element method.

Boundary conditions.

Because of the geometry of the plate and the domain the physical boundary conditions are easy to implement. The tangency condition on the plate (plane z = 0) reads w = 0 and the symmetry condition is v = 0 on y = 0.

On the contrary the boundary conditions at the limits of the domain are not so easy especially downstream, where we must allow the vortex to go through the exit plane. In particular we cannot impose a constant value of the pressure. With the unsteady model one generally takes a non reflecting boundary condition or sometimes impose a mean value for the pressure [4]. We have made attempts to extrapolate the pressure or both the pressure and the velocity from inside the domain in the direction of the flow, but we found that global extrapolation in the main flow direction eliminates the vortex downstream.

Finally the flow variables have been set to the freestream values at incoming boundaries (x = x_0, z = z_0) and no condition was imposed at exit boundaries (x = x_1, z = z_1), so that the values there are computed from the variational formulation. It corresponds to requiring that the first order equations be satisfied on those boundaries.

At y = y_1, we use the far field condition v = 0.

Numerical aspects.

Use is made of the structured nature of the mesh system (i,j,k). Thus the global matrix of the discrete linearized equations has a symmetric block structure. However it is too large to be stored on a CRAY-1S with 1 million words of memory and a relaxation method is used. The degrees of freedom are numbered plane by plane in the x direction so that the corresponding matrix is a block matrix where each block corresponds to a plane i = const. A block Gauss-Seidel iterative method is used and the blocks are inverted by means of an Incomplete Cholesky Conjugate Gradient algorithm.

The plate is contained in a double plane z = 0. On the plate the field variables are double valued, and single valued outside and at the tip of the plate, the first single valued point has 10 neighbors instead of 9 in a plane i = const. (Fig. 2).

At the exit plane, i = IMax, the conditionning of the block matrix is poor and the ICCG method is stabilized upon adding the main diagonal in the Gauss-Seidel algorithm, which can be interpreted as an extra time derivative term.

In order to take advantage of the performance of the CRAY, a particular effort is made to vectorize the matrix assembly and the solution algorithm which account for at least 70 % of the CPU time.

For matrix assembly, the interaction of a degree of freedom with a neighboring one is computed for all the elements of the plane, instead of assembling the elementary matrix corresponding to one element. Thus the innermost vectorizable loop is of length JMax × KMax (320 and 900 in our application).

In a plane, all the degrees of freedom related to $\tilde{p}$, $\tilde{u}$, $\tilde{v}$ and $\tilde{w}$ are numbered successively. Thus the matrix can be subdivised in 4×4 blocs. This allows vectorization of the Cholesky factorization when it operates on different blocks.

Numerical results.

The first case corresponds to an incidence of 15°. Two different mesh systems are used with respectively 28×18×19 ≅ 11000 points (medium mesh) and 50×30×31 ≅ 50000 points (fine mesh). The plate tip is embedded in a uniformly spaced grid and away from the tip, the grid is stretched (Fig. 3). The computing time for 30 fixed point iterations in the medium grid is 1 hour CPU, and for 50 iterations in the fine grid is 15 hours.

The vortex structure develops with a more or less conical growth in x as can be seen on the cross-flow velocity plots at the various locations in x of 70 %, 90 % and 110 % (Fig. 4-6). The last figure corresponds to the exit plane which is crossed by the vortex. The spreading of the vortex structure is strongly dependent on the mesh steps. The fine grid gives a better capture of the vortex roll-up (Fig 7-8).

The second case corresponds to an incidence of 30°. Only the medium mesh is used. The presence of the plate creates a very large perturbation to the flow and near the trailing edge the local mach number exceeds unity. The fixed point algorithm does not converge under supersonic flow condition and the flow variables oscillate with the iterations. The qualitative behaviour of the flow is very similar to the 15° case, but the vortex structure is stronger and lifts off more rapidly behind the plate, according to the angle of incidence (Fig. 9-10).

Conclusion.

The first applications of the finite element least square method in three dimension to the computation of the separated flows past a flat plate at incidence indicate that the vortex structure is well captured by the numerical scheme without the help of artificial viscosity. The quality of the steady solution is related to the total number of mesh points in the discretization, which is limited by the cost of the computation (CPU time). Progress in the solution procedure for the linear system is expected using different techniques available with the present memory size of computer (multigrid or frontal methods), or large extended memory size.

References.

[1] Ch.H.Bruneau, J.J.Chattot, J.Laminie, R.Temam : Numerical solutions of the Euler equations with separation by a finite element method. 9th ICNMFD Saclay 1984, Lecture Notes in Physics n° 218, Springer-Verlag.

[2] Ch.H.Bruneau, J.J.Chattot, J.Laminie, J.Guiu-Roux : Finite element least square method for solving full steady Euler equations in a plane nozzle. 8th ICNMFD Aachen 1982, Lecture Notes in Physics n° 170, Springer-Verlag.

[3] S.H.Chiang, G.M.Johnson : An embedding method for the steady Euler equations. J. Comp. Phys., n° 63, 1986.

[4] Ch.Koeck : Computation of three dimensional flow using the Euler equations and multiple-grid scheme. Int. J. Num. Meth. Fluids n° 5, 1985.

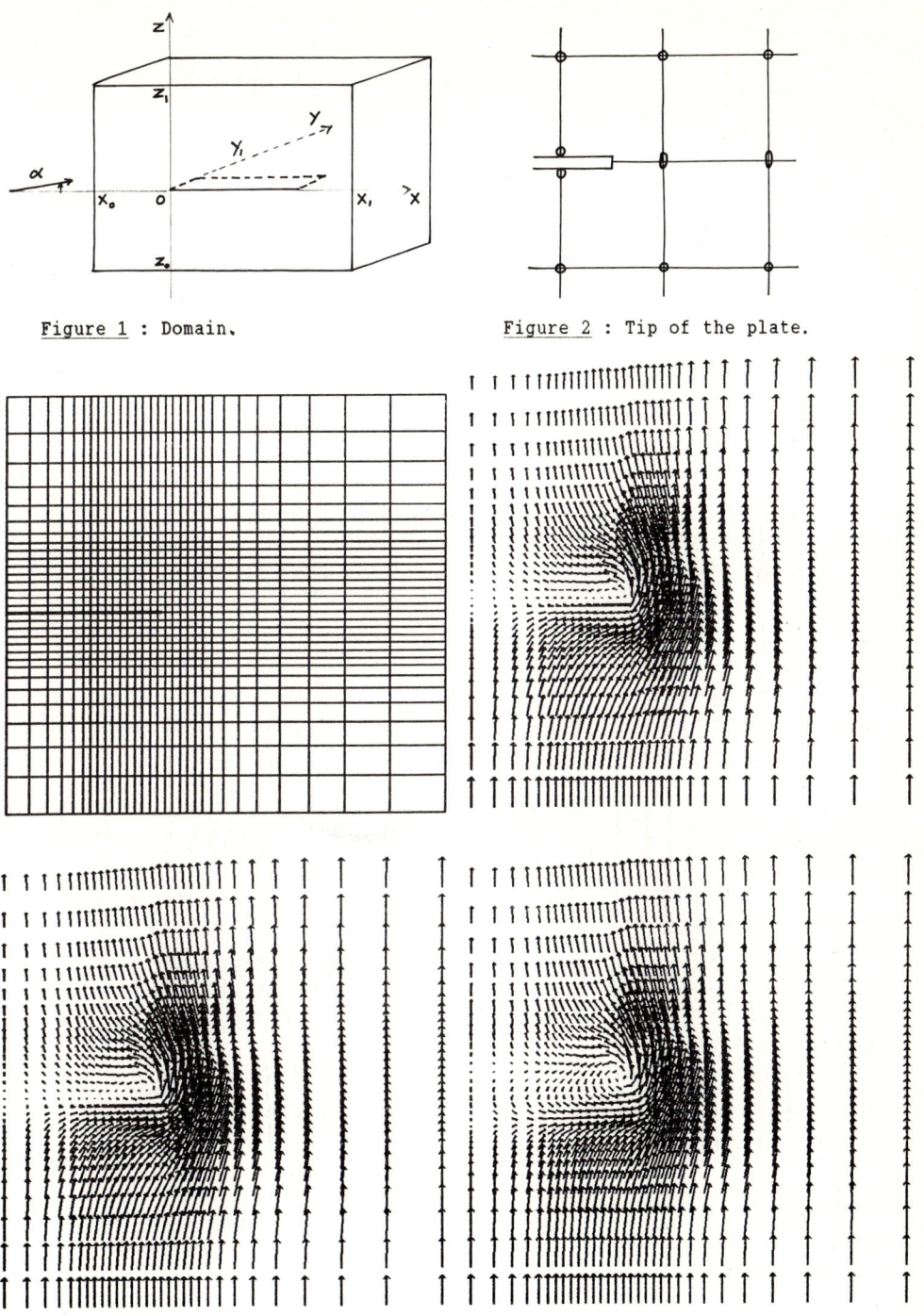

Figure 1 : Domain.

Figure 2 : Tip of the plate.

Figures 3 to 6
Mesh and cross - flow velocities at 70 %, 90 % and 110 % of the plate.

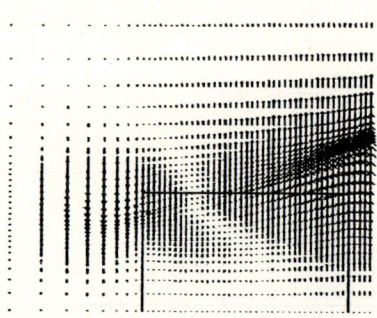

Figure 7
Isobars in a cross plane at the trailing edge.

Figure 8
Component v of the velocity on the upper plate.

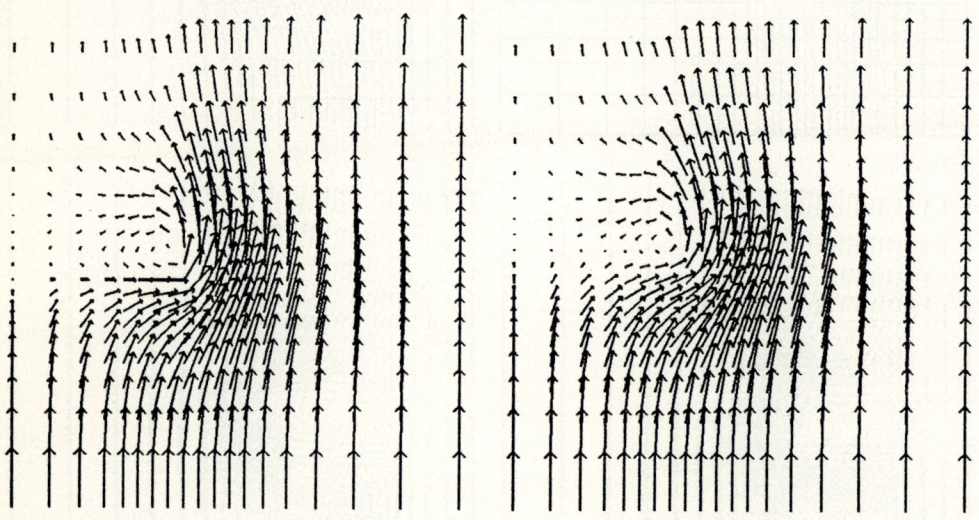

Figure 9
Cross-flow velocities at 90 % of the plate.

Figure 10
Cross-flow velocities at 110 % of the plate.

SPECTRAL SIMULATION OF THERMAL CONVECTION IN COMPLEX GEOMETRIES

E.T. Bullister*, T. Cartage+, M. Deville+, A.T. Patera*

+Université Catholique de Louvain, Louvain-la-Neuve, Belgium
*Massachusetts Institute of Technology, Cambridge, U.S.A.

ABSTRACT

This paper presents two- and three-dimensional spectral calculations of thermal convection flows. Two algorithms are employed : a mono-domain Chebyshev Tau approximation, and a spectral element method. The numerical methods are compared for flows occurring in parallelpipeds, and are evaluated in terms of stability, vectorization, resolution of thin boundary layers, generality, and computational complexity. To illustrate the flexibility of multidomain spectral decompositions, we present a spectral element simulation of thermal convection in a complex geometry arising in float glass processing.

1. NUMERICAL ALGORITHMS

We briefly describe our numerical algorithms here : further details can be found in the references. We solve the Navier-Stokes equations coupled with the energy through the Boussinesq approximation,

$$\rho_o \frac{D\tilde{v}}{Dt} = - \text{grad } p + \rho_o \tilde{g} [1 - \alpha(T-T_o)] + \mu \Delta \tilde{v} \quad , \tag{1}$$

$$\rho_o c_p \frac{DT}{Dt} = k\Delta T \quad , \tag{2}$$

$$\text{div } \tilde{v} = 0 \quad , \tag{3}$$

where $\tilde{v}$ denotes the velocity field, p the pressure, T the temperature, $\tilde{g}$ the gravity, c_p the specific heat coefficient, μ the dynamic viscosity, ρ_o the mass density at the reference temperature T_o, k the thermal conductivity, and α the coefficient of thermal expansion.

Both the Chebyshev Tau (1-3) and the spectral element methods (4-7) use similar semi-implicit time-stepping procedures, starting with an explicit convective step, followed by implicit treatment of the Stokes problem. In particular, the Chebyshev Tau method uses Adams-Moulton predictor-corrector convective step, with a reduced boundary divergence

O(dt**2) scheme for the pressure (8). The spectral element method uses a third-order Adams-Bashforth technique for the non-linear terms, followed by an O(dt) splitting scheme for the Stokes problem (8). The stability of both schemes is determined by a Courant condition associated with the explicit treatment of the convective terms. As regards efficiency, the critical point is that these time-stepping methods reduce the Navier-Stokes equations to a series of decoupled (scalar) hyperbolic and elliptic problems.

It is primarily in the spatial treatment that our two schemes are different. Whereas both methods use higher-order polynomial expansions and achieve convergence by increasing the order of the polynomial representation, the Chebyshev Tau method uses collocation (hyperbolic) and Galerkin-Chebyshev (elliptic) projection operators, while the spectral element method employs Legendre-Galerkin (hyperbolic) and classical variational (elliptic) projection operators. The use of "finite element" variational projection operators (9) in the spectral element methods allows for construction of C° continuous approximation spaces (piecewise high-order polynomials) that still yield exponential convergence (10). By breaking up the flow domain into macro-elements, the spectral element method can be applied to a wide variety of interesting fluid flow problems.

Both the spectral and spectral element techniques use as bases tensor product representations, the former in terms of Chebyshev expansions, the latter in terms of Lagrangian interpolants through Gauss-Lobatto Chebyshev or Legendre collocation points. The tensor product form is essential in constructing efficient (iterative and direct) solvers for high order methods. For example, the Stokes solver for the Chebyshev Tau method (for parallelpipeds) is based on multiple diagonalizations using eigenfunction decompositions (11); similar techniques can be used in certain cases in spectral element discretizations as well. It is also the tensor product form that allows for efficient implementation of conjugate gradient iteration (matrix vector products) for high-order methods; this technique is used in the three-dimensional spectral element calculations reported in Section 3.

2. TWO-DIMENSIONAL FLOW IN COMPLEX GEOMETRY

We consider first the case of natural convection in the configuration shown in Fig.1. The Grashof number of the flow is $Gr = \rho^2 g\alpha\Delta TD^3/\mu^2 = 10,000$, while the Prandtl number is taken to be $Pr = \mu c_p/k = 0.1$. No-slip boundary conditions are applied at all boundaries. The spectral element mesh for the problem is shown in Fig.2, illustrating the geometric flexibility and accuracy of the macro-elemental decomposition. As regards resolution, not only is there natural crowding of points at elemental boundaries, but there is also the option of clustering elements near regions of expected rapid function variation.

We plot in Figs.3 and 4 the (steady) velocity vectors and isotherms obtained using the spectral element method. The complex geometry results in the formation of several distinct recirculation regions.

3. THREE-DIMENSIONAL FLOW

We now consider fully three-dimensional flow in the large aspect ratio parallelpiped shown in Fig.5. The two endplates are at different (uniform) temperatures, with the remainder of the boundaries thermally insulated. No-slip boundary conditions are imposed on all walls. This problem is one for which there is analytical and experimental results as regards steady flow patterns and instability (12).

We show in Figs.6.a and 6.b the spectral and spectral element meshes in the "A" cross-section indicated in Fig.5. The corresponding velocity vectors in this plane obtained by the Tau method, are shown in Fig.7.a for the case of $Gr = \rho^2 \alpha g\Delta TD^4/L\mu^2 = 7,100$, $Pr = .026$. The flow is found to reach a steady-state in agreement with experiment (12); as can be seen in Fig.7.b, the form of the velocity field agrees with the asymptotic result given in (12). The three-dimensional effects are limited here to boundary layers at the side walls, as shown in Fig.8 by the plot of x,z components of the velocity in the "B" plane obtained with the Tau method. These boundary layers are accurately resolved by the natural clustering of Chebyshev collocation points near the boundaries.

At a Grashof number of $Gr = 1.8 \; 10^4$ both experiment and theory predict that the flow should be unstable to oscillatory modes. We demonstrate that our schemes accurately resolve this change in behavior by plotting in Fig.9 the y-component of the velocity as a function of time obtained using the spectral element method. The period of oscillation is approximately 40% lower than the theory (which assumes periodicity in the horizontal planform), and 50% lower than the experimental result. Tests at higher resolution are required to determine the source of the discrepancy. Velocity vectors are shown in Fig.10.

As regards the relative efficiency of the two algorithms, for this three-dimensional problem the Tau method and the spectral element method require approximately $5.7 \, 10^{-4}$ and $2.3 \, 10^{-3}$ seconds/timestep-node respectively on a CRAY X-MP (using a single processor), the Chebyshev Tau method using 6 times greater time steps (Courant condition of 4 against 0.7 for the spectral element method). As both codes are highly vectorized, the discrepancy is due to the conjugate gradient iterations required at each time step for the spectral element solution. As might be expected, the specialized direct solution procedure is more efficient for this (mono-domain) problem than the more general conjugate gradient algorithm, however there may be difficulties in extending the eigenfunction method to more complex three-dimensional configurations.

The results indicate that accurate solutions can be obtained with moderate resolution for three dimensional flows dominated by thermal convection. The efficiency of the algorithms and the flexibility of macro-elemental decomposition demonstrates the capacity of the codes to solve complex problems in general configuration.

ACKNOWLEDGEMENTS

Authors are grateful to C. Schneidesch for the graphical treatment of their results. T. Cartage would like to thank IRSIA for continuous financial support and FNRS for providing computer time on CRAY. Patera's research was partially supported by the ONR and DARPA, contract N00014-85-k-0208. Authors benefited from NATO grant NOSA.5-2-05 RG (035/84) for international collaboration.

REFERENCES

1. Gottlieb, D., Orszag, S.A., "Numerical analysis of spectral methods : theory and applications", SIAM monograph n°26, CBMS-NSF, Philadelphia, 1977.
2. Cartage, T., Demaret, P., Deville, M., "Chebyshev spectral and pseudospectral solutions of the Navier-Stokes equations", 9th Int.

Conf. on Num. Meth. in Fluid Dyn., Saclay, 1984, Lecture notes in Physics, 218, Springer Verlag, Berlin, pp. 127-132, 1985.
3. Cartage, T., Deville, M., "Improved time marching schemes for Navier-Stokes equations using Chebyshev-Tau approximation", 6th GAMM Conf. on Num. Meth. in Fluid Mech., Notes on Num. Fluid Mech., 13, Vieweg Verlag, pp. 47-54, 1986.
4. Patera, A.T., "A spectral element method for fluid dynamics : Laminar flow in a channel expansion", JCP, 54, pp. 468-488, 1984.
5. Korczak, K.Z., Patera, A.T., "An isoparametric spectral element method for solution of the Navier-Stokes equations in complex geometry", JCP, 62, pp.361-382, 1986.
6. Karniadakis, G.E., Bullister, E.T., Patera, A.T., "A spectral element method for solution of the two-and three-dimensional time-dependent incompressible Navier-Stokes equations", in Proc. Europe-US Conf. on Finite Element Methods for non-Linear Problems, Springer, pp. 803-817, 1986.
7. Bullister, E.T., Karniadakis, G.E., Ronquist, E.M., and Patera, A.T., "Solution of the unsteady Navier-Stokes equations by spectral element methods", in Proc. 6th Int. Symp. on Finite Element Methods, Antibes, to appear.
8. Orszag, S.A., Israeli, M., Deville, M., "Boundary conditions for incompressible flows", to appear.
9. Strang, G. and Fix, G., "An analysis of the Finite Element Methods", Prentice Hall, New Jersey, 1971.
10. Babuska, I. and Dorr, M.R., "Error estimates for the combined h- and p-versions of the finite element method", Numer. Meth., 37, 257, 1981.
11. Haldenwang, P., Labrosse, G., Abboudi, S., Deville, M., "Chebyshev 3-D and 2-D pseudospectral solvers for the Helmholtz equation", JCP, 55, pp. 115-128, 1984.
12. Hart, J., "A note on the stability of low-Prandtl-number Hadley circulations", JFM, 132, pp. 271-281, 1983.

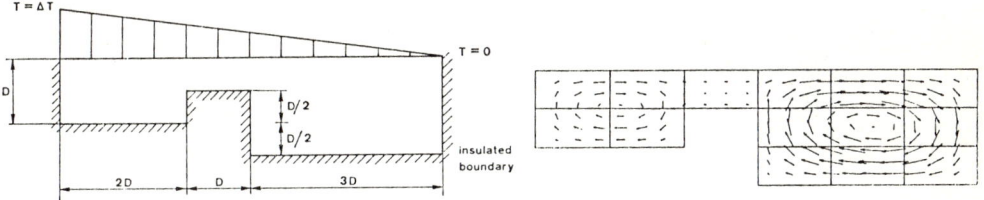

Fig.1 : Geometry and thermal boundary condition of the 2-D problem.

Fig.3 : Velocity pattern at steady state.

Fig.2 : Spectral element mesh.

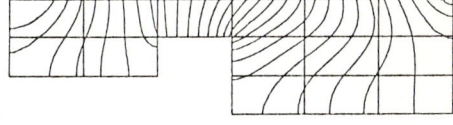

Fig.4 : Isotherms at steady state.

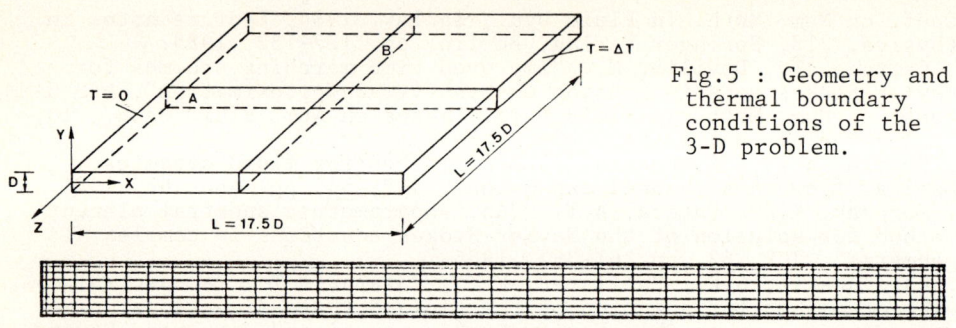

Fig.5 : Geometry and thermal boundary conditions of the 3-D problem.

Fig.6.a : Spectral mesh in the "A" plane 41x11.

Fig.6.b : Spectral element mesh in the "A" plane 21 x 9.

Fig.7.a : Velocity vectors in the "A" plane for $Gr = 7.1 \; 10^3$.

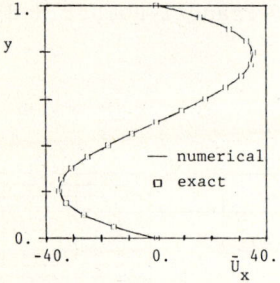

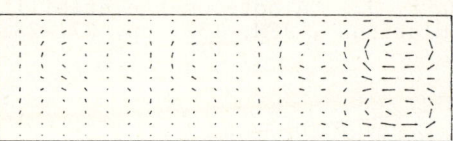

Fig.8 : Velocity vectors near the boundary in the "B" plane for $Gr = 7.1 \; 10^3$.

Fig.7.b : Basic $\bar{U}_x$ velocity profile in the "A" plane.

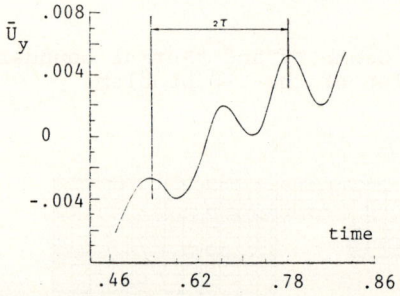

Fig.9 : y-component of the velocity versus time for $Gr = 1.8 \; 10^4$.

$2\tau = 0.265$
$\omega = 47.5$
$\omega_{ex} = 92 \pm 5$

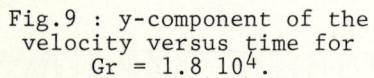

Fig.10 : Velocity pattern in the "A" plane for $Gr = 1.8 \; 10^4$.

THREE-DIMENSIONAL SEPARATED VISCOUS FLOW ANALYSES

J. E. Carter, R. L. Davis and D. E. Edwards
United Technologies Research Center
East Hartford, CT 06108/USA

M. M. Hafez
University of California
Davis, CA 95616/USA

INTRODUCTION

The analysis of complex three-dimensional flows is a problem that continues to challenge fluid dynamicists. Present day design techniques are primarily based on 2-D methodology; however, it is becoming increasingly clear that advanced aerodynamic designs will need to make greater use of the third dimension for flow control to achieve desired aerodynamic performance. In recent years significant advances have been made in the development of viscous flow analyses for 2-D separated flow. These methods are based on approaches ranging from numerical solutions of the Navier-Stokes equations to solutions of the governing equations of Interacting Boundary Layer Theory (IBLT). The latter approach has been demonstrated to be particularly attractive due to its efficiency and capability to capture the fine length scales in the viscous region while simultaneously solving the global inviscid region. Numerous problems have been solved (e.g. see Refs. 1-2) with IBLT such as transitional separation bubbles, transonic shock induced separated flow, and subsonic boattail separated turbulent flow. Despite these advances for 2-D flow, the more physically realistic occurrence of separation in three dimensions is only now beginning to be addressed. In this paper, which is a condensed version of Refs. 3 and 4, results of recent research efforts on the development of analyses for 3-D separated flows are presented. This work is being pursued through the development of an analysis based on IBLT as well as a more generalized technique based on a stream function-vorticity formulation. The latter approach is an attempt to develop an efficient Navier-Stokes solution technique using, where appropriate, IBLT concepts. Numerical comparisons are presented between these two techniques for a closed 3-D laminar separation bubble.

3-D INTERACTING BOUNDARY LAYER THEORY

In this approach to the analysis of 3-D separated flow, the 3-D boundary layer equations, expressed in finite difference form, are solved iteratively through displacement thickness coupling with a small disturbance 3-D representation of the outer inviscid flow. The boundary layer procedure is based on that described in Ref. 5 in which a detailed numerical study was conducted to determine under what conditions the inverse form of the 3-D boundary layer equations can be solved as an initial value problem. With this methodology now in place, our efforts have turned to the development of a complete viscous-inviscid interaction analysis. The numerical coupling method used to iteratively solve the viscous and inviscid equations is the three-dimensional extension of the Veldman[6] quasi-simultaneous procedure with the modifications introduced by Davis and Werle[7]. At the present time this overall interaction method is limited to low speed laminar flow over flat surfaces with small geometrical protuberances, as shown schematically in Fig. 1, that result in separated flow. Despite these simplifications, it is anticipated that this work will play the same fundamental role in the development of procedures for more complex 3-D flows as the work presented in Ref. 8 did for the subsequent development of

2-D IBLT. In particular the focus of the present research is to determine proper differencing procedures in the boundary layer equations to account for the local 3-D flow direction, efficient coupling strategies for iteration of the governing sets of viscous and inviscid equations, and the overall stability and robustness of such an approach to properly describe 3-D separated flow. The details of this procedure are discussed in a paper by Edwards, Ref. 3.

3-D STREAM FUNCTION-VORTICITY FORMULATION

IBLT provides a very efficient approach to describe strongly interacting viscous and inviscid flows; nonetheless, increases in configuration complexity for either internal or external flows result in a displacement surface that becomes increasingly difficult to track and hence, IBLT techniques become very tedious to implement. In addition, many 3-D flows such as corner or wing tip vortex flows are dominated by axial vortex effects and thus the role of the displacement body concept is anticipated to be of secondary importance in the analysis of such 3-D flows. It is therefore not surprising that most current research on 3-D strongly interacting viscous and inviscid flows is based on numerical solutions of the Navier-Stokes equations which are the fundamental governing equations of fluid motion. Most of these methods are relatively inefficent though, particularly for practical flows in which high Reynolds number viscous effects introduce a wide range of diverse length scales. Clearly, an effort is needed to build on what has been learned about high Reynolds number interacting flows in the development of new formulations and numerical methods for the Navier-Stokes equations. Some work[9-10] in this area has been initiated in the development of efficient generalized viscous-inviscid solution procedures which can be eventually extended to 3-D flows.

In the present effort IBLT concepts have been incorporated into a Navier-Stokes procedure in order to take advantage of the efficiency of IBLT techniques but retain the generality of a Navier-Stokes approach. This approach has been pursued in the present effort by casting the 3-D Navier-Stokes equations into a stream function-vorticity formulation. Lighthill[11] has presented an extensive discussion that shows that vorticity is a very useful quantity which provides a complete description of most fluid dynamic phenomena. It is particularly attractive for describing low speed flows since vorticity is produced only at solid surfaces due to viscosity and hence is non-zero only in viscous regions. A brief description of the governing equations will now be presented along with some remarks as to the relationship of this approach to that of IBLT.

The governing equations in the present formulation are based on the vorticity transport equation

$$\vec{V} \cdot \nabla \vec{\omega} = \vec{\omega} \cdot \nabla \vec{V} + \nu \nabla^2 \vec{\omega} \tag{1}$$

the definition of vorticity $\vec{\omega} = \nabla \times \vec{V}$, and the continuity equation $\nabla \cdot \vec{V} = 0$. The continuity equation is automatically satisfied by the introduction of two streamlike functions which are related to the Cartesian velocity components by

$$u = \frac{\partial \psi}{\partial y}, \quad v = -\frac{\partial \psi}{\partial x} - \frac{\partial \theta}{\partial z}, \quad w = \frac{\partial \theta}{\partial y} \tag{2}$$

This relationship is not unique; however, it is a simple procedure that has been found to work very well in the present study. The overall governing equations consist of two Poisson type equations for ψ and θ, which are easily derived from the definition of vorticity, and the thin layer form of two vorticity transport equations for the Cartesian vorticity components ω_x and ω_z shown in the schematic diagram in Fig. 1. It is unnecessary to consider the third component of vorticity as is discussed in Ref. 4. The governing equations, expressed in body sheared coordinates, are iteratively solved with Gauss-Seidel line relaxation which is marched in the mainstream direction. The complete details of the formulation and numerical approach are presented in Ref. 4.

This formulation retains some of the advantages of IBLT in which the inviscid and viscous flow regions are solved separately with subsequent coupling through displacement thickness. In the stream function-vorticity formulation the equations reduce to elliptic equations for ω and θ in the inviscid region since the vorticity is zero outside of the viscous layer. In the inner or viscous flow region, the flow is represented by the vorticity transport equations which describe the diffusion and convection of the vorticity produced at the surface due to the imposition of the no-slip condition. These equations have a parabolic or initial value character, similar to the boundary layer equations, in that they are solved independent of a downstream boundary condition on vorticity. The interaction effect of the viscous flow on the inviscid region takes place through the vorticity terms in the Poisson equations which act as "forcing functions" to set the streamline pattern in the outer flow. A final point to be noted on the close relationship between the present stream function-vorticity approach and IBLT is that if, on the scale of the inviscid flow, the vorticity field is assumed to collapse to an infinitesimal thickness (e.g. at the surface) then this approach reduces precisely to the viscous-inviscid interaction coupling identified by Lighthill. In a pioneering paper[12] in 1958, Lighthill showed that viscous effects can be introduced into the inviscid flow through one of three equivalent methods: a) solid body displacement thickness, b) surface transpiration, which is generally used in IBLT due to its simplicity, or c) surface vorticity.

RESULTS AND DISCUSSION

A number of comparisons have been made[3,4] between the present IBLT and stream function-vorticity formulation of the thin layer Navier-Stokes equations. One case is the Re=8000 laminar flow over a 3-D trough shown schematically in Fig. 2 and given by the function $y(x,z)=-t\operatorname{sech}(4x-10)\operatorname{sech}(4z)$. Along the plane of symmetry, z=0, this configuration is identical with the 2-D shape used by Carter and Wornom[8] and others to study 2-D separated flow. In the present calculations, which are for t=-0.06, a uniform 81x21 mesh was used in both computations in the x- and z-directions, respectively. The x-z computational domain was $1 \leq x \leq 4$ and $0 \leq z \leq 1.5$. In the IBLT calculation 21 grid points, distributed nonuniformly, were placed across the boundary layer whereas in the stream function-vorticity formulation 81 points were distributed nonuniformly with at least 32 points in the boundary layer region. Both solutions have been tested with finer y-meshes with little change in the computed results.

Figure 3 shows the contours from both analyses for the axial component of skin friction with a more detailed comparison shown at two z-locations in Fig. 4. The agreement is excellent between these solutions for this case in which separation occurs since the main stream velocity component undergoes flow reversal. Figure 3 shows that the disturbance in the flow caused by the trough vanishes as the z-farfield is approached. Similar agreement is shown in Figs. 5 and 6 for the spanwise component of skin friction. Both solutions show spikes at the upstream boundary which is probably due to the assumption that the v- and w-velocities are zero at this boundary. These numerical disturbances quickly dampen, as shown in Fig. 6, with good quantitative results demonstrated between the two analyses. The spanwise skin friction is qualitatively similar to that computed by Duck and Burggraf[13] for laminar, incompressible flow over a 3-D trough at infinite Reynolds number using triple deck theory. Figures 7 and 8 show a comparison of a viscous integral thickness (defined the same as the 2-D displacement thickness) deduced from the two approaches. The agreement is good although a slight difference exists near the trough center position. Favorable agreement is also observed in the u- and w-velocity profiles as shown in Ref. 4.

CONCLUDING REMARKS

Research efforts have been initiated in the development of two procedures for the analysis of 3-D strongly interacting viscous and inviscid flows. One method is based on Interacting Boundary Layer Theory whereas the other is a thin layer Navier-Stokes approach based on a 3-D stream function-vorticity formulation. Favorable comparison between the two solutions for a 3-D laminar closed separation bubble provide strong encouragement that further development of both methods should be continued. The IBLT approach provides an excellent method by which localized interactions can be analyzed, whereas the stream function-vorticity approach should provide a general method which can address flows with strong axial vorticity effects.

ACKNOWLEDGEMENT

The authors express their gratitude for the support of this work to the following technical monitors and agencies: Dr. James D. Wilson, Air Force Office of Scientific Research (Contract F49620-84-C-0032) and Dr. R. E. Whitehead, Office of Naval Research (Contract N00014-81-C-0381).

REFERENCES

1. Carter, J. E. and V. N. Vatsa: Eighth International Conference on Numerical Methods in Fluid Dynamics, Springer-Verlag, Vol. 170, pp. 167-174, Aachen, Germany, 1982.
2. Carter, J. E., D. E. Edwards, R. L. Davis, and M. Hafez: Ninth International Conference on Numerical Methods in Fluid Dynamics, Springer-Verlag, Vol. 218, pp. 133-137, Saclay, France, 1984.
3. Edwards, D. E.: IUTAM Conference on Boundary Layer Separation, University College, London, August, 1986.
4. Davis, R. L., J. E. Carter and M. Hafez: AFOSR-TR-86, May, 1986.
5. Edwards, D. E. and J. E. Carter: AIAA Paper No. 85-1499, July, 1985.
6. Veldman, A. E. P.: NLR TR 79023 U, 1979.
7. Davis, R. T. and M. J. Werle: Proceedings of the First Symposium on Numerical and Physical Aspects of Aerodynamic Flows, Long Beach, 1981.
8. Carter, J. E. and S. F. Wornom: NASA SP-347, 1975.
9. Whitfield, D. L.: Proceedings of the Third Symposium on Numerical and Physical Aspects of Aerodynamic Flows, Long Beach, 1985.
10. Steger, J. L. and W. R. Van Dalsem: Proceedings of the Third Symposium on Numerical and Physical Aspects of Aerodynamic Flows, Long Beach, 1985.
11. Lighthill, M. J.: <u>Laminar Boundary Layers</u>, ed. L. Rosenhead, pp. 46-109, 1963.
12. Lighthill, M. J.: J. Fluid Mechanics, Vol. 4, pp. 383-392, 1958.
13. Duck, P. W. and O. R. Burggraf: J. Fluid Mechanics, Vol. 162, pp. 1-22, 1986.

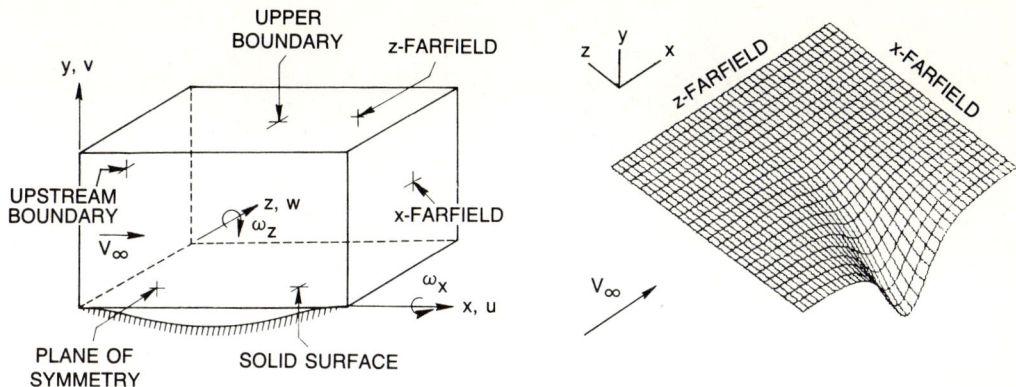

Fig. 1 Computational region around a localized, symmetric disturbance on flat plate

Fig. 2 Schematic of 3-D trough

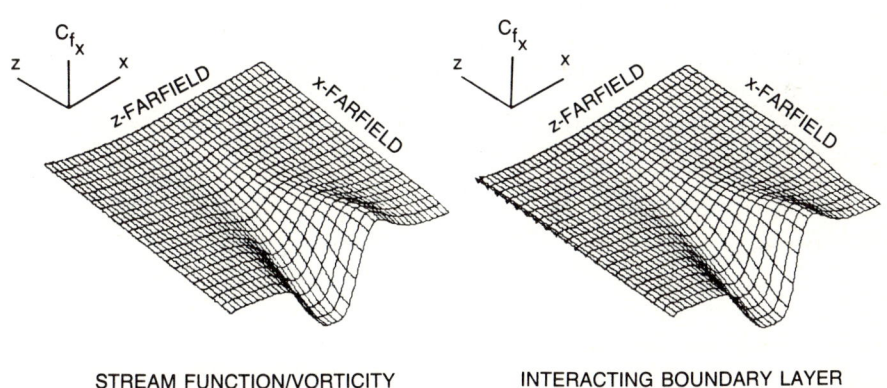

STREAM FUNCTION/VORTICITY

INTERACTING BOUNDARY LAYER

Fig. 3 Axial skin friction contours for 3-D trough

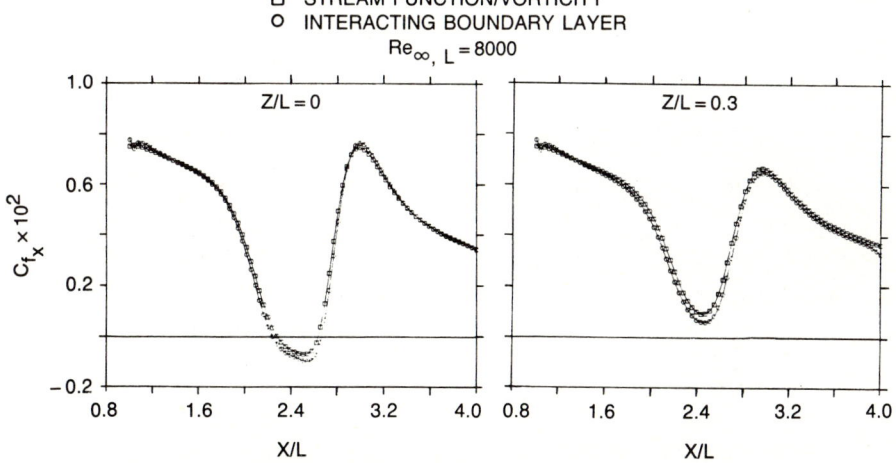

Fig. 4 Predicted axial skin friction distributions for 3-D trough

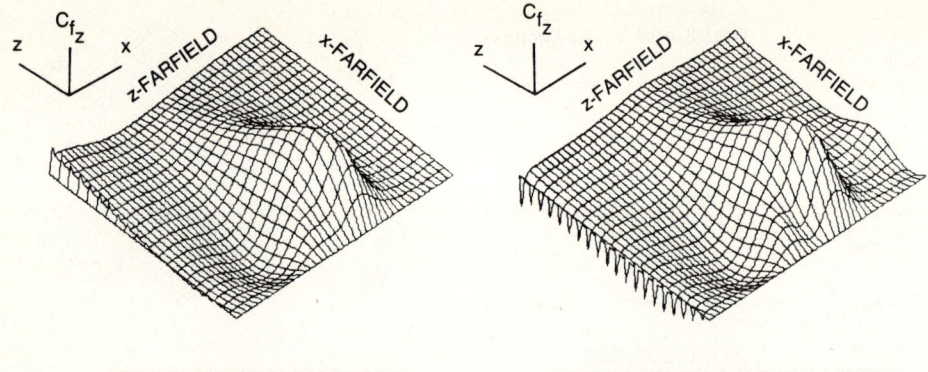

Fig. 5 Spanwise skin friction contours for 3-D trough

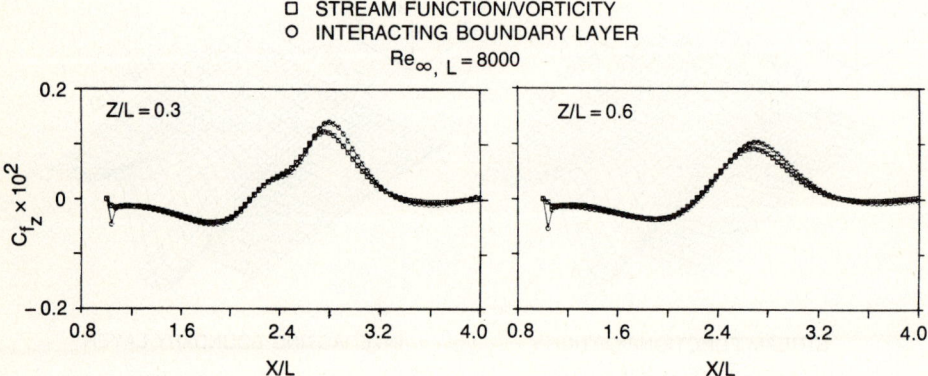

Fig. 6 Predicted spanwise skin friction distributions for 3-D trough

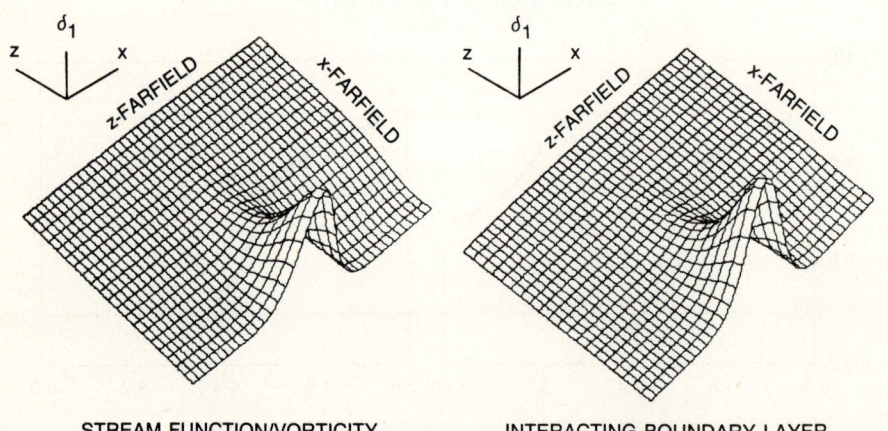

Fig. 7 Integral thickness contours for 3-D trough

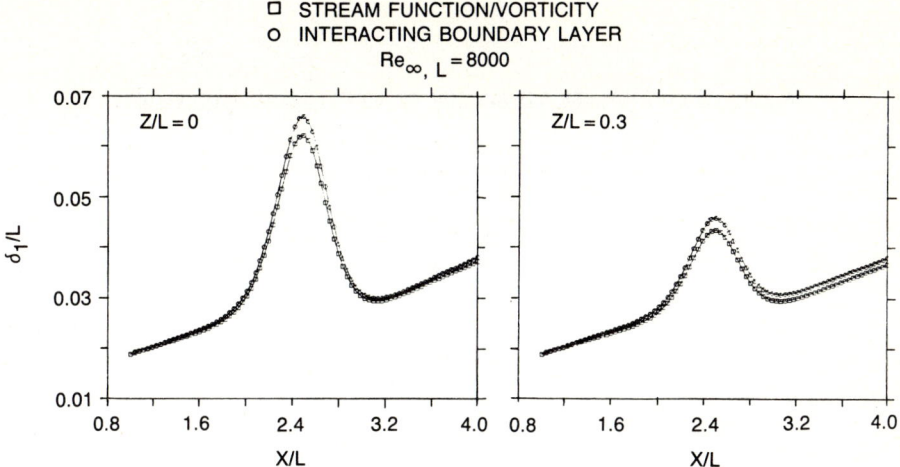

Fig. 8 Predicted integral thickness distributions for 3-D trough

A FULL NAVIER-STOKES SIMULATION OF COMPLEX INTERNAL FLOWS

James L. C. Chang and Ruey-Jen Yang
Rockwell International/Rocketdyne Division
Canoga Park, California USA

Dochan Kwak
NASA/Ames Research Center
Moffett Field, California USA

I. Introduction

In the past two decades, or so, significant advancements have been made in external flow simulations. Effort in interval flow problems has been limited by computer capacity and computing speed. With the fast advances of the modern supercomputers, it is now feasible to extend the knowledge accumulated in external flows to complex internal flow problems.

Problems of internal flows are of great interest in flow device engineering, such as the design of nozzles, diffusers, ducts, turbines, and turbopumps in propulsion systems. Internal flows differ from external ones in many ways. For instance, the boundary layer in external flow is usually very thin compared with the dimensions of the moving object, while in internal flow it may occupy the entire flow field. Since flow is confined by the boundary of the device, the concept of far-field boundary conditions does not apply in these problems. Flow geometries are generally three-dimensional and likely complex. The complex internal flows in the Space Shuttle Main Engine (SSME) powerhead provide a good example of these problems.

Figure 1 illustrates the current arrangement of the SSME powerhead components. Inside the Hot Gas Manifold (HGM), hot gas discharged from the turbine enters the annular turnaround duct (TAD), and experiences a 180-degree turn before it diffuses into the fuel bowl. Then it flows into the main injector through three transfer tubes. Because of the high gas temperature, the Mach number is less than 0.12. The flow is turbulent and practically incompressible.

Figure 2 demonstrates cross-sectional views of the three-dimensional grid representing the flow path for a newly proposed two-elliptical-duct HGM. The grid is generated by using algebraic functions. The grid generation code is written with high degree of flexibility; changes of the geometric configurations can readily be obtained.

II. Description of the Method

Method of Pseudocompressibility

To achieve an efficient computation in solving steady-state incompressible flows, Chorin [1] proposed the use of artificial compressibility. This method was later studied by Steger and Kutler [2]. It was further developed into a method of pseudocompressibility by Chang and Kwak [3.4]. By this method, the incompressible Navier-Stokes equations in dimensionless form are modified as follows:

$$\frac{1}{\beta} \frac{\partial p}{\partial t} + \frac{\partial u_i}{\partial x_i} = 0 \tag{1a}$$

$$\frac{\partial u_i}{\partial t} + \frac{\partial u_i u_j}{\partial x_j} = -\frac{\partial p}{\partial x_i} + \frac{\partial T_{ij}}{\partial x_j} \tag{1b}$$

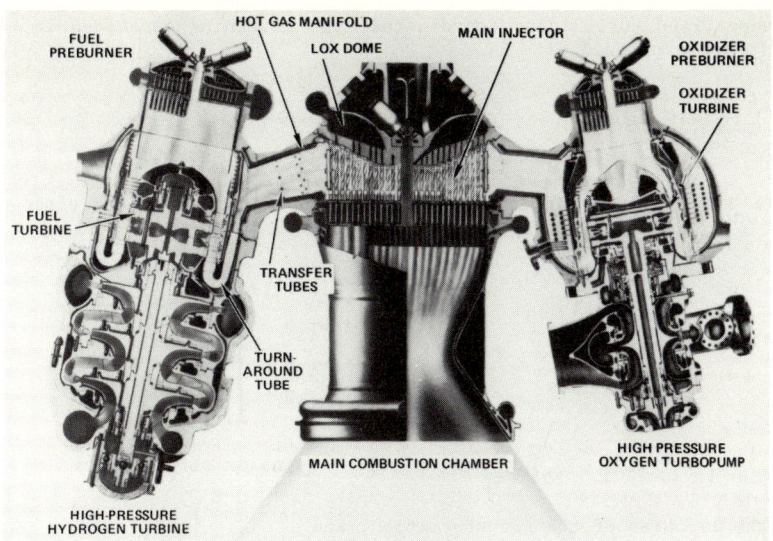

Fig. 1. SSME Powerhead Component Arrangement

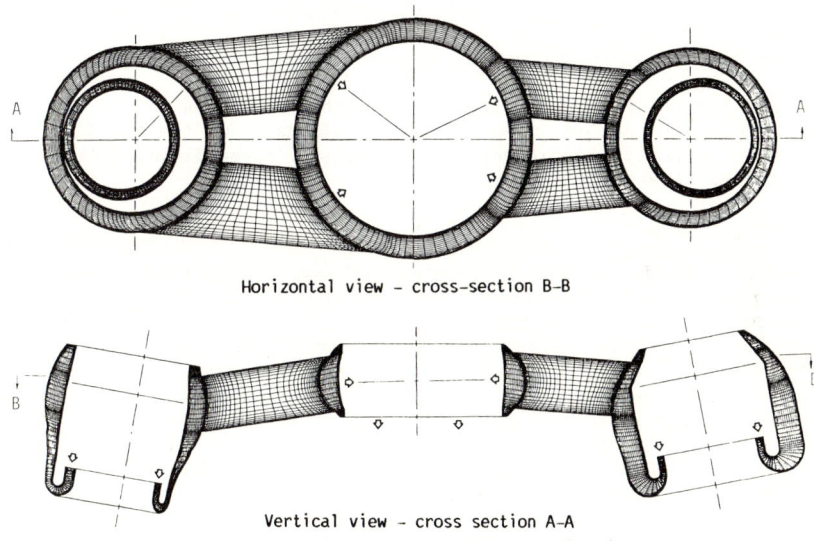

Horizontal view – cross-section B-B

Vertical view – cross section A-A

Fig. 2. Grid for the SSME Powerhead

As the computation converges to a steady state, the effect of pseudocompressibility vanishes, yielding an incompressible solution.

Numerical Algorithm

To accommodate arbitrary three-dimensional geometries, the coordinates are transformed by using the following independent variables:

$$\begin{aligned} \tau &= t \\ \xi_i &= \xi_i(x,y,z,t) \end{aligned} \qquad (2)$$

In these generalized curvilinear coordinates, the governing equations in conservation-law form are expressed as

$$\frac{\partial \hat{Q}}{\partial \tau} + \frac{\partial (\hat{E}_i - \hat{\Gamma}_i)}{\partial \xi_i} = 0 \qquad (3a)$$

The pressure and velocity vector $(\hat{Q})$, the flux vector $(\hat{E})$, and the viscous diffusion vector $(\hat{\Gamma})$, are described by

$$\hat{Q} = \frac{Q}{J} = \frac{1}{J}\begin{bmatrix} p \\ u \\ v \\ w \end{bmatrix}, \quad \hat{E}_i = \frac{1}{J}\begin{bmatrix} \beta U_i + L_{i0}(p-\beta) \\ u U_i + L_{i1} p \\ v U_i + L_{i2} p \\ w U_i + L_{i3} p \end{bmatrix} \qquad (3b)$$

$$\hat{\Gamma}_i = \frac{\nu}{J}(\nabla \xi_i \cdot \nabla \xi_j)\frac{\partial}{\partial \xi_j}[0,u,v,w]^T + \text{Reynolds stress terms}$$

where J is the Jacobian of the transformation, and

$$U_i = L_{i0} + L_{i1} u + L_{i2} v + L_{i3} w$$
$$L_{i0} = (\xi_i)_t, \quad L_{i1} = (\xi_i)_x, \quad L_{i2} = (\xi_i)_y, \quad L_{i3} = (\xi_i)_z \qquad (3c)$$

are the contravariant velocities and the metrics of transformation, respectively.

An implicit finite-difference algorithm together with the ADI or approximate-factorization scheme [5,6] are used to advance solution to the Eq. (3a) in time. These procedures have been implemented to develop a three-dimensional incompressible Navier-Stokes flow solver (INS3D) [7].

Method of Multiple Zones

A large number of mesh points is required to resolve the three-dimensional viscous flow in geometrically complicated problems such as the HGM. The restrictions of the computational speed and storage are still the main factor in the Navier-Stokes flow simulation. To overcome the limitation of the computer in-core memory, a method of multiple zones is introduced [8]. By this method, the domain of interest is divided into several zones as shown in Fig. 3. Solutions to these zones are linked by zonal interpolations at each time-step of advancement in the computation. This method effectively utilizes external memories such as the fast input-output solid state device (SSD) of a Cray XMP to increase the allowable number of mesh points. It makes the computation for highly complex geometries possible.

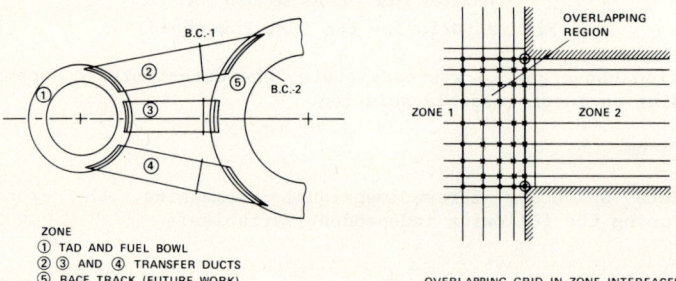

Fig. 3. Computational Zones

Boundary Conditions

At the solid wall, no-slip conditions are used. The pressure at the wall can be obtained simply by applying a zero pressure gradient at the no-slip boundaries. Upstream inlet conditions can be specified as desired or using experimental data. In this paper, a constant and uniform static pressure is applied at the inlet. The primary velocity is specified by using a profile corresponding to a uniform inlet flow with thin boundary layers at the walls for the laminar solution. For the turbulent case, a fully developed channel flow profile is used.

In this paper, only the fuel-side HGM, which includes the turnaround duct, the fuel bowl, and the transfer tubes, is used in the formulation. The downstream boundary is the outlet of the transfer tubes. A mass and momentum weighted extrapolation method [8] is applied to provide stable free-flow conditions. The flow in the HGM is taken to be symmetric about the plane through the centerlines of the fuel bowl and the main injector. Symmetric boundary conditions can be easily obtained for this plane.

Turbulence Models

Several levels of turbulent models have been implemented into the code. These include a two-equation $K-\epsilon$ model with low Reynolds number corrections for the near-wall effects. However, for many engineering applications, economy is still an overriding factor. Simple algebraic models are therefore desirable. The model used in the present paper is an extended Prandtl-Karman mixing length theory [9]. The length scale is based on the vorticity thickness, δ_ω, defined as the distance between the wall and the location of minimum vorticity. The expression is given as

$$\frac{\ell}{\delta_\omega} = \kappa^2 (1 - e^{-\tilde{y}/K}) , \quad \tilde{y} = \frac{y}{\delta_\omega} \tag{4}$$

and the eddy viscosity is given by

$$\nu_t = \ell^2 |\omega| \tag{5}$$

where $|\omega|$ is the absolute value of the three-dimensional vorticities. In the inner sublayer, the Van Driest (1956) damping function is applied to make corrections for the viscous effects.

III. Computed Results

The steady-state laminar solution for a three-circular-duct HGM is shown in Fig. 4. The Reynolds number for this flow is 1000 based on the width of the turnaround duct entrance. In Fig. 4a, 4b, and 4c, velocity vectors are shown in the horizontal and vertical cross sections. Downstream of the 180-degree bend along the inner walls of the turnaround duct, an extended separation bubble is revealed. The flow in the center transfer tube, as illustrated in Fig. 4b, is highly nonuniform. A large recirculation zone is formed just downstream of the tube entrance. It extends to the exit of the tube. The predicted mass flow through the center tube is 9.8% of the total flow which agrees with the test data.

Figures 4d and 4e show the swirling flow patterns at the entrance to the outer and the center tubes, respectively. The three-dimensional velocity profiles in the fuel bowl are illustrated in Fig. 4f and 4g at two unwrapped planes. Swirl patterns are predicted in the vicinity of the entrance to the transfer tubes. Figure 4h is a photograph that indicates similar swirls obtained by means of surface streak (shear pattern) visualization in the air flow test.

Computed velocity distribution for turbulent flow in a two-elliptical-duct HGM at $Re=10^5$ are shown and explained in Fig. 5. The computed results agree satisfactorily with the air flow test data conducted in a slightly different configuration.

IV. Concluding Remarks

This paper summarizes the further development of the INS3D code. This code has been applied to a variety of internal as well as external flow problems, and has been demonstrated to be accurate and efficient in obtaining steady-state solutions to the three-dimensional Navier-Stokes equations. In particular, it has been applied successfully as a design tool to analyze a variety of new HGM configurations. The validity of the computational results has been confirmed by experiments.

References

1. Chorin, A. J., "A Numerical Method for Solving Incompressible Viscous Flow Problems," Journal of Computational Physics, Vol. 2, 1967, pp. 12-26.

2. Steger, J. L. and P. Kutler, "Implicit Finite-Difference Procedures for the Computation of Vortex Wakes," AIAA Journal, Vol. 15, No. 4, April 1977, pp. 581-590.

3. Chang, J. L. C. and D. Kwak, "On the Method of Pseudo Compressibility for Numerically Solving Incompressible Flows," AIAA Paper No. 84-0252, Reno, Nevada, January 1984.

4. Kwak, D., J. L. C. Chang, and S. P. Shanks, "A Solution Procedure for Three-Dimensional Incompressible Navier-Stokes Equation and Its Application," Ninth International Conference on Numerical Methods in Fluid Dynamics, C.E.N. Saclay, France, 25-29 June 1984.

5. Beam, R. M. and R. F. Warming, "An Implicit Finite-Difference Algorithm for Hyperbolic Systems in Conservation-Law Form," Journal of Computational Physics, Vol. 22, Sept. 1976, pp. 87-110.

6. Briley, W. R. and H. McDonald, "Solution of the Three-Dimensional Compressible Navier-Stokes Equations by an Implicit Technique," Proceedings of Fourth International Conference on Numerical Method in Fluid Dynamics, Lecture Notes in Physics, Vol. 35, Springer-Verlag, New York, 1975, pp. 105-110.

7. Kwak, D., J. L. C. Chang, S. P. Shanks, and S. Chakravarthy, "An Incompressible Navier-Stokes Flow Solver in Three-Dimensional Curvilinear Coordinate Systems Using Primitive Variables," AIAA Paper 84-0253, Reno, Nevada, January 1984; AIAA Journal, Vol. 24, No. 3, March 1986; H. Julian Allen Award 1985, NASA/Ames Research Center.

8. Chang, J. L. C., D. Kwak, S. C. Dao, and R. Rosen, "A Three-Dimensional Incompressible Flow Simulation Method and Its Application to the Space Shuttle Main Engine, Part I - Laminar Flow," AIAA Paper 85-0175, Reno, Nevada, January 1985.

9. Chang, J. L. C., D. Kwak, S. C. Dao, and R. Rosen, "A Three-Dimensional Incompressible Flow Simulation Method and Its Application to the Space Shuttle Main Engine, Part II - Turbulent Flow," AIAA Paper 85-1670, Cincinnati, Ohio, July 1985.

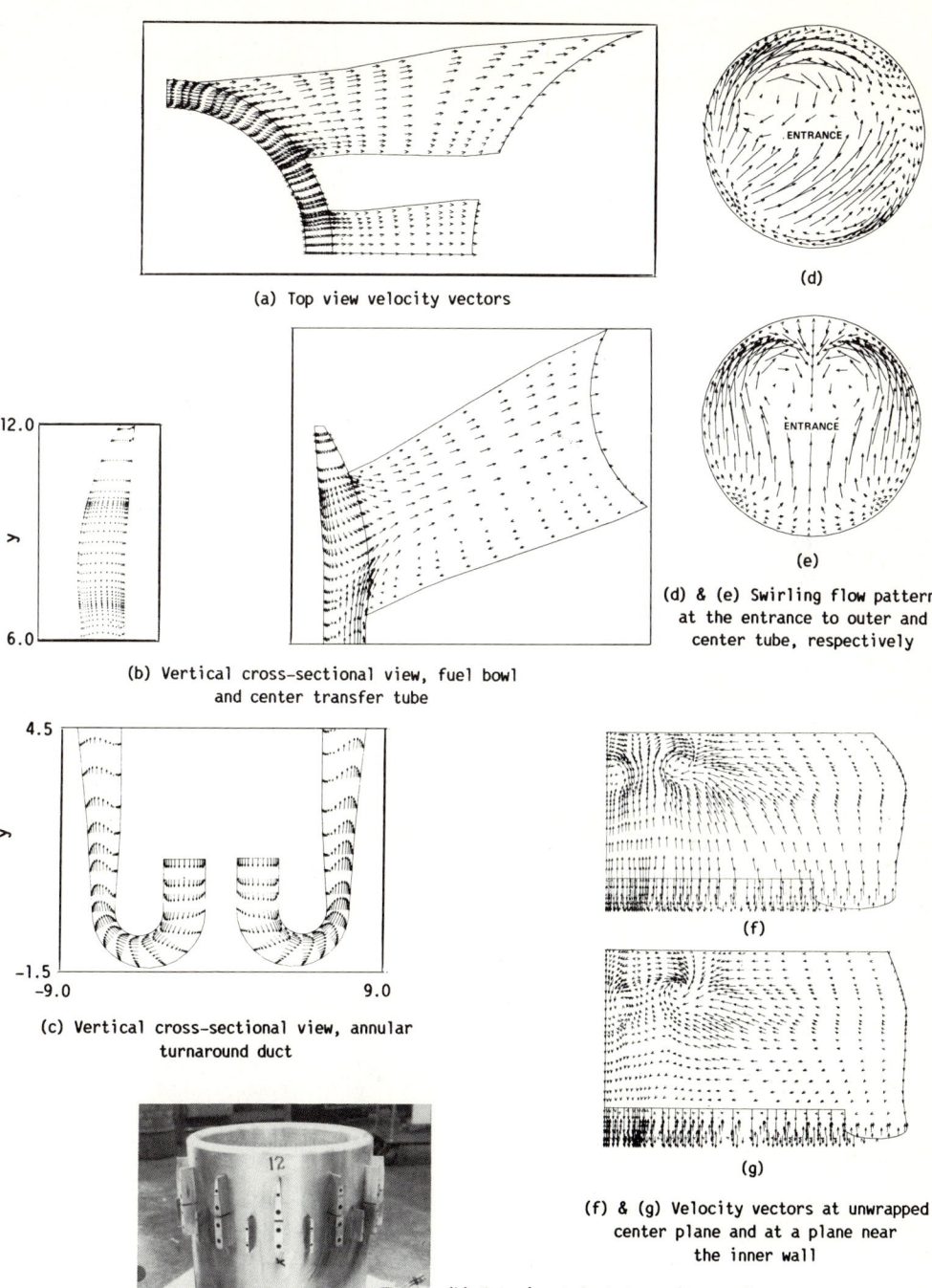

(a) Top view velocity vectors

(b) Vertical cross-sectional view, fuel bowl and center transfer tube

(c) Vertical cross-sectional view, annular turnaround duct

(d) & (e) Swirling flow patterns at the entrance to outer and center tube, respectively

(f) & (g) Velocity vectors at unwrapped center plane and at a plane near the inner wall

(h) Experimental photograph - surface streak visualization

Fig. 4. Computed Velocity Distributions for Three-Duct HGM, Laminar Flow, Re = 1000

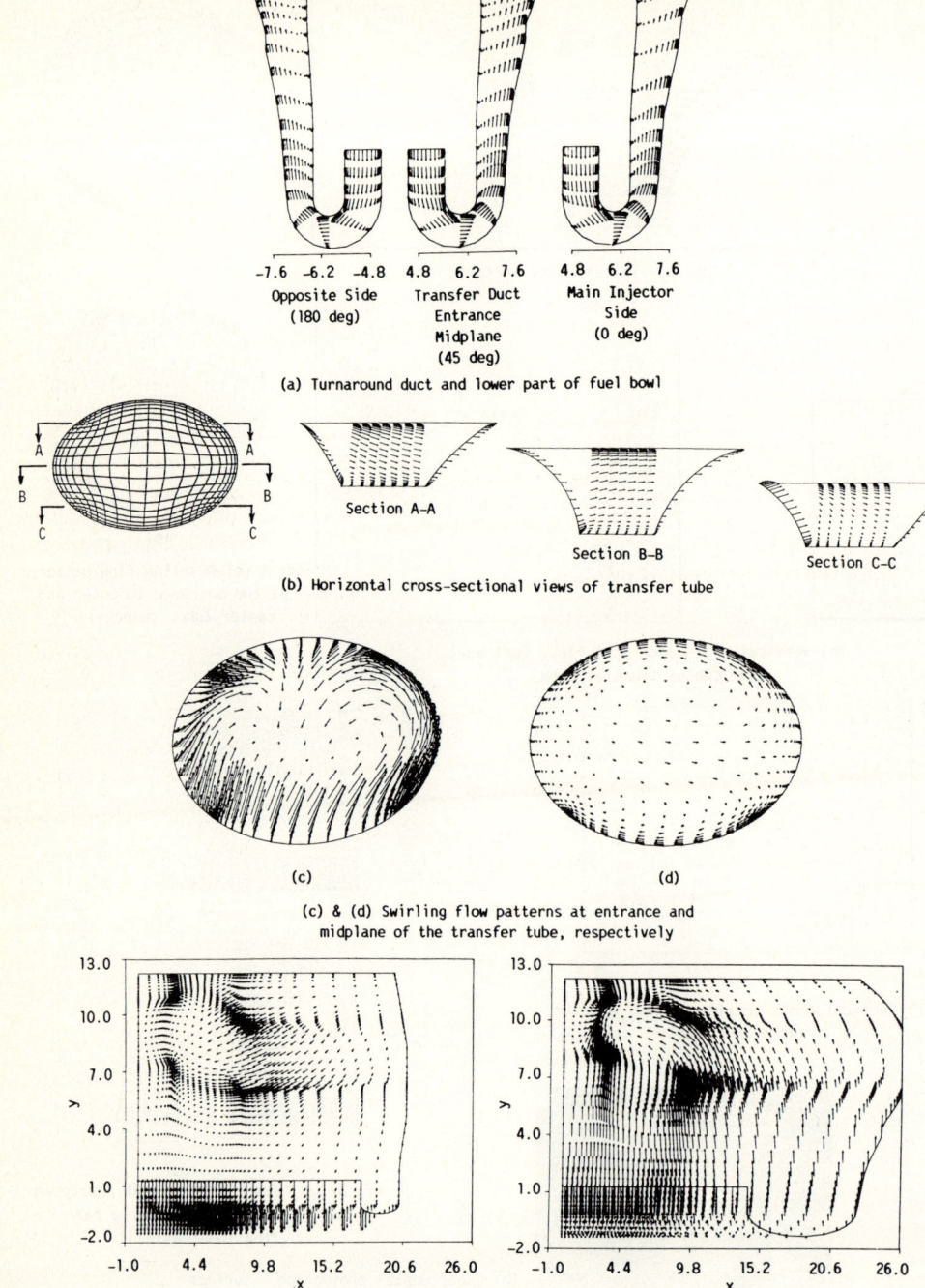

Fig. 5. Velocity Distributions for Two-Elliptical-Duct HGM, Turbulent Flow, Re = 10^5

A STRONG INVISCID-VISCOUS INTERACTION SOLUTION
OF A PLANE TRANSONIC CASCADE FLOW

Y.W.Chen   Y.K.Zhang
M.Y.Shen   D.T.Huang

(Dept. of Engineering Mechanics, Tsinghua Univ., Beijing China)

INTRODUCTION

For designing high efficiency and high load turbomachines it is necessary to take into account of the effect of viscosity in aerodynamic computation. In cases of no separation or small separation region the flow field can be divided into two regions: the viscous thin layer and the inviscid flow region, the former includes the boundary layer on the blade surface and the wake behind the blade. So the flow field in these two regions can be determined by use of Euler equations and boundary layer equations respectively. The interaction between inviscid and viscous flow is taken into account by an iterative method.

An inviscid calculation is carried out for the given blade geometry and aerodynamic parameters, and the resulting blade surface Mach number distribution is inputted to a boundary layer calculation to produce the first estimation of the blade surface boundary layer thickness. That information is used to modify the blade geometry and this cycle is then repeated until convergence is achieved.

In present paper the inviscid flow is calculated by a time marching finite area method[1,2] and the viscous flow by integral methods for laminar and turbulent boundary layer[3]. The location of transition point is predicted by an empirical correlations[4]. After convergence is achieved a mixing calculation is then carried out to determine the pressure loss coefficient[5].

INVISCID MAIN-STREAM CALCULATION

For plane adiabatic flow of an inviscid perfect gas with constant specific heat the governing equations can be taken as follows[2]

$$\frac{\partial}{\partial t}\iint_A \rho \, dxdy + \oint_S \rho \vec{v} \cdot \vec{n} \, ds = 0 \tag{1}$$

$$\frac{\partial}{\partial t}\iint_A \rho u \, dxdy + \oint_S [(p+\rho u^2)\vec{i} + (\rho uv)\vec{j}] \cdot \vec{n} \, ds = 0 \tag{2}$$

$$\frac{\partial}{\partial t}\iint_A \rho v \, dxdy + \oint_S [(\rho uv)\vec{i} + (p+\rho v^2)\vec{j}] \cdot \vec{n} \, ds = 0 \tag{3}$$

$$\frac{\gamma}{\gamma-1} \cdot \frac{p}{\rho} + \frac{1}{2}(u^2 + v^2) = i_o = \text{const} \tag{4}$$

As shown in figure 1, ABCDEFGH is the computational region, $\widehat{GPF}$ and $\widehat{BSC}$ represent respectively the equivalent boundaries of the pressure surface and suction surface of two neighbouring blades. The inlet and exit boundaries are parallel to the Y axis. Parallelogram ABGH is the periodic region in front of the blade, the angle between AB(or HG) and X axis is equal to the inlet flow angle β_1. In the first inviscid calculation FE and CD are oriented to the estimated exit flow angle and in the following calculations FE, CD are the lower and upper wake boundaries determined from the last viscous calculation and to be modified in the present calculation.

The equations (1),(2) and (3) are discretized using Denton's basic scheme[1] and correction scheme developed in (2).

Various kinds of boundary conditions for inviscid flow are applied. If $U \leq C$ (C is the speed of sound), P_0, T_0 and β_1 are given on AH and if $U < C$, P_2 on DE. The periodic conditions are applied in front of the cascade. The slip condition is applied on the equivalent blade surfaces.

In actual flow the difference of pressure across the wake is small and can be neglected. Therefore in a converged solution the pressure at the corresponding points on the both sides of the wake must be equal. In the present paper this is realized by adjustment of base line of the wake.

The original base line of the wake is determined as the streamline through the trailing edge of the blade and in following iteration it will be regulated in accordance with the relation of Prandtl-Meyer function with consideration of the pressure equality mentioned above.

VISCOUS FLOW CALCULATION

The viscous flow consists of four parts: laminar, transition, turbulent boundary layer and wake.

The laminar boundary layer is calculated by Thwaite's (Loitsianskii's) method after performing Illingworth-Stewartson transformation. The step length for the integration is taken as constant at 1% arc length of the blade surface. After each step a check is made on whether the transition criteria are satisfied. There are two kinds of transition: bubble transition and natural transition. In this computation the bubble length or the length of transition is neglected, in both cases the momentum thickness is assumed to be unchanged after transition.

The critical Reynolds number and the thickness ratio after transition are determined by empirical correlations.

The turbulent boundary layer and wake are calculated by Green's Lag-Entrainment method [2],[6]. In the computation it has been found that in some cases the flow is under such a strong acceleration, that near the trailing edge the empirical correlations given by Green et.al. are not valid and calculation can not proceed. In this case a modified Nash-McDonald method which takes the flow upstream history effects into account, presented by Chen[7], is used to solve the turbulent boundary layer.

CONVERGENCE CRITERIA

At the end of each cycle of the inviscid-viscous interaction procedure, a check is made on convergence. The convergence criteria are:
    (a) The maximum difference or averaged difference in axial velocity component between two successive time steps of the inviscid calculations is less than 0.02% or 0.005% respectively.
    (b) The maximum difference in resulting blade surface Mach number distribution between last and present boundary layer calculations is less than 0.5%.
    (c) The maximum difference between Y coordinates of the equivalent blade surfaces (i.e. the Y coordinates produced by adding to the blade surfaces the boundary layer displacement thickness calculated within the cycle) and the corresponding Y coordinates of the blade surfaces actually used throughout the cycle is less than 0.3% of the blade pitch.

If any of these criteria is not satisfied, then the flow geometry is revised by applying under-relaxation to the predicted changes, and the cycle is repeated.

SAMPLE CALCULATIONS

(a) Transonic flow in a 624 gas turbine cascade
The details of the geometric and aerodynamic parameters and its experimental results are taken from [8].
The comparison between the predicted and measured surface dimensionless pressure distributions is shown in Fig.2.

The result shows that in this case the inviscid solution is improved by the inviscid-viscous interaction method.

(b) Transonic flow in RA gas turbine cascade
The details of the geometric and aerodynamic parameters and its experimental data are taken from [9].

The comparison between the predicted and measured surface Mach number distributions is shown in Fig. 3.

REFERENCES

1. Denton, J. D., ARC R&M 3775 (1975)
2. Zhang Yaoke, Shen Mengyu, Gong Zengjin, Mathematica Numerica Sinica, NO.4, (1978)
3. Green, J. E., Weeks, D.J., Brookman, J.W.F., ARC R&M 3791 (1973)
4. Singh, U. K., VKI Lecture Series 1980-8.
5. Stewart, W.L., NACA TN 3515 (1955)
6. East, l. F., Smith, P. D., Merryman, P. J., RAE TR 77046 (1977)
7. Chen Yunwen, Journal of Engineering Thermophysics, 5, 4, (1985)
8. Liu Shaozhong, Master Thesis, Tsinghua University, Jan., (1981)
9. C.G.Graham & F.H.Kost, ASME 79-GT-37.

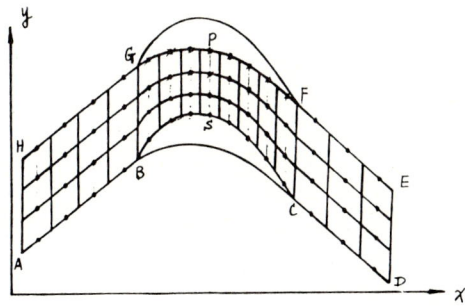

Fig. 1

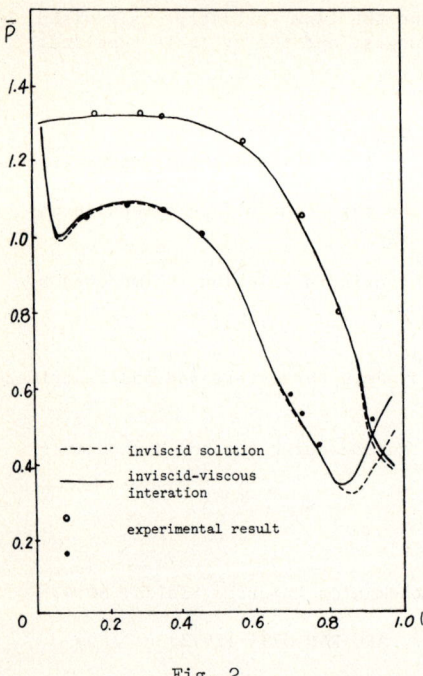

Fig. 2

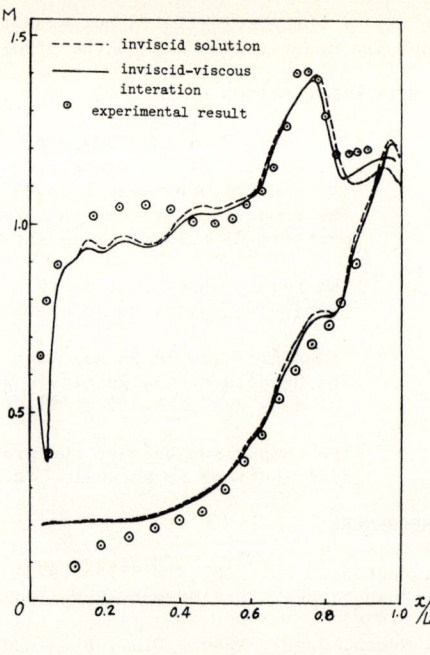

Fig. 3

COMPUTATION OF TURBULENT SPOT EVOLUTION

Sin-I Cheng
Princeton University
Princeton, New Jersey 08544/USA

The temporal evolution of a large disturbance in an incompressible shear flow from a laminar to a turbulent state is found to be three dimensional, abrupt and local.[1] The random occurrence of such events is the predecessor of flow transition. We report here our computational simulation of such evolutions in a uniform shear flow at Re = 3000, based on the Navier stokes equations (N.S) and its implications.

Computational solution of time dependent N.S. in three spatial dimensions with iterative pressure corrections $\delta p = -(\Delta/\varepsilon)\delta t$ based on residual velocity divergences, has been attempted with various choices of $\varepsilon(\Delta)$, explicitly in primitive variables [2] or implicitly through vorticity [3] through some discrete analogs of:

$$\varepsilon \frac{\partial p}{\partial t} + \frac{\partial u_i}{\partial x_i} = 0 \tag{1}$$

$$\rho \frac{\partial u_i}{\partial t} + \rho u_j \frac{\partial u_i}{\partial x_j} + \frac{\partial p}{\partial x_i} = \frac{1}{Re} \frac{\partial^2 u_i}{\partial x_j \partial x_j} + F_i \tag{2}$$

which reduces formally to the N.S. with $\varepsilon = F_i = 0$. F_i is the momentum carried by the mass influx of the residual divergence Δ. It was hoped that "converging" iterants of u_i and p at small ε, F_i and Δ would give reasonable approximate solutions.

We multiply (1) by p and (2) by u_i to construct an energy integral for $E = (\rho u^2 + \varepsilon p^2)/2$:

$$\frac{\partial}{\partial t}\int_V E dv = -\frac{1}{Re}\int_V \rho \frac{\partial u_i}{\partial x_j}\frac{\partial u_i}{\partial x_j} dv + \int_V u_i(\rho u_i \Delta/2 - F_i) dv$$

$$+ \frac{1}{Re}\int_S \rho \frac{\partial}{\partial x_i}(\frac{u^2}{2}) ds_i + \int_S -(p + \rho u^2/2) u_i ds_i \tag{3}$$

over the computational volume V with boundary S. If both surface integrals in (3) should vanish for periodic or other trivial boundary value problems, the choice $F_i = \rho u_i \Delta/2$ guarantees a negative definite energy integral so that solution sequences from (1) and (2) may converge as ε, $\Delta \to 0$.

For nontrivial initial boundary value problems of flows containing drag bodies, the last surface integral remains positive and nonzero as $\Delta \to 0$. Therefore for any choices of F_i and ε that vanish as $\Delta \to 0$, the integral (3) remains positive, at sufficiently large Re. Iterative solution sequences would diverge as $\Delta \to 0$ if not kept bounded by repeated smoothing and filtering. While impressive results can be obtained for certain types of problem, such artifacts are not useful in the present problem since we cannot distinguish oscillations of physical and numerical origin, to be kept or removed.

The energy integral (3) suggests strongly that iterative methods of solution of N.S. for nontrivial problems should limit the velocity iterants in the solenoidal subspace. When the N.S. is linearized, we can find the solenoidal velocity and the pressure fields. The nonlinear N.S. can then be solved through quasi-linearization by successively updating the solenoidal convective velocity. One may avoid the tedious quasi-linear iteration with intelligent guesses of the convective velocity, known or preconceived, and possibly including random fields locally or globally. Unable to outguess the physical system in creating the ever more complex flows for linearization, we would rather guess the nonlinear dynamic mechanism of turbulence based on first principles. With proper dynamics and solenoidal velocity iterants, quasi-linear iteration may give "converging" approximations to the bifurcated state at each time step if not the details of bifurcating processes in the interim.[4] The nonlinear dynamic mechanism must be able to restore any disturbed flow field to a solenoidal state.

Equation (3) suggests that if F_i were Dirac functions its volume integral could cancel the surface integral of the outflux of total pressure in the limit of $\Delta \to 0$. Then some iterative procedure might converge to the true solenoidal state. With singular F_i, convergence is to be accepted in some weak sense. This is not damaging since we are calculating an "unstable" system through "stable" discrete algorithm at finite resolutions through "quasi-stable" approximation. The Dirac functions F_i represent impulsive forces which relax instantly to a smooth pressure field while generating a velocity potential ϕ to maintain or restore solenoidal velocity fields. [5] They are generated by the nonsolenoidal disturbances, either imposed externally as initial boundary data or generated internally by physical and numerical processes. The evolution of such "converged" flow field is thus "Determined" by the internal physical process of nonlinear transport and by the external influences as initial-boundary data, presuming minimal numerical errors.

The above "deterministic view" is in sharp contrast with the conventional "statistical view" that subscale modeling is necessary and that only statistical information are meaningful. We feel that the classical statistical theory of turbulence is deficient fundamentally.[6] According to the modern theory of nonlinear dynamics, chaos result from repeated solution bifurcations. The successive bifurcations create ever more complex topological structures within the confines of earlier bifurcated states. Thus, later bifurcations do not alter the global features of turbulent spots resulting from the first or first few bifurcations at large scales. With adequate resolution of such large scales a "converging" computational simulation of such global features should be possible with the proper nonlinear dynamic mechanism. Whatever numerical errors, organized or chaotic, should be confined to smaller scales comparable to mesh size and pertaining to structural details of the spot interior. Accordingly, a successful simulation of the experimentally observed global features of spot evolution tends to support the deterministic nature of fluid turbulence and the proposed dynamic mechanism of turbulence production.

In the following, we present our computational algorithm based on the vorticity transport equations:

$$\frac{\partial}{\partial t} \underline{\omega} = \nabla \times (\underline{u} \times \underline{\omega}) + \frac{1}{Re} \nabla^2 \underline{\omega} \tag{4}$$

with $\underline{\omega} = \nabla \times \underline{u}$, rather than the momentum equations. This is to bring out the dynamic mechanism of turbulence production in addition to passive vorticity transport. Vorticity is by definition solenoidal; but advanced vorticity $\underline{\omega}_r$ from (4), especially with approximate boundary formulation, will not remain solenoidal. It calls for potential correction ψ (equal and opposite impulsive force pairs F_i) to restore solenoidal vorticity $\underline{\omega}$.

From this solenoidal vorticity field $\underline{\omega}$, we can advance the velocity field from the Poisson equations.

$$\nabla^2 \underline{u} = - \nabla \times \nabla \times \underline{u} = - \nabla \times \underline{\omega} \tag{5}$$

The advanced velocity field from (5) $\underline{u}_r$, need not be solenoidal and a potential field ϕ (as a result of impulsive forces F_i) is needed to restore an advanced solenoidal velocity field. Thus, we have

$$\underset{\sim}{u} = \underset{\sim r}{u} + \nabla \phi$$

$$\underset{\sim}{\omega} = \underset{\sim r}{\omega} + \nabla \psi \qquad (6)$$

with $\quad \nabla^2 \phi = - \text{div } \underset{\sim r}{u}$

$$\nabla^2 \psi = - \text{div } \underset{\sim r}{\omega} \qquad (7)$$

The potential fields ϕ and ψ are to be solved from (7) under Neumann conditions on the boundary. In terms of the new solenoidal velocity, some quasi-linearization format (say arithmetic average) for iterative correction of the convective velocity in advancing $\underset{\sim}{\omega}$ from (4) is adopted. This iterative process repeats until some norm of the differences between successive iterants becomes less than a prescribed small quantity (say 10^{-5}). The advanced pressure field can then be evaluated from the latest available $\underset{\sim}{u}$ or $\underset{\sim}{\omega}$ from the Poisson equation (8):

$$\nabla^2 p = - \rho \frac{\partial u_j}{\partial x_j} \frac{\partial u_j}{\partial x_i} \qquad (8)$$

We discretize equation (5)-(8) with forward time and centered space differencing schemes, using the latest available values while sweeping over the field of computation in a specific order. The details of the differencing schemes are unimportant so long as they are consistent, linearly stable and without artificial smoothing and filtering. We computed with resolutions (15-10) x 15 x 15 on IBM 3081 at Princeton University and then with resolutions (30-45) x (15-25) x (15-25) on CRAY 1 at NCAR, Boulder, Colorado. We altered our program for IBM 3081, by using a different Poisson solver and other details to achieve 17 times speed advantage on CRAY 1. For the test case of a set of unit streamwise impulsive velocity introduced into a uniform shear flow near the stationary plate at flow Re = 3000, we compared the local and temporal results from both machines at comparable points within the developing turbulent spot. They agree to 10^{-4} with at least one significant figure. Our results are "insensitive" to the computational details on different machines. They are moreover, "reproducible" and "deterministic." We then computed on CRAY 1, at 45 x 25 x 25 the development of impulsive disturbances of different velocity and vorticity components with magnitudes 0.1 to 10, up to 50 - 100 Δt. Each time step Δt is 0.1 of the reference transit time.

Generally speaking, larger disturbances generate in 40-50 Δt local turbulent regions with computationally sharp boundary propagating into neighboring unstable laminar region. The r.m.s. vorticity drops precipitously (i.e. within 1-2 meshes) across the boundary. For some cases, the spots promptly assume the heart or kidney shape observed in physical experiments. Smaller disturbances generate similarly disturbed, almost chaotic local region initially, but without developing any sharp boundary. A dominant horse shoe vortex soon emerges, stretching and lifting into higher velocity region while diffusing and decaying globally. The flow fields apparently remain or will be laminar. For the smallest few disturbances, undisturbed uniform shear flow was restored at 40 ~ 50Δt. The last results suggest minimal effects of "numerical turbulence" on our computed results. The large scale, organized turbulent structures in transitional flow fields, despite our coarse resolution, are well simulated.

The computed pressure field over a transverse section across the developing turbulent spot at 10Δt is illustrated.
Sharp peaks and valleys of magnitudes 10^{-2} of the reference dynamic head are prominent. The pressure elevations over longitudinal sections both horizontal and vertical are similar. Such pressure elevations propagate with the local turbulent region. The steep pressure gradients

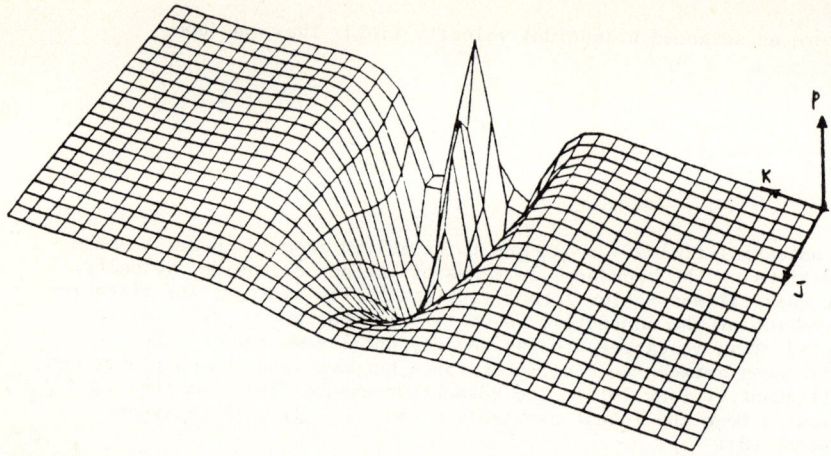

generate streamwise vortices (in the transverse plane) along its path of propagation, leading to a dominant "horse-shoe" vortex within the propagating spot, hugging onto the plate surface. This horseshoe vortex is distinct from the lifting, diffusing and decaying horse shoe vortex generated by those initial disturbances that failed to produce propagating local turbulent regions. The former is sustained by continuous generation while the latter is passively transported and dissipated.

The pressure distribution produced by different initial impulses are similar, differing largely in magnitudes. The particular pressure distribution generates the horseshoe vortices. Whether a propagating local turbulent region or a self-sustaining pressure disturbance and horseshoe vortex will result depends on the disturbance magnitude and location. Other structural details or secondary features reminiscent of experimental observations such as bursts, ejections etc are also noted. Thus, we suggest that:
1. "Deterministic" and "reproducible" computational simulation of transitional flows can provide global characteristics of practical interest as the solutions of initial boundary value problems of N.S. without hypothetical statistical closure or turbulence modeling.
2. Sufficiently large pressure disturbances associated with or generated by nonsolenoidal velocity disturbances is a crucial physical mechanism of laminar-turbulent transition.

This work is sponsored by NSF under NSM 8312094. Details of the computational procedures and sample results are contained in Ref.7

REFERENCES

1. Hinze, H.O. <u>Turbulence</u>, 2nd Edition., McGraw Hill, Inc. (1975).
2. Nichols, B.D. and Hirt, C.W. <u>Jour.Comp.Phys.</u> Vol 12,No.2.,(1973).
3. Leonard, A. NASA TM. 78579. (1979).
4. Cheng, S.I. "Computational Fluid Dynamics Convergent, or Asymptotic". Proc. of 11th IMACS World Congress. (1985). To be published by North Holland Pub.
5. Lamb, H. <u>Hydrodynamics.</u> 6th Edition, Dover Pub. (1932).
6. Cheng, S.I. "Fluid Turbulence, Deterministric or Statistical". Proc. of IUTAM Conference on "Shock Wave and Turbulent Boundary Layer Interaction". To be published by Springer-Verlag.
7. Roy.S. Ph.D. Thesis. 1986. Princeton Univ. Princeton, N.J. U.S.A.

COMPUTATION OF LOW MACH NUMBER FLOWS WITH BUOYANCY

Y.-H. Choi and C. L. Merkle
Department of Mechanical Engineering
The Pennsylvania State University
University Park, PA 16802/USA

INTRODUCTION

The difficulties of computing flows at low Mach numbers are well known. Low Mach number conditions can be encountered as local regions in high Mach number flowfields, or as flows that are low speed throughout. When the entire flow is low speed, the incompressible equations represent the preferable system, but when only a local region is low speed, or when heat addition is present, the compressible equations must be used. Such flows also can exhibit buoyancy effects. A primary example of problems of this type is combustion applications although our interest lies in laser-gasdynamic interactions[1]. The specific purpose of the present paper is to apply contemporary time-iterative algorithms to flowfields with arbitrarily low Mach numbers and buoyancy. As a test problem, we consider low speed flow with specified heat addition. Because the major difficulties with such problems stem from the inviscid terms, we treat the two-dimensional Euler equations.

The difficulties encountered by either implicit[2-4] or explicit[5,6] algorithms are somewhat different, but they arise from the stiff eigenvalues. One-dimensional experiments and multi-dimensional stability analyses[7] show implicit schemes are unaffected by Mach number unless approximate factorization is used. Approximate factorization is nearly as restrictive as explicit differencing. The stiffness can be removed by preconditioning[7-9] that rescales the acoustic speeds to lower values, but this is only effective to Mach numbers of about 0.01.

The presence of source terms is likewise detrimental to numerical algorithms. Source terms give rise to exponential growth that results in numerical instability, and buoyancy is particularly troublesome in this regard.

PROBLEM FORMULATION

The non-dimensional Euler equations in generalized coordinates are,

$$\frac{\partial Q}{\partial t} + \frac{\partial E}{\partial \xi} + \frac{\partial F}{\partial \eta} = H \tag{1}$$

where,

$$Q = J^{-1}[\rho,\ \rho u,\ \rho v,\ e]^T \tag{2a}$$

$$E = J^{-1}[\rho U,\ \rho U u + \xi_x p/\varepsilon,\ \rho U v + \xi_y p/\varepsilon,\ (e+(\gamma-1)p)U]^T \tag{2b}$$

$$H = J^{-1}[0, -\rho/F_{Rx}, -\rho/F_{Ry}, S-\rho(\Gamma_x u + \Gamma_y v)]^T \tag{2c}$$

with an analogous relation for F.

Here, the standard symbols have their normal meaning and the new symbols express the non-dimensional parameters,

$$\varepsilon = \gamma M_R^2; \quad F_R = \frac{u_R^2}{gL}; \quad S = \frac{(\gamma-1)Lq}{p_R u_R}; \quad \Gamma = \frac{(\gamma-1)\rho_R gL}{p_R} \tag{3}$$

where M is the Mach number, F_R is the Froude number, and the subscripts x and y refer to appropriate Cartesian components while R denotes a reference state.

To extend the computations to arbitrarily low Mach numbers, we apply a perturbation expansion in the small parameter, $\varepsilon = \gamma M_R^2$, as,

$$p = p_0 + \varepsilon p_1 + \ldots \tag{4}$$

with similar expansions for ρ, u and v. To order $1/\varepsilon$ the system reduces to grad $p_0 = 0$. For steady solutions, this implies p_0 is a constant.

To order ε^0, the system becomes,

$$Q_0 = J^{-1}[\rho, \rho u, \rho v, p_1/\beta]^T \tag{5a}$$

$$E_0 = J^{-1}[\rho U, \rho U u + \xi_x p_1, \rho U v + \xi_y p_1, \gamma p_0 U]^T \tag{5b}$$

$$H_0 = J^{-1}[0, -\rho F_{Rx}, -\rho F_{Ry}, S-\rho(\Gamma_x u + \Gamma_y v)]^T \tag{5c}$$

with F_0 analogous to E_0. (We have dropped the subscript o on all quantities except p_0.) The time derivative in the energy equation is an artificial term that has been added by analogy with Chorin's[10] artificial compressibility procedure. The perturbation expansion removes the rapidly propagating acoustic waves. The arbitrary time derivative replaces them by pseudo-acoustic waves that travel at speeds given by,

$$c' = [U^2 + 4\beta(\xi_x^2 + \xi_y^2) \gamma p_0/\rho]^{1/2} \tag{6}$$

Choosing $\beta = 1/4\gamma$ ensures c' is of the order of U so the equations are no longer stiff. To solve rescaled equations, we use implicit, approximately-factored methods combined with central differencing in space. A small amount of explicit fourth order viscosity[4] was added only for the strongest heat addition cases.

The eigenvalues for the perturbed system are,

$$\lambda_\xi = [U, U, (U \pm c')/2] \tag{7}$$

Calculations with this system converge at the same rate for all Mach numbers attempted (down to 10^{-6}) so long as gravity is omitted. With gravity present, rapid convergence is observed when the Froude number is greater than unity (Mach numbers of 10^{-3} for the present problem), but for Froude numbers below unity, the system becomes unstable and divergence results. This has been shown theoretically by stability analyses and experimentally by computations.

Representative stability calculations in the presence of buoyancy are shown in Fig. 1 for upward flow in a constant area duct at a Froude number of 0.4. Only the largest of the four eigenvalues of the amplification matrix is shown. The eigenvalues exceed unity near the maximum and minimum wavenumbers in ξ. Even calculations for uniform flow (without heating) diverged for these conditions. In the absence of buoyancy, all eigenvaues are less than unity and the solution converges rapidly[11].

CONTROL OF THE SOURCE TERM INSTABILITY

One physical reason for low Froude number instability is because the pressure gradient induced by buoyancy exceeds that due to convection. To include buoyant effects in our perturbation expansion, we consider the one-dimensional momentum equation. For low Froude numbers, a more proper scaling is,

$$\frac{\partial}{\partial t}\rho u + \frac{\partial}{\partial x}\rho u^2 + \frac{1}{\delta}\left(\frac{1}{\varepsilon'}\frac{\partial p}{\partial x} + \rho\right) = 0 \tag{8}$$

where both δ and ε' represent small parameters. Here, δ is the Froude number, and $\varepsilon' = \rho_R g L / p_R = \varepsilon/\delta$. This suggests the double expansion,

$$p = p_0 + \varepsilon'[p_{10} + \delta p_{11} + \ldots] + \ldots \tag{9}$$

Substituting this into the conservation equations again indicates p_0 is constant, but now the pressure gradient (p_1) in Eqn. 5 contains two parts. One part, p_{10}, is the hydrostatic part due to buoyancy, while the other part, p_{11}, balances convective effects.

There are two ways to incorporate this scaling. An approximate way is to rescale the problem by the new reference velocity, $u_R = \sqrt{gL}$, instead of by the convective velocity. If we express this new reference velocity as $u_R = K u_\infty$ (where u_∞ is the previous reference velocity), we need only replace F_R in Eqn. 5 by F_R/K^2. The more complete way is to incorporate the two parameter expansion directly into the equations to obtain the (5x5) system:

$$Q'_0 = J^{-1}[\rho, \rho u, \rho v, p_{11}/\beta_1, p_{10}/\beta_0]^T \tag{10a}$$

$$E'_0 = J^{-1}[\rho U, \rho U u + \xi_x p_{11}, \rho U v + \xi_y(p_{10}+p_{11}), \gamma p_0 U, -\xi_x p_{10}]^T \tag{10b}$$

$$F'_0 = J^{-1}[\rho V, \rho V u + \eta_x(p_{10}+p_{11}), \rho V v + \eta_y(p_{10}+p_{11}), \gamma p_0 V, p]^T \tag{10c}$$

$$H'_0 = J^{-1}[0, 0, -\rho F_{Ry}, S-\rho(\Gamma_x u + \Gamma_y v), \rho F_{Rx}]^T \tag{10d}$$

Thus far, only results with the approximate procedure have been obtained. Figure 2 shows stability results for the $F_R=0.4$ case with the reference velocity increased by a factor of 5 (K=5). This artifice removes the instability at this Froude number (compare Fig. 1) and provides a stable algorithm.

Convergence rates for heat addition in a constant area duct (using the heating distribution given in Ref. 11) are given on Fig. 3 for K=1 and K=3. The divergence encountered with K=1 is replaced by relatively good convergence at K=3. The converged velocity and temperature contours for this case are shown on Fig. 4

This approximate procedure allows us to extend computations about an order of magnitude lower in Froude (Mach) number. Calculations at lower Mach numbers will require the more complete procedure. The difficulty with the K-scaling is manifest by the eigenvalues of the system. They are identical to Eqn. 7, except that,

$$c' = \sqrt{U^2 + K^2(\xi_x^2 + \xi_y^2)p_0/\rho} \tag{11}$$

As the Froude (Mach) number is reduced, we require larger values of K to ensure stability. This increases c' and makes the system stiff again. This is the original difficulty we sought to avoid. In this regard, it is interesting to note that the complete Euler equations remain stable in the presence of buoyancy because the stiffness in the original equations suppresses the effects of the buoyant source as it does in the present K-modified equations. This should be avoided by the scheme detailed in Eqn. 10.

REFERENCES

1. Merkle, C. L., AIAA Journal, Vol. 22, No. 8, August 1984, pp. 1101-1107.
2. Briley, W. R. and McDonald, H., J. of Comp. Physics, Vol. 34, 1980, pp. 54-73.
3. Warming, R. F. and Beam, R. M., SIAM-AMS Proceeding, Vol. 11, 1978, pp. 85-129.
4. Pulliam, T. H. and Barton, J. T., AIAA Paper 85-0018, AIAA 23rd Aerospace Sciences Meeting, Reno, NV, 1985.
5. MacCormack, R. W., AIAA Paper 69-354, 1969.
6. Jameson, A., AIAA Paper 81-1259, AIAA 14th Fluid and Plasma Dynamics Conference, June 23-25, 1981, Palo Alto, CA.
7. Merkle, C. L. and Choi, D., AIAA Journal, Vol. 23, No. 10, Oct. 1985, pp. 1518-1524.
8. Turkel, E., NASA Contractor Report 172543, February 1985.
9. Briley, W. R., McDonald, H., and Shamroth, S. J., AIAA Journal, Vol. 21, No. 4, Oct. 1983, pp. 1467-1469.
10. Chorin, A. J., J. of Comp. Physics, Vol. 2, 1967, pp. 12-26.
11. Merkle, C. L. and Choi, Y.-H., AIAA Paper 86-0351, AIAA 24th Aerospace Sciences Meeting, Jan. 6-9, 1986, Reno, NV.

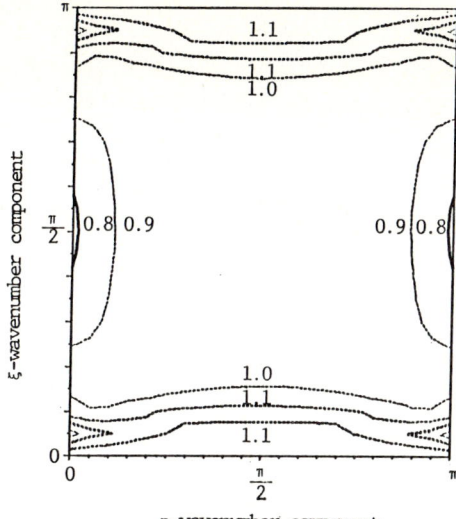

Fig. 1 Contour plot of maximum eigenvalues from vector stability analysis for low Mach number equations with buoyancy, with approximate factorization: $M_x=10^{-3}$, $M_y=0$, CFL=5, $F_r=0.4$, K=1.

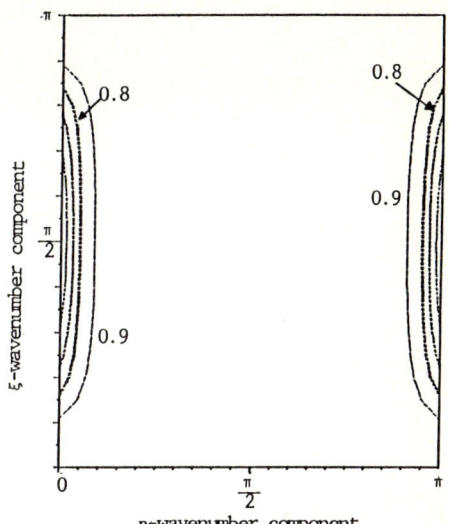

Fig. 2 Contour plot of maximum eigenvalues from vector stability analysis of low Mach number equations with buoyancy, with approximate factorization: $M_x=1\times10^{-3}$, $M_y=0$, CFL=5, $F_r=0.4$, K=5.

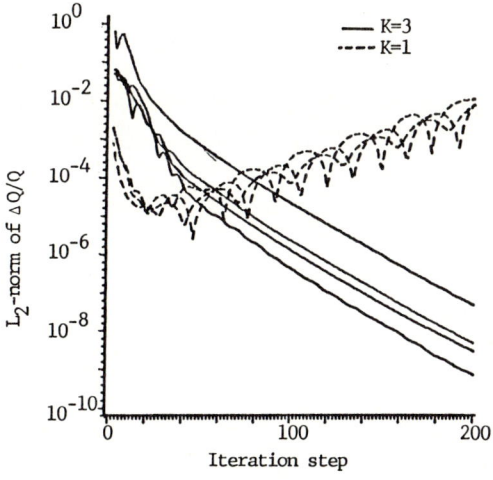

Fig. 3 Rate of convergence of calculations at Mach number 1.0×10^{-3} in terms of L_2 norms; $F_r=0.4$.

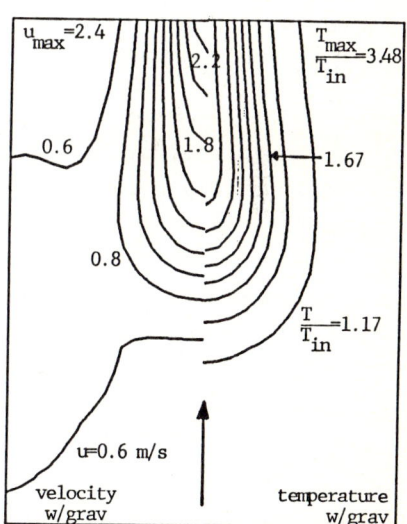

Fig. 4 Temperature and velocity contours for an inlet Mach number of 5×10^{-4}; $p=5\times10^6$ w/m^3, $F_r=0.1$, K=30.

VORTEX MULTIPOLE METHODS FOR
VISCOUS INCOMPRESSIBLE FLOWS

S. Choudhury and R. A. Nicolaides
Department of Mathematics
Carnegie Mellon University
Pittsburgh, PA 15213, USA

1. INTRODUCTION

Recently there has been a resurgence of interest in point vortex methods stemming, at least partly, from the theoretical work done in [1],[3],[4],[9],[12],[13],[15] and elsewhere. These references deal with inviscid flows in two or three space dimensions with boundaries at infinity. The procedure used is to make a point vortex approximation on a uniform mesh to a given smooth initial vorticity distribution. The evolution of the point vortices is then determined by the Euler equations. Induced velocities are computed using a smoothed Green's function in order to avoid the singularity at the inducing particle position. Error estimates in terms of the uniform mesh spacing h and the smoothing radius ρ have the form

$$O(\rho^k) + O(h^m/\rho^{m-1}) \qquad (1.1)$$

where k can be made arbitrarily large in principle, and m is the number of derivatives of the initial vorticity distribution. The unexpectedly highly accurate second term results from the use of a uniform mesh. Use of a nonuniform mesh, necessary for example when there are bodies in the flow field, reduces m to at most 2. In Sections 2 and 3 we introduce a method which achieves high order on general meshes, along with a technique for incorporating viscous effects; Section 4 contains numerical results.

2. INVISCID ALGORITHMS

We will define the algorithms for the two dimensional Navier-Stokes equations expressed in vorticity/velocity form. The three dimensional case is mentioned later. The equations are the usual ones, namely the two dimensional vorticity transport equation

$$D\omega/Dt = \nu\Delta\omega \qquad (2.1)$$

together with the velocity equations

$$\text{div } u = 0 \qquad (2.2)$$
$$\text{curl } u = \omega \qquad (2.3)$$

and the initial and boundary conditions

$$\omega(x,0) = \omega_0(x) \qquad x \in \mathbb{R}^2 \qquad (2.4)$$

$$u \to u_\infty \qquad |x| \to \infty. \qquad (2.5)$$

As usual, D/Dt is the material derivative operator, $u = (u_1, u_2)$ is the velocity field, and in (2.3) curl denotes the operator $(\partial/\partial x, -\partial/\partial y)$. (2.4) is associated with (2.1) and (2.5) with (2.2)-(2.3). Here and below it is assumed that ω_0 is zero outside of some bounded domain. We will first consider the inviscid case of (2.1) with $\nu = 0$. A uniform mesh with spacing h in both the x and y directions with the coordinate axes as mesh lines will be used to define the algorithm. The modifications for a nonuniform mesh will be self evident. Let B_{ij} denote the mesh cell whose upper right corner is at (ih, jh), and let ω_{ij} denote the corresponding value of ω_0. The basic point vortex method approximates $\omega_0(x)$ of (2.4) by the point vortex array

$$\omega_0^h = \Sigma_{ij} h^2 \omega_{ij} \delta(x - x_{ij}) \qquad (2.6)$$

and then, using the fact that the vortices are convected by the flow, approximates $\omega(x,t)$ by

$$\omega^h = \Sigma_{ij} h^2 \omega_{ij} \delta(x - X_{ij}) \qquad (2.7)$$

where X_{ij} is obtained by solving the trajectory equation

$$dX/dt = u^h(X,t) \qquad (2.8)$$
$$X(0) = x_{ij}.$$

In (2.8) u^h denotes an approximation to the velocity field u. This approximation is computed using an approximate Green's function G_ρ for (2.2)-(2.3). G_ρ is defined as the convolution

$$G_\rho(x) = \iint G(x-z) f_\rho(z) dz \qquad (2.9)$$

where

$$G(x) = 1/(2\pi |x|^2)(-x_2, x_1) \qquad (2.10)$$

is the exact Green's function, and $f_\rho(x) = (1/\rho^2) f(x/\rho)$ where the smoothing kernel f satisfies $\iint f = 1$. One reason for the use of the smoothing kernel is to avoid the arbitrarily large values which can occur in (2.10) when $|x| \to 0$. We will not further consider the choice of f in this paper since it is already well documented in the previous references. However it is helpful to point out another interpretation of the smoothing operation, which follows from the definition of $u^h(x,t)$ as

$$u^h(x,t) = \iint G_\rho(x-y) \omega^h(y,t) dy + u_\infty. \qquad (2.11)$$

Substituting (2.9) and rearranging shows that the numerical velocity field u^h is computed as the exact velocity field, not of the point vortex array, but of an array of smoothed vortices having the form $h^2 \omega_{ij} f_\rho$. This is the origin of the name "vortex-blob" method due to [6].

To motivate our generalization of the above algorithm, we write the induced velocity using (2.11), as

$$u^h(x,t) = \Sigma_{ij} G_\rho(x-X_{ij}) h^2 \omega_{ij} + u_\infty. \qquad (2.12)$$

At $t=0$ (2.12) is a numerical integration of $\iint G_\rho(x-y)\omega(y,t)dy$. By the incompressibility of the flow this property is preserved at all later times. By using more information about the initial distribution ω_0 in (2.6) we can expect to obtain greater accuracy from this numerical integration. Thus, we generalize (2.6) to

$$\omega_0^h = \Sigma_{ij} a_{ij} \delta(x-x_{ij}) + b_{ij} \delta_x(x-x_{ij}) + c_{ij}\delta_y(x-x_{ij}) \qquad (2.13)$$

where the new subscripts denote partial differentiations and the coefficients are to be determined. In [14] several different methods for choosing these are examined. Here we consider one of these which is such that when (2.13) is substituted into (2.11) a composite integration formula results which is exact on each B_{ij} for cubic polynomials. As an integration rule, this will be 4th order accurate in contrast to the simpler midpoint rule used in (2.6), which in this context is only 2nd order accurate.

The integration formula, exact for cubic polynomials, on which the composite rule is based is the following: for a box centered at the origin of side lengths h_x and h_y with vertices P,Q,R,S listed anticlockwise from the upper right corner:

$$\iint g\,dx\,dy \cong (h_x h_y/4)(g_P + g_Q + g_R + g_S) + (h_x^2 h_y/24)(-g_{xP} + g_{xQ} + g_{xR} - g_{xS})$$
$$+ (h_x h_y^2/24)(-g_{yP} - g_{yQ} + g_{yR} + g_{yS}).$$

The weights a_{ij}, b_{ij}, c_{ij} in (2.13) can be easily computed from this for any given mesh.

To complete the description of the algorithm note that vortex configurations of the form (2.13) are not simply convected in the flow unaltered in form as was the case for (2.6). In fact, the solution of (2.1) with the initial condition (2.13) turns out to be

$$\omega^h(x,t) = \Sigma_{ij} a_{ij}\delta(x-X_{ij}) + b'_{ij}\delta_x(x-X_{ij}) + c'_{ij}\delta_y(x-X_{ij})$$

where, as previously, X_{ij} is found from the trajectory equations (2.8). b'_{ij} and c'_{ij} satisfy $(b'_{ij},c'_{ij})^T = M(b_{ij},c_{ij})^T$ where M is the 2×2 Jacobian matrix of the flow map which may be found by solving the ode system

$$dM/dt = \nabla u M \qquad M(0) = 1. \qquad (2.14)$$

Essentially, the inviscid algorithm reduces to the integration of (2.8) and (2.14), with computation of the velocity field and its derivatives from (2.11).

Details of the analysis leading to (2.14) are given in [7] and [14]. In [7] a 3 dimensional version of this algorithm is defined and rigorously analyzed. It is proved that under smoothness conditions, the error in both the 2 and 3 dimensional cases is

$$O(\rho^k) + O(h^3/\rho^2).$$

Observe that (2.14) required computation of the gradient of the velocity vector. The extra work required here is the price paid for the increase in order of accuracy.

3. VISCOUS ALGORITHMS

Let $H(x,t)$ denote the diffusion kernel

$$H(x,t) = \exp(-|x|^2/4\nu t)/4\pi\nu t$$

which is the solution of the two dimensional diffusion equation with initial condition $\delta(x)$. Given an array of diffusion sources and dipoles at $t = 0$, say

$$\Sigma_{ij} \, a_{ij}\delta(x-x_{ij}) + b_{ij}\delta_x(x-x_{ij}) + c_{ij}\delta_y(x-x_{ij}) \qquad (3.1)$$

the solution at time $t = t_0$ at x_{kl} is given by integration as

$$\Sigma_{ij} \, a_{ij}H(x_{kl}-x_{ij}, t_0) - b_{ij}H_x(x_{kl}-x_{ij}, t_0) - c_{ij}H_y(x_{kl}-x_{ij}, t_0) \qquad (3.2)$$

By differentiation of (3.1) and similar integrations, expressions for the derivatives of the solutions at x_{kl} can be obtained.

The approach to solving the viscous problem is via splitting of the convection and viscous steps over a time interval δt. For the convection step, we use the method of section 2. For the diffusion step, we simply modify the coefficients a_{ij} according to (3.2) with $t_0 = \delta t$ and b_{ij}, c_{ij} using the corresponding derivative formulas. The low order case of this algorithm, with $b_{ij} = c_{ij} = 0$, was implemented in [8]. An analysis for the linear problem is given in [10]. In [2] it is proved that the splitting error for the Navier-Stokes equations is $O(\nu\delta t)$, so that the splitting algorithm becomes more effective as $\nu \to 0$.

The major part of the computer time in vortex methods is spent evaluating the inter vortex distances for the two kernels G_ρ and H. Very great savings in these computations may be made in the future using a technique in [11].

4. COMPUTATIONS

Some preliminary computations to verify the theoretical results have been carried out. One series of runs was used to check that the gain in accuracy from the higher order scheme was worthwhile allowing for the extra work that has to be done. Using a kernel for which $k = 4$, and a known radially symmetric C^6 function as the exact solution, Table 1 gives some sample results. The upper figure in each box represents a time averaged maximum spatial error in velocity over the vortex positions, and the lower figure is the relative run time. N denotes the number of point vortices used to approximate the initial vorticity distribution. Roughly speaking, the accuracy of the higher order scheme with a given number of vortices is seen to be comparable with the accuracy of the low order scheme using 4 times the number of vortices. However, the cost of the high order scheme is only 0.2-0.25 of the low order scheme. With an increase in the number of vortices this advantage would become more pronounced.

Some viscous computations are shown in Figures 2-4. The same kernel with $k = 4$ was used. The kinematic viscosity $\nu = 1/1000$. The figures show the vorticity contours at the shown times for 1) a pair of corotating point vortices of unit strength initially at (.14,0) and (-.14,0), 2) a pair of counterrotating point vortices of unit strength at the same initial positions and 3) three corotating point vortices of unit strength initially equally spaced on a circle of radius .16 with one at (0,.16). The contours are at intervals of $\omega = 2$. In the corotating cases, the merging of the vortices is already apparent by $t = 10$. In all cases, 4th order Runge-Kutta was used as the ODE solver, and 121 (first order) vortices were used.

REFERENCES

[1] C. ANDERSON and C. GREENGARD, "On vortex methods," SIAM J. Num. An., 22 (1985) pp. 413-440.
[2] J. T. BEALE and A. J. MAJDA, "Rates of convergence for viscous splitting of the Navier-Stokes Equations," Math. Comp., 37 (1981) pp. 243-259.
[3] J. T. BEALE and A. J. MAJDA, "Vortex methods 1: Convergence in three dimensions," Math. Comp., 39 (1982) pp. 1-27.
[4] J. T. BEALE and A. J. MAJDA, "Vortex methods 2: higher order accuracy in two and three dimensions," Math. Comp., 39 (1982) pp. 29-52.
[5] J. T. BEALE and A. J. MAJDA, "Higher order accurate vortex methods with explicit velocity kernels," Jnl. Comp. Phys. 58 (1985) pp. 188-208.
[6] A. J. CHORIN, "Numerical study of slightly viscous flow," J. Fluid Mech., 57 (1973) pp. 785-796.
[7] C. CHIU and R. A. NICOLAIDES, "Convergence of a vortex method of higher order for the two and three dimensional Euler equation," submitted to Math. Comp.
[8] J. P. CHOQUIN and S. HUBERSON, "Particles simulation of viscous flow," to appear.

[9]   G. H. COTTET, "Methodes particulaires pour l'equation d'Euler dans le plan," These de 3e cycle, Univ. P. et M. Curie, Paris (1982).
[10] G. H. COTTET and S. GALLIC, "A particle method to solve transport-diffusion equations," Report 115, Centre de Math. Appl., Ecole Polytechnique (1985).
[11] L. GREENGARD and V. ROKHLIN, "A fast algorithm for particle simulations," Yale University research report YALEU/DCS/RR-459 (1986).
[12] O. HALD and V. M. DELPRETE, "Convergence of vortex methods for solving Euler's equations," Math. Comp., 32 (1978) pp. 791-809.
[13] O. HALD, "Convergence of vortex methods II," SIAM J. Num. An., 16 (1979) pp. 726-755.
[14] R. A. NICOLAIDES, "Construction of higher order accurate vortex and particle methods," to appear in Applied Num. Math.
[15] P. A. RAVIART, "An analysis of particle methods," CIME course, Numerical Methods in fluid dynamics, Como (1983).

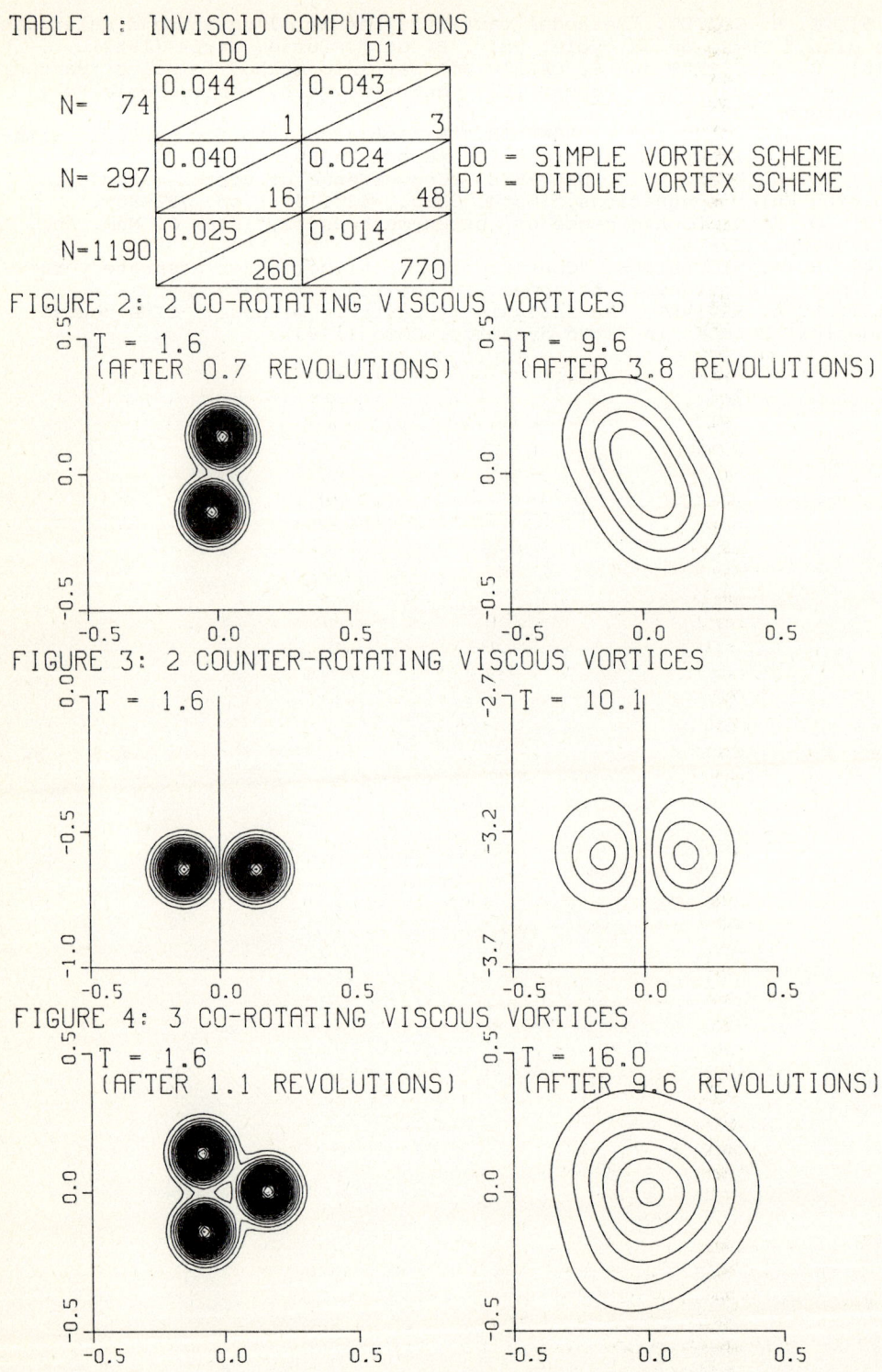

TABLE 1: INVISCID COMPUTATIONS

	D0	D1
N= 74	0.044 / 1	0.043 / 3
N= 297	0.040 / 16	0.024 / 48
N=1190	0.025 / 260	0.014 / 770

D0 = SIMPLE VORTEX SCHEME
D1 = DIPOLE VORTEX SCHEME

FIGURE 2: 2 CO-ROTATING VISCOUS VORTICES
T = 1.6 (AFTER 0.7 REVOLUTIONS)
T = 9.6 (AFTER 3.8 REVOLUTIONS)

FIGURE 3: 2 COUNTER-ROTATING VISCOUS VORTICES
T = 1.6
T = 10.1

FIGURE 4: 3 CO-ROTATING VISCOUS VORTICES
T = 1.6 (AFTER 1.1 REVOLUTIONS)
T = 16.0 (AFTER 9.6 REVOLUTIONS)

FREE-LAGRANGIAN HYDRODYNAMICS USING MASSLESS TRACER POINTS

Robert A. Clark
Computational Physics Group, X-7
Los Alamos National Laboratory
Los Alamos, New Mexico, USA

I. Introduction

The partial differential equations (PDEs) describing the time evolution of compressible fluid flow in two and three dimensions are usually solved numerically in the Eulerian frame of reference as opposed to the Lagrangian frame. This is especially true of flows involving large distortions. Lagrangian codes based upon Lagrangian cells do not run very long when the cells become distorted. In order to avoid the problem of cell distortion we have developed a Lagrangian code named "HOBO" based upon massless tracer points. These tracer points can be thought of as being embedded in, and moving with the fluid. The PDEs which we are trying to solve at each of the points are equations 1, 2, 3, and 4 which represent respectively conservation of mass, momentum and energy, and an equation of state.

$$\frac{D\rho}{Dt} = - \rho \vec{\nabla} \cdot \vec{U} \qquad (1) \qquad\qquad \frac{D\vec{U}}{Dt} = - \frac{1}{\rho} \vec{\nabla} P \qquad (2)$$

$$\frac{De}{Dt} = - \frac{P}{\rho} \vec{\nabla} \cdot \vec{U} \qquad (3) \qquad\qquad P = P(\rho, e) \qquad (4)$$

The basic method is as follows: to update the kth point we first select a set of representative neighbors near the point k. We then make a finite difference approximation to the spacial derivatives $\vec{\nabla} P$ and $\vec{\nabla} \cdot \vec{U}$. Then, from equations 1, 2, and 3 we have the time rate of change of the density, velocity, and internal energy at the point k which allows us to calculate the changes in the state variables from time t to time $t + \delta t$. We then move the point a distance $\vec{U} \delta t$ and we have completed a cycle. The code is classified as "Free-Lagrangian" because there is no fixed or permanent connectivity between points.

The neighbors used to approximate $\vec{\nabla} P$ and $\vec{\nabla} \cdot \vec{U}$ at point k can change from cycle to cycle. The changing of neighbors does not require any sort of rezoning since the variables being calculated are associated only with the points and not with any sort of Lagrangian cell.

The details of how the code selects representative neighbors and how it approximates $\vec{\nabla} P$ and $\vec{\nabla} \cdot \vec{U}$ can be found in Reference [1]. In this paper we will concentrate on two recent improvements to the method and present an example calculation which makes use of these improvements.

II. The Independent Time Step

The possibility of using an independent time step for each point in the problem was suggested by Eltgroth in Reference [2]. The standard technique would be to run all points in the problem at the same δt with this being the minimum (δt_{min}) of all the points. Suppose a point k could be advanced at 10 δt_{min} with accuracy and stability maintained. One thing we could do would be to advance point k from t to t+10 δt_{min} and then, if values of the variables at k for times between t and t+10 δt_{min} were needed for the calculation of another point, we could interpolate in time.

There is an easier way to accomplish the same thing. Every point in the calculation is advanced every cycle at the time step δt_{min}, but the time derivatives $\dot{\rho}_k$, $\dot{\vec{U}}_k$ and $\dot{e}_k$ at the point k are recalculated based on a local time step δt_k. To put it another way, if local stability and accuracy allow the point k to advance 10 δt_{min} we freeze the values of $\dot{\rho}_k$, $\dot{\vec{U}}_k$ and $\dot{e}_k$ for 10 cycles. The advantage to this is that while it takes on the order of a hundred floating point operations to calculate $\vec{\nabla} P$ and $\vec{\nabla} \cdot \vec{U}$, it takes only 2 floating point operations to evaluate $\rho_k(t+\delta t_{min}) = \rho_k(t) + \delta t \cdot \dot{\rho}_k$.

A further consideration is that the local time step δt_k may change as information is propogated between points. The obvious example is a shock propogating toward the kth point which the kth point has no knowledge of when its local δt_k is calculated and its time derivatives frozen. To solve this problem we allow active points to reduce the local δt of its neighbors, be they active or inactive (frozen). Thus, as a shock is propogated the inactive points in front of the shock are turned on by the active points approaching them. In the example calculation in Section IV the average ratio of inactive to active points throughout the entire calculation is almost 15 to 1 resulting in a more than 10 to 1 speedup in the calculation.

III. The Sliding Surface Treatment

In certain calculations it is desirable to have one material essentially slide over another material; in effect, simulating a boundary layer which is much thinner than the point separation. In the problem in Section IV, we are propelling a plate out of a cylindrical barrel and we use the sliding surface treatment between the plate and the cylinder. If materials A and B are declared to be a sliding surface, then whenever A has one or more neighbors of type B (or the other way around), we separate the calculation of $\vec{\nabla} P$ and $\vec{\nabla} \cdot \vec{U}$, into a tangential and normal component. One of the materials is designated as the controlling material. Each point of the controlling material has a unit vector pointing normal to the surface associated with it. This vector is rotated as the surface moves. The rotation is determined by the movement of the neighbors of the point. To calculate the tangential components, we draw an imaginary line through the point k and contruct a local reflective boundary. Velocity components in the normal direction are set to zero. This guarantees that $\vec{\nabla}P$ will be in the tangential direction and $\vec{\nabla}\cdot\vec{U}$ will be due only to flow in the tangential direction. Then we do a second calculation in the normal direction. We use the actual points and the actual pressures and velocities except that the tangential velociites are set to zero. Also, we keep only the normal component of $\vec{\nabla}P$. The results of the tangential and normal calculations of $\vec{\nabla}P$ and $\vec{\nabla}\cdot\vec{U}$ are summed.

IV. Example Calculation

The following problem has been proposed and studied extensively by McCall [3] in one-dimension. We are using HOBO to study the two-dimensional effects. The intent is to accelerate a solid metal plate to a hypervelocity, on the order of 20 km/sec, while keeping the plate intact in order to experimentally study impact physics at these velocities. The setup of the experiment is shown in Figure 1.

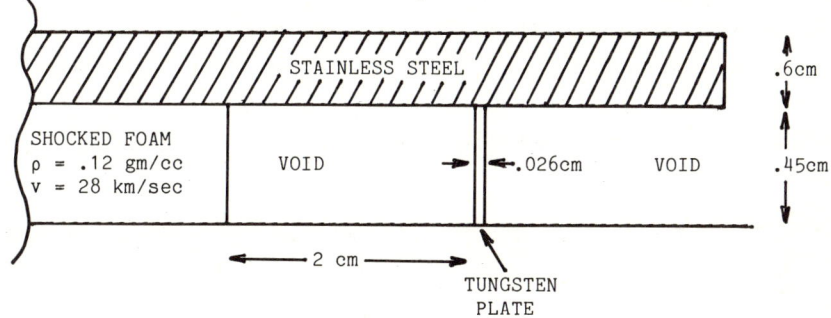

FIGURE 1

A strong shock is driven through the CH foam. The 28 km/sec particle velocity driving the shock is produced by an axial shock tube driven by PBX 9501 high explosive. A detailed description is given in Reference [3]. When the foam is shocked a free expansion moves through the void region and the high-velocity, low-density foam strikes the tungsten (Tn) plate. As the pressure builds behind the plate, a shockless acceleration is produced and the Tn plate exits the cylinder with a kinetic energy much higher than its vaporization energy.

The one-dimensional calculation of McCall could not examine the Raleigh-Taylor instability at the foam-Tungsten interface or the effect of the expansion of the cylinder as the pressure builds up behind the plate. We used the HOBO code to evaluate three test cases. In each case, we used the exact analytic solution to the shock and free expansion of the foam to the point where the foam initially contacts the plate as our initial condition.

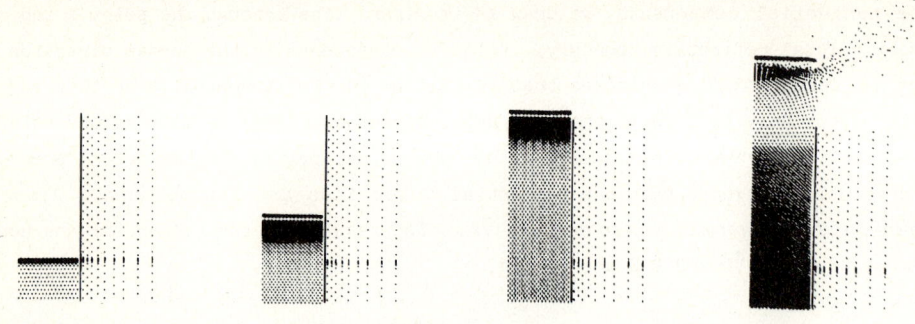

FIGURE 2

In the first problem we gave the cylinder an artificially high density so that it would not move. This was to test whether under ideal conditions, using the sliding surface treatment between Tn and the steel wall and between the foam and the steel wall, we could accelerate the plate out of the cylinder. As can be seen in Figure 2 the plate is intact after it exits the cylinder. The plate velocity of 19 km/sec agrees with the one-dimensional calculation of McCall. The internal energy of the plate is less than .01% of the kinetic energy.

In the second problem, see Figure 3, we eliminated the artificially high density in the cylinder and made it stainless steel. The cylinder begins to bulge, perturbations in the plate are initiated and the plate breaks up completely; well before it exits the cylinder.

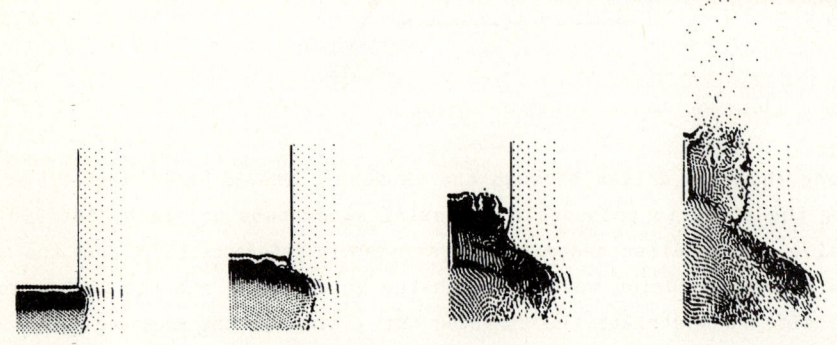

FIGURE 3

In the third problem, see Figure 4, we tested the idea of reducing the plate instability by placing a beryllium (Be) cushion behind the plate. The Be density of 1.845 gm/cc is less than 1/10 that of Tn, 19.237 gm/cc. The Be has the same thickness as the Tn so the total mass has changed less than 10% while the density ratio at the foam interface has been reduced by a factor of 10. In addition, the

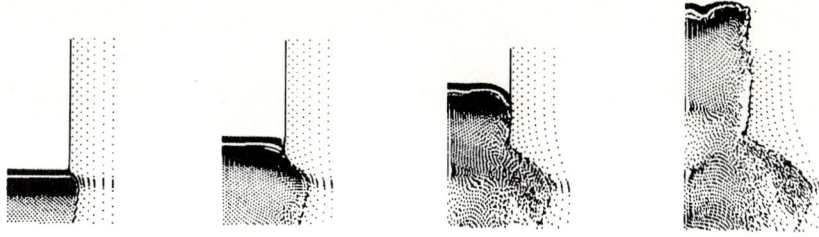

FIGURE 4

sound speed in Be is 8.1 km/sec as opposed to 3.5 km/sec in Tn. The results were remarkable. Although the plate is not perfectly plane as it exits the cylinder, it still looks essentially like a plate.

V. Summary

The free-Lagrangian Hydrodynamics using massless tracer points used in the HOBO code, is able to calculate highly distorted fluid-flow. Because it is Lagrangian, we are able to concentrate the calculational points at the region of interest. The method of independent time steps greatly increase the efficiency of the code. In problem 1, we used 42,187,201 point cycles, i.e., we advanced that many points that many cycles. However, there were only 2,877,707 active point cycles which involved updating $\vec{\nabla}P$ and $\vec{\nabla}\cdot\vec{U}$. The "grind time" was 85 μsec per point cycle on a CRAY1 computer; slightly over half of which was used in testing for neighbors. When the independent time step is operational for testing neighbors, we anticipate a grind time of 40 μsec per point cycle.

The sliding surface treatment in HOBO allows us to simulate the effects of a thin boundary layer without having to resolve it. We would not have had much confidence in the calculations of problems 2 and 3 if we could not have calculated problem 1.

REFERENCES

1. Clark, R. A. (1986) "Compressible Lagrangian Hydrodynamics Without Lagrangian Cells," Numerical Methods for Fluid Dynamics II (The Institute of Mathematics and its Applications Confeerence Series. New Series: No. 7), 255-271, Oxford University Press, New York.

2. Eltgroth, P. E. (1983), "The Independent Time Step Method for Hydrodynamics," Report UCRL-89853, Lawrence Livermore National Laboratory, Livermore, California.

3. McCall, G. H. (1984), "A Method for Producing Shockless Acceleration of Masses to Hypervelocities Using High Explosives," LJI-TM-84-106, La Jolla Institute, La Jolla California.

IMPACT OF TURBULENCE MODELING ON NUMERICAL ACCURACY AND EFFICIENCY OF COMPRESSIBLE FLOW SIMULATIONS

Thomas J. Coakley
NASA Ames Research Center
Moffett Field, CA 94035

The basic objective of this work is to compare and evaluate the performance of various turbulence models which are used in the numerical simulation of complex turbulent flows. The approach utilizes the Reynolds-averaged compressible Navier-Stokes equations in which the Reynolds stresses and heat fluxes are mathematically modeled by suitable turbulence models. This paper focuses on the simulation of transonic flows about a 2-D airfoil and the estimation of performance characteristics such as lift and drag. Attention is given to separated flows where differences between model predictions are generally greater than those observed for unseparated flows.

The turbulence models used in the present study are eddy viscosity models which include the family of zero-, one-, and two-equation models. Zero- and one-equation models are numerically the simplest of the eddy viscosity models but lack the generality of the two-equation models because their length scales must be determined algebraically rather than from a field equation. However, two-equation models are more complicated than zero- and one-equation models, and these complications can sometimes lead to numerical difficulties.

Six turbulence models are studied in this paper. They are the zero-equation models of Cebeci and Smith[1], Baldwin and Lomax[2], and Johnson and King[3], and the two-equation k-ε and q-ω models respectively of Chien[4] and the present author, Refs. 5 and 6. These models are listed in Table I. All of these models utilize the procedure of integration-to-the-wall in which no-slip boundary conditions are applied at solid surfaces. For the zero-equation models, a common formulation is used in the airfoil wake region which is due to Cebeci[7]. The location of transition in the computations was taken from the experimental boundary-layer trip locations except in the cases of the two-equation q-ω model predictions where the transition locations were allowed to occur naturally.

The numerical differencing method consists of a second-order implicit upwind differencing algorithm combined with a finite volume discretization technique which produce accurate resolution of shock waves and discontinuities[8]. Additional features of the method include the use of spatially varying time steps to speed convergence to a steady state and an implicit treatment of boundary conditions. Inviscid boundary conditions at far field boundaries are based on the method of characteristics and account for circulation due to lift.

The airfoil investigated in the present study was the RAE 2822 airfoil which has extensive experimental documentation[9]. Three cases were investigated, Cases 1, 9, and 10, which correspond respectively to unseparated subcritical flow, unseparated supercritical flow and separated supercritical flow. The experimental conditions of Mach number, Reynolds number and geometric angle of attack for each of the three cases are shown in Table I. Since the calculations were done in free air, the angle of attack must be changed from the experimental or geometric angle of attack to account for wind tunnel wall effects. The angles of attack used in the calculations are shown in Table I and were obtained from the recommendations in Ref. 9.

The numerical grid used in the computations is shown in Fig (1). It consists of an algebraically generated 240x60 C-grid. The mesh spacing in the y direction at the surface was such that y^+ at the first mesh point was less than one everywhere on the airfoil. The spacing in x over the central portion of the airfoil was $\Delta X/C = .013$ with the spacings reduced at the leading and trailing edges. The far field boundary was placed approximately 20 chords from the airfoil surface.

Surface pressure and upper surface skin friction and displacement thickness distributions are compared for Case 1 in Fig. (2). It is apparent that the surface pressure distributions predicted by the turbulence models are in good agreement with one another and with experiment. Larger differences between model predictions and experiment are indicated by the skin friction and displacement thickness distributions. Models which are in best agreement with experiment on skin friction, especially near the trailing edge, are the C-S, B-L and q-ω2 models. The best models in predicting displacement thickness are the J-K and q-ω2 models. The lift and drag predictions for this case are shown in table I. It appears that all three zero-equation models give better predictions of drag than the two-equation models while the J-K and q-ω2 models give the best predictions of lift.

The second case studied was the unseparated supercritical case 9. Computed Mach contours for this case obtained using the J-K model are shown in Fig. (3). Surface pressure and upper surface skin friction and displacement thickness distributions are shown in Fig. (4). Computations of surface pressure indicate that all models are in close agreement with experiment except near the shock wave where the J-K and q-ω2 models predict shock locations which are slightly upstream of the locations given by experiment and the other models. With regard to skin friction, the C-S, B-L and q-ω2 models give the best predictions downstream of the shock wave while the J-K, q-ω2 and k-ε models are closest in their predictions of displacement thickness. The k-ε model shows an unrealistically large increase in skin friction downstream of the shock wave. From Table I it is evident that the models giving the best overall predictions of lift and drag are the J-K, q-ω2 and k-ε models.

Computed Mach contours and surface distributions for Case 10 are shown in Figs. (5) and (6). In this case the flow downstream of the shock wave is separated for some distance. The corresponding model predictions show more differences in surface pressure distributions and especially shock wave locations than in the previous cases and this gives rise to larger differences in lift and drag. It is evident that the J-K model provides the best overall predictions of surface pressure, skin friction and displacement thickness distributions as well as lift and drag (Table I). It should be noted in this case that the calculations using the J-K and q-ω2 models were unsteady with the shock wave undergoing small periodic oscillations about its mean position. The corresponding oscillations in lift were from 2% (J-K) to 4% (q-ω2) of their mean values. The k-ε model showed an anomalous behavior (or weak instability) in skin friction for this case which is believed to result from numerical stiffness associated with the low-Reynolds-number damping terms of the model (see Ref. 5).

Numerical efficiency of the various turbulence models was measured by the computing time required to achieve steady state, or in the case of unsteady flows, the time required to achieve a periodic state. For the unseparated cases 1 and 9 about 1000 steps were needed for convergence (to approximately 3 significant figures in C_L). Computing time for the zero-equation models was about 480 seconds (on the NASA Ames Cray XMP/48) and about 600 seconds for the two-equation models. For case 10, the models which predicted steady solutions (i.e., C-S, B-L, k-ε, q-ω1) required approximately the same time as cases 1 and 9. The unsteady solutions (J-K and q-ω2 models) required about 2.5 times more computing time than the other cases and a constant (instead of spatially varying) time step was used to maintain time accuracy.

The primary conclusions resulting from this work are that differences between experiment and turbulence model predictions which are relatively small for subcritical unseparated transonic flows, become greater for supercritical separated flows. The principal differences between model prediction of lift and drag for the supercritical cases appear to be the result of differences in predicted shock location. The best overall model in predicting the three cases was the J-K model.

References

[1] Cebeci, T. and Smith, A. M. O., <u>Analysis of Turbulent Boundary Layers</u>, Academic Press, 1974.

[2] Baldwin, B. S. and Lomax, H., "Thin Layer Approximation and Algebraic Model for Separated Turbulent Flows," AIAA Paper 78-257, 1978.

[3] Johnson, D. A., "Predictions of Transonic Separated Flow with an Eddy-Viscosity/Reynolds-Shear-Stress Closure Model," AIAA Paper 85-1683, Cincinatti, OH, Jul. 1985.

[4] Chien, K. Y., "Predictions of Channel Boundary-Layer Flows with a Low-Reynolds-Number Turbulence Model," <u>AIAA Journal</u>, Vol. 20, Jan. 1982, pp. 33-38.

[5] Coakley, T. J., "Turbulence Modeling Methods for the Compressible Navier-Stokes Equations," AIAA Paper 83-1693, Danvers, MA, Jul. 1983.

[6] Coakley, T. J., and Hsieh, T., "A Comparison Between Implicit and Hybrid Methods for the Calculation of Steady and Unsteady Inlet Flows," AIAA Paper 85-1125, Monterey, CA, Jul. 1985.

[7] Mehta, U., Chang, K. C., and Cebeci, T., "A Comparison of Interactive Boundary Layer and Thin-Layer Navier-Stokes Procedures," Numerical and Physical Aspects of Aerodynamic Flows III, Ed. T. Cebeci, Springer-Verlag, 1986.

[8] Coakley, T. J., "Implicit Upwind Methods for the Compressible Navier-Stokes Equations, AIAA Journal, Vol. 23, No. 3, Mar. 1985, p. 374.

[9] Cook, P. H., McDonald, M. A., and Firmin, M. C. P., AEROFOIL RAE 2822 Pressure Distributions, and Boundary Layer and Wake Measurements," AGARD Advisory Report No. 138, 1979.

Table I - Turbulence Models and Experimental/Computational Characteristics - RAE2 822 Airfoil

Model	Originator	Case 1		Case 9		Case 10	
		c_L	c_D	c_L	c_D	c_L	c_D
Zero-equation models							
C-S	Cebeci-Smith[1]	.610	.0086	.833	.0172	.815	.0269
B-L	Baldwin-Lomax[2]	.620	.0087	.861	.0185	.859	.0298
J-K	Johnson-King[3]	.573	.0087	.787	.0159	.745	.0243
Two-equation models							
$k-\varepsilon$	Chien[4]	.599	.0094	.821	.0179	.776	.0268
$q-\omega\ 1$	Coakley[5]	.607	.0095	.846	.0181	.836	.0284
$q-\omega\ 2$	Coakley[6]	.576	.0093	.783	.0159	.726	.0267
Experiment[9]		.566	.0085	.803	.0168	.743	.0242
α(geom), α(comp):		2.40 ,	1.93	3.19 ,	2.80	3.19 ,	2.80
M_∞ , $R_e \times 10^{-6}$:		.676 ,	5.70	.730 ,	6.50	.750 ,	6.20

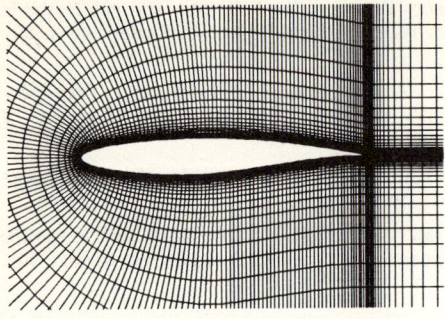

Fig. 1 240x60 C-Mesh, RAE 2822

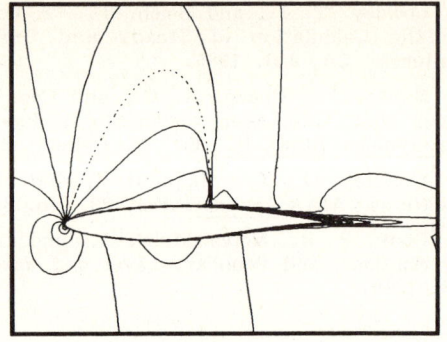

Fig. 3 Mach Contours, Case 9

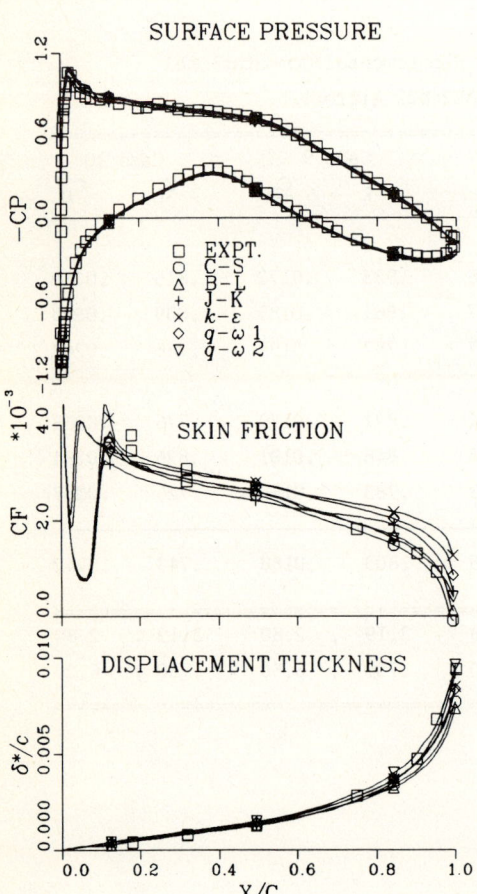

Fig. 2 Surface pressure, upper surface skin friction and displacement thickness distributions, Case 1

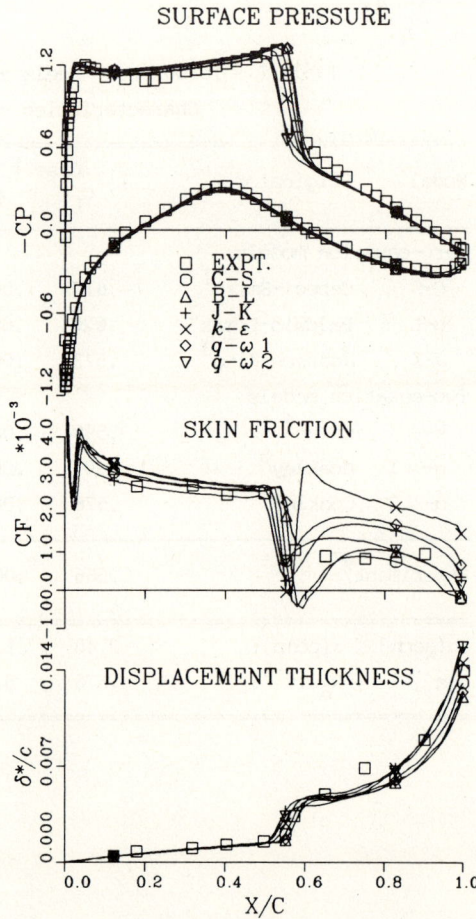

Fig. 4 Surface pressure, upper surface skin friction and displacement thickness distributions, Case 9

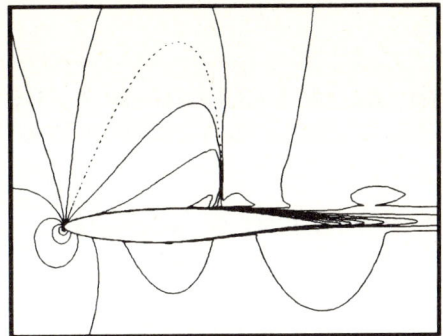

Fig. 5 Mach Contours, Case 10

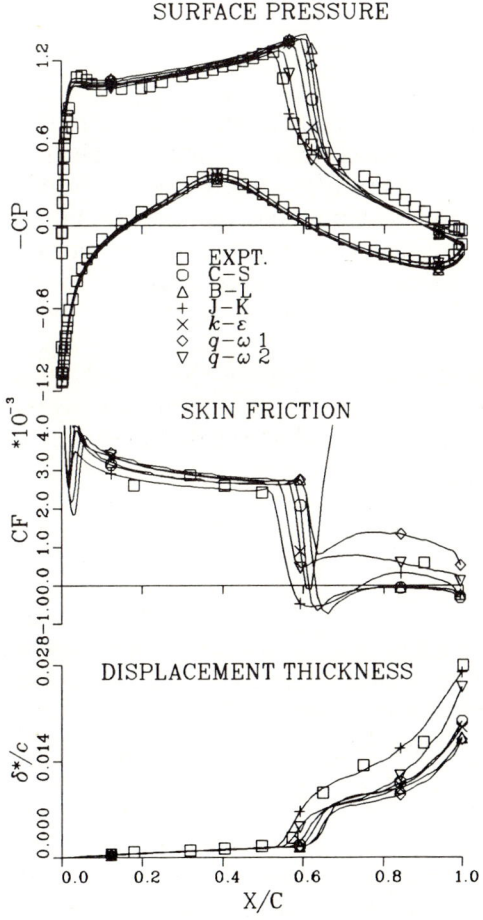

Fig. 6 Surface pressure, upper surface skin friction and displacement thickness distributions, Case 10

NUMERICAL MODELLING OF A BUBBLE RISING THROUGH VISCOUS FLUID

B. Couët, G. S. Strumolo
Schlumberger-Doll Research, Ridgefield, CT 06877-4108, U.S.A.

A. E. Dukler
Department of Chemical Engineering,
University of Houston, Houston, TX 77004, U.S.A.

INTRODUCTION

A problem that has been studied for many decades is that of the so-called Taylor bubbles rising in vertical tubes. These bubbles occupy a significant cross-section of the tube and are often many tube-radii long. The bubble itself is a free surface with a constant pressure maintained along its length while the fluid runs around the outside of the bubble and remains as a layer down the surface of the tube, falling freely under gravity. A variety of numerical techniques have been used in an attempt to describe both the bubble shapes and the dynamics of the liquid region [1,2]. In this paper, unlike previous modelling in the literature, we determine the bubble shape by computing it simultaneously with the development of a boundary layer along the wall of a two-dimensional duct. A comparison to physical experiments is also presented.

ANALYSIS

If U is the velocity with which the gas bubble penetrates the liquid and b the duct's half-width, the flow can be brought to a steady motion by giving the whole system an equal downward velocity. With a viscous fluid, a boundary layer will develop along the duct walls. We assume that it starts across from the bubble nose as shown in Fig. 1. A mass balance condition yields:

$$\int_{b-m}^{b} \tilde{u}\, dy = U b \,, \tag{1}$$

where $\tilde{u} = \tilde{u}(x,y)$ is the velocity of the liquid in coordinates moving with the bubble. A momentum balance can also be made on a differential element of liquid film situated at a distance x from the nose of the bubble:

$$-\frac{\partial}{\partial x}\int_{b-m}^{b} \tilde{u}^2\, dy + \nu \frac{\partial \tilde{u}}{\partial y}\bigg|_{y=b} + g\, m = 0 \,. \tag{2}$$

The terms in (2) take into account liquid acceleration and the forces due to viscous wall shear and gravity, respectively. The force due to surface tension is assumed to be negligible. To follow the development of the boundary layer in the liquid region, we adopt the approximate method due to von Karman and Pohlhausen [3]. We begin with the momentum equation for the boundary layer:

$$\frac{\tau_0}{\rho} = u_\infty^2 \frac{d\delta_2}{dx} + (2\delta_2 + \delta_1)u_\infty \frac{du_\infty}{dx}. \tag{3}$$

δ_1 and δ_2 are the displacement and momentum thicknesses, respectively, and τ_0 is the wall shear stress.

Define the following dimensionless variables:

$$u = \frac{\tilde{u}}{U}, \quad X = \frac{x}{b}, \quad F_r = \frac{U}{(gb)^{1/2}}, \quad u_0 = \frac{u_\infty}{U},$$

$$\eta = \frac{b-y}{\delta}, \quad \beta = \frac{m}{b}, \quad R_e = \frac{Ub}{\nu}, \quad \gamma = \frac{\delta}{b},$$

where δ is the boundary-layer thickness.

Following Pohlhausen [3], we assume a polynomial of the fourth degree for the velocity profile in the dimensionless distance η, i.e.,

$$u(X,\eta) = \frac{\tilde{u}}{U} = a + b\eta + c\eta^2 + d\eta^3 + e\eta^4. \tag{4}$$

This velocity profile is subject to the following boundary conditions at $y = b$ (duct wall, $\eta = 0$) and at $y = b - \delta$ (outer region of boundary layer, $\eta = 1$):

$$\tilde{u}(x,b) = U, \quad \nu \left. \frac{\partial^2 \tilde{u}}{\partial y^2} \right|_{y=b} + g = 0,$$

$$\tilde{u}(x, b-\delta) = u_\infty(x), \quad \left. \frac{\partial \tilde{u}}{\partial y} \right|_{y=b-\delta} = 0, \quad \left. \frac{\partial^2 \tilde{u}}{\partial y^2} \right|_{y=b-\delta} = 0.$$

These conditions enables us to solve for the coefficients a through e in (4), giving:

$$u = 1 + (2u_0 - 2 + A)\eta - 3A\eta^2 + (3A - 2u_0 + 2)\eta^3 + (u_0 - 1 - A)\eta^4,$$

where $A = \frac{R_e}{6}\left(\frac{\gamma}{F_r}\right)^2$. Using this velocity profile, the dimensionless form of equations 1, 2 and 3 becomes

$$\beta = \frac{1 - \gamma \left[\frac{3}{10}(1-u_0) + \frac{R_e \gamma^2}{120 F_r^2}\right]}{u_0}, \tag{5}$$

$$\frac{du_0}{dX} = (1-2u_0\gamma f_2)^{-1}\left[\frac{\beta}{F_r^2} + \frac{1}{\gamma R_e}\left[2(u_0-1) + \frac{R_e\gamma^2}{6F_r^2}\right]\right] + u_0^2\frac{d(\gamma f_2)}{dX}, \quad (6)$$

$$\frac{d\gamma}{dX} = \frac{1}{\gamma R_e f_2^2}\left[\frac{f_3 - f_4(2 + f_1/f_2)}{u_0} - f_2\gamma^2 R_e\frac{df_2}{dX}\right], \quad (7)$$

where $f_i = f_i(u_0, \gamma, R_e, F_r), i=1,\ldots,4$. This is a *coupled* set of algebraic and differential equations for the bubble shape, the boundary-layer thickness and the liquid velocity outside the boundary layer. In other words, the development of the boundary layer is dependent upon the development of the bubble shape, and vice-versa.

NUMERICAL RESULTS

These equations were solved using a variant of the GEAR package which allows for the automatic switching between stiff and nonstiff methods of solution. This is needed since, near $x=0$, the ODE describing the boundary-layer development exhibits a stiff behavior while, further downstream, it becomes nonstiff. A relative and absolute error tolerance of 10^{-12} was maintained throughout the calculations. By specifying a value of the Reynolds number, R_e, and the Froude number, F_r, we can solve equations 5 through 7 for the bubble shape and the boundary layer. One such calculation is illustrated in Fig. 2 for the case $R_e = 2000$ and $F_r = 0.25$. The computation is stopped when the developing boundary layer reaches the bubble. At this point, we conjecture that the film has reached its limiting thickness. A previous paper by the authors [2] showed a similar calculation; in that paper, the bubble shape was determined by a potential analysis and the boundary layer developed in this fixed flow domain. In this paper, the bubble and the boundary layer interact with each other to determine their ultimate shapes.

While our numerical calculations will produce bubbles for any values of R_e and F_r, physical experiments provide for a single or at most a narrow range of F_r given any R_e [4]. These experiments have been conducted on two-dimensional bubbles between parallel plates of high aspect ratio so as to approximate a two-dimensional flow. The digitization of a typical experimental bubble is illustrated by the dots in Fig. 3. This case corresponds to $R_e = 24,000$ and $F_r = 0.36$. Using these values, we solve equations 5-7 and the resulting numerical bubble is shown by the solid line in Fig. 3. The agreement is quite good. The boundary layer can also be seen close to the wall, as expected for such a high value of the Reynolds number.

SUMMARY

In this paper, we have presented a numerical model for the dynamics of a two-dimensional bubble rising through a viscous fluid. Unlike previous attempts in the literature, we have combined the growth of the boundary layer with the evolution of the bubble shape. This results in a coupled set of algebraic and differential equations. The computed shape from the numerical solution agrees well with physical experiments.

REFERENCES

[1] Couët, B., G. S. Strumolo & A. E. Dukler 1986, Modelling two-dimensional large bubbles in a rectangular channel of finite width. *Phys. Fluids* (to appear).

[2] Couët, B., G. S. Strumolo & A. E. Dukler 1985, Modelling of two-dimensional bubbles in vertical tubes. *Lecture notes in physics*, **218**.

[3] Schlichting, H. 1979, *Boundary Layer Theory*, McGraw-Hill.

[4] Maneri, C. C. 1970, The motion of plane bubbles in inclined ducts. Ph. D. Thesis, Polytechnic Institute of Brooklyn, New York.

Figure 1.

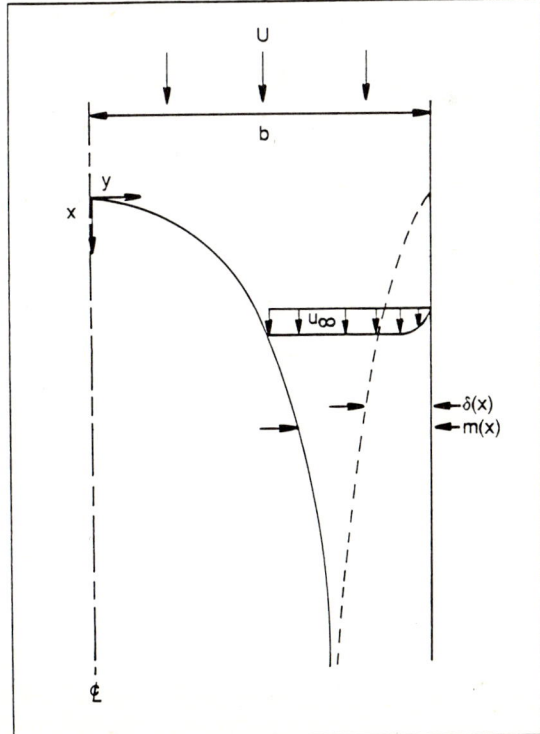

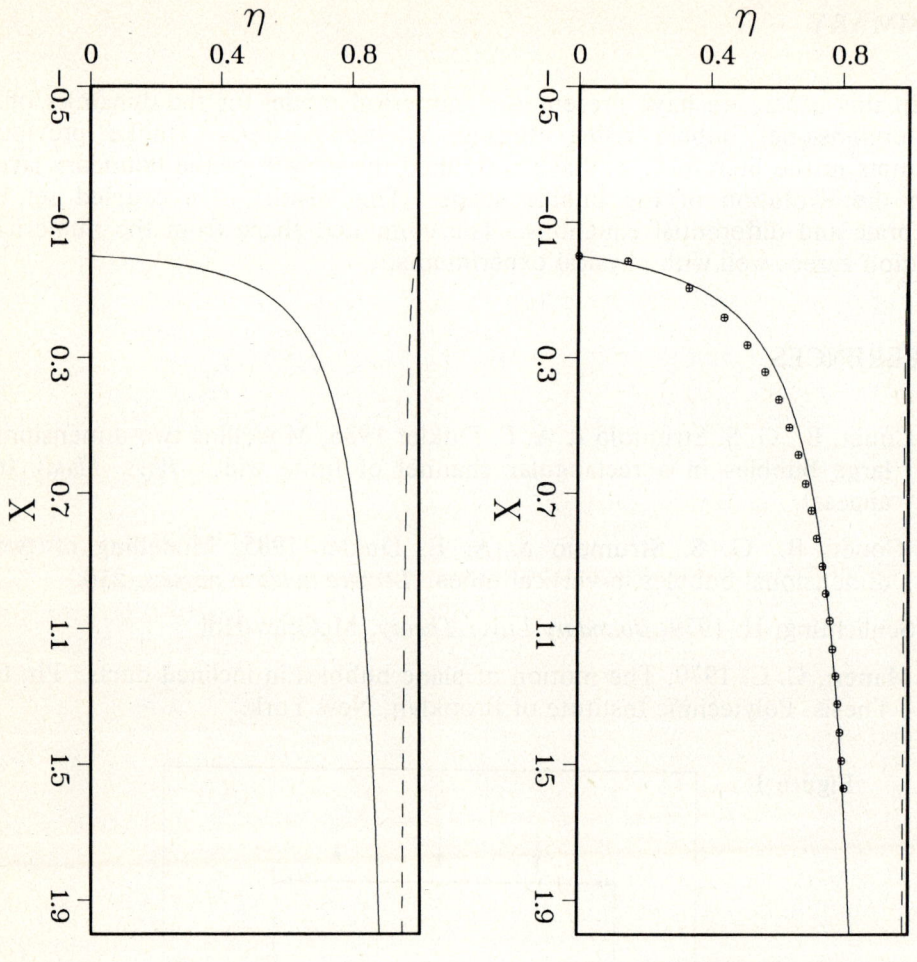

Figure 2.   Figure 3.

THERMOCAPILLARY FREE BOUNDARIES IN CRYSTAL GROWTH

C. CUVELIER   J. M. DRIESSEN [1]

Delft University of Technology
Dept. of Mathematics and Informatics
Delft
The Netherlands

[1] present address:
Koninklijke/Shell-Laboratorium
Shell Research bv
Amsterdam
The Netherlands

There are many important technological and engineering-science applications in which capillary free boundaries play a dominant part, for example in lubrication, electrochemical plating, corrosion, coating, polymer technology, separation processes, metal and glass forming processes, crystal growth and aerospace technology, there are abundant technologically important fluid flows with capillary free boundaries
    A situation of special interest, both from the technological as well as from the purely mathematical point of view, is the study of the behavior of liquids in an open container in a low-g environment, where capillary free boundaries must be taken into account.

In this paper we shall focus our attention to the (finite element) analysis of thermocapillary free boundaries in crystal growth processes. Crystal growth, by definition, is concerned with the formation of a single crystal by which we mean a solid in its most uniform structure that can be attained. This uniform structure of crystals, which can be modified by the controlled addition of impurities, allows, for instance, the transmission without scattering of electromagnetic waves and charged particles and forms the basis of the electronic-industrial fabrication of devices like chips, transistors, semiconductors etc.

In crystal growth techniques there are, in general, two types of free boundaries. The first is the crystal-melt interface. The determination of this interface is reminiscent to the so-called Stefan problem, because heat transfer in the crystal and in the melt is coupled by requirements of temperature continuity and energy conservation. In the heat flux balance on this interface, the latent heat must be taken into account since it determines the growth-rate of the crystal into the melt. The second kind of free boundary is the liquid-air interface. According to the basic principles of hydrodynamics the condition of a balance of forces must be fulfilled on this free boundary.

Surface tension and the temperature dependence of the surface tension coefficient must be taken into account, since they influence the flow pattern of the melt and consequently the temperature distribution near the crystal. This temperature distribution near the crystal in turn affects the crystal morphology and growth-rate.

The possibility nowadays of using the space shuttles to carry a manned space laboratory has intensified the interest in the concept of crystal manufacturing in space. Prolonged periods of micro-gravity environment of about $10^{-3}g$ to $10^{-6}g$ can be achieved in a near-earth orbit, where g denotes the terrestrial gravity. The interest of a low-g environment for the crystal growth is the fact that natural convection (i.e. convection due to density variations in the melt) can be reduced considerably. Moreover, it was found that in a low-g situation more homogeneous and less striated crystals can be produced. Although it is not certain whether there are crystals for use in commercial electronic devices which can only be manufactured in space and whose quality is unattainable on Earth, crystal growth experiments or, in general, the behavior of liquids in space can help us to understand fundamental physical processes, in particular when these experiments are complemented by terrestrial experiments, mathematical modelling and numerical analysis.

Although, as mentioned above, a low-g environment can reduce substantially the phenomenon of natural convection due to density variation in the melt, other types of convection processes become important. The presence of a free melt-gas interface can influence the motion of the melt when the coefficient of surface tension varies from point to point. In a non-isothermal melt there will be a surface tension gradient on the melt-gas interface, because of the temperature dependence of the surface tension coefficient. This surface tension gradient acts like a shear stress on the melt-gas interface and thereby generates a surface flow from the region of low surface tension to that of high surface tension, which usually means from hot to cold. Owing to the viscosity of the melt this surface motion penetrates into the melt and induces a bulk flow. This type of convection is called Marangoni or thermocapillary convection.

The aim of this paper is to investigate stationary (melt-gas) free boundary problems in the context of crystal growth from the numerical point of view. As we have mentioned above, the crystal growth system is subjected to two types of convection. The first is the convection due to gravitational buoyancy forces (which we call Grashof convection). The second type of convection is induced by a gradient of the coefficient of surface tension (called Marangoni convection).

We shall study the relative influence of these two types of convection on the shape of the melt-gas free boundary. The melt-crystal interface shape is not considered and is assumed to be constant. Since one needs to solve the general hydrodynamic equations based on the Navier-Stokes equations, we shall restrict ourselves to a model problem of open boat type which has some relevance from the crystal growth point of view and which is general enough to admit a realistic study of free boundary shapes. Although some limitations exist when comparing with real crystal growth techniques, the open boat situation to be discussed can serve as a model for various crystal growth configurations.

We formulate the open boat model problem and write the partial differential equations (Navier-Stokes equations coupled with the heat-conduction equation in a variable domain) governing the melt flow field. The equations are written in dimensionless form and characteristic numbers (Reynolds, Grashof, Prandtl, Marangoni, Biot-cooling, Biot-radiation, Bond, Ohnesorge) are introduced. Some iterative methods for the computation of the free boundary shape will be considered and the finite element discretization of the system will be discussed. Finally, we present the results of numerical experiments.

References:

C. Cuvelier, J.M. Driessen.
  Thermocapillary free boundaries in crystal growth.
  To appear in J. of Fluid Mechanics, 1986.

C. Cuvelier, A. Segal, A.A. van Steenhoven.
  Finite-element methods and Navier-Stokes equations.
  Reidel Publishing Company, 1986.

A QUASI-CONSERVATIVE COIN LAMBDA FORMULATION

A. Dadone
Universita' di Bari, Bari, Italy

INTRODUCTION

The present paper is concerned with the numerical simulation of transonic inviscid flows. Among the several numerical methods developed for such computations, the so-called lambda formulation (e.g., Ref. 1) has several, very desirable features. In order to enhance the accuracy of this formulation, Moretti et al.(e.g., Ref. 2) have developed a very accurate numerical scheme, while Dadone and Napolitano /3/ have shown that a significant improvement is obtained by employing a perturbative formulation.

The classical lambda formulation has its drawbacks when transonic flows have to be computed: the supersonic region is decoupled from the shocked subsonic one, once the shock wave is established. On the contrary such a coupling is allowed by the "flux difference splitting" methodologies, by splitting the waves containing a "sonic point". By inserting such a basic mechanism into the classical lambda formulation, Dadone and Magi /4/ obtained a modified lambda formulation, characterized by a supersonic flow region coupled with the shocked subsonic one and apt to compute transonic flows with weak shocks.

The present paper provides the extension of the previous quasi conservative methodology to higher shock strengths, by modifying the correction terms used in the shock transition region. Moreover, a COIN variant is presented which has been obtained by modifying the perturbative formulation /3/. Transonic flows can thus be accurately and efficiently computed.

GOVERNING EQUATIONS

The classical lambda formulation equations in a general orthogonal curvilinear coordinate system are reported in Ref. 2; such equations allow to evaluate the time derivatives of the two velocity components, of the speed of sound and of the entropy as function of the space derivatives of these same variables. Alternatively, the derivatives of the speed of sound can be eliminated in favour of the derivatives of the logarithmic pressure, as here done.

Let us now consider a one-dimensional transonic nozzle flow /4/. Following the technique suggested in /5/, a flux difference splitting approach leads to the solution of the Riemann problems sketched in Fig. 1 for the mesh HI, bounded by supersonic flow conditions on the left and subsonic flow conditions on the right, and for the two neighbouring meshes GH and IL; the corresponding characteristic lines pertaining to the classical lambda formulation are plotted in Fig. 2. Let us now compare Figs. 1 and 2; the wave shape is different because the converging fan wave 1' has no corresponding wave in Fig. 2; the classical lambda formulation cannot account for such a wave and for its capability of coupling the last supersonic meshpoint with the shocked subsonic region.

When dealing with a two-dimensional flow where only one component of the velocity of the fluid can be supersonic, the flux difference splitting method suggested in Ref. 6 indicates that only one of the four Riemann problems to be solved for evaluating the flux differences can contain a converging fan wave (simulating a shock) with a vertical characteristic. Such a wave is neglected by the classical lambda formulation, also for two-dimensional flows. Accordingly, the Euler equations can be rewritten, as reported in Ref. 4 for the case of a supersonic first component of the velocity of the fluid, by separating the flux terms due to this wave. Such equations can be rearranged to obtain the lambda formulation equations with the addition of correction terms f_{c1}, f_{c2}, f_{c3} corresponding to the previously outlined flux terms.

The COIN variant can be obtained as follows. Let us define an appropriate "incompressible" flow, characterized by the following governing equations:

$$\text{div } \underline{V}' = \text{curl } \underline{V}' = 0 \quad ; \quad a'^2 + \delta(u'^2 + v'^2) = a_o^2 \tag{1; 2}$$

$$P' = \gamma/(\gamma - 1) \ln(a'^2/\gamma) \tag{3}$$

where $\underline{V}'$, u', v', a', P' are the "incompressible" flow velocity vector, velocity components, speed of sound and logarithmic pressure, while a_o is the stagnation speed of sound of the compressible flow and $\delta = (\gamma - 1)/2$. The following COIN variables are then defined:

$$\bar{u} = u - u' \quad ; \quad \bar{v} = v - v' \quad ; \quad \bar{P} = P - P' \tag{4}$$

where u, v, P are the compressible velocity components and logarithmic pressure.

In terms of the COIN variables with the addition of the correction terms, the final equations are given as:

$$2\bar{u}_t = -\lambda_{1x}(a/\gamma \bar{P}_\xi + \bar{u}_\xi) + \lambda_{2x}(a/\gamma \bar{P}_\xi - \bar{u}_\xi) - 2\lambda_{3y}\bar{u}_\eta + 2\beta v + 2f_{p1} + f_{c1} \tag{5}$$

$$2\bar{v}_t = -\lambda_{1y}(a/\gamma \bar{P}_\eta + \bar{v}_\eta) + \lambda_{2y}(a/\gamma \bar{P}_\eta - \bar{v}_\eta) - 2\lambda_{3x}\bar{v}_\xi - 2\beta u + 2f_{p2} + f_{c2} \tag{6}$$

$$2a/\gamma \bar{P}_t = -\lambda_{1x}(a/\gamma\bar{P}_\xi+\bar{u}_\xi)-\lambda_{2x}(a/\gamma\bar{P}_\xi-\bar{u}_\xi)-\lambda_{1y}(a/\gamma\bar{P}_\eta+\bar{v}_\eta)-\lambda_{2y}(a/\gamma\bar{P}_\eta-\bar{v}_\eta)-2F+2f_{p3}+f_{c3} \tag{7}$$

being:

$$\beta = (h_{2\xi}\bar{v} - h_{1\eta}\bar{u})/(h_1 h_2) \quad ; \quad F = a(h_{2\xi}\bar{u} + h_{1\eta}\bar{v})/(h_1 h_2) \tag{8; 9}$$

$$f_{p1} = -\{k_1\bar{u} + k_2\bar{v} + k_3(a^2 - a'^2)\} \quad ; \quad f_{p2} = -\{k_4\bar{u} + k_5\bar{v} + k_6(a^2 - a'^2)\} \tag{10; 11}$$

$$f_{p3} = -a(k_3 u + k_6 v) \quad ; \quad f_{c1} = 2(u f'_{1x} - f'_{2x})/(\rho h_1) \tag{12; 13}$$

$$f_{c2} = 0 \quad ; \quad f_{c3} = (\gamma-1)\{(v^2-u^2) f'_{1x} + 2(u f'_{2x} - f'_{3x})\}/(\rho a h_1) \tag{14; 15}$$

$$\lambda_{1x} = (u + a)/h_1 \quad ; \quad \lambda_{2x} = (u - a)/h_1 \quad ; \quad \lambda_{3x} = u/h_1 \tag{16}$$

$$\lambda_{1y} = (v + a)/h_2 \quad ; \quad \lambda_{2y} = (v - a)/h_2 \quad ; \quad \lambda_{3y} = v/h_2 \tag{17}$$

$$k_1 = u'_\xi/h_1 \; ; \; k_2 = v'_\xi/h_1 \; ; \; k_3 = P'_\xi/(\gamma h_1) \; ; \; k_4 = u'_\eta/h_2 \; ; \; k_5 = v'_\eta/h_2 \; ; \; k_6 = P'_\eta/(\gamma h_2) \tag{18}$$

$$f'_{1x} = (\rho u)'_\xi \quad ; \quad f'_{2x} = (p + \rho u^2)'_\xi \quad ; \quad f'_{3x} = \{u(p + e)\}'_\xi \tag{19}$$

In eqns. (5-19) h_1 and h_2 are the scale factors while the subscripts ξ, η, t indicate derivatives with respect to the orthogonal curvilinear coordinates and time. Moreover f'_{1x}, f'_{2x}, f'_{3x} represent the flux terms corresponding to the wave 1' in Fig. 1.

Finally a steady flow condition has been enforced, namely the constancy of the total temperature, instead of the classical time-dependent entropy equation. The numerical technique and the boundary conditions reported in Ref. 2 have been used for the present computations.

Numerical experiments have shown that the previously used perturbative formulation /3/ is able to compute accurate results with a small increase in the total temperature along the body; such an error has shown a tendency to increase when the undisturbed flow Mach number is increased. Consequently, in the present COIN variant, the "incompressible" flow velocity components have been scaled on the basis of

$$V'_u = (\rho/\rho_o)_u V_u \qquad (20)$$

where V_u and $(\rho/\rho_o)_u$ are the velocity and the ratio of the density to the total density in undisturbed compressible flow conditions. With such a scaling the compressible and incompressible flows in the leading edge region of a symmetric body at no incidence are practically coincident, while minor variations occur in other cases (the isoMach lines of the compressible and "incompressible" flows near the leading edge of a circular cylinder with an undisturbed Mach number equal to .38 are plotted in Fig. 3). Consequently the COIN variant allows to perform accurate computations: in the present flow case the maximum entropy deviation, or the equivalent total temperature error for a time-dependent formulation, are of the order of .0001.

Numerical experiments have also shown that the conservation of mass, momentum and energy is better achieved if a flux-difference splitting technique (for the corresponding Riemann problem) is used to compute the flow conditions in the first shocked meshpoint; as a consequence the governing equations pertaining to this meshpoint must be slightly modified.

RESULTS

The present approach has been tested at first by means of some one-dimensional numerical experiments. Transonic one-dimensional flows in a converging-diverging nozzle have been computed; the nozzle outlet pressure has been varied between .4 and .9 (corresponding to a maximum Mach number ranging approximately from 1.3 up to 2.4) and the computations have been carried out using 21, 41 and 81 meshpoints: the computed shock was always correctly located. To demonstrate the accuracy of the present approach, the entropy (S) and Mach number (M) values, corresponding to the nozzle outlet section and computed by means of 21 meshpoints, are plotted as symbols in Fig. 4 versus the outlet pressure; for comparison the exact values are also plotted as lines in this figure.

The present approach has also been successfully applied to two-dimensional transonic external as well as internal flows. A plane channel with a circular bump /7/ has been considered at first. The steady state pressure coefficient results at the bump surface, corresponding to a downstream isentropic Mach number equal to .85, have been computed using 30x10 gridpoints and are plotted (symbols) in Fig. 5 together with the results computed by Montagne' (continuous line) using a flux difference splitting methodology and 72x21 gridpoints /8/.

The flow past a circular cylinder with an undisturbed Mach number equal to .5 has then been computed. Fig. 6 shows the steady state pressure coefficient results at the cylinder surface obtained using 51x21 gridpoints and performing the computation on a half plane; a recirculation bubble of approximately a diameter in width was found at the rear of the cylinder. The computed pressure coefficient results and recirculation bubble dimensions are in good agreement with published data /9/.

Finally the transonic flow past a NACA 0012 airfoil with an undisturbed Mach number equal to .85 at 1 degree angle of attack has been computed by using a conformal C-grid with 128x32 mesh intervals and 83 meshpoints on the airfoil. The Mach number distribution on the airfoil surface is plotted in Fig. 7. A very accurate solution obtained by Jameson, using 320x64 mesh intervals and 320 gridpoints on the airfoil surface, is reported in Ref. 10. A comparison between the present computed solution and Jameson's results shows the following features:

- shock position on the two sides: .643 and .866 versus .646 and .862;
- maximum Mach number on the two sides: 1.27 and 1.43 versus 1.28 and 1.44;
- lift coefficient: .3580 versus .3584.

Finally it must be remarked that the Rankine-Hugoniot conditions are correctly satisfied in all the three considered transonic flow cases. Moreover, the entropy errors along the body upstream of the shock are always very small, ranging from .0001 to .001; such a remarkable accuracy is due to the COIN variant, which has significantly increased the already good accuracy of the lambda formulation coupled with the numerical technique suggested in Ref. 2.

By taking into account the one- and two-dimensional flow results previously outlined, it can be concluded that the present formulation shows a good accuracy and a correct shock "capturing" capability.

ACKNOWLEDGEMENTS

This research has been supported by M.P.I. and C.N.R..

REFERENCES

1. Moretti G., "The λ- Scheme", Computers and Fluids, Vol. 7, 1979, pp.191-205.
2. Moretti G., "Numerical Studies of Two-Dimensional Flows", NASA CR 3930, 1985.
3. Dadone A. and Napolitano M., "A Perturbative Lambda Formulation", AIAA Journal, Vol. 24, March 1986, pp. 411-417.
4. Dadone A. and Magi V., "A Quasi-Conservative Lambda Formulation", AIAA Journal, to appear.
5. Pandolfi M., "A Contribution to the Numerical Prediction of Unsteady Flows", AIAA Journal, Vol. 22, May 1984, pp. 602-610.
6. Pandolfi M.,"On the Flux-Difference Splitting Method in Multidimensional Unsteady Flows", AIAA paper 84-0166.
7. Rizzi A. and Viviand H. (Eds.), "Numerical Methods for the Computation of Inviscid Transonic Flows with Shock Waves", Notes on Numerical Fluid Mechanics, Vol. 3, Vieweg Verlag, 1981.
8. Montagne' J. L., "A Second-Order Accurate Flux Splitting Scheme in Two-Dimensional Gasdynamics", Lecture Notes in Physics, Vol. 218, 1985, pp. 406-411.
9. Salas M. D., "Recent Developments in Transonic Euler Flow over a Circular Cylinder", Mathematics and Computers in Simulation, Vol. 25, 1983, pp. 54-58.
10. AGARD AR 211, May 1985, pp. 6.23-6.29.

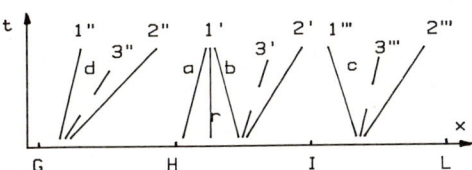

Fig. 1 - Riemann problems

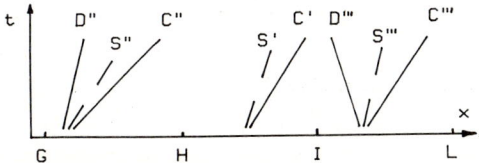

Fig. 2 - Characteristic lines

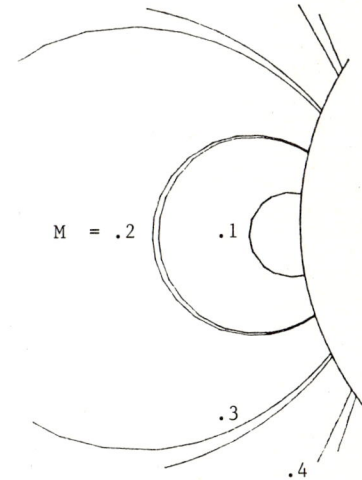

Fig. 3 - Leading edge region

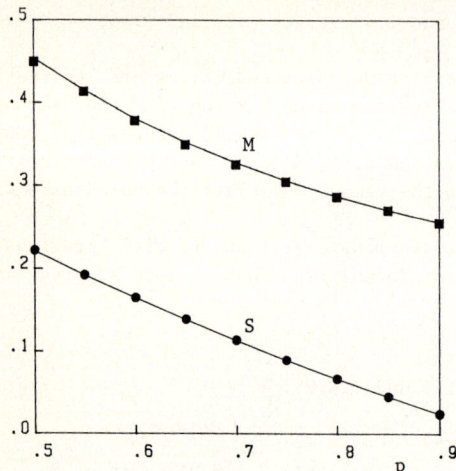

Fig. 4 - Onedimensional flow results

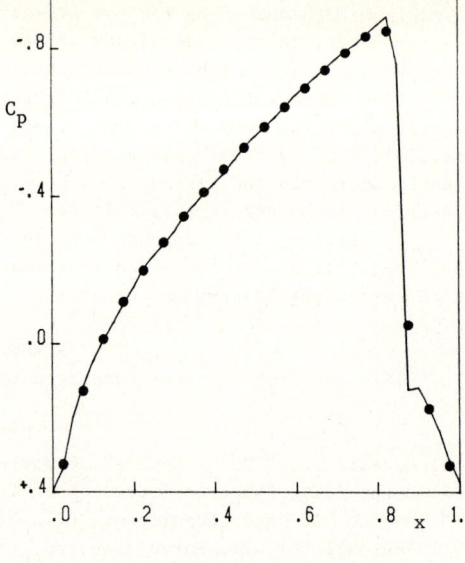

Fig. 5 - Circular bump results

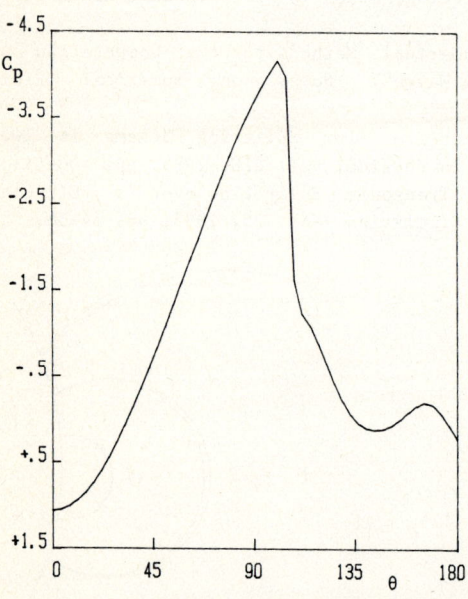

Fig. 6 - Circular cylinder results

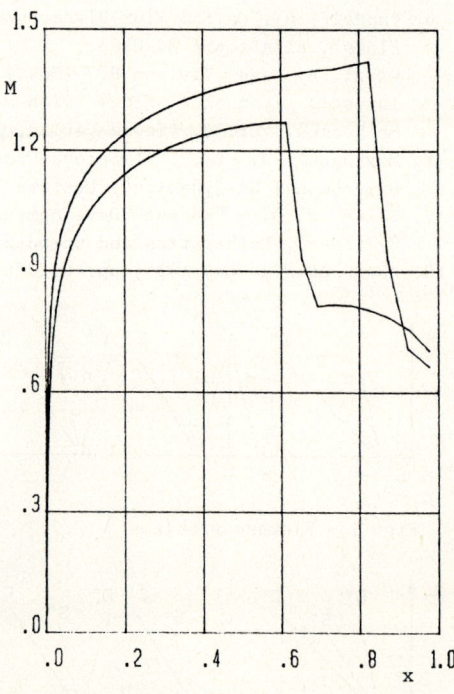

Fig. 7 - NACA 0012 airfoil results

AN IMPLICIT TIME-MARCHING METHOD FOR SOLVING THE 3-D COMPRESSIBLE EULER EQUATIONS

Hisaaki DAIGUJI, Yasuo MOTOHASHI and Satoru YAMAMOTO
Department of Mechanical Engineering
Tohoku University, Sendai, Japan

INTRODUCTION

The compressible Euler equations for arbitrary shaped flow fields have been solved using a number of explicit and implicit time-marching finite-difference methods and finite-volume methods. As far as we know, in these existing methods the Euler equations of physical velocities are used. But in the previous paper[1] the Euler equations of contravariant velocities are employed. The Beam-Warming delta-form approximate-factorization scheme[2][3], the uncoupled diagonal form by Pulliam-Chaussee[4], and the upstream-difference scheme by Steger-Warming[5] can also be applied to such Euler equations. The employment of the Euler equations of contravariant velocities makes the treatments of solid wall boundary conditions easy.

In the present paper, we first extend the previous method[1] developed for the two-dimensional Euler equations to a three-dimensional method, and then propose some explicit and implicit versions.

FUNDAMENTAL EQUATIONS

The fundamental equations of unsteady three-dimensional inviscid compressible flows in general curvilinear coordinates $\xi_1 \xi_2 \xi_3$ used in the existing methods are

$$\hat{q}_t + \hat{E}_\xi + \hat{F}_\eta + \hat{G}_\zeta \equiv \hat{q}_t + \partial \hat{E}_i / \partial \xi_i = 0 \tag{1}$$

where

$$\hat{q} = J \begin{pmatrix} \rho \\ \rho u_1 \\ \rho u_2 \\ \rho u_3 \\ e \end{pmatrix}, \quad \hat{E}_i = J \begin{pmatrix} \rho U_i \\ \rho u_1 U_i + (\partial \xi_i / \partial x_1) p \\ \rho u_2 U_i + (\partial \xi_i / \partial x_2) p \\ \rho u_3 U_i + (\partial \xi_i / \partial x_3) p \\ (e + p) U_i \end{pmatrix} \tag{2}$$

Jacobian J and the contravariant velocities U_i are expressed as

$$J = \partial(x_1, x_2, x_3) / \partial(\xi_1, \xi_2, \xi_3) \tag{3}$$

$$U_i = (\partial \xi_i / \partial x_k) u_k, \quad (i = 1, 2, 3) \tag{4}$$

Eqs. (1) and (2) are used in most of the existing methods as the fundamental equations in curvilinear coordinates, but involve some troubles in the treatments of solid wall boundary conditions.

In the present method, the velocities $u_i (i = 1, 2, 3)$ in $\hat{q}$ of Eq. (1) are replaced by the velocities $U_i (i = 1, 2, 3)$, that is,

$$\tilde{q}_t + \tilde{L}(\tilde{q}) = 0$$

or

$$\tilde{q}_t + \partial \tilde{E}_i / \partial \xi_i + \tilde{R} = 0 \tag{5}$$

where

$$\tilde{q} = J \begin{pmatrix} \rho \\ \rho U_1 \\ \rho U_2 \\ \rho U_3 \\ e \end{pmatrix}, \quad \tilde{L}(\tilde{q}) = \begin{pmatrix} \hat{e}_{i0} / \partial \xi_i \\ (\partial \xi_1 / \partial x_j)(\partial \hat{e}_{ij} / \partial \xi_i) \\ (\partial \xi_2 / \partial x_j)(\partial \hat{e}_{ij} / \partial \xi_i) \\ (\partial \xi_3 / \partial x_j)(\partial \hat{e}_{ij} / \partial \xi_i) \\ \hat{e}_{i4} / \partial \xi_i \end{pmatrix} \tag{6}$$

$$\tilde{E}_i = J \begin{Bmatrix} \rho U_i \\ \rho U_1 U_i + g_{1i}p \\ \rho U_2 U_i + g_{2i}p \\ \rho U_3 U_i + g_{3i}p \\ (e+p)U_i \end{Bmatrix}, \quad \tilde{R} = - \begin{Bmatrix} 0 \\ \hat{e}_{ij}(\partial/\partial \xi_i)(\partial \xi_1/\partial x_j) \\ \hat{e}_{ij}(\partial/\partial \xi_i)(\partial \xi_2/\partial x_j) \\ \hat{e}_{ij}(\partial/\partial \xi_i)(\partial \xi_3/\partial x_j) \\ 0 \end{Bmatrix} \quad (6)$$

$$(i = 1,2,3)$$

The first and fifth equations in Eq.(5) are those of Eq.(1), but the second to fourth equations are their linear combinations. $\hat{e}_{ij}$ is the (j+1)th component of $\hat{E}_i$, and

$$g_{ij} = (\partial \xi_i/\partial x_k)(\partial \xi_j/\partial x_k), \quad (i,j = 1,2,3) \quad (7)$$

$\tilde{R}$ is an additional term introduced to realize the linearization, the diagonalization and the upstreaming of $\tilde{E}_i$, and the values are generally small for smooth grids.

The equation of state is

$$p = \tilde{k}e - \rho\phi^2, \quad \phi^2 = \tilde{k}u_i u_i/2, \quad \tilde{k} = k - 1 \quad (8)$$

The fundamental equations in the present method are Eqs.(5) to (8), and here the initial value problem of Eq.(5) is solved using an implicit time-marching method for steady flows.

COMPUTATIONAL METHOD

The flux vectors $\tilde{E}_i (i = 1,2,3)$, which are homogeneous functions of degree one in $\tilde{q}$, can be expressed as

$$\tilde{E}_i = \tilde{A}_i \tilde{q} \quad (i = 1,2,3) \quad (9)$$

where

$$\tilde{A}_i = \begin{Bmatrix} 0 & \delta_{1i} & \delta_{2i} & \delta_{3i} & 0 \\ -U_1 U_i + g_{1i}\phi^2 & U_i - g_{1i}\alpha_1 + \delta_{1i}U_1 & -g_{1i}\alpha_2 + \delta_{2i}U_1 & -g_{1i}\alpha_3 + \delta_{3i}U_1 & \tilde{k}g_{1i} \\ -U_2 U_i + g_{2i}\phi^2 & -g_{2i}\alpha_1 + \delta_{1i}U_2 & U_i - g_{2i}\alpha_2 + \delta_{2i}U_2 & -g_{2i}\alpha_3 + \delta_{3i}U_2 & \tilde{k}g_{2i} \\ -U_3 U_i + g_{3i}\phi^2 & -g_{3i}\alpha_1 + \delta_{1i}U_3 & -g_{3i}\alpha_2 + \delta_{2i}U_3 & U_i - g_{3i}\alpha_3 + \delta_{3i}U_3 & \tilde{k}g_{3i} \\ -\psi^2 U_i + U_i\phi^2 & -U_i\alpha_1 + \delta_{1i}\psi^2 & -U_i\alpha_2 + \delta_{2i}\psi^2 & -U_i\alpha_3 + \delta_{3i}\psi^2 & kU_i \end{Bmatrix},$$

$$(i = 1,2,3) \quad (10)$$

$$\alpha_i = \tilde{k}(\partial x_k/\partial \xi_i)u_k \quad (i = 1,2,3) \quad (11)$$

$$\psi^2 = (c^2 + \phi^2)/\tilde{k}$$
$$\phi^2 = \alpha_i U_i/2 \quad (12)$$

c is the velocity of sound. Therefore, applying the Beam-Warming delta-form approximate-factorization scheme to Eq.(5), we get

$$(I + \Delta t \theta \delta_\xi \tilde{A}_1)(I + \Delta t \theta \delta_\eta \tilde{A}_2)(I + \Delta t \theta \delta_\zeta \tilde{A}_3)\Delta \tilde{q}^n = \text{RHS} \quad (13)$$

where

$$\tilde{q}^{n+1} = \tilde{q}^n + \Delta \tilde{q}^n \quad (14)$$

$$\text{RHS} = -\Delta t \tilde{L}^*(\tilde{q}^n) \quad (15)$$

$\theta = 1$ or $1/2$, δ_x is the second-order central-difference operator, and $\tilde{L}^*$ is an approximate operator of $\tilde{L}$.

Next, applying the theory of characteristics, we get

$$\tilde{A}_i = \tilde{S}_i^{-1} \Lambda_i \tilde{S}_i \quad (i = 1,2,3) \quad (16)$$

where Λ_i is the diagonal matrix of eigenvalues of $\tilde{A}_i$,

$$\Lambda_1 = \begin{pmatrix} \lambda_1^1 & & & & 0 \\ & \lambda_1^4 & & & \\ & & \lambda_1^1 & & \\ & & & \lambda_1^1 & \\ 0 & & & & \lambda_1^5 \end{pmatrix}, \quad \begin{aligned} \lambda_i^1 &= U_i \\ \lambda_i^4 &= U_i + c\sqrt{g_{ii}} \\ \lambda_i^5 &= U_i - c\sqrt{g_{ii}} \end{aligned} \tag{17}$$

and $\tilde{S}_i$ is the matrix composed of the eigenvectors of $\tilde{A}_i$,

$$\tilde{S}_1 = \begin{pmatrix} 1 & 0 & 0 & 0 & -1/c^2 \\ 0 & 1 & 0 & 0 & \sqrt{g_{11}}/c \\ 0 & -g_{12}/g_{11} & 1 & 0 & 0 \\ 0 & -g_{13}/g_{11} & 0 & 1 & 0 \\ 0 & -1 & 0 & 0 & \sqrt{g_{11}}/c \end{pmatrix} \begin{pmatrix} 1 & 0 & 0 & 0 & 0 \\ -U_1 & 1 & 0 & 0 & 0 \\ -U_2 & 0 & 1 & 0 & 0 \\ -U_3 & 0 & 0 & 1 & 0 \\ \phi^2 & -\alpha_1 & -\alpha_2 & -\alpha_3 & \tilde{k} \end{pmatrix} \tag{18}$$

Then, by splitting the matrices according to the positive or negative sign of eigenvalues

$$\Lambda_i = \Lambda_i^+ + \Lambda_i^- \qquad (i = 1,2,3) \tag{19}$$
$$\tilde{A}_i = \tilde{A}_i^+ + \tilde{A}_i^- \qquad (i = 1,2,3) \tag{20}$$
$$\tilde{E}_i = \tilde{E}_i^+ + \tilde{E}_i^- \qquad (i = 1,2,3) \tag{21}$$

where $\tilde{S}_i^{-1} \Lambda_i^\pm \tilde{S}_i = \tilde{A}_i^\pm$, $\tilde{A}_i^\pm \tilde{q} = \tilde{E}_i^\pm$.

$$\tilde{E}_i^\pm = \tilde{q} \lambda_i^{1\pm} + Jp \begin{pmatrix} 0 \\ g_{1i} \\ g_{2i} \\ g_{3i} \\ U_i \end{pmatrix} \frac{\lambda_i^{a\pm}}{c\sqrt{g_{ii}}} + \frac{Jp}{c^2} \begin{pmatrix} 1 \\ U_1 \\ U_2 \\ U_3 \\ \psi^2 \end{pmatrix} \lambda_i^{b\pm}, \qquad (i = 1,2,3) \tag{22}$$

$$\lambda_i^a = (\lambda_i^4 - \lambda_i^5)/2, \quad \lambda_i^b = (\lambda_i^4 + \lambda_i^5)/2 - \lambda_i^1.$$

Using Eqs.(16) to (22), Eq.(13) can be rewritten in the uncoupled diagonal form with the upstream-difference schemes according to sign of the characteristics speeds λ_i^j ($i = 1,2,3$; $j = 1,4,5$).

$$\tilde{S}_1^{-1}\{I + \Delta t\theta(\Lambda_1^+ \nabla_\xi + \Lambda_1^- \Delta_\xi)\} \tilde{M}_1 \{I + \Delta t\theta(\Lambda_2^+ \nabla_\eta + \Lambda_2^- \Delta_\eta)\}$$
$$\tilde{M}_2 \{I + \Delta t\theta(\Lambda_3^+ \nabla_\zeta + \Lambda_3^- \Delta_\zeta)\} \tilde{S}_3 \Delta \tilde{q}^n = \text{RHS} \tag{23}$$

where $\tilde{M}_1 = \tilde{S}_1 \tilde{S}_2^{-1}$, $\tilde{M}_2 = \tilde{S}_2 \tilde{S}_3^{-1}$, and ∇_x and Δ_x are backward- and forward-differences. Eq.(23) can be easily solved by dividing into the following seven steps.

(1) $\Delta q^1 = \tilde{S}_1 \text{RHS}$ \hfill (24a)

(2) $\{I + \theta \Delta t (\Lambda_1^+ \nabla_\xi + \Lambda_1^- \Delta_\xi)\} \Delta q^2 = \Delta q^1$ \hfill (24b)

(3) $\Delta q^3 = \tilde{M}_1 \Delta q^2$ \hfill (24c)

(4) $\{I + \theta \Delta t (\Lambda_2^+ \nabla_\eta + \Lambda_2^- \Delta_\eta)\} \Delta q^4 = \Delta q^3$ \hfill (24d)

(5) $\Delta q^5 = \tilde{M}_2 \Delta q^4$ \hfill (24e)

(6) $\{I + \theta \Delta t (\Lambda_3^+ \nabla_\zeta + \Lambda_3^- \Delta_\zeta)\} \Delta q^6 = \Delta q^5$ \hfill (24f)

(7) $\Delta \tilde{q}^n = \tilde{S}_3^{-1} \Delta q^6$ \hfill (24g)

In order to capture the shock wave stably, the calculation of RHS, i.e., $L^*(\tilde{q})$ is performed using the following formulas

$$(u_x)_0 = (u_{1/2} - u_{-1/2})/\Delta x \tag{25}$$

where

$$u_{1/2} = \{\mp \alpha u_{-1} + (6 \pm 3\alpha)u_0 + (6 \mp 3\alpha)u_1 \pm \alpha u_2\}/12 \tag{26a}$$
$$u_{1/2} = \{-(1 \pm \alpha)u_{-1} + (7 \pm 3\alpha)u_0 + (7 \mp 3\alpha)u_1 - (1 \mp \alpha)u_2\}/12 \tag{26b}$$

$$u_{1/2} = \{-(1 \pm \alpha)u_{-1} + (9 \pm 3\alpha)u_0 + (9 \mp 3\alpha)u_1 - (1 \mp \alpha)u_2\}/16 \tag{26c}$$

The terms containing α are the fourth-order artificial dissipation term, and the upper and lower signs are taken for the waves of u propagating to the positive and negative directions of x, respectively. Eq.(26a) for $\alpha = 0$ is the second-order central-difference scheme, and Eq.(26b) for $\alpha = 0$ the fourth-order central-difference scheme. Eq.(26b) for $\alpha = 3$ turns into the Kawamura-Kuwahara scheme[6], and Eq.(26c) for $\alpha = 1$ the QUICK scheme[7].

TREATMENTS OF SOLID WALL BOUNDARY CONDITIONS

The condition on this boundary is basically only one condition that the contravariant velocity component across the boundary be zero,

$$U_i = 0, \quad \text{on } \xi_i\text{-constant boundary} \tag{27}$$

These conditions must not only be used in the coefficients of Eq.(24), but also considered immediately after the right hand side computation and in the left hand side computation implicitly. That is,

$$(RHS)_{i+1} = 0, \quad \text{on } \xi_i\text{-constant boundary} \tag{28}$$

And for the ζ-constant boundary, since $U_3 = (\Delta \tilde{q}^n)_4 = 0$, Eq.(24g) becomes

$$\Delta q^6 = \tilde{S}_3 \Delta \tilde{q}^n = \begin{pmatrix} \Delta \tilde{q}^n - \Delta q^*/c^2 \\ -U_1 \Delta \tilde{q}_1^n + \Delta \tilde{q}_2^n \\ -U_2 \Delta \tilde{q}_1^n + \Delta \tilde{q}_3^n \\ \Delta q^* \sqrt{g_{33}}/c \\ \Delta q^* \sqrt{g_{33}}/c \end{pmatrix} \tag{29}$$

where $\Delta q^* = \phi^2 \Delta \tilde{q}_1^n - \alpha_1 \Delta \tilde{q}_2^n - \alpha_2 \Delta \tilde{q}_3^n + \tilde{k}\Delta \tilde{q}_5^n$. From this equation we can deduce the simple relation

$$\Delta q_4^6 = \Delta q_5^6, \quad \text{on } \zeta\text{-constant boundary} \tag{30}$$

Eq.(30) is an equivalent condition to Eq.(27).

The computations of sweep along the boundary are applicable also to the boundary straightaway, and the computations of sweep across the boundary can be executed by using equations like Eq.(30). Wave equations propagating towards the boundary from the outside are unreasonable ones at the solid wall boundary points. Therefore, these equations in Eq.(24f) must be replaced by the condition(30). Consequently, the fourth and fifth sets of linear equations (24f) are coupled as

$$\begin{pmatrix} B_1 & C_1 & & & & 0 \\ & B_2 & C_2 & & & \\ & & \ddots & \ddots & & \\ & & & \ddots & \ddots & \\ 0 & & & & B_{N-1} & C_{N-1} \\ C_N & & & & & B_N \end{pmatrix} \begin{pmatrix} (\Delta q^6{}_5)_{k=1} \\ \vdots \\ (\Delta q^6{}_5)_{K-1} \\ (\Delta q^6{}_4)_K \\ \vdots \\ (\Delta q^6{}_4)_2 \end{pmatrix} = \begin{pmatrix} (\Delta q^5{}_5)_{k=1} \\ \vdots \\ (\Delta q^5{}_5)_{K-1} \\ (\Delta q^5{}_4)_K \\ \vdots \\ (\Delta q^5{}_4)_2 \end{pmatrix} \tag{31}$$

where $N = 2(K-1)$.

The solid wall boundary condition is reasonably considered in the computation by these treatments, and in the case of steady flow the values of residual, RHS, become sufficiently small. It is not possible to carry out such treatments unless Eq.(6) is employed as the fundamental equations.

EXPLICIT SCHEMES

The fundamental equations for compressible flows, Eq.(5), can also be solved by using some explicit schemes. For example, using the Euler forward-difference method, the fourth-order Runge-Kutta method and the Adams-Bashforth method, we get

$$\Delta \tilde{q}^n = -\Delta t \tilde{L}^*(\tilde{q}^n) \qquad (32a)$$

$$\Delta \tilde{q}^n = (k_1 + 2k_2 + 2k_3 + k_4)/6 \qquad (32b)$$

where
$$k_1 = -\Delta t \tilde{L}^*(\tilde{q}^n), \quad k_2 = -\Delta t \tilde{L}^*(\tilde{q}^n + k_1/2),$$
$$k_3 = -\Delta t \tilde{L}^*(\tilde{q}^n + k_2/2), \quad k_4 = -\Delta t \tilde{L}^*(\tilde{q}^n + k_3)$$

$$\Delta \tilde{q}^n = -\Delta t \left[\tilde{L}^* + \frac{1}{2}\nabla \tilde{L}^* + \frac{5}{12}\nabla^2 \tilde{L}^* + \frac{3}{8}\nabla^3 \tilde{L}^* + \cdots \right]^n \qquad (32c)$$

where
$$\nabla f^n = f^n - f^{n-1}, \quad \nabla^2 f^n = f^n - 2f^{n-1} + f^{n-2},$$
$$\nabla^3 f^n = f^n - 3f^{n-1} + 3f^{n-2} - f^{n-3}, \quad \cdots$$

The computations of $\tilde{L}^*$ in Eq.(32) are performed using Eqs.(25) and (26). These equations are central-difference schemes added an artificial dissipation term for stabilization, and can be interpreted as the upstream-difference schemes. Hence, $\tilde{L}^*$ is expressed as follows.

$$\tilde{L}^* = \nabla_\xi \tilde{E}_1^+ + \Delta_\xi \tilde{E}_1^- + \nabla_\eta \tilde{E}_2^+ + \Delta_\eta \tilde{E}_2^- + \nabla_\zeta \tilde{E}_3^+ + \Delta_\zeta \tilde{E}_3^- + \tilde{R} \qquad (33)$$

NUMERICAL EXAMPLE

The transonic flows with a shock wave in a converging-diverging nozzle as shown in Fig.1 were computed. The area ratio of throat to inlet is 1:2. The computational grid had $37 \times 9 \times 9$ grid points, and was generated analytically. The computation was performed using Eqs.(24), (25), (26c) and (31). Fig.2 shows the computed pressure distributions on the solid wall and along the centre line as compared with the results of the one dimensional flow theory. And Fig.3 shows the Mach number contours on a plane containing the centre line.

CONCLUSIONS

An implicit time-marching method for analysing steady three-dimensional inviscid transonic flow problems has been proposed. This method is based on the well-known Beam-Warming delta-form approximate-factorization scheme, and improved in the following points. (i) In order to treat the solid wall boundary condition without difficulty, the momentum equations of contravariant velocity as the fundamental equations in curvilinear coordinates are used. (ii) To save the computer time and to increase the stability, the existing techniques of diagonalization and upstreaming are applied. Some explicit and implicit versions of the present method have been proposed. These methods are easily extended to the viscous flow problems.

REFERENCES

(1) Daiguji, H. and Yamamoto, S., "An Efficient Time-Marching Method for Solving the Compressible Euler Equations," Int. Symp. on Computational Fluid Dynamics—Tokyo, (1985-9), Vol.I, 217-228.
(2) Beam, R.M. and Warming, R.F., "An Implicit Factored Scheme for the Compressible Navier-Stokes Equations," AIAA J., 16-4(1978-4), 393-402.
(3) Steger, J.L., "Implicit Finite-Difference Simulation of Flow about Arbitrary Two-Dimensional Geometries," AIAA J., 16-7(1978-7), 679-686.
(4) Pulliam, T.H. and Chaussee, D.S., "A Diagonal Form of an Implicit Approximate-Factorization Algorithm," J. Comp. Phys., 39-2(1981-2), 347-363.
(5) Steger, J.L. and Warming, R.F., "Flux Vector Splitting of the Inviscid Gasdynamic Equations with Application to Finite-Difference Methods," J. Comp. Phys., 40-2(1981-4), 263-293.
(6) Kawamura, K. and Kuwahara, K., "Computation of High Reynolds Number Flow around a Circular Cylinder with Surface Roughness," AIAA Paper 84-0340, (1984-1).
(7) Leonard, B.P., "A Survey of Finite Differences with Upwinding for Numerical Modeling of the Incompressible Convective Diffusion Equations." Computational Techniques in Transient and Turbulent Flows, Taylor, C. and Morgan, K. ed., (1981), 1-35, Pineridge Press.

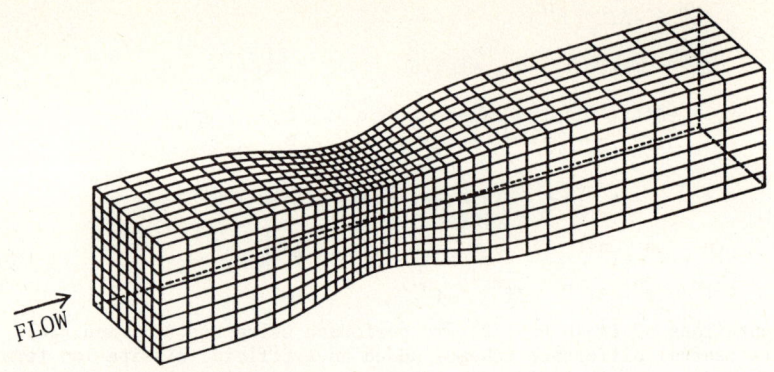

Fig.1　Converging-diverging nozzle and computational grid

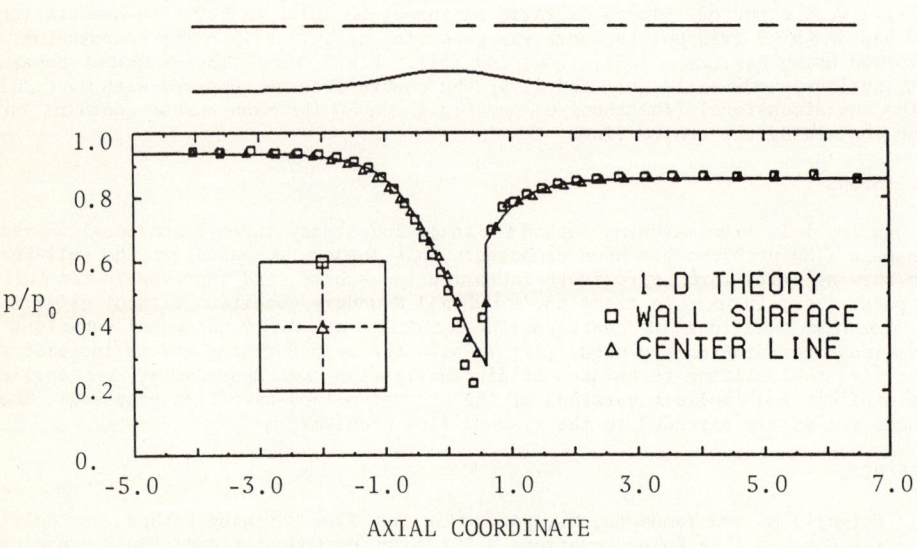

Fig.2　Comparison of pressure distributions

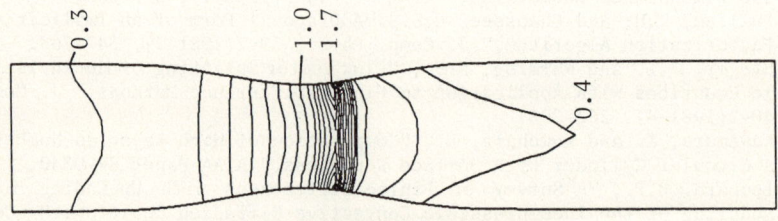

Fig.3　Mach number contours

LARGE EDDY SIMULATION OF A NARROW SOURCE OF PASSIVE SCALAR
IN HOMOGENEOUS STRAINED TURBULENCE

K. DANG & Y. F. MORCHOISNE

O.N.E.R.A. 29 Av de la Division Leclerc
BP 72, 92322 Chatillon Cedex - France.

1. Introduction

32^3 Large eddy simulations of a passive scalar in 3-D homogeneous turbulence submitted to constant mean velocity gradient flows have been performed on a Cray 1 computer with the aim of obtaining useful data for turbulence modelling.
As opposed to direct and large eddy simulations achieved by other authors, where the fluctuating scalar field is homogeneous (the scalar is present everywhere in the turbulent field at the initial time), we are interested here by narrow sources (small compared to the scale of the flow) which develop with time into an inhomogeneous scalar wake as sketched in figure 1. This type of diffusion is found in many engineering fluid flows and environment pollution situations.

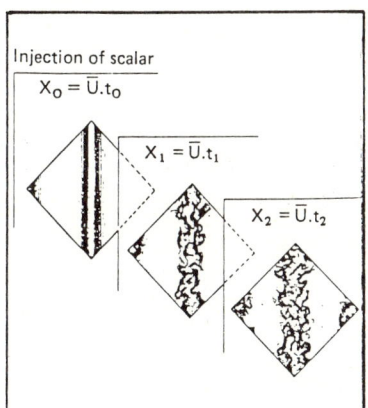
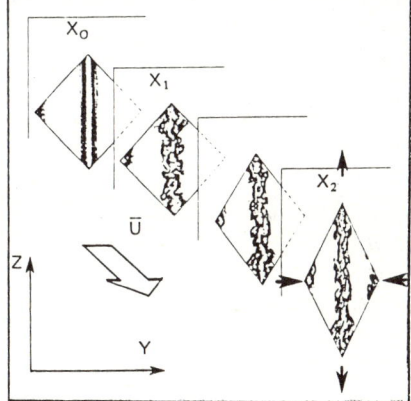

- Fig 1.a. Scalar wake in decaying   Fig 1.b. Scalar wake submitted to a plane
-         isotropic turbulence   strain (compressed wake)
-         Fig 1. Iso-scalar contours. Time evolution of the scalar wake
-                 in homogeneous strained turbulence.

2. Numerical method.

The semi-Lagrangian method used to solve the Navier-Stokes equations (spectral-collocation Fourier method in space with a second order finite-difference time scheme) has been described in (9). It has been extended to simulate the turbulent mixing of a passive scalar by solving, in addition the diffusion equation describing the scalar field (4).
Due to the low resolution of the 32^3 code, the drain of turbulent energy to the subgrid scales has to be modelled. This is done by an eddy viscosity model evaluated and used with succes in (2) for the simulation of homogeneous turbulence submitted to two successive plane strains and in (5) (6) for the simulation of the effect of rotation on turbulence with a 64.128^2 code running on a parrallel system. In this model, the eddy viscosity ν_u depends on the wave number k. It is defined in spectral space, in the frame following the distortion imposed by the mean flow: $\nu_u = C\sqrt{E(k_c)/k_c}(1+\lambda(k/k_c)^\mu)$ where k_c is the cut-off wave number of the sharp filter used, and $E(k_c)$ the turbulent kinetic energy at the cut-off.
The drain of the scalar variance is modelled in a similar way, via an eddy diffusivity ν_θ related

to ν_u by an eddy Prandtl number $Pt=\nu_u/\nu_\theta$ (4)(7).
The constants of the subgrid model have been evaluated (in particular Pt) to match the experimental results of Yeh & Van Atta (10) in the case of a heated isotropic grid turbulence. Results obtained for the decay rates of velocity and scalar are similar to those of Domis (7), the eddy Prandtl number has been found to be equal to $Pt=0.6$ as suggested by EDQNM analyses of Chollet & Lesieur (1).

3. Narrow line sources of scalar in homogeneous turbulence.

The aim of the present work is to detail the influence of imposed strains (plane compression or expansion) on statistical moments of the scalar field: scalar variance $\overline{\theta^2}$, velocity-scalar covariance function $\overline{u_i\theta}$ and correlations involved in their conservation equations.
Results will be compared mainly with the experimental work of Polychronidis & Marechal (8). The figure 1 visualizes by means of iso-scalar contours (in a cross section $X=Cte$ of the computational box) the time evolutions of the scalar (or temperature) wake in two of the three cases studied. A Gaussian initial temperature distribution is "injected" at time t_0 and diffuses under the action of the isotropic turbulent velocity field (fig 1.a). At time t_1, a plane compression (fig 1.b) or expansion (not shown here) of rate of strain D can be imposed to both velocity and scalar fields. The total strain $D(t-t_1)$ is kept inferior to 0.7 in order to limit the numerical problems resulting from the distortion of the computational domain (2).
Table 1 gives characteristic time and length scales of the velocity field at the time t_0 of injection of scalar and at time t_1 of imposition of the strain (note the big length scale of the initial scalar distribution compared to the experimental thermal source). Results are given here for one value $|D|=20$ of the rate of strain.

	$t_0=0.05$	$t_1=0.075$	$experiment(8)$
$\tau_u=q^2/\epsilon$	0.07	0.11	0.118
$\tau_D=1/D$	0.05	0.05	0.03
R_λ	17	19	48
σ/L	0.68	0.68	~ 0

Table1. Time,length scales of simulations. σ=wake's width. L=integral length scale

3.1. Mean Flow Quantities.

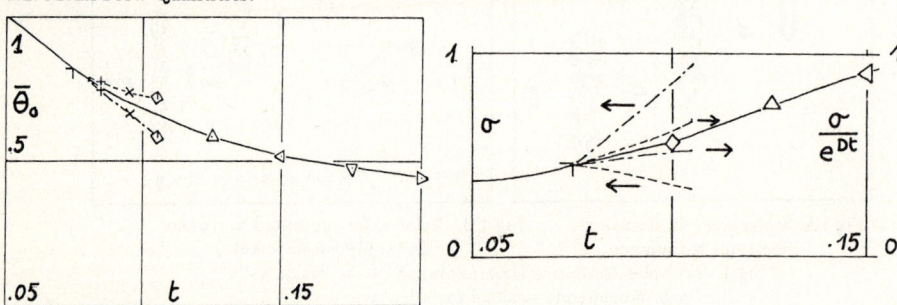

-Fig 2.a Temperature on wake's axis   Fig 2.b Width of the wake
-Fig 2. Influence of plane strains on mean flow quantities —— unstrained wake, - - - compressed wake,-- - -- expanded wake.

The mean temperature profiles $\overline{\Theta}(Y)$ (not shown here, and obtained in the unstrained case over 3 turnover times of the velocity field) remain Gaussian without and with imposed strain. The mean temperature value $\overline{\Theta}_0$ on the wake's axis decreases more (less) rapidly for the compressed (expanded) wake (figure 2.a), showing an enhanced (reduced) diffusion.
The width of the wake (measured here by the standard deviation σ of the Gaussian profile $\overline{\Theta}(Y)$) is greatly influenced by the deformation ratio e^{Dt} imposed to the mean streamlines (fig 2.b). The relative width σ/e^{Dt} (shown in the same figure) characteristic of the effective diffusion is greater (smaller) for the compressed (expanded) wake as compared to the case without strain.

3.2 Scalar Fluctuations and Moments.

Averaged profiles $\sqrt{\overline{\theta^2}}/\Theta_0$ of temperature fluctuations are shown on figure 3 for $0.05 < t < 0.2$ without strain (fig 3.a) and $0.05 < t < 0.08$ with strain (fig 3.b, 3.c).
Temperature fluctuations are much more important within the compressed wake and the small depression on the wake's axis is comparatively less pronounced due to the enhanced diffusion in that case. The expanded wake seems to reach a self similar state more rapidly than the unperturbed wake.

The same trends can be observed more quantitatively on the time evolution of the global quantity $I_\theta = \int \overline{\theta^2} \, dY / \int \overline{\Theta}^2 \, dY$ (fig 3.d). These results are in good qualitative agreement with experiment.

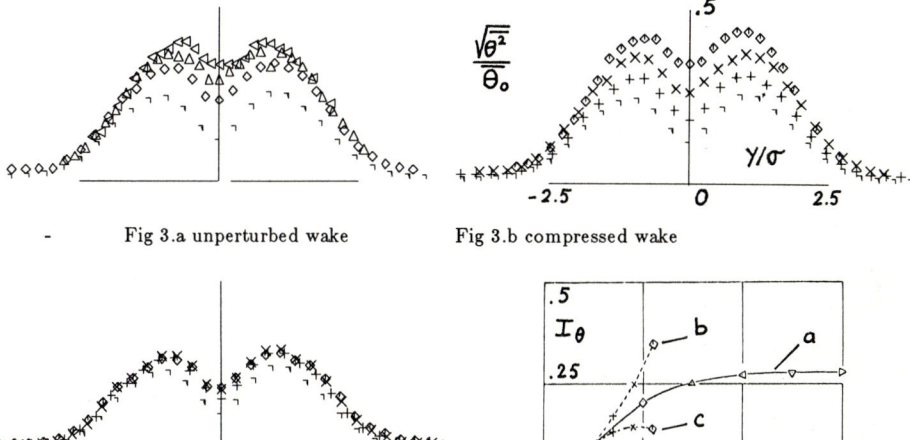

Fig 3.a unperturbed wake   Fig 3.b compressed wake

Fig 3.c expanded wake   Fig 3.d evolution of I_θ
Fig 3. Influence of plane strains on fluctuating scalar quantities

These results can be related to the fine structure of the temperature wake. Probability density functions (Pdf) of the temperature signal on the wake's axis are shown on figure 4 for the two perturbed wakes, near the end of the strains ($Dt = 0.6$). As commented in (8) for experimental Pdf, the probability to find on the axis, the ambient temperature ($\overline{\Theta} = 0$) is much greater inside the compressed wake. This greater intermittency of the compressed thermal wake is clearly evidenced by the instantaneous iso-scalar contours of figure 5, where the compressed wake has been redrawn in the expanded grid to make comparisons easier.

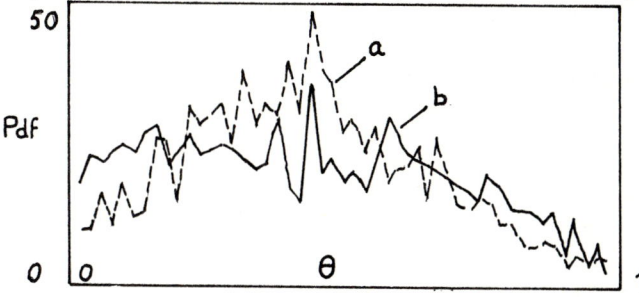

- Fig 4 Probability density functions of Θ. a.-compressed wake, b.-expanded wake.

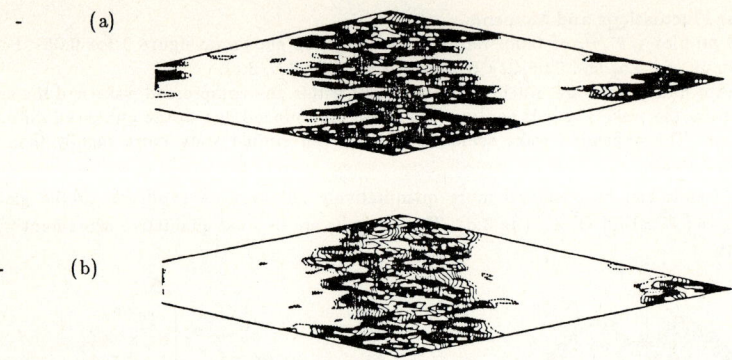

- (a)

- (b)

- Fig 5. Instantaneous iso-scalar contours. a.-compressed wake, b.-expanded wake

3.3 Scalar-velocity Correlation Functions
The turbulent flux $\overline{u_i\theta}$ is important only in the direction Y orthogonal to the line source ($\overline{u_2\theta} \gg \overline{u_1\theta}, \overline{u_3\theta}$). The time evolutions of the correlation coefficient $\overline{u_2\theta}/\sqrt{\overline{u_2^2}}\sqrt{\overline{\theta^2}}$ as functions of Y/σ are given in figure 6. These coefficients are rapidly decreased by the expansion, whereas they are sustained at higher values by the compression.

4. Data concerning turbulence modelling
Second and higher order moments involved in the conservation equations of $\overline{\theta^2}$ and $\overline{u\theta}$ have been evaluated. We present here only two examples of results interesting turbulence modelling.

1)- the ratio τ_θ/τ_u of characteristic times for the fluctuating scalar ($\tau_\theta = \overline{\theta^2}/\epsilon_\theta$) and velocity ($\tau_u = q^2/\epsilon$) are given in figure 7. In the central part of the three wakes ($-2. < Y/\sigma < 2.$), this ratio does not depend on X as found experimentally in heated grid experiments and varies with Y about a mean value of 0.5 measured in several equilibrium flows and used in current models.

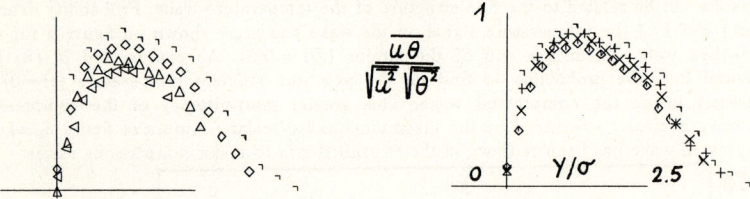

- Fig 6.a unperturbed wake   Fig 6.b compressed wake

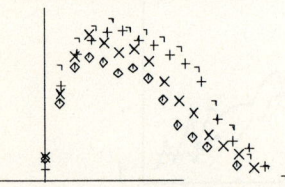

- Fig 6.c expanded wake   Fig 6. Correlation coefficient
                                                $\overline{u\theta}/\sqrt{\overline{u^2}}\sqrt{\overline{\theta^2}}$

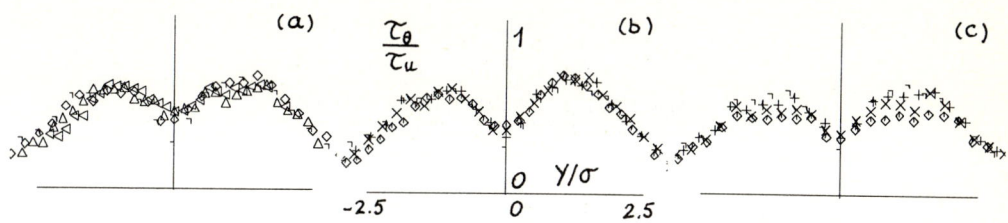

- Fig 7.a. unperturbed wake   Fig 7.b. compressed wake   Fig 7.c. expanded wake
- Fig 7. Ratio of Characteristic Times τ_θ/τ_u

2)- Production terms ($-\overline{u_2 u_k} \partial \overline{\Theta}/\partial X_k$), pressure-scalar terms ($\overline{(p/\rho)\, \partial \theta/\partial X_2}$) appearing in the conservation equation for $\overline{u_2 \theta}$ are shown on figure 8 for the unperturbed and compressed wakes. In the expansion case, the production decreases rapidly in time and the pressure-scalar correlation is too small to give precise results. As can be seen, production and pressure scalar correlation have opposite signs: the effect of the latter is to decrease $\overline{u_2 \theta}$, which confirms the model currently used for this term: $\overline{p/\rho \theta}_{,2} \sim -\overline{u_2 \theta}/\tau_\theta$.

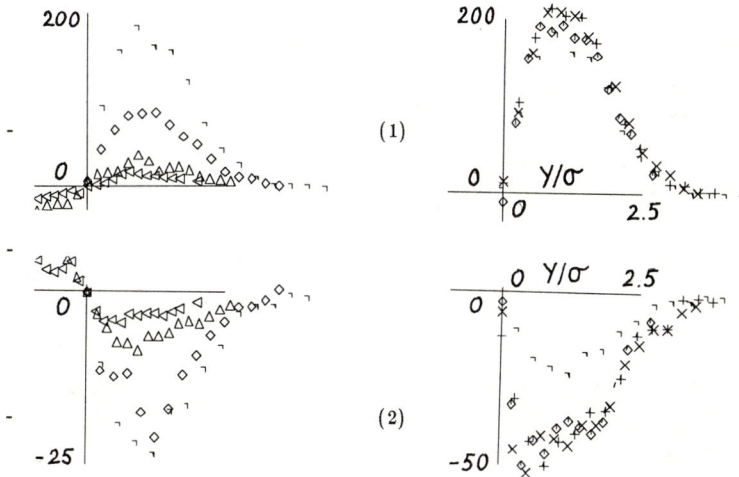

- Fig 8.a unperturbed wake   Fig 8.b compressed wake
- Fig 8. Time evolutions of 1. Production of $\overline{u_2 \theta}$: $-\overline{u_2 u_k}\,\overline{\Theta}_{,k}$ - 2. Pressure scalar correlation $\overline{(p/\rho)\theta}_{,2}$

5. Conclusion

Turbulent transport fluxes have been evaluated by large eddy simulation for an inhomogeneous thermal wake in homogeneous strained velocity fields. Results obtained are in very good qualitative agreement with available experimental ones. Evaluation of current models for these fluxes are underway, based on the numerical data.

6. References

1. Chollet J.P. - NCAR TN-206-STR, 1983.
2. Dang K. - AIAA J. Vol 23, Nb 2, Feb 1985, p221.
3. Dang K. - Euromech 180,Karlsruhe, RFA 4-6 July 1984. TP ONERA 1984-54
4. Dang K. - Laminar & Turbulent Flow, Pineridge Press 1985. ONERA TP 1985-73.
5. Dang K., Roy Ph. - Springer Verlag 1985 vol 230. ONERA TP 1985-14.
6. Dang K., Roy Ph. -TSF V 1985,TP ONERA 1985-88.
7. Domis A.M.- J. Fluid Mech (1981),Vol 104,pp 55-790 ,
8. Polychronidis H.C., Marechal J.- JMTA , Vol 2, N0 6,1983 pp 1013-1055.
9. Roy Ph. - 8th ICMNFD Springer Verlag, Lectures Notes in Physics Vol 170.
10. Yeh T.T., Van Atta C.W. - J. Fluid Mech 1973, Vol 58, Part 2, pp223-261.

CHARACTERISTIC DECOMPOSITION METHODS FOR THE MULTIDIMENSIONAL EULER EQUATIONS

H. Deconinck, Ch. Hirsch and J. Peuteman
Vrije Universiteit Brussel
Brussels, Belgium

INTRODUCTION

Various computational methods for solving the Euler equations in one space dimension make use of a decoupling of the system to a set of scalar simple wave transport equations. By decoupling the equations, the simple waves are separated and resolved independently by taking into account their individual character such as propagation direction and speed.

In two space dimensions, a simple decoupling of the equations is not possible since the Jacobian matrices in the x- and y-direction cannot be diagonalized simultaneously. This corresponds to the presence of an infinite number of simple wave solutions connected to the infinite set of directions present in a multidimensional space. The problem of identifying a limited number of physically relevant simple waves and their propagation direction was treated for the first time by Roe in 1985 [1], [2]. Roe proposes to decompose the local flow variations in contributions from 4 acoustic waves, an entropy wave and a shear wave. The parameters in the model are the orientation and strength of these waves, which are determined by identification to the local flow gradients.

In the present paper, a different approach is used for the determination of the physically relevant simple waves and their orientation, which is based on characteristic theory. This leads to a straightforward and algebraically simple decomposition consisting of two acoustic waves, one shear wave and one entropy wave. The propagation direction of the shear wave is parallel with the local pressure gradient and the propagation direction of the acoustic waves depends on the local strain rate tensor.

In the second part of the paper it is shown how the decomposition can be used to construct a genuinely two-dimensional upwind-differencing scheme.

CHARACTERISTIC DECOMPOSITION IN TWO SPACE DIMENSIONS

Consider the system of Euler equations in two space dimensions in conservative variables

$$\frac{\partial U}{\partial t} + \frac{\partial F(U)}{\partial x} + \frac{\partial G(U)}{\partial y} = 0 \quad \text{or} \quad \frac{\partial U}{\partial t} + A\frac{\partial U}{\partial x} + B\frac{\partial U}{\partial y} = 0 , \tag{1}$$

with A and B the Jacobian matrices in the x- and y-direction and F and G the fluxes of U in the x- and y-direction:

$$U = \begin{vmatrix} \rho \\ \rho u \\ \rho v \\ \rho E \end{vmatrix}, \quad F = \begin{vmatrix} \rho u \\ \rho u^2 + p \\ \rho uv \\ \rho uH \end{vmatrix}, \quad G = \begin{vmatrix} \rho v \\ \rho uv \\ \rho v^2 + p \\ \rho vH \end{vmatrix}. \tag{2}$$

The locally linearized system (1) is transformed to a set of characteristic variables W, such that $\partial U = L^{-1} \partial W = \Sigma r^\alpha \partial W^\alpha$, where the particular form of ∂W and L^{-1} (with columns r^α) will be determined in the following. Substituting in the quasi-linear Euler equations, one obtains

$$\frac{\partial W}{\partial t} + LAL^{-1} \frac{\partial W}{\partial x} + LBL^{-1} \frac{\partial W}{\partial y} = 0. \tag{3}$$

A simple decoupling as in the 1-D case, by constructing the matrix L such that the Jacobians LAL^{-1} and LBL^{-1} are diagonal, is not possible since A and B do not commute, i.e. they have distinct eigenvalues. However, it might be possible to define L and W such that

$$LAL^{-1} = D^x + C^x, \quad LBL^{-1} = D^y + C^y, \quad \text{with} \quad C^x \frac{\partial W}{\partial x} + C^y \frac{\partial W}{\partial y} = 0, \tag{4}$$

and where D^x and D^y are diagonal matrices with diagonal elements λ_x^α resp. λ_y^α. If such a set of variables W with corresponding matrix L can be constructed, the Euler system eq. (3) is reduced to the decoupled form

$$\frac{\partial W}{\partial t} + D^x \frac{\partial W}{\partial x} + D^y \frac{\partial W}{\partial y} = 0 \quad \text{or} \quad \frac{\partial W^\alpha}{\partial t} + \lambda_x^\alpha \frac{\partial W^\alpha}{\partial x} + \lambda_y^\alpha \frac{\partial W^\alpha}{\partial y} = 0, \quad \alpha=1,4. \tag{5}$$

This equation expresses the transport of the simple wave W^α along a characteristic curve with slope $(\lambda_x^\alpha, \lambda_y^\alpha, 1)$ in the x-y-t-space.

A guideline for the construction of the variables W is provided by characteristic theory: the 2D Euler equations are written as a set of compatibility relations on characteristic surfaces with normal $\vec{N}=(N_x, N_y, N_t)$ given by

$$(\vec{V}.\vec{N})^2 [(\vec{V}.\vec{N})^2 - c^2] = 0, \tag{6}$$

where $\vec{N}$ has been scaled such that $N_x^2 + N_y^2 = 1$ and $\vec{V} = (u,v,1)$ is directed along the pathline in the x-y-t-space. The first factor in eq.(6) describes the streamsurfaces and the second factor the wavesurfaces with the Mach conoid as envelope (fig.1). The set of compatibility relations equivalent to the Euler equations is then given by

$$\begin{array}{ll}
\partial_t W^1 + \vec{u}.\vec{\nabla} W^1 = 0 & \partial W^1 = \partial \rho - \frac{1}{c^2} \partial p \\
\partial_t W^2 + \vec{u}.\vec{\nabla} W^2 + \frac{1}{\rho} \vec{S}.\vec{\nabla} p = 0 & \partial W^2 = \vec{S}.\partial \vec{u} \\
\partial_t W^3 + (\vec{u} + c\vec{N}).\vec{\nabla} W^3 + c\vec{S}.(\vec{S}.\vec{\nabla})\vec{u} = 0 & \partial W^3 = \vec{N}.\partial \vec{u} + \frac{1}{\rho c} \partial p \\
\partial_t W^4 + (\vec{u} - c\vec{N}).\vec{\nabla} W^4 + c\vec{S}.(\vec{S}.\vec{\nabla})\vec{u} = 0 & \partial W^4 = -\vec{N}.\partial \vec{u} + \frac{1}{\rho c} \partial p
\end{array} \tag{7}$$

Here, $\vec{S}$ is a unit vector along the intersection of the characteristic surface with

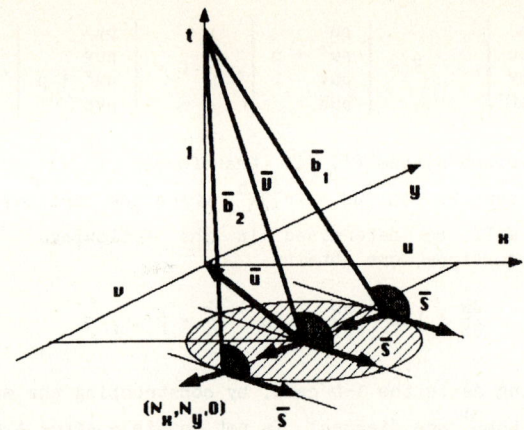

Figure 1 : characteristic surfaces and Mach cone

the x-y-plane, $\vec{S} = (N_y, -N_x, 0)$. Hence, $\vec{S}$ lies in the wave front surface and $(N_y, N_x, 0)$ is the direction of propagation of the wave front in the x-y plane.

System (7) can be identified to (5) provided that the coupling terms in (7) can be made to vanish. This is obtained for a particular choice of the characteristic surfaces such that locally

$$\vec{S} \cdot \vec{\nabla} p = 0 \qquad \text{for the second compatibility equation and}$$
$$\vec{S} \cdot (\vec{S} \cdot \vec{\nabla}) \vec{u} = 0 \qquad \text{for the third and fourth compatibility equation.} \qquad (8)$$

Hence, the characteristic variables are defined as in (7), where $\vec{S}$ (or N_x and N_y) are determined from (8a) for ∂W^2, and from (8b) for ∂W^3, ∂W^4 and the corresponding compatibility relations. The decomposed system, eq. (5) is now completely defined by the following relations

$$\partial W = \begin{vmatrix} \partial \rho - \frac{1}{c^2} \partial p \\ \vec{S}_1 \cdot \partial \vec{u} \\ \vec{N}_2 \cdot \partial \vec{u} + \frac{1}{\rho c} \partial p \\ -\vec{N}_2 \cdot \partial \vec{u} + \frac{1}{\rho c} \partial p \end{vmatrix} \qquad (9)$$

$$D^x = \begin{vmatrix} u & & & \\ & u & & \\ & & u+cN_{2x} & \\ & & & u-cN_{2x} \end{vmatrix}, \quad D^y = \begin{vmatrix} v & & & \\ & v & & \\ & & v+cN_{2y} & \\ & & & v-cN_{2y} \end{vmatrix}, \qquad (10)$$

with the unit vectors $\vec{S}_1$ and $\vec{S}_2$ obtained from

$$\vec{S}_1 \cdot \vec{\nabla} p = 0, \qquad \vec{S}_2 \cdot (\vec{S}_2 \cdot \vec{\nabla}) \vec{u} = 0 \qquad (11)$$

and $\vec{N}_1$, $\vec{N}_2$ unit normals to $\vec{S}_1$ resp. $\vec{S}_2$ in the x-y plane. Figure 2 illustrates the intersection of the three characteristic surfaces with the x-y plane. $\vec{S}_1$ is the wave front for the entropy and shear waves, eq. (7a) and (7b), and the two vectors $\vec{S}_2$ on opposite sides of the Mach cone are the wave fronts for the two acoustic waves, eq. (7c) and (7d).

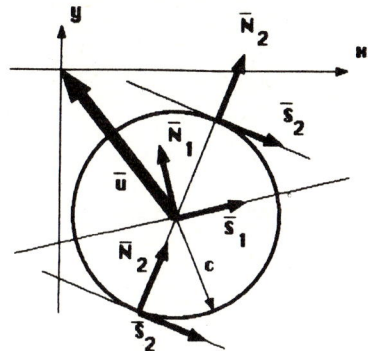

Figure 2 : Simple wave fronts and propagation direction for two-dimensional flow.

The transformation matrix L^{-1} with columns r^α is readily obtained from the transformation $\partial U = L^{-1} \partial W = \Sigma r^\alpha \partial W^\alpha$, with $P = N_{1x} N_{2x} + N_{1y} N_{2y} = \vec{N}_1 \cdot \vec{N}_2$:

$$L^{-1} = \begin{vmatrix} 1 & 0 & \dfrac{\rho}{2c} & \dfrac{\rho}{2c} \\ u & \rho \dfrac{N_{2y}}{P} & \dfrac{\rho}{2}[\dfrac{u}{c} + \dfrac{N_{1x}}{P}] & \dfrac{\rho}{2}[\dfrac{u}{c} - \dfrac{N_{1x}}{P}] \\ v & -\rho \dfrac{N_{2x}}{P} & \dfrac{\rho}{2}[\dfrac{v}{c} + \dfrac{N_{1y}}{P}] & \dfrac{\rho}{2}[\dfrac{v}{c} - \dfrac{N_{1y}}{P}] \\ \dfrac{u^2+v^2}{2} & \rho \dfrac{uN_{2y} - vN_{2x}}{P} & \dfrac{\rho H}{2c} + \rho \dfrac{uN_{1y} + vN_{1x}}{2P} & \dfrac{\rho H}{2c} - \rho \dfrac{uN_{1y} + vN_{1x}}{2P} \end{vmatrix} \quad (12)$$

The characteristic wave vectors $\vec{N}_1$ and $\vec{N}_2$ have a clear physical interpretation in terms of the local flow variables : obviously, $\vec{N}_1$ is in the direction of the local pressure gradient. On the other hand, the relation for $\vec{S}_2$, eq. (11b), can be written explicitly in terms of the local velocity gradient (or strain rate) tensor ϵ :

$$I = \vec{S}_2 \cdot (\vec{S}_2 \cdot \vec{\nabla})\vec{u} = S_{2x}^2 \epsilon_x + 2 S_{2x} S_{2y} \epsilon_{xy} + S_{2y}^2 \epsilon_y = 0$$
$$\text{with } \epsilon_{ij} = \frac{1}{2}(\partial_i u_j + \partial_j u_i) \quad (13)$$

and in a local orthogonal coordinate system with axes X and Y directed along the principal axes of the strain rate tensor :

$$I = \vec{S}_2 \cdot (\vec{S}_2 \cdot \vec{\nabla})\vec{u} = S_{2X}^2 \epsilon_X + S_{2Y}^2 \epsilon_Y = 0, \quad (14)$$

with ϵ_X and ϵ_Y the eigenvalues of ϵ_{ij} corresponding to the principal directions X resp. Y. Hence, ϵ_X and ϵ_Y are the minimum and maximum normal strain rates in the fluid.

Equation (13) or (14) allows two real solutions for the direction of $\vec{S}_2$, provided that the discriminant $\epsilon_{xy}^2 - \epsilon_x \epsilon_y = -\epsilon_X \epsilon_Y \geq 0$. If the discriminant is negative, a complete decoupling is not possible and the direction of $\vec{S}_2$ is determined such that the coupling term $|cI|$ is minimal. Hence, the following two cases are to be considered :

<u>CASE 1</u> : $-\epsilon_X \epsilon_Y = \epsilon_{xy}^2 - \epsilon_x \epsilon_y \geq 0$

one obtains for the solution of eq. (14)

$$\tan\theta = \frac{S_{2Y}}{S_{2X}} = \pm \sqrt{-\epsilon_X/\epsilon_Y} \qquad \text{and} \qquad |I| = 0 \qquad (15)$$

This situation corresponds to a complete decoupling of the Euler equations and $\vec{S}_2$ makes an angle θ with the principle axis X, (fig. 3).

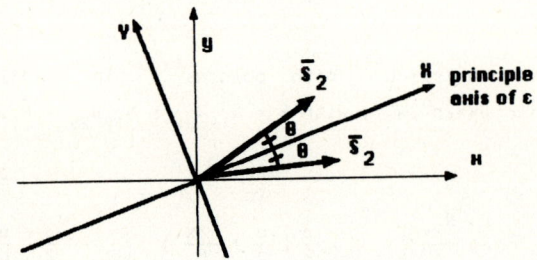

Figure 3 : Determination of wave front vector $\vec{S}_2$

In particular, for incompressible flow, one obtains $\theta = \pi/4$ and $\vec{S}_2$ is the direction of maximum shearing stress in the fluid.

<u>CASE 2</u> : $-\epsilon_X \epsilon_Y = \epsilon_{xy}^2 - \epsilon_x \epsilon_y < 0$

A complete decoupling is not possible. A maximal decoupling is obtained if the direction of $\vec{S}_2$ is chosen along the principal axis corresponding to the minimum strain rate in the fluid. The remaining coupling term in the third and fourth wave equation is then minimal, equal to $c.\min(|\epsilon_X|, |\epsilon_Y|)$

A further investigation of this condition for complete decoupling has been performed in the case of stationary flow. It follows from this analysis that complete decoupling is always possible if the local flow is subsonic. For supersonic flow, a bounded region for the parameter v_y/u_x can be defined where complete decoupling is not possible. If this parameter lies out of the interval, complete decoupling is also possible for supersonic flow.

Extension to three dimensional flows, although not presented here, has been examined and is straightforward, requiring no more than two propagation directions [3] : the two shearwaves are propagated in the pressure gradient direction and the two acoustic waves in directions connected to the strain rate tensor.

EXPLICIT CHARACTERISTIC DECOMPOSITION SCHEME

The above decomposition can be used in various ways to construct genuinely upwind schemes for the linearized Euler equations in two space dimensions. As an illustration, consider the following first order upwind scheme on an orthogonal Cartesian mesh

$$U_i^{n+1} - U_i^n = -\frac{\Delta t}{\Delta x}(F_{i+1/2,j}^n - F_{i-1/2,j}^n) - \frac{\Delta t}{\Delta y}(G_{i,j+1/2}^n - G_{i,j-1/2}^n) \qquad (16)$$

where the numerical fluxes are defined in a true upwind way as follows, e.g. for the x-component [2]:

$$F_{i+1/2,j} = F(U_{i+1/2,j})$$

where

$$U_{i+1/2,j} = \frac{U_{i,j} + U_{i+1,j}}{2} - \frac{1}{2} \delta_x U_{i+1/2,j}^+ + \frac{1}{2} \delta_y U_{i+1/2,j}^- \qquad (17)$$

with

$$\delta_x U^+ = \sum_{\alpha+} r^{\alpha+} \delta_x W^{\alpha+}, \qquad \delta_x U^- = \sum_{\alpha-} r^{\alpha-} \delta_x W^{\alpha-}, \qquad \delta_x [\,]_{i+1/2,j} = [\,]_{i+1,j} - [\,]_{i,j}$$

Here, $\alpha+$ denotes summation over all characteristics with positive speeds λ_x^α and $\alpha-$ over all characteristics with negative speed λ_x^α. The characteristic speeds and the characteristic variations are obtained from the decomposition, eq. (9) and (10), and r^α are the colums of matrix L^{-1}, eq. (12). These quantities are intrinsic to the flow. In a classical approach, e.g. based on the Steger and Warming flux vector splitting [4], λ_x^α and r^α are the eigenvalues and eigenvectors of the Jacobian A, which is a one-dimensional splitting in the x-direction.

Implicit and second order extensions of the above scheme in a finite volume approach for arbitrary grids can be considered along the same principles as for classical upwind schemes.

CONCLUSION

An algebraically simple decomposition of the Euler equations for two-dimensional flows has been presented. This decomposition is based on characteristic theory and makes use of two acoustic waves with orientation depending on the local strain rate tensor, and one entropy and shear wave with orientation parallel to the local pressure gradient.

REFERENCES

[1] P. L. ROE, J. Comp. Phys., Vol. 63, 458-476, 1986
[2] P. L. ROE, 'A Basis for Upwind Differencing of the Two-Dimensional Unsteady Euler Equations', Proceedings of CFD Conference, Reading University, april 1985.
[3] CH. HIRSCH, 'A diagonalization procedure for the multidimensional Euler equations', Proc. 1985 Seminar on High Speed Aerodynamics, T.H. Aachen, Ed. by A. Nastase, 1986
[4] J. L. STEGER AND R.F. WARMING, J. Comp. Phys., Vol. 40, 263-293, 1981

GENERALIZED FINITE DIFFERENCES FOR OPERATORS OF NAVIER-STOKES TYPE

S.C.R. Dennis and   Q. Wing
University of Western Ontario
London, Canada

INTRODUCTION

In this presentation we consider the approximation of operators of the form $L = \partial^2/\partial x^2 - u\partial/\partial x$ by a generalized finite-difference procedure. The operator L is of Navier-Stokes type and satisfies an equation of the form

$$L\phi = \phi'' - u\phi' = r(x), \qquad (1)$$

where x is a space variable and a prime denotes differentiation with respect to x. Once we have set up a method to deal with (1) it is straightforward to write down approximations to equations in one space dimension and time t such as

$$\partial\phi/\partial t = \partial^2\phi/\partial x^2 - u\partial\phi/\partial x \qquad (2)$$

and in two space dimensions such as

$$\partial\phi/\partial t = \partial^2\phi/\partial x^2 + \partial^2\phi/\partial y^2 - u\partial\phi/\partial x - v\partial\phi/\partial y, \qquad (3)$$

or in more space dimensions. Some simple examples of approximations to equations of these types will be mentioned which illustrate results of different orders of accuracy, depending on the order of approximation.

The basis of the generalized finite differences is simply a procedure which follows the structure of the formal exact solution of (1), considered as an ordinary differential equation. Such a solution can be written as

$$\phi(x) = Af(x) + Bg(x) + F(x) \qquad (4)$$

where A and B are constants. The functions f, g and F must be found from the differential equation (1) and then if boundary conditions, e.g. two-point conditions, are given one can determine A and B and hence the solution for $\phi(x)$.

Exactly the same expression (4) may be used locally over a restricted domain of x which forms part of the grid structure used in a numerical solution, e.g. over the three grid points at $x = x_o - h$, x_o, $x_o + h$. The new point is that now the functions f, g and F will be some form of approximating functions to a given order of accuracy, rather than exact solutions. If (4) is fitted to values of $\phi(x)$ at $x = x_o - h$, x_o, $x_o + h$ one obtains the three-point approximation

$$c(x_o - h)\phi(x_o - h) - c(x_o)\phi(x_o) + c(x_o + h)\phi(x_o + h) = G(x_o) \qquad (5)$$

where the coefficients $c(x_o - h)$, $c(x_o)$, $c(x_o + h)$ can be identified in terms of grid values of f, g and F. To simplify the notation we denote

values of quantities at $x = x_0 - h$, x_0, $x_0 + h$ by subscripts 3, 0, 1 (in accordance with the Southwell notation, cf. Smith, 1978, p. 219). Then
$$c_0 = f_1 g_3 - f_3 g_1, \quad c_1 = f_0 g_3 - f_3 g_0, \quad c_3 = f_1 g_0 - f_0 g_1 \tag{6}$$
and
$$G_0 = F_1 c_1 - F_0 c_0 + F_3 c_3. \tag{7}$$

With a given choice of functions f, g and F equation (5) is a generalized three-point difference equation. It is natural to think of the functions f, g, F as power series in x since this forms the basis of most finite-difference methods of approximation. This is not, however, necessary as long as the functions in (4) model in some way the differential equation (1). A previously published example of obtaining an approximation to (1) using non-polynomial functions in (4) is the method of Allen and Southwell (1955). In this the equation (1) is modelled approximately by assuming that $\phi(x)$ satisfies
$$\phi'' - u_0 \phi' = r_0 \tag{8}$$
in $x_0 - h \leq x \leq x_0 + h$. The exact solution gives
$$f(x) = 1, \quad g(x) = \exp(u_0 x), \quad F(x) = -r_0 x / u_0 \tag{9}$$
and from this it is readily deduced that, after division by a suitable factor, (5) takes the form
$$E^{-1} \phi(x_0 - h) - (E + E^{-1}) \phi(x_0) + E \phi(x_0 + h) + r_0 (E - E^{-1}) h / u_0 = 0 \tag{10}$$
where $E = \exp(-\tfrac{1}{2} h u_0)$. Nevertheless, methods using power series are often extremely accurate and effective so we now develop the method far enough to give a compact h^4-accurate approximation to (1) involving only ϕ and not its derivative ϕ'.

APPROXIMATION USING POWER SERIES

As a simple illustration of the power series method, we temporarily make x_0 the origin of x and put
$$f = 1, \quad g = x + \tfrac{1}{2} u_0 x^2, \quad F = \tfrac{1}{2} r_0 x^2 \tag{11}$$
to evaluate the coefficients (6) and (7). On substitution in (5) and division by a factor we obtain the h^2-accurate central-difference approximation
$$(1 + \tfrac{1}{2} u_0 h) \phi(x_0 - h) - 2\phi(x_0) + (1 - \tfrac{1}{2} u_0 h) \phi(x_0 + h) - h^2 r_0 = 0. \tag{12}$$
We note that in this case (4) is nothing more than the second-degree Maclaurin polynomial with ϕ_0'' eliminated in terms of ϕ_0' using the differential equation (1). In forming (12) the first derivative ϕ_0' is eliminated completely at the start. This is basic in this method of

approximation, which is simply a finite-difference adaptation of the Taylor-Maclaurin series method of solving differential equations.

An h^4-accurate approximation to (1) is obtained by taking, with x_o as origin of x,

$$f = 1, \quad g = x + \tfrac{1}{2}u_o x^2 + (u_o^2 + u_o')x^3/6 + (u_o^3 + 3u_o u_o' + u_o'')x^4/24,$$

$$F = \tfrac{1}{2}r_o x^2 + (u_o r_o + r_o')x^3/6 + [(u_o^2 + 2u_o')r_o + u_o r_o' + r_o'']x^4/24. \tag{13}$$

The coefficients obtained from (6) and (7) are then found to be

$$c_1, c_3 = 1 \mp \tfrac{1}{2}u_o h + (u_o^2 + u_o')h^2/6 \mp (u_o^3 + 3u_o u_o' + u_o'')h^3/24 \tag{14a}$$

$$c_o = 2 + (u_o^2 + u_o')h^2/3 \tag{14b}$$

$$G_o = r_o h^2 + [(u_o^2 + 4u_o')r_o - u_o r_o' + r_o'']h^4/12, \tag{14c}$$

where in (14a) the upper sign on the right-hand side applies to c_1 and the lower to c_3. The essential point about the approximation (5) with the coefficients (14) is that it remains compact, involving only the three points $x_o - h$, x_o and $x_o + h$, even when $u(x)$ is a function whose derivatives must be expressed using finite differences. No further accuracy is lost if central differences are used. For example the terms in $u_o' h^2$ in (14a,b) appear on the left-hand side of (5) as

$$u_o' h^2[\phi(x_o - h) - 2\phi(x_o) + \phi(x_o + h)]/6 \tag{15}$$

which is of order h^4. The central-difference formula for u_o' is $u_o' = [\phi(x_o + h) - \phi(x_o - h)]/2h + O(h^2)$ so neglect of the error term in this result still leaves the corresponding approximation to (15) within the required order of accuracy for (5), with coefficients defined by (14), to give an h^4-accurate approximation to (1).

This h^4-accurate approximation is of quite similar form to that given by Dennis (1985), Dennis and Hudson (1985). It is not precisely the same and will naturally give somewhat different results when applied to ordinary differential equations of type (1) or time-dependent problems of type (2). However, by suitable expansions of terms in (14c) we can identify (5) and (14) with the approximation given by Dennis (1985, Eqs. (21)-(23)). The steps in the identification are to express the derivatives r_o', r_o'' in (14c) in central differences and to write

$$(u_o^2 + 4u_o')r_o = (u_o^2 + 4u_o')(\phi_o'' - u\phi_o') \tag{16}$$

using (1) and then express the derivatives ϕ_o', ϕ_o'' in central differences. After simplification of terms in (5) we arrive at Eqs. (21)-(23) of Dennis (1985)

This reduction is not given specifically in the present section since (5) and (14) adequately give an h^4-accurate representation of (1) as they stand. The reduction is considered in the following section since it occurs quite naturally there when the case of a two-dimensional partial differential equation is considered. For the present case of (1) we can continue to obtain approximations of h^6 and higher accuracy by present methods in the same compact form (5), except that derivatives of u higher than the second at $x = x_0$ start to enter the coefficients c_0, c_1, c_3. If $u(x)$ and $r(x)$ in (1) are given as analytical expressions we can evaluate the derivatives exactly and thus find truly three-point approximations (5) of high accuracy. If $u(x)$ is given numerically then accuracy of order higher than h^4 requires more points than three to approximate higher derivatives of $u(x)$. At any rate the method of procedure is clear and we shall not pursue it further here.

POWER SERIES APPROXIMATIONS IN TWO DIMENSIONS

Equations of time-dependent type (3) can easily be considered once we have approximated the operator on the right-hand side. In this section we consider the equation

$$\partial^2\phi/\partial x^2 + \partial^2\phi/\partial y^2 - u\partial\phi/\partial x - v\partial\phi/\partial y = 0 \tag{17}$$

by means of methods of the previous section. Equation (17) is split, following the method used by Dennis (1985), Dennis and Hudson (1985) into the two equations

$$\partial^2\phi/\partial x^2 - u\partial\phi/\partial x = r(x,y); \quad \partial^2\phi/\partial y^2 - v\partial\phi/\partial y = -r(x,y) . \tag{18a,b}$$

Along the grid line $y = y_0$ of constant y an approximation to (18a) over the three points $(x_0 - h, y_0)$, (x_0, y_0) and $(x_0 + h, y_0)$ can be written in the slightly more generalized form of (5) given by

$$c_3\phi(x_0 - h, y_0) - c_0\phi(x_0, y_0) + c_1\phi(x_0 + h, y_0) = G(x_0, y_0). \tag{19}$$

Again we can use any approximating function of type (4) to determine the coefficients c_n and G in (19) along the line $y = y_0$. In the present section we shall consider only the h^4-accurate method using power series; in this case the corresponding coefficients are defined by (14), where the prime denotes partial differentiation with regard to x along the line $y = y_0$ and subscript o refers to values at (x_0, y_0). The h^2-accurate method is a much simpler application and will not be considered.

Equation (18b) is now approximated in a similar way along the grid line $x = x_0$ over the three points $(x_0, y_0 - h)$, (x_0, y_0) and $(x_0, y_0 + h)$ using the same methods of the previous section. The approximation yields a similar formula to (19) given by

$$c_4\phi(x_0, y_0 - h) - c_0^*\phi(x_0, y_0) + c_2\phi(x_0, y_0 + h) = G^*(x_0, y_0). \tag{20}$$

Again we can use any modelling procedure for (18b), but we shall here restrict ourselves to the h^4-accurate power series method. The coefficients c_2, c_o^* and c_4 denote values associated with the three points $(x_o, y_o + h)$, (x_o, y_o) and $(x_o, y_o - h)$ respectively. For the h^4-accurate modelling using power series they are defined by reference to (14) to be

$$c_2, c_4 = 1 \mp \tfrac{1}{2} v_o h + (v_o^2 + v_o')h^2/6 \mp (v_o^3 + 3v_o v_o' + v_o'')h^3/24 \qquad (21a)$$

$$c_o^* = 2 - (v_o^2 + v_o')h^2/3 \qquad (21b)$$

$$G_o^* = -r_o h^2 + [(v_o^2 + 4v_o')r_o - v_o r_o' + r_o'']h^4/12 . \qquad (21c)$$

Here the upper sign in (21a) again applies to c_2 and the lower sign to c_4. The subscript o still refers to values at (x_o, y_o) but the prime in (21) now refers to partial differentiation with regard to y.

There is nothing basically new in the method of setting up the simultaneous approximations (19) and (20) corresponding respectively to (18a,b). It was used by Dennis (1985), Dennis and Hudson (1985) and previously by Allen and Southwell (1955) in setting up their exponential method of approximation and again by Dennis (1960), Dennis and Hudson (1979) in deriving more generalized exponential approximations. However, the formulae (14) and (21) appropriate to the h^4-accurate power series approximation do appear to be new. In order to obtain the final approximation to (17) in this case, we add (19) to (20). The terms in $r_o h^2$ cancel leaving only correction terms of order h^4 on the right-hand side and a combination of terms involving values of ϕ with associated coefficients at the points (x_o, y_o), $(x_o \pm h, y_o)$, $(x_o, y_o \pm h)$ on the left-hand side. If we neglect the h^4 terms on the right-hand side and terms in h^2 and higher powers in (14a,b) and (21a,b) we obtain the usual h^2-accurate approximation to (17).

On the other hand, if we retain all terms in (14) and (21) we obtain an h^4-accurate approximation. To do this we follow the principle of the method used by Dennis and Hudson (1985). The derivatives $r_o' = (\partial r/\partial x)_o$, $r_o'' = (\partial^2 r/\partial x^2)_o$ in (14c) are evaluated by partial differentiation of $r(x,y)$ defined by (18b). The derivatives $r_o' = (\partial r/\partial y)_o$, $r_o'' = (\partial^2 r/\partial y^2)_o$ in (21c) are likewise obtained by partial differentiation of $r(x,y)$ defined by (18a). All partial derivatives of ϕ resulting from this process are then approximated using central differences; this brings in values of ϕ at $(x_o + h, y_o + h)$, $(x_o - h, y_o + h)$ $(x_o - h, y_o - h)$ and $(x_o + h, y_o - h)$ which are denoted respectively by ϕ_5, ϕ_6, ϕ_7 and ϕ_8. The term of order $h^4 r_o$ in (14c) is replaced by the right-hand side of (16) and the partial derivatives of ϕ with regard to x on this right-hand side are expressed in central differences. Similarly, the term

r_o in the expression $(v_o^2 + 4v_o')r_o h^4/12$ on the right-hand side of (21c) is written as $r_o = -(\partial^2 \phi/\partial y^2)_o + v_o(\partial \phi/\partial y)_o$ using (18b) and the derivatives of ϕ expressed in terms of central differences. After simplifications of various similar terms which arise, the equation (17) is approximated by an h^4-accurate formula

$$\sum_{n=1}^{8} d_n \phi_n - d_o \phi_o = 0 \tag{22}$$

where

$$d_o = 4o + 2h^2(u_o^2 + v_o^2) - 4h^2[(\partial u/\partial x)_o + (\partial v/\partial y_o)] \tag{23a}$$

$$d_1, d_3 = 8 \mp 4hu_o + h^2[u_o^2 - 2(\partial u/\partial x)_o]$$
$$\pm \tfrac{1}{2}h^3[u_o(\partial u/\partial x)_o + v_o(\partial u/\partial y_o) - (\nabla^2 u)_o] \tag{23b}$$

$$d_2, d_4 = 8 \mp 4hv_o + h^2[v_o^2 - 2(\partial v/\partial y)_o]$$
$$\pm \tfrac{1}{2}h^3[u_o(\partial v/\partial x)_o + v_o(\partial v/\partial y)_o - (\nabla^2 v)_o] \tag{23c}$$

$$d_5, d_7 = 2 \mp h(u_o + v_o) + \tfrac{1}{2}h^2(u_o v_o - H_o) \tag{23d}$$

$$d_6, d_8 = 2 \pm h(u_o - v_o) - \tfrac{1}{2}h^2(u_o v_o - H_o). \tag{23e}$$

In (23b-e) the upper and lower signs on the right-hand side again correspond respectively to the first and second terms on the left, $H_o = (\partial u/\partial y)_o + (\partial v/\partial x)_o$ and $\nabla^2 = \partial^2/\partial x^2 + \partial^2/\partial y^2$. The derivatives in (23) can be approximated by central-difference formulae without any loss of h^4 accuracy. The terms in the last bracket in (23a) vanish by continuity if (u,v) are the velocity components of an incompressible fluid. The inclusion of a term $R(x,y)$ on the right-hand side of (17) requires only that a set of terms

$$h^2[8R_o + (1-\tfrac{1}{2}hu_o)R_1 + (1-\tfrac{1}{2}hv_o)R_2 + (1+\tfrac{1}{2}hu_o)R_3 + (1+\tfrac{1}{2}hv_o)R_4] \tag{24}$$

be added to the right-hand side of (22).

NUMERICAL ILLUSTRATIONS

It may be noted that the formulae (23) which define the h^4-accurate approximation (22) to (17) are exactly those given by Dennis and Hudson (1985) by considerably different methods. The terms (24) which must be added to the right-hand side of (22) when $R(x,y)$ is present on the right-hand side of (17) are also identical. Thus the illustrations of two-dimensional steady-state problems given by Dennis and Hudson (1985) also serve to illustrate the present method. In order to illustrate the approximation (5) to (1) with the coefficients (14) we have taken the time-dependent problem (2) considered by Dennis and Hudson (1985) in which $\phi = \phi(x,t)$ with

$$u = -2x; \quad \phi(x,0) = 0, \quad x > 0; \quad \phi(0,t) = 1, \quad \phi(\infty,t) = 0, \quad t > 0. \quad (25)$$

This problem was solved numerically using a Crank-Nicolson procedure in time with small enough time steps to give an accurate time integration, as described by Dennis and Hudson (1985). Solutions were obtained using the h^4-accurate spatial approximations derived from (14) and also with an h^6-accurate spatial model obtained by extending the functions g and F in (13) up to terms including powers of x^6.

In Table 1 we give values of three approximations ϕ_A, ϕ_B and ϕ_C to the steady-state solution $\phi(x,\infty)$ to the above problem obtained after a large number of time steps. The approximations are given at three separate stations of x together with the exact solution $E = 1 - \text{erf}(x)$ at these same three stations of x. The spatial grid size was $h = 0.2$ and the value $x_\infty = 5$ was taken as the value for which $\phi(x_\infty,t) = 0$. The approximation ϕ_A is the h^4-accurate approximation given by Dennis and Hudson (1985), ϕ_B is the h^4-accurate approximation derived from (14) and, finally, ϕ_C is the h^6-accurate approximation. It has also been found that when $h = 0.1$ ϕ_C agrees with the exact solution to almost 9 decimal places and when $h = 0.05$ to almost 10.

x	0.2	0.8	1.6
ϕ_A	0.777296	0.257910	0.023671
ϕ_B	0.777313	0.257933	0.023644
ϕ_C	0.7772973	0.2578989	0.0236517
E	0.7772974	0.2578990	0.0236516

Table 1: Approximations to $\phi(x,\infty)$ defined by Eqs. (2) and (25).

Several other ways of choosing $f(x)$, $g(x)$ and $F(x)$ in (4) have been considered, such as using Fourier series and Chebyshev series and these results will be reported later.

REFERENCES

Allen, D.N. de G. and Southwell, R.V. 1955 Quart. J. Mech. Appl. Math. **8**, 129.
Dennis, S.C.R. 1985 Lecture Notes in Physics **218**, 23.
Dennis, S.C.R. and Hudson, J.D. 1985 submitted to J. Computational Physics.
Dennis, S.C.R. 1960 Quart. J. Mech. Appl. Math. **13**, 487.
Dennis, S.C.R. and Hudson, J.D. 1979 J. Inst. Math. Applics. **23**, 43.
Smith, G.D. 1978 Numerical Solution of Partial Differential Equations: Finite Difference Methods, Second Edition, Oxford University Press.

ON THE COUPLING OF INCOMPRESSIBLE VISCOUS FLOWS AND
INCOMPRESSIBLE POTENTIAL FLOWS VIA DOMAIN DECOMPOSITION

Q. V. Dinh, J. Periaux and G. Terrasson
AMD/BA Industries
78 Quai Carnot
St. Cloud, France

and

R. Glowinski
University of Houston
Department of Mathematics
4800 Calhoun
Houston, TX 77004 U.S.A.
and
INRIA
78153, le Chesnay, France

1. Generalities.

The main goal of this paper is to present a preliminary discussion of a computational method for the coupling of two distinct mathematical models describing the unsteady flow motion of an incompressible viscous fluid in two adjacent regions. Since one of the regions contains an obstacle, the viscous effects cannot be neglected there; on the other hand, we suppose that the second region is far enough from the obstacle that we can suppose the flow in it is inviscid, more precisely potential. The main motivation for considering this problem is to take advantage of situations for which simplified models can be used in some regions of the flow domain. Another advantage of such an approach is to split a global problem into simpler ones which can be solved independently on parallel machines; nevertheless these partial solutions have to be coupled later on and it is one of the objectives of this paper to present such a coupling method, together with the results of some preliminary numerical experiments.

2. Formulation of a model flow problem. Basic simplifications.

Using the notation of Figure 2.1, let us consider the flow of an incompressible viscous fluid around the obstacle B; we denote by ∂B the boundary of B and by Ω the flow domain, i.e. $\Omega = R^N - B$. Such a flow is modelled by the unsteady Navier-Stokes equations (with classical notation):

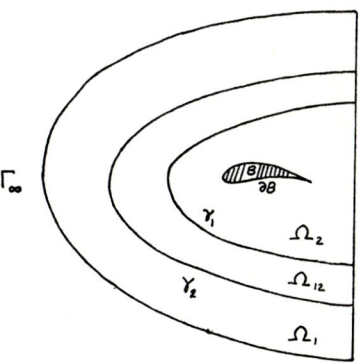

Figure 2.1

$$u_t - \nu \nabla^2 u + (u \cdot \nabla)u + \nabla p = 0 \quad \text{in } \Omega, \tag{2.1}$$

$$\nabla \cdot u = 0 \quad \text{in } \Omega \text{ (incompressibility condition)}, \tag{2.2}$$

$$u(x,0) = u_0(x) \text{ on } \Omega \text{ (initial condition; } \nabla \cdot u_0 = 0), \tag{2.3}$$

$$u = 0 \quad \text{on } \partial B \text{ (no-slip condition)} \tag{2.4}$$

$$u = u_\infty \quad \text{at infinity.} \tag{2.5}$$

In view of the numerical solution of (2.1) - (2.5), we introduce the following simplifications: (a) We replace Ω by a sufficiently large bounded computational domain - still denoted by Ω - whose external boundary will be denoted by Γ_∞; (b) we suppose that the viscosity ν is sufficiently small so that viscous effects are concentrated close to B and in its wake. We take advantage of these simplifications by splitting Ω into the two overlapping subdomains Ω_1 and Ω_2 - with Ω_2 containing B - and by assuming that the flow is inviscid potential in Ω_1, and still modelled by (2.1), (2.2), (2.4) in Ω_2. The boundary conditions on Γ_∞ will be the following:

$$u = u_\infty \quad \text{on} \quad \partial \Omega_2 \cap \Gamma_\infty; \text{ and} \tag{2.6}$$

$$\frac{\partial \phi}{\partial n} = u_\infty \cdot n \quad \text{on} \quad \partial \Omega_1 \cap \Gamma_\infty, \tag{2.7}$$

where ϕ is the flow potential, such that $\nabla \phi = u$ in Ω_1 and n is the unit outward normal vector. Finally, we define $\Omega_{12} = \Omega_1 \cap \Omega_2$ and γ_1 (resp. γ_2) as the interface of Ω_{12} and Ω_2 (resp. Ω_1), (see Figure 2.1).

3. Coupling the inviscid and viscous flows.

We can think of several methods for coupling the viscous and inviscid flows a according to the domain decomposition described in Section 2. The Schwarz's alternating method can be one of them; however, from previous calculations, (cf. e.g., [1]-[3]), we shall force the coupling on Ω_{12} through a least squares criterion; the numerical results of Section 6 justify such an approach.

From a mathematical point of view, the coupled problem can be formulated as follows:

Find a triple $\{\phi, u, p\}$ such that for each time t, we have

$$\begin{cases} u_t - \nu \nabla^2 u + (u \cdot \nabla)u + \nabla p = 0 & \text{in } \Omega_2, \quad \nabla \cdot u = 0 \quad \text{in } \Omega_2, \\ u(x,0) = u_0(x) \quad \text{in } \Omega_2, \\ u = 0 \quad \text{on } \partial B, \quad u = u_\infty \quad \text{on } \partial \Omega_2 \cap \Gamma_\infty, \end{cases} \tag{3.1}$$

$$\nabla^2 \phi = 0 \text{ in } \Omega_1, \frac{\partial \phi}{\partial n} = u_\infty \cdot n \text{ on } \partial \Omega_1 \cap \Gamma_\infty, \tag{3.2}$$

$$\int_{\Omega_{12}} |\nabla \phi - u|^2 \, dx \text{ is minimal.} \tag{3.3}$$

Remark 3.1: The above least squares methods for matching viscous and potential solutions has been introduced in the context of time dependent problems; in fact, it can also be used to capture steady state solutions by integrating on a large time interval; this will be validated by the numerical results of Section 6.

4. <u>Time discretization by operator splitting and application to the matching</u>

problem.

The time discretization of the Navier-Stokes equations by operator splitting methods (see [4], [5]) can be applied to the solution of the matching problem (3.1)-(3.3); in fact, we can take advantage of the operator splitting to require the optimal matching (3.3) for the solutions of linear subproblems defined on a sequence of discrete times. The time discretization will be done through the Peaceman-Rachford operator splitting scheme; however, the calculations presented in Section 6 have been performed through the closely related θ-schemes, discussed in [4], [5], which are more efficient, but whose description is more complicated.

Description of the discrete time scheme:

For Δt (>0) a time discretization step we have

$$\underline{u}^0 = \underline{u}_0 \quad \text{in } \Omega_2; \tag{4.1}$$

for $n \geq 0$, $\underline{u}^n$ being known on Ω_2, we look for $\{\phi^{n+\frac{1}{2}}, \underline{u}^{n+\frac{1}{2}}, p^{n+\frac{1}{2}}\}$ solution of

$$\begin{cases} \dfrac{\underline{u}^{n+\frac{1}{2}} - \underline{u}^n}{\Delta t/2} - \dfrac{\nu}{2} \nabla^2 \underline{u}^{n+\frac{1}{2}} + \nabla p^{n+\frac{1}{2}} = \dfrac{\nu}{2} \nabla^2 \underline{u}^n - (\underline{u}^n \cdot \nabla)\underline{u}^n \quad \text{in } \Omega_2, \\ \nabla \cdot \underline{u}^{n+\frac{1}{2}} = 0 \quad \text{in } \Omega_2, \\ \underline{u}^{n+\frac{1}{2}} = \underline{0} \quad \text{on } \partial B, \quad \underline{u}^{n+\frac{1}{2}} = \underline{u}_\infty^{n+\frac{1}{2}} \quad \text{on } \partial\Omega_2 \cap \Gamma_\infty, \end{cases} \tag{4.2}_1$$

$$\nabla^2 \phi^{n+\frac{1}{2}} = 0 \quad \text{in } \Omega_1, \quad \dfrac{\partial \phi^{n+\frac{1}{2}}}{\partial n} = \underline{u}_\infty^{n+\frac{1}{2}} \cdot \underline{n} \quad \text{on } \partial\Omega_1 \cap \Gamma_\infty, \tag{4.2}_2$$

$$\int_{\Omega_{12}} |\nabla\phi^{n+\frac{1}{2}} - \underline{u}^{n+\frac{1}{2}}|^2 \, dx \text{ is minimal, and then } \underline{u}^{n+1} \text{ solution of} \tag{4.2}_3$$

$$\begin{cases} \dfrac{\underline{u}^{n+1} - \underline{u}^{n+\frac{1}{2}}}{\Delta t/2} - \dfrac{\nu}{2} \nabla^2 \underline{u}^{n+1} + (\underline{u}^{n+1} \cdot \nabla)\underline{u}^{n+1} = \dfrac{\nu}{2} \nabla^2 \underline{u}^{n+\frac{1}{2}} - \nabla p^{n+\frac{1}{2}} \quad \text{in } \Omega_2, \\ \underline{u}^{n+1} = \underline{0} \quad \text{on } \partial B, \quad \underline{u}^{n+1} = \underline{u}_\infty^{n+1} \quad \text{on } \partial\Omega_2 \cap \Gamma_\infty, \quad \underline{u}^{n+1} = \underline{u}^{n+\frac{1}{2}} \quad \text{on } \gamma_2 \end{cases} \tag{4.3}$$

(with $\underline{u}_\infty^s = \underline{u}_\infty(s\Delta t)$).

Solution methods for problems such as (4.3) have been already discussed in [4]-[6]; we shall therefore concentrate on the solution of the subproblems (4.2). Omitting the subscripts in (4.2), this last problem is clearly a particular case of

$$\begin{cases} \alpha \underline{u} - \nu \nabla^2 \underline{u} + \nabla p = \underline{f} \quad \text{on } \Omega_2, \quad \nabla \cdot \underline{u} = 0 \quad \text{in } \Omega_2, \\ \underline{u} = \underline{g}_2 \quad \text{on } \partial B \cup (\partial\Omega_2 \cap \Gamma_\infty), \end{cases} \tag{4.4}_1$$

$$\nabla^2 \phi = 0 \text{ in } \Omega_1, \quad \dfrac{\partial \phi}{\partial n} = g_1 \quad \text{on } \partial\Omega_1 \cap \Gamma_\infty \tag{4.4}_2$$

$$\int_{\Omega_{12}} |\nabla\phi - \underline{u}|^2 \, dx \text{ is minimal.} \tag{4.4}_3$$

In view of solving the above matching problem (4.4), it is very convenient to take as master variables the trace of ϕ on γ_1 and the trace of $\underline{u}$ on γ_2. Once these traces have been specified ϕ, $\underline{u}$ and p are uniquely defined (p is in fact defined within an arbitrary additive constant and the trace $\underline{z}$ of $\underline{u}$ on γ_2 has to satisfy

$$\int_{\gamma_2} \underline{z} \cdot \underline{n} \, d\gamma_2 + \int_{\Gamma_\infty \cap \partial\Omega_2} \underline{g}_2 \cdot \underline{n} \, d\Gamma_\infty = 0).$$

Problem (4.4) has in fact the structure of an optimal control problem whose solution will be briefly discussed in the following Section 5.

5. <u>Solution methods for the matching problem (4.4).</u>

We formulate problem (4.4) as an optimal control problem by

$$\text{Min}_{\eta,\underline{z}} \ J(\eta,\underline{z}); \quad \{\eta,\underline{z}\} \in V_1 \times V_2, \text{ where} \tag{5.1}$$

$$V_1 \text{ is a space of suitable functions defined over } \gamma_1, \text{ and} \tag{5.2}$$

$$V_2 = \{\underline{z} | \underline{z} \text{ satisfies } \int_{\gamma_2} \underline{z} \cdot \underline{n} \, d\gamma + \int_{\Gamma_\infty \cap \partial \Omega_2} \underline{g}_2 \cdot \underline{n} \, d\Gamma_\infty = 0 \}, \tag{5.3}$$

with

$$J(\eta,\underline{z}) = \frac{1}{2} \int_{\Omega_{12}} |\underline{u} - \underline{\nabla}\phi|^2 \, dx, \tag{5.4}$$

where ϕ and $\underline{u}$ are solutions of the following problems

$$\underline{\nabla}^2\phi = 0 \text{ in } \Omega_1, \frac{\partial \phi}{\partial n} = \underline{u}_\infty \cdot \underline{n} \text{ on } \Gamma_\infty \cap \partial \Omega_1, \quad \phi = \eta \text{ on } \gamma_1, \text{ and} \tag{5.5}$$

$$\begin{cases} \alpha \underline{u} - \nu \underline{\nabla}^2 \underline{u} + \underline{\nabla} p = \underline{f} \text{ in } \Omega_2, \ \underline{\nabla} \cdot \underline{u} = 0 \text{ in } \Omega_2, \\ \underline{u} = \underline{g}_2 \text{ on } \partial B \cup (\partial \Omega_2 \cap \Gamma_\infty), \ \underline{u} = \underline{z} \text{ on } \gamma_2; \end{cases} \tag{5.6}$$

ϕ and $\underline{u}$ appears therefore as the solutions of a Poisson problem and of Stokes problem, respectively. Efficient solvers exist now for such problems (see e.g. [4]-[7]).

Using variational principles and finite element approximations like those discussed in [1]-[7], we can reduce problem (4.4), (5.1) to a finite dimensional linear quadratic control problem, which can be solved by a conjugate gradient algorithm. Such an algorithm requires at each iteration the solution of two discrete Poisson problems on Ω_1, and of two discrete Stokes problems on Ω_2.

The lengthy technical details associated with the above procedure are fully described in [8], [9]. Let's mention, however, that a key step is to obtain the adjoint (co-state) equations associated to (5.5), (5.6); this is done through classical arguments of Control Theory for Partial Differential Equations (see, e.g., LIONS [10], [11]).

6. <u>Numerical Experiments.</u>

To test the above methodology, we consider as a test problem the flow of an incompressible viscous fluid around a circular cylinder at Re=50; at Re=50 there exists a stable steady flow.

Figures 6.1, 6.2 show the streamlines and vorticity contours, for the steady flow obtained at Re=50, through the time dependent solution of the Navier-Stokes equations on the whole domain $\Omega_1 \cup \Omega_2$, using as initial flow $\underline{u}_0 = \underline{u}_\infty$.

Figures 6.3(a), 6.3(b) (resp. 6.4) show the streamline distribution in Ω_1 and Ω_2 (resp. the vorticity contours in Ω_2).

We observe the very good agreement between the global results, and those obtained by the matching.

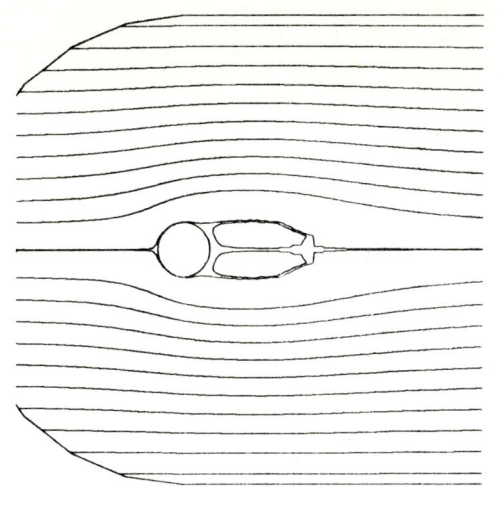

Figure 6.1
Global solution (streamlines)

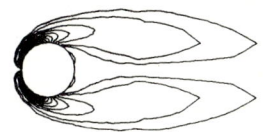

Figure 6.2
Global solution (vorticity contours)

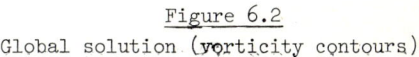

Figure 6.3(b)
Viscous flow streamlines

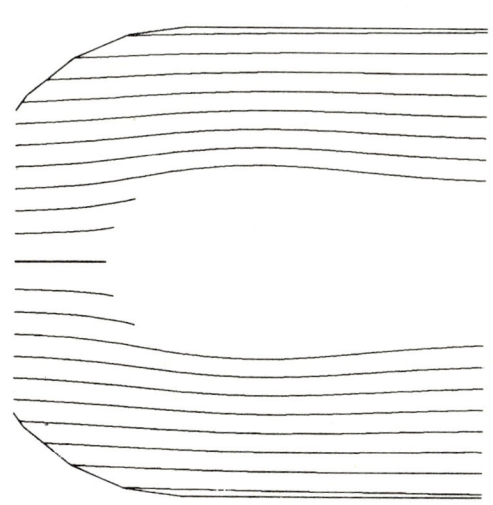

Figure 6.3(a)
Potential flow streamlines

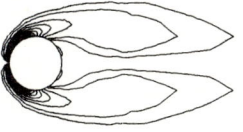

Figure 6.4
Vorticity contours in Ω_2 after matching

7. Conclusion.

Least squares matching seems to provide a robust methodology for coupling multi-models description of flow problems; its efficient implementation on multi-processor machines requires further studies in order to benefit from the advantage of the domain decomposition.

Acknowledgements: This work was partly supported by DRET under contract 85/175

References

[1] DINH, Q.V., Simulation numerique en elements finis d'ecoulements de fluides visqueux incompressibles par une methode de decomposition de domaines sur processeurs vectoriels, These 3eme Cycle, L'Universite Pierre et Marie Curie, Paris, 1982.

[2] DINH, Q.V., GLOWINSKI, R., PERIAUX, J., Domain decomposition for elliptic problems, in Finite Elements in Fluids, vol. v, R.H. Gallagher, J.T. Oden, O.C. Zienkiewicz, T.Kawai, M. Kawahara ets., Wiley, Chichester, 1984, pp. 45-106.

[3] DINH, Q.V., GLOWINSKI, R., PERIAUX, J., Domain Decomposition Methods for Nonlinear Problem in Fluid Dynamics, Comp. Meth. Appl. Mech. Eng., 40,(1983), pp.27-109.

[4] BRISTEAU, M.O., GLOWINSKI, R., MANTEL, B., PERIAUX, J., PERRIER, P., Numerical Methods for Incompressible and Compressible Navier-Stokes Problems, in Finite Elements in Fluids, vol. 6, R.H. Gallagher, G. Carey, J.T. Oden, O.C. Zienkiewiz eds., J. Wiley, Chichester, 1985, pp. 1-40.

[5] GLOWINSKI, R., Viscous Flow simulations by finite element methods and related numerical techniques, in Progress and Supercomputing in Computational Fluid Dynamics, E.M. Murman, S.S. Abarbanel eds., Birkhauser, Boston, 1985, pp.173-210.

[6] GLOWINSKI, R., Numerical Methods for Nonlinear Variational Problems, Suringer-Verlag, New York, 1984.

[7] HAUGEL, A., CAHOUET, J., Finite element method for incompressible Navier-Stokes equations and for shallow water equations, EDF report 41/86.03, Electricite de France, Direction des Etudes et Recherches, 1986.

[8] GLOWINSKI, R., PERIAUX, J. (to appear).

[9] TERRASSON, G., Thesis (to appear).

[10] LIONS, J. L., Controle Optimal des Systemes Gouvernes par des Equations aux Derivees Partielles, Dunod-Gauthier Villars, Paris, 1968.

[11] LIONS, J.L., Controle des Systemes Distribues Singuliers, Dunod-Gauthier Villars, Paris, 1983.

FREE MASS-LUMP METHOD FOR TWO-DIMENSIONAL COMPRESSIBLE FLOW
S.S.Dong, Z.X.Wang and H.Lee
Institute of Applied Physics and Computational Mathematics
Beijing, China

1. Introduction

   The Free-Lagrange Method was put forward by W.P.Crowley in 1970 [1]. The First International Conference on Free-Lagrange Method was held last year and many valuable papers were collected in the Conference Proccedings [2]. The ordinary Lagrangian methods with their plenty good computer programs are powerful for solving many problems of compressible multi-materials in one, two and three dimensions. But they have a fundamental difficulties associated with large distortion . To overcome the difficulties such procedures as rezonning, reconnection and introduction of slide lines are used. In Free-Lagrange methods the neighbors of a fluid element may be changed from cycle to cycle. There is nothing to be afraid of mesh distortions,because there is not any fixed mesh at all.

   In this paper an algorithm of Free-Lagrange Method is given. We call it "Free Mass-Lump Method". In this method the fluid is considered as a set of small fluid mass lumps. Every small mass lump has its own integral index, definite mass and material number. The intensive variables associated with the fluid, i.e. density,velocity, pressure and specific internal energy are defined at the centroids of the mass lumps. To determine the domain and the neighbors of each lump, we adopt the Voronoi mesh [3],[4]. The transfers of momentum and energy between neighboring mass lumps are calculated by the contour integrations along the borders of Voronoi mesh. On these borders velocity and pressure are interpolated, and artificial viscosity is introduced. The algorithm can keep all of the conservation law being satisfied. If time steps are chosen properly the mass lumps will change their neighbors partly and gradually in one cycle. For an unneighboring pair of mass lumps to become new neighbors with each other, the necessary condition is that they must be located at the tops of two "back to back" triangles respectively before this cycle.

   In the next section, we will describe the algorithm of this method. The results from several test problems are showed in the third section. Finally we will discuss the method briefly.

2. Description of the Algorithm

We define all hydrodynamic quantities at the centroids of the mass lumps. Each mass lump (Lagrangian cell) has its lump number N and material number m (they are integers). We adopt Voronoi mesh to determine and reconnect the neighboring lumps. The Voronoi mesh is defined as the subdivision of space, associated with a random set of Lagrangian points, into a set of convex polygons such that all space inside a polygon is closer to the enclosed point than to any other point. The faces of the polygon are segments of the perpendicular bisectors of the lines joining neighboring Lagrangian points (Fig.1).

We also **choose** the Voronoi mesh crudely as the control volume of each lump, about that we will discuss in the last section.

We consider both plane geometry and cylinder geometry denoted by $r=0$ and $r=1$ respectively.

Conservative Difference Scheme

$$u_N^{n+1} = u_N^n - \frac{(2\pi Y_N^n)^r dt}{M_N} \sum_{N' \in B_N} (p+q)_{NN'}^n \, \ell_{NN'}^n \cdot (\vec{n}_{NN'}, \vec{i}) + g \, dt$$

$$v_N^{n+1} = v_N^n - \frac{(2\pi Y_N^n)^r dt}{M_N} \sum_{N' \in B_N} (p+q)_{NN'}^n \, \ell_{NN'}^n \cdot (\vec{n}_{NN'}, \vec{j})$$

$$X_N^{n+1} = X_N^n + (u_N^{n+1} + u_N^n) dt/2$$

$$Y_N^{n+1} = Y_N^n + (v_N^{n+1} + v_N^n) dt/2$$

$$e_N^{n+1} = e_N^n + [(u_N^n)^2 + (v_N^n)^2 - (u_N^{n+1})^2 - (v_N^{n+1})^2]/2 - \frac{dt}{M_N} \sum_{N' \in B_N} (p+q)_{NN'}^n \, \ell_{NN'}^n (2\pi Y)_{NN'}^r \cdot (\vec{n}_{NN'} \cdot \vec{u}_{NN'}^n)$$
$$+ g(X_N^{n+1} - X_N^n)$$

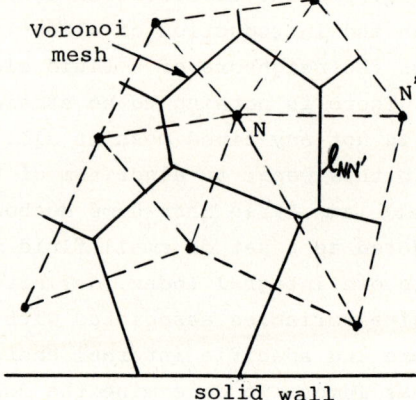

Fig. 1

V_N^{n+1} equals the volume of the Voronoi mesh N at time t^{n+1}

$$\rho_N^{n+1} = M_N / V_N^{n+1}$$

$$p_N^{n+1} = f_m(e_N^{n+1}, \rho_N^{n+1})$$

where B_N is the set of lumps neighboring to lump N, $l_{NN'}$ is the length of the border between lumps N and N', g is the gravity acceleration towards to the X axis.

$$\vec{n}_{NN'} = (\vec{X}_{N'} - \vec{X}_N)/|\vec{X}_{N'} - \vec{X}_N|$$

is the normal unit vector at the border. $p_{NN'}^n$, $\vec{u}_{NN'}^n$ are the interpolated values of pressure and velocity, and $q_{NN'}^n$ is the artificial viscosity. We use the following formulas to calculate them

$$p_{NN'}^n = \frac{p_N^n (\rho c)_{N'}^n + p_{N'}^n (\rho c)_N^n}{(\rho c)_{N'}^n + (\rho c)_N^n}$$

$$\vec{u}_{NN'}^n = \frac{\vec{u}_N^n + \vec{u}_{N'}^n}{2}$$

$$q_{NN'}^n = \begin{cases} 0 & U_{NN'}^n \leq 0 \\ \dfrac{c_1^2 (\rho c)_N^n (\rho c)_{N'}^n U_{NN'}^n}{(\rho c)_N^n + (\rho c)_{N'}^n} + \dfrac{2 c_2^2 \rho_N^n \rho_{N'}^n (U_{NN'}^n)^2}{\rho_N^n + \rho_{N'}^n} & U_{NN'} > 0 \end{cases}$$

where

$$U_{NN'}^n = U_{N'N}^n = (\vec{u}_N^n - \vec{u}_{N'}^n) \cdot \vec{n}_{NN'}$$

c_N is the sound speed of lump N, c_1 and c_2 are constants evaluated by 0.7~2. Here for $\vec{u}_{NN'}^n$ we have tried different ways of interpolation, the simplest is as above.

Boundary Treatments

There are two kinds of boundaries that have been treated by us.
a. Solid Boundaries. They can be treated easily, because they are the borders of near boundary Voronoi meshes and the normal velocity is zero on them.
b. Free Surfaces. To calculate the location of free surfaces we defined a set of surface mass lumps. These lumps as well as inner lump have their own dynamical quantities. But the position of the representative point for a surface lump is set at the free surface.
As concerns the neighbor selection we give a special rule. At first we require that the surface lumps must have at least one inner neighbor. Otherwise we shall interpolate a new inner lump E for surface lump C (Fig. 2). If such process is repeated several times for the same point C and the time step is rather small, then we cut point C off and set it fully free. We call such lump a splashed lump.

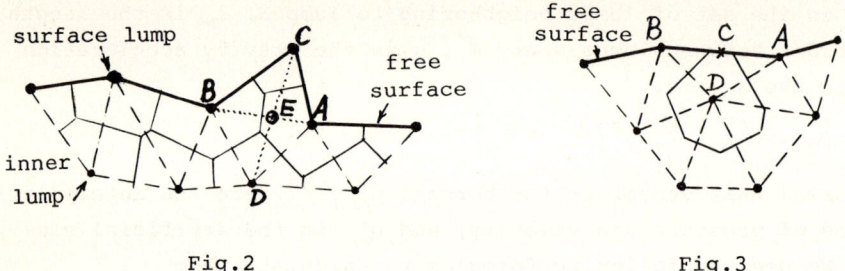

Fig.2　　　　　　　　　　Fig.3

Secondly, when the inner lump approaches to contact the free surface as shown in Fig.3. We shall interpolate a new surface lump C and the position of C is at the middle of two previous neighboring surface lumps A and B. Using the procedures described above we have calculated the penetration problem as showed in the next section.

3. Results from Several Test Problems

3.1 First we use two-dimension code to calculate one-dimensional piston impulsion problem. The calculation results represent rigorously **one-dimensional and are in basic agreement** with the analytic solution of the one-dimensional flow (Fig.4).

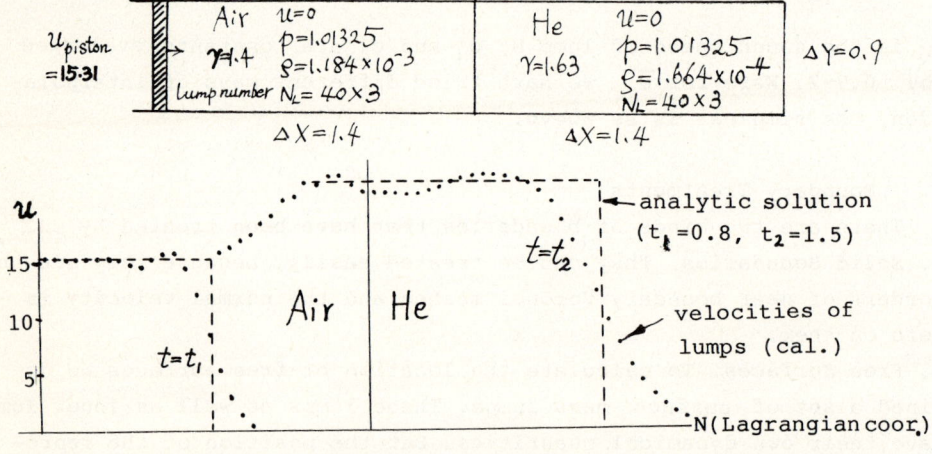

Fig. 4

3.2 Secondly, we have calculated the Meshkov-Richtmyer Instability [5]. Clark [6] has already done this problem by his code. In our calculation the initial conditions are showed in Fig.5.

The Fig.6 shows several snapshots of forty air lumps originally close to the interface between air and helium. Under the impulsion of a shock wave, the interface amplitude reverses and increases rapidly.

It is approximately the same as Clark's result.

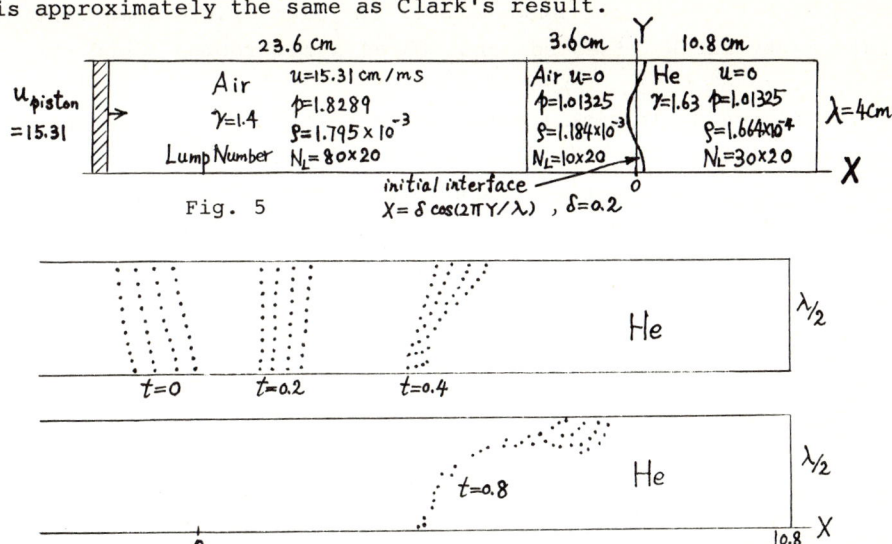

Fig. 5

Fig. 6

3.3 Referring to the Rayleigh-Taylor instability model that was considered by Crowley [7], We have calculated the following similar problem. Initially there are two different polytropic gases with $\gamma=2$. They are put in a gravitational field and the heavy gas is over the light one. Both gases are isentropic. According to the static equilibrium condition the initial densities are varied in X direction and $\rho(X=+0,t=0)= 4$, $\rho(X=-0,t=0)= 8$, $p(X=0,t=0)= 300$. The initial interface is $X=0.1\cos(2\pi Y/\lambda)$, where $\lambda=2.55$. The four boundaries are solid walls.

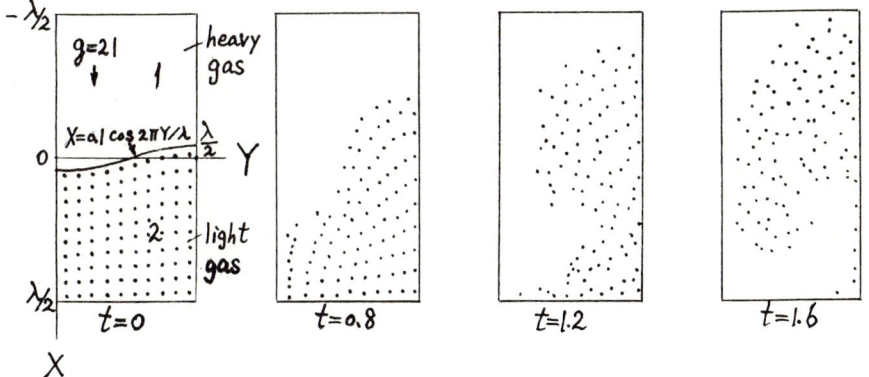

Fig. 7

The Fig. 7 shows four snapshots of the light fluid lump centroids. The light fluid rises up and forms a mushroomed cloud gradually.

3.4 We have dealt with a penetration problem to examine the free surface calculation. In this problem a steel cylinder is chosen as a projectile, an aluminium plate as a target. The initial size is shown in Fig.8. The initial impact velocity is 0.07cm/μs. The shapes of the free surface and interface at four different times are given.

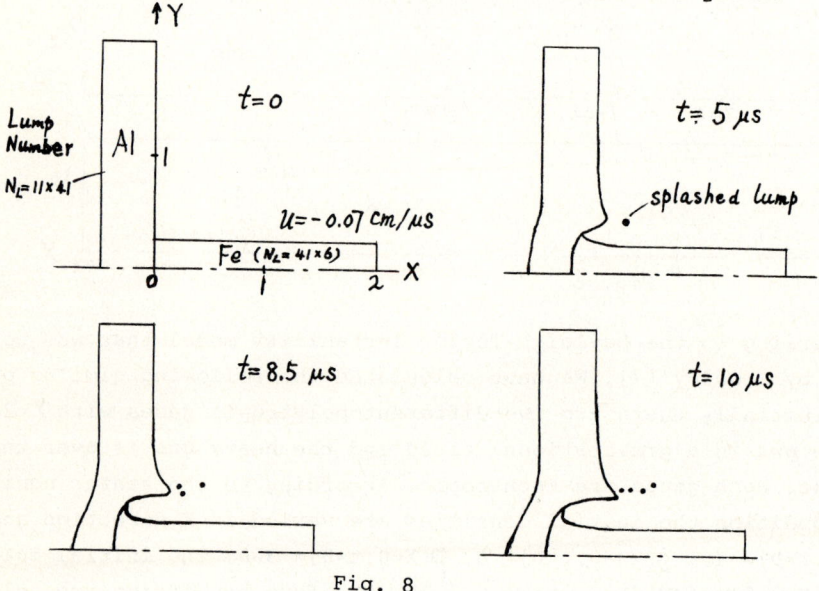

Fig. 8

4. Discussion

For Free Lagrangian Methods, the choice of control volume is one of the most important problems. Dukowicz [4] uses the Voronoi-Delaunay mesh in Lagrangian Fluid Dynamics. In [4] he has used a median-based control volume fomed by median lines of all the Delaunay triangles surrounding a cell point, as illustrated in Fig.9.

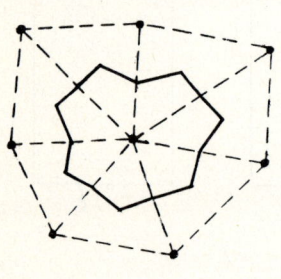

Fig. 9

In this paper we use directly the Voronoi mesh as the control volume of Lagrangian cell. This method has an advantage in space separation, it can keep the interface strictly one dimensional in one dimensional problems. But the method requires approximately the same scale of neighboring lumps. Otherwise larger error will be introduced. To improve the accuracy combining or redividing lumps is needed. Taking Voronoi mesh as control volume will bring us some error, but this **in**accuracy gives us the freedom of calculating the large distortion problems. We think for different problems different methods must be used. Our work is only at the start. The results of above test examples are qualitatively fine.

Finally we thank Prof. D.Y.Lee for his help and guidence.

References

1. W.P.Crowley, "FLAG: A Free-Lagrange Method for Numerically Simulating Hydrodynamic Flows in Two Dimensions", Proceedings of the Second International Conference on Numerical Methods in Fluid Dynamics, Springer-Verlag, pp.37-43.
2. "The Free-Lagrange Method", Proceedings of the First International Conference on Free-Lagrange Methods(1985), Edited by M.J.Fritts, et. al. Springer-Verlag.
3. G.Voronoi, Z.reine Angew. Math., 134, p.198 (1908)
4. J.K.Dukowicz, "Lagrangian Fluid Dynamics Using the Voronoi Delaunay Mesh", Numerical Methods for Coupled Problems. Pineridge Press, Swansea, U.K. (1981)
5. E.E.Meshkov, JETP, Vol.44, No.2, p.424 (1976)
6. R.A.Clark, "Compressible Lagrangian Hydrodynamics Without Lagrangian Cells", in [2].
7. W.P.Crowley, "Some Numerical Experiments with Rayleigh-Taylor Instability for a compressible Inviscid Fluid", UCRL-50845.

SPECTRAL METHODS FOR MODELING CHEMICALLY REACTING FLOW FIELDS

J. Philip Drummond
NASA Langley Research Center
Hampton, Virginia 23665-5225, USA

Introduction

A research effort is underway at the NASA Langley Research Center to achieve a detailed understanding of important physical phenomena present in supersonic chemically reacting flow fields. Important phenomena include fuel-air mixing and ignition, flameholding and stability, fluid-chemistry coupling, and the physical structure of the resulting flame. Such flows are governed by the Navier-Stokes equations coupled to an additional set of continuity equations describing each species present due to chemical reaction. Chemical reaction in a supersonic flow is controlled by both kinetics effects and the turbulent mixing of the fuel and air making up the flow field. The purpose of this paper will be to describe current work to develop an improved technique for solving the equations governing a chemically reacting flow field; this approach should offer both a high degree of accuracy as well as an acceptable level of computational efficiency.[1,2]

Mathematical Formulation and Analysis

Since the present research centers around the development of a numerical algorithm, geometric complexities have been simplified by considering quasi-one-dimensional flows. Also, since all flows considered are supersonic, streamwise diffusive effects are small, and therefore only the inviscid form of the governing equations have been considered. With these assumptions, the governing equations can be written in conservation law form as

$$\frac{\partial \vec{U}}{\partial t} + \frac{\partial \vec{F}}{\partial x} + \vec{H} = 0 \qquad (1)$$

where U is the solution vector for density, velocity, internal energy, and species mass fraction; F is a vector containing the corresponding fluxes; and H is a source vector for momentum and species production and loss. The production terms for the various species present when hydrogen and air undergo chemical reaction are determined using a global finite-rate chemistry scheme.[3] This model assumes four active species that are distributed according to the following two reactions.

$$H_2 + O_2 \rightleftarrows 2OH \quad (a) \qquad\qquad H_2 + 2OH \rightleftarrows 2H_2O \quad (b)$$

The model describes reasonably well the reaction of hydrogen and air. The scheme also well represents the property of numerical stiffness present in most coupled fluid-kinetics systems. Reaction (a) takes place on a time scale which is several orders of magnitude smaller than reaction (b). Likewise reaction (b) has a scale smaller than the scale of the fluid dynamics. The presence of such disparate time scales in a single coupled system; i.e., stiffness, presents additional complications in the solutions of equation (1).

The system of equations (1) have typically been solved in the past by discretizing in both space and time using either explicit or implicit finite-difference techniques. The equations are solved in this work by using a spectral discretization in space. Spectral methods are based on the representation of the solution to a problem by a finite series of global functions. The derivatives of the solution are then approximated by the corresponding derivatives of the finite series expansion. With proper application, these high-order approximations can produce extremely accurate numerical solutions, resulting in a significant reduction in the required number of computational nodes needed relative to a finite-difference solution.[4]

The Chebyshev spectral collocation method is chosen for this work, and the vector F is expanded in a truncated Chebyshev series

$$F(x) = \sum_{n=0}^{N} \hat{F}_n T_n(x) \quad (2)$$

where the $\hat{F}_n$ are the expansion coefficients of the series and N is one less than the number of grid points. Next, the change of variables $x = \cos\theta$ is introduced, and a set of collocation points, x_j, is defined by

$$x_j = \cos \frac{\pi j}{N}. \quad (3)$$

The discrete form of equation (2) then becomes

$$F_j = F(x_j) = \sum_{n=0}^{N} \hat{F}_n \cos \frac{n\pi j}{N} \quad (4)$$

where

$$\hat{F}_n = \frac{2}{N \bar{c}_n} \sum_{j=0}^{N} \bar{c}_j^{-1} F_j \cos \frac{n\pi j}{N}, \quad \bar{c}_j = \begin{cases} 2 & j = 0 \text{ or } j = N \\ 1 & 1 \leq j \leq (N-1) \end{cases}. \quad (5)$$

To find the spatial derivatives of F, required by equation (1), equation (2) is differentiated with respect to x giving

$$F'(x) = \sum_{n=0}^{N} \hat{F}_n T'_n(x) \quad (6)$$

A form of equation (6) without derivatives of the Chebyshev polynomials is preferred, so equation (6) is written in terms of another series

$$F'(x) = \sum_{n=0}^{N} \hat{F}_n^{(1)} T_n(x) \tag{7}$$

and then the coefficients of the two series are related. The following recursion relation exists between Chebyshev polynomials and their derivatives.

$$\frac{T'_{n+1}}{n+1} - \frac{T'_{n-1}}{n-1} = \frac{2}{c_n} T_n, \qquad c_n = \begin{cases} 2 & n = 0 \\ 1 & n \geq 1 \end{cases} \tag{8}$$

Putting equation (8) into (7) and algebraically manipulating the resulting expression gives

$$2n \hat{F}_n = c_{n-1} \hat{F}_{n-1}^{(1)} - \hat{F}_{n+1}^{(1)} \tag{9}$$

an expression for the $\hat{F}_n^{(1)}$ given the $\hat{F}_n$. The procedure for finding the $\hat{F}_n^{(1)}$ is initialized by setting $\hat{F}_{N+1}^{(1)} = \hat{F}_N^{(1)} = 0$ and then the remaining $\hat{F}_{N-1}^{(1)}$ through $\hat{F}_0^{(1)}$ are calculated from equation (9) by back substitution. Then, knowing all the $\hat{F}_n^{(1)}$, the derivatives of F can be calculated from equation (7). It should be noted that the operations required by equations (5) and (7) can be done quite efficiently using the fast Fourier transform.

Once values for $\partial F/\partial x$ and H are determined as described above, a system of kinetically stiff ordinary differential equations in time remains to be solved. To relax the severe stability restriction imposed by that stiffness, the kinetic source terms H are computed implicitly at the new time level while the spatial derivatives, $\partial F/\partial x$, are computed explicitly.[5] Using this approach, the discretized form of equation (1) becomes

$$U_i^{n+1} = U_i^n - \Delta t \left[\left(\frac{\partial F}{\partial x}\right)_{i_{sp}}^n + H_i^{n+1} \right] + O(\Delta t)^2 \tag{10}$$

where n is the old time level, (n+1) is the new time level, and the subscript sp denotes a spectral evaluation. The source term H is then linearized by expanding it in a Taylor series in time. After some algebra, equation (10) is reduced to

$$[I + \Delta t\, J_i^n] \Delta U_i^{n+1} = -\Delta t \left[\left(\frac{\partial F}{\partial x}\right)_{i_{sp}}^n + H_i^n \right] \tag{11}$$

where I is the identity matrix, J is the Jacobian of H with respect to U, and $\Delta U_i^{n+1} = U_i^{n+1} - U_i^n$. Equation (11) is then solved by applying a classic two-stage Runge-Kutta technique, yielding the following predictor-corrector formulae:

$$\overline{\Delta U_i^{n+1}} = -\Delta t \left[I + \Delta t J_i^n \right]^{-1} \left[\left(\frac{\partial F}{\partial x}\right)_{i_{sp}}^n + H_i^n \right], \qquad \overline{U_i^{n+1}} = U_i^n + \overline{\Delta U_i^{n+1}} \qquad (12)$$

$$\Delta U_i^{n+1} = -\Delta t \left[I + \Delta t \overline{J_i^{n+1}} \right]^{-1} \left[\left(\frac{\partial F}{\partial x}\right)_{i_{sp}}^{\overline{n+1}} + \overline{H_i^{n+1}} \right], \qquad U_i^{n+1} = U_i^n + \frac{1}{2}\left(\overline{\Delta U_i^{n+1}} + \Delta U_i^{n+1} \right) \qquad (13)$$

Results

Using the method described above, the chemically reacting flow field in several nozzle/diffuser configurations has been computed.[1,2] One such rapid expansion diffuser configuration is shown in figure 1. The diffuser is 2 units long, has an initial cross-sectional area of 0.79 and a final cross-sectional area of 3.14. A three-tenths stoichiometric mixture of hydrogen fuel and air is introduced at Mach 1.4, a temperature of 1900K and a pressure of 0.081 MPa. Figure 2 gives a time history to steady state of evolution of the OH radical and water at the center of the diffuser. The solid line describes results from a benchmark calculation using an Adams-Moulton implicit finite-difference procedure with 101 grid points. The symbols give results from the partial implicit spectral calculations for the same flow field, but with only 17 grid points. Note that the agreement between the spectral calculation and the benchmark is excellent. Figure 3 shows the spatial distribution of the OH radical and water in the diffuser at steady state using the same symbol key that was used in the previous figure. The agreement between the benchmark and the spectral calculations are again excellent, even when only 9 nodes are used in the spectral calculation. A final comparison of the methods is given in figure 4, which shows the rate of reduction of steady-state residual, $\left[\left(\frac{\partial F}{\partial x}\right)^n + H^n \right]$, with iteration count at the first interior grid point. The residual reduction rate is significantly greater than that provided by the finite-difference code. The maximum residual is reduced with the spectral code by ten orders of magnitude in only 2400 iterations, whereas the finite-difference code requires 6000 iterations to achieve the same level of residual reduction.

Concluding Remarks

The above results indicate that the spectral technique yields a highly resolved picture of a chemically reacting flow field. Because of its potential for accurately and efficiently modeling combustion phenomena, the method developed in this work is currently being extended to study the details of a supersonic reacting mixing layer.

References

1. Drummond, J. P.; Hussaini, M. Y.; and Zang, T. A.: Spectral Methods for Modeling Supersonic Chemically Reacting Flow Fields. AIAA Paper No. 85-0302, January 1985. Also accepted for publication in AIAA Journal.
2. Drummond, J. P.; Hussaini, M. Y.; and Zang, T. A.: Spectral Methods for Modeling Supersonic Chemically Reacting Flow Fields. NASA CR-172578, March 1985.
3. Rogers, R. C.; and Chinitz, W.: On the Use of a Global Hydrogen-Air Combustion Model in the Calculation of Turbulent Reacting Flows. AIAA Journal, April 1983.
4. Hussaini, M. Y.; Salas, M. D.; and Zang, T. A.: Spectral Methods for Inviscid, Compressible Flows. Advances in Computational Transonics, W. G. Habashi, ed., Pineridge Press, Swansea, U.K., 1983.
5. Widhopf, G. F.; and Victoria, K. J.: On the Solution of the Unsteady Navier-Stokes Equations Including Multicomponent Finite-Rate Chemistry. Computers and Fluids, v. 1, pp. 159-184, 1973.

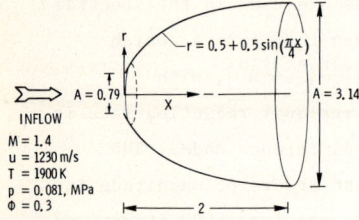

Figure 1.- Rapid expansion supersonic diffuser test case.

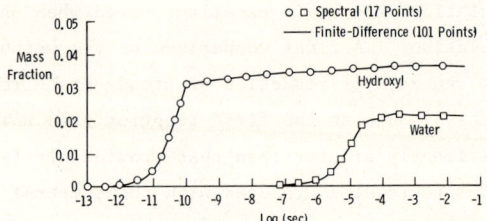

Figure 2.- Comparison of time histories of hydroxyl and water mass fractions.

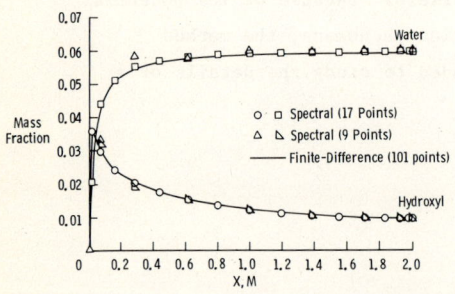

Figure 3.- Comparison of axial hydroxyl and water mass fraction profiles.

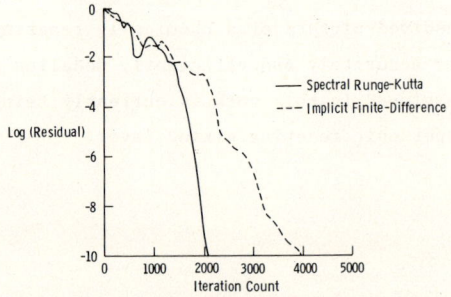

Figure 4.- Comparison of steady-state residual reduction rates of the methods.

TIME ACCURATE SOLUTIONS OF THE NAVIER-STOKES EQUATIONS FOR REACTING FLOWS

H. S. Dwyer - Professor
M. Soliman - Research Assistant
M. Hafez - Professor

Department of Mechanical Engineering
University of California, Davis
Davis, California 95616, USA

A method of solving the Navier-Stokes equations for variable density flows with chemical reactions has been developed and applied to the time-dependent ignition and burning of a hydrocarbon fuel droplet. The most important parts of the method are the following three features: (1) A pressure correction algorithm that is time accurate and allows for varible density, (2) A direct solution of the pressure correction algorithm with the use of a LU-decomposition of a banded matrix obtained from a Poisson equations, (3) A predictor/corrector method of the ADI sweeping type for the momentum and transport equations. The algorithm to determine the pressure field is of the low Mach number type and is a variation of an incompressible method developed by Chorin, Reference [1]. The pressure correction method can be utilized with both staggered and non-staggered grids and is also not directly dependent on the numerical method used to solve the momentum and transport equations. However, it is extremely important that the resulting Poisson equation be solved accurately, and this is achieved very efficiently with the use of the banded solver. This high efficiency is accomplished because of the fact that the LU-decomposition only has to be carried out at the beginning of the calculation and then saved for future use with a back solve.

For all of the problems that have been solved, this new method has shown itself to be a significant improvement over iterative techniques to solve the Poisson equation for the pressure field, References [2] and [3]. Whether the flow is incompressible or variable density, time dependent or steady, the use of the direct solver for the pressure Poisson equation improves the time dependent accuracy and accelerates the steady state convergence.

The Problem and Method of Approach

The physical problem that will be solved is the time dependent ignition and burning of a spherical decane fuel droplet in a hot air environment. The flow is assumed to be described by the variable density Navier-Stokes equations and the finite rate chemical kinetics and transport properties have been taken from

Reference [4]. For these types of problems the maximum velocities are of the order of ten meters per second and there is no influence of Mach number. Under these circumstances there are substantial economies to be gained by neglecting acoustic wave effects, and the limit of an infinite sound speed has been taken. The basic equations solved are:

$\partial \rho / \partial t + \partial(\rho u_i)/\partial x_i = 0$ \hfill CONTINUITY

$\partial(\rho u_i)/\partial t + \partial(\rho u_i u_j)/\partial x_j$
$= \partial p/\partial x_i + \partial \tau_{ij}/\partial x_j$ \hfill MOMENTUM EQUATIONS

$\partial(\rho e)/\partial t + \partial(\rho u_j h)/\partial x_j$
$= \partial q_j/\partial x_j + \sum \Delta h_k w_k$ \hfill ENERGY EQUATION

$\partial(\rho Y_k)/\partial t + \partial(\rho u_j Y_k)/\partial x_j$
$= \partial J_{kj}/\partial x_j + \sum C_k w_k$ \hfill SPECIES EQUATIONS

$p_R = \rho R_k T$ \hfill EQUATION OF STATE

(Note p_R is not equal to ρ in the momentum equations.)
These equations were then transformed to generalized coordinates (ξ, η), so that a more general formulation can be employed.

The solution of the above system of equations is of the split type with the momentum and transport equations being solved first and a pressure correction applied. The time differencing for the transport equations was a second order three point backward implicit formula, and central differences are applied for all spatial derivatives in a conservative fashion. The momentum and transport equations are solved along one of the generalized coordinate directions (for example-η) in a block-tridiagonal form with the old pressure field and the use of Newton's method. For the derivatives along the other direction (ξ) and the cross derivatives the old values at time n are utilized, and this step is called the predictor. Next a block-tridiagonal sweep (corrector) in the other generalized coordinate direction (ξ) is applied with all the cross derivative terms, nonlinear terms and η-direction terms evaluated with the predicted values. Of course this method is iterative and would converge as an implicit solution if the continuity equation was included in the procedure. In a practical time dependent problem a global iteration is not necessarily applied, but the method still has the advantage of updating on the second sweep all of the terms which cannot be included in the block-tridiagonal method.

The density is determined from the equation of state and the background reference pressure (p_R) is essentially constant and "not" influenced by the dynamic pressure of th flow field. The pressure correction α is then determined in the following fashion:

Two Step Operator Splitting

Step I. Solution of the Momentum Equations

$$[(\rho u_i)^{n+1} - (\rho u_i)^n]/\partial t = -(\partial p/\partial x_i)^n + RHS \quad (1.)$$

Step II. Pressure Correction

$$[(\rho u_i)^{n+1} - (\underline{\rho u_i})^{n+1}]/\partial t = -\partial \alpha/\partial x_i \quad (2.)$$

Definition
$$p^{n+1} = p^n + \alpha \quad (3.)$$

Step III. Apply Continuity Equation

$$\partial(\rho u_i)^{n+1}/\partial x_i = -\partial \rho/\partial t \quad (4.)$$

A. Take the divergence of eq. (2.)
B. Substitute eq. (4.) into the divergence of eq. (2.)

Result

$$+\partial(\underline{\rho u_i})^{n+1}/\partial x_i + \partial \rho/\partial t = (\partial^2 \alpha/\partial x_i^2)\delta t \quad (5.)$$

Step IV. Solve the Poisson equation (5.) for α with LU-decomposition that has been saved for a back solve.

Step V. Iterate or take the next time step.

Note: α is a potential function and does not change the vorticity, and addition of Steps I. and II. is a consistent approximation to the Navier-Stokes equations.

The above presentation is for a non-staggered grid but it could easily be modified for a staggered grid. Or, if one likes, the velocity correction could be applied at the cell boundaries and interpolated to the cell centers. With the use of the exact LU-decomposition many of the reported problems for the pressure correction algorithm take on much less importance. Also, for variable density flows it is impossible to avoid interpolation to the cell flux boundaries since the fluid density is only known at the cell centers.

RESULTS

The calculation of a burning hydrocarbon fuel droplet is a complex one and it is difficult to find good experimental data to compare against. However, for incompressible flows and for steady variable density flow there is considerable data available for comparison, Reference [5]. A comparison of the Drag coefficient for Reynolds numbers between 2 and 100 gives very good agreement with all data and previous calculations for both incompressible and variable density flows. Also, there is good agreement with other calculations involving surface mass transfer[5]. The time dependent calculations could not be compared with experiments directly, and were therefore compared against an unsteady incompressible streamfunction-vorticity computer program. The two methods were found to give the same results.

The resulting steady state isotherm patterns for a burning droplet are shown in Figures (1) and (2) for Reynolds numbers of 10 and 100. For these calculations the mass transfer from the surface was determined from the surface heat transfer, and this surface condition was the controlling factor in determining the time step for a stable calculation. The isotherm patterns show that ignition and burning can only be stabilized in the wake of the droplet for the high Reynolds number flow, while the burning region or flame completely surrounds the droplet for the lower Reynolds number flow.

The above solutions were obtained by integration of the equations in a time accurate fashion until steady state was achieved. Due to a lack of space in this paper only a limited amount of the time-dependent results can be shown, and these are exhibited in Figures (3) and (4). Shown in the figures is the ignition and flame propagation process for the Re=10 case, and it can be seen from Figure (3) that the ignition process has begun at the front stagnation point of the droplet. A short time later in the process the flame completely wraps itself around the droplet as shown in Figure (4), and the final steady state is approached slowly as given previously in Figure (1). For the calculations which will be presented in the future the droplet radius will change as a function of time as well as the free stream velocity.

CONCLUSIONS

The major conclusions of the investigations are the following:
1. A significant improvement in the numerical simulation of gas phase combustion processes at low Mach number has been achieved with the use of a pressure correction type method. The transport equations have been solved with a predictor/corrector algorithm of the ADI type.
2. The use of a direct banded solver for the Poisson equation for the pressure correction algorithm considerably improves the efficiency and accuracy of the method.
3. The pressure correction method is a time consistent approximation to the momentum equations which effectively decouples the velocity change between time steps into a vortical and potential component.

REFERENCES

1. Chorin, A., Mathematics Computations, 22, 745-762, 1968.
2. Patnaik, G., et. al., Proc. 10th Inter. Coll. on the Dynamics of Explosions and Reactive Systems, Berkeley, CA (1985).
3. Vaughn, H.R., et. al., JFM, Vol. 150, 1985, pp. 121-138.
4. Westbrooke, C., et. al., Combustion Science and Technology, 27, 31-43 (1981).
5. Dwyer, H.A., et. al., Twentieth Symposium on Combustion, p. 1743, The Combustion Institute, 1984.

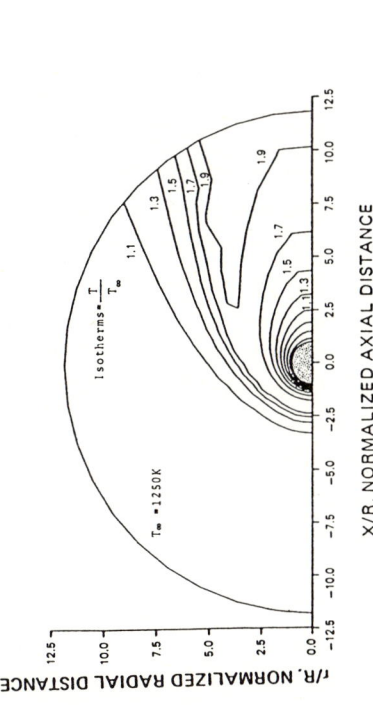

FIGURE (1) STEADY ISOTHERMS Re = 10

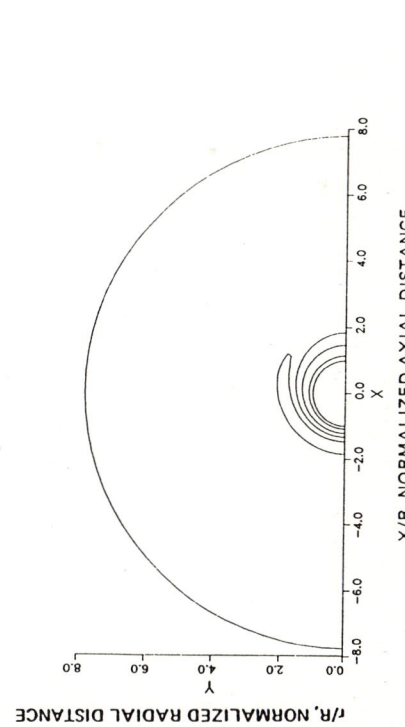

FIGURE (2) STEADY ISOTHERMS Re = 100

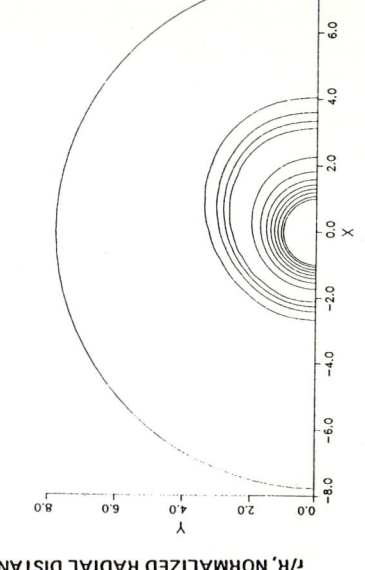

FIGURE (3) UNSTEADY ISOTHERMS Re = 10

FIGURE (4) UNSTEADY ISOTHERMS Re = 10

A FINITE ELEMENT METHOD FOR SIMULATION OF UNSTEADY FLOWS

Yuzuru Eguchi and Laszlo Fuchs
Department of Gasdynamics,
The Royal Institute of Technology,
S-10044, Stockholm, Sweden.

SUMMARY

The application of a Conjugate-Residual method to the solution of the time-dependent incompressible Navier-Stokes equations, is studied. This method is applied to the system of equations obtained by a finite element approximation to the differential equations, in each (semi-implicit) time step. Both steady and unsteady flows can be computed, with a rather acceptable efficiency. The computer code has been applied to the simulation of several flow problems, such as the flow in the lid-driven cavity and the steady and pulsatile flow past valve models.

INTRODUCTION

Finite element methods are less used in computational fluid dynamics compared to finite-differences. One of the main reasons is that the resulting (linearized) system of equations that has to be solved, is, in general, more irregular than corresponding finite-difference formulations. In addition, the system corresponding to the incompressible Navier-Stokes (N-S) equations, is indefinite (and unsymmetric). Often used methods, for the primitive variable form, are based on the separation of the velocity and the pressure, where the incompressibility constraint is indirectly satisfied by solving the Poisson equation for the pressure or by introducing a potential and solving a Poisson equation for that variable [1-4]. These types of uncoupled methods use the lumped mass matrix, and as a result the solutions may loose accuracy in cases of flows dominated by convective phenomena [5]. Furthermore, these methods do not satisfy, strictly, the tangential velocity boundary conditions [1], unless one uses such an iterative correction as the preconditioned Uzawa algorithm [6]. This discrepancy is likely to occur if the flow is developing from rest, or if the boundary conditions themselves are time-dependent. Since our goal is to study unsteady flows, methods that use the separation of the velocity and the pressure are not adopted. Instead, we solve the coupled equation system simultaneously. For time integration we use the semi-implicit scheme for the momentum equations, as proposed by Gresho and Chan [4], whereas the continuity equations (and the pressure itself) are updated directly. The resulting system of equations, in each time step, could be solved by the direct solvers. However, such a procedure would require much more additional computer storage, and lead to a major restriction on the applicability of the computer code to complex problems. Most iterative methods (such as SOR) for the coupled system would fail because of the diagonal zero entries. The indefiniteness and the lack of symmetry of the system exclude the use of the Conjugate-Gradient (CG) method.

Led by these facts, we adopted the Conjugate-Residual (CR) method (or the so-called ORTHOMIN [7,8]) which could be used to solve unsymmetric and indefinite matrix equations, although no rigorous proof of the convergence can be given in general. The use of this iterative method allows a considerable reduction in required computer memory since only non-zero entries of the matrix are stored.

THE NUMERICAL SCHEME

We assume that the fluid is incompressible and Newtonian and that the motion of the fluid is laminar and two-dimensional. The equations which govern the motion of the fluid are the following conservation equations of momentum and mass:

$$\dot{u} + u_j u_{i,j} + p_{,i} - (\nu_{kl} u_{i,k})_{,l} = 0 \qquad (1)$$

$$u_{i,i} = 0 \qquad (2)$$

where p is the pressure divived by the density and ν_{kl} is the viscosity tensor. The other symbols are subject to conventional notation. The suffixes i and j run from 1 to 2 corresponding the x- and y-directions, respectively. On the boundaries, we prescribe either the velocity vector or the components of the traction.

The Galerkin finite element formulation for equations (1) and (2) leads to a system of algebraic equations. Here, we approximate the velocity and the pressure by bilinear and piecewise-constant interpolation functions, respectively.

$$\begin{pmatrix} M & 0 \\ 0 & 0 \end{pmatrix} \begin{pmatrix} \dot{U} \\ \dot{P} \end{pmatrix} + \begin{pmatrix} N(U)+K & -C \\ C^T & 0 \end{pmatrix} \begin{pmatrix} U \\ P \end{pmatrix} = \begin{pmatrix} F \\ 0 \end{pmatrix} \qquad (3)$$

where U and P are the unknown vectors of the discretized velocity components and the pressure, respectively. $\mathbf{M}$, $\mathbf{N}(U)$, $\mathbf{K}$, $\mathbf{C}$ and $\mathbf{C}^T$ are the consistent mass-, convection-, diffusion-, gradient- and divergence-matrices, respectively. F is the boundary traction vector.

The momentum equations in (3) are integrated in time via the semi-implicit method proposed by Gresho and Chan [4]: The diffusion terms are integrated by using the trapezoidal rule and the convection terms are integrated via the explicit Euler scheme. The artificial balancing tensor viscosity is also added to the physical viscosity (for details see [4, 9]). The pressure gradient terms must be integrated in fully implicit manner. Using a Crank-Nicolson type discretization for the pressure gradient results in temporal oscillations in the pressure field (unless the pressure is exact initially). Even in case of oscillatory pressure field, however, the velocity field does not contain such oscillations. We solve the equations of the fully coupled system simultaneously. Therefore, the matrix equations to be solved in each time step $t=t^{n+1}$, with U^n and P^n being known, are as follows:

$$\begin{pmatrix} \mathbf{M}/\Delta t + \mathbf{K}/2 & -\mathbf{C} \\ \mathbf{C}^T & 0 \end{pmatrix} \begin{pmatrix} U^{n+1} \\ P^{n+1} \end{pmatrix} = \begin{pmatrix} F^{n+1} + (\mathbf{M}/\Delta t - \mathbf{N}(U^n) - \mathbf{K}/2)U^n \\ 0 \end{pmatrix} \quad (4)$$

Since the matrix in equation (4) is unsymmetric and indefinite, the CR method is applied on (4). The algorithm can to be found in [8]. The CR method consists, in fact, of a family of schemes denoted by CR(m) depending on the number (m) of the former searching direction vectors. Some numerical experiments [10,11] reveal that the larger the number the searching direction vectors is, the faster is the convergence of the CR(m) algorithm. However, the larger m is, the larger computer storage is required. In our computations we have used the CR(2) scheme. For 3-D computations, one is more limited to the CR(1) scheme, as Robichaud and Tanguy [12] did for computing the steady-state solution. The CR(m) scheme can be improved considerably by using proper preconditioning. Such a method is not straightforward for the fully coupled system, since simple preconditioners, such as SOR, cannot be used. Other types of preconditionings are currently under consideration.

The storage requirement of the CR(2) including the nodal co-ordinates is estimated by
$$M = 18*NODE + 8*NELM$$
where NODE and NELM stand for the number of nodes and the number of elements, respectively. In addition, the matrix itself must be stored. The storage requirement of the matrices can be minimized by using the hybrid finite element technique whose effectiveness has already been demonstrated for the computation of some 2-D steady flows using the Bi-Conjugate Gradient (BCG) method [13].

NUMERICAL RESULTS

We consider first the classical problem of the lid-driven cavity. A 43X43 regular uniform-element mesh for the approximation of the square domain is used. The Reynolds number is set to 1000. The time increment is fixed at 0.02 throughout the time integrations. Although the artificial viscosity is estimated to be 10 times as large as the physical viscosity, at worst case, it does not seem to reduce the accuracy seriously since the tensor viscosity is added only in the streamline direction [9]. The solutions of streamlines and pressures at steady-state are illustrated in Figures 1 (a) and (b), respectively. These results show good agreements with other numerical results (e.g. [1]). The storage requirement to solve this problem is about 0.4 MB in double precision computations.

The flows controlled by a valve are seen in wide ranges of engineering applications from gigantic power plants to elaborate artificial hearts. Here we consider a flow past a typical valve model whose finite element mesh is shown in the Figure 2. The velocities are assumed to vanish on the walls and on the valve surface. While the traction at the outlet is fixed at zero for all time $t \geq 0$, a time-dependent traction may be applied at the inlet. Two cases of steady-state results are shown in Figures 3. A more detailed study of the unsteady flow past different valve configurations is to be published elsewhere [14]. It should be noted that the storage requirement for this computation is only 0.5 MB in double precision owing to the efficiency of the hybrid mesh and the usage of the CR iterative solver.

CONCLUDING REMARKS

The application of the CR(2) method to the fully coupled time-dependent incompressible N-S equations have been studied. Some numerical experiments reveal that the CR method is applicable to the solution of the indefinite matrix system and allows a substantial reduction of the storage requirements. The proposed time integration strategy allows us to solve rather complex problems, with the time-dependent boundary conditions, with reasonable accuracy and CPU costs. The scheme can be, probably, improved by using an adaptive time step control and by using a proper preconditioning of the CR scheme.

ACKNOWLEDGEMENT

The authors wish to thank the Swedish National Board for Technical Development (STU) for its financial supports.

REFERENCES

1. J.Donea, S.Giuliani, H.Laval and L.Quartapelle, 'Finite Element Solution of the Unsteady Navier-Stokes Equations by a Fractional Step Method', Comp. Meth. Appl. Mech. & Engng., 30, 53-73 (1982).
2. P.M.Gresho, R.L.Lee and R.L.Sani, 'On the Time-dependent Solution of the Incompressible Navier-Stokes Equations in Two and Three Dimensions', in Recent Advances in Numerical Methods in Fluids Volume 1, Pineridge Press Ltd., Swansea, 27-79 (1980).
3. A.Mizukami and M.Tsuchiya, 'A Finite Element Method for the Three-dimensional Non-steady Navier-Stokes Equations', Int. J. Numer. Meth. Fluids, 4, 349-357 (1984).
4. P.M.Gresho and S.T.Chan, 'A New Semi-implicit Method for Solving the Time-dependent Conservation Equations for Incompressible Proc. 4th Int. Conf. Numer. Meth. Laminar and Turbulent Flow, Swansea, 3-21 (1985).
5. P.M.Gresho, R.L.Lee and R.Sani, 'Advection-dominated Flows with Emphasis on the Consequences of Mass Lumping', in Finite Elements in Fluids, Volume 3, Wiley, Chichester 1978.
6. J.P. Gregoire, J.P.Benque, P.Lasbleiz and J.Goussebaile, '3D Industrial Flows Calculations by Finite Element Method', Lecture Notes in Physics 218, Springer-Verlag, 245-249 (1985).
7. P.K.W.Vinsome, 'ORTHOMIN --- An Iterative Method for Solving Sparse Sets of Simultaneous Linear Equations', Proc. 4th SPE Symp. Reservoir Simulation, Los Angeles, 149-160 (1976).
8. Y.Saad, 'The Lanczos Biorthogonalization Algorithm and Other Projection Methods for Solving Large Unsymmetric Systems', SIAM J. Numer. Anal. 19, 485-506 (1982).
9. P.M.Gresho, S.T.Chan, R.L.Lee and C.D.Upson, 'A Modified Finite Element Method for Solving the Time-dependent, Incompressible Navier-Stokes Equations. Part 1: Theory', Int. J. Numer. Meth. Fluids, 4, 557-598 (1984).
10. C.P.Jackson and P.C.Robinson, 'A Numerical Study of Various Algorithms Related to the Preconditioned Conjugate Gradient Int. J. Numer. Meth. Engng., 21, 1315-1338 (1985).
11. K.Ito, 'An Iterative Method for Infinite Systems of Linear Equations', NASA ICASE Rept. No.84-13 (1984).
12. M.Robichaud and P.Tanguy, 'Incomplete Factorization in 3-D Incompressible Fluid Flow Problems', Proc. 4th Int. Conf. Numer. Meth. Laminar and Turbulent Flow, Swansea, 1783-1793 (1985).
13. Y.Eguchi and L.Fuchs, 'Conjugate-Gradient Methods Applied to the Finite-Element Approximation of the Navier-Stokes Equations', Proc. Int. Conf. Comput. Mech., Tokyo, (1986).(to appear)
14. L.Fuchs and Y.Eguchi, 'Numerical Simulation of Steady and Unsteady Flows Past Heart-Valve Models', submitted to J. Biomechnics.

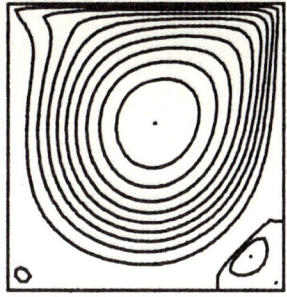 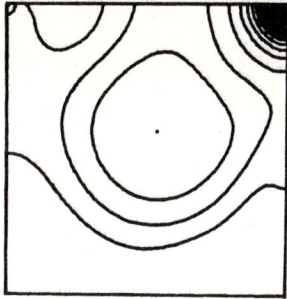

Fig. 1 Cavity flow solution (Re=1000);
(a) streamline pattern: (b) pressure distribution

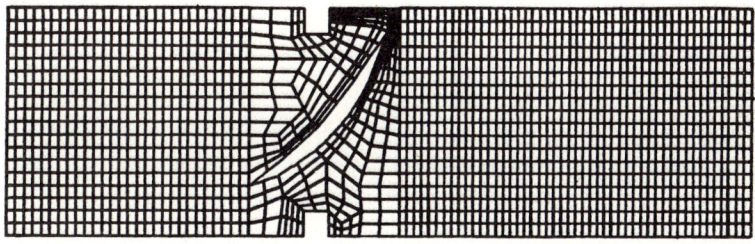

Fig. 2 Finite element mesh for a valve configuration

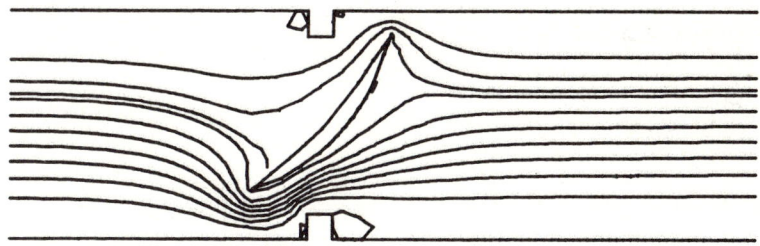

Fig. 3 (a) Streamline pattern at steady-state; θ≈40°

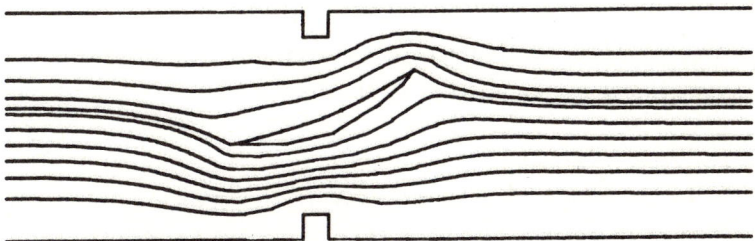

Fig. 3 (b) Streamline pattern at steady-state; θ≈70°

ALTERNATING DIRECTION ADAPTIVE GRID
GENERATION FOR THREE-DIMENSIONAL REGIONS

Peter R. Eiseman
Department of Applied Physics and Nuclear Engineering
Columbia University
New York, NY 10027

INTRODUCTION

An alternating direction adaptive procedure is a process of moving grid points to adapt to a disturbance by performing a sequence of coordinate directional sweeps. In each direction, points are moved along coordinate curves that are taken one at a time in a sequence that covers all grid points. The primary motion comes from the equidistribution of a weight function that in general can be applied to curves on a surface over our physical region. Such surfaces have been called "monitor surfaces" [1] and are merely a convenient way of expressing the adaptive data as a single relatively simple object. The basic alternating direction method and its application to monitor surfaces is given in [2] where also a historical account of work leading up to the general methods developed therein is presented. In the present work, we split the directional sweeps into two phases, we consider three-dimensional grids, and we establish orthogonality controls in three dimensions.

THE STRUCTURE OF A SWEEP

The adaptive sweep in a given direction consists of an active phase from which the primary grid point motion is imparted and a passive phase that produces a secondary corrective motion. Altogether, a directional sweep appears in the predictor-corrector format which is given by

 (1) The active phase
     Move points to equidistribute a weight function along
     each coordinate curve of a given direction.

 (2) The passive phase
     Relax the grid with a low pass filter to remove any
     wiggles or abrupt changes in spacing created by the

active phase.

In the active phase, the coordinate curves are taken in a sequential order that is consistent with a lexigraphic ordering of all grid points. Along each curve in the sequence, a weight w is created with respect to the curve arc length s. This is formed by expressing the desired objects of attraction as masses M_1, M_2, ..., M_m that compete with each other in the general linear weight

$$w = 1 + c_1 M_1 + \ldots + c_m M_m \tag{1}$$

Typical attracting objects are gradients (when curves are not on monitor surfaces), curvature, cell areas or volumes, angles, conformality, etc. The positive coefficients c_1, c_2, ..., c_m give the level of importance of the respective objects.

In the absence of such objects, the weight is unity and the grid points will become equally distributed along the arc length of the curve. This comes directly from the differential equidistribution statement

$$w \, ds = A \, dt \tag{2}$$

where A is a proportionality constant and t is the desired curvilinear variable for which dt is held constant. With a constant right-hand side, the spacing ds shrinks to adjust for large values of w. For grids, the corresponding integral statement is that a distribution $0 = s_1 < s_2 < \ldots < s_N = s_{max}$ is determined such that

$$T(s_{i+1}) - T(s_i) = \text{constant}$$

where $\tag{3}$

$$T(s) = \int_0^s w(r) \, dr$$

for $i = 1, 2, \ldots, N-1$.

The equidistribution process for the active phase starts with the curve given by a sequence of grid points $\mathbf{P}_1$, $\mathbf{P}_2$, ..., $\mathbf{P}_N$ and ends with a new sequence of grid points $\mathbf{Q}_1$, $\mathbf{Q}_2$, ..., $\mathbf{Q}_N$ determined by the following steps:

(1) Evaluate the integral of Eq. 3 at the arc length positions a_i of each $\mathbf{P}_i$ by using

$$T(a_i) = \sum_{k=1}^{i-1} w_{k+1/2} ||\mathbf{P}_{k+1} - \mathbf{P}_k||$$

to form the array of values

$$T(a_1), T(a_2), \ldots, T(a_N)$$

(2) For interior points $j = 2, 3, \ldots, N-1$, establish the uniform spacing for T by setting

$$T_j = \left(\frac{j-1}{N-1}\right) T(a_N)$$

(3) Find the interval locations $k = k(j)$ such that

$$T(a_k) \leq T_j < T(a_{k+1})$$

where $<$ is replaced by $\leq$ if $k+1 = N$.

(4) For $j = 2, 3, \ldots, N-1$ with $k = k(j)$, linearly interpolate to find

$$\mathbf{Q}_j = \mathbf{P}_k + \frac{T_j - T(a_k)}{T(a_{k+1}) - T(a_k)} (\mathbf{P}_{k+1} - \mathbf{P}_k)$$

(5) For endpoints, set $\mathbf{Q}_1 = \mathbf{P}_1$ and $\mathbf{Q}_N = \mathbf{P}_N$.

In the passive phase, a "low pass filter" is applied to remove high frequency variations while leaving low frequency properties intact. The low frequency properties are essentially the pointwise distributions that would have appeared in the absence of numerical approximations for the equidistribution process. The high frequency variations come primarily from the approximations and manifest themselves in the grid as local wiggles or abrupt changes in spacing. The removal of these local irregularities is a smoothing operation that permits us to obtain finite difference estimates of coordinate derivatives which are continuous over the grid. Derivative continuity is an essential ingredient in both the numerical solution of equations over the grid and in the application of grid controls of any following sweep. Such controls include grid orthogonality and curvature clustering.

The smoothing operation for the passive phase comes from the Gauss-Seidel relaxation of

$$\mathbf{Q}_{ijk} = \mathbf{P}_{ijk} + \frac{1}{12} \{\mathbf{P}_{i+1,jk} + \mathbf{Q}_{i-1,jk} + \mathbf{P}_{i,j+1,k} + \mathbf{Q}_{i,j-1,k} + \mathbf{P}_{ij,k+1} + \mathbf{Q}_{ij,k-1} - 6\mathbf{P}_{ijk}\} \quad (4)$$

where $\mathbf{Q}$ and $\mathbf{P}$ are respectively new and old values of the position vector for grid points. The subscripts using i, j, and k give the logical space position with lexigraphic ordering. Suitable truncations of Eq. 4 are applied at the boundaries. The smoothing rate can be estimated in the same manner that Brandt [3] employed for his grid management decisions. The properties as a "low pass filter" can be derived by following the discussion in Hamming [4]. For our purposes, we found that three complete Gauss-Seidel passes through the grid produced the desired effect for an impulsive start. In situations where the grid movement is less dramatic, the number of passes can be reduced.

A further consequence of the passive phase is that the equidistribution process becomes more stable and robust. The growth of small perturbations in pure equidistribution schemes was examined in one dimension by Flaherty et al. [5] and was found to decay only if the weight increased in time. In our context, a similar restriction on the weights would have occurred had there been no smoothing filter. The presence of the filtering operation has alleviated this restrictive condition.

ORTHOGONALITY

Points along a given coordinate curve in a three-dimensional grid can be orthogonally aligned with the grid points on adjacent curves in either of the two coordinate sheets containing the given curve. In the active phase of a directional sweep, the chosen adjacent curves are those which have already been updated and thus are a consistent part of the new grid. The consistency is particularly important when significant grid point motion is being created. With lexographic ordering, the adjacent curves are the previous curves directly behind and below the current one. From each of these previous curves, orthogonal trajectories are computed to the current curve, inverse equidistribution is applied to find the weight which would reproduce the orthogonally aligned points, and then this weight is inserted as a mass within the general linear weight of Eq. 1. There it competes with other attracting masses on an equal basis. Altogether, the process starts with a previous curve grid P_1, P_2, ..., P_N and a current curve grid Q_1, Q_2, ..., Q_N and ends with weighting values at each Q_i. In stepwise fashion, it is outlined as follows:

(1) At internal points P_i of the previous curve, define the curve normal plane by computing the curve tangent
$$\tau_i = P_{i+1} - P_{i-1}.$$

(2) As j is increased, find the <u>first</u> positive value of $(Q_{j+1} - P_i) \cdot \tau_i$ to get the index j of the orthogonally aligned interval.

(3) Get the arc length position s_i at the orthogonally aligned point $R_i = Q_j + t(Q_{j+1} - Q_j)$ by solving $(R_i - P_i) \cdot \tau_i = 0$ for t, and then adding the distance $t||Q_{j+1} - Q_j||$ to the arc length at Q_j.

(4) Apply the inverse equidistribution step to get the weights $w_{i+1/2} = 1/(s_{i+1} - s_i)$ and $w_{i-1/2} = 1/(s_i - s_{i-1})$ at the midpoint locations $s_{i+1/2} = \frac{1}{2}(s_{i+1} + s_i)$ and $s_{i-1/2} = \frac{1}{2}(s_i + s_{i-1})$.

(5) Use linear interpolation from $s_{i-1/2}$ to $s_{i+1/2}$ to get the weight w_i at s_i corresponding to $\mathbf{R}_i$.

(6) Extrapolate to get boundary weights.

(7) Obtain the weights at each $\mathbf{Q}_k$ by local arc length interpolation from the s_i at $\mathbf{R}_i$.

(8) Normalize the weights to have a maximum value of unity.

AN APPLICATION

To examine the basic movement scheme in a nontrivial three-dimensional situation, an artificial disturbance was created about a pair of ellipsoids that intersected both the boundary of a fixed Cartesian volume and each other. Moreover, the initial grid was a Cartesian 21 x 21 x 21 system that covered the volume and thus represented an impulsive start.

The disturbance was created by defining a fourth coordinate u that smoothly varies from 0 outside of the ellipsoids to 1 inside of them. With the coordinates (x,y,z,u), a three-dimensional monitor surface [1] is embedded in four-dimensional space.

The algorithm was applied to generate a grid on the surface to obtain an equidistribution of its arc length. This was done by evaluating Eq. 3 with each distance $||\mathbf{P}_{k+1} - \mathbf{P}_k||$ being the square root of a sum of the four squares corresponding to increments in x, y, z, and u. The physical space grid was obtained by setting u = 0 or simply taking only (x,y,z). The scheme was converged for all practical purposes in just one step consisting of one sweep in each direction. In Figures 1, three coordinate surfaces through a given point in space are viewed. The object in the field that represents the disturbance passes through one corner of the Cartesian volume and is indicated by faint dots. By observing the boundaries of the dotted object together with the grid surfaces, we can clearly see the adaptive action of the grid.

ACKNOWLEDGEMENT

This research was supported by grants AFOSR-82-0176 and NASA NAG-1-479.

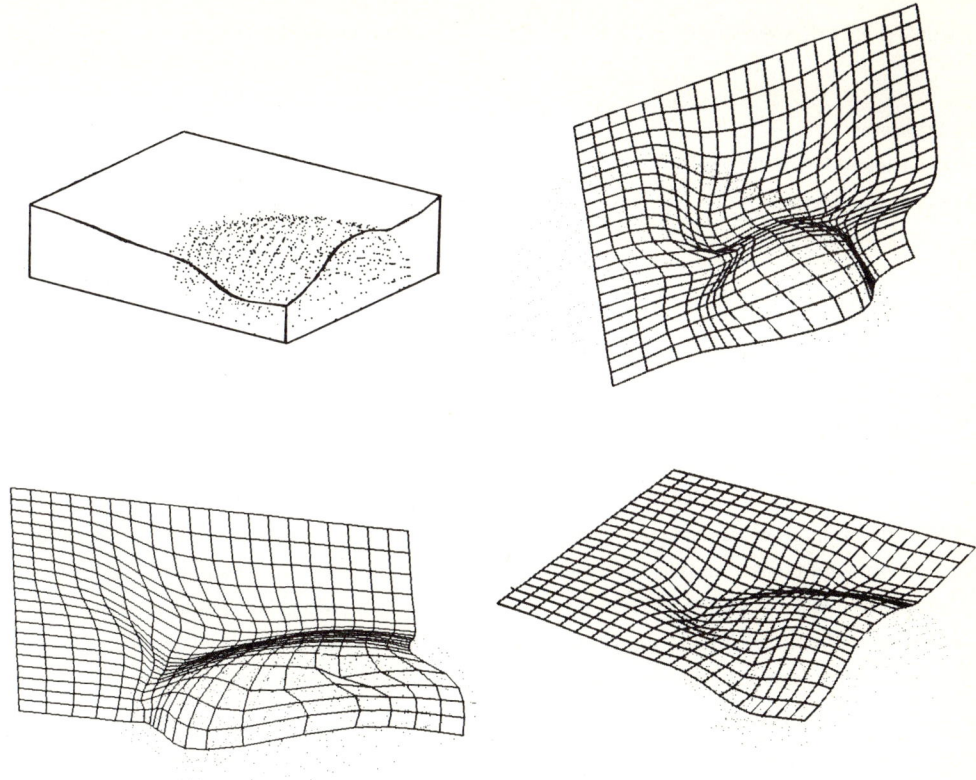

Figure 1: Three coordinate surfaces of the adapted grid
which pass through a given point in space

REFERENCES

[1] Eiseman, P.R., "Grid Generation for Fluid Mechanics Computations," <u>Annual Review of Fluid Mechanics</u>, Vol. 17, 1985, pp. 487-522.

[2] Eiseman, P.R., "Alternating Direction Adaptive Grid Generation," <u>AIAA Journal</u>, Vol. 23, No. 4, 1985, pp. 551-560.

[3] Brandt, A., "Multi-Level Adaptive Solutions to Boundary-Value Problems," <u>Mathematics of Computation</u>, Vol. 31, No. 138, 1977, pp. 333-390.

[4] Hamming, R.W., <u>Numerical Methods for Scientists and Engineers</u>, McGraw-Hill, 2nd Edition, 1973.

[5] Flaherty, J.E., Coyle, J.M., Ludwig, R. and Davis, S.F., "Adaptive Finite Element Methods for Parabolic Partial Differential Equations," <u>Adaptive Computational Methods for Partial Differential Equations</u>, ed. by I. Babuska, J. Chandra and J.E. Flaherty, SIAM, 1983, pp. 123-164.

TRANSITION PHENOMENA OVER A FLAT PLATE FOR COMPRESSIBLE FLOWS

G. Erlebacher
NASA Langley Research Center
Hampton, Va. 23669

1. Introduction

Although the predominant interest in the stability and transition to turbulence lies in compressible flows, research undertaken over the past 15 years has mainly focused on incompressible phenomena because of their comparative simplicity. As a result, a fairly extensive collection of theoretical [1], experimental [2] and numerical [3] data has been accumulated and cross-correlated for incompressible channel and Blasius boundary layer flows. Detailed studies of several instability mechanisms has led to a fairly comprehensive picture of the incipient stages of transition in incompressible flows. One such mechanism involves the interaction of a two-dimensional primary unstable wave with two skewed waves. This secondary instability leads to what is commonly referred to as K-type breakdown whose signature is a peak-valley vortical structure[1]. These structures have been successfully simulated numerically for incompressible boundary-layer flows [3]. Moreover, results from these simulations agree very well with theoretical and experimental findings.

As a step towards a better understanding of the instability mechanisms in compressible flows, the K-type breakdown of laminar flow at high Mach numbers is studied in this report. To this end, a three-dimensional, fully-spectral compressible Navier-Stokes code capable of direct simulation of parallel boundary-layer flows over a flat plate has been developed. The code is validated in the linear regime against the unstable eigenfunctions of the compressible linear stability eigenvalue problem. Then the temporal evolution of a triad of waves superimposed on a parallel boundary-layer is followed up to the incipient breakdown.

2. Algorithm

The full, time-dependent, three-dimensional, compressible Navier-Stokes equations are solved in conservative form under the parallel flow assumption [5]. Periodicity boundary-conditions in the streamwise(x) and spanwise(y) directions permit a Fourier representation of the primitive variables (velocity, pressure and density). For example, the double Fourier decomposition of the streamwise velocity, u, is

$$u(x,y,z,t) = \sum_{m=-\frac{N_x}{2}}^{\frac{N_x}{2}-1} \sum_{n=-\frac{N_y}{2}}^{\frac{N_y}{2}-1} u_{i,j}(z,t)\, e^{i\,(m\alpha x + n\beta y)} \qquad (1)$$

where N_x and N_y are the total number of nodes in x and y directions. The periods of the physical domain in the streamwise and spanwise directions (L_x and L_y) are related to the wave numbers α and β by $L_x = \frac{2\pi}{\alpha}$ and $L_y = \frac{2\pi}{\beta}$. The code has the option to evaluate normal spatial derivatives either by finite-difference approximation or by a Chebyshev collocation method, in which case $u_{ij}(z,t)$ has the series representation

$$u_{ij}(z,t) = \sum_{k=1}^{N_z} u_{ijk}(t) \, T_k(z)$$

where $T_k(z)$ is the Chebyshev polynomial of order k. The number of nodes in the normal direction is N_z. In the absence of discontinuities in the solution, spectral collocation methods are far more accurate than typical finite-difference methods for a specified distribution of a fixed number of nodes.

In all the mean flow calculations considered thus far, the critical layer is always found to lie between δ^* and $1.5\delta^*$ from the wall and special care is excercised to resolve it. The physical domain $(0, z_{max})$ in the direction normal to the plate is mapped onto the computational domain with an algebraic mapping. Letting one half the nodes lie between the wall and z=2 was found to be optimal for resolving the relevant flow in the chosen parameter range.

Currently, the code is fully explicit. A third order low-storage Runga-Kutta method is used for time discretization. Theoretically the scheme is unconditionally stable for a CFL below 0.55; however it is empirically found that above a CFL of 0.2 that the algorithm is unstable. More details are given in reference [5].

3. Initial and Boundary Conditions

The initial conditions consist of a triad of waves superimposed on a mean flow. Mean flow profiles are generated from the solution to the similar compressible boundary layer equations with zero pressure gradient and zero heat transfer at the wall. The 2-D and 3-D perturbation waves are solutions to the linearized compressible stability equations [4].

Under the assumption of parallel flow, all the variables are periodic in the streamwise and spanwise directions. No slip conditions are applied to the velocities at the wall which is adiabatic. In the far-field, all the variables are frozen at their initial values.

4. Results

The coordinate system and nomenclature used for the flat plate geometry is illustrated in figure 1.

Comparisons of the growth rates predicted by the Navier-Stokes code against linear results are made using both Chebyshev collocation and finite-differences in the direction normal to the wall. A single 3-D perturbation wave is added to the mean flow. Input parameters are summarized in table I. A value of .001 for the 2-D perturbation amplitude is sufficient to insure the absence of non-linear interactions of the fundamental mode with its higher harmonics. Figures 2-3 summarize the results of the linear test. Plotted in these figures is $log(E_K)$ versus time, where E_K is the perturbed kinetic energy. At Mach 0.5, the growth curves of E_K for a 33 collocation point normal distribution and a distribution with 129 finite-difference nodes are almost indistinguishable from each other and from the growth predicted by the linear eigenvalue code. However, 65 Chebyshev collocation nodes are required to match the predicted linear growth curves at Mach 4.5. In this case, 129 finite-difference nodes are not sufficient to resolve the structure of the eigenfunctions near the wall. The need for extra resolution at the higher Mach number is explained by the more complicated structure of the eigenfunctions presented in figures 5 and 6b. At Mach 4.5, the displacement thickness is almost an order of magnitude greater than at Mach 0.5, but in units of δ^*, the distance of the critical layer from the wall always lies between 1 and 1.5.

For the non-linear simulation, the parameters are: $M_\infty=4.5$, $R_e=10000$, $\overline{T}_\infty=110.85°R$, $\alpha=0.6$ and $\beta=1.03923$. The spanwise wavenumber is chosen to maximize the growth rate of the 3-D mode in order to accelerate the onset of the instability. The initial wave angle of the three-dimensional wave is 60°.

Starting amplitudes for the 2-D and 3-D wave are respectively $\varepsilon_{2D}=.054$ and $\varepsilon_{3D}=.012$. The 2-D growth rate predicted by linear theory is $\omega_{2D}=.5011+.00203i$ with a period of $T=12.54$. The 3-D growth rate, $\omega_{3D}=.09765+.01098i$, is substantially larger than its 2-D counterpart, a property typical of compressible flows as the Mach number increases. Above Mach 3, the phase angle that produces maximum growth is between 55° and 65°. As stated earlier, the primitive variables are expanded in Fourier series in the x and y directions. The (i,j) mode of u is defined by the coefficient $u_{i,j}$ in (1). Figure 4 shows the evolution of the kinetic energy of selected Fourier modes (integrated along the normal direction) as a function of time. The energy content of the (1,0) mode remains fairly constant. The spanwise modes (0,1) and (0,2) continuously grow until their strength is apparently sufficient to trigger the growth of the (1,1) mode, which is largely responsible for the presence or absence of the secondary instability. Prior to the rapid growth of the (1,1) mode, it goes through a stable interval starting at about 3 periods. Once the (0,1) mode is within 0.5 decades of the fundamental 2-D mode, the (1,1) mode suddenly becomes unstable again. At this stage, non-linear interactions between the 2-D and the 3-D mode begin to dominate the evolution of the flow.

Spanwise vorticiy contours are illustrated in figure 5 in the peak spanwise plane (normal to the spanwise direction) after 5.5 periods. In this and subsequent figures, the normal coordinate extends to $z=2$. The initial formation of the high-shear layer (computed on a 36x16x65 grid) is quite apparent. Its downward curving of the shear layer near its maximum is similar to the structure found in incompressible flow in the early stages of transition [3]. Streamwise and spanwise vorticity contours in streamwise planes (the normal to the plane is in the streamwise direction) are shown in figures 6-7 after 5.5 periods in planes at 0.0, 0.25, 0.5 and 0.75 wavelengths along the streamwise direction. Saddle points in the spanwise vorticity correspond to maximum streamwise vorticity. If these maxima are followed in figures 6a-d, the uplifting of the vortex tube becomes apparent, although it is rather flat.

5. Conclusions

A fully spectral, 3-D, compressible Navier-Stokes code specialized to parallel flow over a flat plate has been presented. It has been demonstrated that in the linear regime an initial eigenfunction grows at the same rate as predicted by a linear eigenvalue code in both subsonic and supersonic flow. As a test of the non-linear behaviour of the code at high Mach numbers, initial conditions are set up to numerically generate, for the first time, a high Mach number peak-valley vortical structure similar to that observed and computed for incompressible flows.

Case	Mach nb.	α	β	R_e	T_∞ (°R)
I	0.5	.24933	.20944	1200	520.00
II	4.5	.4652	.8057	10000	110.85

Table I: parameters input to the linear eigenvalue code

6. References

[1] Herbert, T.: Secondary Instability of Plane Channel Flow to Subharmonic Three-Dimensional Disturbances. Phys. Fl., Vol. 26, No. 4, 1983, pp. 871-874.

[2] Kovaznay, L.S., Komoda, H. and Vasudeva, B.R., Proc. of the 1962 Heat Transfer and Fluid Mech. Inst., pp. 1-26, Stanford University Press, 1962.

[3] Wray, A. and Hussaini, M.Y.: Numerical experiments in Boundary-layer Stability. Proc. R. Soc. Lond. 392, pp. 373--389, 1984.

[4] Malik, M.R.: Finite-Difference Solution of the Compressible Stability Eigenvalue Problem. NASA Contractor Report 3584, 1982.

[5] Erlebacher, G. and Hussaini, M.Y.: Incipient Transition Phenomena in Compressible Flows over a Flat Plate. ICASE Report No 86-39, 1986.

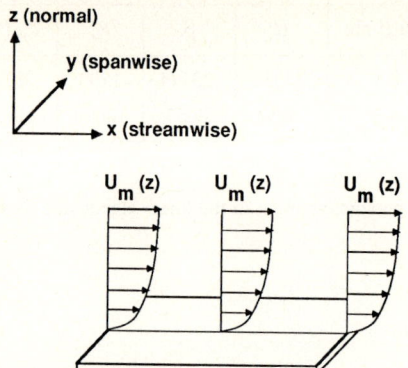

Figure 1: Geometry

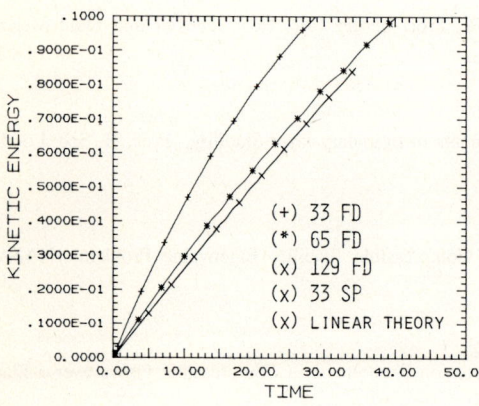

Figure 2: M=0.5 linear growth curves

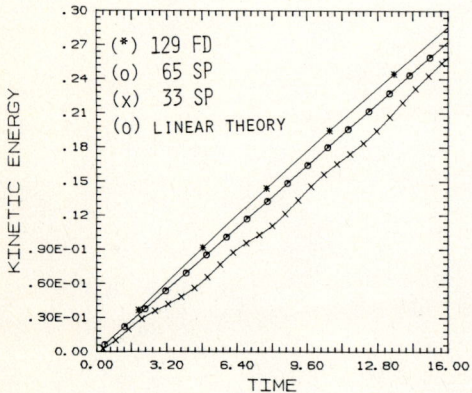

Figure 3: M=4.5 linear growth curves

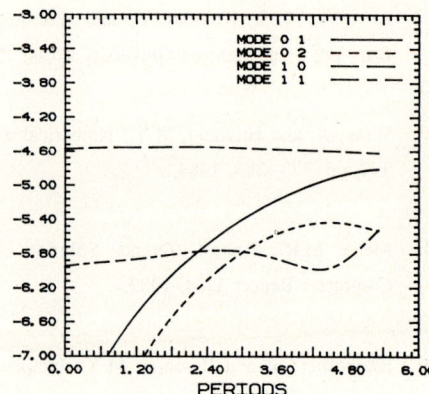

Figure 4: Time evolution of E_K

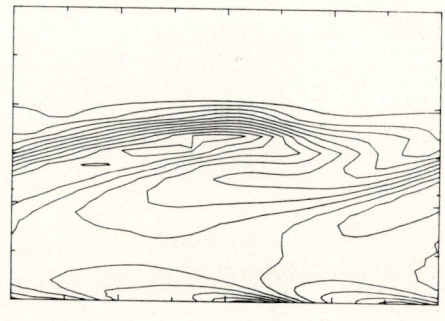

Figure 5: Spanwise vorticity in the peak plane

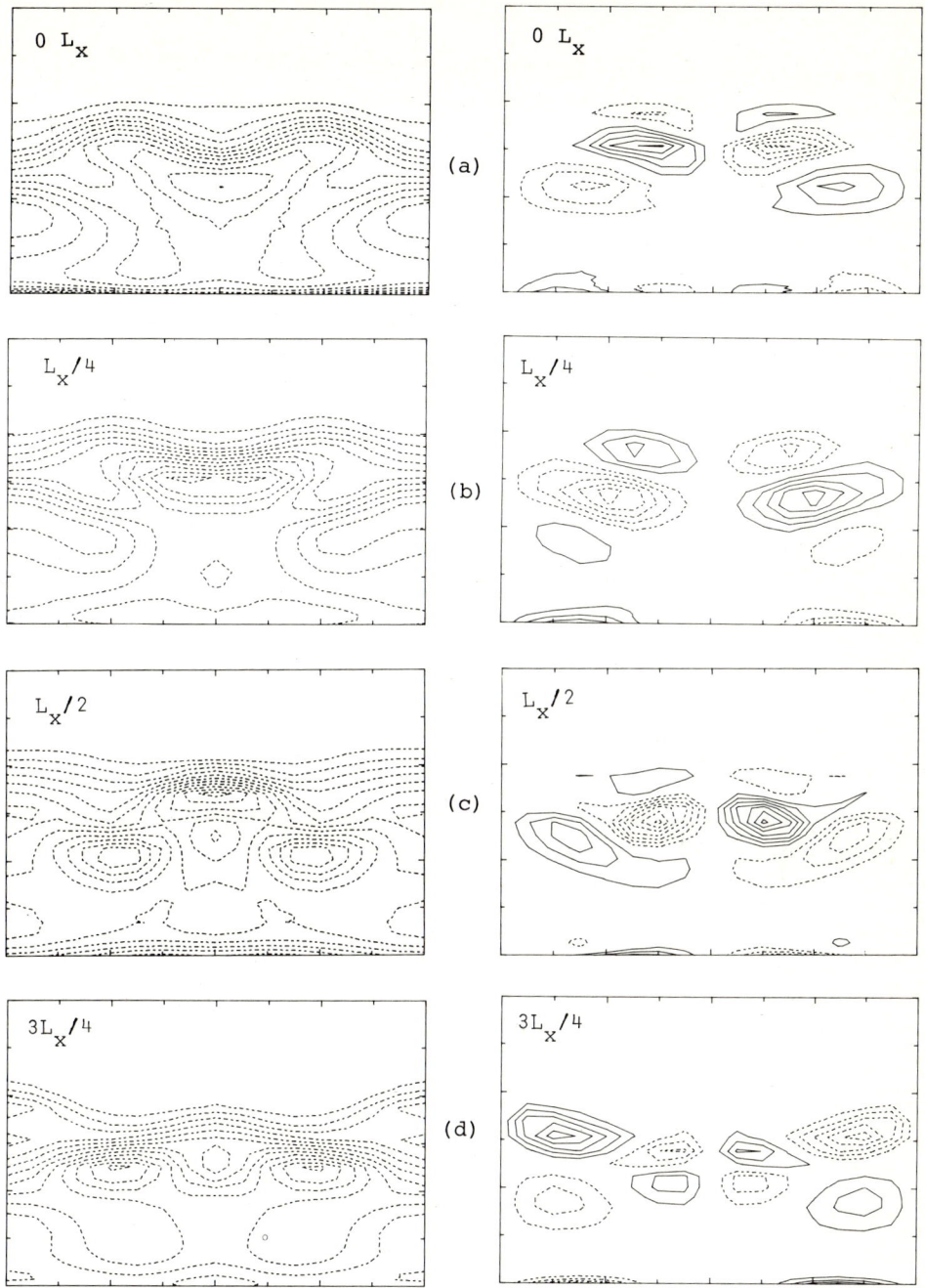

Figure 6: Spanwise vorticity contours in streamwise planes

Figure 7: Streamwise vorticity contours in streamwise planes

ON CONSERVATIVE PROPERTIES AND NON-CONSERVATIVE FORMS OF EULER SOLVERS

B. Favini[+], L. Zannetti[++]

SUMMARY

A new non-linear scheme for the solution of Euler equations is developed. The equations are solved in non-conservative form in the smooth regions of the flows, while across the discontinuities the conservation form is retained. The switching from the non-conservation to the conservation form is performed in an exact and rigorous way. The Jacobian is splitted in a positive and negative part and the spatial derivatives are approximated by upwind differences. The solution is advanced in time by an explicit two-steps procedure, and the scheme turns out to be second order accurate almost everywhere.

INTRODUCTION

In this paper we consider the numerical solution of unsteady, compressible inviscid flows, which are governed by the Euler equations (for simplicity of exposition we will consider here the one dimensional case, only):

$$w_t + F(w)_x = 0 \qquad (1)$$

with suitable initial and boundary values. Here the variables $w(x,t)$ and the fluxes $F(w)$ are the two vectors:

$$w = [\rho;\ \rho u;\ e]^T \qquad (2.a)$$

$$F(w) = [\rho u;\ \rho u^2 + p;\ u(e+p)]^T \qquad (2.b)$$

The system (1) is of hyperbolic type, that is, the Jacobian matrix $A = \partial F/\partial w$ has only real eigenvalues.

In virtue of the Lax's theorem [1] the conservative methods have enjoyed a widespread acceptance. Weak solutions of Euler's equations could be approximated by an "automatic capture" of discontinuities, that is, without having to detect and treat

---

+ Research Assistant, Dipartimento di Meccanica e Aeronautica, Rome, Italy.
++ Professor, Dipartimento di Ingegneria Aeronautica e Spaziale, Torino, Italy.

such discontinuities explicitly. To develop a shock-capturing scheme, which satisfies the design property of sharp and monotone solution of steady shocks, we have to find a suitable approximation of the spatial derivative of the flux such that the variation of the flux on each mesh can be splitted in two parts, one carrying the information moving to the left and the other to the right along the characteristics.

For instance, following Roe [2] we have to define a matrix A such that for each interval,

$$\Delta F = \hat{A} \Delta w \tag{3}$$

and the matrix $\hat{A}$ has three distinct eigenvalues, that is the associated eigenvectors are linearly independent.

Once the matrix $\hat{A}$ is constructed we can split it in a positive and negative part analysing the sign of the eigenvalues,

$$\hat{A}^+ = \frac{\hat{A} + |\hat{A}|}{2} \qquad \hat{A}^- = \frac{\hat{A} - |\hat{A}|}{2} \tag{4}$$

with

$$|\hat{A}| = R|\Lambda|L \quad \text{and} \quad \hat{A} = \hat{A}^+ + \hat{A}^- \tag{5}$$

where R, L, Λ are the matrices of the right eigenvectors, of the left eigenvectors and of the eigenvalues. After multiplication by the vector Δw we obtain

$$\Delta F = \hat{A}^+ \Delta w + \hat{A}^- \Delta w = \Delta F^+ + \Delta F^- \tag{6}$$

At this stage anyway, the scheme satisfies only half of the previous design requirement: indeed, as found by Godunov [3], a linear upwind scheme is monotone only if accurate to the first order. A first-order accurate scheme, if used over the entire flow field, is unsatisfactory. Upwind conservative methods have, therefore, evolved into non-linear schemes which gain higher order spatial accuracy at the expenses to self-adapt to the solution which they generate according to the local nature of the flow [4,5]. Impressive results have been obtained [6]; unfortunately, some advantages of conservative methods - automatism and simplicity, mostly - have been lost in the process.

Since it is necessary to change the integration scheme according to the local nature of the solution to be approximated, we investigate the possibility of developing schemes which are not consistent with the conservation form overall. Indeed, knowing the topology of the flow, that is the location of the shock and of the other discontinuities, there are no mathematical or physical reason to be consistent with the conservation form (1) everywhere in the domain, but in the smooth regions of the flow it is correct to be consistent with the equations in non-conservative form,

$$w_t + A w_x = 0 \tag{7}$$

Moreover we can adopt whatever set of variables we prefer. For instance we can choose the primitive variables - sound speed, flow velocity and entropy - which have the properties to be associated with a Jacobian matrix whose eigenvectors are practically constant.

CONSTRUCTION OF THE SCHEME

The construction of a scheme which switches from the non-conservation to the conservation form based on the local nature of flow, can be easily accomplished, exploiting the property that the fluxes $F(w)$ are homogeneous function of order one with respect to the conserved variables w,

$$F = A w \tag{8}$$

By differentiation we obtain

$$F_x = A w_x + A_x w \tag{9}$$

and for the definition of Jacobian matrix the second term in the right hand side is identically zero. We can obtain an approximation of the previous expression starting from the (8):

$$\Delta F = F_R - F_L = \Delta(Aw)$$
$$= \frac{A_R + A_L}{2}(w_R - w_L) + (A_R - A_L)\frac{w_R + w_L}{2} \tag{10.a}$$

or also, using non-symmetric expressions,

$$\Delta F = A_L(w_R - w_L) + (A_R - A_L)w_R \tag{10.b}$$

$$\Delta F = A_R(w_R - w_L) + (A_R - A_L)w_L \tag{10.c}$$

A scheme in non-conservative form will neglect the second term, what is completely justified in the absence of discontinuities as stated before. In the presence of a discontinuity, instead, this term is not negligible: it is, indeed, what provides the property of shock-capturing to a numerical scheme written in conservation form.

The scheme has been implemented in the following way. First of all we split the Jacobian matrix in a positive and a negative part - using (4) and (5),

$$\Delta F = F_R - F_L = A_R w_R - A_L w_L = (A_R^+ + A_R^-) w_R - (A_L^+ + A_L^-) w_L \qquad (11)$$

then we decompose the difference of the fluxes in two parts – using (10.b) and (10.c),

$$\Delta F = A_R^+(w_R - w_L) + (A_R^+ - A_L^+) w_L + A_L^-(w_R - w_L) + (A_R^- - A_L^-) w_R \qquad (12)$$

The first two terms will contribute to the variation in time for the node on the right of the mesh and on the left the other two terms.

Therefore the approximation of the Euler's equation (1), for a generic node i, is:

$$w_t|_i + A_i^+ \Delta_{i-1/2} w + A_i^- \Delta_{i+1/2} w + (\Delta_{i-1/2} A^+) w_{i-1} + (\Delta_{i+1/2} A^-) w_{i+1} = 0 \qquad (13)$$

In the smooth regions of the flow the last two terms are neglected and the scheme coincide with the extension of the Courant-Isaacson-Rees scheme [7] for non-linear hyperbolic system, developed by Moretti et al. [8,9]. Across a discontinuity the two terms are reintroduced to correctly compute the variation in time, and the scheme turns out to be the Flux-Vector Splitting scheme proposed by Steger and Warming [10].

NUMERICAL EXPERIMENTS

The numerical scheme has been structured as a predictor-corrector algorithm, and this provides a very simple way of switching from second order non conservative lambda scheme to first order upwind conservative scheme, described in the previous section. No particular attention has been directed to the detection of the discontinuities: the supersonic-subsonic shocks are located checking the change of sign of the characteristics.

Three preliminary examples are described in the following. The first one refers to a quasi-onedimensional flow in a°convergent divergent nozzle. Fig. 1 shows the theoretical (solid line) and computed (squares) Mach number (multiplied by ten) along the nozzle. Fig. 2 shows the theoretical and computed entropy distribution (multiplied by ten) along the nozzle. Fig. 3 shows the isoMach lines on the space-time plane. It describes the transient occurring in the nozzle when starting with a gas at rest and imposing the theoretical pressure at the outlet. The vertical straight lines are the theoretical steady isoMach lines. The second example refers to the computation of the transonic 2D flow field past a Joukowsky airfoil with $M_\infty = .85$ and $\alpha = 2°$ angle of attack. The computation is performed on a 120 x 15 orthogonal O-grid. In Fig. 4 the isoMach lines, and in Fig. 5 the constant entropy lines, are shown, respectively. Fig. 6 shows the Cp diagram (multiplied by ten). The third example is the calculation of an inviscid, compressible flow past a circle, with a free stream Mach number equal to 0.5. The flow is unsteady. Shock waves are generated and destroyed on the upper and lower surface, alternatively. Such shocks produce a strong rotationality and regions of very low stagnation pressure [11]. Consequently, recirculation occurs and vortices are shed. The effect is similar to the shedding of vortices due to boundary layer separation but it is totally inviscid. Fig. 7 is an example of isoMach lines at a randomly chosen instant of time.

CONCLUDING REMARKS

We have shown, in a rigorous way, the possibility to develop a non-linear scheme which changes from non-conservative to conservative forms according to the local nature of the solution to be approximated. Therefore each approximation scheme to the Euler equations in non-conservative form can be modified in a shock-capturing method simply by adding a "correction term", whose role is to restore the entire differences of the fluxes.

AKNOWLEDGEMENT

We would like to thank prof. G. Moretti and prof. M. Freedman for the helpful conversations and valuable suggestions.

REFERENCES

1. P.D. Lax, Comm. Pure and Appl. Math., 7, 1954, 159.
2. P.L. Roe, J. Comp. Phys., 43, 1981, 357.
3. S.K. Godunov, Mat. Sb., 47, 1959, 271.
4. P.L. Roe, M.J. Baines, Proceedings IV GAMM Conference, Vieweg, 1982.
5. A. Harteri, SIAM J. Numer. Anal., 21, 1, 1984, 1.
6. P. Woodward, P. Colella, J. Comput. Phys., 54, 1984, 115.
7. R. Courant, E. Isaacson, M. Rees, Comm. Pure and Appl. Math, 5, 1952, 243.
8. G. Moretti, Comp. Fluidsn 7, 1979, 191.
9. L. Zannetti, G. Colasurdo, AIAA Journal, 19, 7, 1981, 852.
10. J.L. Steger, R.F. Warming, J. Comput. Phys., 40, 1981, 263.
11. P.G. Burning, J.L. Steger, AIAA Paper 80-0971, 1982.

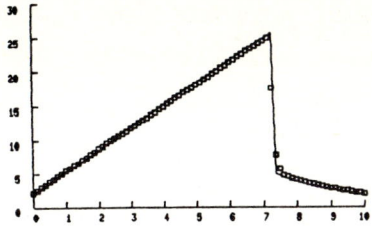

Fig. 1 - Mach distribution.

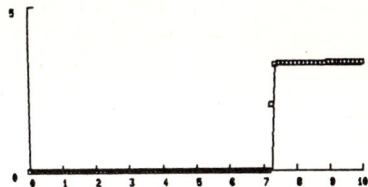

Fig. 2 - Entropy distribution.

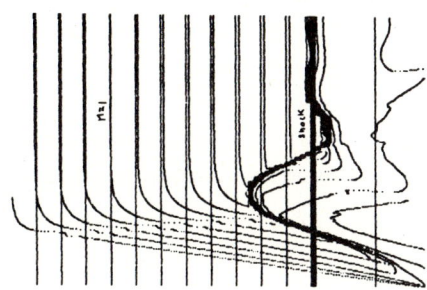

Fig. 3 - Time history of the isoMach lines.

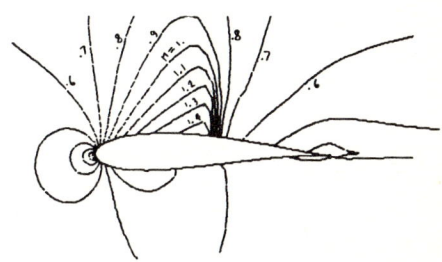

Fig. 4 - IsoMach lines.

Fig. 5 - Constant entropy lines.

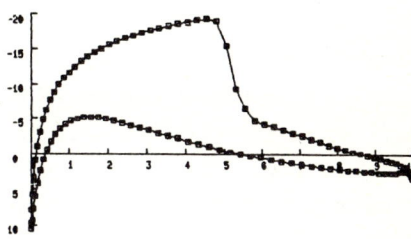

Fig. 6 - Pressure coefficient distribution.

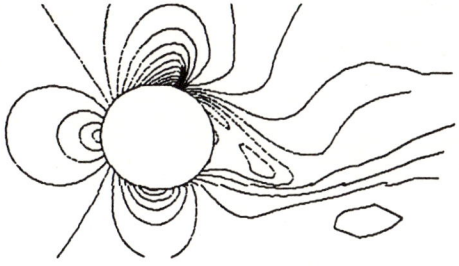

Fig. 7 - IsoMach lines.

A NUMERICAL SCHEME FOR THE UNSTEADY TRANSONIC FLOW AROUND AN OSCILLATING AIRFOIL

K. Förster and F.W. Li
Institut für Aerodynamik und Gasdynamik
Universität Stuttgart

Introduction

The computation of the flow around an unsteadily moving airfoil, especially in the transonic regime, is still not generally mastered in spite of its importance to problems as wing flutter or flow through rotors of helicopters and windturbines. The few methods for the fully nonlinear case suffer from difficult time-dependant boundary conditions at the contour and at the far-field boundary which result in either only approximate or else very clumsy and time-consuming implementations.

The new concept

The present paper tackles this problem in a physically and mathematically thorough manner by extending the computational domain over the complete infinite x,y-plane. Then the computational grid can be fixed to the contour of the airfoil, resulting in simple and physically real boundary conditions there. On the other hand, the boundary conditions at infinity consist simply in one set of variables: the onflow quantities. If now the airfoil is being moved in a prescribed but else arbitrary way, the grid moves with it and the said boundary values at infinity become exactly known functions of time.

This basic concept is verified by a chain of transformations. First, the airfoil is mapped into a near-circle by the Kármán-Trefftz method (accounting correctly for the trailing edge angle). The second step, an application of Theodorsen's scheme, transforms the near-circle into a circle and then into the unit circle:

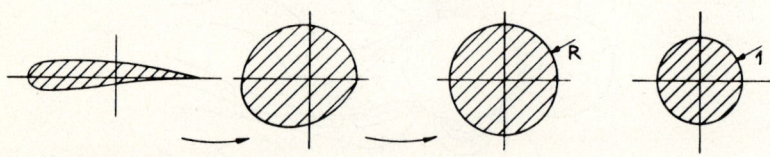

Kármán-Trefftz   Theodorsen

The third transformation brings in the motion of the airfoil which we take here as an oscillation around the origin (midpoint of airfoil) plus a translation to shift the center of rotation arbitrarily:

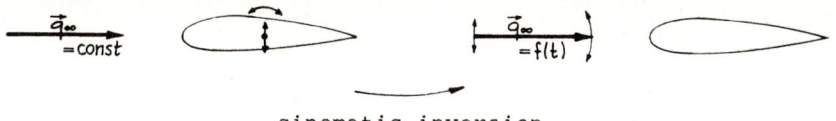

cinematic inversion

The next step is an inversion bringing the infinite point into the origin, and then the inner region of the unit circle is plotted as a rectangle mapping the origin into one of its sides:

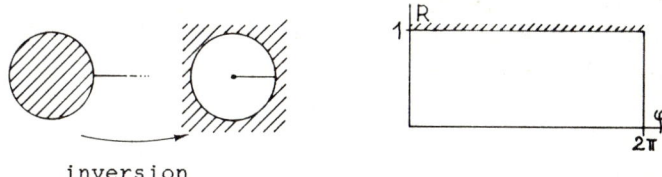

inversion

Additional transformations are used to condense the grid lines in those regions where steep gradients of the flow variables (e.g. shocks) are expected:

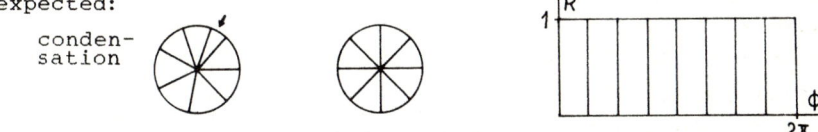

condensation

Details

All the transformations mentioned above except the second one are analytical and so the corresponding transformation coefficients entering the differential equations will be exact, too. The Theodorsen mapping, however, is only approximate, because it is based upon an infinite series which must be truncated and thus produces a truncation error. To keep this error small - more precisely: as small as possible with a given number of nodes around the airfoil - the Kármán-Trefftz mapping should produce a near-circle whose difference to a true circle is, in terms of the Fourier spectrum, characterized by long-wave components only. This is achieved by choosing the second Kármán-Trefftz parameter such that the nose radii of the given airfoil and of the corresponding Kármán-Trefftz airfoil are equal. The figure shows sample distributions of the polar coordinates of the near-circle for a NACA 0012 airfoil depending from this parameter.

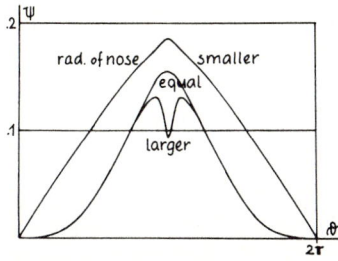

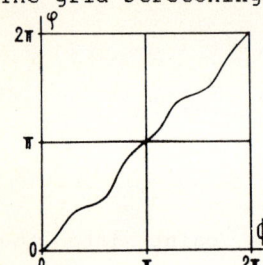

The grid stretching (which in our paper was only applied to the circumferential coordinate) can be done by suitable superposition of sine- and hyperbolic-tangent-functions, e.g. $\varphi = \alpha\phi + \beta \tanh(\gamma(\phi-\phi_0)) + \delta \sin(\pi\phi/\phi_1)$. The large number of undetermined coefficients gives sufficient freedom to fit any situation. The figure shows a distribution used in a real case.

The rest of the paper is mostly standard technique. The sequence of transformations described above is applied to the conservation laws of gasdynamics and the resulting system of nonlinear partial differential equations is solved by a finite difference scheme for which we have chosen the explicit, second order accurate Lax-Wendroff-Richtmyer scheme. At the solid wall boundary we used the one-sided scheme given by Rusanov and Nazhestkina /1/ for the computation of values which are regarded as preliminary only. Namely, to secure the long-time stability of the computational procedure as a whole (it might be necessary to march through several thousand steps in time) we found it indispensable to treat the boundary in accordance with the theory of characteristics (see also /2/). We therefore applied the characteristic correction principle of deNeef /3/ to a plane normal to the contour and used the resulting compatibility equation together with the boundary condition to correct the computed preliminary boundary values.

One more unconventional feature is the computation of the grid point at the trailing edge. In unsteady flow, the Kutta condition is not valid and we replaced it (following Theilemann /4/) by considering this point as a field point and treating it by a special algorithm which secures the appropriate influence from a cluster of neighbouring points at the upper and lower side of the airfoil. This physically inspired treatment has yielded correct results already in a number of similar unsteady flow problems.

Tests

A novel scheme such as the described one must be tested. Fortunately there exists a number of careful calculations for the steady case. So we set the oscillatory amplitude to zero and compared the results of our computation with those of
- a very accurate special method for subcritical flow /5/ and with two

other papers /6/,/7/ and
- the results of the well-known workshop by Rizzi and Viviand /8/ about the supercritical case.

Though we used only 60 meshes around the airfoil and 10 from the contour to infinity, the results from these comparisons, laid down in the diagrams 1 and 2, are very encouraging and, in our opinion, justify the application to the unsteady case.

Application to an oscillating NACA 0012 airfoil

The reduced frequency $f=\omega c/U_\infty$ was chosen as 2.5. We will show three computed samples which are distinct in type:

a) subcritical flow $\quad M_\infty = 0.63$, small amplitude $\alpha = \pm 2°$
b) supercritical flow $\quad M_\infty = 0.80$, small amplitude $\alpha = \pm 2°$
c) supercritical flow $\quad M_\infty = 0.80$, large amplitude $\alpha = \pm 10°$

The results are shown in diagram 3. In case a) one can easily imagine the rolling of the unsteady pressure waves from the numbering of the different curves, and in case b) the undulating motion is small, the shock being nearly stationary. Case c) however, can only be conceived by taking into account some more curves for intermediate time levels which would have obscured the diagram, so we give a verbal description of what happens:
the shocks (dot-accompanied portions of curves) are generated just about the trailing edge (curves 1 and 2). They gain strength (curve 3) and sweep upstream quickly (curves 3 to 0), then loose strength and speed (curves 0 to 1), become stationary (curve 2) and disappear (curve 3). This process takes more than one-and-a-half period so that always two generations of shocks are present around the airfoil.
- A much-better-than-verbal impression is of course given by cinematographic pictures we have taken from some cases.

References

/1/ Rusanov,V. and E.Nazhestkina, Boundary conditions in difference schemes for hyperbolic systems.
In: Notes Num. Fluid Mech. vol.2, Vieweg 1980

/2/ Pandolfi,M. and L.Zannetti, Some tests on finite differnce algorithms for computing boundaries in hyperbolic flows.
In: Notes Num. Fluid Mech. vol.1, Vieweg 1978

/3/ deNeef,T., Treatment of boundaries in unsteady inviscid flow computations. Delft University, Rep.LR-262, 1978

/4/ Theilemann,L., Ein gitterfreies Differenzenverfahren.
Dissertation Stuttgart 1983

/5/ Jäger,H., Singularitätenverfahren höherer Ordnung zur Berechnung der ebenen Unterschallströmung.
Dissertation Stuttgart 1984

/6/ Sells,C.C.L., Plane subcritical flow past a lifting airfoil.
    Proc.Roy.Soc.A. 308, 1968

/7/ Labrujere,T.E., W.Loeve and J.W.Slooff, An approximate method for
    the determination of the pressure distribution on wings in the
    lower critical speed range. AGARD CP 35, 1968

/8/ Rizzi,A. and H.Viviand, Numerical methods for the computation of
    inviscid transonic flows with shock waves.
    Notes Num. Fluid Mech., vol.3, Vieweg 1981

Details about the presented method are contained in
/9/ Li,F., Ein Beitrag zur Berechnung der instationären, transsonischen
    reibungsfreien Strömung um ein Tragflügelprofil.
    Dissertation Stuttgart 1986

Diagram 1

Steady subcritical flow around NACA 0012

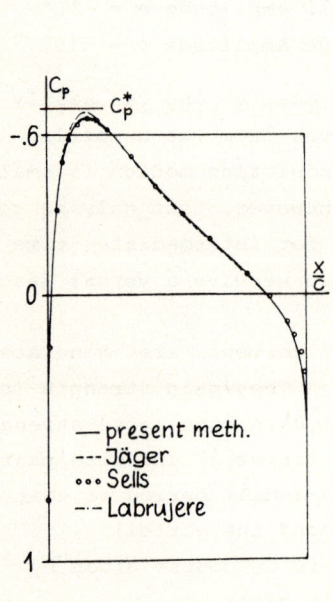

$M_\infty = 0.72$, $\alpha = 0°$

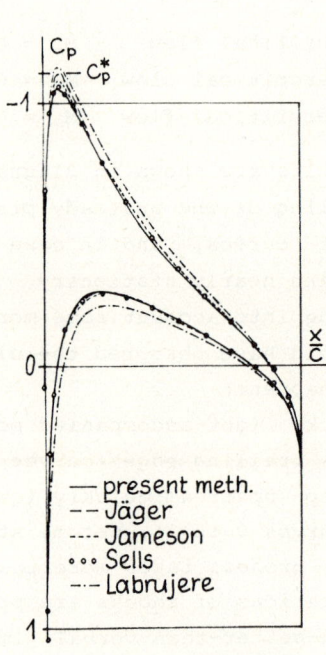

$M_\infty = 0.63$, $\alpha = 2°$

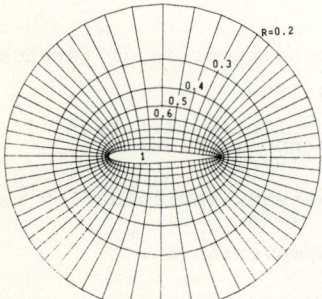

Typical discretization
of physical domain.

60 meshes around airfoil,
10 (8 shown) from contour to ∞.

Slightly condensed grid around
nose and trailing edge.

Diagram 2
Steady supercritical flow around NACA 0012

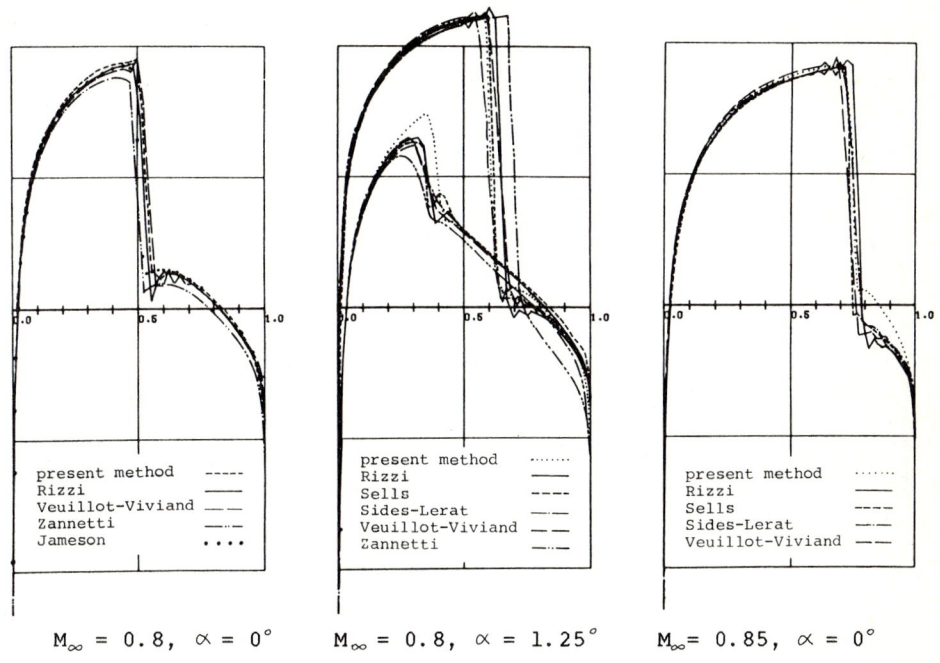

$M_\infty = 0.8$, $\alpha = 0°$ $M_\infty = 0.8$, $\alpha = 1.25°$ $M_\infty = 0.85$, $\alpha = 0°$

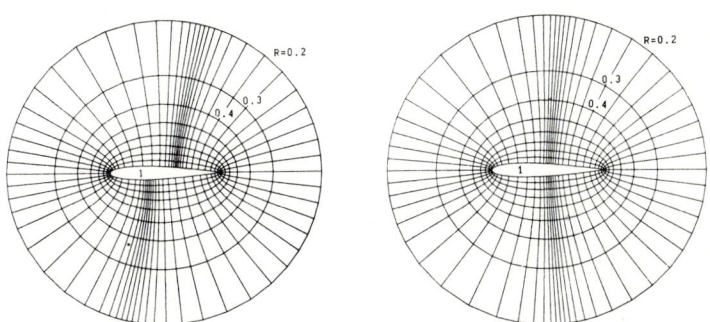

Typical discretization for flow with shocks
with   without angle of attack

Diagram 3

Unsteady transonic flow around NACA 0012,
reduced frequency f = 2.5

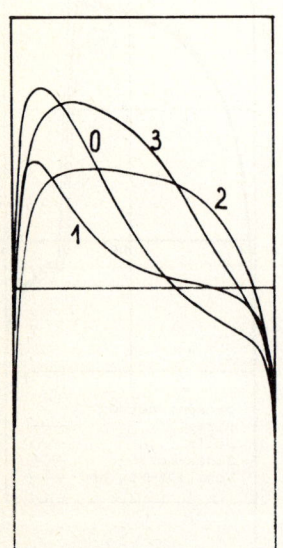
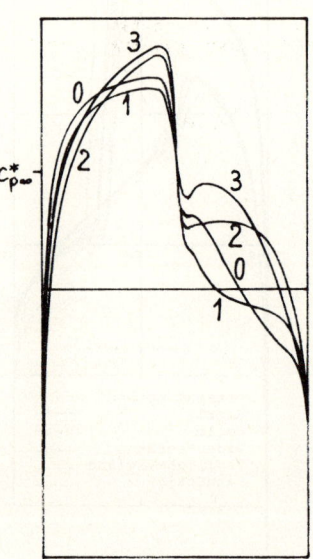

$M_\infty = 0.63$, $\alpha = \pm 2°$
subcritical

$M_\infty = 0.8$, $\alpha = \pm 2°$
supercritical

$M_\infty = 0.8$, $\alpha = \pm 10°$

line #	0	1	2	3	
upper side	2	3	0	1	*T/4
lower side	0	1	2	3	*T/4

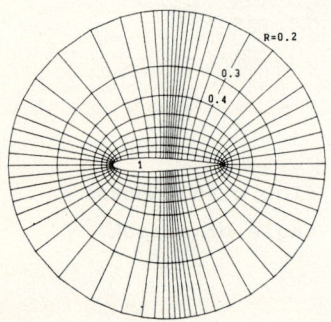

Typical discretization for nearly stationary shocks

FREE SURFACE CALCULATION OF CAPILLARY SPREADING

J. E. FROMM
IBM Almaden Research Center
San Jose, CA 95120 USA

ABSTRACT: A second generation finite difference method[1,2] for treating free surface problems is described. Basically, a streamfunction-vorticity formulation for incompressible flow is used, but because of the free surface, a velocity potential function is also introduced. Surface momentum equations are integrated to provide time dependent surface boundary conditions for the streamfunction and velocity potential. The numerical method is applied to spreading of liquids over no-slip surfaces and includes the full surface stress conditions.

I. THE GOVERNING EQUATIONS

An appropriate choice of scale for capillary flows[3] leads to the nondimensional parameters

$$\frac{W}{R} = \left(\frac{\rho \nu^2}{\sigma r}\right)^{1/2} \quad \text{and} \quad B = \frac{\rho r^2 a}{\sigma}. \tag{1}$$

W/R is a ratio of Weber number to Reynolds number and B is the Bond number. The former is a measure of viscous to capillary forces. The latter is a measure of a body force to capillary forces. Here ρ is the fluid density, σ the surface tension coefficient, ν the kinematic viscosity of the liquid, r is a characteristic length and "a" is a uniform and constant acceleration.

The equations governing the flow are the time dependent vorticity equation

$$\frac{\partial \omega}{\partial t} + \frac{\partial u \omega}{\partial x} + \frac{\partial v \omega}{\partial y} = \frac{W}{R}\left(\frac{\partial^2 \omega}{\partial x^2} + \frac{\partial^2 \omega}{\partial y^2}\right), \tag{2}$$

and the streamfunction equation

$$\frac{\partial^2 Q}{\partial x^2} + \frac{\partial^2 Q}{\partial y^2} = -\omega. \tag{3}$$

The velocities u and v may be extracted from the streamfunction Q with

$$u = \frac{\partial Q}{\partial y} \quad \text{and} \quad v = -\frac{\partial Q}{\partial x}. \tag{4}$$

Because of the nature of a free surface boundary, (2) and (3) are not sufficient for solution. The problem is that the free surface streamfunction boundary condition is unknown. Our approach to this is to divide the velocities into irrotational and solenoidal parts (see Ref. 4). Thus, let

$$u = u_i + u_s = \frac{\partial \phi}{\partial x} + \frac{\partial \psi}{\partial y} \quad \text{and} \quad v = v_i + v_s = \frac{\partial \phi}{\partial y} - \frac{\partial \psi}{\partial x}. \tag{5}$$

Here ϕ is a velocity potential which satisfies

$$\frac{\partial^2 \phi}{\partial x^2} + \frac{\partial^2 \phi}{\partial y^2} = 0, \tag{6}$$

and ψ is a secondary streamfunction which satisfies

$$\frac{\partial^2 \psi}{\partial x^2} + \frac{\partial^2 \psi}{\partial y^2} = -\omega. \tag{7}$$

The latter is distinguished from Q by having different boundary conditions. Now (2), (6), and (7) are complete for obtaining solutions, but as it turns out, it is just as convenient and more intuitive physically to use ϕ and ψ only to obtain boundary conditions for Q. It is necessary to solve (6) in interior regions but not (7).

Now consider local surface variables (τ, η) as indicated in Fig. 1. Through rotational transformation, we have

$$u_\tau = u \cos \alpha + v \sin \alpha \quad \text{and} \quad u_\eta = - u \sin \alpha + v \cos \alpha, \tag{8}$$

where α is the angle of τ with x. Follow-up transformations of required equations may be obtained by noting that, for example,

$$\frac{\partial}{\partial x} = \frac{\partial \tau}{\partial x} \frac{\partial}{\partial \tau} + \frac{\partial \eta}{\partial x} \frac{\partial}{\partial \eta} = \cos \alpha \frac{\partial}{\partial \tau} - \sin \alpha \frac{\partial}{\partial \eta}. \tag{9}$$

The appropriate form of the momentum equations required at the surface follows from the invarient form of the global equations. Thus in the (τ, η) frame

$$\frac{\partial u_\tau}{\partial t} + \frac{1}{2} \frac{\partial}{\partial \tau} \left(u_\tau^2 + u_\eta^2 \right) - u_\eta \omega = - \frac{\partial \left(\frac{P}{\rho} \right)}{\partial \tau} - \frac{W}{R} \frac{\partial \omega}{\partial \eta} + B_\tau$$

$$\frac{\partial u_\eta}{\partial t} + \frac{1}{2} \frac{\partial}{\partial \eta} \left(u_\tau^2 + u_\eta^2 \right) + u_\tau \omega = - \frac{\partial \left(\frac{P}{\rho} \right)}{\partial \eta} + \frac{W}{R} \frac{\partial \omega}{\partial \tau} + B_\eta. \tag{10}$$

The appropriate forms for continuity and vorticity are respectively

$$\frac{\partial u_\eta}{\partial \eta} + u_\tau \frac{\partial \alpha}{\partial \eta} + \frac{\partial u_\tau}{\partial \tau} - u_\eta \frac{\partial \alpha}{\partial \tau} = 0, \tag{11}$$

$$\omega = \frac{\partial u_\eta}{\partial \tau} + u_\tau \frac{\partial \alpha}{\partial \tau} - \frac{\partial u_\tau}{\partial \eta} + u_\eta \frac{\partial \alpha}{\partial \tau}. \tag{12}$$

Similarly, it may be shown that the full global forms of the normal and tangential stress conditions (see Ref. 5) become

$$\text{Normal stress:} \quad \frac{P}{\rho} - \left(\frac{P}{\rho} \right)_g = - \frac{\partial \alpha}{\partial \tau} + 2 \frac{W}{R} \left(\frac{\partial u_\eta}{\partial \eta} + u_\tau \frac{\partial \alpha}{\partial \eta} \right), \tag{13}$$

$$\text{Tangential stress:} \quad \frac{\partial u_\eta}{\partial \tau} + u_\tau \frac{\partial \alpha}{\partial \tau} + \frac{\partial u_\tau}{\partial \eta} - u_\eta \frac{\partial \alpha}{\partial \eta} = 0. \tag{14}$$

$\left(\frac{P}{\rho} \right)_g$ in (13) is the ambient pressure in the gas (see Fig. 1); it may here be set to zero. The first term on the right of (13) is the single curvature in rectangular coordinates. It has a coefficient of unity in our chosen scaling.

With the transformed equations in hand, we proceed next by writing the surface velocities as combinations of irrotational and solenoidal parts

$$u_\tau = u_\tau^i + u_\tau^s = \frac{\partial \phi}{\partial \tau} + \frac{\partial \psi}{\partial \eta} = \frac{\partial Q}{\partial \eta}, \quad u_\eta = u_\eta^i + u_\eta^s = \frac{\partial \phi}{\partial \eta} - \frac{\partial \psi}{\partial \tau} = - \frac{\partial Q}{\partial \tau}. \tag{15}$$

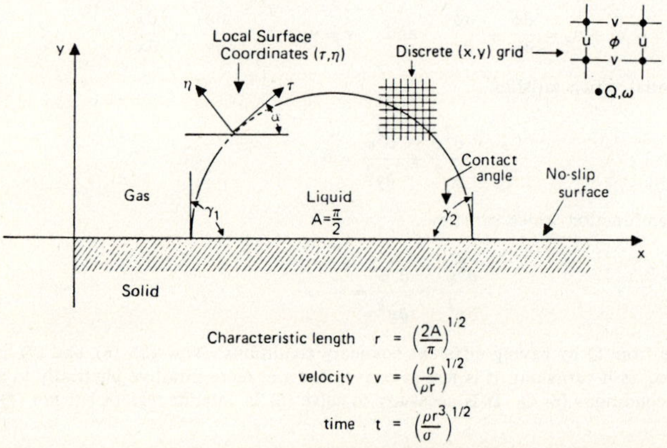

Figure 1. Problem geometry and scale.

Now from (10) we are led to a double set of equations for $\dot\phi = \dfrac{\partial \phi}{\partial t}$ and $\dot\psi = \dfrac{\partial \psi}{\partial t}$. These are

$$\frac{\partial \dot\phi}{\partial \tau} = - \frac{\partial \left(\dfrac{P}{\rho}\right)}{\partial \tau} + B_x \frac{\partial x}{\partial \tau} + B_y \frac{\partial y}{\partial \tau} - \frac{1}{2}\frac{\partial}{\partial \tau}\left(u_\tau^2 + u_\eta^2\right),$$

$$\frac{\partial \dot\psi}{\partial \tau} = - \frac{W}{R}\frac{\partial \omega}{\partial \tau} + u_\tau \omega,$$

(16)

and

$$\frac{\partial \dot\phi}{\partial \eta} = - \frac{\partial \left(\dfrac{P}{\rho}\right)}{\partial \eta} + B_x \frac{\partial x}{\partial \eta} + B_y \frac{\partial y}{\partial \eta} - \frac{1}{2}\frac{\partial}{\partial \eta}\left(u_\tau^2 + u_\eta^2\right),$$

$$\frac{\partial \dot\psi}{\partial \eta} = - \frac{W}{R}\frac{\partial \omega}{\partial \eta} + u_\eta \omega.$$

(17)

Equations (17) are not used and are included only to show the redundancy that follows in the invarient formulation. The $u_\tau \omega$ term of (16) compared to the $u_\eta \omega$ term of (17) simply reflects a property of the streamfunction.

It is clear then that with (16) we may forward march $\dot\phi$ and $\dot\psi$ and hence ϕ and ψ to obtain surface boundary conditions. With these quantities in hand and with solid-liquid interface boundary conditions (to be discussed in the following), we may obtain ϕ and ψ in the interior with (6) and (7). Instead, however, we obtain only ϕ in the interior so that u_η^i may be extracted. u_η^s may be obtained directly from ψ at the surface so that with (15) we can obtain free surface boundary conditions for Q. Solving (3) in the interior permits us to obtain u_τ at the surface and with the inverse transformation of (8), the global surface velocities u and v.

Local (τ,η) reference frames are transported to new positions in a Lagrangian sense with

$$\frac{dx}{dt} = u \quad \text{and} \quad \frac{dy}{dt} = v.$$

(18)

The angles α relating the global and local frames of reference are, of course, changed at each modification of the surface by (18). The local vorticities and pressures required in (16) follow from (12) and (13). We avoid normal derivatives by using (14) in (12) and (11) in (13). Thus, the surface vorticity and pressure equations become

$$\omega = 2\left(\frac{\partial u_\eta}{\partial \tau} + u_\tau \frac{\partial \alpha}{\partial \tau}\right),$$

(19)

and

$$\frac{P}{\rho} = - \frac{\partial \alpha}{\partial \tau} - 2\frac{W}{R}\left(\frac{\partial u_\tau}{\partial \tau} - u_\eta \frac{\partial \alpha}{\partial \tau}\right).$$

(20)

Our final consideration is the no-slip boundary condition at the solid surface. Clearly we must have

$$v_b = \left(\frac{\partial \phi}{\partial y}\right)_b - \left(\frac{\partial \psi}{\partial x}\right)_b = -\left(\frac{\partial Q}{\partial x}\right)_b = 0,$$

(21)

and

$$u_b = \left(\frac{\partial \phi}{\partial x}\right)_b + \left(\frac{\partial \psi}{\partial y}\right)_b = \left(\frac{\partial Q}{\partial y}\right)_b = 0.$$

(22)

To satisfy the former, we require

$$\left(\frac{\partial \phi}{\partial y}\right)_b = \left(\frac{\partial \psi}{\partial x}\right)_b = 0,$$

(23)

and for the latter

$$\left(\frac{\partial \psi}{\partial y}\right)_b = -\left(\frac{\partial \phi}{\partial x}\right)_b = -(u_i)_b.$$

(24)

While (24) is not used in calculation, it is important to see in what respect ψ and Q boundary conditions differ here.

With consideration of (24), we have

$$\omega_b = -\left(\frac{\partial^2 Q}{\partial y^2}\right)_b = -\left(\frac{\partial^2 \psi}{\partial y^2}\right)_b - \left(\frac{\partial u_i}{\partial y}\right)_b. \tag{25}$$

This is the usual "afterthought[6]" determination of the boundary vorticity at a no-slip surface that follows from (3). If we were to use ψ as a calculation variable, we would specify ψ as constant on the no-slip surface in accordance with (23), but the "afterthought" vorticity would require a contribution obtained from ϕ also.

II. THE FINITE DIFFERENCE EQUATIONS AND ORDER OF CALCULATION

Here we use a uniform rectangular grid for calculation of internal fields. A square grid is preferred whenever possible, because any stretching is accompanied by larger truncation errors. We regard the surface as being composed of Lagrangian particles to which are attached the local (τ,η) coordinate frames. Generally, 3 to 10 particles cover a mesh distance of the internal grid. One assumes that the accuracy is governed by the Eulerian (internal) grid, and hence the more dense surface grid is only a means of providing sufficiently adequate particles that motion does not lead to intergrid regions, vacant of associated surface particles. This would affect the logic by which the program determines the appropriate approximations to be used near the free surface. Local interpolation of surface variable values occurs often throughout calculation in order to align information along grid lines. Where information is required for such alignment, linear interpolation is used. On the other hand, we also need to transfer information from interior points to the surface when normal derivatives at the surface are required. This involves taking one sided derivatives at nearest surface approach to selected interior points and then distributing this information to all surface particle positions. In these cases Lagrange interpolation up to fourth order is used, except near contact points (lines).

Let us consider a case where the half cylinder fluid configuration is placed on a solid surface, Fig. 1. Motion is initiated through the input contact angles, if they differ from 90°, causing a nonuniform curvature $(\partial \alpha/\partial \tau)$ near the contact lines. This is reflected as a reduced pressure which manifests itself in a spatial change of ϕ, the time derivative of the velocity potential. In the given case, we choose $\dot\phi$ and ϕ to be forever zero at the crest of the initial configuration (Fig. 1). Integrating both directions from this point and recognizing the conservative form of (16), we may write

$$\dot\phi_k = \left(\frac{P}{\rho}\right)_0 - \left(\frac{P}{\rho}\right)_k + B_x(x_k - x_0) + \frac{1}{2}\left\{(u_\tau^2 + u_\eta^2)_0 - (u_\tau^2 + u_\eta^2)_k\right\}, \tag{26}$$

where k represents the k'th particle relative to the reference 0'th particle at the initial lens crest. If the surface velocities and derivatives are stored quantities, we have

$$\left(\frac{P}{\rho}\right)_k = -\left(\frac{\partial \alpha}{\partial \tau}\right)_k + 2\frac{W}{R}\left\{(u_\eta)_k \left(\frac{\partial \alpha}{\partial \tau}\right)_k - \left(\frac{\partial u_\eta}{\partial \tau}\right)_k\right\}, \tag{27}$$

where derivatives are evaluated locally by first order differences.

With a time dependent free surface boundary condition

$$\phi_k^{n+1} = \phi_k^n + \Delta t\, \dot\phi_k, \tag{28}$$

and a solid surface condition (23) in the form

$$\phi_{i,\, j_0 - 1/2} = \phi_{i,\, j_0 + 1/2}, \tag{29}$$

we are in a position to solve (6) in the fluid interior. Here we are using (i,j) as the (x,y) grid index (j=j_0 at the solid boundary), and n as an explicit time index. Note in Fig. 1 that ϕ is defined at cell centers.

An analogous expression to (26) for ψ is

$$\dot\psi_{k+1} = \dot\psi_k + \frac{W}{R}(\omega_k - \omega_{k+1}) + \frac{1}{2}\left\{(u_\tau\omega)_{k+1} + (u_\tau\omega)_k\right\}(\tau_{k+1} - \tau_k) \tag{30}$$

where

$$\psi_k^{n+1} = \psi_k^n + \Delta t\, \dot\psi_k. \tag{31}$$

We arbitrarily take $\dot\psi$ and ψ to be forever zero at the solid-fluid boundary and take a mean of outward integrations from the contact lines to reduce truncation errors. Note that $\dot\psi$ is here treated locally because of the $u_\tau \omega$ term.

Assume that u_η^i is available to us (as it will be after solution of (6)), and we obtain u_η^s directly from ψ at the surface.

Now by integrating from the contact lines (Q forever zero there), we obtain Q at the surface with

$$Q_{k+1} = Q_k + \left(u_\eta^i + u_\eta^s\right)\Delta\tau_{k+\frac{1}{2}}. \tag{32}$$

With Q also zero at the entire solid-liquid interface, we may solve (3).

Now equations (3) and (6) are integrated numerically in identical form for the Laplacian, except for displacement of ϕ and Q on the discrete grid (see Fig. 1). Here we will discuss only Q, in terms of the forms that the difference equations take. In the case of (3), the right-hand side must be determined.

The time dependent vorticity equation (2) provides us with interior vorticity values. It is integrated in flux form, as described in Ref. 7. Near the free boundary, vorticity values are updated if they lie greater than a half mesh distance from the surface. If less than a half mesh distance, they are not required in (3) and may take on interpolated values, simply for the sake of completing the field. Surface vorticity follows from (19) and no-slip surface vorticity from (25).

This "floating" nature of mesh points less than a half mesh distance from the surface is crucial in solutions to (3) and (6). We write for a "somewhat" general difference form of (3) (see Fig. 2) the "Diamond" formula

$$Q_0 = (a_1 Q_1 + b_1 Q_2 + c_1 Q_3 + d_1 Q_4 + \omega_0)/\left(\frac{2}{ac} + \frac{2}{bd}\right) \tag{33}$$

where

$$a_1 = \frac{2}{a(a+c)}, \quad c_1 = \frac{2}{c(a+c)},$$

$$b_1 = \frac{2}{b(b+d)}, \quad d_1 = \frac{2}{d(b+d)},$$

and

$$\tfrac{1}{2}\Delta x \leq a \leq \tfrac{3}{2}\Delta x, \quad \tfrac{1}{2}\Delta x \leq c \leq \tfrac{3}{2}\Delta x$$

$$\tfrac{1}{2}\Delta y \leq b \leq \tfrac{3}{2}\Delta y, \quad \tfrac{1}{2}\Delta y \leq d \leq \tfrac{3}{2}\Delta y.$$

Equation (33) is not, in fact, as general as we may need. When concave curvatures are very small, a special form must be devised in order to satisfy the limitations on a, b, c, or d. The same is true for very large concave curvatures, but there is no corresponding difficulty with convex surfaces. We would have no problem even with the former, if a diagonal equivalent of (33) could be used. Unfortunately the diagonal form is relevant only to a square grid, otherwise an extra grid point is required to remove the cross term from the approximation! Fortunately the "Star 1" layout of Fig. 2 is a simple modification of (33). That is, require $a=\Delta x$, then take

$$Q_1 = \{\Delta y Q_5 + (b' - \Delta y)Q_8\}/b'$$

where

$$\Delta y \leq b' \leq \tfrac{3}{2}\Delta y.$$

The reason for this modification is that otherwise we may have $a \gg (3\Delta x/2)$, and hence have large truncation errors.

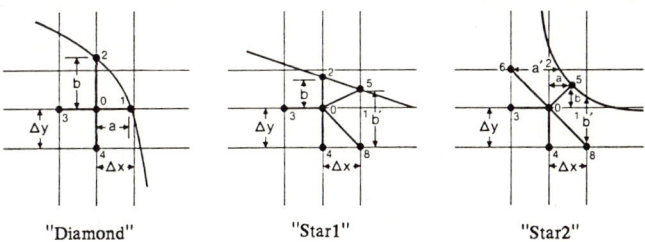

"Diamond"   "Star1"   "Star2"

Figure 2. Grid layout for special near surface elliptic difference equation formulas.

The "Star 2" formula (layout sketched in Fig. 2) is more complicated and is not included here.

Only the single particle in contact with the solid no-slip surface (at the contact line) is permitted *free* slip. This is the simplest assumption one can make and is perhaps as far as one should go on the basis of continuum mechanics alone. For all other particles on the free surface the global (u,v) velocities are obtained by rotational transformation, but at the contact point the inverse transformation (Eq. 8) is required because u and v are the given quantities while u_τ and u_η are not given.

Finally the best assumption concerning curvature at contact lines seems to be an extrapolation of the input contact angle beyond contact. This gives the proper angular gradient ($\partial \alpha/\partial \tau$) and satisfies equilibrium in that it provides an ultimately *constant* curvature surface.

III. RESULTS

We illustrate layer spreading with an example in which W/R=10, B=0, and $\gamma_1 = \gamma_2 = 45°$. This is a case which is in the "viscous" domain; the transition where nonlinearity becomes important is at W/R~0.5. Figure 3 gives a series of five successive states of time dependent spreading. Streamlines and lines of constant velocity potential define the flow. The former are solid lines with negative values including tick marks. Constant potential lines are dashed lines with a characteristic normal trend toward the solid surface. Flow throughout the sequence is symmetrically downward toward contact lines. The flow is initially rapid and diminishes as equilibrium is approached. Experience indicates that the surface energy minimum is broad.

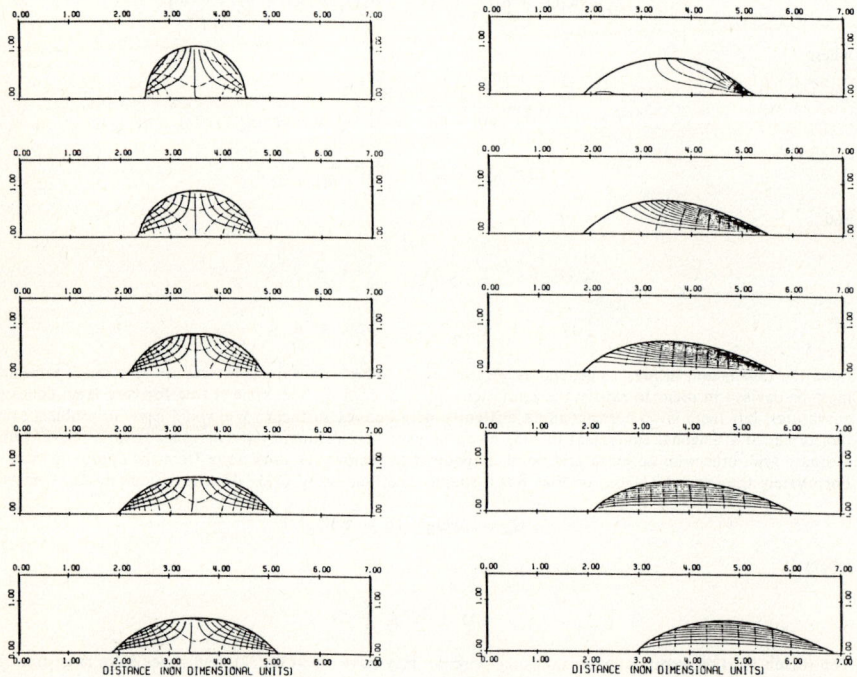

Figure 3. Time sequence of spreading for W/R=10, B=0, $\gamma_1 = \gamma_2 = 45°$.

Figure 4. Time sequence of spreading into a higher surface energy region: W/R=0.05, $\gamma_1 = 45°$, $\gamma_2 = 30°$.

The meniscus is barely visible at the earlier times, but we know that the low pressure there is immediately felt throughout the liquid. The meniscus curvature diminishes with time, passes through zero curvature, and ultimately meets with the changing curvature of the surface of the bulk of the liquid. Thus, the early part of the flow is driven by the low pressure at the meniscus, but as the sign of the curvature changes, the driving pressure comes from other regions of the surface in such a way as to finally equalize the pressure throughout at the lower value consistent with the equilibrium contact angle.

An interesting aspect of the flow is less apparent in this rather viscous example. That is, the effect of the no-slip surface. For less viscous cases, one can observe slow growth of the boundary layer, allowing faster spreading at early times.

For an additional example, we include Fig. 4, a case in which W/R=0.05, B=0, $\gamma_1=45°$ and $\gamma_2=30°$. This simulates a situation in which surface energies differ at the contact lines. One might think of this as introducing a blotter at the edge of a stationary region of fluid on a flat surface. We begin the calculation with equilibrium for $\gamma_1=\gamma_2=45°$, but then modify γ_2 to be 30° to initiate the motion. A history plot reveals that the right contact line moves into the higher surface energy region at a near constant velocity at first, while the left contact line remains stationary. Once the motion has progressed like a rarefaction wave to the left contact line, one has an exponential build up of the velocity of the layer (except for the moderating effect of the no-slip surface). Note the sparseness of streamlines at the solid surface.

REFERENCES

1. J. E. Fromm, "Finite Difference Computation of the Capillary Jet, Free Surface Problem," Lecture Notes in *Physics* 141, 188-193, Seventh International Conference on Numerical Methods in Fluid Dynamics, Springer-Verlag, Berlin 1981.
2. J. E. Fromm, "A Numerical Study of Drop-on-Demand Ink Jets," Proceedings of the Second International Colloqium on Drops and Bubbles, NASA JPL Publication 82-7, California Institute of Technology, Pasadena (1982).
3. J. W. S. Rayleigh, *The Theory of Sound* Vol. 2, Dover, p. 351 (1945).
4. N. E. Kochin, I. A. Kibel, and N. V. Roze, "Theoretical Hydrodynamics," (Interscience Publishers John Wiley and Sons, Inc., New York 1964).
5. G. K. Batchelor, "An Introduction to Fluid Dynamics" (Cambridge University Press, Cambridge 1967).
6. J. Gazdag, Y. Takao, and J. Fromm, "Rigorous Numerical Treatment of the No-Slip Conditions in a Vorticity Formulation," Proceeding of the NASA Symposium on Numerical Boundary Condition Procedures, NASA Conference Publication 2201, Moffett Field, California, October 1981.
7. D. B. Bogy, J. E. Fromm, and F. E. Talke, "Exit Region Central Source Flow Between Closely Spaced Parallel Co-rotating Disks," *The Physics of Fluids* Vol. 20, No. 2, 176-186 (1977).

A COMBINED NUMERICAL SCHEME FOR TRANSONIC FLOWS

Laszlo Fuchs
Department of Gasdynamics,
The Royal Institute of Technology,
S-100 44 Stockholm, Sweden.

SUMMARY

A combined algorithm using potential/Euler solvers, for transonic flow computations, is described. This combined scheme is substantially more efficient than the basic Euler solver. The potential equations are solved very effeciently by a Multigrid or an AF scheme. These solver require also considerably less computer memory than the corresponding Euler solver. The potential solver is used to provide initial approximation to the Euler solver, to determine the regions where the flow is vortical and where the Euler solver has to be used. Thus, the full Euler equations are solved only in some parts of the computational domain, and the potential equations are used elsewhere. Since the potential solver provides a relatively good approximation to the Euler solution, few (one or two) coupled iterative steps are adequate. The coupled scheme, is faster and requires less computer storage than current Euler solvers.

INTRODUCTION

The potential equations are an inviscid and irrotational approximation to the Navier-Stokes equations. These equations can be solved numerically, with good efficiency and accuracy. In several transonic flows, however, the (non-linear) potential theory is not a good enough approximation of real flows. These types of compressible flows are treated, in general, by solving the Euler equations. The only underlying assumption for these equations beside continuum, is that the fluid is inviscid. The potential approximation is obtained from the Euler equations by assuming irrotational motion. For external airfoil problems, the flow is irrotational if the inflow velocity profile is irrotational <u>and</u> if there are no shocks (assuming that the fluid is inviscid and that the vorticity due to the circulation around the airfoil can be represented by a singular surface, i.e. a vortex sheet).

Numerical methods for solving the potential equations require less computer capacity. The potential model consist of a single PDE (both in 2-D and 3-D), whereas the Euler equations consist of 2+d equations, where d is the dimensionality of the problem. (One PDE may be replaced by an algebraic equation if one assumes that the total enthalpy is constant). Furthermore, the character of the potential PDE and the Euler system of PDE's are different. The former is (locally) elliptic whenever the flow is subsonic, and hyperbolic at regions of supersonic flows. The Euler equations are always hyperbolic, independent of the flow. As a result of the different types, these two approximations require different boundary conditions with regard to number and type. The iterative methods for solving the two models are also different. The potential equations can be solved very efficiently by Multi-Grid (MG) [1-3,5] and by Approximate-Factorization (AF) methods [4,5]. The convergence rates achieved by MG methods can be very close, even for supercritical cases, to those attained by MG methods for the Laplace equation [5]. In the case of the Euler equations, most iterative methods use a pseudo time-marching techniques (e.g. [11]). These schemes, even in a MG configuration [6,7], do not attain the efficiency level attained by the potential solvers. MG solvers for the steady-state problem, using flux splitting methods [8] have been reported to be faster, but these methods seem to have low order of accuracy (i.e. large artificial viscosity). All these factors together make Euler solvers to be an 'expensive' alternative to the potential solvers. For these reasons there have been different attempts to improve the potential model, while keeping its numerical efficiency.

One of the early attempts of non-isentropic potential flow modelling were made by Fuchs [1], who used the transonic small perturbation equation, and included in it a correcting term, that accounted for the increase in entropy behind the shock. Methods for correcting the full potential equations was suggested by Klopfer and Nixon [9] who included in the potential model the conservation of mass and momentum or momentum and energy across shocks. Hafez and Lovell [10] solved, beside the equation for the potential an additional equation, with a similar structure, for the streamfunction. These models are, however, more limited in applications than the complete set of Euler equations. In this paper we use the potential solution to improve the efficiency of the Euler solver. This is done by providing a good initial approximation to the Euler solver and by allowing a significant reduction of the computational domain. In most applications one does not have to iterate between the potential and the Euler solvers, and a single step of a local Euler correction yields very good results. The combined scheme results in a considerable improvement of numerical efficiency: shorter computational times and reduction in computer memory requirements.

In the following we describe shortly the Euler scheme that we use and how it is modified to accommodate the potential results. Finally, we give some computational examples on the effects of the different modifications on the numerical efficiency.

THE NUMERICAL SCHEME

Both the potential and the Euler approximations have been applied for solving the flow past airfoils. A body fitted mesh is generated, such that the spacing near the airfoil, and especially close to the leading and trailing edges is finer. The potential equation is written in terms of the of the transformed coordinates (in conservative form). The equations are discretized by centered finite differences, and artificial viscosity is added to the density in supersonic flow regions (see [1,4,5]). The discrete equations are solved by an Approximate Factorization (AF) scheme [4,5]. Our Euler solver is based on the code of Rizzi [12]. The basic code uses a finite-volume technique with second and fourth order artificial damping terms. The integration of the equations is done by a multi-stage Runge-Kutta technique. This code has been modified to accommodate MG processing, local mesh refinements and includes the new additional features of the combination with the potential solver.

Improved efficiency is obtained by combination of both methods. By improved numerical efficiency we mean shorter computational times and less computer memory for the solution of a given problem. Improved efficiency may be attained by one or a combination of the following measures:
 * Modified modelling (reduction in number of unknowns).
 * Improved discretization techniques, so that the number of degrees of freedom can be reduced without hamperring accuracy.
 * Faster iterative methods.
In the present work we address the first possibility. The other options are being at different stages of testings.

To improve the efficiency of the Euler solver we use the potential solver. The potential equation is solved initially. The potential results are interpolated (averaged) to the cell centers, where they are defined in the current finite-volume scheme. The potential equation provides the velocity vector, the density and the pressure. The last two quantities are computed by using isentropic relations. The potential solution is used by the Euler solver in two ways: It is used as an initial approximation for the Runge-Kutta steps and it provides boundary condition on regions where the potential approxmiantion is not good enough. Once the Euler equations, in the reduced domain, are solved, one has to update the potential solver (i.e. correct for the rotational part of the flow field). This can be done by adding a correction to the isentropic density formula. Our solution algorithm has the following form:

Step i. Solve the potential equation for ϕ and ρ_p: $\nabla \cdot (\rho_p \nabla \phi^{(1)}) = 0$

Set n=1. Define the 'Euler domain' to include at least the regions downstream of the shocks.

Step ii. Interpolate the correction due to the potential solution (i.e. the differences between the current potential and Euler solutions) to the Euler mesh.

Step iii. Solve the Euler problem in the 'Euler domain' (possibly, a subdomain of the full potential domain), to certain accuracy, using the modified initial approximation (of step ii.).

Step iv. Compute the corrections for the potential solver: $\Delta\rho_p = \rho_E^{(n)} - f(\mathbf{u}_E)$ where f is the value of the density using the isentropic relation and the Euler velocity vector, $\mathbf{u}_E$. Note that the right hand side of the potential problem should also be modified:

Step v. Solve the modified potential equation: $\nabla \cdot (\bar{\rho}_p \nabla \phi^{(n+1)}) = R^{(n+1)}$

where $\bar{\rho}_p = f(u_p, v_p) + \Delta\rho_p$

and $R^{(n+1)} = \nabla \cdot (\bar{\rho}_p \nabla \phi^{(n)} - \rho_E \mathbf{u}_E^{(n)})$

Step vi. If the changes in the density are too large set n←n+1 and go to Step ii.

The scheme described above can be regarded as a defect correction scheme for the potential equation (with the defects correcting for the entropy rise due to the rotational character of the flow). The better the potential model in approximating the Euler equations is, the less number of defect correction steps one has to make. The number of steps that one should make, can be estimated in terms of the deviation from isentropic flow (i.e. in terms of entropy production). It is well known that the entropy production across a shock is proportional the shock strength to the third power. Thus, for transonic flows the number of defect correction cycles, in the iterative process above, can be as low as 1 or 2. That is, in most cases it is enough if both the potential and the Euler equations are solved once or at most twice. The amount of additional computational work in solving the potential equation is only a small fraction compared to that required for the solution of the full Euler problem. This is so since the potential problem consists of a single PDE and the iterative process has higher rate of convergence than that of the corresponding Euler solver. It should be noted that when the above iterative scheme converges, the solutions (the density) of the potential and the Euler equations are identical.

COMPUTATED EXAMPLES

The algorithm that is described above, have been applied for the solution of the flow past airfoils (NACA-0012 and RAE-2822). These problems have been solved by many different schemes and therefore are good candidates for studying our scheme.

A typical mesh is shown in Figures 1. For subsonic flows the potential solver provides avery good approximation to the Euler solver. The only errors in the initial approximation are due to the difference in the truncation errors and due to interpolating the potential solution to the Euler grid. For supercritical cases, the errors in the initial approximation are larger, and depend on the strength of the shock (i.e. freestream Mach number, M_∞, and the angle of attack, α).

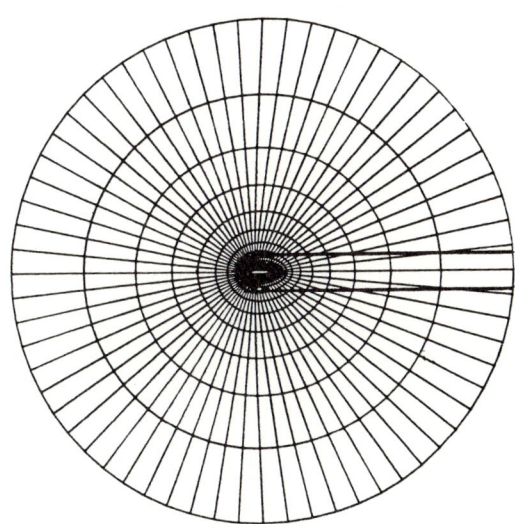

Figure 1.a: The computational grid for the airfoil problem.
The potential problem (3) is solved in the whole domain, while the Euler equations (2) are solved only in a limited (typically a rectangular domain as shown in the figure for case a).

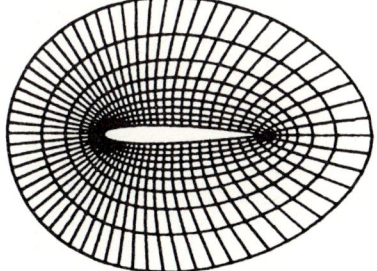

Figure 1.b: Enlargement of the computational mesh near the airfoil.

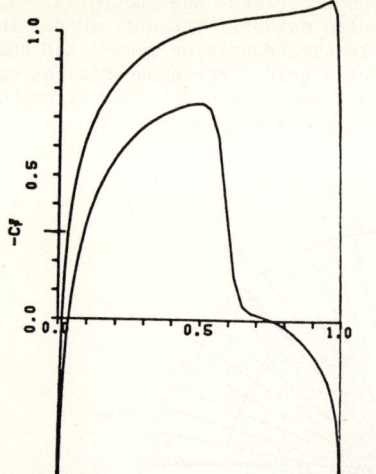

Figure 2.a: The pressure coefficient, C_p, on the airfoil. Full grid potential solution. $M_\infty=0.85$, $\alpha=1°$.

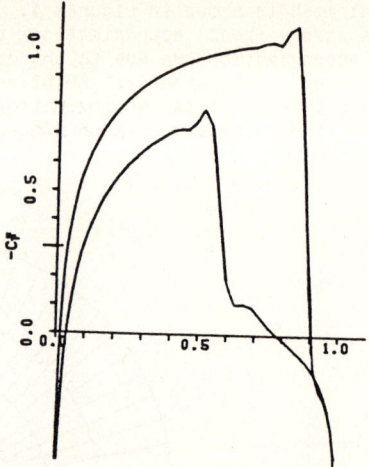

Figure 2.b: The pressure coefficient, C_p, on the airfoil. Full grid Euler solution. $M_\infty=0.85$, $\alpha=1°$. Mean convergence factor, $\theta=0.984$ (not fully converged).

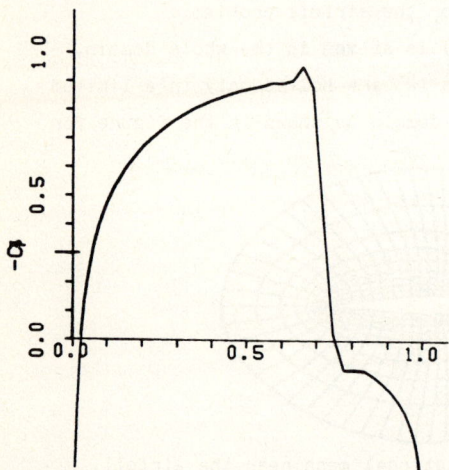

Figure 3.a: C_p on the airfoil. Full Euler solution using interpolated potential solution as initial approximation. $M_\infty=0.85$, $\alpha=0°$. $\theta=0.958$

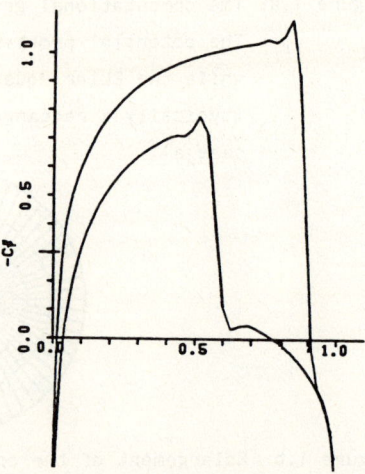

Figure 3.b: C_p on the airfoil. Full Euler solution using interpolated potential solution as initial approximation. $M_\infty=0.85$, $\alpha=1°$. $\theta=0.962$

First, we have used the original potential and the Euler solvers for two supercritical cases; a: $M_\infty = 0.8$ and $\alpha = 0°$, and : $M_\infty = 0.85$ and $\alpha = 1°$. The potential and the Euler solutions differ considerably with increasing shock strength (see Fig. 2.a and 2.b, respectively). The convergence factor of the unmodified Euler solver for cases a and b are 0.980 and 0.984, respectively. The convergence factors are defined as the mean reduction of the error for a computational effort equivalent to one (three staged) time step in the Euler solver on the full (finest) grid. The slow asymptotic rates of convergences are not improved by using different initial approximations (e.g. by coarse grid approximation or by potential solution). However, the use of good initial approximations does improve the initial rate of convergence (to a level better than that due to discretization errors). When the initial approximation is computed by interpolating a coarse grid solution, the number of time steps may be halved (to about 450). When we use the interpolated potential solution as initial approximation, the Euler solver converges well, in less than 110 iterations; giving a mean convergence factor of 0.958 (Fig 3.a) and 0.962 (Fig. 4.a) for cases a and b respectively. When the Euler equations are solved only in a limited, 'Euler region' (see Fig. 1) the convergence rate improves further. The convergence factors for the two cases, are 0.888 (Fig. 3.b) and 0.929 (Fig. 4.b), respectively. It should be noted that in these computations the coupled algorithm was used only once (n=1). That is, the local Euler solution was not used to improve the outer (potential) solution. Therefore, the extent of the Euler region in case b had to be extented compared to case a, leading to a reduction in convergence rate. For comparison, one may consider one of the best MG-Euler solver as reported by Jameson [13]. That scheme uses a four stage scheme (and therefore the 'work unit' in that case is larger by about 33% compared to the current three stage scheme). The convergence factors reported in [13] are somewhat larger than 0.9. That is, the current scheme can provide a solution, using much simpler schemes, at less programming and computational effort (and requires less computational memory, since the Euler equations are solved only on parts of the domain).

The limited 'Euler domain' solutions (as shown in Fig. 3.b and 3.c) have been compared to the full region Euler solutions. In all the computed cases, the differences could be made as small as one wishes by enlarging the 'Euler domain'. Compared to the results published in the literature it was found that for practicle applications, it was enough to limit the 'Euler domain' to include the shocks and the regions behind these shocks. Thus, in most cases the computational domain could be reduced by a factor between 2 and 4, without loss of (graphically noticable) accuracy. For the RAE airfoil the 'Euler domain' had to be relatively larger than required only from vorticity production point of view. It is believed that the larger truncation errors, especially near the leading edge play an important role in determining the size of the Euler domain.

ACKNOWLEDGEMENT

This work was supported by the Swedish National Board for Technical development (STU) (Grant No. AU#2040:3).

REFERENCES

[1] Fuchs, L.: "Finite-difference methods for plane steady inviscid transonic flows" The Royal Institute of Technology Report. TRITA-GAD-2, ISSN 0281-7721 (1977).
[2] Fuchs, L.: "Transonic flow computation by a multi-grid method", GAMM-Workshop on Numerical Methods for the computation of Inviscid Transonic Flows with Shock Waves, Rizzi, A. and Viviand, H. (eds.) Vieweg & Sohn, 1979, pp. 58-65.
[3] Jameson, A.: "A multi-grid scheme for transonic potential calculations on

arbitrary grids", Proc. 4-th AIAA CFD conference, pp. 122-146 (1979).
[4] Holst, T. L.: "Implicit algortihm for the conservative transonic full potential equation using an arbitrary mesh", AIAA paper 78-1113 (1978).
[5] Gu, C-Y., Fuchs, L.: "Numerical computation of transonic airfoil flows", in Num. Meth. Laminar and Turbulent Flow-IV, Taylor, C., Olson, M.D., Gresho, P.M., Habashi, W.G. (eds.), Pineridge Press, Swansea, 1985, pp. 1501-1512. (Details in: Gu, Y-C.: Transonic flow computations. TRITA-GAD-8, ISSN 0281-7721, 1985).
[6] Schmidt, W., Jameson, A.: "Application of multiple-grid methods for transonic flow calculations". in 'Lecture Notes in Mathematics', Hackbusch, W. and Trottenberg, U. (eds.), Springer Verlag, Berlin, 1982, pp. 599-613.
[7] William, J., Usab, Jr., Murman, E.M.: "Embedded mesh solutions of the Euler equations using a multiple-grid method", in 'Advances in computational transonics' Habashi W.G. (ed.), Pineridge Press, Swansea 1985, pp. 447-472.
[8] Mulder, W. A.: "Computation of the quasi-steady gas flow in a spiral galaxy by means of a multigrid method", 2nd Copper-Mountain Multigrid Conference, 1985.
[9] Klopfer, G. and Nixon, D.: "Non-isentropic potential formulation for transonic flows", AIAA paper 83-375 (1983).
[10] Hafez, M., Lovell, D.: "Entropy and vorticity corrections for transonic flows", AIAA paper 83-1926 (1983).
[11] Jameson, A., Schmidt, W., Turkel, E.: "Numerical solution of the Euler equations by finite volume methods using Runge-Kutta time- marching schemes", AIAA Paper 81-1259 (1981).
[12] Rizzi, A. W.: "Damped Euler-equation method to compute transonic flow around wing-body combinations", AIAA J. 10, (1982), pp. 1321-1328.
[13] Jameson, A.: A multi-grid solution method for the Euler equations, Proc. ICFD conference, Morton K.W. and Bains, M.J. (eds.), Oxford University Press, 1986.

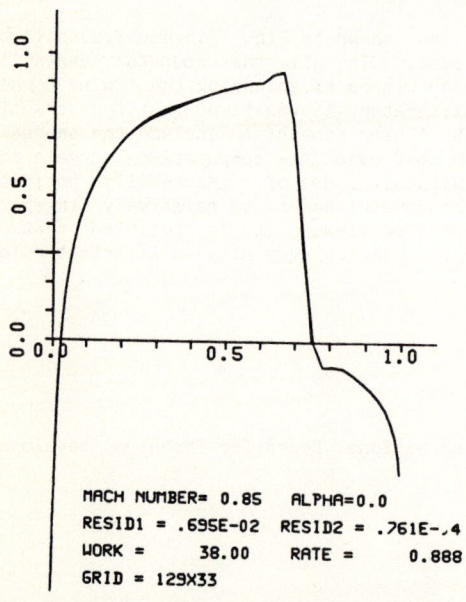

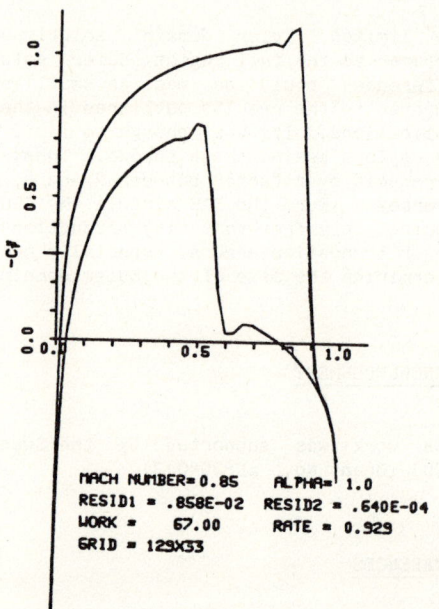

Figure 4.a: C_p on the airfoil. Euler solution only in a limited 'Euler domain'. $M_\infty=0.85$, $\alpha=0°$. $\theta=0.888$

Figure 4.b: C_p on the airfoil. Euler solution only in a limited 'Euler domain'. $M_\infty=0.85$, $\alpha=1°$. $\theta=0.929$

FULLY DEVELOPED PULSATILE FLOW IN A CURVED PIPE

C.C. Hamakiotes and S.A. Berger
Department of Mechanical Engineering
University of California
Berkeley, California 94720

INTRODUCTION

Curved pipe flows arise frequently in many engineering disciplines. Although steady flows are more common in industry than unsteady ones, unsteady flows do occur due to intentional or unintentional transient processes. In some cases the pressure gradient driving the flow is of interest in order to design an efficient piping system, thus avoiding catastrophic results in case of malfunction. In other cases it is the mixing properties of such flows which are of interest. Because the secondary currents in curved pipe flows enhance mixing, chemical and biomedical engineers take advantage of this property by using helical coils in their applications. In biomedical applications, knowing the velocity profiles and shear stress distributions would be useful for purposes like designing a proper artificial heart valve or explaining phenomena such as fatty deposition on the aortic walls.

Although curved pipe flows in general have been studied extensively, little is yet known about unsteady flows in such geometries. We have attempted to simulate the fully developed region of flows subject to a pulsatile volumetric flow-rate in tightly curved pipes. The experiments of Talbot and Gong (1979, 1983) show that pulsating flows entering a curved pipe become fully developed, in the sense that they become periodic in time at any cross section and independent of the axial direction, at about 180° downstream. Our problem is therefore two-dimensional, applicable to any cross-section in the fully-developed region. The geometric configuration and the toroidal coordinate system used is shown in Figure 1; the pulsatile volumetric flow-rate we have imposed, $Q = Q_{DC} + Q_{AC} \cos(\omega t)$, chosen to model the flow of blood through the aortic arch, is shown in Figure 2.

MATHEMATICAL DESCRIPTION
Governing equations

The governing equations are the fully-developed, time-dependent Navier-Stokes equations, written below in dimensionless form:

Continuity $\quad \nabla \cdot V = 0 \quad$ (1)

Momentum $\quad \frac{\partial V}{\partial t} + St_m^{-1}((V \cdot \nabla)V + V(\nabla \cdot V)) = - St_m^{-1} \nabla P + \frac{1}{St_m \cdot Re_m} \nabla^2 V \quad$ (2)

where

$$St_m = \frac{a\omega}{W_m}, \quad Re_m = \frac{aW_m}{\nu}, \quad \text{and} \quad W_m = \frac{Q_{DC}}{\pi a^2}.$$

In the above equations the velocity field has been non-dimensionalized using W_m, the mean axial velocity, the pressure by ρW_m^2, the radial coordinate using the radius of the pipe, and the time using the frequency of the volumetric flow-rate, ω.

The boundary conditions for these equations are:

No-slip $V(r = 1, \phi) = 0$ (3a)

Fully-Developed Condition $\partial/\partial\theta = 0$ (3b)

Symmetry $\frac{\partial u}{\partial \phi} = \frac{\partial w}{\partial \phi} = 0, \quad v = 0 \quad \text{at} \quad \phi = 0, \pi$ (3c)

Incompressibility $Q(t) = 2 \int_{\phi=0}^{\pi} \int_{r=0}^{1} wr\,dr\,d\phi = \pi(1 + \frac{Q_{AC}}{Q_{DC}} \cos \omega t).$ (3d)

Dimensional analysis indicates that unsteady flows through curved pipes are determined by three parameters: (i) the frequency parameter, α; (ii) the Dean number, κ; and (iii) the amplitude ratio, γ. Their definitions and the values used in the case presented here are:

$$\alpha = a\sqrt{\omega/\nu} = 15,$$

$$\kappa = \frac{2W_m a}{\nu}\sqrt{\frac{a}{R}} = 1134,$$

$$\gamma = \frac{Q_{AC}}{Q_{DC}} = 1.$$

The curvature ratio $\delta = a/R = 1/7$ is small, yet large enough to require solution of the complete equations. The Reynolds number, $Re_m = W_m a/\nu = 1500$, is large enough as to play a crucial role in the choice of a numerical scheme and in the introduction of artificial errors, i.e., artificial viscosity, as we shall see in the next section.

Numerical method

To solve this set of equations we use the Projection Method (1968), which consists of two steps. In the first step the momentum equation is solved without the pressure gradient, to obtain an auxilliary velocity field, V^*:

$$\frac{v^*-v^n}{\Delta t} + \frac{1}{St_m}((v^* \cdot \nabla)v^* + v^*(\nabla \cdot v^*)) - \frac{1}{Re_m \cdot St_m}\nabla^2 v^* = 0 \ . \qquad (4)$$

We implement this step using an ADI method, developed by Douglas and Gunn (1964), which is second order accurate in time and space. In the second step we obtain v^{n+1} by considering the equations,

$$\frac{v^{n+1}-v^*}{\Delta t} + St_m^{-1} \nabla P^{n+1} = 0 \ , \qquad (5a)$$

$$\nabla \cdot v^{n+1} = 0. \qquad (5b)$$

Taking the divergence of (5a) and using (5b), we obtain the Poisson equation,

$$\nabla^2 P^{n+1} = \frac{St_m}{\Delta t} \nabla \cdot v^* \ , \qquad (6a)$$

which we solve using successive overrelaxation iteration, thus obtaining the pressure field. The proper boundary condition for the pressure in (6a) is a Neumann-type condition obtained by projecting the vector equation (5a) on the outward normal unit vector N to the boundary Γ,

$$(\frac{\partial P}{\partial N})_\Gamma^{n+1} = - \frac{St_m}{\Delta t} (v_\Gamma^{n+1} - v_\Gamma^*) \cdot N \ . \qquad (6b)$$

Due to the two-dimensionality of the problem, we cannot obtain w^{n+1}, the axial component of the velocity, directly from (5a). Instead, we integrate the axial component of (5a) over the area, A, of the cross section obtaining

$$\int_A (w^{n+1} - w^*) \, r dr d\phi = - \frac{\delta \cdot \Delta t}{St_m} \frac{\partial P^{n+1}}{\partial \theta} \int_A \frac{r}{B} dr d\phi \qquad (7)$$

where $B = 1 + \delta r \cos\phi$. We note that w^{n+1} must integrate over the cross-sectional area to the specified volumetric flow-rate, and $\partial P^{n+1}/\partial \theta$ is uniform over the cross section. The integral of w^* is evaluated numerically. Equation (7) can then be used to solve for $\partial P^{n+1}/\partial \theta$, which in turn is used in the axial component of (5a) to obtain w^{n+1}.

We have used a 14x19 staggered mesh which is finer near the wall and the inner bend, and coarser everywhere else. The grid points are distributed using a general transformation proposed by Roberts (1971).

To initiate the calculation we used as the initial velocity field an inviscid vortex-like axial velocity distribution with its origin at the center of curvature, and zero secondary velocities. This choice is motivated by the results of Singh (1974) and Agrawal et al. (1978) which show that flows entering curved pipes develop into such an inviscid vortex.

Finally, we use centered differencing in space, which makes the algorithm second order accurate, and in addition, in the case of insufficient spatial resolution, has the advantage of keeping the results bounded, if oscillating, unlike forward or backward differencing.

Results and conclusions

The velocity profile obtained at the end of the 9th cycle and used as input for the 10th cycle is shown in Figure 3, while in Figure 4 we show the flow-field development over the 10th cycle. Because the flow is symmetric over the center plane we plot the secondary velocity vectors only on the upper half of the the cross section, and the isovelocity contours for the axial velocity on the lower half. The inner bend is located on the left hand side, the outer bend on the right hand side.

Although periodicity of the flow has not been established by the end of the 10th cycle, the results show very good qualitative agreement with the experimental results of Talbot and Gong. A strong vortex appears at the inner bend at the beginning of the deceleration period, while the main secondary flow consists of a jet-like flow from the inner to the outer bend. As the deceleration continues, this vortex becomes weaker as it moves towards the center, while at $\omega t = 3\pi/4$ we observe negative axial flow. When the acceleration period starts, the fluid gains enough momentum to be swept downstream, thus making any existing vortex weaker. During the end of the cycle an additional weak vortex appears close to the wall at about $60°$ from the outer bend; we also observe that the maximum axial velocity is located off center towards the outer bend at all times.

During cycles prior to the 10th we observed a kind of oscillation of the flow around the center. Whereas the flow field everywhere else seems to be converging to the periodic state, the flow around the center undergoes a bounded oscillation. Such behavior, we believe, is attributable to the size of the Reynolds number. It is known that ADI methods are sensitive to the size of the Reynolds number as it introduces artificial viscosity. Preliminary investigation of this matter indicates that finer spatial resolution might be required to compensate for the diffusive errors due to the large Reynolds number.

The fully developed results of this investigation will help us in the study of the entering pulsatile flow in a curved pipe, the ultimate goal of our research in this area.

ACKNOWLEDGEMENT

This work was supported by the National Science Foundation (Grant No. MEA-8116360 and ECE-8417852).

REFERENCES

Y. Agrawal, L. Talbot, and K. Gong, "Laser anemometer study of flow development in curved circular pipes," J. Fluid Mech. v. 85, pp. 497-518, 1978.

A.J. Chorin, "Numerical solution of the Navier-Stokes equations," Math. Comp. v. 22, pp. 745-762, 1968.

J. Douglas and J.E. Gunn, "A general formulation of alternating direction implicit methods, Part I, Parabolic and hyperbolic problems," Numer. Math. v. 6, pp. 428-453, 1964.

K.O. Gong, "Experimental study of unsteady entrance flow in a curved pipe," Ph.D. thesis, University of California, Berkeley, 1979.

G.O. Roberts, "Computational meshes for boundary layer problems," Proc. Second Int. Conf. Num. Methods Fluid Dyn., Lecture Notes in Physics, v. 8, Springer-Verlag, New York, pp. 171-177, 1971.

L. Talbot and K.O. Gong, "Pulsatile entrance flow in a curved pipe," J. Fluid Mech. v. 127, pp. 1-25, 1983.

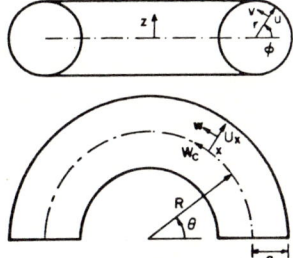 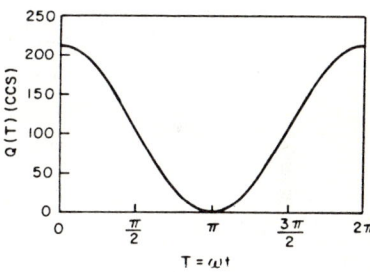

Figure 1. Toroidal coordinate system.　　Figure 2. Pulsatile, volumetric flow rate.

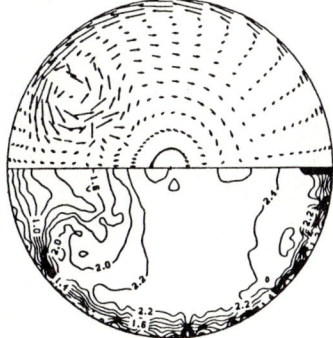

Figure 3. Velocity profile and isovelocity contours at the end of the 9th cycle.

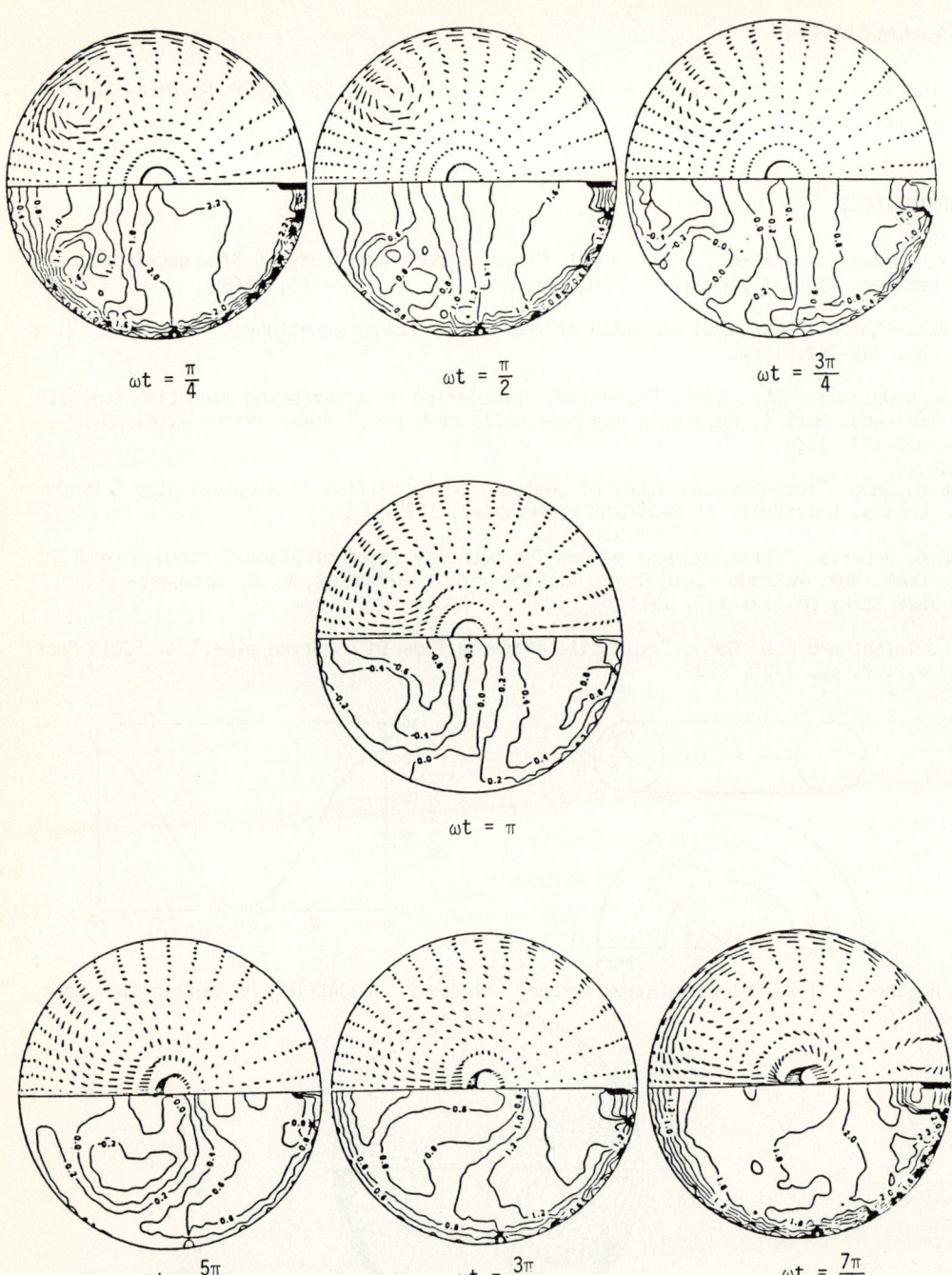

Figure 4. Velocity profiles and isovelocity contours at various times during the 10th cycle.

Implicit Hybrid Schemes for the Flux-Difference Split, Three-Dimensional Navier-Stokes Equations

Peter-M. Hartwich, Chung-Hao Hsu, and Chen-Huei Liu
Analytical Methods Branch
NASA Langley Research Center
Hampton, Virginia 23665-5225, U.S.A.

INTRODUCTION

Upwind schemes for the solution of the Euler equations have gained considerable popularity [1] during recent years. Upwinding leads to naturally dissipative finite-difference schemes, alleviating the necessity to add and to tune dissipative terms for numerical stability and accuracy as in schemes with central differencing. Implicit algorithms benefit from the diagonal dominance of the coefficient matrices ensured by the one-sided differences. Additionally, upwinding allows for the construction of faster and simpler implicit schemes. These features led to the extension of split methods to the Navier-Stokes equations (see, for example, Refs. 2-4). Except for Ref. 4, these efforts concentrated on supersonic and transonic flows. With decreasing Mach number, the governing equations become stiffer, and tremendous difficulties are encountered in converging the solution for M<0.3. The stiffness can be partly relaxed by neglecting compressibility altogether and solving the incompressible Navier-Stokes equations. For this purpose, a series of hybrid and of 2-factor AF schemes for three-dimensional flow have recently been developed. With the hybrid schemes, sufficient experience has been accumulated to assess their relative merits, which will be reported after a brief description of the schemes.

ANALYSIS

The governing equations are formulated in general coordinates and written in conservation law form:

$$\bar{Q}_t + (\bar{E}-\bar{E}_v)_\zeta + (\bar{F}-\bar{F}_v)_\xi + (\bar{G}-\bar{G}_v)_\eta = 0 \tag{1}$$

with
$$\begin{bmatrix} \bar{E}-\bar{E}_v \\ \bar{F}-\bar{F}_v \\ \bar{G}-\bar{G}_v \end{bmatrix} = \begin{bmatrix} \zeta_x & 0 & 0 \\ \xi_x & \xi_y & \xi_z \\ \eta_x & \eta_y & \eta_z \end{bmatrix} \begin{bmatrix} \hat{E}-\hat{E}_v \\ \hat{F}-\hat{F}_v \\ \hat{G}-\hat{G}_v \end{bmatrix} \quad \text{and} \quad \begin{aligned} \hat{E} &= J^{-1}(u, u^2+p, uv, uw)^T \\ \hat{F} &= J^{-1}(v, uv, v^2+p, vw)^T \\ \hat{G} &= J^{-1}(w, uw, vw, w^2+p)^T \end{aligned}$$

where the Jacobian of the coordinate transformation is given by $J^{-1} = x_\zeta (y_\xi z_\eta - y_\eta z_\xi)$. With the assumption of constant density and viscosity, the viscous flux vectors are written like

$$\hat{E}_v = (Re \cdot J)^{-1}(0, 2u_x, u_y+v_x, u_z+w_x)^T, \text{ etc.}$$

The solution vector $\bar{Q}$ is defined as $\bar{Q} = J^{-1}(p/\beta, u, v, w)^T$. The time-derivative of the pressure p_t permits the integration of the incompressible Navier-Stokes equations like a conventional parabolic time-dependent system of equations. However, for $\beta \gg 1$, or for $p_t \to 0$ with $t \to \infty$, the continuity equation is still satisfied [5]. Note that the assumption has been made that all ζ = constant planes coincide with x = constant planes. This allows a simplified treatment of wing and wing-body geometries without affecting the generality of the three-dimensional spatial-differencing scheme or the implicit algorithms.

NUMERICAL PROCEDURE

Differencing of the Inviscid Fluxes

Pressure and convective terms are upwind differenced using flux-difference splitting. This splitting was originally developed by Roe [6] as a Riemann solver for compressible flow. Since this splitting does not require flux vectors which are homogeneous of degree one in the solution vector $\bar{Q}$, it allows to construct a fully conservative scheme for the incompressible Navier-Stokes equations. The details of the implementation of the flux-difference splitting in a finite-difference code using general coordinates are given in Ref. 4. Through the operator split concept, the three-dimensional conservation laws are split into three quasi one-dimensional problems, i.e.,

$$\bar{Q}_t + \bar{H}_\chi = \bar{Q}_t + (\tilde{D}\bar{Q})_\chi = 0 \tag{2}$$

where χ stands for ζ, ξ, or η, $\bar{H} = \bar{E}$, $\bar{F}$, or $\bar{G}$, and $\tilde{D} = \partial \bar{H}/\partial \bar{Q}$. Instead of finding an approximate solution to Eq. (2), an exact solution to an approximate problem is constructed:

$$Q_t + D\Delta_\chi Q = 0 \quad (Q: \text{discrete representation of } \bar{Q}) \tag{3}$$

The locally frozen matrix D has the same eigenvalues as $\tilde{D}$ if for $Q_{i-1} \to Q_i$, $D(Q_i,Q_{i-1}) = \tilde{D}(Q_i,Q_{i-1})$. Moreover, Eq. (3) is conservative if $D(Q_i,Q_{i-1}) \times (Q_i - Q_{i-1}) = H_i - H_{i-1}$. A first-order accurate, semidiscrete formulation of Eq. (3) reads

$$(Q_t)_i - D^-_{i+1/2}(Q_{i+1}-Q_i) + D^+_{i-1/2}(Q_i - Q_{i-1}) = 0 \tag{4}$$

with $D^\pm = X\Lambda^\pm X^{-1}$ and $\Lambda^\pm = (|\Lambda| \pm \Lambda)/2$. The diagonal matrix $\Lambda^+(\Lambda^-)$ contains the positive (negative) eigenvalues. The columns of X consist of the right eigenvectors of Λ $(= \Lambda^+ - \Lambda^-)$.

<u>Hybrid Algorithms</u>

The finite-difference soluton to Eq. (1) is advanced in pseudotime using Euler-implicit time-differencing. To avoid the steady-state solution to be time-step size dependent, the two level iteration procedure is cast in delta-formulation:

$$N\Delta Q^n_{ijk} + L(Q^n_{ijk}) = 0 \tag{5}$$

where $\Delta Q^n = Q^{n+1} - Q^n$ and the subscripts i,j,k indicate the location in ζ-, ξ-, and η-direction, respectively. The operator N defines the type of iterative procedure and is an approximation to L. The residual operator L gives the discretized steady part of the Navier-Stokes equations. The inviscid fluxes in L are constructed by applying the Riemann solver separately in each coordinate direction. The overall discretization is obtained via adding up all independent discretizations of the flux derivatives in each direction. The viscous fluxes are centrally differenced in the usual manner [7]. For the operator N, a hybrid approach has been adopted. Choosing ζ as the coordinate direction for the relaxation, one possible choice for N is written as

$$[M-(B^-+S)_{j+1/2}\Delta_{j+1/2} + (B^++S)_{j-1/2}\Delta_{j-1/2}][M^{-1}] \times$$
$$[M-(C^-+T)_{k+1/2}\Delta_{k+1/2} + (C^++T)_{k-1/2}\Delta_{k-1/2}]\Delta Q^n = -L(Q^n, Q^{n+1}) \tag{6}$$

with $\quad M = I/\Delta tJ + (A^++R)_{i-1/2} + (A^-+R)_{i+1/2}$

where I = identity matrix. The symbol $\Delta_{j-1/2}$, for instance, indicates the difference $\Delta Q^n_j - \Delta Q^n_{j-1}$. The matrices R, S, and T stem from the viscous shear fluxes E_v, F_v, and G_v, respectively, and they are derived like in Ref. 7. Sweeping back and forth in ζ-direction, a symmetric planar Gauss-Seidel (SPGS) relaxation is recovered, indicated by the non-linear updating of the residuals in Eq. (6). Since each factor in Eq. (6) has the same block-tridiagonal structure as in the three-dimensional, 3-factor AF scheme, let this approach be termed AF-SPGS to distinguish it from the second hybrid scheme, given by

$$[M-(B^-+S)_{j+1/2}\Delta_{j+1/2} - (C^-+T)_{k+1/2}\Delta_{k+1/2}][M^{-1}] \times$$
$$[M+(B^++S)_{j-1/2}\Delta_{j-1/2} + (C^++T)_{k-1/2}\Delta_{k-1/2}]\Delta Q^n = -L(Q^n, Q^{n+1}) \quad (7)$$

This formulation leads to an approximate LU-factorization of the left-hand side. Let this approach be termed LU-SPGS. The inversion of the upper and lower block-triangular matrices requires only 37% of the grid point operations which are necessary for the inversion of the block-tridiagonal matrices in Eq. (6). The drawback of LU-SPGS is that only 56% of these operations can be vectorized, compared with 100% of the AF-SPGS method.

Boundary Conditions

Unknown values of Q along the boundaries are updated explicitly, and ΔQ^n is set to zero. The boundary conditions consist of freestream conditions on the outer boundary, except for the outflow crossplane, reflection conditions at the plane of symmetry, and no-slip conditions along the wing surfaces. Values for p on the wing were simply extrapolated. Averaging extrapolates from above and below the wing extension (branch cut) give the flow variables there. Zeroth-order extrapolation give the values in the outflow cross-section.

COMPUTATIONAL RESULTS

The finite-difference solutions are computed on C-H-type grids which are generated using a recently developed grid generation code [8]. The grids envelop a slender, sharp-edged delta wing (AR=1) with zero thickness. Fig. 1 illustrates the integration domain which consists of 51x101x72 grid points in radial, circumferential, and chordwise direction, respectively. Figs. 2 and 3 show the vortex flow (Re=$9 \cdot 10^5$) around the delta wing at 20.5 degrees incidence. The crossplane velocity vector plot in Fig. 2a (Fig. 2b) clearly indicates the primary (secondary) vortex at 70% root chord station. The corresponding surface pressure coefficient distribution in Fig. 3 is compared with experimental data by Hummel [9] and with numerical results by Fujii and Kutler [10]. The performance of the two hybrid schemes is documented in Figs. 4a and 4b. The computations were done in half precision on the CDC-VPS 32 vector computer at NASA Langley. Both schemes reduce the L_2-norm of all residuals to machine zero in about the same CPU-time and they have almost the same spectral radius. The lift develops so similarily that the C_L-traces in Figs. 4a and 4b cannot be distinguished from each other. Using the same parameter as above, except for Re(=10^4), these computations were repeated on a coarser grid with 51x51x72 grid points. Then a difference in the performance of the two hybrid schems is revealed in Figs. 5a and 5b. Compared to the AF-SPGS scheme, the LU-SPGS saves about 16% of the CPU-time. This indicates that the LU-SPGS scheme is a viable method when used on sequentially operating computers or on vector machines which do not so heavily depend on long vector lengths for optimum performance like the VPS 32. In all computations, the β-parameter was set to unity, and local time stepping (CFL=10) has been used. Converging results for the LU-SPGS scheme are only obtained when the first approximate factor in Eq. (7) is chosen such that the inversion procedure of the block-triangular matrices marches from the outer boundary towards the wing. That is most probably due to the boundary conditions for the intermediate variables (they are set to zero along all boundaries).

CONCLUDING REMARKS

Basing on flux-difference splitting, an upwind scheme has been developed for finite-difference solutions to the three-dimensional, incompressible Navier-Stokes equations. Using implicit hybrid algorithms, vortex flows around delta wings have been efficiently computed. The hybrid scheme with block-tridiagonally structured coefficient matrices (AF-SPGS) is faster on vector computers which achieve optimum performance with very long vector lengths. The hybrid scheme with block-triangular coefficient matrices (LU-SPGS) is superior on sequentially operating machines and on vector computers which operate with short vector lengths (ca 50 words). Future work will assess the merits of two-factor, three-dimensional AF schemes. The extension of the present first-order accurate scheme to a higher-order, monotone scheme is another interesting endeavor.

REFERENCES

1. Enquist, B. J. et al. (eds.), Lect. Appl. Math., Vol. 22, 1985.
2. Lombard, C. K. et al., AIAA 83-1895-CP, July 1983.
3. Fujii, K. and Obayashi, S., AIAA 86-0513, Jan. 1986.
4. Hartwich, P.-M. and Hsu, C.-H., AIAA 86-1839-CP, June 1986.
5. Steger, J. L. and Kutler, P., AIAA J., Vol. 15, No. 4, 581-590, April 1977.
6. Roe, P. L., Proc. 7th Intl. Conf. Num. Meth. Fluid Dyn., 1981.
7. Pulliam, T. H. and Steger, J. L., AIAA J., Vol. 18, No. 2, 159-167, Feb. 1980.
8. Hartwich, P.-M., AIAA 86-0430, Jan. 1986.
9. Hummel, D., AGARD-CP 247, Paper No. 15, 1978.
10. Fujii, K. and Kutler, P., AIAA 84-1550, June 1984.

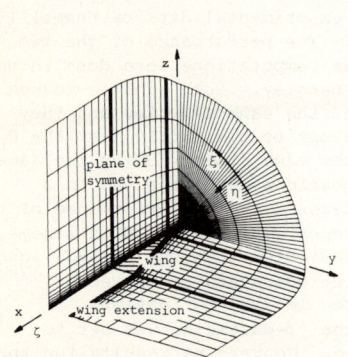

Fig.1 Perspective view of the three-dimensional finite-difference grid for a thin delta wing (AR=1) with 51x101x72 grid points in radial, circumferential, and chordwise direction, respectively.

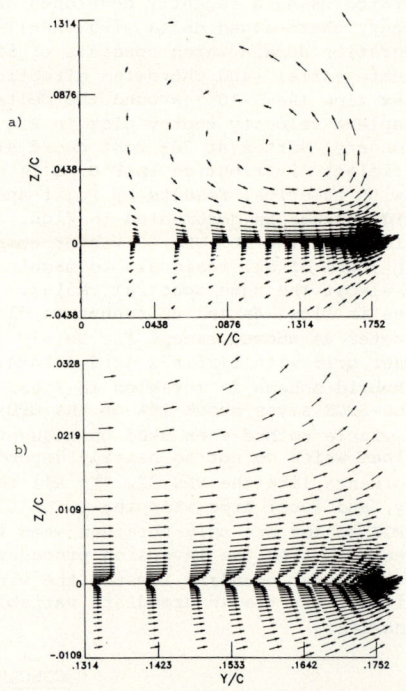

Fig. 2 Crossflow velocity vector plot at 70% root chord station, $\alpha = 20.5°$, $Re = 9 \times 10^5$:
a) overall flow pattern; b) detail at leading edge.

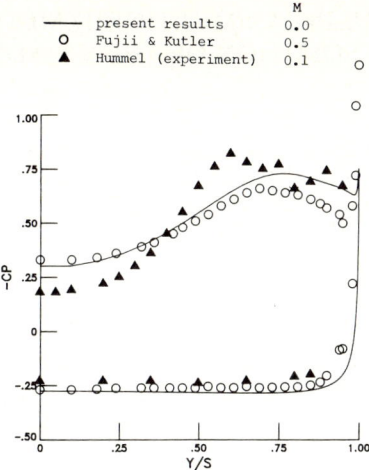

Fig.3 Spanwise surface pressure coefficient distribution at 70% root chord station, $\alpha=20.5°$, $Re=9\cdot 10^5$ (s: local half span).

	present results	M
—	present results	0.0
○	Fujii & Kutler	0.5
▲	Hummel (experiment)	0.1

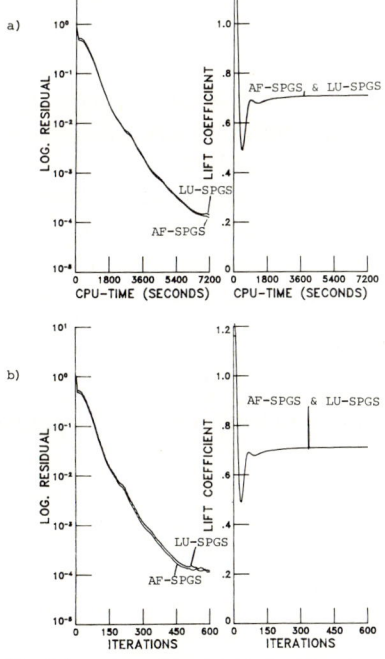

Fig. 4 Comparison of two hybrid algorithms for a vortex flow around a thin delta wing of AR=1, $\alpha=20.5°$, $Re=9\times 10^6$ (51x101x72 grid points) : a) convergence rate and lift evolution versus CPU-time; b) convergence rate and lift evolution versus the number of iterations.

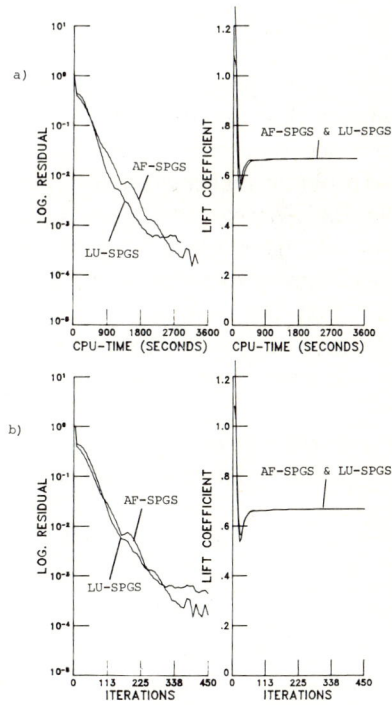

Fig. 5 Comparison of two hybrid algorithms for a vortex flow around a thin delta wing of AR=1, $\alpha=20.5°$, $Re=10^4$ (51x51x72 grid points) : a) convergence rate and lift evolution versus CPU-time; b) convergence rate and lift evolution versus the number of iterations.

A NONLINEAR MULTIGRID METHOD FOR THE EFFICIENT SOLUTION OF THE STEADY EULER EQUATIONS

P.W.Hemker, B.Koren and S.P.Spekreijse
CWI, Centre for Mathematics and Computer Science
P.O.Box 4079, 1009 AB Amsterdam, The Netherlands

ABSTRACT

An efficient iterative method has been developed for the accurate solution of the non-isenthalpic steady Euler equations for inviscid flow.

First, the system of conservation laws is space-discretized by a first order finite-volume Osher-discretization. Without time stepping, the steady equations are solved by iteration with nonlinear multiple grid cycles, where a Symmetric Gauss-Seidel method is used as a relaxation. Initial estimates are obtained by the Full Multigrid method. In the pointwise relaxation, the equations corresponding to each cell are kept in block-coupled form, i.e. a Collective Symmetric Gauss-Seidel relaxation is used. In this relaxation local linearization of the equations and the boundary conditions is applied, and one (or a few) step(s) of a Newton iteration is (are) used for the approximate solution of these small nonlinear systems. The first order Osher-discretization has many good properties which foster the efficiency of multigrid iteration. It appears that for all meshsizes the discrete system is solved up to truncation error accuracy in only a few (1 to 3) iteration cycles (3 to 8 work units).

To obtain higher accuracy, we use second order finite volume schemes (e.g. the newly developed superbox scheme [3]), again based on Osher's approximate Riemann solver. The more accurate discretizations are less stable, and hence harder to solve by relaxation iteration. Therefore, we make use of the fact that the solution of the first order scheme can be computed very efficiently, and we solve the second order system (up to truncation error) by one or a few cycles of a defect correction process.

1. INTRODUCTION

For a 2-D domain Ω^*, we solve the system of non-isenthalpic *Euler equations*

$$\frac{\partial}{\partial t} q + \frac{\partial}{\partial x} f(q) + \frac{\partial}{\partial y} g(q) = 0, \tag{1.1}$$

$$q = \begin{pmatrix} \rho \\ \rho u \\ \rho v \\ \rho e \end{pmatrix}, \quad f = \begin{pmatrix} \rho u \\ \rho u^2 + p \\ \rho uv \\ \rho u H \end{pmatrix}, \quad g = \begin{pmatrix} \rho v \\ \rho vu \\ \rho v^2 + p \\ \rho v H \end{pmatrix};$$

where ρ, u, v, e, p and $H = e + p/\rho$ represent density, velocity component in x- and y-direction, specific energy, pressure and specific enthalpy. For a perfect gas, (1.1) is completed by

$$p = (\gamma - 1)\rho(e - \tfrac{1}{2}(u^2 + v^2)),$$

in which γ is the ratio of specific heats. In symbolic form we write (1.1) as

$$q_t + N(q) = 0. \tag{1.2}$$

The *steady* equations are obtained by the assumption $q_t = 0$.

To construct a nested sequence of discretizations for our multigrid solution procedure, we use the *finite volume* technique. We divide the domain Ω^* in quadrilateral cells Ω_{ij}, such that a mapping is introduced from a regular and rectangular "computational domain" to the irregular "physical domain". By regular refinement of the computational domain, this mapping generates the coordinates for the cell vertices in a sequence of refining irregular grids. To prove the accuracy of the resulting schemes, we take this mapping non-singular and sufficiently smooth.

The discrete approximation q_h of $q(x,y)$ is represented by a (vector-) quantity q_{ij} for each Ω_{ij}. Each q_{ij} is associated with the mean value of q over Ω_{ij}. The space discretization now requires the approximation of $\int_{\Gamma_{ijk}} (f.n_x + g.n_y)\, ds$, $k = N,E,S,W$, at the four walls Γ_{ijk} of cell Ω_{ij}. Each wall Γ_{ijk} may be either a common boundary with a neighbouring cell Ω_{ijk} or a part of the boundary $\partial\Omega^*$. In both cases the integral is approximated by $f^k(q_{ij}^k, q_{ijk}^k) \cdot \text{meas}(\Gamma_{ijk})$, i.e. at each Γ_{ijk} we approximate $fn_x + gn_y$ by a constant value, only depending on q_{ij}^k and q_{ijk}^k, which are approximations to $q(x,y)$ at Γ_{ijk} in Ω_{ij} and Ω_{ijk} respectively. First and second order schemes are obtained by different choices for these approximations.

Thus, the discretization of the steady equation (1.2) is the set of nonlinear equations

$$N_h(q_h)|_{i,j} := \sum_{k=N,E,S,W} f^k(q_{ij}^k, q_{ijk}^k)\, \text{meas}(\Gamma_{ijk}) = 0. \tag{1.3}$$

for all (i,j) with $\Omega_{ij} \subset \Omega^*$.

By the rotation invariance of the Euler equations, we can relate $f^k(.,.)$ to a local coordinate system, rotated such that it is aligned with Γ_{ijk}. Then we find $f^k(q_{ij}^k, q_{ijk}^k) = T_{ijk}^{-1} f(T_{ijk} q_{ij}^k, T_{ijk} q_{ijk}^k)$. Here, the operator T_{ijk} takes care of the local rotation of the coordinate system at Γ_{ijk} and $f(.,.)$ is a *numerical flux function*, independent of the orientation of Γ_{ijk}. For $f(.,.)$ we use the numerical flux function as proposed by Osher [6]. For details see [4].

2. THE FULLY IMPLICIT NONLINEAR MULTIGRID METHOD

Most methods developed so far for the solution of the steady equations (1.1) are based on integrating the equation (1.2) in time until a steady state is reached. We disregard the time-dependence, and assume that a suitable space discretization takes into account the proper characteristic directions in Ω^*, and that for $h \to 0$ the discrete solution q_h approaches an (existing) steady solution $q(x,y)$ that satisfies the entropy condition. Hence, we restrict ourselves to the direct solution of the nonlinear system

$$N_h(q_h) = 0. \tag{2.1}$$

For this solution we apply the *nonlinear multiple grid* (FAS-) algorithm.

We construct the nested set of refining grids, such that each set of 2×2 cells in a fine mesh forms a single cell in the next coarser mesh.

Slightly generalising the equation (2.1) to

$$N_h(q_h) = r_h, \tag{2.2}$$

where r_h denotes a possible correction term, we select a (nonlinear) relaxation procedure

$$q_h^{(n+1)} := S_h(q_h^{(n)}, r_h) \tag{2.3}$$

for its iterative solution.

The coarser grids are used to accelerate this basic procedure. For this a coarse grid correction is used: starting with an approximation $q_h^{(k)}$ on the fine mesh and some approximation q_{2h}^{old} on the next coarser, an approximate solution for the coarse grid problem

$$N_{2h}(q_{2h}^{new}) = N_{2h}(q_{2h}^{old}) - \overline{R}_{2h,h}(N_h(q_h^{(k)}) - r_h); \tag{2.4}$$

is computed. Then the value $q_h^{(k)}$ is updated by

$$q_h^{(k+1)} = q_h^{(k)} + P_{h,2h}(q_{2h}^{new} - q_{2h}^{old}). \tag{2.5}$$

The equations (2.4) and (2.5) describe the *coarse grid correction* step.
Our FAS-cycles for the solution of (2.2) consist of the following steps:
(0) Start with an approximate solution q_h.
(1) Improve q_h by application of a (pre-) relaxation sweep (2.3).
(2) If the present grid is the coarsest, skip to (3); otherwise improve q_h by application of one coarse-grid-correction step, where the approximate solution of (2.4) is effected by application of a single FAS-cycle to this coarser grid problem.
(3) Improve q_h by another (post-) relaxation sweep (2.3).

For the FAS-procedure, we obtain an initial estimate by the *Full Multi-Grid* (FMG-) technique [2] : the initial estimate is obtained by interpolation from the approximate solution on the next coarser grid. For many problems this process gives very good results, even if one starts with rough approximations on a really coarse grid [5].

With a particularly simple restriction $\overline{R}_{2h,h}$ and prolongation $P_{h,2h}$ as transfer operators between the coarse and fine grids, the coarse discrete operator N_{2h} is a *Galerkin approximation* to the (first-order) fine grid discretization N_h. Viz. with $P_{h,2h}$ the piecewise constant interpolation over cells, and $\overline{R}_{2h,h}$ the summation of the residual over 2×2 fine mesh cells to form a residual on the corresponding coarse cell, we find

$$N_{2h}(q_{2h}) = \overline{R}_{2h,h} N_h(P_{h,2h} q_{2h}). \tag{2.6}$$

This formula has an interesting implication for a coarse grid correction. Viz. if (2.4)-(2.5) transform the approximation q_h into $\tilde{q}_h$, the residual of $\tilde{q}_h$ generally satisfies

$$\overline{R}_{2h,h}\left[r_h - N_h(\tilde{q}_h)\right] = \mathcal{O}(||q_h - \tilde{q}_h||^2).$$

This means that after the coarse grid correction step the residual mainly contains high frequency components.

A necessary property of a *relaxation method* in a multiple grid context is the capability to damp high frequency components in the residual. To ensure this, the discretization should be sufficiently dissipative. For the first order scheme, well-known and simple nonlinear relaxation procedures such as Collective Symmetric Gauss-Seidel work well. ("Collective" means that the 4 variables corresponding to a single cell are relaxed simultaneously.) In most applications we use CSGS in one diagonal direction as pre- and CSGS in the other diagonal direction as post-relaxation. The smoothing behaviour of the relaxations can be analyzed by local mode analysis. If we study plots of reduction factors of Fourier components (spectral radii, or norms for the error or residual amplification operator), we see that two CSGS-sweeps are usually sufficient for a significant reduction of the high frequencies (Hemker, unpublished results). For second order schemes the smoothing rates are not satisfactory.

Wanting at least *second order accuracy*, we start with a first order approximation $q_h^{(1)}$, obtained by a single sweep of the FMG-process, and improve the accuracy by a *defect correction process* (DCP) [1, 3]

$$N_h^1(q_h^{(n+1)}) = N_h^1(q_h^{(n)}) - N_h^2(q_h^{(n)}). \tag{2.7}$$

Here N_h^p, p = 1,2, denotes the p-th order discretization. For smooth solutions a single step of (2.7) is sufficient to obtain the higher order of accuracy [2]. But also, for solutions with discontinuities (where the formal order of convergence has no practical meaning) it is seen that one or a few steps of (2.7) improve the accuracy of the solution significantly, even if the new iterands are approximated by only a few FAS-cycles [3, 5]. The first iterand $q_h^{(2)}$ of (2.7) can also be approximated by application of one additional FMG-sweep that reduces the error by another factor h. This means that only two FMG-sweeps may solve the second order equations sufficiently accurate.

3. RESULTS

As standard testcases we consider the NACA0012-airfoil at $M_\infty = 0.63$, $\alpha = 2^o$ (subsonic flow), and at $M_\infty = 0.8$, $\alpha = 1.25^o$ (transonic flow with shock). As a finest grid we use a 128×32 O-type mesh with an outer boundary at approx. 100 chord lengths away from the airfoil (fig.1). At the outer boundary we impose unperturbed flow. As 2nd order scheme we use the superbox scheme [3,5]. In fig.2 and 3 we present results and make a comparison with solutions from [7].

In fig.2a and 3a, the convergence histories of the lift and drag coefficent are shown. As starting point, $q_h^{(1)}$, we use a single-FAS FMG-approximation of the first order scheme. The lift and drag as published in [7] are spread over the shaded areas. Clearly visible is the excellent improvement of the drag which is obtained in the first DCP-cycle. Taking the results from [7] as a standard, we see that we need 3 DCP-cycles for the subsonic flow, and only 1 DCP-cycle for the transonic flow with a shock.

In fig.2b and 3b, the left graphs show the pressure distributions obtained after the 3rd DCP-cycle, the right graphs are taken from [7]. For the subsonic flow the good agreement is evident. Due to scattering in shock position, this agreement is less for the transonic flow with shock. For the latter the superbox scheme yields solutions of good quality in the smooth parts of the flow, but (being non-TVD) it introduces some spurious non-monotonicity.

For the multigrid computation of airfoil flows with the steady Euler equations, DCP is found to be an efficient solution method for stable 2nd order discretizations. It appears that it is sufficient to perform only a few DCP-cycles in which all sub-problems (2.2) are solved by a single FAS-cycle.

ACKNOWLEDGEMENT

The investigations were supported in part by the Netherlands Technology Foundation.

REFERENCES
1. K. BÖHMER and P. HEMKER, and H.J. STETTER (1984). The Defect Correction Approach. *Computing Suppl.5*, 1-32.
2. W. HACKBUSCH (1985). *Multigrid Methods and Applications*, Springer Series Comp. Mathematics, 4, Springer Verlag, Berlin, Heidelberg.
3. P.W. HEMKER (1985). *Defect correction and higher order schemes for the multigrid solution of the steady Euler equations,* CWI Report NM-R8523, To appear in Proceedings Multigrid Conference, Cologne, Oct. 1985.
4. P.W. HEMKER and S.P. SPEKREIJSE (1985). *Multiple Grid and Osher's Scheme for the Efficient Solution of the Steady Euler Equations,* CWI Report NM-R8507, To appear in Appl. Num. Math. 1986.
5. B. KOREN (1986). *Evaluation of second order schemes and defect correction for the multigrid computation of airfoil flows with the steady Euler equations,* To appear.
6. S. OSHER (1981). Numerical solution of singular perturbation problems and hyperbolic systems of conservation laws. O. AXELSSON, L.S. FRANK AND A. VAN DER SLUIS (eds.). *Analytical and Numerical Approaches to Asymptotic problems in Analysis*, Springer Series Comp. Mathematics, North Holland Publ. Comp..
7. A. RIZZI and H. VIVIAND (1981). *Numerical Methods for the computation of inviscid transonic flows with shock waves,* Springer Series Comp. Mathematics, Vieweg Verlag, Proceedings GAMM Workshop, Stockholm, 1979.

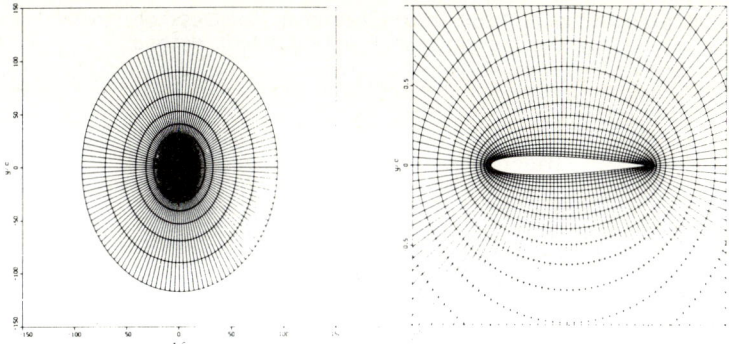

Fig.1: 128×32-grid NACA0012-airfoil.

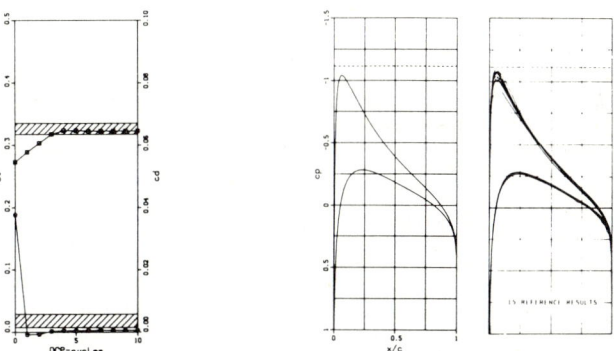

a. Convergence history lift (square) and drag (circular) coefficient

b. Surface pressure distributions

Fig.2: Results for NACA0012-airfoil at $M_\infty = 0.63$ and $\alpha = 2^o$.

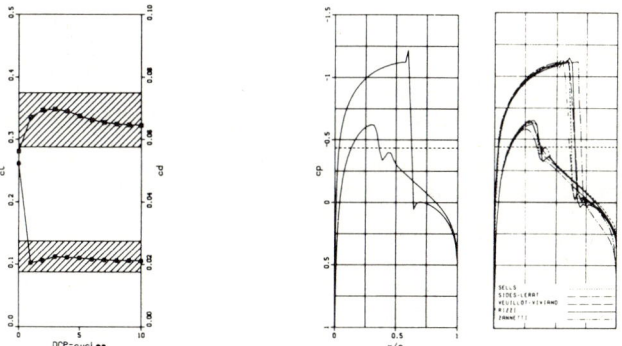

a. Convergence history lift (square) and drag (circular) coefficient

b. Surface pressure distributions

Fig.3: Results for NACA0012-airfoil at $M_\infty = 0.8$ and $\alpha = 1.25^o$.

Calculation of Flow in a Supersonic Compression Corner
by the Dorodnitsyn Finite Element Method

by

Maurice Holt and Christopher Pace

Department of Mechanical Engineering
University of California
Berkeley, CA 94720

Abstract

The calculation of laminar boundary layer flow in two dimensions, using the Dorodnitsyn Method of Integral Relations, was successfully extended to separated regions by Holt and others (Holt, Numerical Methods in Fluid Dynamics, Springer 1984). The extension requires incorporation of a square root term in the representation of the local shearing stress as a function of the streamwise velocity component. This limits the order of approximation that can be conveniently carried out for various flow configurations in plane flow and presents obstacles to the generalisation of M.I.R. for three dimensional flow. The same difficulties arise in developing the orthonormal version of M.I.R. both for laminar and turbulent boundary layers.

In a recent paper Fletcher and Fleet (Int. J. for Num. Meths. in Fluids, $\underline{4}$, 399-419, 1984) treat laminar boundary layer flow in two dimensions by solving the Dorodnitsyn integral form of the equations of motion using a Finite Element Method. In the present paper this approach is extended to boundary layer flows dominated by positive pressure gradients. Free interaction couples the viscous and inviscid regions in which no iteration between these regions is required.

Introduction

Calculation of laminar incompressible boundary layers with a known external pressure gradient has been successfully achieved with a Dorodnitsyn Finite Element Method (Fletcher and Fleet 1984). This paper extends this method to supersonic laminar compressible boundary layers in an unknown (*a priori*) positive pressure gradient up to separation.

Free interaction has been successfully used in M.I.R. which is the method used to couple the viscous and inviscid regions of boundary layer flows (Holt, 1966). Free interaction is a process which requires no iteration between these regions and therefore, the steady solution can be marched directly downstream. The initial shear stress profile is represented by Blasius flow. The inviscid region is given an initial positive pressure perturbation to start the marching process, which will cause the flow to separate downstream due to the free interaction process.

Mathematical Formulation

Dorodnitsyn (1962) originally applied his transformation to the laminar incompressible boundary layer equations. The resulting equation were solved by Fletcher and Fleet using a Galerkin Finite Element method with a known, *a priori*, outside velocity (Falkner-Skan profiles). Pavlovskii (1962) extended Dorodnitsyn's transformation to include compressible terms in the laminar boundary layer equations using the Stewartson transformation (Stewartson, 1949). The resulting equation is identical to that of Dorodnitsyn. The following is Pavlovskii's transformed equation with the appropriate boundary conditions (see Holt, 1966).

$$\frac{\partial}{\partial \xi} \int_0^1 \frac{uf}{z} du = \frac{\dot{U}_1}{U_1} \int_0^1 \frac{(1-u^2)}{z} \frac{df}{du} du + \int_0^1 \frac{df}{du} \frac{\partial z}{\partial u} du \quad (1)$$

$$z = \frac{\partial u}{\partial \eta}, \quad \dot{U}_1 = \frac{dU_1}{d\xi}$$

$$u = \frac{U}{U_1}, \quad U = \bar{u} \frac{a_0}{a_1}$$

0 – denote free stream
1 – denotes edge of boundary layer
$\bar{u}$ – streamwise velocity in x-y space
ξ – streamwise coordinate

with boundary conditions

$$f = 0, \ z = 0, \ \text{at} \ u = 1$$

The boundary condition on f is needed to decouple the transverse velocity from the equation.

The benefit of the change of independent variable from y to u is two-fold. Firstly the infinite domain $0 \leq y \leq \infty$ is transformed into a finite domain $0 \leq u \leq 1$, being more adaptable to fixed grids, and, secondly, a greater resolution of the solution is obtained near the wall.

The domain $0 \leq u \leq 1$ is separated into finite elements on a variably spaced grid. The Galerkin Finite Element method is applied to Equation (1). The interpolation functions for z and 1/z are the following.

$$z = \sum_{i=1}^{N} (1 - u) \, \psi_i(u) \, T_i(z)$$

$$\frac{1}{z} = \sum_{i=1}^{N} \frac{1}{(1-u)} \, \psi_i(u) \, \theta_i(z)$$

$$T_i(z) = \frac{z_i}{1 - u_i} \qquad \theta_i(z) = \frac{1 - u_i}{z_i}$$

In this form the interpolation functions automatically satisfy the boundary conditions. The shape functions $\psi_i(u)$ are linear or quadratic. The exact relationship between θ_i and T_i, $(\theta_i = 1/T_i)$ is satisfied only at the nodal points. As $N \to \infty$ the exact relationship will be satisfied everywhere (Fletcher and Fleet 1984).

The weighting function f(u) follows from the Galerkin method

$$f(u) = \sum_{i=1}^{N} (1 - u) \, \psi_i(u)$$

which also explicitly satisfies the boundary condition.

The term $(1 - u_i)$ appears in e_i and T_i, so we recover z_i at each node while decoupling $(1 - u_i)$ from the shape functions $\psi(u)_i$. Substituting these into Equation (1) we obtain

$$\sum_{j=1}^{N} A_{ij} \frac{de_j}{d\xi} - \frac{d}{d\xi}(\log U_1) \sum_{j=1}^{N} B_{ij} e_j = \sum_{j=1}^{N} C_{ij} T_j \qquad (2)$$

$$A_{ij} = \int_0^1 u\, \psi_i(u)\, \psi_j(u)\, du$$

$$B_{ij} = \int_0^1 (1 + u)\, \psi_j((1 - u)\frac{d\psi_i}{du} - \psi_i)\, du$$

$$C_{ij} = \int_0^1 [((1 - u)\frac{d\psi_i}{du} - \psi_i)((1 - u)\frac{d\psi_j}{du} - \psi_j)]\, du$$

<u>Free Interaction Process</u>

The free interaction process applies to supersonic flow. The velocity at the edge of the boundary layer U_1 undergoes an isentropic compression as the boundary layer grows downstream. Therefore the magnitude of the edge velocity can be related to its direction relative to the free stream direction thru the Prandtl-Meyer relationship. The local edge velocity direction is approximated by the local tangent to the curve traced out by the displacement thickness. Then we have

$$\frac{d\delta^*}{dx} = \tan \phi \qquad \qquad \phi - \text{angle between free stream and edge velocity}$$

$$\delta^* = \int_0^\infty (1 - \frac{\rho \bar{u}}{\rho_1 \bar{u}_1})\, dy$$

After some manipulation (see Nielsen 1964), we obtain

$$\delta^* = \frac{\rho_o a_o}{\rho_1 a_1} \frac{\sqrt{U_o \ell \nu_o}}{U_1} \int_0^\infty (1 + m_1)(1 - u) + m_1 u (1-u)\, d\eta \qquad m_1 = \frac{\gamma + 1}{2} M_1^2$$

$$M = \text{Mach number}$$

Transforming to u-space and substituting for $1/z$, we find

$$\delta^* = \frac{\rho_o a_o}{\rho_1 a_1} \frac{\sqrt{U_o \ell \nu_o}}{U_1} \sum_{i=1}^{N} \int_0^1 ((m_1 + 1) + m_1 u)\, \psi_i\, e_i\, du$$

Further manipulation and consideration of the transformed planes yields

$$\frac{d\delta^*}{d\xi} = \frac{dx}{dX} \frac{dX}{d\xi} \frac{d\delta^*}{dx} = \frac{dx}{dX} \frac{dX}{d\xi} \tan \phi$$

We have

$$\frac{d}{d\xi}(\log U_1)[\sum_{i=1}^{N} \theta_i [D_i (\frac{\gamma+1}{\gamma-1} \frac{m_1}{(1+m_1)}] + 2m_1 E_i] + \sum_{i=1}^{N} D_i \frac{d\theta_i}{d\xi} \quad (3)$$

$$= (\frac{1+m_1}{1+m_0}) \sqrt{\frac{U_o \ell}{\nu_o}} \tan \phi$$

$$D_i = \int_0^1 ((1+m_1) + m_1 u) \psi_i \, du$$

$$E_i = \int_0^1 (1+u) \psi_i \, du$$

Numerical Procedure

Now, to build an implicit downstream march, equation (2) is linearized about the nth level (Briley and McDonald, 1976). Equation (3) is left explicit, so that equation (2) becomes

$$\sum_{j=1}^{N} P_{ij} \Delta k_j^{n+1} = \Delta \xi [\beta S_i^{n+1} + (1-\beta) S_i^n] \quad (4)$$

$$S = \sum_{j=1}^{N} C_{ij} T_j \qquad S_i^{n+1} = S_i^n + (\frac{\partial S_i}{\partial k_j})^n \Delta k_j^{n+1} + \theta(\Delta\xi^2)$$

$$\Delta k_j^{n+1} = \Delta \theta_j^{n+1} = \theta_j^{n+1} - \theta_j^n \qquad j = 1 \to N$$

$$\Delta k_{N+1}^{n+1} = \Delta(\log U_1)^{n+1} = (\log U_1)^{n+1} - (\log U_1)^n$$

Substituting for S^{n+1}, equation (4) becomes

$$\sum_{j=1}^{N} (A_{ij} + \Delta\xi\beta C_{ij} T_j^2) \Delta\theta_j^{n+1} - \sum_{j=1}^{N} B_{ij}\theta_j \Delta(\log U_1)^{n+1} = \Delta\xi \sum_{j=1}^{N} C_{ij} T_j$$

The degree of implicitness was set to $\beta = .5$ for all runs.

Results and Discussion

Figures 1-4 show the u, z, δ^*, and P_1/P_o profiles for various ξ values including the point of separation $\xi = 1.24$. As the flow approaches separation the gradient of $\theta = 1/z$ near the wall becomes very large. A fixed exponentially spaced grid near the wall was required to get a stable solution near separation.

The initial conditions correspond to free stream Mach number = 2.7 and pressure perturbation P_1/P_∞ = 1.0215. A total of 11 nodes were used across the boundary layer with the minimum and maximum step size in u equal to 8.3×10^{-5} and .25 respectively. The step size $\Delta\xi$ was .001. The solution was repeatable for $N \geq 11$ and $\Delta\xi \leq .001$ but unstable for $\Delta\xi > .001$.

Some effort was required to get the solution to converge near separation. The grid spacing was varied to produce a smooth transition between near wall nodes and adjacent nodes. A purely explicit scheme with $\beta = 0$ was unstable and diverged rapidly within a few $\Delta\xi$ steps regardless of the size.

A boundary condition can be obtained from the initial equations of motion at u = 0. However its imposition sacrifices overall accuracy for local accuracy (Fletcher and Fleet, 1984).

Conclusions

The Galerkin Finite Element method was applied to Pavlovskii's integral equation (1). Free interaction was used to couple the viscous and inviscid regions. The method worked well in a positive pressure gradient but encountered some trouble near separation due to steep gradients at the wall. Imposition of an exponentially spaced grid near the wall overcame this problem. This method could be applied to bodies of various shapes where the external conditions are unknown.

This work was supported by the Air Force Office of Scientific Research, Mechanics Branch, Dr. James D. Wilson, Program Manager.

References

1. W.R. Briley and H. McDonald, J. Comp. Phys., 24, 372-397, 1977.

2. A.A. Dorodnitsyn, Advances in Aeronautical Sciences. 3, 207-219, Macmillan, New York, 1962.

3. C.A.J. Fletcher and R.W. Fleet, Int. J. for Num. Meths. in Fluids, 4, 399-419, 1984.

4. M. Holt, AGARD Conference Proceedings, No. 4, Part 1, 69-87, 1966.

5. M. Holt, Numerical Methods in Fluid Dynamics, Springer Verlag, 1984.

6. J.N. Nielsen et al., AIAA 2nd Aerospace Sciences Meeting, Paper No. 65-50, New York, January 1965.

7. Y.N. Pavlovskii, Zh. Vych. Mat, i Mat. Fiz., 2, 884-901, 1962.

8. K. Stewartson, Proc. Roy. Soc., Series A, Vol. 200, 84-100, 1949.

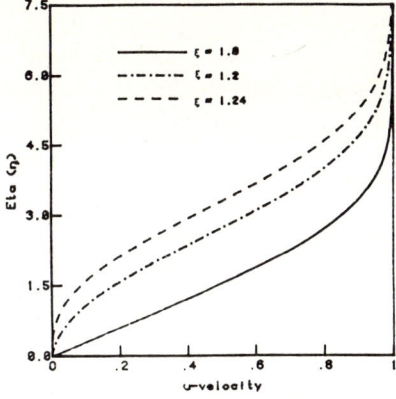

FIGURE 1. Velocity Profile

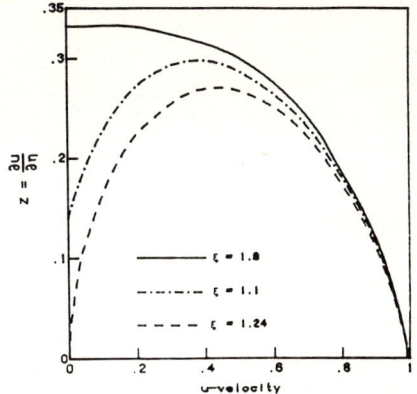

FIGURE 2. Shear Stress Profile

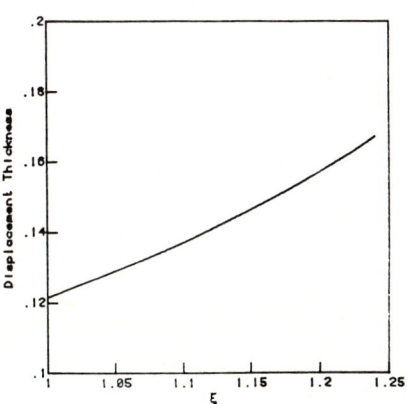

FIGURE 3. Displacement Thickness

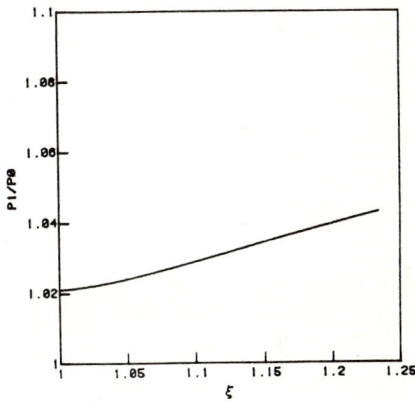

FIGURE 4. Outer Flow Pressure Ratio

THE SOLUTION OF SYSTEM OF NON-LINEAR ALGEBRAIC EQUATIONS GENERATED IN BOUNDARY POINTS CALCULATION

T. X. Hou

Beijing Meteorological Institute, Beijing, China

Abstract

In the numerical algorithm based on the multi-characteristic method there is a system of complicated non-linear algebraic equations. The purpose of this paper is to derive an analytic solution of this system. The uniqueness of solution of this system is shown theoretically in some sense.

1. Introduction. In numerical computations of initial-boundary value problems of systems of hyperbolic equations by difference methods the scheme of boundary points calculation has great influence on the stability of computation and the accuracy of results. The modified characteristic method is one of the most accurate algorithm[1], but there are derivatives of unknowns involved in the right hand side of compatible equations, it can not be computed a priori. One can only compute the terms which do not contain the derivatives of unknowns and average them. This can increase the accuracy greatly, but the method is still not a genuine second-order scheme. For the purpose to obtain a full second-order boundary algorithm we have presented a multi-characteristic method[2] which corresponds to the L-W or the three-level schemes. By the multi-characteristic method all derivatives of unknowns in the $t + \Delta t$ time level are cancelled, so the boundary scheme has the genuine second-order accuracy. But in this scheme will occur a system of complicated non-linear algebraic equations with the components of velocity vector as the unknowns. In the three dimensional case for the solution of these non-linear equations the iteration procedure is very complicated. This fact will greatly limit the application of the method.

In this paper we give an analysis about the system of equations and simplify it, so that the exact solution is given. At the same time the uniqueness of solution is shown.

2. A concise description about multi-characteristic method for body points. Now we consider the gas dynamics equations in Eulerian coordinates. The sphere coordinate system (r,θ,φ) is adopted as the basic coordinate. The components of velocity vector in the directions of r, θ and φ are denoted by u, v and w. Let R denote the natural logarithm of density, S denote the entropy. In the case of supersonic flow the equations of body surface and the bow shock may be written as

$$r - b(\theta) = 0 \tag{1}$$

$$r - s(\theta,\varphi,t) = 0 . \tag{2}$$

By the known transformation of sphere coordinate to computational system ζ,ψ,ϕ,T, the body surface is transformed to $\zeta = 0$ and the bow shock to $\zeta = 1$. The basic equation has the form

$$\frac{\partial U}{\partial T} + \Pi_1 \frac{\partial U}{\partial \zeta} + \Pi_2 \frac{\partial U}{\partial \psi} + \Pi_3 \frac{\partial U}{\partial \phi} + f = 0 \qquad (3)$$

Here $U = (u, v, w, R, S)^T$; $\Pi_1, \Pi_2,$ and Π_3 are 5×5 matrices whose elements are the known functions of s, b and U; the components of vector f are known functions of U, s, b too. The matrices Π_1, Π_2 and Π_3 have five linearly independent left eigenvectors, denoted by ℓ_i, m_j, n_k respectively. Let Q denote the point on body surface where the unknowns at time $t + \Delta t$ need to be computed and all quantities at time t are known and denoted by index 0. The index h denotes the quantities at time $t + \Delta t$, and we suppose that all functions belong to C^3 in the computational domain. By the Taylor series we can prove the following relation

$$U_h - U_0 + \frac{\Delta t}{2}\left[(\Pi_1 \frac{\partial U}{\partial \zeta} + \Pi_2 \frac{\partial U}{\partial \psi} + \Pi_3 \frac{\partial U}{\partial \phi} + f)_h + (\Pi_1 \frac{\partial U}{\partial \zeta} + \Pi_2 \frac{\partial U}{\partial \psi} + \Pi_3 \frac{\partial U}{\partial \phi} + f)_0\right] = 0(\Delta t^3) \qquad (4)$$

Now we consider the (t,ζ)-plane. The quantities at the point where the i-th characteristic line issuing from point Q intersects the ζ-axis at time t are denoted by index i, then again by the Taylor series it can be shown that

$$\ell_{i,0}(U_h - U_i) + \ell_{i,0}\left[(\Pi_2 \frac{\partial U}{\partial \psi} + \Pi_3 \frac{\partial U}{\partial \phi} + f)_h + (\Pi_2 \frac{\partial U}{\partial \psi} + \Pi_3 \frac{\partial U}{\partial \phi} + f)_i\right]\frac{\Delta t}{2} = 0(\Delta t^3)$$
$$i = 1, 2, \cdots, 5 \qquad (5)$$

where
$$\ell_{i,0} = \frac{1}{2}[(\ell_i)_h + (\ell_i)_i].$$

For the planes (t,ψ) and (t,ϕ) the similar relations can be obtained. Because we only consider the second order scheme then the terms $0(\Delta t^3)$ may be omitted. From these equations the space derivatives of unknowns at time $t + \Delta t$ can be cancelled. In what follows we give the final form of equations:

$$(\frac{D}{r})_{1,0} u_h + v_h + [-(\frac{D}{r})_{1,0} \frac{v_h^2 + w_h^2}{r_h} - A_h u_h + B_h w_h^2 \cos\psi] \frac{\Delta t}{2} + \sigma_1 = 0 \qquad (6)$$

$$-(BE)_{2,0} u_h + w_h + [(BE)_{2,0} \frac{v_h^2 + w_h^2}{r_h} + \frac{u_h w_h}{r_h} - v_h w_h B_h \cos\psi]\frac{\Delta t}{2} + \sigma_2 = 0 \qquad (7)$$

$$S_h + \sigma_3 = 0 \qquad (8)$$

$$-(\frac{1}{aZ})_{5,0}(u_h - u_5^i) - (\frac{K}{GaZ})_{5,0}(v_h - v_5^i) - (\frac{BE}{aZ})_{5,0}(w_h - w_5^i) + R_h - R_5^i + \frac{1}{\gamma}(s_h - s_5^i)$$
$$+ [(\frac{1}{aZ})_{5,0} \frac{v_h^2 + w_h^2}{r_h} - (\frac{k}{GaZ})_{5,0}(-A_h u_h + B_h w_h^2 \cos\psi) - (\frac{BE}{aZ})_{5,0}(\frac{u_h w_h}{r_h} - v_h w_h B_h \cos\psi)$$
$$+ \frac{2u_h}{r_h} - v_h B_h \cot\psi] \frac{\Delta t}{2} + \sigma_4 = 0 \qquad (9)$$

Here A, B, D $\cdots$ are known functions and $\sigma_1, \sigma_2, \cdots$ denote the sum of all terms which involve only the known quantities at time t and the sound speed a at time $t + \Delta t$. The signs g^i_{α}, $\alpha=1,2,\cdots,5$ denote the value of g at that points, where the α-th characteristic line issuing from point Q at time $t + \Delta t$ intersects with the ζ-axis in plane (t,ζ). Similarly the signs g^j_{α}, g^k_{α} can be defined. If the initial value of the body surface entropy is given as constant S_c, then the equation (8) can be simplified to $S_h = S_c$.

3. The solution of equations (6), (7) and (9). Since $\zeta=0$ on body points it can be verified that

$$\left(\frac{D}{r}\right)_{1,0} = -\frac{1}{b}\frac{db}{d\psi}, \quad (BE)_{2,0} = 0$$

The equation (6) can be written in the form

$$\frac{1}{b}\frac{\Delta t}{2}\left(\frac{\cos\psi}{\sin\psi} + \frac{1}{b}\frac{db}{d\psi}\right) w^2 + \nu^2 v + \Sigma_1 = 0 \tag{10}$$

where $\nu^2 = 1 + \left(\frac{1}{b}\frac{db}{d\psi}\right)^2$, Σ_1 denotes the sum of known terms. By the same way equation (7) can be written in the following form

$$-\frac{1}{b}\frac{\Delta t}{2}\left(\frac{\cos\psi}{\sin\psi} + \frac{1}{b}\frac{db}{d\psi}\right) vw + w + \Sigma_2 = 0 \tag{11}$$

and equation (9) in the form

$$c_1 v^2 + c_2 w^2 + c_3 vw + c_4 v + c_5 w + \Sigma_3 = 0 \tag{12}$$

where $c_1 - c_5$, Σ_3 denote the sum of terms containing the a_h and the t-quantities. The procedure of solution of equations (10)-(12) is the following. First the trial value of a_h is given as $a^{(0)}$, then the coefficients of (10), (11) are known. Since $\nu > 0$, from (10) the unknown v may be solved and can be substituted into (11) to obtain the following equation

$$\left(\frac{q}{\nu}\right)^2 w^3 + \left(1 + \frac{q\Sigma_1}{\nu^2}\right) w + \Sigma_2 = 0 \tag{13}$$

where

$$q = \frac{1}{b}\frac{\Delta t}{2}\left(\frac{\cos\psi}{\sin\psi} + \frac{1}{b}\frac{db}{d\psi}\right).$$

For the non-zero meridian planes the points on the spherical coordinate axis are not necessary to compute, hence $\psi \geq \Delta\psi > 0$ and $q \leq$ const. $< \infty$. From this fact we have $q = O(\Delta t)$. When Δt is small enough the coefficient of w will be greater than zero. According to the nature of cubic equations one can conclude that equation (13) has only one real root, and it can be denoted by the known formula. Now we substitute the w into (10) then the v can be obtained. Finally, we substitute the v and w into (12) then an equation with variable $a^{(0)}$ will be obtained. This equation can be solved easily. In the zero-meridian plane the axis points need to be computed, but in that plane $w \equiv 0$, hence from (10) the unknown v can be obtained immediately, and one can verify that $\Sigma_2 = 0$, hence equation (11) reduces to an equality.

4. Calculated examples. We have computed two examples, a sphere-cone and a rectangular body with zero angle of attack. Here the shape of bow shock and the pressure distribution on the surface of rectangular body are presented.

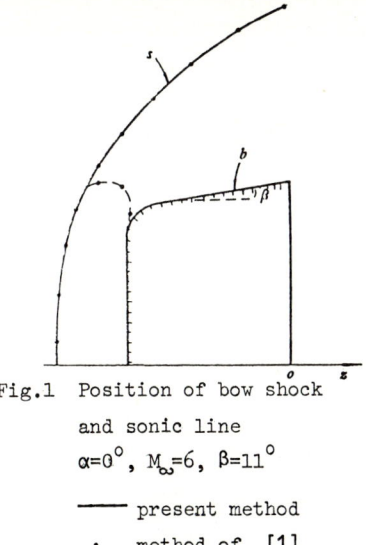

Fig.1 Position of bow shock
and sonic line
$\alpha=0°$, $M_\infty=6$, $\beta=11°$

——— present method
· method of [1]

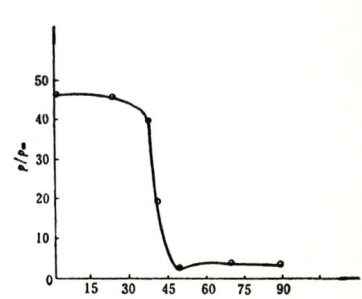

Fig.2 Pressure distribution on
body surface
$\alpha=0°$, $M_\infty=6$

——— present method
· method of [1]

Literature references

[1] Moretti. G. and Bleich, G., Three-dimensional flow around blunt bodies, AIAA J. No 9, 1967.

[2] Hou Tianxiang. The multi-characteristic method for bound points calculation. Proceedings of second Conference of Hydrodynamics, 1983 China. pp. 13-15.

A TEST PROBLEM FOR UNSTEADY SHOCK WAVE CALCULATION

D. Huang

Institute of Mathematics, Peking University, Beijing China

I. INTRODUCTION

This work is a sequel to a paper presented at the 7th conference [1], and is a brief overall description of a series of Chinese papers [2-13]. They serve a dual purpose: to understand complex unsteady flow phenomena with highly nonuniform entropy distribution, and to provide a complex test problem in developing numerical schemes for the solution of one-dimensional gasdynamic problems. One dimensional schemes are useful by themselves, and if time splitting technique is used, they serve as constituent parts of multidimensional schemes.

As well known, test runs can clarify many features of a scheme. The test problem suggested in this paper is not self-similar and thus is more complex than the well known and frequently used Riemann initial value or shock tube problem. Test runs with this problem help us to understand the effects of variation of shock strength, mesh size, boundary condition involving vanishing of particle velocity, artificial viscosity, scheme viscosity, etc., of several schemes. It is highly desirable to have some knowledge about these effects in order to have judicious understanding of the numerical results of many complex problems in science and engineering.

II. A BLAST WAVE REFLECTION PROBLEM   FORMULATION AND FEATURES

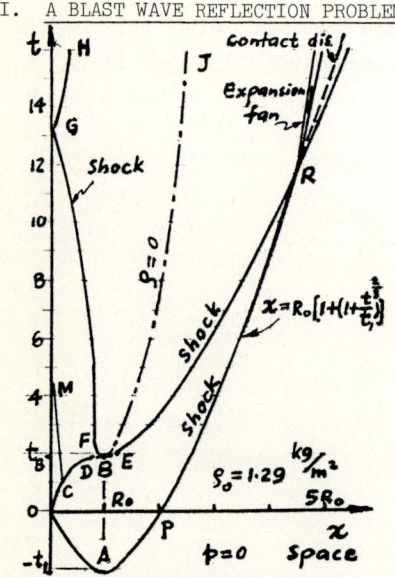

Fig. 1. x-t diagram

The x-t diagram facilitates the explanation of the formulation [1,6] and fluid physics of our 1-D time-dependent Eulerian gasdynamic problem.

Let a finite amount of energy E_0 be released [14] at $(R_0, -t_1)$ in the x-t diagram. Two planar shocks AO and AP propagate into still air with finite density $\rho = \rho_0$, and zero (neglegible) pressure, p=0. The unsteady wave AO reflects at the rigid wall located at x=0. We adjust E_0, R_0, t_1 & ρ_0, so that the pressure at x=0 just after reflection is about 800 atm [1]. The curve OCDBER denotes the reflected shock wave, whose strength varies.

In the numerical solution of the above problem using shock-capturing technique the region near the origin (Fig.1.) should be treated with care. There are three values of pressure p, say, zero, 100 atm. and 800 atm.; three values of density, say, ρ_0, $6\rho_0$ and $21\rho_0$. Also the pressure on the wall drops rapidly with time (see data tabulated in the following). This unsteady normal reflection of plane shock wave is quite different from the reflection of a shock wave with constant flow variables behind the shock wave. The steep gradients of the density ρ, entropy and pressure p of the impinging blast wave require very fine mesh near the origin.

More complex is the region near point B, called a high temperature singularity [5,6,8,9]. It relates to the singular behavior of the symmetric plane of the point explosion [14], denoted by the line AB in Fig.1. Near the line AB, say, for small $|R_0-x|/R_0$, density decreases rapidly, while temperature T and sound speed a increase rapidly as can be described by the following asymptotic formulae:

$$p \sim t^{-2/3}, \qquad \rho \sim t^{-5/3} |R_0-x|^{5/2}$$
$$a \sim \sqrt{T} \sim t^{\frac{1}{2}}|R_0-x|^{-5/4}, \quad u \sim t\ |R_0 - x|$$

In this paper the line AB is the model for the small region of very high temperature due to explosion in an inviscid fluid. When reflected shock wave OCD moves to about $0.8R_0$, its strength decreases while its speed increases rapidly, as if it were sucked by the line AB. This can be easily understood, the shock wave is supersonic and the sound speed a in front of it increases rapidly.

At the moment t=(1.99355/2.17182) t_1 [11,9] the reflected shock wave OCDB hits the line AB, this high temperature line starts to move to the right. The curve BJ is its trace. Accurate results about BJ is obtained in [11]. This gives insight to the motion of the high temperature gas particle for our particular set of values of E_0, R_0 and Tt_1.

One may obtain the trace of the high temperature gas particle with another formulation of the problem, i.e. to choose a narrow region of high constant temperature as Brode [16] for the case of spherical symmetry. On the basis of the research since 1955 we can anticipate that the above two formulations have close results. We adopt the model of point explosion [14], since it is more suitable for construction of a standard test problem for numerical methods; that is, the curves AO and APR have simple analytic expressions and in the region AOCDBERPA finite analytical expressions are available [14], and we can obtain very accurate standard numerical results under this formulation.

Another feature of the flow in the neighbourhood of point B is that, the left going characteristics of the same family, starting from a segment of the reflected shock DB ($x_B = R_0$, $x_D \sim 0.95R_0$), converge and cause the formation of a new shock wave FG propagating into the region of un-steady flow near the wall (x = 0). GH is its reflection.

Our typical problem shows that Euler's equation possesses a large class of solutions, some of which may be complex and can not be solved without the use of computer.

III. METHOD USED AND SOME STANDARD DATA

To ensure high accuracy of the results we used analytical method to find OC and DBE [1, 5, 9] . An accurate modified version of singularity-separating method was developed to calculate ODBER, [6,11]. The data for OD obtained in [8] was again checked in [7,9] by the method of characteristics with characteristic mesh and variable step size to ensure high accuracy. The variable steps used in [9] to calculate DB are much finer than those used in [6], i.e., adaptive to the features of the physical phenomena found in [5]. Accurate data can be found in [8,9,11]. Some typical values are tabulated below, with typical profiles shown in Fig.2.

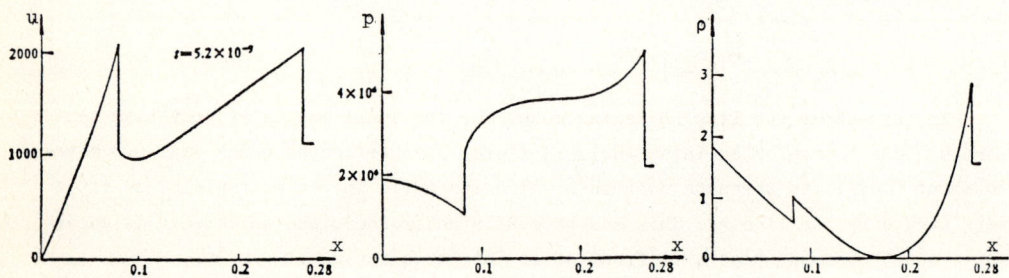

Fig.2. Velocity, pressure and density profiles at $t=5.2\times\kappa \times 10^{-5}$

TIME 10^{-5} sec. t/κ	POSITION OF REFLECTED SHOCK R_3/R_0	SHOCK MACH NUMBER M_s	VALUES JUST BEHIND REFLECTED SHOCK OCDB		PRESSURE ON RIGID WALL AT $x = 0$ 10^7 n/m^2
			u_2 m/s	ρ_2/ρ_0	
0	0	$\sqrt{7}$	0	21	8.103424
0.06	0.006555				6.900
0.12	0.01399				5.950
0.3	0.04131				4.063
0.6	0.10279	2.22345	1031.5	4.3602	2.490
1.0	0.21598	1.97740	1607.1	1.8412	1.540
1.4	0.37513	1.73660	2283.7	0.68274	1.067
1.8	0.62808	1.45159	3691.7	0.12053	0.7939
1.9905	0.94702	1.11075	13296	5.6891×10^{-4}	
1.9920	0.96224	1.08863	16442	3.0444×10^{-4}	
1.99355	1	1	∞	0	
2.01					0.6882
3.6	2.15				0.3208
5.2	2.68				0.1857
12.6	REFLECTED SHOCK OVERTAKES BLAST WAVE APR				
13.44	NEW SHOCK FG REFLECTS ON THE LEFT WALL				

TABLE: SOME SELECTED STANDARD ACCURATE RESULTS   [1,7,8,9]:

κ = arbitrary constant, $t_1 = 2.1718193 \times \kappa \times 10^{-5}$ sec., $\rho_0 = 1.29$ kg/m^3, $R_0 = 0.1 \times \kappa$ m.

$E_0 = \kappa \times 2.7349056 \times 10^6$ joules/m^2 $\qquad R_0 = (\frac{E_0}{\rho_0})^{1/3} (t_1)^{2/3}$.

IV. BRIEF DESCRIPTION OF SEVERAL NUMERICAL TESTS

We tested several schemes in addition to those reported in [1]. During his visit to Peking Univ. P.D.Lax sent us a paper on random choice finite difference scheme [17]. Lei [2] showed that if the coefficient of the artificial viscosity term used in [17] is properly modified, one can relax the restrictions, under which the approximate Riemann solver is consistent with entropy condition. Lei then tested that scheme with our test problem. We [4] tested widely used in China Rusanov's early scheme [15] and found that numerical viscosity is excessive for a region near the wall at x=0. This causes about 8% error in density near x=0, while pressure profile is acceptable. The error in density can be greatly reduced when artifical viscosity is set to zero for the region near x=0. Rusanov's artifical viscosity is interesting in that as the strength of reflected shock decreases, the transition zone representing shock becomes narrower. This fact is in contrast to the widening of transition zone obtained by Godunov's scheme or the second order MUSCL of Leer [3]. These facts can not be observed if we test schemes only by the Riemann initial value problem. Cao [13] observed also excessive numerical viscosity near x=0 when anti-diffusion method of Boris and Book is used to solve this test problem. This means that truncation error varies in a wide range in the region from the wall to the reflected shock. Zhang [12] tested MacCormack's scheme published in 1969 [19] and also widely used in China, and found a formula for the main part of the scheme viscosity. Substracting from MacCormack's scheme this term the scheme is still stable while resolution of shock wave becomes better. In the above test we change Δx sometimes even by two orders of magnitude to observe the effect of meshing on accuracy. In short these tests helped us to understand features of several schemes and of meshing. This understanding can help us to be more judicious in estimating possible errors of a given numerical solution with unsteady shock wave of **variable** strength.

Acknowledgements. We review here the efforts of many of my friends, whose papers are cited. To these friends I express my grateful acknowledgements.

References

1. HUANG, D., LI, Y.F., Huang, L.P. and Liu, Y.Z., Lecture Notes in Physics, Vol. 141 pp. 218-223 (1981).
2. LEI, G.Y., Jour. Comp. Math of High Schools (in Chinese with English abstract) No. 1, pp. 81-90 (1982), also Master degree thesis, Peking Univ. (1981).
3. CHOU, N., Jour. Comp. Physics, Vol. 1, No. 1, pp. 21-30 (1984) (in Chinese).
4. HUANG, D., LI, W.X., Explosion and shock waves, Vol. 2, No. 2 pp. 27-38 (in Chinese with English abstracts) (1982).

5. HUANG, D., Rong, S., Proceedings of 2nd Asian Congress of Fluid Mechanics, pp. 302-308, Science Press, Beijing, (1983).
6. WU, X. H., HUANG, D., ZHU, Y.L., Jour. Comp. Math., Vol. 1, No. 3. (China) pp. 247-259 (1983).
7. TING, A.L., Acta Mechanica Sinica, Vol. 17 No.1, pp. 50-55 (1985).
8. WU, X. H., Master Degree thesis, Computing Center, Academia Sinica.(1981).
9. WANG, A.P., Master Degree thesis, Peking Univ. Math. Depart. (1985).
10. ZHU, Y.L., WU, X.H., Ni, L.A., WANG, Y., Lecture Notes in Physics, Vol. 218 pp. 611-612 (1985).
11. WU, X.H., Doctor Degree thesis, Chap. 3. Computing Center, Academia Sinica. (1984).
12. Zhang Nai-xin, Master Degree thesis, Peking Univ. Math. Department (1981).
13. Cao Yi-min, Test of anti-diffusion method of Boris and Book, Report of Computing Center, Academy Sinica (1982).
14. Sedov, L.I., Methods of Similarity and Dimensions in Mechanics (in Russian), 5th. ed., pp.169-171, 228-252 (1965).
15. Rusanov, V. V., Jour. of Comp. Math. & Math. Phy., Vol. 1 (in Russian), pp. 267-279 (1961).
16. Brode, H., Jour. of Applied Physics, Vol. 26, No. 6 (1955), pp. 766-775.
17. A. Harten and P.D. Lax, Report DOE/ER/03077-167, New York Univ., May (1980).
18. Van Leer, J. Comput Phys., 32, 101-136 (1979).
19. MacCormack, AIAA Paper, 69-354 (1969).
20. Boris and Book, J. Comp. Phys., 11, 38-69 (1973).

APPLICATIONS OF NUMERICAL CONFORMAL MAPPING TECHNIQUE
M.K.Huang
Department of Aerodynamics
Nanjing Aeronautical
Institute, China

I. Introduction

The grid generation based on the solution of elliptical partial differential equations is one of the effective methods. However, the use of this method needs to specify in advance the boundary grid distribution. If the boundary has **slope discontinuities, the improper** specification of the boundary grid distribution near the corners often leads to poor grid distribution in the field. The grid generation based on conformal mapping has no such a drawback since the grid points are generated automatically not only in the field but also on the boundary. Furthermore, the orthogonality of the grid generated by conformal mapping often simplifies the form of the equations in the mapped plane for the flow problems. This paper will present a numerical conformal mapping method for mapping an arbitrary polygon with curved sides into a circle, so that the boundary shapes that can be studied cover most of the practical ones. In fact, a similar technique has been mentioned by Ives[1], and the method is actually an extension of the mapping technique used by Jameson[2] in computation of the transonic flow past an airfoil that has only one corner, the trailing edge. Some of the application examples are to be given in this paper.

II. Conformal mapping method

In z plane, we have a polygon with curved sides and M corners. Suppose that the derivative of the mapping function, which maps the exterior of such a polygon onto that of a unit circle in σ plane, is of the form

$$\frac{dz}{d\sigma} = \prod_{j=1}^{M} \left(1 - \frac{e^{-i\theta_j}}{\sigma}\right)^{1-\frac{\epsilon_j}{\pi}} \exp \sum_{n=0}^{N} \frac{c_n}{\sigma^n} \tag{1}$$

where ϵ_j are the included angles of the corners, $e^{-i\theta_j}$ the complex coordinates of the corner vertices in σ plane, $c_n = a_n - ib_n$ are the unknowns as well as θ_j, the term number N should be taken to be larger enough for accuracy consideration. The mapping for closed shape boundary requires that the expansion of the Eq.(1) in Lauren's series has no term in $1/\sigma$, so that c_1 should be taken as

$$c_1 = \sum_{j=1}^{M} (1 - \frac{\epsilon_j}{\pi}) e^{-i\theta_j} \tag{2}$$

The Schwarz-Christoffel factors in Eq.(1) account for the slope discontinuities over the boundary.

To determine c_n, we note that $\sigma = e^{-i\theta}$ and $d\sigma = -ie^{-i\theta}d\theta$ on the circle, and that $dz = e^{i\beta}ds$ on the polygon boundary, where β and s are the tangent angle and the arc length respectively. Substituting them into Eq.(1) and separating the equation into real and imaginary parts, we arrive at two equations

$$\log(e^{-a_0}\frac{dS}{d\theta}) - \sum_{j=1}^{M}(1-\frac{\epsilon_j}{\pi})\log|2\sin\frac{\theta-\theta_j}{2}| = \sum_{n=1}^{N}(a_n\cos n\theta + b_n\sin n\theta) \tag{3}$$

$$(\beta + \theta - \frac{3}{2}\pi) - \sum_{j=1}^{M}(1-\frac{\epsilon_j}{\pi})\frac{(\theta-\theta_j) + 2\pi H(\theta_j - \theta) - \pi}{2}$$
$$= -b_0 + \sum_{n=1}^{N}(a_n\sin n\theta - b_n\cos n\theta) \tag{4}$$

Here $H(\theta_j - \theta)$ is the unit step function that takes on the value of zero when $\theta > \theta_j$ and one when $\theta < \theta_j$. The unknowns c_n and θ_j can be determined by the equations (3) and (4) through the following iteration steps:

Step 1:

We first give an initial guess on s as a function of θ. Having $s(\theta)$, β as a function of θ is then known since $\beta(s)$ is known for a given polygon. The required θ_j can be evaluated by interpolation.

Step 2:

With the known $\beta(\theta)$, the LHS of the Eq.(4) becomes a known function of θ, and it can be expanded into a Fourier series by means of FFT technique. Thus all the Fourier coefficients except a_0 are obtained. At this step, it is easy to see that the function representing the LHS of the Eq.(4) is continuous everywhere. According to the theory of Fourier series, the corresponding Fourier coefficients, a_n and b_n, go to zero quickly as increasing n.

Step 3:

The value of c_1 is frozen by the formula (2). By taking the use of FFT technique, the summation of the Fourier series on the RHS of the Eq.(3) leads to

$$\frac{dS}{d\theta} = e^{a_0} f(\theta) \tag{5}$$

where $f(\theta)$ is known. Integration of this equation yields the new guess of $s(\theta)$, and a_0 is calculated by the condition that $s(2\pi)$ should be equal to the total arc length of the given shape boundary. Then, turn back to the step 1 for iteration until convergence to a certain tolerance. In the previous iteration procedure, the under-relaxation factor is usually needed if the polygon considered is of complex form.

It should be indicated now that the great care must be taken in the numerical integration of the Eq.(5) because, from Eq.(3), $f(\theta)$ behaves at the corner as

$$f(\theta) \sim |\theta - \theta_j|^\mu \tag{6}$$

where $\mu = 1 - \epsilon_j/\pi$. It is seen that θ_j is an infinite but integrable singularity point of the function $f(\theta)$ when $\pi < \epsilon_j < 2\pi$. Suppose that θ_j is within a small subinterval (θ_i, θ_{i+1}). We approximate $f(\theta)$ in (θ_i, θ_j) by

$$f(\theta) = f(\theta_i) \frac{(\theta_j - \theta)^\mu}{(\theta_j - \theta_i)^\mu}$$

A similar approximation is taken in (θ_j, θ_{i+1}). Thus the integration can be performed analytically. It yields

$$\int_{\theta_i}^{\theta_{i+1}} f(\theta) d\theta = \frac{f(\theta_i)(\theta_j - \theta_i) + f(\theta_{i+1})(\theta_{i+1} - \theta_j)}{2 - \frac{\epsilon_j}{\pi}} \tag{7}$$

For other subintervals that contain no singularity point, a simple trapezoidal rule may be taken for the quadrature. The expression (7) also shows that the method fails when $\epsilon_j = 2\pi$.

III. Applications

The first application is the grid generation which may be used in solving the flow problems by finite difference methods. In fact, with the use of the polar coordinate system, the orthogonal mesh is first generated in σ plane with equal step size in θ. The desirable conformal grid is then formed by mapping back to the physical z plane.

The second application is the estimation of the apparent mass coefficients, which are useful in prediction of the aerodynamic derivatives for missiles and aircrafts. To begin with, we introduce $\zeta = \exp(c_0)\sigma$ to map σ plane onto ζ plane in order to have $dz/d\zeta$ taking on the value of one at infinity. Then, the integration of the Lauren's series expansion of $dz/d\zeta$ yields

$$Z = \zeta + \sum_{n=0}^{\infty} \frac{A_n}{\zeta^n} \tag{8}$$

where

$$A_1 = -\exp(2C_o)\left[C_2 - \frac{C_1^2}{2} - \sum_{j=1}^{M} \frac{(1-\frac{\epsilon_j}{\pi})\frac{\epsilon_j}{\pi}}{2} e^{-2i\theta_j} + \sum_{j=1}^{M-1} \sum_{K=j+1}^{M} (1-\frac{\epsilon_K}{\pi})(1-\frac{\epsilon_j}{\pi}) e^{-i(\theta_K + \theta_j)}\right] \tag{9}$$

The formula (10-83) in the text by Nielsen[3] together with the value of A_1 given by Eq.(9) can be applied to predict the required apparent mass coefficients for cross-sections.

The third application is the computation of the incompressible potential flow over a cross-section. This is well known so we have no need to give further interpretation here.

Four typical cross-section shapes are considered as examples, such as a half circle having two corners, a regular triangle having three corners, regular and irregular hexagons that have six corners, and a wing-body cross-section having also six corners. Some of the grids generated by the present method are shown in Figs. 1 through 3. The estimated apparent mass coefficients with the notation used by Nielsen[3] are presented in table I together with the exact ones[3,4] for comparison. In the table, ρ is the density of the fluid, a represents the height of the triangle, and it also denotes the radius for the half circle and the circumscribed circle of the hexagon. Fig. 4 shows the predicted pressure distribution of the incompressible flow past the wing-body section. In our computations, the term number N is taken to be 32 and 64. It is shown in all cases that the agreement with the exact solutions is excellent.

The experience with the use of the present method indicates that there is no difficulty if all the included angles θ_j are less than π. However, the example of the wing-body section indicates a difficult case because of the presence of the θ_j greater than π. In such a case, a small under-relaxation factor of 0.15 was taken and more than 50 iterations were required to reach the result shown in Fig.4.

Table I. Apparent mass coefficients.

		$m_{11}/2\pi\rho a^2$	$m_{22}/2\pi\rho a^2$
half circle	present	0.5314	0.1579
	exact[4]	19/36=0.5278	17/108=0.1574
regular triangle	present	0.1471	0.1464
	exact[3,4]	0.1453	0.1453
regular hexagon	present	0.4399	0.4396
	exact[3]	0.434	0.434

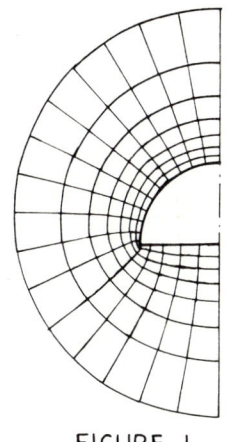

FIGURE 1

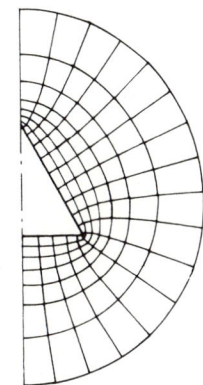

FIGURE 2

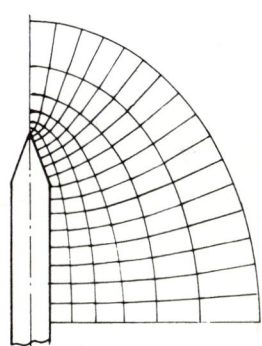

FIGURE 3

References

1. Ives,D.C., AIAA paper no. 83-1906,(1983)
2. Jameson,A., Symposium Transonicum II (1975).
3. Nielsen,J.N., Missile Aerodynamics, McGram-Hill (1960).
4. Huang,M.K. and Chow,C.Y., J. of Aircraft,20,9(1983),810-816.

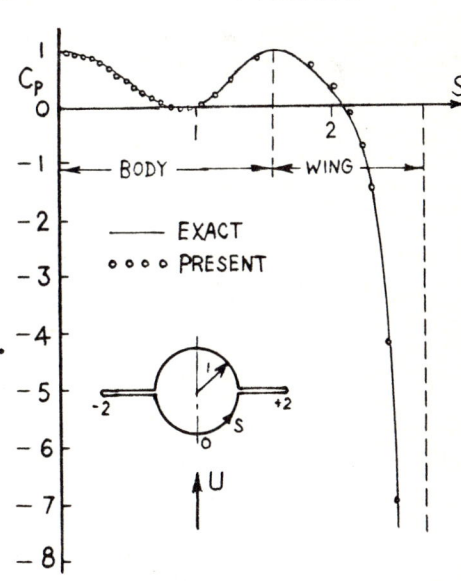

FIGURE 4

EULER CALCULATIONS FOR A COMPLETE AIRCRAFT

A. Jameson and T.J. Baker
Department of Mechanical and Aerospace Engineering
Princeton University
Princeton, New Jersey 08544
U.S.A.

Abstract

We describe a new finite element method for solving the Euler equations, and present the results of a transonic flow calculation for a commercial aircraft with pylon mounted engines. The finite element method uses a tetrahedral mesh, and establishes conservation of mass, momentum and energy in polyhedral control volumes by summing fluxes through the faces of the tetrahedra. The tetrahedra are generated by using a DeLaunay triangulation to connect a swarm of mesh points surrounding the aircraft.

Introduction

During the last two decades the science of aerodynamics has been transformed by the widespread introduction of computational methods to treat previously intractable problems, such as the calculation of transonic flows. Improvements in high speed electronic computers have made it feasible to attempt numerical calculations of progressively more complicated mathematical models of fluid flow, and to apply these methods to increasingly elaborate geometric configurations. Following the introduction of panel methods for subsonic flow in the sixties [1-2], and major advances in the simulation of transonic flow by the potential flow approximation in the seventies [3-6], the eighties have seen rapid developments in methods for solving the Euler and Navier Stokes equations [7-12].

A major pacing item of the emergence of a capability to treat a complete aircraft has been the development of a suitable method of mesh generation. For simple wing body combinations it is possible to generate rectilinear meshes without too much difficulty [9]: for more complicated configurations containing, for example, pylon mounted engines, it becomes increasingly difficult to produce a structured mesh which is aligned with all solid surfaces. An alternative is to use tetrahedral cells in an unstructured mesh which can be adapted to conform to the complex surface of an aircraft. Finite element methods of solving the potential flow and Euler equations on triangular and tetrahedral meshes have been developed by several authors [6,13,14]. Bristeau, Glowinski, Periaux, Perrier, Pironneau and Poirer achieved a striking success in solving the potential flow equation to predict the flow around a Falcon 50. Nevertheless, the generation of a tetrahedral mesh around configurations of such complexity remains a formidable problem. This paper describes a new finite element method for solving the Euler equations on a tetrahedral mesh, and its application to the calculation of transonic flow around a complete commercial aircraft with pylon mounted engines. Preliminary results were presented at the AIAA 24th Aerospace Sciences Meeting in January [15].

The finite element approximation is obtained by directly approximating the integral equations for conservation of mass, momentum and energy in polyhedral control volumes. The scheme can also be regarded as a Galerkin method in which the test function space is the set of piecewise linear tetrahedral elements: this can be shown to be equivalent to a flux balance based on polyhedral control volumes formed by the union of tetrahedra meeting at a common vertex. It turns out that each face is associated with precisely two such control volumes. It is therefore possible to reformulate the calculation in a particularly elegant way, in which the fluxes are evaluated in a single main loop over the faces. This novel decomposition leads to a substantial reduction in computational complexity. Steady state solutions are obtained by integrating the time dependent equations with a multistage time stepping scheme.

Convergence is accelerated by the use of locally varying time steps, residual averaging and enthalpy damping.

A new method is used to generate the tetrahedral mesh. Separate meshes are first generated around the individual aircraft components to create a cluster of points surrounding the whole aircraft. We do not require any regularity in this initial point distribution, only that a reasonable point density is created corresponding to the anticipated variation in the flowfield. The swarm of mesh points is then connected together to form tetrahedral cells which provide the basis for a single finite element approximation for the entire domain. This use of triangulation to unify separately generated meshes bypasses the need to devise interpolation procedures for transferring information between overlapping grids. The triangulation of a set of points to form disjoint tetrahedra is in general nonunique: our procedure is to generate the Delaunay triangulation [16-19]. This is dual to the Voronoi diagram that results from a division of the domain into polyhedral neighborhoods, each consisting of the subdomain of points nearer to a given mesh point that any other mesh point. The implementation of this method and the need to maintain the integrity of solid surfaces present a number of interesting problems. Although the Delaunay triangulation and associated Voronoi diagram has been exploited by others as a natural setting for calculations involving irregularly spaced points [16-17], we believe that the use of the Delaunay criterion as an explicit method of generating meshes for complex shapes is a new departure.

2. Finite Element Approximation

Let p, ρ, u, v, w, E and H denote the pressure, density, Cartesian velocity components, total energy and total enthalpy. For a perfect gas

$$E = \frac{p}{(\gamma-1)\rho} + \frac{1}{2}(u^2 + v^2 + w^2), \quad H = E + p/\rho$$

where γ is the ratio of specific heats. The Euler equations for flow of a compressible inviscid fluid can be written in integral form as

$$\frac{\partial}{\partial t}\iiint_\Omega w\, d\Omega + \iint_{\partial\Omega} \underline{F} \cdot d\underline{s} = 0 \tag{1}$$

for a domain Ω with boundary $\partial\Omega$ and directed surface element $\underline{dS}$. Here w represents the conserved quantity and $\underline{F}$ is the corresponding flux. For mass conservation

$$w = \rho, \quad \underline{F} = (\rho u, \rho v, \rho w)$$

For momentum conservation

$$w = \rho u, \quad \underline{F} = (\rho u^2 + p, \quad \rho uv, \quad \rho uw)$$

with y and z momentum quantities similarly defined, and for energy conservation

$$w = \rho E, \quad \underline{F} = (\rho Hu, \quad \rho Hv, \quad \rho Hw)$$

Consider the differential form of equation (1)

$$\frac{\partial w}{\partial t} + \nabla \cdot \underline{F} = 0$$

Multiplying by a test function ϕ and integrating by parts over space leads to

$$\frac{\partial}{\partial t}\iiint_\Omega \phi w\, d\Omega = \iiint_\Omega \underline{F} \cdot \nabla\phi\, d\Omega - \iint_{\partial\Omega} \phi\underline{F} \cdot \underline{dS} \tag{2}$$

Suppose now that we take ϕ to be the piecewise linear function with the value unity at one node (denoted by 0 in Figure 1), and zero at all other nodes. Then the last term vanishes except in the case when 0 is adjacent to the boundary. Also $\nabla\phi$ is constant in every tetrahedron, and differs from zero only in the tetrahedra with a common vertex at node 0. Since ϕ_x is constant in a tetrahedron it may be evaluated as

$$\phi_x = \frac{1}{V}\iiint \phi_x\, dxdydz = \frac{1}{V}\sum_k S_{x_k} \bar{\phi}_k$$

where V is the cell volume, S_{x_k} and $\bar{\phi}_k$ are projected area of the kth face in the x direction and the average value of ϕ on the kth face, and the sum is taken over the faces of the tetrahedron. For the given test function $\bar{\phi} = 1/3$ on the faces 012, 023, and 031 and zero on the face 123. Also the projected area S_x on face 123 is equal and opposite to the sum of the projected face areas of the other three faces. Using the same procedure to evaluate ϕ_y and ϕ_z, it follows that

$$\nabla \phi = -\underline{S}/3V \tag{3}$$

where $\underline{S}$ is the directed area of the face opposite vertex 0. Now treat $\underline{F}$ as piecewise linear and use equation (3) to evaluate the volume integral on the right side of equation (2). Then each tetrahedron meeting at node 0 introduces a contribution $(\bar{\underline{F}} \cdot \underline{S})/3$ where $\bar{\underline{F}}$ is the average value of $\underline{F}$ in the cell. For the cell illustrated in Figure 1, for example,

$$\bar{\underline{F}} = \frac{1}{4}(\underline{F}_0 + \underline{F}_1 + \underline{F}_2 + \underline{F}_3)$$

Summing over all cells meeting at node 0 leads to the total contribution

$$\frac{1}{3} \sum_k \bar{\underline{F}}_k \cdot \underline{S}_k$$

Since the control volume is closed, however,

$$\sum_k \underline{S}_k = 0$$

Therefore the contribution of F_0 to $\bar{F}_k$ can be discarded, leading to a sum over the faces multiplied by a constant. Thus if we write

$$\tilde{\underline{F}} = \frac{1}{3}(\underline{F}_1 + \underline{F}_2 + \underline{F}_3)$$

for the average value of $\underline{F}$ on the face opposite vertex 0 we find that the right-hand side of equation (2) can be replaced by

$$-\frac{1}{4} \sum_k \tilde{\underline{F}}_k \cdot \underline{S}_k$$

On the left-hand side of equation (2) we take w to be constant inside the control volume. Since ϕ is piecewise linear, the volume average value is $\bar{\phi} = 1/4$. The factor $1/4$ cancels on each side and the approximation to equation (2) can therefore be written as

$$\frac{d}{dt}\left(\sum_k V_k\right) w + \sum \tilde{\underline{F}}_k \cdot \underline{S}_k = 0 \tag{4}$$

Referring to Figure 2, which illustrates a two dimensional mesh, it may be seen that with a triangular or tetrahedral mesh, each face is a common external boundary to exactly two control volumes. Therefore each internal face can be associated with a set of 5 mesh points consisting of its three corners 1, 2 and 3, and the vertices 4 and 5 of the two tetrahedra based on the face, as illustrated in Figure 3. Vertices 4 and 5 are the centers of the two control volumes influenced by the face. It is now possible to generate the approximation (4) by presetting the flux balance at each mesh point to zero, and then performing a single loop over the faces. For each face one first calculates the fluxes of mass, momentum and energy across the face, and then one assigns these contributions to the vertices 4 and 5 with positive and negative signs respectively. Since every contribution is transferred from one control volume into another, all quantities are perfectly conserved. Mesh points on the inner and outer boundaries lie on the surface of their own control volumes, and the accumulation of the flux balance in these volumes has to be correspondingly modified. At a solid surface it is also necessary to enforce the boundary condition that there is no convective flux through the faces contained in the surface.

3. Dissipation

Equation (4) represents a nondissipative approximation to the Euler equations. Dissipative terms may be needed for two reasons; to eliminate the occurrence of undamped or lightly damped nodes, and to prevent oscillations near shock waves.

The simplest form of dissipation is to add a term generated from the difference between the value at a given node and its nearest neighbors. That is, at node 0, we add a term
$$D_0 = \sum_k \varepsilon^{(1)}_{k0} (w_k - w_0) \tag{5}$$
where the sum is over the nearest neighbors, as illustrated in Figure 4. The contribution $\varepsilon^{(1)}_{k0}(w_k - w_0)$ is balanced by a corresponding contribution $\varepsilon^{(1)}_{k0}(w_0 - w_k)$ at node k, with the result that the scheme remains conservative. The coefficients $\varepsilon^{(1)}_{k0}$ may incorporate metric information depending on local cell volumes and face areas, and can also be adapted to gradients of the solution. It is shown in reference 15 that the addition of properly controlled differences along edges can be used to assure a positivity condition on the coefficients of the semi-discrete scheme, which will prevent growth in the maximum norm and inhibit oscillations is the solution.

Formula (5) is no better than first order accurate unless the coefficients are proportional to the mesh spacing. A more accurate scheme is obtained by recycling the edge differencing procedure. After first setting
$$E_0 = \sum_k (w_k - w_0) \tag{6}$$
at every mesh point, one then sets
$$D_0 = - \sum_k \varepsilon^{(2)}_{0k} (E_k - E_0) \tag{7}$$
An effective scheme is produced by blending formulas (5) and (7), and adapting $\varepsilon^{(1)}_{0k}$ to the local pressure gradient. This is accomplished by calculating
$$P_0 = \sum_k \left| \frac{P_k - P_0}{P_k + P_0} \right|$$
at every mesh point, and then taking $\varepsilon^{(1)}_{0k}$ proportional to max (P_0, P_k). Formulas of this type have been found to have good shock capturing properties, and the required sums can be efficiently assembled by loops over the edges.

4. Integration to a Steady State

The discretization procedures of Sections 2 and 3 leads to a set of coupled ordinary differential equations, which can be written in the form
$$\frac{dw}{dt} + R(w) = 0 \tag{8}$$
where w is the vector of the flow variables at the mesh points, and R(w) is the vector of the residuals, consisting of the flux balances defined by equation (4), together with the added dissipative terms. These are to be integrated until they reach a steady state.

For this purpose we use a multistage time stepping scheme of the same type which has proved effective in calculations on rectilinear meshes. Let w^n be the result after n steps. To advance one step Δt with an m stage scheme we set
$$w^{(0)} = w^n$$
$$w^{(1)} = w^{(0)} - \alpha_1 \Delta t \, R^{(0)}$$
$$\cdots$$
$$w^{(m-1)} = w^{(0)} - \alpha_{m-1} \Delta t \, R^{(m-2)}$$
$$w^{(m)} = w^{(0)} - \Delta t \, R^{(m-1)}$$
$$w^{n+1} = w^{(m)}$$
The residual in the (q+1)-st stage is evaluated as
$$R^{(q)} = \frac{1}{V} \sum_{r=0}^{q} \{ \beta_{qr} Q(w^{(r)}) - \gamma_{qr} D(w^{(r)}) \}$$

where $Q(w)$ is the approximation to the Euler equations and $D(w)$ represents the dissipative terms, and the coefficients β_{qr} and γ_{qr} satisfy the consistency condition that

$$\sum_{r=0}^{q} \beta_{qr} = \sum_{r=0}^{q} \gamma_{qr} = 1$$

In practice a three stage scheme in which the dissipative terms are evaluated only once has proved effective. For this scheme

$$\alpha_1 = .6, \quad \alpha_2 = .6$$
$$\beta_{qq} = 1, \quad \beta_{qr} = 0, \quad q > r$$
$$\gamma_{q0} = 1, \quad \gamma_{qr} = 0, \quad r > 0$$

Convergence to a steady state is accelerated by using a variable time step close to the stability limit at each mesh point. The scheme is accelerated further by the introduction of residual averaging [9]. At the mesh point 0 the residual R_0 is replaced by $\widetilde{R}_0$ where $\widetilde{R}_0$ is an approximation to the solution $\overline{R}_0$ of the equation

$$\overline{R}_0 + \sum_k \varepsilon(\overline{R}_0 - \overline{R}_k) = R_0 \qquad (9)$$

in which the sum is over the nearest neighbors. This is similar to the weighted average appearing in the Galerkin method, but with the opposite sign for the coefficient ε, leading to an increase in the permissible time step instead of a reduction. In practice it has been found effective to obtain $\overline{R}$ by using two steps of the Jacobi iteration

$$\widetilde{R}_0^{(m)} + \sum_k \varepsilon(\widetilde{R}_0^{(m)} - \widetilde{R}_k^{(m-1)}) = R_0 \qquad (10)$$

starting from $\widetilde{R}_0^{(0)} = R_0$.

5. Mesh Generation

The triangulation procedure will connect an arbitrary collection of points to form a tetrahedral mesh. If the aircraft surface is adequately defined, we can introduce the aircraft into some pre-defined cloud of points, remove all points lying inside the aircraft structure, and then connect up the remaining points including a prescribed set of points lying on the aircraft surface. In the present version of our code we have chosen to make use of existing mesh generation techniques to create a cloud of points around the wing/body/tail/fin combination, and a further cloud of points around each nacelle.

A mesh for the wing/body/tail combination is generated by the procedure used in FLO59 [9,20]. This starts with a C-mesh around the wing which is generated by the introduction of sheared parabolic coordinates. This is accomplished in two stages. First we define a parabolic mapping which unwraps the wing to a shallow bump above a half plane. Let $\underline{X} = (X,Y,Z)$ be a point in the mapped space corresponding to $\underline{x}$ in the physical space. The unwrapping transformation is

$$\underline{X} = P_w \underline{x}$$

where P_w is defined by

$$x - x_0(z) = X^2 - Y^2$$
$$y - y_0(z) = 2XY$$
$$z = Z$$

and $\underline{x}_0$ is a point just inside the wing leading edge. The bump is then removed by a shearing transformation. Let $Y_w(X,Z)$ be the surface of the wing in mapped space. We define a shearing S_w taking $X' = (X', Y', Z')$ to $\underline{X}$ by the transformation

$$\underline{X} = S_w \underline{X}'$$

where S_w is defined by

$$X = X'$$

$$Y = Y' + Y_w(X,Z)$$
$$Z = Z'$$

This maps the half space $Y' \geq 0$ onto the region in $\underline{X}$ - space above Y_w. The C mesh is then generated by introducing rectangular coordinates in the half-space, and reversing the transformations.

The mesh around a combination of a wing plus body is generated by introducing a further transformation which maps an arbitrary shaped body into the symmetry plane, $z = 0$. This mapping can be constructed as a combination of a Joukowski mapping plus a shearing. This sequence of operations will generate a mesh that conforms with the body surface but such that the crest line of the body is not necessarily aligned with any mesh line. This deficiency is rectified by deforming the mesh lines in the mapped space to ensure that the resulting mesh is completely aligned with the body surface. The extension of these ideas to include a tail and fin follows the same principle of first utilizing a mapping to simplify the configuration, fitting a mesh in the mapped space and then mapping back to obtain the mesh in physical space.

Separate meshes are generated for each nacelle, again using a combination of unwrapping plus shearing. In this case we define a mapping P_N by the conformal transformation

$$\xi = z - e^{-z}$$

where $z = x + iy$ and $\xi = X + iY$. Here x is a coordinate aligned with the nacelle axis, and y is the radial coordinate corresponding to a cylindrical coordinate system such that $y = 0$ is the nacelle axis. If the nacelle is not axisymmetric we take the axis of the cylindrical coordinate system to be an approximate center line through the nacelle. The z coordinates are scaled so that $y = \pi$ corresponds to a cut inside the nacelle section with the point $(0, i\pi)$ just inside the section leading edge. The mapping defined above is applied to each nacelle section, transforming the space around the nacelle onto the space inside a deformed cylinder with the nacelle surface mapped to the cylinder surface. A shearing transformation can now be combined with the inverse of the mapping to generate a mesh that is aligned with the nacelle surface. A straightforward extension of the sequential mapping procedure can be used to accommodate a center body. Finally we can generate points around the pylons by treating each pylon as an isolated wing and using the mapping sequence that has previously been described.

6. Delaunay Triangulation

If the set of points is denoted by $\{P_i\}$, the Voronoi neighborhood of the point P_i is defined as the region of space

$$V_i = \{x \mid d(x,P_i) < d(x, P_j) \text{ for all } i \neq j\}$$

Here x is a point in three dimensional Euclidean space and d is the Euclidean metric. Each such region V_i is the intersection of the open half spaces bounded by the perpendicular bisectors of the lines joining P_i to each of the other P_j. The regions are thus convex polyhedra and, in general, four such regions meet at each vertex of the Voronoi diagram. We refer to regions that have common boundary faces as contiguous and likewise denote the points associated with two such regions as contiguous points. For each vertex of the Voronoi diagram we can join the four contiguous points, which have that vertex in common, by four planes to form a tetrahedron. The aggregate of tetrahedra forms the unique triangulation of the convex hull of points $\{P_i\}$ known as the Delaunay triangulation. Each Voronoi vertex is the circumcenter of the tetrahedron with which it is associated, and no point lies within the sphere that circumscribes the tetrahedron. This property ensures that the aspect ratio of the tetrahedra is reasonable and, in some sense, leads to an optimum triangulation for a given distribution of points.

The computation of the Voronoi diagaram and its associated triangulation has received considerable attention recently [18-19]. The algorithm used here is based on Boywer's method [19]. As Boywer notes, it is possible to record the structure of the triangulation by constructing two lists for each vertex in the structure. Each list has four entries: the first contains the forming points of the tetrahedron associated with the vertex and the second list holds the addresses of the neighboring vertices. The process is sequential: each new point is introduced into the existing structure which is broken and then reconnected to form a new Delaunay triangulation. When a new point is introduced into the existing triangulation, it is first necessary to identify a vertex of the Voronoi diagram that will be deleted by the new point. As the vertex at the circumcenter of the tetrahedron in which the point lies must necessarily be deleted, we are assured that at least one deleted vertex can be identified. Next we look at the neighbors of the deleted vertex for other vertices of the Voronoi diagram that may be deleted. We continue the tree search, creating a list of deleted vertices until all deleted vertices have been identified. From the list of deleted Voronoi vertices, we can determine the neighboring contiguous vertices in the undeleted set. Each point lying on the interface with the deleted region is joined to the new point. The deleted region is necessarily simply connected and star shaped. The new tetrahedra thus formed will exactly fill the deleted region and, moreover, will also satisfy the Delaunay criterion. It remains to label the new Voronoi neighborhoods and revise the lists that record the data structure.

Our strategy is to triangulate the entire space including the interior of the aircraft as well as the exterior. It is then important to identify interior tetrahedra correctly, as these must be removed before carrying out the flow calculation. Furthermore, it is necessary to prevent connections from exterior points breaking through the aircraft surface. We start the triangulation by introducing the outer boundary and then the aircraft surface points, component by component. After all the surface points have been introduced the interior tetrahedra are identified. Subsequently, if the insertion of a new point would cause a reconnection penetrating the surface, that point is rejected from the triangulation. This will occur if the point lies inside the DeLaunay sphere of an interior tetrahedron. To allow the introduction of points close to the surface it is therefore essential to make sure that the DeLaunay spheres of all the interior tetrahedra are sufficiently small. After the initial triangulation of the surface points we check the size of the DeLaunay spheres. Then, if any of these exceed a predetermined threshold, we introduce additional surface points until no excessively large spheres remain before proceeding to the introduction of the flow field points.

7. Results

In Figure 5 we show the result of a transonic flow calculation for a Boeing 747-200 flying at Mach .84 and an angle of attack of 2.73 degrees. The result is displayed by computed pressured contours on the surface of the aircraft. Flow is allowed through the engine nacelles which are modelled as open tubes. The mesh contains 24685 points and 132793 tetrahedra. The calculation was performed at Cray Research on a Cray XMP 216: the complete calculation took 3924 seconds. Of these 1448 seconds were spent in generating the mesh points and triangulating them. The remaining 2476 seconds were spent in the flow computation, which was performed with 400 cycles of the three stage scheme. Implicit smoothing with a smoothing parameter $\varepsilon = 1$ allowed the use of time steps corresponding to a nominal Courant number of 5. The number of supersonic points was frozen after 200 cycles, and the average residual was reduced from $.335 \times 10^2$ to $.161 \times 10^{-3}$ after 400 cycles. Although the mesh is fairly coarse, the significant features of the flow are evident, including the interference effects of the wing and tail on the body, and the mutual interference of the wing, nacelle and pylon.

Calculations with this number of mesh points require slightly more than 8 million words of memory. Within the limit of 16 million words available on a Cray XMP 216 it should be possible to introduce nearly twice as many mesh points to produce a mesh with about 1/4 million tetrahedrons. This should be sufficient to resolve the main features of the flow over the complete configuration. Eventually, in order to provide a detailed representation of the aircraft, we anticipate the need to increase the number of mesh points by a factor of between five and ten. This will require access to machines with a much larger memory, such as the Cray 2.

8. Conclusion

The results for the Boeing 747 clearly establish the feasability of our approach. We are now pursing the development of a variety of **improvements and extensions** of the method. These include:

(1) Vectorization

Vectorization of the main loops has already been achieved by separating the cells, faces and edges into groups such that no vertex at which contributions are being accumulated is referred to more than once in each group. Using this procedure, rates of computation ranging from 17-38 megaflops have been realized on a Cray XMP computer, depending on the mesh. These variations stem from variations in the sizes of the groups and the associated vector lengths. The efficiency can be improved by making sure that no group is too small. The analysis of the associated sorting problems leads to some general map coloring problems: for example, what is the minimum number of colors needed to color the tetrahedra in such a way that tetrahedra meeting at the same vertex do not have the same color.

(2) Improved Distribution of Mesh Points

The DeLaunay triangulation procedure connects an arbitrary cluster of points to form a tetrahedral mesh. It can be anticipated, however, that the accuracy will be improved by ensuring a favorable distribution of the points, with sufficient concentration in the neighborhood of the surface, and particularly in critical regions such as the pylon wing intersection. The present mesh generating procedure needs to be improved to provide better control of the size and aspect ratio of the tetrahedra.

(3) Adaptive Mesh Refinement

The unstructured tetrahedral mesh provides a natural setting for the introduction of an adaptive mesh refinement procedure in which additional mesh points are inserted in regions where there are rapid variations in the flow, or an indication of relatively large discretization error. This provides a method of reducing the thickness, for example, of a computed shock layer. The promise of this approach has already been demonstrated in the work of Lohner, Morgan and Peraire [21], and Holmes and Lamson [22].

(4) Multigrid Acceleration

It should be possible to make a further reduction in the cost of the flow calculation by using multiple grids to accelerate the convergence to a steady state. Since the meshes are unstructured, no simple relationship can be assumed between a coarse and a fine mesh, and rather complex procedures must be used to transfer data between the meshes.

(5) Extension to Navier Stokes Equations

By using the weak form, equation (2), the viscous terms of the Navier Stokes equations can rather easily be approximated within the present framework. Then, as a result of the integration by parts, only first derivatives of the velocities are needed to evaluate the rate of strain and stress tensors. These may be taken as constant in each tetrahedron, consistent with the assumption of linear variation in each element. A new version of the program containing additional subroutines to evaluate the viscous terms is currently under development.

(6) Simulation of Engine Power Effects

The present model allows free flow through the engine nacelles. A more realistic simulation can be achieved by introducing source terms to represent the engine power effects.

Acknowledgments

Most of our work on the airplane computer program has been carried out on computers belonging to the Cray Research Corporation. We should like to thank Cray Research for providing us with access to their computers and we are particularly grateful to Kent Misegades for his help and support. Substantial financial support for our work has been provided by the IBM Corporation, the Office of Naval Research, and the NASA Langley Research Center.

References

1. Hess, J. L. and Smith, A. M. O., "Calculation of Non-Lifting Potential Flow About Arbitrary Three-Dimensional Bodies", Douglas Aircraft Report, ES 40622, 1962.

2. Rubbert, P. E. and Saaris, G. R., "A General Three Dimensional Potential Flow Method Applied to V/STOL Aerodynamics", SAE Paper 680304, 1968.

3. Murman, E.M. and Cole, J.D., "Calculation of Plane Steady Transonic Flows", AIAA Journal, Vol. 9, 1971, pp. 114-121.

4. Jameson, Antony, "Iterative Solution of Transonic Flows Over Airfoils and Wings, Including Flows at Mach 1", Comm. Pure. Appl. Math, Vol. 27, 1974, pp. 283-309.

5. Jameson, Antony and Caughey, D. A., "A Finite Volume Method for Transonic Potential Flow Calculations", Proc. AIAA 3rd Computational Fluid Dynamics Conference, Albuquerque, 1977, pp. 35-54.

6. Bristeau, M. O., Pironneau, O., Glowinski, R., Periaux, J., Perrier, P., and Poirier, G., "On the Numerical Solution of Nonlinear Problems in Fluid Dynamics by Least Squares and Finite Element Methods (II). Application to Transonic Flow Simulations", Proc. 3rd International Conference on Finite Elements in Nonlinear Mechanics, FENOMECH 84, Stuttgart, 1984, edited by J. St. Doltsinis, North Holland, 1985, pp. 363-394.

7. Jameson, A., Schmidt, W., and Turkel, E., "Numerical Solution of the Euler Equations by Finite Volume Methods Using Runge-Kutta Time Stepping Schemes", AIAA Paper 81-1259, AIAA 14th Fluid Dynamics and Plasma Dynamics Conference, Palo, Alto, 1981.

8. Ni R.H., "A Multiple Grid Scheme for Solving the Euler Equations", Proc. AIAA 5th Computational Fluid Dynamics Conference, Palo Alto, 1981, pp. 257-264.

9. Jameson, Antony, and Baker, Timothy J., "Solution of the Euler Equations for Complex Configurations", Proc. AIAA 6th Computational Fluid Dynamics Conference, Danvers, 1983, pp. 293-302.

10. Jameson, Antony, "Multigrid Algorithms for Compressible Flow Calculations", Second European Conference on Multigrid Methods, Cologne, October 1985, Princeton University Report MAE 1743.

11. Pulliam, T.H., and Steger, J.L., "Recent Improvements in Efficiency, Accuracy and Convergence for Implicit Approximate Factorization Algorithms", AIAA Paper 85-0360, AIAA 23rd Aerospace Sciences Meeting, Reno, January 1985.

12. MacCormack, R.W., "Current Status of Numerical Solutions of the Navier-Stokes Equations", AIAA Paper 85-0032, AIAA 23rd Aerospace Sciences Meeting, Reno, January 1985.

13. Lohner, R., Morgan, K., Peraire, J. and Zienkiewicz, O.C., "Finite Element Methods for High Speed Flows", AIAA Paper 85-1531, AIAA 7th Computational Fluid Dynamics Conference, Cincinnati, Ohio, July 1985.

14. Jameson, A., and Mavriplis, D., "Finite Volume Solution of the Two Dimesional Euler Equations on a Regular Triangular Mesh, AIAA Paper 83-0435, AIAA 23rd Aerospace Sciences Meeting, Reno, January 1985.

15. Jameson, A., Baker, T.J. and Weatherill, N.P., "Calculation of Inviscid Transonic Flow over a Complete Aircraft", AIAA Paper 86-0103, AIAA 24th Aerospace Sciences Meeting, Reno, Nevada, January, 1986.

16. Augenbaum, J.M., "A Langrangian Method for the Shallow Water Equations Based on a Voronoi Mesh-One Dimensional Results", J. Comp. Physics, Vol. 53, No. 2, February 1984.

17. McCartin, B., "Discretization of the Semiconductor Device Equations", in

New Problems and New Solutions for Device and Process Modeling, pp.72-82, Bode Press, 1985.

18. Watson, D.F., "Computing the n-Dimensional Delaunay Tessellation with Application to Voronoi Polytopes", The Computer Journal, Vol. 24, No. 2, pp. 162-166.

19. Bowyer, A., "Computing Dirichlet Tessellations", The Computer Journal, Vol. 24, No. 2, pp. 162-166.

20. Baker, T.J., "Mesh Generation by a Sequence of Transformations", to appear in Applied Numerical Mathematics, December, 1986, Princeton University Report MAE 1739.

21. Lohner, R., Morgan, K., and Peraire, J., "Improved Adaptive Refinement Strategies for Finite Element Aerodynamic Configurations", AIAA Paper 86-0499, AIAA 24th Aerospace Sciences Meeting, Reno, January 1986.

22. Holmes, D.G., and Lamson, S.H., "Adaptive Triangular Meshes for Compressible Flow Solutions", First International Conference on Numerical Grid Generation in Computational Dynamics, Landshut, W. Germany, July 1986.

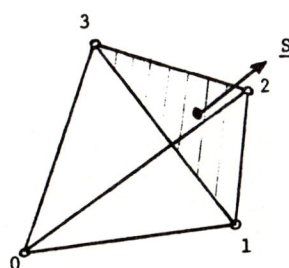

Figure 1. One tetrahedron of the control volume centered at node 0.

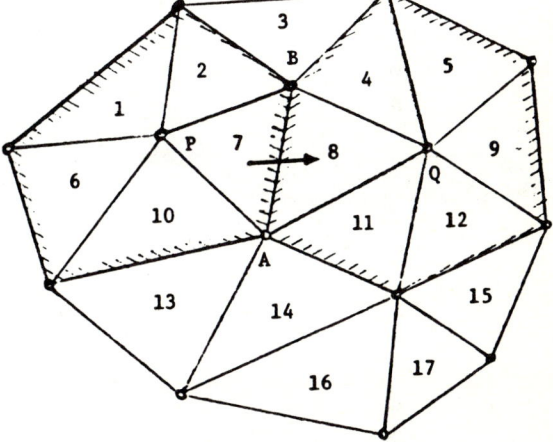

Figure 2. A triangular mesh in 2 dimensions: The control volume at P is the union of triangles 1, 6, 10, 7 and 2, while that at Q is the union of triangles 4, 8, 11, 12, and 9. The flux across the edge AB is from the control volume at P to the control volume at Q.

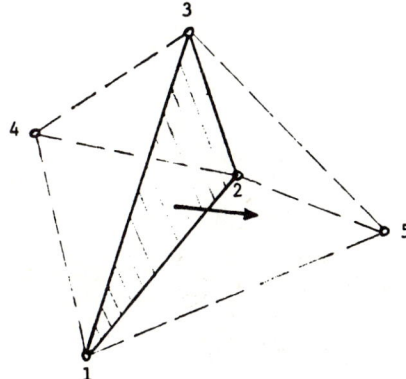

Figure 3. Flux through face defined by nodes 1, 2 and 3 is out of the control volume centered at node 4 and into the control volume centered at node 5.

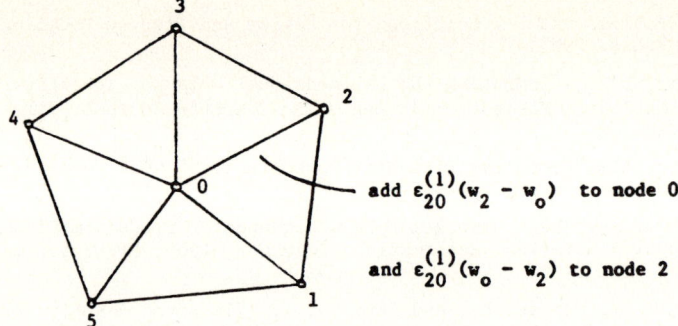

Figure 4. Construction of dissipation from differences along edges in a two dimensional mesh.

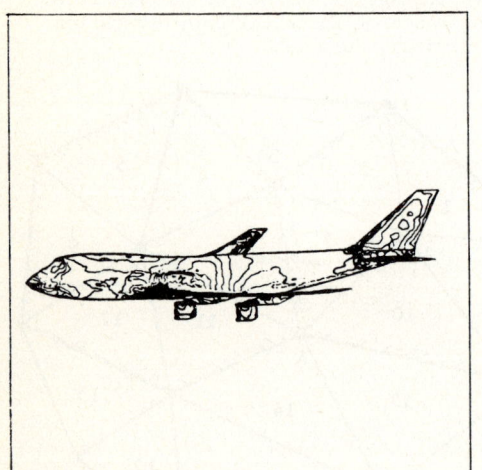
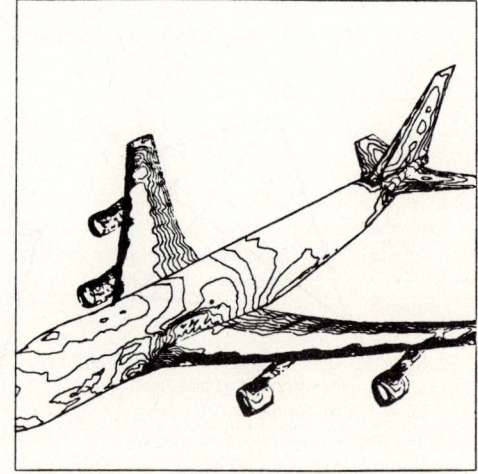

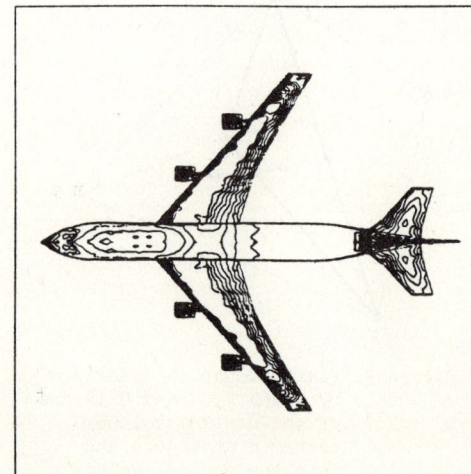

Figure 5. Surface Pressure Contours

ON THE CONVERGENCE OF PARTICLE METHODS APPLIED TO
THE EULER AND FREE SURFACE EQUATIONS

A. JAMI, M. KERMAREC
ENSTA - CNRS*
Chemin de la Hunière
91120 PALAISEAU - France

1. INTRODUCTION

The purpose of particle methods is the approximate solution of some hyperbolic systems. They are currently used to solve the equations for incompressible fluid flows and, in this case, high Reynolds number unsteady flows can be modelized which is a still open field in many respects ...

The wake behind a cylinder has been one of the first problems investigated by a particle method [1] ; however, a more academic problem, that is the evolution of a tangential velocity line of discontinuity is still studied [2]. More sophisticated methods, on the basis of the original work, have been worked out to simulate more realistic situations [3, 4, 5].

Only recently these particle methods have been considered as reliable approximations of the Euler and Navier-Stokes equations and some theoretical results [6, 7, 8] are now available which support previously acknowledged numerical difficulties. Nevertheless when looking over the large amount of papers dedicated to this subject, very few numerical studies of particle methods can be found.

On the one hand, this paper is an attempt to fill this lack and to give an example of a choice between various licit particle approximations. On the other hand, in the case of the 2-D Euler equations, we introduce the coupling method between finite elements and an integral representation formula [9]. This coupling method brings together the advantages of the F-E-M (account for boundary conditions, order of approximation, modular programming ...) and those of Boundary Integral Equation Methods (account for behavior at infinity) while avoiding any singular kernel.

For the sake of simplicity and to compare numerical convergence with theoretical estimations, we shall only consider the case of free space flows.

Moreover a new formulation for free surface flows is shown ; the numerical solution, using particles distributed on the free surface and Fast Fourier Transform, is compared with a Localized Finite Element Solution for a steadily moving pressure patch on the free surface [10].

2. EULER EQUATIONS AND THE PARTICLE FORMULATION

  2.1. Formulation of the Problem

Let $\underline{x} = (x, y)$ denote the coordinates in the plane and t the time variable. The stream-function ψ and vorticity ω formulation of the Euler equations is :

---

* Groupe Hydrodynamique Navale, associé à l'Université PARIS VI.

$$\text{(1)} \quad \begin{cases} \Delta\psi = -\omega \\ \dfrac{d\omega}{dt} = \dfrac{\partial\omega}{\partial t} + u\dfrac{\partial\omega}{\partial x} + v\dfrac{\partial\omega}{\partial y} = 0 \\ \omega(\underline{x}\,;\,0) = \omega_o(\underline{x}) \\ \psi(\underline{x}\,;\,t) \to \psi_\infty(\underline{x}) \text{ as } |\underline{x}| \to \infty \end{cases}$$

where

$$\text{(2)} \quad u = \dfrac{\partial\psi}{\partial y} \text{ and } v = -\dfrac{\partial\psi}{\partial y}.$$

The Poisson equation in the plane yields the integral representation formula :

$$\text{(3)} \quad \psi(\underline{x}\,;\,t) = \psi_\infty(\underline{x}) - \int_{R^2} \omega(\underline{x}'\,;\,t)\, E(\underline{x}',\underline{x})\, dx'\, dy'$$

where $\underline{x}' = (x', y')$ and $E(\underline{x}', \underline{x}) = 1/2\pi \, \text{Log}\,|\underline{x}' - \underline{x}|$ is the fundamental solution of the Laplace operator at $\underline{x}$ in the plane.

By differentiating (3), and using (2), the well-known Biot and Savart formula is obtained for the velocity field $\underline{V} = (u, v)$.

From the hyperbolic transport equation for ω one defines the characteristics network by :

$$\text{(4)} \quad \begin{cases} \dfrac{\partial X}{\partial t}(\underline{a}\,;\,t) = \underline{V}(\underline{X}(\underline{a}\,;\,t)\,;\,t) \\ \underline{X}(\underline{a}\,;\,0) = \underline{a} \end{cases} ;$$

where $\underline{a}$ is a Lagrange coordinate. We thus have the explicit solution

$$\text{(5)} \quad \omega(\underline{X}(\underline{a}\,;\,t)\,;\,t) = \omega_o(\underline{a})$$

In order to solve numerically the system (1), one builds an iterative scheme which determines successively :

- the velocity field $\underline{V}$ by (2) and (3),
- the characteristics network by (4),
- the vorticity ω by (5).

2.2. Variational Formulation for the Poisson Equation

We introduce this formulation as an alternative to the use of the Biot and Savart formula ; the latter being much time-consuming when dealing with a large number of particles.

Let the time t be fixed and the vorticity function $\omega(.;t)$ be given, nonvanishing on a bounded domain 0_t and square integrable on this domain. We choose an open set Ω with regular boundary Σ, such that for example (but not necessarly) $0_t \subset \Omega$.

Denote $R : H^{1/2}(\Sigma) \to H^1_o(\Omega)$ an arbitrary extension operator and

$$a(u, v) = \int_\Omega (\text{grad } u/\text{grad } v)\, dx'\, dy' \quad ;$$

We set the variationnal problem :

$$\text{(6)} \quad \begin{cases} \text{Find } u \in H^1_o(\Omega) \text{ such that } \forall\, v \in H^1_o(\Omega) \\ a(u, v) = \int_{0_t} \omega.v\, dx'\, dy' - a(R\{(\psi_\infty - \omega * E)\big|_\Sigma\}, v) \end{cases}$$

where

$$(\omega * E)(\underline{x}) = \int_{O_t} \omega(\underline{x}'; t) \, E(\underline{x}', \underline{x}) \, dx' \, dy'$$

From the wellknown coerciveness of the bilinear form a, it is easily verified [9], that the unique solution u of this problem is such that, for any extension R, the function

$$u + R\{(\psi_\infty - \omega * E)|_\Sigma\}$$

is the restriction to Ω of the solution ψ of equation (1).

The space discretization of (6) is a classical application of a F-E-M in a bounded domain.

2.3. The Blob-Particle Approximation of the Vorticity

We introduce the set ζ_p of polynomial cut-off functions of degree p and ε the size of the vortex blob ; the initial vorticity is approximated by :

$$\omega_o(\underline{x}) = \frac{1}{\varepsilon^2} \sum_{i=1}^{N} \omega_o(\underline{X}_i) \, \zeta_p\left(\frac{|\underline{x} - \underline{X}_i|}{\varepsilon}\right)$$

where N is the number of particles with positions $(X_i)_{i=1,N}$ regularly filling 0_o.

The quality of a particle method then lie on an ideal choice of 7 parameters ;namely

. N and ε for the number and size of the blobs, p for the type of cut-off function,

. h and k for the size and order of the finite-elements,

. Δt and ℓ for the time step and order of the time discretization of the system (4).

2.4. Convergence results

These are presented, among many others, for the evolution of an initial vorticity uniformly distributed on a unit radius disk ; we use a first order FEM and an $O(\Delta t)$ time discretization of the transport equation ; thus $k = 1$ and $\ell = 1$.

The numerical study shows that

- the speed of convergence is closely related to the cut-off function used, and
- there exists an optimal choice of the parameters which gives a better level of convergence.

The first figure illustrates these two points where the error is evaluated in the norm $\ell^2(\Omega_h)$ on the FEM triangulation for the stream function.

The theoretical estimations are recovered and confirmed by the numerical results when using the Biot and Savart formula ; with the finite element solution of the Poisson equation, we can also verify that the convergence is achieved only if the mesh size h is reduced faster than the blob size ε. This can be deduced from figure 2.

Multiple other applications have been performed mainly in the case where the initial vorticity is distributed on a line, (and thus is singular) because, in those cases, the theoretical estimations are not applicable.

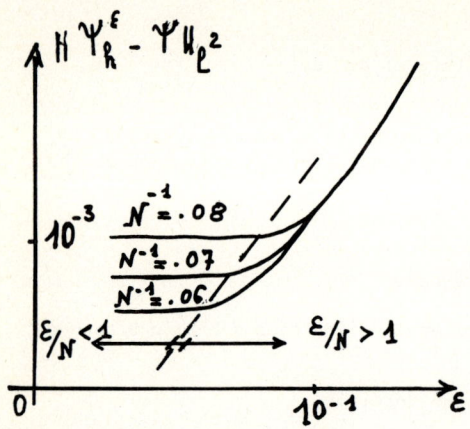 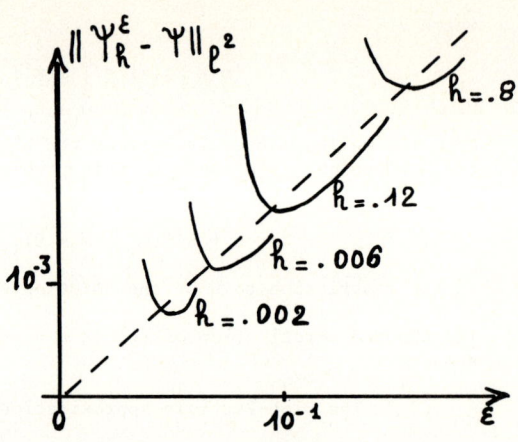

Fig. 1 : Convergence with the blob sizes for different initial discretizations.

Fig. 2 : Convergence with the blob sizes for different finite-element sizes.

3. FREE SURFACE EQUATIONS AND THE PARTICLE FORMULATION

3.1. Formulation of the Problem

We seek for the solution of a 2D irrotationnal flow of an incompressible fluid below a free surface along which the pressure is prescribed. We refer to a system of coordinates (x, z) with z upward vertical and time t. The fluid domain is unlimited in the x direction and limited by the unknown free surface $z = \eta(x ; t)$ above and a bottom $z = -H$. It is then well-known that the equations for our problem are

- the Laplace equation in the fluid domain,
- the slip condition on the bottom,
- the dynamic (Bernoulli) and kinematic conditions on the free surface.

In the sequel we denote u the perturbation of the velocity potential, ϕ its trace and χ its normal derivative on the free surface at any time ; these two functions are only dependent on the x and t variables.

From the solution of the Laplace equation in the time varying fluid domain, we know that ϕ and χ are related by the operator $K(t)$; the latter is defined in the Fourier plane (with respect to x) by

$$\hat{\chi}(\zeta ; t) = \hat{K}(\zeta ; t) \cdot \hat{\phi}(\zeta ; t)$$

when the free surface is approximated by a fixed plane (say $z = 0$), i.e. in the linearized case and where

$$\hat{K}(\zeta ; t) = \begin{cases} 2\pi\zeta \, \text{th}(2\pi H\zeta) & \text{for finite depth } H \\ 2\pi|\zeta| & \text{for infinite depth.} \end{cases}$$

Accordingly, we can denote $\psi(x ; t)$ the tangential velocity on the free surface which, in the linear case, is related to ϕ by $\hat{\psi}(\zeta ; t) = -2i\pi\zeta\hat{\phi}(\zeta ; t)$.

3.2. The Particle Formulation

The free surface conditions have to be written in term of material derivatives ; let this upper boundary be filled with particles which coordinates are $[X(x ; t); Z(x ; t)]$. From Bernoulli's dynamic condition, each particle is carrying the weight ϕ satisfying

$$\frac{d\phi}{dt} = - P/\rho + g Z + \frac{1}{2} [\psi^2 + \chi^2]$$

where ρ is the fluid density, g the gravity constant and P the prescribed pressure. For the sake of simplicity we now give only the linear kinematic conditions which read

$$\frac{dX}{dt} = \psi , \frac{dZ}{dt} = \chi .$$

Now if a set of particles are given on an initial free surface, the previous system may be solved using a time-marching algorithm.

Even in the general non-linear case, the operators K and L must be approximated in order to avoid solving the Laplace equation in a different fluid domain at each time step. The procedure we use is very simple ; first, at each time step, a projection of the potential known on the particles is made on a regular grid of the plane z = 0 using blob particle cut-off functions ; then the Fast Fourier Transform is used to evaluate ψ and χ on the grid and a reverse projection is performed to give the r.h.s. of the system.

The numerical results given below on Fig. 3 have been obtained with only 64 particles, 128 grid-points, first order approximation of the time derivative and a linear cut-off function with support on 5 grid points.

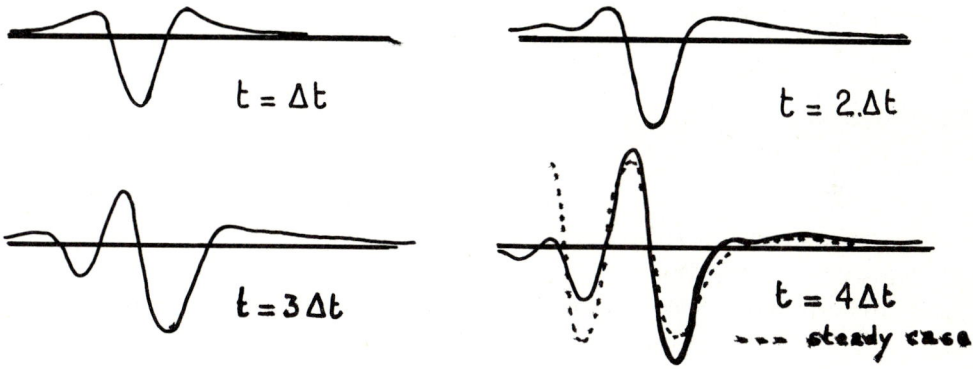

Fig. 3 : Time-history of a free surface of a 2-D canal with a flat bottom perturbated by a constant moving pressure. Froude number is 0.6 .

REFERENCES

[1] L. ROSENHEAD - Proc. Soc. of London, Ser. A, vol. 127, 1930.
[2] R. KRASNY - Courant Institute Report, to appear in the Jal of Fluid Mechanics
[3] C. REHBACH - Proc. AIAA Aer. Sc. Meet., AIAA paper 78-111
[4] J. CHORIN - SIAM Jal of Sc. Stat. Comp., vol. 1, 1980.
[5] A. LEONARD - Jal of Comp. Phys., vol. 37, 1980.
[6] O. HALD - SIAM Jal of Num. Anal., vol. 16, 1979.
[7] J.T. BEALE, A. MAJDA - Math. of Comp., vol. 32, 1982.
[8] P.A. RAVIART - CIME Course on Num. Meth. in Fluid Mech., Como (Italy), 1983
[9] M. LENOIR, A. JAMI - Comp. Meth. in Appl. Mech. and Eng., vol. 16, 1978.
[10] J. POUSIN - Workshop on Water Waves, Washington, Fév. 1986.

MULTITASKED EMBEDDED MULTIGRID FOR THREE-DIMENSIONAL FLOW SIMULATION

Gary M. Johnson, Julie M. Swisshelm,
Daniel V. Pryor and Johnny P. Ziebarth
Institute for Computational Studies
PO Box 1852
Fort Collins, Colorado 80522
U.S.A.

SUMMARY

An efficient algorithm designed to be used for Navier-Stokes simulations of complex flows over complete configurations is described. The algorithm incorporates a number of elements, including an explicit three-dimensional flow solver, embedded mesh refinements, a model equation hierarchy ranging from the Euler equations through the full Navier-Stokes equations, multiple-grid convergence acceleration and extensive vectorization and multitasking for efficient execution on parallel-processing supercomputers. Results are presented for a preliminary trial of the method on a problem representative of turbomachinery applications. Based on this performance data, it is estimated that a mature implementation of the algorithm will yield overall speedups ranging as high as 100.

INTRODUCTION

It is generally recognized that a comprehensive approach to the simulation of flows involving both complex geometries and complex physics will require powerful advanced-architecture supercomputers with very large memories. Machines capable of producing solutions to Reynolds-averaged Navier-Stokes flows over complex geometries within computing times short enough to be of design interest are expected to be available by the end of this decade. In order to use these parallel-processing supercomputers effectively, algorithms must be adapted to focus the power of multiple processing units on a single flow simulation. The purpose of this work is to contribute to the development of such algorithms. The approach selected enhances the efficiency of a robust and flexible solution procedure by implementing it on a collection of local meshes embedded in a global mesh. Either the Euler, thin-layer Navier-Stokes or full Navier-Stokes equations are solved on each mesh. The choice of model equations is determined by the nature of the flow physics to be resolved on a particular mesh. When the requirement for time accuracy is relaxed, a convergence acceleration procedure is applied simultaneously to all meshes and all model equations. The entire algorithm is explicit and is designed to perform well on computers consisting of multiple processing units, each having vector processing capability. Examples of such machines are the Cray X-MP and Cray 2.

EQUATIONS OF MOTION

The nondimensional equations of motion may be written in conservation-law form as

$$q_t = -(F_x + G_y + H_z)$$

where, for the Reynolds-averaged Navier-Stokes equations,

$$F = f - \text{Re}^{-1} p \qquad G = g - \text{Re}^{-1} r \qquad H = h - \text{Re}^{-1} s$$

while, for their thin-layer version,

$$F = f \qquad G = g \qquad H = h - \text{Re}^{-1} d$$

and, for the Euler equations,

$$F = f \qquad G = g \qquad H = h$$

where:

$$q = \begin{bmatrix} \rho \\ \rho u \\ \rho v \\ \rho w \\ E \end{bmatrix} \quad f = \begin{bmatrix} \rho u \\ \rho u^2 + p \\ \rho uv \\ \rho uw \\ (E+p)u \end{bmatrix} \quad g = \begin{bmatrix} \rho v \\ \rho uv \\ \rho v^2 + p \\ \rho vw \\ (E+p)v \end{bmatrix} \quad h = \begin{bmatrix} \rho w \\ \rho uw \\ \rho vw \\ \rho w^2 + p \\ (E+p)w \end{bmatrix}$$

$$p = \begin{bmatrix} 0 \\ \tau_{xx} \\ \tau_{yx} \\ \tau_{zx} \\ \beta_x \end{bmatrix} \quad r = \begin{bmatrix} 0 \\ \tau_{xy} \\ \tau_{yy} \\ \tau_{zy} \\ \beta_y \end{bmatrix} \quad s = \begin{bmatrix} 0 \\ \tau_{xz} \\ \tau_{yz} \\ \tau_{zz} \\ \beta_z \end{bmatrix} \quad d = \begin{bmatrix} 0 \\ \mu u_z \\ \mu v_z \\ (\lambda + 2\mu) w_z \\ \gamma \kappa \text{Pr}^{-1} e_z + (\lambda + 2\mu) w w_z \end{bmatrix}$$

$$\tau_{xx} = \lambda(u_x + v_y + w_z) + 2\mu u_x \qquad \beta_x = \gamma \kappa \text{Pr}^{-1} e_x + u\tau_{xx} + v\tau_{xy} + w\tau_{xz}$$

$$\tau_{yy} = \lambda(u_x + v_y + w_z) + 2\mu v_y \qquad \beta_y = \gamma \kappa \text{Pr}^{-1} e_y + u\tau_{yx} + v\tau_{yy} + w\tau_{yz}$$

$$\tau_{zz} = \lambda(u_x + v_y + w_z) + 2\mu w_z \qquad \beta_z = \gamma \kappa \text{Pr}^{-1} e_z + u\tau_{zx} + v\tau_{zy} + w\tau_{zz}$$

$$\tau_{xy} = \tau_{yx} = \mu(u_y + v_x), \qquad \tau_{xz} = \tau_{zx} = \mu(u_z + w_x), \qquad \tau_{yz} = \tau_{zy} = \mu(v_z + w_y)$$

Here ρ, u, v, w, p and E are respectively density, velocity components in the x-, y- and z-directions, pressure and total energy per unit volume. This final quantity may be expressed as

$$E = \rho \left[e + \frac{1}{2}(u^2 + v^2 + w^2) \right]$$

where the specific internal energy, e, is related to the pressure and density by the simple law of a calorically-perfect gas

$$p = (\gamma - 1)\rho e$$

with γ denoting the ratio of specific heats. The coefficient of thermal conductivity, κ, and the viscosity coefficients, λ and μ, are assumed to be functions only of temperature. Furthermore, λ is expressed in terms of the dynamic viscosity μ by invoking Stokes' assumption of zero bulk viscosity. Re and Pr denote the Reynolds and Prandtl numbers, respectively. Although, for simplicity, the equations of motion are presented here written in Cartesian coordinates, it is well known that their strong conservation law form may be maintained under an arbitrary space- and time-dependent transformation of coordinates.

SOLUTION METHODOLOGY

The integration scheme used here is the forward predictor - backward corrector version of the two-step Lax-Wendroff method due to MacCormack [1]. This version of MacCormack's scheme is used for convenience. Any of its many variants could also be used, as could any other one- or two-step Lax-Wendroff scheme [2]. In fact, the class of fine-grid methods with which the convergence acceleration technique described below may be applied appears to be quite large, including schemes not of Lax-Wendroff type [3]. The advantages of MacCormack's method, in the present context, are its explicit nature, simplicity and low operations count. A disadvantage is its conditional stability and the severe time-step size limitation which this imposes for viscous flows, in particular. The ill effects of conditional stability are mitigated through the use of embedded grid refinements and convergence acceleration.

The embedded-mesh technique developed for the present application is a generalization of that employed in [4] to obtain two-dimensional Euler solutions. The computational domain is divided into regions requiring grids of differing fineness and the resolution of different flow physics. At present, for simplicity, this partitioning is done *a-priori*. However, solution-adaptive gridding based on this technique is possible. Fig. 1 shows typical locations for the mesh regions employed in the computations described subsequently in this paper. Note that, where mesh lines are illustrated, their spacing is much coarser than that employed in the computations. Mesh 3, the coarsest mesh in Fig. 1, covers the entire computational domain. The Euler equations are solved on it. Mesh 2, finer than mesh 3 in all directions by a factor of two, contains the regions near the walls and the blade surface where flow can be modeled by the thin-layer form of the equations of motion. The finest mesh shown, mesh 1, contains the regions of the domain near the juncture of surfaces where all viscous terms have been retained. From this specific example, it is easy to see that quite general collections of embedded meshes may be constructed in this manner. The embedded meshes are not disjoint. Rather, given a mesh labelled m, all coarser meshes from m+1 through the coarsest mesh used in the computation underlie it. This property, together with the coarsening factor of 2, facilitate the use of the multiple-grid acceleration techniques described in [5]. The flowfield updating begins with mesh 1. After one timestep on mesh 1, mesh 2 is updated exterior to mesh 1 while convergence acceleration is applied at the mesh-2 points interior to mesh 1. Next, mesh 3 is updated exterior to mesh 2 while convergence acceleration is applied at the mesh-3 points interior to mesh 2. Updating proceeds in this fashion until the global mesh has been advanced by one timestep. Then, convergence acceleration is applied to coarsenings of the global grid. This cycle is repeated until the desired measure of convergence is satisfied. Observe that when both the basic integration scheme and the coarse-grid convergence accelerator are explicit, the algorithm is particularly easy to vectorize. Additionally, a parallel coarse-grid algorithm has been developed for more efficient execution on both vector- and parallel-processing computers, as described in [5]. Further, note that, while in this paper multiple-grid convergence acceleration is applied only to steady flow simulations, it appears that the technique may extend to time-accurate computation of some unsteady flows [6, 7].

When implementing an algorithm on a multiple instruction-multiple data machine, we are concerned with multitasking overhead and algorithm granularity. By granularity we mean the time required to execute a multitaskable code segment on a single processor. For a given overhead, the best speedup is obtained when algorithm granularity is maximal. Large granularity is usually introduced by top-down programming which exploits global parallelism in the algorithm. Bottom-up programming exploits algorithm parallelism at a low level by making many partitionings, each on small code segments, such as DO loops containing independent statements. The sequential multigrid algorithm contains many opportunities for creating small granularity parallelism but relatively few for the sort of large granularity necessary to produce good speedup in the face of non-trivial overhead. This observation, together with the desirability of non-sequential multigrid schemes for reasons of algorithm flexibility, led to the construction of the parallel multigrid algorithms mentioned above. In these algorithms, grids which are independent of one another may be updated simultaneously on separate processors. In fact, such a simple strategy may result in a poor load balance across processors because of the differing amounts of work inherent in updating grids of different coarseness. However, more refined strategies are possible. Grids may be grouped together into tasks of approximately equal work, or they may be melded into tasks with other large-grained code segments in order to equalize processor loading. Notice futher that, by multitasking large-grained structures, the vectorization potential of code within these structures remains intact.

NUMERICAL SIMULATION

As the algorithm described in this paper is designed to efficiently simulate complex flows over complete configurations, it should be tested under conditions which fully exercise its capabilities. On the other hand, excessive complexity would serve no useful purpose in the initial testing phases of the algorithm. With these considerations in mind, three-dimensional computations are being carried out for the geometry illustrated in Fig. 2, a rectilinear cascade of finite-span, swept blades mounted between endwalls. The sweep angle ranges from 0 to 26 degrees. The blade thickness to chord ratio ranges from 0.0 to 0.2. The subcritical computations are performed at an isentropic inlet Mach number of 0.5. The Mach number for the supercritical computations is 0.675. In the viscous cases, the Reynolds numbers, based on cascade gap and critical speed, span the approximate range from 8.4×10^3 to 2.0×10^5. The mesh structure on which the computations are being performed is illustrated in Fig. 1. The full Navier-Stokes equations are solved on mesh 1. The thin-layer Navier-Stokes equations are solved on mesh 2. The Euler equations are solved on mesh 3. Only steady flows are computed and convergence acceleration, as described previously, is applied. The entire algorithm is vectorized and multitasked to run on a four-processor Cray X-MP or Cray 2. Sample results for a subcritical flow over a swept blade are shown in Fig. 3.

Comparison of the embedded-mesh algorithm with a single-mesh algorithm yields the following conclusion: the accuracy of the embedded-mesh results is essentially that of a global finest mesh, while the convergence rate is like that of a global coarsest mesh. Thus far, in two-dimensional computations using the Euler and thin-layer Navier-Stokes equations and three mesh regions, embedding speedups as high as 30 have been obtained. Three-dimensional embeddings using Euler, thin-layer and full Navier-Stokes regions should produce substantially larger speedups.

Multiple-grid convergence acceleration applied to three-dimensional cases, in the absence of mesh embedding, has yielded speedups ranging from 2.5 to 4.7. It is expected that there will be some tradeoff between embedding and multigrid speedup in the complete algorithm. Vectorization of the three-dimensional algorithm without embedded meshes results in speedups ranging from 3.6 to 5.7. This range should remain about the same in the final algorithm.

Using a top-down multitasking approach, the parallel coarse-grid algorithm has been implemented on a four processor Cray X-MP, for two-dimensional cases without mesh embedding. Initially, only the coarse grids were multitasked so that the performance of parallel grids on a multiprocessor could be evaluated. Then the fine-grid computations were partitioned and multitasked, and the resultant code was integrated with the parallelized coarse grids.

For the multitasking results, performance measures are based on a comparison of multitasked code segments with their unitasked analogs. The parallel coarse-grid scheme results were obtained with a five-grid multigrid sequence length. An efficiency of nearly 90% has been obtained using two processors, but that efficiency deteriorates to 77% when four processors are used. This deterioration is a result of distributing multigrid structures containing unequal amounts of work across four processors. Results obtained from multitasking the fine-grid scheme show that the fine-grid tasks are fairly evenly balanced, and this code segment performs well on both two and four processors. Processor utilization of 90% or better is achieved. The fully multitasked two-dimensional multigrid algorithm attains efficiency levels ranging from 94% on two processors to 83% on four processors. For the three-dimensional Navier-Stokes code, the bottom-up approach is taken by using microtasking software on the Cray X-MP. Microtasking incurs relatively low overhead, which allows parallelization of very fine-grained code segments and alleviates the need for careful *a-priori* load balancing. The resulting fully microtasked three-dimensional code performance ranged from 98% efficiency on two processors to 89% on four processors.

Given that the speedups from the various categories described above are generally multiplicative in effect, it is to be conservatively estimated that a mature implementation of the algorithm will produce overall speedups ranging as high as 100.

CONCLUSIONS

An efficient algorithm designed to be used for Navier-Stokes simulations of complex flows over complete configurations has been presented.

The algorithm makes use of several elements: a robust explicit basic flow solver, locally-embedded mesh refinements, a flow simulation hierarchy ranging from the Euler equations through the full Navier-Stokes equations, an explicit multiple-grid convergence acceleration technique, and both vectorization and multitasking for efficient execution on parallel-processing supercomputers.

Results are presented for a problem representative of turbomachinery applications. These results provide grounds for optimism regarding the algorithm's future application to more challenging internal and external flows. Based on the performance data presently available, this algorithm is expected to reduce simulation times by as much as two orders of magnitude.

ACKNOWLEDGEMENTS

This research is funded in part by the NASA Ames Research Center (NCC-2-344) and the Air Force Office of Scientific Research (AFOSR-85-289). Some of the computing time has been contributed by Cray Research, Inc. All of this support is gratefully acknowledged.

REFERENCES

1. MacCormack, R.W.: AIAA Paper 69-354, 1969.
2. Johnson, G.M.: NASA TM-82843, 1982.
3. Stubbs, R.M.: AIAA Paper 83-1945, 1983.
4. Usab, W.J.: Doctoral Dissertation, Aero and Astro Dept., MIT, 1983.
5. Johnson, G.M., Swisshelm, J.M. and Kumar, S.P.: AIAA Paper 85-1508, 1985.
6. Stubbs, R.M.: Private Communication, 1983.
7. Jespersen, D.C.: AIAA Paper 85-1493, 1985.

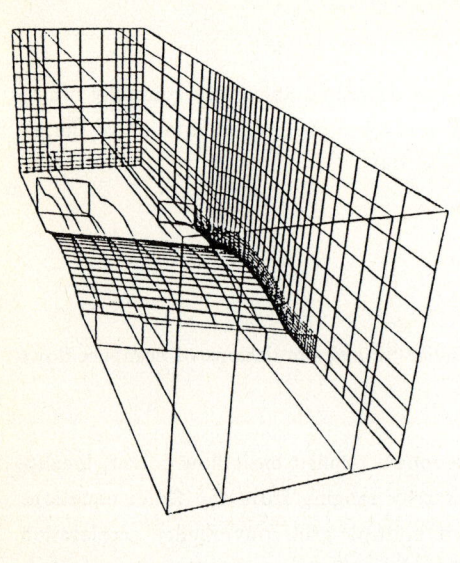

Figure 1. Typical Locations for Embedded Mesh Regions

Figure 2. Computational Domain

Flow Direction

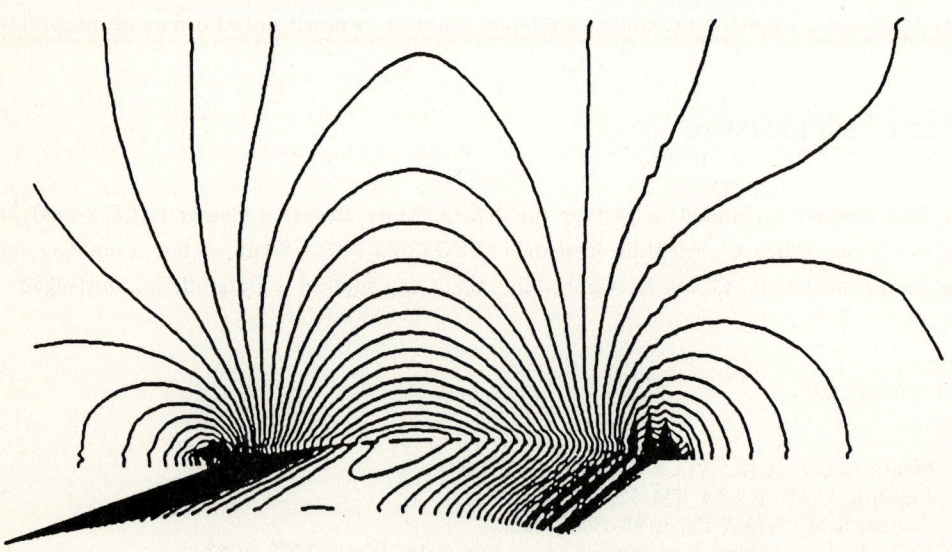

Figure 3. Isobars for Subcritical Flow over Swept Blade

NUMERICAL SIMULATION OF SOME SEPARATED FLOWS

V.F.Kamenetsky, L.I.Turchak

(117333 Moscow, 40 Vavilova str.
U.S.S.R. Academy of Sciences, Computing Centre)

Abstract. Some separated flows which arise as a result of the turbulent boundary layer interaction with shocks and expansion waves are considered. Models are formed on the basis of the inviscid approach using the Euler equations. Turbulent boundary layer effects are taken into account by means of introduction of an appropriate oncoming stream velocity profile. This model was used for the numerical study of the flow in front of the two-dimensional step generated by a supersonic stream. The calculation was carried out with the front shock fitting, and thus the subsonic part of the nonuniform velocity profile was thrown off. A separated flow behind the two-dimensional ledge was also studied. Here the profile of the oncoming stream velocity on the upper surface was given. The results obtained agree with the experimental data.

Introduction. Experimental studies of the interaction of the turbulent boundary layer with shocks or expansion waves [1,2] have shown that the flow parameters are independent of the Reynolds number. On the basis of this property numerical simulation of such flows might be realized using the Euler equations. Such indications were given in ref.[3-6] et al. In ref.[6] supersonic flows about bodies in a nonuniform oncoming stream were considered, and the results of the flow simulation ahead of the step with the boundary layer separation were also given. These preliminary results indicated the possibility of such aproach to be used for separated flows modelling.

Flow in front of the step. The oncoming stream of inviscid gas with a nonuniform velocity profile interacts with a plane step. The flow sketch is shown in Fig. 1. The velocity profile $V_\infty(y)$ corresponds to the familiar one for the turbulent boundary layer, which is taken from experiments,

$$V_\infty = V_0 \left[1 + (y/\delta)^n \right], \quad y \leq \delta.$$

The velocity value at the wall ($y = 0$) is supersonic, i.e. $V_0 > c$. The subsonic part is thrown off to use the calculation scheme with the front shock fitting. The oncoming flow pressure p_∞ = const, the enthalpy might be prescribed. The step height $h = 1$, the boundary layer thickness $\delta < 1$. The velocity value at $y > \delta$ is constant: $V_\infty = V_1$.

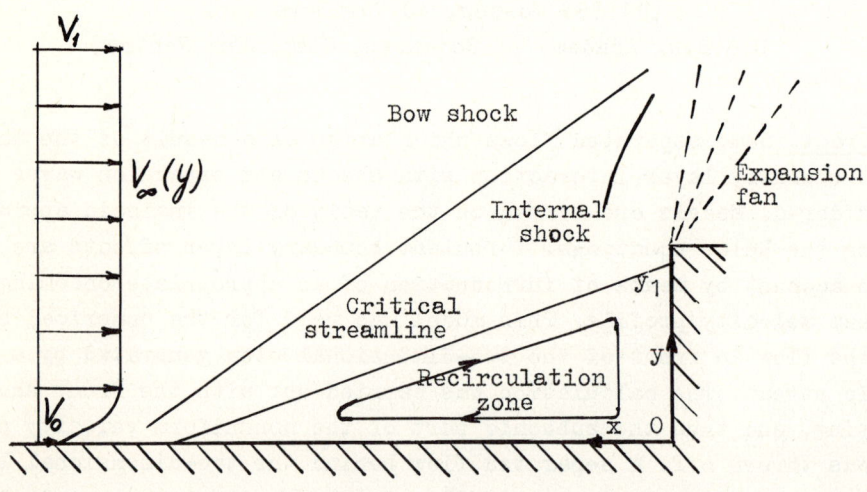

Fig. 1

The problem is solved on the basis of the Euler equations. The boundary conditions are formulated at the bow shock (Hugonio relations) and the wall (the normal velocity component $V_n = 0$). The Grid-Characteristic (GC) method [7] is used for the problem solution.
This method had been successfully used for studies of different flows, and the results were presented at some previous meetings (see [8-10]). We apply here this method for the one-dimensional gasdynamic equations such as

$$u_t + B u_x = 0. \qquad (1)$$

Here $u = (u_1, u_2, \ldots, u_N)^T$ - the unknown vector, B - a square matrix NxN with real eigenvalues.
Let us denote: Ω - the eigenvalue matrix; Λ^+, Λ^- - diagonal matrices with positive and negative eigenvalues, accordingly. Then the GC-method difference scheme might be written as follows:

$$u_k^{n+1} - u_k^n + \sigma \Omega^{-1} \Lambda^+ \Omega (u_k^n - u_{k-1}^n) + \sigma \Omega^{-1} \Lambda^- \Omega (u_{k+1}^n - u_k^n) = 0,$$
$$\sigma = \tau/h. \qquad (2)$$

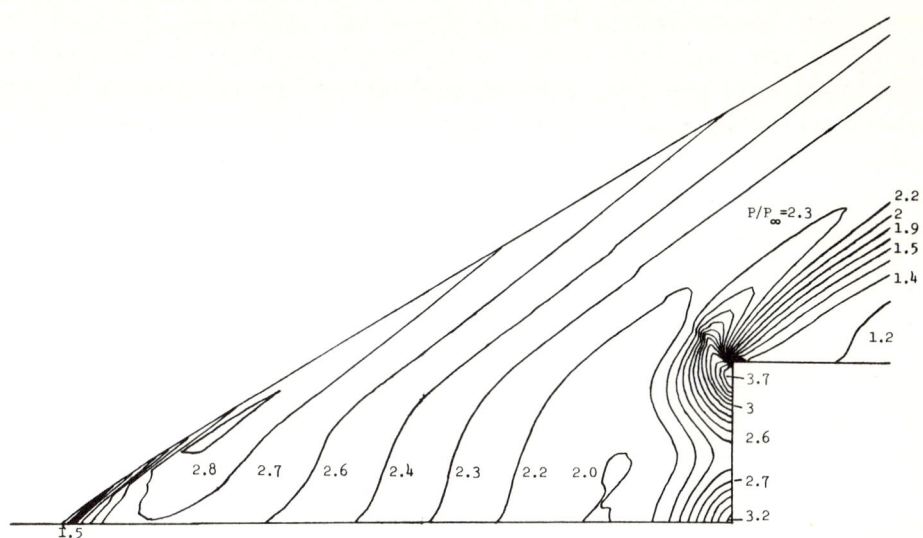

Fig.2

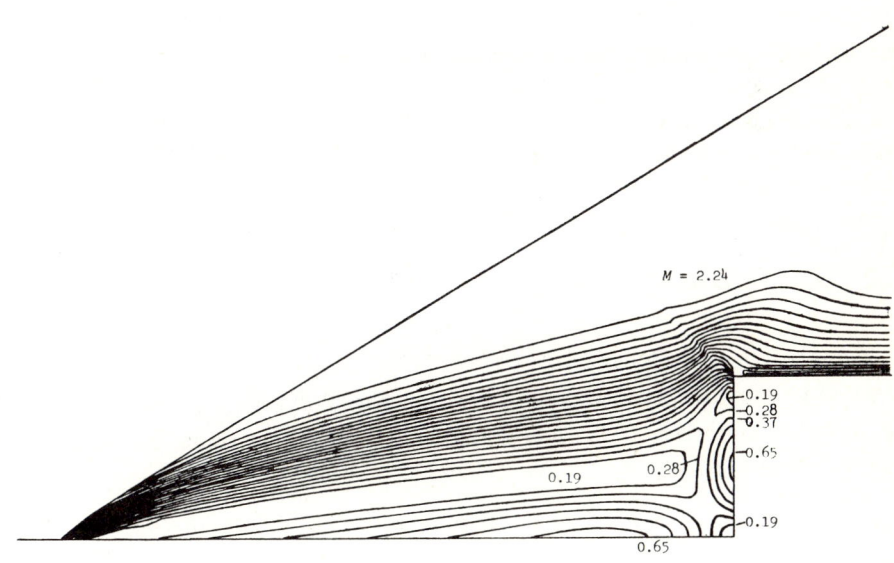

Fig.3

This scheme can be easily written for a multidimensional case.
Flow about the right-angle step was given the following parameters:
$M_0 = 1.01$, $M_1 = 3$, $n = 1/7$. The isobars pattern is shown in Fig.2. The bow shock position practically coincides with the experimental one. A recirculation zone was obtained by the calculation. We can also see expansion fan near the upper corner point. At the same time an internal (hanging) shock is expressed feebly, it is smeared out.
Lines of constant Mach number values are shown in Fig.3.
The pressure distributions along the lower surface y = 0 (a) and on the vertical wall of the step (b) are constructed in Fig.4. The most important features in p(x) distribution are the pressure plateau p_p and the unmonotonous behaviour.
Experimental data [1,2] helped to find out that the p_p value in the developed turbulent boundary layer is independent of Reynolds number and the step size. Some empiric dependences on the basis of experimental data were generated; in particular, in ref.[1] is given

$$p_p/p_\infty = 0.5 \, M_1 + 1.$$

Some additional property of such flows was determined experimentally: the distance between the point of separation of the turbulent boundary layer and the step basis is independent of M_1. It means that the shock distance can be fixed, and it helps to correct the nonuniform stream parameters to improve the suggested model. Thus this model can be considered as semiempiric.

<u>Flow behind the ledge.</u> A conservative version of the GC-method [7] might be used for the through solving of some nonlinear problems with internal discontinuities. To generate such a scheme we rewrite Eq.(2) in the form

$$u_k^{n+1} = u_k^n - \sigma B(u_{k+1/2}^n - u_{k-1/2}^n) + 0.5 \Omega^{-1} |\Lambda| \Omega \, \sigma \, (u_{k+1}^n - 2u_k^n + u_{k-1}^n),$$
$$u_{k\pm 1/2}^n = 0.5(u_{k\pm 1}^n + u_k^n). \qquad (3)$$

As we can see, an additional term in the form cu_{xx}, where $c = 0.5 h \Omega^{-1} |\Lambda| \Omega$, was put into system (1). The original equations must be presented in the divergent form,

$$u_t + F_x = 0, \quad B = F_u.$$

Then scheme (3) might be written as follows:

$$u_k^{n+1} = u_k^n - \sigma (F_{k+1/2}^n - F_{k-1/2}^n) + \sigma [c_{k+1/2}^n (u_{k+1}^n - u_k^n)/h -$$

$- c_{k-1/2}^{n}(u_k^n - u_{k-1}^n)/h]; \; F_{k\pm 1/2}^{n} = 0.5(F_{k\pm 1}^{n} + F_k^n).$

As it was shown in ref. [7], scheme (4) is stable provided the CFL condition is satisfied. This scheme ensures the absolute approximation and its computational viscosity is minimum.

Numerical simulation of the flow behind the plane edge using the two-dimensional version of scheme (4) was realized. Parameters of the velocity profile on the upper surface are: $M_o = 0$, $M_1 = 1.97$. The bottom pressure distribution is presented in Fig.5. Here curve 1 corresponds to the uniform flow ($V_o = V_1$); 2 - nonuniform one with $\quad$ = 0.08; the dotted line - experimental data from ref.[11].

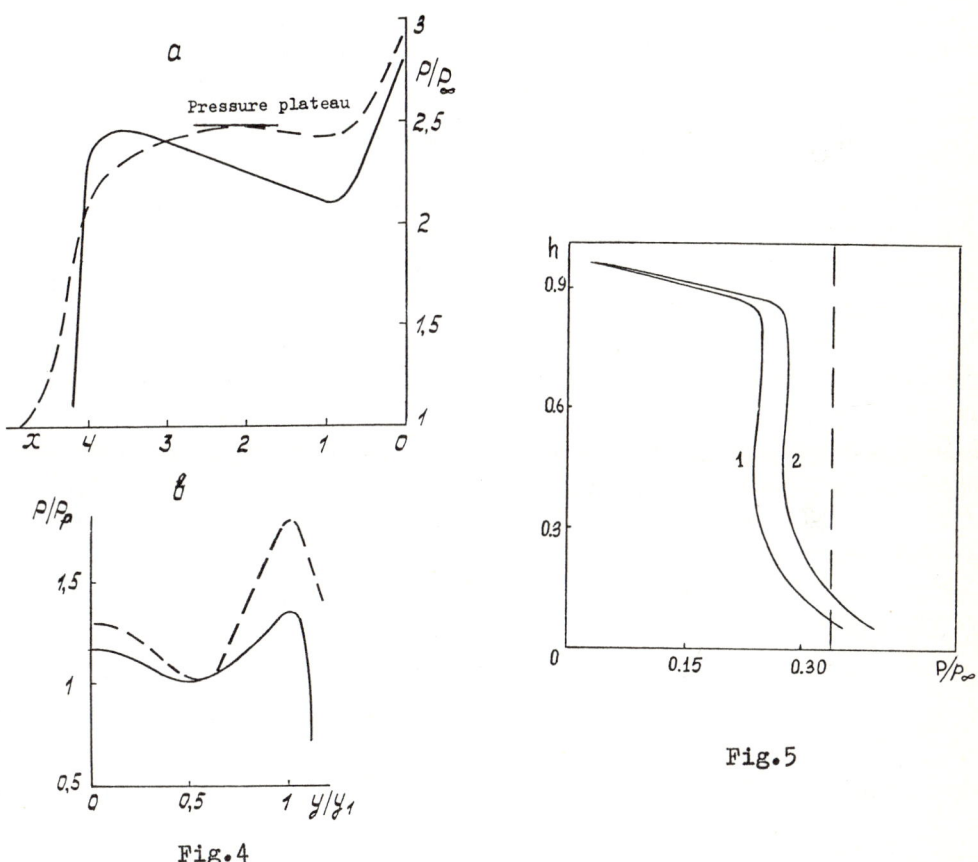

Fig.4

Fig.5

<u>Conclusion.</u> We have presented here some preliminary results of the numerical solution of certain separated flows on the basis of inviscid flow model. The model proposed is rather reasoable. It accounts for the main features of separated flows which arise by the interaction of the turbulent boundary layer with shocks.

References.

1. Zukoski E.E. Turbulent boundary-layer separation of a forward-facing step. AIAA Journal, 1967, v.5, No.10.
2. Желтоводов А.А., Шилейн Э.Х., Яковлев В.Н. Развитие турбулентного пограничного слоя в условиях смешанного взаимодействия со скачками уплотнения и волнами разрежения. ИТПМ СО АН СССР, Препринт № 28, Новосибирск, 1983.
3. Belotserkovskii O.M. New computational models in continuum mechanics. Technical Report No. AE-79-1. University of Maryland, College Park, Dept. Aerospace Engineering, 1979.
4. Белоцерковский О.М. Прямое численное моделирование переходных течений газа и задач турбулентности. В сб.: "Механика турбулентных потоков", М., "Наука", 1980.
5. Петров Г.И. Система скачков уплотнения и волн разрежения при обтекании тел сложной формы. В кн.: "Гидроаэродинамика и космические исследования", М., "Наука", 1985.
6. Каменецкий В.Ф., Турчак Л.И. Сверхзвуковое обтекание тел неоднородным потоком идеального газа. М., ВЦ АН СССР, 1982.
7. Белоцерковский О.М. Численное моделирование в механике сплошных сред. М., "Наука", 1984.
8. Belotserkovskii O.M., Kholodov A.S. Numerical investigation of some gas dynamics problems by net-characteristic method. Lecture Notes in Physics, No.90, Springer-Verlag, 1979.
9. Turchak L.I. Investigation of unsteady supersonic flows about blunt bodies. Lecture Notes in Physics, No.90, Springer-Verlag, 1979.
10. Turchak L.I., Kamenetsky V.F. Numerical simulation of unsteady flowfields near bodies in nonuniform oncoming stream. Lecture Notes in Physics, No.218, Springer-Verlag, 1985.
11. Тагиров Р.К. Экспериментальное исследование отрывных течений за плоским уступом при $M_1 = 1.97$, Известия АН СССР, МЖГ, 1969, № 4.

NUMERICAL STUDY OF THE ENTRANCE FLOW OF A CIRCULAR PIPE

    Hidesada KANDA
        Science Institute, IBM Japan, Ltd.
    and Koichi OSHIMA
        The Institute of Space and Astronautical Science

Summary: Two-dimensional, time-dependent computational scheme has been devised for determining the flow development and the corresponding pressure drop in the entrance region of a circular pipe at the Reynolds numbers based on the pipe diameter of 10, 100, 2000 and 10000. An iterative, stream-function vorticity formulation was applied, utilizing a mesh system in which the axial grid size is nearly proportional to the Reynolds number. The velocity field are compared with the previous analysis. The effects of the Reynolds number and the superposed disturbances on the transition from laminar to turbulent flow were numerically simulated. Moreover, the transition was experimentally confirmed to occure only in the entrance region.

#1. Entrance Length and Transition Length

The entrance length zep is defined as the distance from the inlet to the point where the flow develops fully and the velocity distribution becomes perfectly parabolic. The measured values of zep have been observed to be proportional to the Reynolds number (1). The dimensionless entrance length, which is denoted as Lep, is defined as the entrance length divided by the diameter of the pipe and by the Reynolds number based on the pipe diameter: Lep=zep/(D*Re), where D is the diameter of the pipe and the subscript p means 'parabolic'. The dimensionless entrance length is supposed to be within the range of 0.02875 by L. Shiller (2), 0.058 by H. L. Langhaar (3) and 0.075 by A. K. Mohanty (4).

The transition length zet is defined as the distance from the inlet to the point where the transion from laminar to turbulent flow occurs. The measured values of zet are about 20 to 30 times diameters of the tube diameter by O. Reynolds (5), and less than 17.5 to 35 by M. Arakawa (6). The dimensionless transition length Let is defined in the same way of Lep: Let=zet(D*Re), where t means 'transition'. The dimensionless transition length for zet=30 diameters is 0.013 at Re=2300 and is fairly shorter than the dimensionless entrance length.

#2. Numerical Simulation

2.1 Mesh system and basic equations

For the entrance region described above we consider the 2-dimensional, unsteady flow of an incompressible Newtonian fluid with constant viscosity and density. We neglect gravity and external forces. The dimensionless forms of the stream-function vorticity equation and the Poisson equation are written in the cylindrical coordinates (7). The whole fluid particles start moving downstream with uniform velocity in a pipe at the initial time. The no-slip boundary condition is on the wall and any vorticity does not exist at the inlet and on the center line of a circular pipe. Moreover, the outlet condition is given by extrapolation.

Fig. 1 shows the mesh system formed by the orthogonal lines separated by a constant space increment Δz on the z-axis and Δr on the r-axis, where z and r are the axial and radial coordinates, respectively. The axial point i takes an integer value between 1 (inlet) and I0 (outlet) and the radial point j is between 1 (center line) and J0 (wall).

Table 1 gives I0, J0, the values of the time increment and the number of time step N at which the steady state of the flow is obtained, with respect to the Reynolds number.

As the aspect ratio increases with the Reynolds number, the derivatives with respect to z decreases and become ineffective in the basic equations. However, in the range of 1 and 500, the derivatives with respect to z were meaningfully effective (7). Both of the stream-function vorticity and the pressure drop are calculated by the Gauss-Seidel iterative method, respectively. The results of the numerical simulationin are shown in the dimensionless axial distance from the inlet: ($z^* = z/(D*Re)$).

Table 1
Values of Mesh System

Re	I0	J0	$\Delta z/\Delta r$	Δt	N(xΔt)
10	100	21	1	0.00278	2000
100	100	21	10	0.02621	8000
2000	150	21	100	0.04762	8000
10000	150	21	500	0.09901	20000

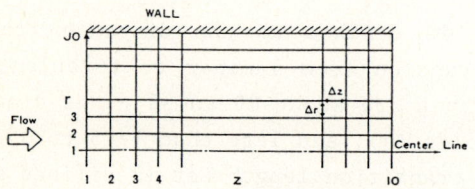

Fig. 1 Mesh System

2.2 Velocity profile

Figs. 2-4 show the results for the velocity distribution in a steady state. At Reynolds number of 10, in the central $0 < r < 0.4$ and $Z^* < 0.05$ region, the axial dimensionless velocities are fairly smaller than those of Reynolds numbers 100 and 10000. However, the flow field in the entrance region develops completely fully near $z^*=0.1$ and far downstream. At the center line $r=0$, $u/u0$ is 1.9977 (99.89% of fully developed value) at Re=10, 1.9988 (99.94%) at Re=100, 1.9929 (99.65%) at Re=2000 and 1.9846 (99.23%) at Re=10000. In Fig. 2● and x are the results of the numerical solution on the center line by J.S. Vrentas (8) and Y. Koyari (9), respectively. J.S. Vrentas calculated in the range of Reynolds numbers 0, 1, 50, 150, and 250. Y. Koyari calculated at Reynolds number of 60. The comparison shows that the axial velocity distributions and the velocity development are nearly same for Reynolds numbers more than 50. Table 2 shows the some results of the numerical solution. We can say approximately that the dimensionless entrance length for 98% and 99% velocity development are 0.045 and 0.055, respectively, for Reynolds numbers above 50.

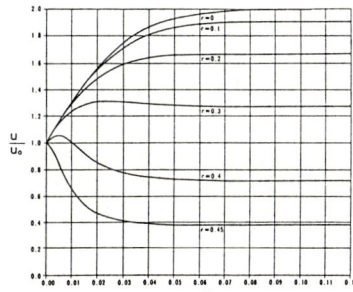

Fig. 2 Velocity Distribution Re=10, steady state

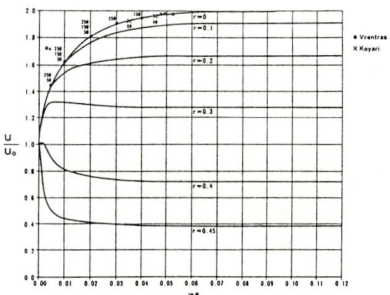

Fig. 3 Velocity Distribution Re=100, steady state

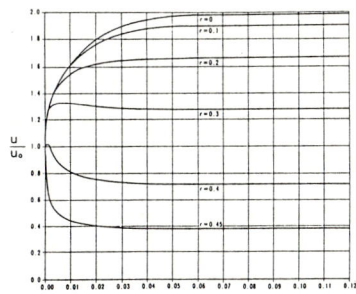

Fig. 4 Velocity Distribution Re=10000, steady state

Table 2   Entrance Length

Re	percent of fully developed value		Bibliography
	98%	99%	
1		0.33	Vrentas
10	0.0583	0.0681	*Kanda
50		0.047	Vrentas
60	0.045		Koyari
100	0.0437	0.0544	*Kanda
150		0.048	Vrentas
250		0.0535	Vrentas
2000	0.0440	0.0546	*Kanda
10000	0.0438	0.0555	*Kanda
		0.058	Langhaar

2.3 Effect of disturbance upon velocity distribution

In this study, 1.2 times singular stream-function were given at two points, $z=(0*\Delta z, 1*\Delta z)$ and $r=0.25$. Then, it was simulated whether the singularity of the velocity distribution would be amplified or would be damped as the time step develops. Figs. 5-6 are the results of the numerical simulation at Re=2000; Figs. 7-8 are the results at Re=10000. The singularity of the velocity distribution seems to be damped smoothly at Re=2000. At Re=10000, the singularity seems to be damped after $z*=5*\Delta z$. However, the strong deformation of the velocity distribution at $z=2*\Delta z$ still remains.

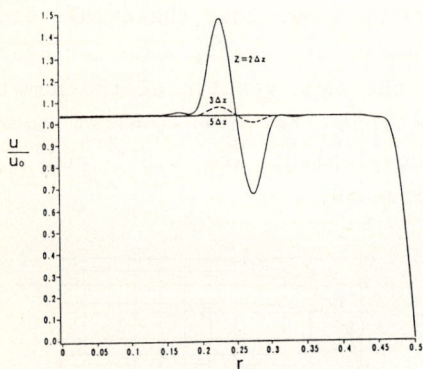

Fig. 5 Velocity Distribution with Disturbances Re=2000, T=10*.t

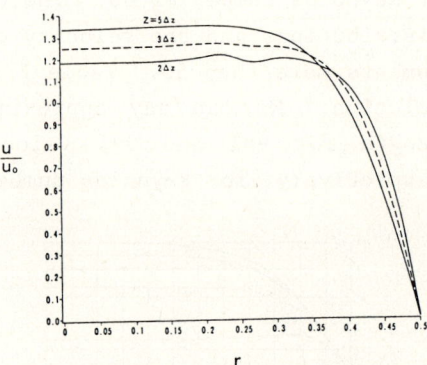

Fig. 6 Velocity Distribution with Disturbances Re=2000, T=500*.t

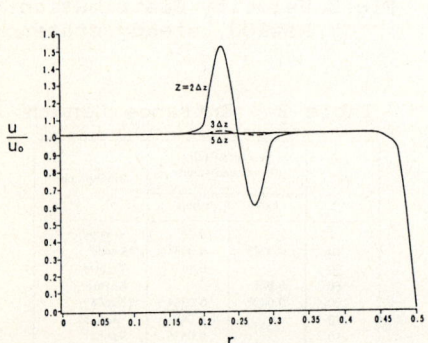

Fig. 7 Velocity Distribution with Disturbances Re=10000, T=10*.t

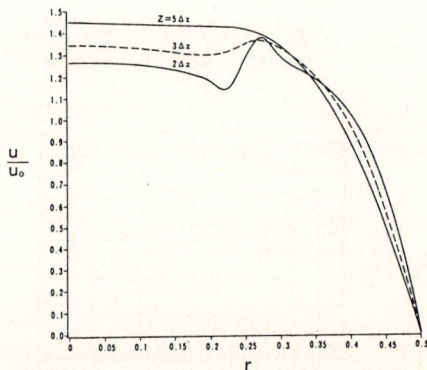

Fig. 8 Velocity Distribution with Disturbances Re=10000, T=4500*.t

#3. Experimental Apparatus and Experimental Results

The experimental apparatus is in Figs. 9-10. Fluid enters into a smooth circular pipe of 3 cm in diameter and 300 cm in length from a very large container. The diameter of the inlet of the bellmouth is 8.7 cm and its central length is 4 cm. The experiments were carried out in two cases: (a) without the bellmouth, (b) with the bellmouth. The objectives of the experiments are to measure the transition length by color-dye method and to determine conditions under which the transition occurs. Fig. 11 shows the experimental results. The measured data were for 858 ≦ Re ≦ 40530. In the case without the bellmouth, the transition length is about 19 cm for Re=2702 (Let=0.0023) and 7 cm for Re=3766 (Let=0.00062). Below Re=1930, no transion appears even under vibration of the container. In the case with the bellmouth, the transion length is about 170 cm for Re=7000 (Let=0.0081), between 80 and 120 for Re=11500 (Let=0.0023--0.0035) and 70 for Re=15440 (Let=0.0015). Below Re=5404, no transition appears even under vibration of the container. It was difficult to determine the precise transition point because it moves considerably upstream and downstream and is no clearly distinguishable. No special disturbance were given in order to make the transition occur.

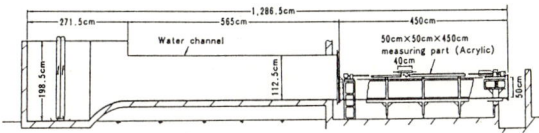

Fig.9 Water Channel

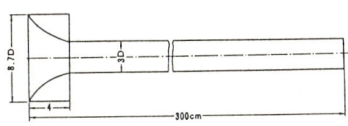

Fig.10 Circular Pipe and Bellmouth

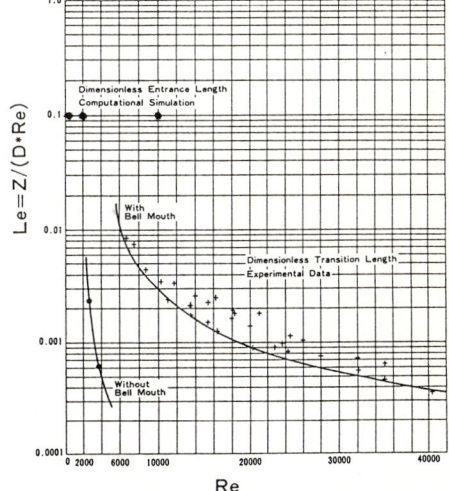

Fig.11 Dimensionless Entrance / Transition Length

#4. Conclusions

1) Smooth, two-dimensional, time-dependent, numerical solution of the Navier-Stokes equations exists regardless of the Reynolds number.
2) The dimensionless entrance length was calculated to be 0.1 for Reynolds numbers above 10. The dimensionless entrance lengths for 98% and 99% velocity development are 0.045 and 0.055, respectively, for Reynolds number above 50.
3) The transition occurs only in the Entrance region.
4) Both of the transition length and the dimensionless transition length become shorter when the Reynolds number is increased.

Acknowledgement

The authors wish to express their thanks to Dr. Y. Oshima and Y. Ishii for their full experimental cooperation.

References

(1) Perry, R.H. and Chilton, C.H.: Chemical Engineer's Handbook (5th edition), 5-34 (1983).
(2) Schiller, L.: ZAMM 2, 96-106 (1922).
(3) Langhaar, H.L.: J. Appl. Mechanics, 9, A55-A58 (1942).
(4) Mohanty, A.K. and Asthana, S.B.L.: J. Fluid Mech., 90, 433-447 (1978).
(5) Prandl, L. and Tietjens, O.G.: Applied Hydro- and Aeromechanics, Dover Publications, 1934.
(6) Arakawa, M. and Mastunobu, Y.: Proc. of the 17th Symposium on Turbulence, Japan Society of Fluid Mechanics, 1985.
(7) Kanda, H., Oshima, K.: Proc. of the Symposium on Mechanics for Space Flight, Institute of Space and Astronautical Scinece, 1986.
(8) Vrentras, J.S., Duda, J.L. and Bargeron, K.G.,: A.I.Ch.E.J, 12, 837 (1966).
(9) Koyari, Y.,: Study of Unsteady Entrance Flow in Pipes (in Japanese), Ph. D. Thesis, Tokyo University, 1980.

A Numerical Method to Assess the Feedback in a Free Shear Layer

Upender K. Kaul*
University of California, Berkeley, CA 94720

Introduction

A free shear layer is now known to be characterized predominantly by the large two-dimensional coherent structures. These structures were observed and studied in detail by Brown and Roshko[1] and Winant and Browand[2]. The shear layer grows downstream as a result of the coalescence of these vortex structures. An extensive study of these shear layers or mixing layers in the laboratory has been carried out by various researchers and considerable amount of information is available about the dynamics of these structures. It has also been shown computationally under controlled conditions by Patnaik, Corcos and Sherman[3], Peltier, Halle and Clarke[4] and Riley and Metcalfe[5] that the same features of these coherent structures hold for a shear layer growing in time, hereafter called the T-layer, as for a shear layer growing in space, hereafter called the S-layer. Also some direct simulations of the S-layer have been carried out, e.g., by Ashurst[6] using the "vortex chasing" methods due to Chorin[7]. In the direct simulation methods, the initial conditions at the origin of the mixing layer have to be prescribed. How the upstream effect of the downstream growing vortex structures on the origin of the flow is accounted for is, therefore, not clear. Although a considerable amount of insight into the dynamics of shear layers acquired through the controlled numerical experiments[3-5] and theoretical analyses, Corcos and Sherman[8,9] and Corcos[10], has helped in understanding the dynamics of the real flow, there is an essential difference between the real shear layers growing in space and those growing in time. The latter problem is parabolic in time whereas the former is elliptic in space. As a result, in the real flow there are some perturbations induced by the downstream vortices at points upstream in the flow. That the effects of the initial conditions on the development of a free shear layer are large was recognised by Bradshaw[11] and others. As observed by Weisbrot et al[12], even for fixed velocity ratios, the shear layer growth is sensitive to the initial conditions at the origin of the flow. They conclude that the shear layer growth is a function of the velocity ratio and the initial Strouhal number based on the initial momentum thickness and the mean velocity. Ho and Huang[13] observed that the most probable passage frequency scaled itself with the initial momentum thickness on the high speed side and the mean velocity. They also conclude that the evolution of coherent structures in a mixing layer appears to be controlled by the global feedback mechanism and the local stability. Miksad[14] showed that the presence of the wake arising out of the boundary layers on two sides of the splitter plate did not affect the most unstable frequency. The wake effect was seen to be present only on the amplification rates of the low frequencies. Since the passage frequency of the coherent structures is determined by the most unstable frequency, the wake effect does not significantly alter the dynamics of the mixing layer. It is therefore important to understand the role the upstream perturbations have in influencing these initial conditions – whether they accentuate or retard the instability mechanism. To quantify these upstream perturbations, a numerical method has been devised which consists in calculating the S-layer vorticity field and then studying the kinematics of the S-layer in time.

---

* Principal Analyst, Sterling Software, Palo Alto, CA; Member, AIAA

Numerical Method

The temporal problem of the nonlinear development of the finite amplitude waves is solved first. Then a set of transformations is contrived which transforms this T-layer problem to the spatially growing mixing layer problem, see Corcos[15] and Kaul[16]. According to the physical basis for this transformation given by Kaul[16], an alternate transformation is conceived and is used in this study. These transformations accurately map the vorticity of the T-layer to that of the S-layer for small values of velocity ratios, $\Delta U/U_m$, where $\Delta U = U_2 - U_1$ and $U_m = (U_1 + U_2)/2$, U_1 and U_2 being the velocities of the low and high speed streams respectively. Due to the nonuniform distribution of vorticity in the streamwise direction in the S-layer, there is a finite net induced effect present at any point upstream in the S-layer. This is also a measure of the ellipticity of the S-layer problem. The difference between this induced velocity and the velocity transformed directly from the T-layer at a given point upstream near the origin will give a measure of the feedback signal. However, in constructing the S-layer vorticity field, the induced normal velocity upstream of the geometric origin of the flow has to be cancelled out to simulate the presence of the splitter plate. A method to do this is devised by Kaul[16]. Other boundary conditions in the S-layer are satisfied by placing vortex sheets of appropriate strength in the flowfield. Having thus completed the specification of the S-layer vorticity field, the feedback signal is measured at various upstream locations.

Frequency and power spectra analyses are carried out to measure the amplitude and phase of the various components of the feedback signal and its effect on the fundamental and the first subharmonic of the transformed velocity itself.

Results

The computations are carried out corresponding to two values of velocity ratios and at a Reynolds number of 50 based on the initial shear layer thickness. The values of $\Delta U/U_m$ are chosen as 2 and 0.666. The initial perturbations are chosen from the linear stability theory and the amplitude ratios between the fundamental and the subharmonics is progressively fixed at 2 so that each 'local fundamental' has a chance to amplify on its own before its subharmonic begins to overtake it. Four different time realizations of the flow pattern shown through the vorticity contours corresponding to $\Delta U/U_m = 0.666$ are shown in Fig. 1(a)-(d). The same realizations are shown through passive scalar contours in Fig. 2(a)-(d). The initial stability, predominantly the fundamental, is seen to roll up. Then the rolled up structures coalesce further downstream. Still further downstream, the once paired structures coalesce again. During this process of pairing, the spacing between and the scale of these structures is seen to approximately double as expected. Same interaction between the structures is seen in Fig. 3, which gives a time realization corresponding to $\Delta U/U_m = 2$. The pattern of passive scalar contours shown in the Fig. 2 above is strikingly similar to that of the structures observed by Brown and Roshko[1], Winant and Browand[2] and later by others.

To quantify the feedback signal in the shear layer, the contribution corresponding to the subharmonic of the initial fundamental is measured upstream of the region where the first subharmonic reaches its peak or saturates. This includes the regions where the fundamental has rolled up into a vortex structure, where the first subharmonic begins to dominate and where it eventually saturates. The relative change in amplitude of the first subharmonic is calculated along with the change in the phase difference between this subharmonic and its fundamental. The change in the phase difference is calculated relative to the phase shift that would inhibit the subharmonic instability. These results are plotted in Fig. 4 versus the downstream distance x, where the relative change in the amplitude and the phase difference are plotted against the crossflow direction, y. The negative values of y correspond to the region of high speed stream.

The feedback signal is measured downstream of the location where the fundamental has rolled up into a vortex structure. It is in this region that the subharmonic begins to dominate since the fundamental has now saturated. Also, the feedback signal can be crucial in altering the subharmonic instability mechanism in this region. In Fig. 4(a), corresponding to $\Delta U/U_m = 2.0$, between the roll-up and the first pairing, i.e., $1.5<x<4.0$, both amplitude and phase difference are altered by about 10 to 15 percent at the mid-plane, y=0, between the high and low speed streams. Away from it, the feedback either increases or decreases, and the maximum alteration is about 30 percent. However, since the perturbations feed on the shear, their effect will be predominant near the mid-plane because of the presence of the maximum mean velocity gradient there. It is, therefore, expected that the subharmonic instability is not altered appreciably even though the feedback away from the mid-plane is not small. In Fig. 4(b), corresponding to $\Delta U/U_m = 0.666$, feedback is negligible everywhere; the maximum value between the roll-up (x=4.5) and the first pairing (x=12.0) is about 10 percent.

Conclusions

A unique method to calculate the spatially growing mixing layer vorticity field has been developed which clearly captures the distribution of vorticity realistically for small values of velocity ratios. The method does not suffer from the ill-posedness of the direct simulation methods for the spatially growing mixing layers in that no arbitrary inflow and outflow conditions have to be imposed. The inflow conditions for the present method derive themselves out of the transformation and include the contribution from the feedback signal. The method offers us the unique ability to measure the feedback signal in the spatially growing mixing layer and thus in having *a priori* knowledge of the appropriate perturbation required to be imposed at the origin of the flow for a given mixing layer growth rate. This does not seem to be possible otherwise. Also, the method can provide the inflow and outflow conditions for a direct simulation of the spatially growing mixing layer, which, in principle, are not available otherwise, e.g., the outflow boundary condition must be of the properly reflecting type rather than of the non-reflecting type.

Although the feedback in the mixing layer is not necessarily small, its effect on the subharmonic instability is seen to be small in the neighborhood of the mid-plane between the high and low speed sides of the mixing layer, even for the largest possible value of $\Delta U/U_m$ (with one stream at rest). Away from the mid-plane, the maximum value of the feedback is about 30 percent for $\Delta U/U_m = 2$ and about 10 percent for $\Delta U/U_m = 0.666$. Since the transformation becomes progressively less accurate as the velocity ratio increases (the error of the mapping is of the order of $\Delta U/6U_m$[16]), the feedback prediction for $\Delta U/U_m = 2$ is not very accurate. However, for small $\Delta U/U_m$, it has been quantitatively shown that the feedback in a forced mixing layer is small. As a result, the proposed transformations offer the novel capability of studying the spatially growing mixing layers[17].

Acknowledgements

The author would like to thank Professor Gilles M. Corcos for helpful discussions during the course of this study at Berkeley, and to Professor Maurice Holt for the insight into Numerical Methods in Fluid Dynamics that the author gained from his courses at Berkeley.

References

[1] Brown, G. R. and Roshko, A., "On the Density Effects and Large Structure in Turbulent Mixing Layer," J. Fluid Mech., Vol. 64, 1974, pp 775

[2] Winant, C. D. and Browand, F. K., "Vortex Pairing: The Mechanics of Turbulent Mixing layer Growth at Moderate Reynolds Numbers," J. Fluid Mech., Vol. 63, 1974, pp 237

[3] Patnaik, P. C., Corcos, G. M. and Sherman, F. S., "A Numerical Simulation of Kelvin-Helmholtz Waves of Finite Amplitude," J. Fluid Mech., Vol. 73, 1976, pp 215

[4] Peltier, W. R., Halle, J. and Clarke, T. L., "The Evolution of Finite Amplitude Kelvin-Helmholtz Billows," Geophys. and Astrophys. Fluid Dynamics, Vol. 10, 1978, pp 53

[5] Riley, J. J. and Metcalfe, R. W., "Direct Numerical Simulation of a Perturbed Turbulent Mixing Layer," AIAA Paper 80-0274, AIAA 18th Aerospace Sciences Meeting, Jan. 14-16, 1980, Pasadena, Calif.

[6] Ashurst, W. T., "Numerical Simulation of Turbulent Mixing Layers via Vortex Dynamics," Turbulent Shear Flows I, ed. by F. Durst et al., Springer-Verlag, Berlin, 1979, pp 402

[7] Chorin, A. J., "Numerical Study of Slightly Viscous Flow," J. Fluid Mech., Vol. 57, 1973, pp 785

[8] Corcos, G. M. and Sherman, F. S., "Vorticity Concentration and the Dynamics of Unstable Free Shear Layers," J. Fluid Mech., Vol. 73, 976, pp 241

[9] Corcos, G. M. and Sherman, F. S., "The Mixing Layer: Deterministic Models of a Turbulent Flow. Part 1. Introduction and the Two-Dimensional Flow," J. Fluid Mech., Vol. 139, 1984, pp 29-65

[10] Corcos, G. M., "The Mixing Layer: Deterministic Models of a Turbulent Flow," Rept No. FM-79-2, University of Calif., Berkeley 1979

[11] Bradshaw, P., "The Effect of Initial Conditions on the Development of a Free Shear Layer," J. Fluid Mech., Vol. 26, 1966, part 2, pp 225-236

[12] Weisbrot, I., Einav, S. and Wygnanski, I., "The Nonunique Rate of Spread of the Two-Dimensional Mixing Layer," Phys. Fluids, Vol. 25, No. 10, 1982, pp 1691-1693

[13] Ho, C.-M and Huang, L.-S., "Subharmonics and Vortex Merging in Mixing Layers," J. Fluid Mech., Vol.119, 1982, pp 443-473

[14] Miksad, R. W., "Experiments on the Nonlinear Stages of Free Shear Layer Transition," J. Fluid Mech., Vol.56, 1972, pp 695-719

[15] Corcos, G. M., "The Deterministic Description of the Coherent Structure of Free Shear Layer," Int. Conf. on Coherent Structure in Turbulent Shear Flow, 1980, Madrid, Spain,

[16] Kaul, U. K., "Do Large Vortices Control Their Own Growth in a Mixing Layer? An Assessment by a Boot-Strap Method," Ph.D. Thesis, University of California, Berkeley, Mech. Engg. Dept., 1982

[17] Kaul, U., K., "A Computational Study of the Subharmonic Instability in Mixing Layers," Proceedings of the Tenth U. S. National Congress of Applied Mechanics, The University of Texas at Austin, Austin, Texas, June 16 - 20, 1986

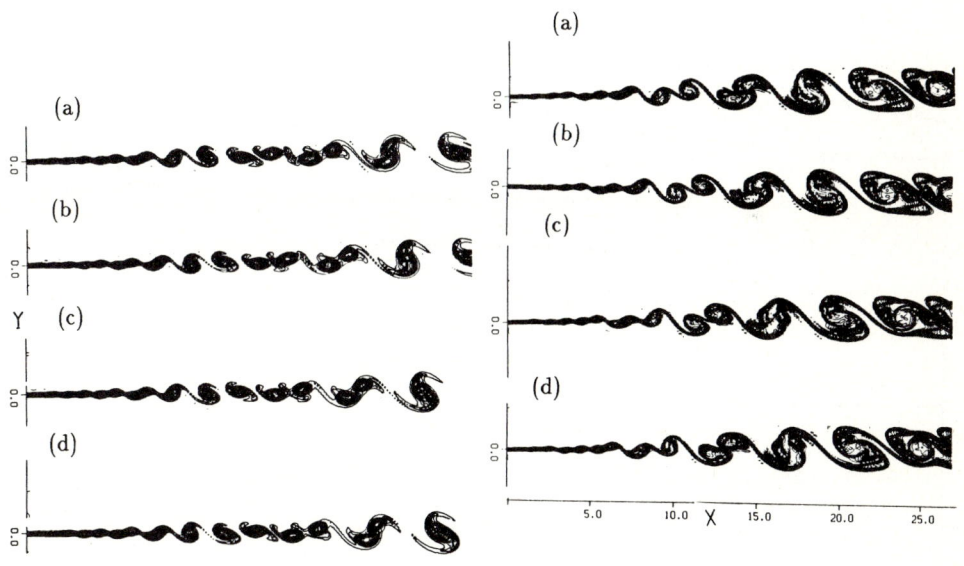

Fig. 1 Vorticity Contours in the S-Layer
$\Delta U/U_m = 0.666$

Fig. 2 Passive Scalar Contours in the S-Layer
$\Delta U/U_m = 0.666$

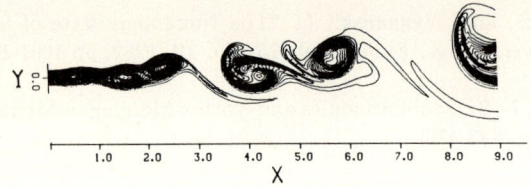

Fig. 3 Vorticity Contours in the S-layer
$\Delta U/U_m = 2.0$

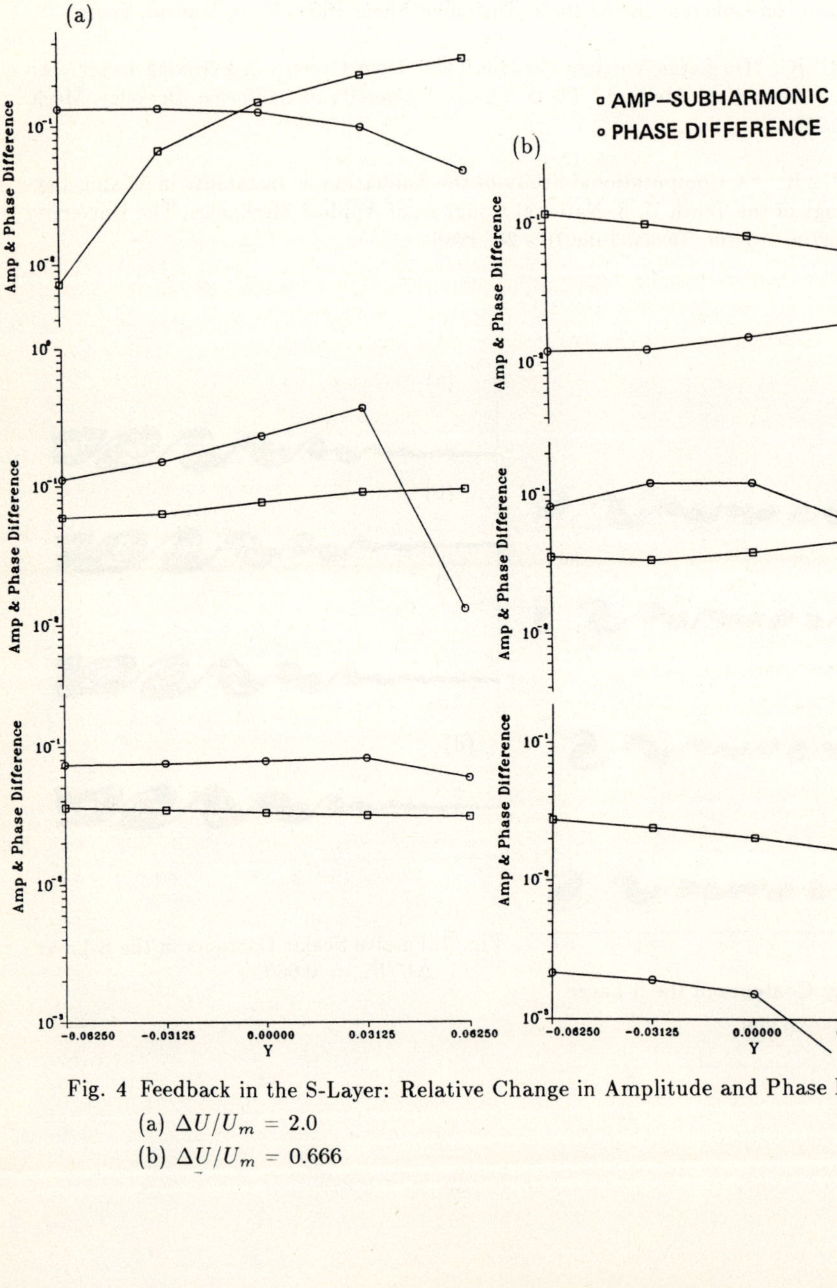

Fig. 4 Feedback in the S-Layer: Relative Change in Amplitude and Phase Difference
(a) $\Delta U/U_m = 2.0$
(b) $\Delta U/U_m = 0.666$

CONSISTENT STRONGLY IMPLICIT ITERATIVE PROCEDURES

P.K. Khosla and S.G. Rubin
University of Cincinnati, Cincinnati, Ohio USA

I. INTRODUCTION

The success of a numerical formulation for the evaluation of two-dimensional time accurate or three-dimensional spatial accurate flows, e.g., reduced Navier-Stokes (RNS) marching or global relaxation, depends largely on the solution algorithm. A non-iterative, unconditionally stable, consistent procedure should provide maximum efficiency. For large time increments (Δt), i.e., steady state calculations, consistency is not critical; rather, the technique should have strong convergence properties. For moderate Δt and transient flows, or moderate Δx for spatial marching, consistency of the numerical scheme plays an important role.

The most generally applied implicit and consistent formulation is the ADI method[1-3]. There are, however, a number of problems that have occasionally been encountered by investigators using ADI factorizations. These are (i) an 'instability' which may be attributable to the choice of intermediate boundary conditions and that generally can only be controlled with smaller values of Δt and (ii) poor rates of convergence for $\Delta t > 1$, i.e., when steady state problems are being considered[4,5]. In spite of these shortcomings, the ADI factorization has been successfully applied for a large number of flow problems, especially where the steady state is not of paramount interest.

Any algorithm that is time consistent, for $\Delta t < 1$, and has strong convergence properties, for $\Delta t \gg 1$, will be optimal for a large class of transient and steady flow problems. For $\Delta t \gg 1$, inconsistent or $O(\Delta t)$ techniques, e.g. the coupled strongly implicit procedure (CSIP), typically possess strong convergence properties[6,7]. However, such algorithms are not time consistent even for moderate values of Δt. Therefore, there is currently no single implicit solution algorithm that has the desirable properties of transient consistency (small Δt), rapid convergence to the steady state (large Δt) and can also be applicable when first or second order accuracy is desirable.

In the present investigation a time-consistent CSIP algorithm that also includes a Sherman-Morrison[8] formula to accelerate convergence is considered. Additional techniques and details are given in reference (13). The formulation is very simple to implement, requires little modification of existing codes and results in less than 5% additional computational effort. Furthermore, it is a single step procedure that can be applied to achieve either $O(\Delta t)$ or $O(\Delta t^2)$ accuracy. Intermediate boundary conditions are not required. The algorithm has been tested successfully on a variety of flow problems.

II. ONE-STEP CONSISTENT COUPLED STRONGLY IMPLICIT PROCEDURE (CSIP)

The consistency of the SIP procedure can be maintained by using a second-order factorization similar to the one originally proposed by Stone[6]. However, for $\Delta t \gg 1$, such a factorization is generally unstable. For $\Delta t < 1$, the technique retains both consistency and stability properties. In the present section, methods for achieving consistency, along with simple remedies to enhance stability for $\Delta t = O(1)$, are discussed.

Given the algebraic system

$$A_1 \phi_{i,j-1} + D_1 \phi_{i-1,j} + B_1 \phi_{i,j} + C_1 \phi_{i,j+1} + E_1 \phi_{i+1,j} = G_1 \tag{1a}$$

Stone[6] had proposed, a sparse second-order factorization that leads to the introduction of the corner values of ϕ_{ij}. For example

$$\phi_{i-1,j+1}^{n+1} = SR1 + \alpha(\phi_{i+1,j}^{n+1} + \phi_{i,j-1}^{n+1} - \phi_{i,j}^{n+1}) \tag{1b}$$

where

$$SR1 = \phi_{i+1,j-1}^{n} - \alpha(\phi_{i+1,j}^{n} + \phi_{i,j-1}^{n} - \phi_{i,j}^{n})$$

A similar expression exists for $\phi_{i-1,j+1}^{n+1}$ and the related SR2, where n+1 denotes the latest time level.

Clearly the terms evaluated at the previous time level n have a spatial truncation error of $O(\Delta x \Delta y)$, when α is chosen to be unity. This method was proposed by Stone for the evaluation of steady heat conduction problems. In order to stabilize the steady-state iteration, Stone used a sequence of values of α between zero and one. The choice of α's was very problem dependent. It has been found by the present authors[7] that the value of $\alpha = 0$ is most suitable for steady state computations, although other values of α might, in some cases, improve the convergence rate marginally. In the present paper, the use of SIP or CSIP implies $\alpha = 0$ unless noted otherwise. Stone did not specify that the SIP technique would be consistent and stable for unsteady problems, with a value of α close to or equal to unity and $\Delta t < 1$. This represents the simplest time-consistent SIP technique. Even in this form, however, Stone's technique is not considered practical by the present authors as the factorization procedure for coupled systems is too complex. However, the following simple implementation has been found to be equivalent and greatly simplifies the matrix factorization. Based on sparsity considerations the SIP solution algorithm has been given by the present authors[7] by:

$$\phi_{i,j} = GM_{i,j} + E_{i,j} \phi_{i,j+1} + F_{i,j} \phi_{i+1,j} . \tag{2}$$

The elimination of the lower triangular terms $\phi_{i-1,j}$ and $\phi_{i,j-1}$ along with the cancellation of the corner points $\phi_{i+1,j-1}$ and $\phi_{i-1,j+1}$ can be carried out in a single step as:

$$\phi_{i,j-1} = (1 - \alpha F_{i,j-1})^{-1} [GM_{i,j-1} + (E_{i,j-1} - \alpha F_{i,j-1}) \phi_{i,j}$$
$$+ \alpha F_{i,j-1} \phi_{i+1,j} + F_{i,j-1} (SR1)] \tag{3}$$

A similar expression is obtained for $\phi_{i-1,j}$. These replace (1) and can be used to eliminate lower diagonal terms from equation (2a). The recurrence relation for GM_{ij}

contains the terms SR1 and SR2 that are calculated explicitly. The effect of the initial conditions on the solution appears through SR1, etc.

For a diagonally dominant algebraic system, as found in fluid flow analysis, $\frac{D_1}{B_1}$ and $\frac{A_1}{B_1}$ are typically $= O(\frac{\Delta t/h^2}{1+4\Delta t/h^2})$, where $\Delta x = \Delta y = h$ is assumed. If $\varepsilon_{ij} = (\phi_{ij}^{n+1} - \phi_{ij}^E)$, where ϕ_{ij}^E is the exact solution to (2a), then the error is given by:

$$\frac{\Delta t/h^2}{1+4\Delta t/h^2} \varepsilon_{ij} \quad \text{for} \quad \alpha = 0 \quad \text{and} \quad \frac{\Delta t/h^2}{1+4\Delta t/h^2} h^2 (\frac{\partial^2 \varepsilon}{\partial x \partial y})_{ij} \quad \text{for} \quad \alpha = 1. \quad (4)$$

For fixed Δt and $h \to 0$, the case $\alpha = 0$ retains an error in the solution which is of order (ε_{ij}), while for $\alpha = 1$, this error tends to zero. Clearly, the consistency requirement is satisfied only for $\alpha = 1$. Where, the values of $GM_{i,j}$ do not depend upon the initial conditions and depend upon the grid spacing to within an error of $O(\Delta x \Delta y)$. A general block version of this algorithm for an arbitrary system of equations has been formulated and developed in a general CSIP code.

II.1. Sherman-Morrison Formula[8]

In order to eliminate the sensitivity of the procedure to the value of α, a rank one improvement of the iterative procedure was found to be quite useful. This is achieved by the application of a Sherman-Morrison formula. The solution is given as:

$$\phi_{n+1} = M^{-1}b - \frac{(M^{-1}U)<\hat{\phi}_n^T M^{-1}b>}{1 + <\hat{\phi}_n M^{-1}U>}, \quad (5)$$

where $<U^T V>$ is the scalar product of the vectors U and V and for simplicity of notation $\hat{\phi}_n \equiv \hat{\phi}_{i,j}^n$. This solution further improves the symmetry property, as well as, the time-consistency of the solution[13].

II.2 Error Analysis

A general first or second order implicit formulation of an arbitrary differential equation can be written as:

$$[I + \theta \Delta t (M+N)]\phi_{n+1} = \phi_n + \Delta t [b - (1-\theta)(M+N)\phi_n] \quad (6)$$

where the index n again corresponds to the time step and (M+N) is the coefficient matrix arising from the spatial terms. The time consistency of the solution procedure, with the Sherman-Morrison update, can be investigated by rewriting the lefthand side of the above equation as:

$$[I + \theta \Delta t M + \theta \Delta t N \hat{\phi}^* \hat{\phi}^{*T}]\phi_{n+1} + \Delta t \theta \{N \hat{\phi}_{n+1} \hat{\phi}_{n+1}^T - N \hat{\phi}^* \hat{\phi}^{*T}\} \phi_{n+1}$$

where $\hat{\phi}$ represents the normalized unit vector defined earlier and ϕ^* is an initial guess for ϕ_{n+1}. For a first-order implicit technique ($\theta=1$), ϕ^* can be chosen to be ϕ_n. For the second order Crank-Nicholson ($\theta=1/2$) procedure, there are two choices for ϕ^*; viz, (i) $\phi^* = \phi_n$ and perform at least one iteration per time step or (ii) $\phi^* = 2\phi_n - \phi_{n-1}$. The Taylor series expansion leads to the error term given as

$$N \hat{\phi}_{n+1} \hat{\phi}_{n+1}^T - N \hat{\phi}^* \hat{\phi}^{*T} = \Delta t^2 [N \hat{\phi}_{tt} \hat{\phi}_{n+1/2}^T + N \hat{\phi}_{n+1/2} \hat{\phi}_{tt}^T$$

$$- \frac{3}{2} \frac{<\hat{\phi}_{n+1/2} \hat{\phi}_{tt}>}{<\hat{\phi}_{n+1/2}\hat{\phi}_{n+1/2}>} N \hat{\phi}_{n+1/2} \hat{\phi}_{n+1/2}] + O(\Delta t^3)$$

Second order temporal accuracy is obtained provided ϕ_t and ϕ_{tt} are of $O(1)$ and N is such that the spatial error vanishes as the grid is refined. In such a case, the limit $\Delta t \to 0$ will render the system consistent.

III. APPLICATIONS

In order to test the validity of the procedure, a variety of model and flow problems have been investigated by the new CSIP method, now termed CSIP1.

III.1 Heat Diffusion Equation:

An exact solution to the 2D diffusion equation is

$$\phi(t,x,y) = e^{\frac{-2\beta^2 t}{Re}} \sin\beta x \sin\beta y + e^{\frac{t}{Re}} e^{-x(1+\beta^2)^{1/2}} \sin\beta y$$

This problem was chosen, because both the boundary conditions and the solution are time dependent. The maximum residual has been computed for 17x17 and 51x51 uniform grids. The results for the latter is depicted in figure 1 for both the consistent SIP1 and the inconsistent SIP procedure. Both techniques include the Sherman-Morrison modification. On coarser grids, the two results are similar; however, on the fine grid there is a significant difference. This reflects the severe inconsistency of the standard SIP procedure. The SIP1 formulation, however, retains an error of the order of the truncation error. Results for second-order time accuracy, as discussed in section II and without any additional iterations, are depicted in figure 2.

III.2. Subsonic Potential Flow Past a Biconvex Airfoil:

As another example of the applicability of the new algorithm, the composite velocity[9] solution past a biconvex heaving airfoil at $M_\infty = 0.65$ has also been computed. The unsteady potential flow is governed by

$$\rho_t + (\rho\phi)_x + (\rho\phi_y)_y = 0 \quad , \quad \rho = [1 + \frac{\gamma-1}{2} M_\infty^2 (1 - \phi_x^2 - \phi_y^2 - \phi_t)]^{1/\gamma-1} \quad (7a,b)$$

The instantaneous thickness of the airfoil is given as

$$F(t,x,y) = \epsilon(t) \, x(1-x) \quad \text{where} \quad \epsilon(t) = [1 - \frac{t}{10} - \frac{t^2}{375}] \, (\frac{t}{15})^3 .$$

ρ_t has been evaluated in terms of ϕ_t from (7b). A simple shearing transformation is used for generating the grid, so that boundary conditions are evaluated at the surface of the heaving airfoil. Computations have been carried out for a variety of time steps on a 75x33 non-uniform grid. For comparison purposes, solutions using a direct solver[10] have also been obtained. The results of these calculations are depicted in figures (3a,b). The accuracy of the new SIP1 procedure is apparent. Calculations have been performed with time steps $\Delta t = 0.1$ and $\Delta t = 0.5$. The results are in close agreement at identical time levels. The pressure time history at the mid-chord has also been compared with the direct solver solution (10). The two solutions are in excellent agreement.

III.3. Unsteady Flow Past a Flat Plate at an Angle of Attack

The algorithm described in section II has been extended to a nine point formulation by Ramakrishna and Rubin[11]. This algorithm has been applied to the reduced form of the Navier-Stokes equations given below:

$$\rho_t + (\rho u)_x + (\rho v)_y = 0 \quad , \quad (\rho u)_t + (\rho u^2)_x + (\rho u v)_y = -p_x + \frac{1}{Re} (\mu u_y)_y$$

$$(\rho v)_t + (\rho uv)_x + (\rho v^2)_y = -p_y \quad , \quad H = C_p T + \frac{u^2+v^2}{2} = \text{const} \quad , \quad p = \rho RT$$

The unsteady laminar flow over a flat plate at 5.7° angle of attack has been computed. The flow Reynolds number is 10^5. The calculations have been performed on a 120×100 H-grid. A time step of $\Delta t \approx 0.002$ has been used in these calculations. Typically, two non-linear iterations are performed at each time step. The streamline pattern at different times are depicted in figure 4. The flow separates at the leading edge. The separation bubble appears at a time level when $t = O(1)$. The bubble moves towards the trailing edge and multiple vortices are predicted at later times.

CONCLUSION

A new time consistent strongly implicit algorithm has been investigated. A single step first or second-order accurate method has been shown to be quite general and applies to both transient and steady flows with moderately large values of the time step Δt. With a sample change of a single parameter this CSIP1 technique reduces to the old CSIP method, which has been found to be quite robust for steady state calculations. The CSIP1 procedure has been applied to a variety of model and flow problems. The algorithm performs exceedingly well for all of the problems considered. For unsteady flow, moderately large time steps can be used to compute transient behavior. For three-dimensional steady spatial marching see reference 13.

ACKNOWLEDGEMENT

This research was supported by the Air Force Office of Scientific Research under Contract No. F49620-85-0027. The United States government is authorized to reproduce and distribute reprints for governmental purposes notwithstanding any copyright notation hereon.

REFERENCES

1. Douglas, J. and Gunn, J.E. (1964), Numerische Mathematik, Vol. 6, pp. 428-453.
2. Briley, W.R. and McDonald, H. (1980), J. Comp. Phys., Vol. 34, pp. 393-402.
3. Beam, R.M. and Warming, R.F. (1978), AIAA Journal, Vol. 16, pp. 393-402.
4. Nietubicz, C.J. (1981), AIAA Paper 81-1262.
5. Osswald, G.A., Ghia, K.N. and Ghia, U. (1983), AIAA Computational Fluid Dynamics Conference, A Collection of Papers, pp. 686-696.
6. Stone, H.L. (1968), SIAM J. Num. Anal., Vol. 5, pp. 530-558.
7. Rubin, S.G. and Khosla, P.K. (1981), Computers and Fluids, Vol. 9, pp. 163-180.
8. Sherman, J. and Morrison, W.J. (1960), Amer. Math. Statistics, Vol. 20, p. 621.
9. Khosla, P.K. and Rubin, S.G. (1983), AIAA Journal, Vol. 21, No. 11, pp. 1546-1551.
10. Bender, E.E. and Khosla, P.K. (1986), under preparation.
11. Ramakrishna, S.V. and Rubin, S.G. (1986), AIAA Paper No. 86-0205.
12. Reddy, D.R. and Rubin, S.G. (1984), AIAA Paper No. 84-1627.
13. Khosla, P.K. and Rubin, S.G. (1986), to appear in Computers and Fluids.

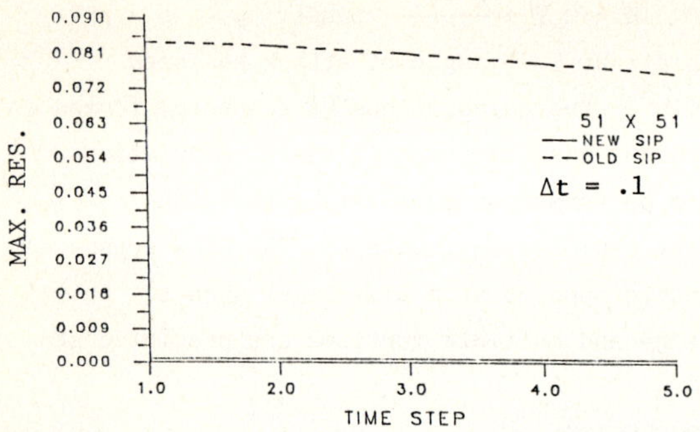

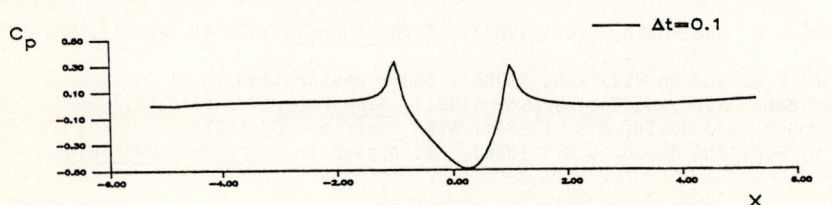

Figure 1 Diffusion Equation, Spatial Consistency

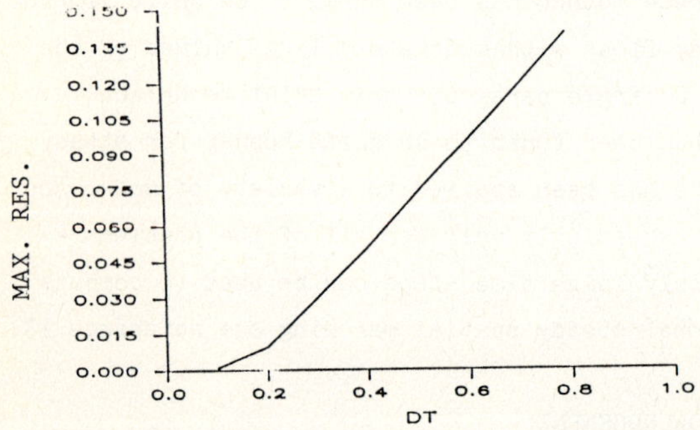

Figure 2 Diffusion Equation, Second Order Temporal Accuracy

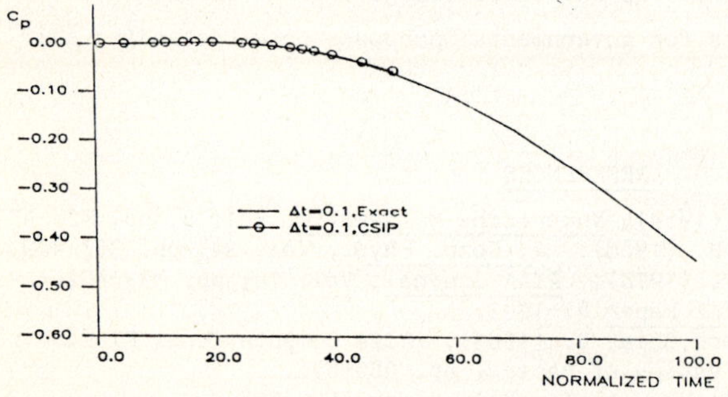

Figure 3a Pressure Time History at Mid-Chord of Biconvex Airfoil, $M_\infty = 0.65$

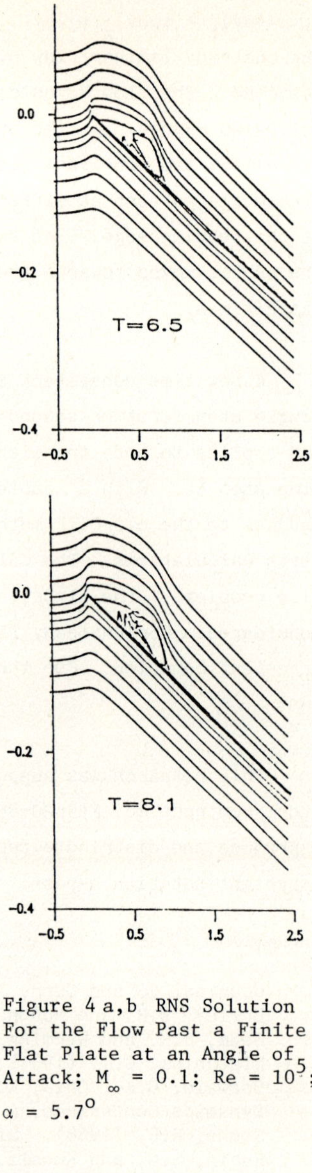

Figure 4 a,b RNS Solution For the Flow Past a Finite Flat Plate at an Angle of Attack; $M_\infty = 0.1$; $Re = 10^5$; $\alpha = 5.7°$

Figure 3b Pressure Coefficient at Normalized Time = 10.0, $M_\infty = 0.65$

AN ISOPARAMETRIC SPECTRAL ELEMENT METHOD IN SIMULATION OF INCOMPRESSIBLE COMPLEX FLOWS

Karol Z. Korczak
Mechanical and Aerospace Engineering, Case Western Reserve University
Cleveland, OH 44106 U.S.A.

ABSTRACT

An isoparametric spectral element method, a high-order finite element technique for solving the incompressible general Navier-Stokes equations, is briefly introduced (general 3-D formulation). The accuracy of the method and its dependence on a choice of collocating points is demonstrated on a Poisson equation case. For inflow-outflow problems, it has been shown that the natural (zero-gradient) boundary conditions at outflow perform reasonably well. The high convergence of this technique, combined with very small numerical dissipation and dispersion errors, allows realistic direct simulation of highly-unsteady and turbulent flows. Here, a numerical study of a turbulent mixing layer (from a splitter plate) is presented. In this application, in addition to the high accuracy, the isoparametric spectral element algorithm permits geometrical singularities in the computational domain, thus allowing a laminar boundary layer flow over the splitter plate, transition, and evolution of turbulent structures to be included in the simulation process.

INTRODUCTION

Among various numerical methods [1,2] implemented for simulations of highly-unsteady flows, high-order methods based on spectral methodology [3] have the potential of allowing realistic direct simulation of turbulent flows by solving the general Navier-Stokes equations. Spectral methods are characterized by rapid convergence with negligible numerical diffusion and dispersion errors - such properties are required to resolve small-scale developments with high gradients. Those techniques have been successfully used for testing turbulence theories [4,5], for validation studies of mixing layers [6,7], chemically reacting turbulent mixing layers [8] and many others. The above simulations were performed in smooth geometries with periodic boundary conditions. In practical applications with geometrical singularities (corners, sharp edges, etc.) or non-periodic discontinuous boundary conditions, the pure spectral formulation cannot be applied successfully. Several attempts have been made to bridge spectral approach with low-order techniques, but only the isoparametric spectral element method [9] has been implemented with success to solve the general incompressible Navier-Stokes equations.
The 2-D isoparametric spectral element method has been demonstrated in detail elsewhere [9,10] and applied to flow simulations in grooved channels [11,12]. This work briefly describes the general 3-D formulation of the algorithm, demonstrating the influence of various computational meshes. Varieties of boundary conditions for pressure and velocity have been investigated with results presented below. In final section, an application of this technique to direct simulation of turbulent structures in a plane mixing layer is demonstrated.

THE NUMERICAL ALGORITHM

The isoparametric spectral element method [9], a high-order finite element technique, brakes the domain into a series of quadrangular elements that are mapped into squares. The geometrical transformation and all functions are represented as tensor-product high-order Lagrangian interpolants through Chebyshev collocating points. The method incorporates a splitting scheme [13] that effectively separates the nonlinear, pressure, and viscous parts of the general incompressible Navier-Stokes equations. An explicit standard collocation approach with third-order Adams-Bashforth is used for the non-linear terms. The remaining pressure and viscous parts are solved implicitly with variational projection operators. The general approach has been described in details elsewhere [9], here we concentrate on the 3-D extension of the formulation. Since the viscous and pressure terms are variations of the Helmholtz equation, we briefly present the variational formulation for that case.

The Helmholtz problem is defined as,

$$\nabla^2 \Theta - \lambda^2 \Theta = f \qquad \text{in } D \qquad (1a)$$

$$\Theta = \Theta_b \quad \text{or} \quad \nabla\Theta \cdot \hat{n} = g(\vec{x}) \qquad \text{on } \partial D. \qquad (1b)$$

The geometrical transformations and all functions are interpolated in the same fashion (according to the isoparametric formula) as,

$$f_N^k = \sum_{i=0}^{N_r} \sum_{j=0}^{N_s} \sum_{m=0}^{N_t} f_{ijm} \, h_i(r) \, h_j(s) \, h_m(t) \quad . \qquad (2a)$$

Here, $h_m(z)$ are Lagrangian interpolants and z_n are Chebyshev collocating points,

$$h_m(z_n) = \delta_{mn} , \qquad (2b)$$

$$z_n = -\cos\frac{\pi n}{N} . \qquad (2c)$$

As the interpolating functions we use Chebyshev polynomials [9], typically of $N \sim 6 \div 14$. In the local (r,s,t) coordinate system (cubical elements), the mesh is trivially determined using (2c) in each direction. The formulation does not require a Chebyshev mesh data distribution in the physical (x,y,z) space. However, the accuracy of the method strongly depends on the type of a chosen mesh as shown in the next section.

The differential equation (1a) is an equivalent to maximization of the functional,

$$I(\Theta) = \int_D [-\frac{1}{2} \nabla\Theta \cdot \nabla\Theta - \frac{1}{2} \lambda^2 \Theta^2 - \Theta f] d\vec{x} + \int_{\partial D} \Theta g \, dS \qquad (3)$$

with restriction to admissible variations associated with the essential boundary conditions (1b). In arbitrary element k the functional in local coordinates can be written as,

$$I^k(\Theta) = \int_{-1}^{1}\int_{-1}^{1}\int_{-1}^{1} [-\frac{1}{2|J|} \tilde{\nabla}\Theta \cdot \tilde{\nabla}\Theta - \frac{1}{2}|J|\lambda^2\Theta^2 - |J|\Theta f] dr\,ds\,dt + F^k \qquad (4)$$

where

$$\tilde{\nabla} \equiv [(y_s z_t - y_t z_s)\frac{\partial}{\partial r} + (-y_r z_t + y_t z_r)\frac{\partial}{\partial s} + (y_r z_s - y_s z_r)\frac{\partial}{\partial t}]\hat{x} + \qquad (5)$$

$$[(-x_s z_t + x_t z_s)\frac{\partial}{\partial r} + (x_r z_t - x_t z_r)\frac{\partial}{\partial s} + (-x_r z_s + x_s z_r)\frac{\partial}{\partial t}]\hat{y} +$$

$$[(x_s y_t - x_t y_s)\frac{\partial}{\partial r} + (-x_r y_t + x_t y_r)\frac{\partial}{\partial s} + (x_r y_s - x_s y_r)\frac{\partial}{\partial t}]\hat{z} ,$$

$$J = x_r y_s z_t - x_r y_t z_s + x_s y_t z_r - x_s y_r z_t + x_t y_r z_s - x_t y_s z_r , \qquad (6)$$

and

$$F^k = \sum_{q=1}^{6} F^{k,q} \quad ; \quad F^{k,q} = \begin{cases} \int_{\partial D} \Theta g\, dS^{k,q} & S^{k,q} \in \partial D \\ 0 & S^{k,q} \notin \partial D \end{cases} . \qquad (7)$$

Here, subscripts refer to differentiation, J denotes the Jacobian of the geometrical transformation, F is a surface integral, and q indicates element's side.

To obtain the discretized equations, we first define the geometric transformation factors as,

$$(x_r)_{pq} = X_{mq} \mathcal{D}_{mp} \quad ; \quad (x_s)_{pq} = X_{pn} \mathcal{D}_{nq} \quad ; \quad \ldots \qquad (8a)$$

$$J_{ijm} = (x_r)_{ijm}(y_s)_{ijm}(z_t)_{ijm} - \ldots \qquad (8b)$$

where X's are the physical mesh coordinates and $\mathcal{D}$ is the derivative matrix [9],

$$D_{pq} = \frac{dh_p}{dz}\bigg|_{z=z_q} \quad . \tag{9}$$

Inserting interpolants and geometric transformations into the functional (4), performing the resulting integrals, and requiring stationarity with respect to variations in the discrete Θ's, results in elemental equations,

$$(A^k_{ijmnop} - \lambda^2 B^k_{ijmnop}) \Theta^k_{nop} = B^k_{ijmnop} f^k_{nop} + F^k_{ijm} \quad . \tag{10}$$

Here, A and B are matrices of coefficients, and F represents the contribution of the surface integral (7). (Note: If a conservation of mass is not preserved explicitly, the total surface integral could be different than zero. In such cases, a corrective constant flux has to be introduced to impose mass conservation.) For a general geometry, the form of these matrices is rather complicated [9]. In case of rectangular elements with local (r,s,t) system parallel to global (x,y,z) axes, they take simple forms,

$$A^k_{ijmnop} = -\frac{L_y L_z}{2L_x} A^r_{in} B^s_{jo} B^t_{mp} - \frac{L_x L_z}{2L_y} B^r_{in} A^s_{jo} B^t_{mp} - \frac{L_x L_y}{2L_z} B^r_{in} B^s_{jo} A^t_{mp} \tag{11}$$

$$B^k_{ijmnop} = \frac{L_x L_y L_z}{8} B^r_{in} B^s_{jo} B^t_{mp} \quad . \tag{12}$$

where L_x, L_y, L_z are the dimensions of element k, and A, B are the integrals [9],

$$A^z_{ij} = \int_{-1}^{1} \frac{dh_i}{dz} \frac{dh_j}{dz} dz \quad ; \quad B^z_{ij} = \int_{-1}^{1} h_i(z) h_j(z) dz \quad . \tag{13}$$

To form the global system of discrete equations, the elemental matrix equations (10) are assembled using direct stiffness [9]. The unknown nodal values Θ are obtained by solving the final set of equations.

In practical applications we use iterative solver (conjugate gradients) with factorization that eliminates the necessity of forming the global system of equations. In 3-D the size of such a system would be prohibitively large. The conjugate gradient solver allows efficient vectorization and solution of the elemental matrix systems in parallel (direct stiffness is applied after matrix multiplications in (10)). The above makes the isoparametric spectral element algorithm to be ideally suited for modern supercomputers.

MESH TYPE VERSUS ACCURACY

As mentioned above, there is some flexibility in a choice of the physical nodal point locations. Our experience indicates that the nodes distribution should resemble a Chebyshev distribution to assure high accuracy [9]. In rectangular elements these locations are easily determined (2c). For general (curvilinear) elements, a "uniform stretching" [10] results in a mesh with a Chebyshev nodes distribution in every direction measured along curves defined by those nodes. Fig. 1 demonstrates three types of meshes (2-D case) used in a solution of the Poisson equation $\nabla^2 u = \sin\Theta\{2\pi/r \cos 2\pi(r-1) - [1/r^2 + (2\pi)^2]\sin 2\pi(r-1)\}$ in the annulus $1 \leq r \leq 2$, $0 \leq \Theta \leq 2\pi$, with boundary conditions $u(r=1,\Theta)=1$, $u(r=2,\Theta)=0$. The error plot indicates that equally-spaced mesh is unacceptable for the spectral element technique. Meshes with Chebyshev distribution and ones determined from Laplace equations ($\nabla^2 x = 0$, $\nabla^2 y = 0$) perform equally well. However, Chebyshev distribution determined by "uniform stretching" [10] is a straightforward procedure while the other requires a solution of additional Laplace equations in every direction (and it seems to cluster nodes too close to boundaries - Fig. 1b).

BOUNDARY CONDITIONS

The isoparametric spectral element formulation requires specification of boundary conditions for pressure and velocities. Natural boundary conditions imposed on pressure [9] introduce an error in the boundary region. However, our experiments with other boundary conditions (pressure predicted based on the interior of the domain, pressure predicted based on several previous steps, etc.) never gave positive results in a long run. We arrived at similar conclusion for velocity boundary conditions at outflow. Solving $\nabla^2\Theta=-32\pi^2\sin(4\pi x)\sin(4\pi y)$, where $0\leq x\leq 3$, $-1\leq y\leq 1$, in a channel with three elements along the channel, with Dirichlet conditions specified at three sides and natural boundary condition at the "outflow", we determined the maximum error along the channel when the imposed "outflow" conditions were opposite to the actual ones. As seen on Fig. 2, the error from incorrect boundary conditions contaminates only the element next to that boundary, without propagating into the rest of the domain. Although this case cannot directly translate to the behavior of the error at outflow for flow equations, it gives some indication on the extend of an influence of outflow conditions.

DIRECT SIMULATION OF TURBULENT MIXING LAYERS

Turbulent mixing layers are common in engineering practice. In addition to many experimental investigations [14,15], attempts have been made to predict turbulent structures by numerical simulations [16]. Realistic direct simulations (spectral methods) have considered only temporal developments due to periodic boundary conditions [7,8]. Using our technique, we have been able to perform a direct simulation of the mixing layer flow in setup similar to an experiment. The flow starts with laminar boundary layers over the splitter plate, undergoes transition to turbulence, and leaves the domain as turbulent. The isoparametric spectral element technique allows to adjust the density of computational nodes proportionally to the complexity of the flow, resulting in efficient computations. The mesh, demonstrated on Fig. 3, has 15 elements in longitudinal direction, 4 in vertical, and 1 element in transverse direction. Every element contains 11-9-9 nodes in appropriate directions. The inflow velocities are determined from the Blasius profile. At the outflow and both vertical sides natural boundary conditions are imposed. In the transverse direction we assume periodicity. The velocity ratio varies in different simulations.

Our investigation [17] of the coherent structures, pairing mechanisms, pressure interactions, influence of forced oscillatory motion at the inflow on the generation or suppression of coherent structures, shows good agreement with experimental data [14,15]. The availability of complete data sets for the entire flow field allows us to study the complex interactions and mechanisms responsible for the formation and development of particular flow structures.

All major computations have been performed on the NASA Lewis Research Center's supercomputer CRAY-XMP.

REFERENCES

1. Peyret R., Taylor T., Comp. Meth. for Fluid Flows, Springer-Verlag, 1983.
2. Anderson D.A., Tannehill J.C., Pletcher R.H., Computational Fluid Mechanics and Heat Transfer, Hemisphere Publishing Corporation, 1984.
3. Gottlieb D.O., Orszag S.A., Num. Anal. of Spectral Meth.: Theory and Appl. NSF-CBMS Mon., No. 26, SIAM, Philadelphia, 1977.
4. Orszag S.A., Paterson G., Statist. Models of Turb., Springer-Verlag, 1972.
5. Schumann V., Patterson G., JFM, vol. 88, 1978.
6. Riley J.J., Metcalfe R.W., Turb. Shear Flows II, Springer-Verlag, 1980.
7. Riley J.J., Metcalfe R.W., AIAA Paper No. 80-0274, 1980.
8. Riley J.J., Metcalfe R.W., NASA-CR-174640, 1984.
9. Korczak K.Z., Patera A.T., An Isoparametric Spectral Element Method for Sol. of the Incomp. N-S Eq. in Complex Geom., J. Comp. Phys., v. 63, 1986.
10. Korczak K.Z., Ph.D. Thesis, Mechanical Engineering, MIT, 1985.
11. Ghaddar N.K., Patera A.T., Mikic B.B., AIAA Paper no. 84-0495, 1984.
12. Ghaddar N.K., Korczak K.Z., Mikic B.B., Patera A.T., JFM, v. 163, 1986.
13. Orszag S.A., Kells L.C., JFM, v. 96, 1980.
14. Jimenez J., JFM, v. 132, 1983.
15. Chin-Ming Ho, Lein-Saing Huang, JFM, v. 119, 1982.
16. Grinstein F.F., Oran E.S., Boris J.P., JFM, v. 165, 1986.
17. Hu Didi, M.S. Thesis, Mech. and Aero. Eng., CWRU., in progress.

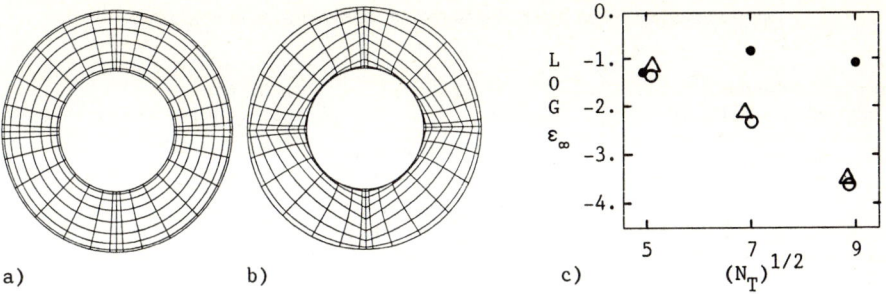

Figure 1. Computational mesh and accuracy (2-D). a) Chebyshev mesh, b) Laplace mesh c) Error L_∞ versus $(N_{Total\ nodes})^{1/2}$, and ● - equally spaced mesh, ○ - a , △ - b.

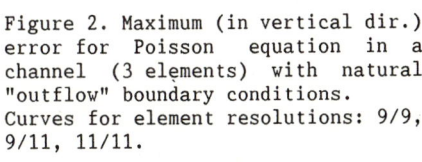
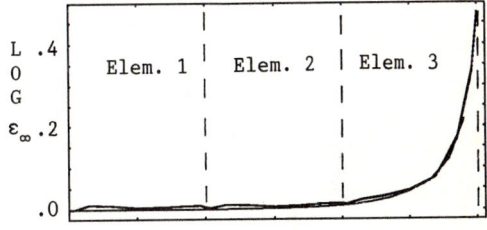

Figure 2. Maximum (in vertical dir.) error for Poisson equation in a channel (3 elements) with natural "outflow" boundary conditions. Curves for element resolutions: 9/9, 9/11, 11/11.

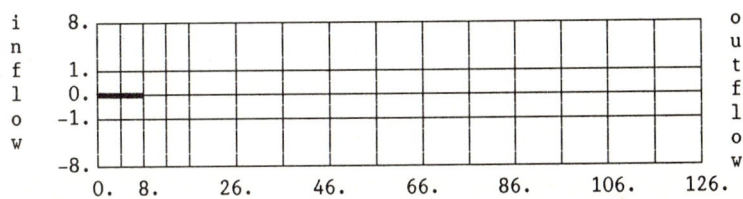

Figure 3. A computational mesh for planar mixing layer (one element in transverse direction). The thick line at inflow indicates the splitter plate.

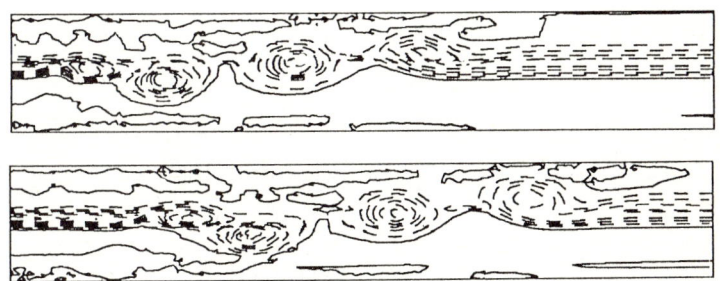

Figure 4. Early evolution of coherent structures in a turbulent mixing layer. Note the merging of vortices (pairing). The dashed lines denote negative vorticity with constant interval between lines.

INITIATION OF BREAKDOWN IN SLENDER COMPRESSIBLE VORTICES

E. Krause, S. Menne
Aerodynamisches Institut
RWTH Aachen, West Germany

C. H. Liu
NASA Langley Research Center
Hampton, Virginia, USA

Statement of the Problem

Analyses of vortex breakdown so far have been restricted to incompressible flows [1,2]. In this investigation, the initiation of the breakdown process for compressible flows is studied with a numerical solution of the conservation equations for mass, momentum, and energy, which were simplified with the following assumptions [3]: 1) the vortex is isolated, with its axis parallel to the direction of the main stream; 2) the flow is axially symmetric, and 3) the core radius R is small compared to the breakdown length L. The breakdown length is the distance between the point of initiation of the vortex, x_0, and the free stagnation point further downstream on the axis of the vortex (see Fig. 1). With these assumptions, the problem can be stated as follows:

Continuity equation

$$\frac{\partial}{\partial x}(\rho u) + \frac{1}{r}\frac{\partial}{\partial r}(\rho v r) = 0 \tag{1}$$

Axial momentum equation

$$\rho u \frac{\partial u}{\partial x} + \rho v \frac{\partial u}{\partial r} = -\frac{\partial p}{\partial x} + \frac{1}{r}\frac{\partial}{\partial r}(r \eta \frac{\partial u}{\partial r}) \tag{2}$$

Circumferential momentum equation

$$\rho u \frac{\partial \Gamma}{\partial x} + \rho v \frac{\partial \Gamma}{\partial r} = \frac{1}{r}\frac{\partial}{\partial r}(\eta r^3 \frac{\partial}{\partial r}(\frac{\Gamma}{r^2})) \tag{3}$$

Radial momentum equation

$$\frac{\rho w^2}{r} = \frac{\partial p}{\partial r} \tag{4}$$

Energy equation

$$c_p(\rho u \frac{\partial T}{\partial x} + \rho v \frac{\partial T}{\partial r}) = u \frac{\partial p}{\partial x} + v \frac{\partial p}{\partial r} + \frac{1}{r}\frac{\partial}{\partial r}(r \lambda \frac{\partial T}{\partial r})$$

$$+ \eta [(r \frac{\partial}{\partial r}(\frac{\Gamma}{r^2}))^2 + (\frac{\partial u}{\partial r})^2] \tag{5}$$

Equation of state

$$\rho = \frac{p}{RT} \tag{6}$$

The quantity Γ denotes the local circulation and all other quantities have the usual meaning.

Initial conditions

$$x = x_0, \quad 0 \leq r: \quad u = f_1(r), \quad w = f_2(r), \quad T = f_3(r) \tag{7}$$

The functions $f_1(r)$, $f_2(r)$, and $f_3(r)$ cannot be prescribed arbitrarily. They must satisfy a compatibility condition of the form

$$a_1(r)\frac{\partial^2}{\partial r^2}\left(\frac{v}{u}\right) + a_2(r)\frac{\partial}{\partial r}\left(\frac{v}{u}\right) + a_3(r)\left(\frac{v}{u}\right) + a_4(r) = 0 \tag{8}$$

in such a way that the slenderness condition

$$\frac{v}{u} = O\left(\frac{R}{L}\right) \ll 1 \tag{9}$$

is not violated. The quantities $a_i(r)$, $i=1,\ldots,4$ depend only on f_j, $j=1,2,3$. The boundary conditions are given by the symmetry condition along the axis

$$r = 0, \quad x_0 \leq x: \quad \frac{\partial u}{\partial r} = v = w = \frac{\partial T}{\partial r} = 0 \tag{10}$$

and the conditions for $r \to \infty$

$$r \to \infty, \quad x_0 \leq x: \quad u = g_1(x), \quad \Gamma = g_2(x), \quad T = g_3(x) \tag{11}$$

Solutions of equations (1) - (6) can be obtained as long as a solution exists for the axial pressure gradient, which is given by the following integro-differential equation

$$\frac{\partial p}{\partial x}(x,r) = \frac{\partial p}{\partial x}(x,r \to \infty) - \int_r^\infty \frac{w^2}{a^2 r'}\frac{\partial p}{\partial x}dr' + \int_r^\infty \bar{a}_1(r')dr' \tag{12}$$

where $\bar{a}_1$ is a known function of r. The solution of Eq. (12) depends on the local flow and on the external boundary conditions. The solution of the system (1) - (6) is said to breakdown at a station x where the solution of equation (12) can no longer be obtained for the given flow conditions. It must be emphasized that the corresponding axial position cannot immediately be identified with the onset of the destruction of the vortex core, normally identified as the actual breakdown process. This is because the axial velocity component of the slender vortex approximation does not necessarily have to reach zero, when the solution of Eq. (12) breaks down, as has often been inferred in earlier investigations of incompressible breakdown in an analogy to boundary-layer separation. The vanishing of the axial velocity component, however, is mandatory for the formation of a stagnation point. Because of conditions (7) and (9), the flow in the immediate vicinity of the stagnation point cannot be computed with the slender-vortex approximation. Finally, it should be noted that only small perturbations in the pressure field are sufficient to lead to a breakdown of the solution. It can be shown that the ratio of the radial to the axial velocity component can be expressed by the following relations

$$\frac{v}{u} = \exp(-I)\left\{\int_0^r \left[(1 - \frac{u^2}{a^2})\frac{1}{\rho u^2}\frac{\partial p}{\partial x}\right] \exp(I)dr' + \int_0^r [F(x,r)\exp(I)]dr'\right\} \quad (13)$$

where I represents the integral

$$I = \int_0^r (1 + \frac{w^2}{a^2})\frac{dr'}{r'}$$

Whenever equation (13) violates condition (9), a solution can no longer be obtained for equation (12). The breakdown of an actual vortex must have occurred before the axial position x corresponding to this violation.

The Finite-Difference Problem

The system of differential equations (1) – (6), and the boundary conditions (10) and (11), can be cast into non-linear or locally linearized implicit finite-difference equations of the form

$$A_{i,j}\vec{F}_{i+1,j-1} + B_{i,j}\vec{F}_{i+1,j} + C_{i,j}\vec{F}_{i+1,j+1} = D_{i,j} \quad (14)$$

where the solution vector $\vec{F}$ is defined in the following way

$$\vec{F}_{i,j} = (u, \Gamma, v, \rho, T)^{-1} \quad (15)$$

The coefficient matrices A, B, and C are 5 × 5 matrices, and the system (14) can be inverted by the recursion

$$\vec{F}_{i+1,j} = X_{i,j}\vec{F}_{i+1,j+1} + Y_{i,j} \quad (16)$$

The quantities $X_{i,j}$ and $Y_{i,j}$ are given by

$$X_{i,j} = -(B_{i,j} + A_{i,j}X_{i,j-1})^{-1}(C_{i,j}) \quad (17)$$

and

$$Y_{i,j} = (B_{i,j} + A_{i,j}X_{i,j-1})^{-1}(D_{i,j} - A_{i,j}Y_{i,j-1}) \quad (18)$$

Since the numerical effort for the solution of the system (14) is rather large, the finite-difference forms of the axial and circumferential momentum equations and of the continuity equation were decoupled from the finite-difference representations of the radial momentum equation and the energy equation. The former equations yield u, Γ and v while the latter yield ρ and T. These result again in a system of the form (14), but now the coefficient matrices are 3 × 3 matrices, and the inversion (16) can be facilitated for compatible initial conditions in an iteration process.

Discussion of Results

Several flowfields were computed with the method of solution just described. Some of the results obtained are shown in Figs. 2 and 3. In Fig. 2, the axial velocity component as computed along the axis is shown as a function of the axial coordinate divided by the reference Reynolds number. The initial conditions correspond to a uniform axial flow ($\alpha = 0$) and a uniform temperature distribution ($\gamma = 0$) at the initial station x_0. The circumferential velocity component at the edge of the core is eight-tenths of the freestream velocity ($\beta = 0.8$). The influence of the freestream Mach number on the "breakdown" length of the slender-vortex approximation is clearly indicated. With increasing Mach number, the breakdown of the solution is shifted further downstream until, finally, breakdown is no longer observed.

Figure 3 shows the influence of the initial temperature distribution on the breakdown length of the solution for the Mach numbers $Ma_\infty = 0.5$ and $Ma_\infty = 1.1$. In the subsonic case ($Ma_\infty = 0.5$), the influence is more pronounced than in the supersonic case. It can be seen that heating of the core ($\gamma > 0$) enhances breakdown, while cooling ($\gamma < 0$) delays it. It can also be seen that the breakdown of the solution always occurs for non-vanishing axial velocity components.

References

1. Leibovich, S.: Vortex Stability and Breakdown: Survey and Extension. AIAA Journal, Vol. 22, Sept. 1984, pp. 1192-1206.

2. Hall, M. G.: Vortex Breakdown. Annual Review of Fluid Mechanics, Vol. 4, (1972), pp. 195-218.

3. Krause, E.: Der Einfluss der Kompressibilität auf schlanke Wirbel, Heft 27, Abh. Aerodynamisches Institut, RWTH Aachen, October 1985.

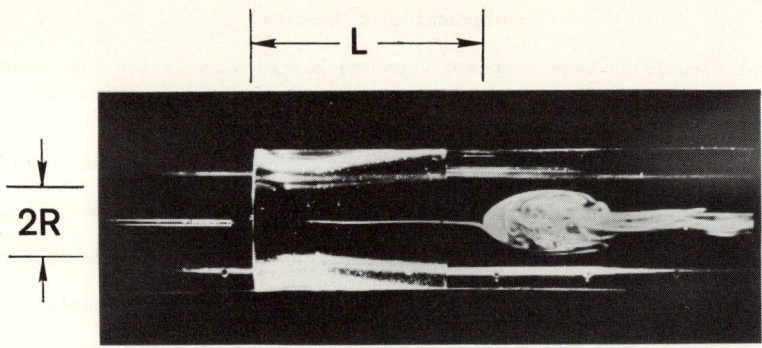

Figure 1. Breakdown length L and core radius R of a slender vortex.

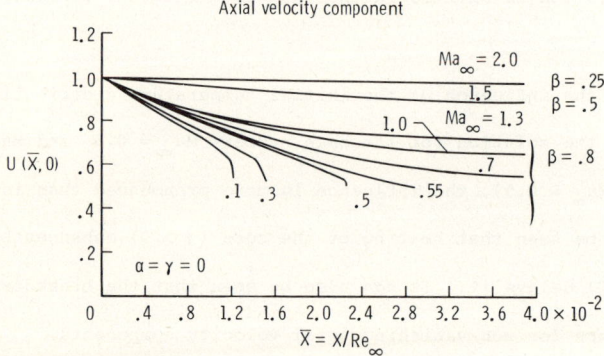

Figure 2. Breakdown length of the slender-vortex approximation for sub- and supersonic flows; axial velocity component u(x,r=0).

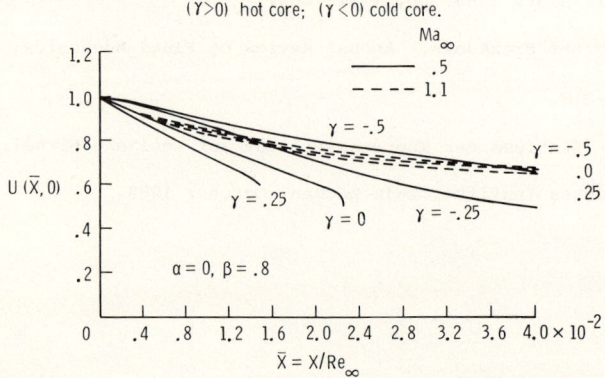

Figure 3. Influence of the temperature on the breakdown length. ($\gamma=0$) uniform temperature at initial station; ($\gamma>0$) hot core; ($\gamma<0$) cold core.

A PSEUDOSPECTRAL METHOD FOR SOLUTION OF THE THREE-
DIMENSIONAL INCOMPRESSIBLE NAVIER-STOKES EQUATIONS

Hwar C. Ku, Richard S. Hirsh, and Thomas D. Taylor
The Johns Hopkins University Applied Physics Laboratory
Laurel, Maryland 20707

INTRODUCTION

A new Chebyshev pseudospectral matrix technique, based on Chorin's projection method, which was previously applied by the authors to the solution of the two-dimensional incompressible Navier-Stokes equations in primitive variables for non-periodic boundary conditions [1] is extended in the present work to solve the three-dimensional Navier- Stokes equations. The crucial point of the method is the requirement that the continuity equation be satisfied everywhere in the domain, on the boundaries as well as in the interior. The key feature of the work presented in this paper is that the computer storage requirement for a matrix inversion resulting from direct solution of the pressure Poisson equation in three dimensions is greatly reduced by considering an eigen- function decomposition. This is accomplished by creating an eigenvalue expansion of the Poisson differential operator in two of the three dimensions, resulting in the need to invert only a simple one-dimensional matrix for the solution of the pressure. The method was tested on a two-dimensional driven cavity flow and the results were compared with those of the most accurate finite-difference calculation. The three-dimensional driven cavity flow was then calculated at the same Reynolds numbers as the two-dimensional cases, i.e., Re = 100, 400, and 1000. In the calculated results, three-dimensional boundary effects were observed in all cases and became more apparent with increasing Reynolds number.

FORMULATION OF THE PSEUDOSPECTRAL MATRIX METHOD

The three-dimensional cavity flow in a cubic box (Figure 1) can most easily be expressed in terms of a primitive variable formulation. Due to the geometric configuration shown in the figure it is only necessary to solve these equations in the domain $0.5 \leq z \leq 1.0$, and assume symmetry about the plane $z = 0.5$. The incompressible Navier- Stokes equations in non-dimensional form can thus be written in tensor notation as:

$$\frac{\partial u_i}{\partial t} + \frac{\partial p}{\partial x_i} = F_i \qquad (1)$$

$$\frac{\partial u_i}{\partial x_i} = 0 \qquad (2)$$

where

$$F_i = \frac{1}{Re}\frac{\partial^2 u_i}{\partial x_j^2} - u_j \frac{\partial u_i}{\partial x_j}$$

The boundary conditions to be imposed on each velocity component are

$$u = v = w = 0, \quad \text{at } x = 0 \text{ and } x = 1$$
$$u = 1, v = w = 0, \quad \text{at } y = 1$$
$$u = v = w = 0, \quad \text{at } y = 0 \qquad (3)$$
$$u = v = w = 0, \quad \text{at } z = 1$$
$$\frac{\partial u}{\partial z} = \frac{\partial v}{\partial z} = \frac{\partial p}{\partial z} = 0, w = 0, \quad \text{at } z = 0.5$$

Note that as in our previous two-dimensional studies [1], no boundary condition is imposed (or required) for the pressure. The incompressibility constraint, Eq. (2), is used in our formulation to close the derived Poisson equation for the pressure.

To solve these equations, Chorin's splitting technique [2] is used with the Chebyshev pseudospectral matrix method. The first step in the method is to split the velocity into a sum of a predicted and corrected value. The predicted velocity field is determined by time integration of the momentum equations without the pressure terms in the form

$$\bar{u}_i^{n+1} - u_i^n = \Delta t \, F_i^n \qquad (4)$$

The second step is to develop the corrected velocity field that satisfies the continuity equation by using the relationships

$$u_i^{n+1} = \bar{u}_i^{n+1} - \Delta t \frac{\partial p}{\partial x_i} \qquad (5)$$

and

$$\frac{\partial u_i^{n+1}}{\partial x_i} = 0 \qquad (6)$$

Instead of using FFT's to form the derivatives necessary in the equations, direct matrix multiplication is used in all cases. For example, to form the derivative $\underline{f}'$ from $\underline{f}$ the simple form

$$\underline{f}' = \hat{\underline{G}} * \underline{f} \qquad (7a)$$

or equivalently

$$f'(x_j) = \sum_{l=1}^{N+1} \hat{G}_{j,l}^{(1)} f_l \qquad (7b)$$

is used, where the matrix $\hat{G}$ is an N x N matrix dependent on the properties of the Chebyshev polynomials and needs to be calculated only once at the outset of the calculation. We have found that it is less effort to calculate the field variables directly in physical space rather than calculate through the transformations between real and spectral space. Similarly the Poisson equation matrix can be formed as described below, and then inverted once at the beginning of the program. Subsequent solutions of the Poisson equation are then accomplished by a simple multiplication of this inverted matrix.

The discretization of Eq. (5) by the aforementioned pseudospectral matrix method takes the form

$$u_{i,j,k}^{n+1} = \bar{u}_{i,j,k}^{n+1} - \Delta t \sum_{m=1}^{NX+1} \hat{G}X_{i,m}^{(1)} \, p_{m,j,k}$$

$$v_{i,j,k}^{n+1} = \bar{v}_{i,j,k}^{n+1} - \Delta t \sum_{l=1}^{NY+1} \hat{G}Y_{j,l}^{(1)} \, p_{i,l,k} \qquad (8)$$

$$w_{i,j,k}^{n+1} = \bar{w}_{i,j,k}^{n+1} - \Delta t \sum_{n=1}^{NZ/2+1} \hat{G}S_{k,n}^{(1)} p_{i,j,n}$$

Note that only half of the grid points, NZ/2 + 1, are needed in the z direction due to the geometric configuration. Hence z-dependent variables u, v, and p are expanded in a symmetric form, while w is expanded in an anti-symmetric form.

By requiring that the velocity components satisfy the continuity equation throughout the whole domain, and with the incorporation of prescribed velocity boundary conditions (which were described in detail in [1,3], the substitution of Eq. (8) into Eq. (6) yields the following general form at interior points of the domain

$$\sum_{m=1}^{NX+1} BX_{i,m}\, p_{m,j,k} + \sum_{l=1}^{NY+1} BY_{j,l}\, p_{i,l,k} + \sum_{n=1}^{NZ/2+1} BZ_{k,n}\, p_{i,j,n} = S_{i,j,k} \qquad (9)$$

$$BX_{i,m} = \sum_{p=2}^{NX} \hat{G}X_{i,p}^{(1)} \hat{G}X_{p,m}^{(1)} \qquad BY_{j,l} = \sum_{q=2}^{NY} \hat{G}Y_{j,q}^{(1)} \hat{G}Y_{q,l}^{(1)} \qquad BZ_{k,n} = \sum_{r=2}^{NZ/2} \hat{G}A_{k,r}^{(1)} \hat{G}S_{r,n}^{(1)} \quad (10)$$

and the $\hat{G}S$ and $\hat{G}A$ are the symmetric and anti-symmetric form of the first derivative matrix [4]. The source term in Eq.(9) contains only known values of velocity derivatives. Since the operators in Eq.(9) involve values of the pressure on the boundary, a supplemental pressure relation can be obtained by imposing the continuity equation at each boundary, i.e., $\partial u/\partial x = -(\partial v/\partial y + \partial w/\partial z)$

$$\sum_{m=1}^{NX+1} BX_{i,m}\, p_{m,j,k} = S_{i,j,k} \qquad i = 1,\, NX+1 \qquad (11)$$

with similar relations on the y and z boundaries

Equation (9), together with supplemental Eq. (11) and others in y and z at the boundaries, constitutes the overall solution for the unknown pressure, with which continuity is satisfied to machine accuracy. What is more important, the pressure Poisson equation is a linear operator, and only an initial matrix inversion is required. The stored matrix inversion coefficients can then be used for all the following time steps to compute the pressure solution. Despite the advantage of the linear operator for the pressure equation, the remaining aspect is the challenge of how to solve the huge three-dimensional matrix without constraints on the size of the computer storage. This question is dealt with below.

THREE-DIMENSIONAL POISSON EQUATION SOLUTION

The three-dimensional Poisson equation can be solved either in real space or spectral space. Gottlieb and Orszag [5], Haidvogel and Zang [6] and Tan [7] employed a spectral space approach that is sometimes called the tau method. For limited expansions the scheme appears to work, but as the modes increase one finds high order expansion terms being driven by the boundary conditions [8]. The real space pseudospectral solution approach seems to be more straightforward in dealing with boundary conditions. As a result, this is the approach adopted here. The method used is similar to the tensor product method [9] used by Murdock [10] for two-dimensional flows in that eigenfunction expansions are used to reduce differential operators in the Poisson equation to algebraic relationships, but differs considerably in the treatment of boundary conditions. In this approach two spacial operators are reduced in the three-dimensional problem and the resulting one-dimensional second order equation is solved using a matrix method that has been previously described [4].

In order to develop an eigenfunction expansion for the operators in Eq. (9) it is necessary to first subtract out or remove boundary pressure terms. This can be accomplished in the x and y direction if one uses supplemental pressure conditions like Eq. (11). From these equations, $p_{1,j,k}$, $p_{NX+1,j,k}$, $p_{i,1,k}$ and $p_{i,NY+1,k}$ can be expressed in terms of field variables in the interior and substituted into Eq. (9) to yield:

$$\sum_{m=2}^{NX} BX^*_{i,m}\, p_{m,j,k} + \sum_{l=2}^{NY} BY^*_{j,l}\, p_{i,l,k} + \sum_{n=1}^{NZ/2+1} BZ_{k,n}\, p_{i,j,n} = S^*_{i,j,k} \qquad 2 \le k \le NZ/2+1 \qquad (12)$$

where BX^*, BY^* and S^* are complicated combinations of the original matrix elements and boundary values of these elements.

Equation (12) together with the z direction equivalent of Eq. (11) at k = 1 form a reduced Poisson problem which has been solved by an eigenfunction expansion.

The key feature of the Poisson solution technique is in the decomposition of the original three-dimensional matrix. Instead of the typical LU decomposition, eigenfunctions and eigenvalues have been used to decompose the matrix. As a result, in the final equation to be solved it is necessary to deal with a matrix of dimensions $(NZ/2 + 1)^2$ rather than one of $(NX)^2 \times (NY)^2 \times (NZ/2 + 1)^2$. The consequence of these steps has been to reduce the minimum storage requirements for direct matrix inversion from $O[(NX)^2(NY)^2(NZ/2+1)^2]$ to $O[(NX)(NY)(NZ/2+1)]$; i.e., the minimum storage requirement is no longer limited by the size of the reduced matrix inversion but by the storage of the field variables themselves, $O(N^3)$. This important size of the result permits use of much smaller computers to perform direct solution of the 3D Poisson equation.

RESULTS AND DISCUSSION

This section discusses the results obtained for the two- and three-dimensional cavity flows. First, computation of the flow in a two-dimensional square cavity using a primitive variable formulation is examined for Re = 100, 400, and 1000, and then results are compared with those of the most accurate finite-difference method in streamfunction vorticity formulation. Next, the results for a cubic cavity flow are presented for the same Reynolds numbers.

Since the technique for the three-dimensional case is essentially the same as that for the two-dimensional case, testing of the method on a two-dimensional square cavity flow for which other results exist makes it possible to ensure the validity of the present algorithm for a three-dimensional cavity flow. For the two-dimensional test 25 x 25 modes were chosen for Re = 100 and 400, and 31 x 31 modes for Re = 1000. Stream lines for the driven cavity at Re = 1000 are shown in Fig. 2. These are in good agreement with the fine grid results (129 x 129) reported by Ghia et al [11].

Computational results for a three-dimensional cavity flow at the highest Reynolds number of 1000 are given in the succeeding figures. Displayed in Figures 3 and 4 are the velocity profiles of the u component on the vertical centerline and the v component on the horizontal centerline of the plane z = 0.5. The two-dimensional results are shown for comparison. These plots clearly indicate that the three-dimensional boundary affects the flow. However, as expected, for the high Reynolds number (Re = 1000) shown this boundary effect is more prominent then at the lower Reynolds numbers.

The yz plane flow patterns for Re = 1000 starting from upstream to downstream at various x are plotted in Figures 5-8. Based on the plotted velocity vectors, one can easily distinguish whether the w component velocity flows inward or outward since without the w component velocity, the v component velocity should be vertical. Note that the flow patterns are not visible in some places due to very small values of both v and w. The vector plots are normalized by the largest vector in the plane displayed. The results not plotted show, for Re = 100 in the downstream yz plane at x = 0.854, a minor outward flow observed at the bottom, while for Re = 400 at x = 0.962 and 0.854, a small recirculation flow is located at the bottom. In the yz plane at x = 0.5, two distinct secondary vortices are found that gradually shift toward the corners with increasing Reynolds numbers. For Re = 1000, the yz plane flow patterns shown in Figs. 5 to 8 show similar structure except that two more small secondary vortices appear at the upper corner of the side walls, and the distortion of v component is more severe. The presence of Taylor-Görtler-like longitudinal vortices for Re = 1000 could not be established. For Re > 1000, their presence remains to be determined; this is a task for future study.

CONCLUSIONS

The three-dimensional Navier-Stokes equations using a primitive variable formulation have been solved by a Chebyshev pseudospectral matrix method for a three-dimensional driven-cavity flow by using a time-splitting technique. In the solution approach, the continuity equation is satisfied everywhere in the interior and on the boundaries, except at the corner singular points. This eliminates the need for momentum-equation-derived Neumann boundary conditions on the pressure.

The key feature of the work presented is that the resulting three-dimensional direct matrix inversion for the pressure Poisson equations is reduced to simple one-dimensional matrix operations by employing eigenfunction expansions. Since only one-dimensional matrices are involved, this formulation avoids the large storage normally associated with three-dimensional solutions of Poisson equations. The approach permitted storage of the overall inverted matrix coefficients that were applied during integration in the time domain using only the limited storage of a VAX 11/780.

Results for a two- and three-dimensional driven cavity flow have been compared for Re = 100, 400, and 1000. For all of the Reynolds numbers studied in this paper, Taylor-Görtler-like longitudinal vortices were not observed, and this topic remains an open question for Re > 1000.

ACKNOWLEDGMENT

This work was partially supported by the Office of Naval Research under U.S. Navy Contract N00024-85-C-5301.

REFERENCES

[1] H. C. Ku, T. D. Taylor, and R. S. Hirsh, to appear in Computer and Fluids.
[2] A. J. Chorin, Math. Comp., Vol. 22, 1968, p. 745.
[3] T. D. Taylor and H. C. Ku, Int. Symp. Comp. Fluid Dynam., Tokyo, 1985.
[4] H. C. Ku, T. D. Taylor, and D. T. Hatziavramidis, Proc. Sixth GAMM Conf., Gottingen, 1985.

[5] D. Gottlieb and S. A. Orszag, SIAM, Philadelphia, 1977.
[6] D. B. Haidvogel and T. Zang, J. Comp. Phys., Vol. 30, 1979, p. 167.
[7] C. S. Tan, J. Comp. Phys., Vol. 59, 1985, p. 81.
[8] R. Peyret and T. D. Taylor, Computational Methods for Fluid Flow, Springer, New York, 1983.
[9] R. E. Lynch, J. R. Rice and D. H. Thomas, Num. Math., Vol. 6, 1964, p. 185.
[10] J. W. Murdock, AIAAJ., Vol. 8, 1977, p. 1167.
[11] U. Ghia, K. N. Ghia, and C. A. Shin, J. Comp. Phys., Vol. 48, 1982, p. 387.

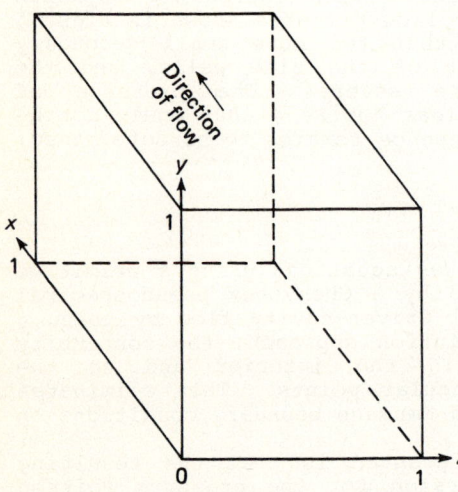

Figure 1
Three dimensional cavity flow configuration and coordinate system

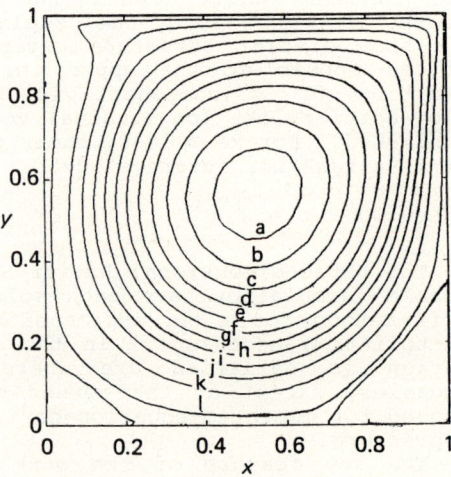

Figure 2
Streamline pattern for Re=1000; ψ_{min} =0.11619; contour levels: a=0.11, b=0.1, c=0.9, etc., l=0.001

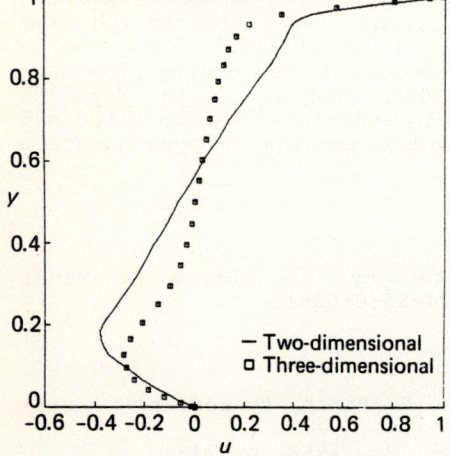

Figure 3
Velocity profiles on vertical centerline of the cubic cavity for Re=1000

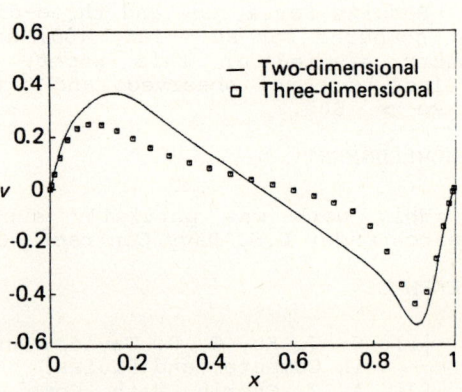

Figure 4
Velocity profiles on horizontal centerline of the cubic cavity for Re=1000

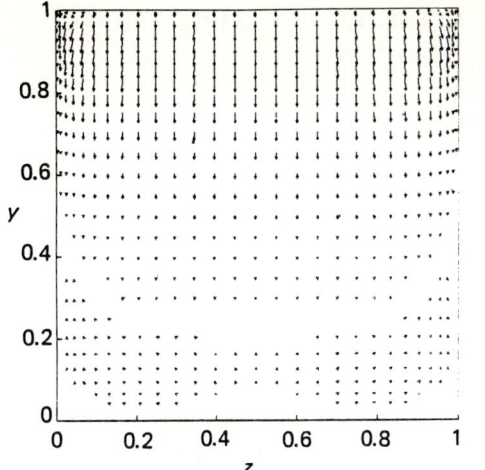

Figure 5
Flow direction vectors for
Re=1000 in the X=0.975 plane

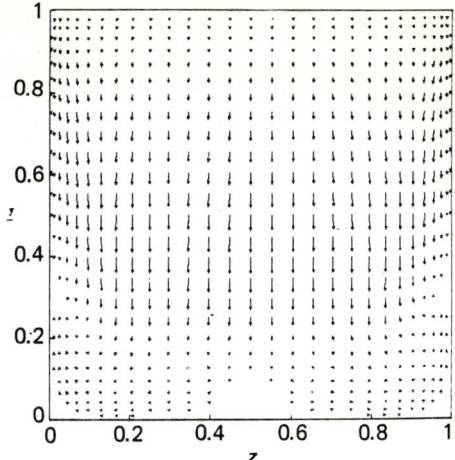

Figure 6
Flow direction vectors for
Re=1000 in the X=0.871 plane

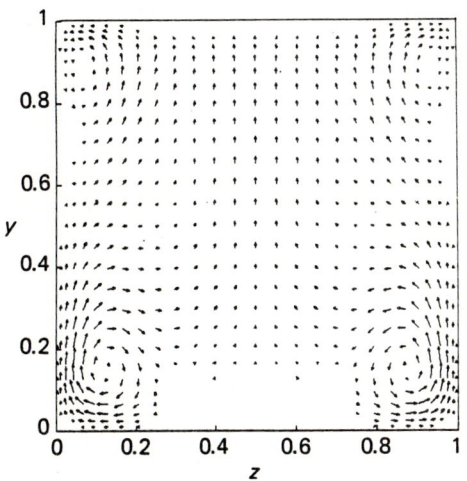

Figure 7
Flow direction vectors for
Re=1000 in the X=0.5 plane

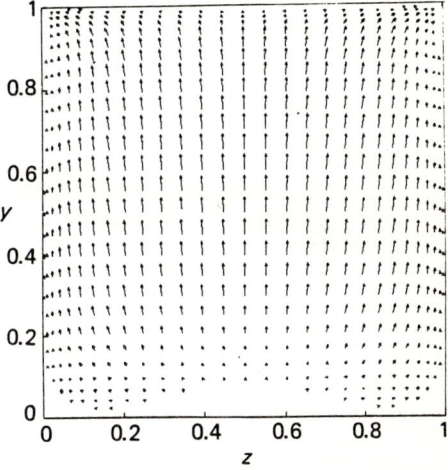

Figure 8
Flow direction vectors for
Re=1000 in the X=0.129 plane

A NUMERICAL STUDY OF INCOMPRESSIBLE JUNCTURE FLOWS

D. Kwak
NASA Ames Research Center, Moffett Field, CA, U.S.A.

S. E. Rogers, U. K. Kaul
Sterling Software, Palo Alto, CA, U.S.A.

and

J. L. C. Chang
Rocketdyne, Rockwell International, Canoga Park, CA, U.S.A.

I. Introduction

The present study is initiated to analyze the local flow around liquid oxygen (LOX) posts in the Space Shuttle Main Engine (SSME). As sketched schematically in figure 1, LOX posts are densely packed in the main injector assembly into which very non-uniform hot gas flows from the transfer ducts [1]. As a first step toward a complete flow analysis around the LOX post, the present study deals with idealized cases of laminar, steady juncture flows around a single and single and double rows of posts mounted between two flat plates.

The flow around a cylinder-plate juncture produces a very interesting viscous phenomena due to the interaction between the boundary layer from the plate and the cylinder. The three-dimensional separation of the boundary layer and subsequent formation of the so-called horseshoe vortex and its development is very challenging to analyze both experimentally and numerically. This type of flow can occur in many practical engineering problems. Flow around a wing-fuselage junction is a simple example of its kind, and the flow near the endwall of turbomachinery blades might be one of the most complicated problems. Most of the earlier studies on juncture flow have been experimental. Baker[2] shows that laminar juncture flow is confined to a very limited region. A similar result has been obtained most recently by Thomas [3]. Eckerle and Langston [4] reported a single primary vortex and saddle point contrary to multiple vortex systems observed earlier by other researchers.Interpretation of the phenomena also varies [3,5].

Computational simulation of these flows involves distinctively different features from that of external aerodynamics. For instance, the thickness of viscous layer for these types of flows is of the same order as the characteristic flow-field dimension, while the viscous region tends to be confined in a thin layer near the body for external flows. Realistic juncture flows under internal flow enviroment are likely to have large deflection as in the case of LOX post regions in the SSME. Separated and recirculating zones need to be resolved in the wake region of the first and the second row of posts in the two row cases. Recently, some numerical studes on this flow have been attempted. Kaul et al [6] reported an extensive numerical study on a single cylinder-plate flow. Subsequently, Rogers et al [7] simulated flow around a multiple post arrangement. Independently, Kiehm et al [8] reported a numerical study of flow around a single post in a channel. These computations reported qualitatively similar phenomena. The present report summarizes the recent numerical study performed at NASA Ames to understand and quantify the laminar juncture flow relevant to the SSME type flow. The turbulence modeling of the juncture flows will constitute the second phase of the study.

II. Solution Procedure

Since the Mach number of the flow is less than 0.12 [1], the flow field is computed using the incompressible Navier-Stokes flow solver, INS3D [9], which utilizes an implicit finite difference procedure in generalized curvilinear coordinates. This code has been verified for both two- and three-dimensional test problems, and has been applied to several realistic problems in aerospace design. Most notably, the application to the SSME flow field analysis has made a significant impact on the Shuttle program.

This code was used as a test code for the Numerical Aerodynamic Simulator (NAS) Facility Cray 2 at NASA Ames Research Center. Most of the numerical results presented here were obtained on the Cray 2.

III. Results

In Fig. 2, the computational domain for a single post on flat plate is illustrated. This is an extreme idealization of the flow around a post in the multiple post assembly in Fig. 1. However, understanding this flow is of considerable value toward a full analysis of the entire assembly. Upstream boundary layer thickness is varied by using a partially and fully developed channel flow profiles. The convergence characteristics of the flow solver are shown in Fig. 3 by the history of RMSDQ, which denotes the root-mean-square value of the change per iteration in the pressure and velocities. The three curves in the figure show three variations of the INS3D code; namely a block tridiagonal, a diagonal, and a diagonal version with fourth order implicit smoothing terms. The flow solver converges fast to about four orders of magnitude reduction in RMSDQ. The computing time per iteration per grid point is 91 μsec for the block tridiagonal version and 32 μsec for the diagonal version of the code. In Fig. 4, particle traces for a single post at Re= 1000 is shown. Saddle point separation and horseshoe vortex can be seen from the traces near the flat plate. The secondary flow in front of the cylinder wraps around toward the wake region and forms a counter-rotating pair of vortex filaments. These spiraling twin vortices demonstrate a striking difference between this type of juncture flow and a two-dimensional cylinder. The vortex filaments are washed upward and then attenuate as they interact and move down stream. In reality, vortex shedding and possible unsteady motion take place at this stage. These tornado shape vortices are very difficult to observe experimentally, and validation of this phenomenon was very much needed. Recently, Schewe [10] made oil flow visualization around a single post which shows a clear evidence of the twin vortex behind the cylinder (Fig. 5). This experimental observation is qualitatively similar to the computed results, shown by the particle traces in Fig. 4 and isobars near the flat plate in Fig. 6. This juncture flow structure will lead to a strong variation in skin friction and pressure along the cylinder and hence significantly affects the overall loading on the post.

Figures 7 through 10 show the computational results for a multiple post arrangement. A C-H type hybrid grid is used for two row computation as shown in Fig. 7 in the horizontal plane. This two-dimensional grid is stacked up vertically to form a three-dimensional grid. A total of 120,000 grid points were used for the calculation on the Cray-XMP. For the computation on the Cray 2, a refined grid with 463,000 mesh points was used. In the y-direction, periodicity is assumed. Comparison for the mass averaged total pressure drop between the single row and the double row of posts is shown in Fig. 8. In the figure, the first and the second row of the two row posts arrangement are designated as double row 1 and 2 respectively. The varying incidence angle corresponds to various relative locations of the posts in Fig. 1(b). Zero angle corresponds to the centerline of the center transfer duct and the angle increases clockwise. The angular dependency shown for the single row case is partially confirmed computationally by Chang and Yang (will be published as a continuation of [1]). In Fig. 9, particle traces near the post plate junction for two rows of posts is shown. Twin vortices behind the first row of posts are still observed and the saddle point of separation for the second row of posts is influenced by the presence of this spiraling vortices behind the first row as shown in Fig. 10 (a). Whether, in reality, the flow maintains steadiness behind the second row at this Reynolds number remains to be verified. The present flow solver is designed for obtaining steady state solutions, and the results can be regarded as an ensemble average in the wake region of the second row of posts. As shown in Fig. 10 (b), another tornado type vortex is generated behind the second row when the incident angle is increased to $78°$. Understanding the vortex dynamics and quantifying the resulting load on posts will be of significant engineering importance as well as an important contribution to the basic understanding the flow physics of juncture flows.

IV. Concluding Remarks

The present study shows the computational capability to study the juncture flow relevant to the SSME LOX posts. The computed flow compares favorably with experimental observation. By comparing the multiple-row post loading with the single-row post loading and varying the angles of flow incidence, the degree of interaction between the front and the rear rows of posts can be quantitatively determined. An improved multiple-row posts arrangement can, therefore, be determined based on the results from the present study. High Reynolds number flows need to be simulated to analyze more realistic cases in the SSME. This will require that special attention be given to the turbulence models for an accurate prediction of the loading on the posts. To complete the juncture flow study, a time-dependent flow solver will be essential.

Acknowledgements

Special thanks to Dr. G. Schewe of DFVLR, West Germany for providing his oil-flow visualization picture. This work is partially sponsored by NASA Marshall Space Flight Center.

References

1. Chang, J. L. C., Yang, R-J, and Kwak, D., A Full Navier-Stokes Simulation of Complex Internal Flows, Tenth International Conference on Numerical Methods in Fluid Dynamics, Beijing, Peoples Republic of China, 23-27, 1986.
2. Baker, C. J., The Laminar Horseshoe Vortex, J. Fluid Mech., vol. 95, part 2, pp346-367, 1979.
3. Thomas, A., Laminar Juncture Flow - A Visualization Study, Lockheed-Georgia Co., in press 1986.
4. Eckerle, W. A. and Langston, L. S., Horseshoe Vortex around a Cylinder, ASME International Gas Turbine Conference, Dusseldorf, West Germany, June 8-12, 1986.
5. Peake, D. J. and Tobak, M., Three-Dimensional Interactions and Vortical Flows with Emphasis on High Speeds, NASA TM 81169, March 1980.
6. Kaul, U. K., Kwak, D. and Wagner, C., A Computational Study of Saddle Point Separation and Horseshoe Vortex System, AIAA Paper 85-0182, 23rd Aerospace Sciences Meeting, Jan. 14-17, 1985.
7. Rogers, S. E., Kaul, U., and Kwak, D., 'A Numerical Study of Single and Multiple LOX Posts and Its Application to the Space Shuttle Main Engine,' AIAA Paper 86-0353, AIAA 24th Aerospace Sciences Meeting, Reno, Nevada, January 6-8, 1986.
8. Kiehm, P., Mitra, N. K. and Fiebig, M., Numerical Investigation of Two- and Three-Dimensional Confined Wakes behind a Circular Cylinder in a Channel, AIAA paper 86-0035, 24th Aerospace Sciences Meeting, Jan. 6-9, 1986.
9. Kwak, D., Chang, J. L. C., Shanks, S. P., and Chakravarthy, S., A Three-Dimensional Incompressible Navier-Stokes Flow Solver Using Primitive Variables, AIAA J, vol 24, No. 3, 390-396, Mar. 1986.
10. Schewe, G., Private communication, 1985. DFVLR, West Germany.

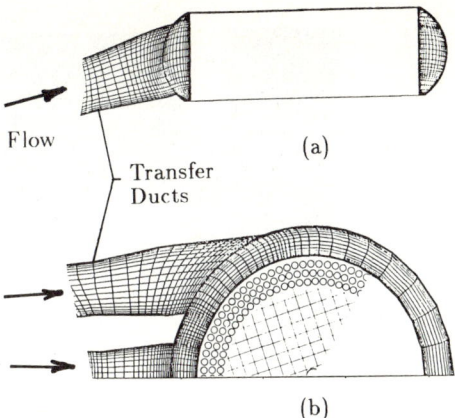

Fig. 1 Schematic of the liquid oxygen post in the Space Shuttle Main Engine:
(a) vertical cross section
(b) horizontal cross section

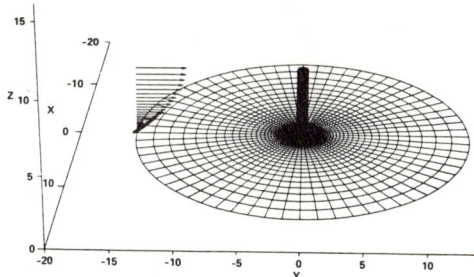

Fig. 2 Grid for a single post.

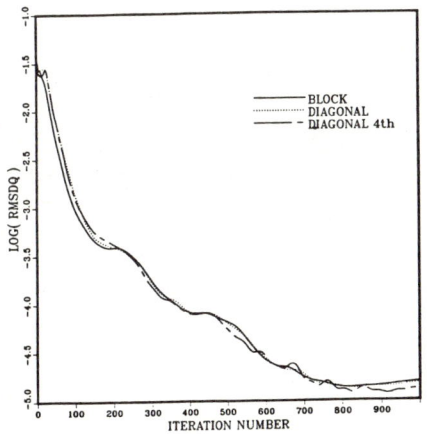

Fig. 3 Convergence history for a sigle post.

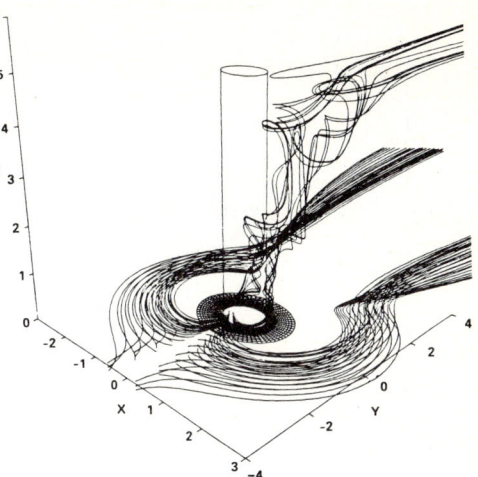

Fig. 4 Particle traces for a single post: Re=1000

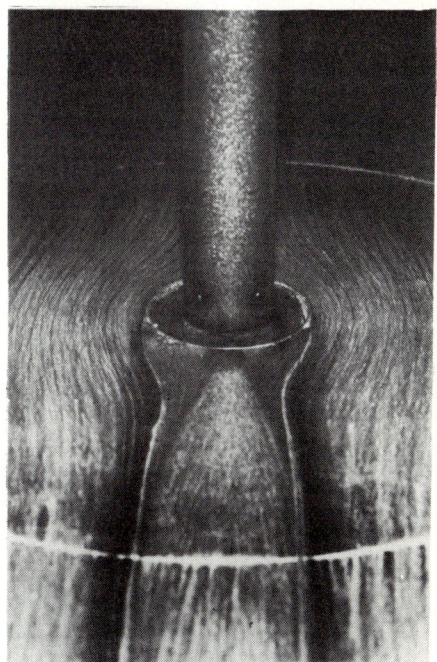

Fig. 5 Flow around a sigle post at $Re=1.85 \times 10^5$: oil-flow visualization by Schewe (DFVLR).

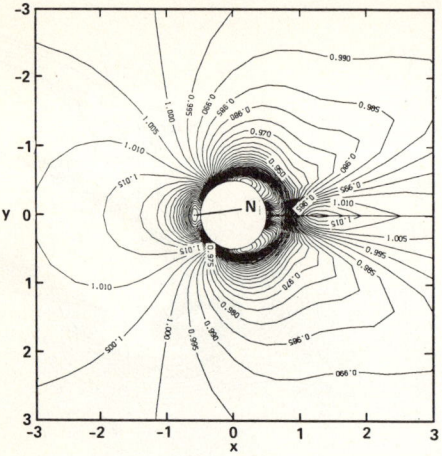

Fig. 6 Isobars in a horizontal plane at z/D=0.01.

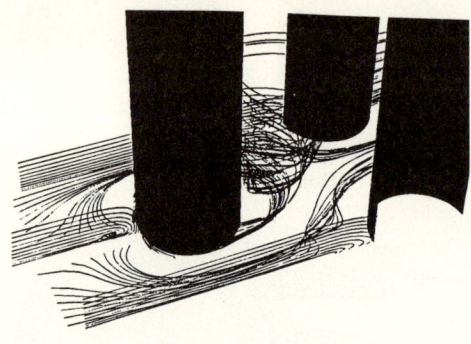

Fig. 9 Particle traces near the post-plate junction for two rows of posts.

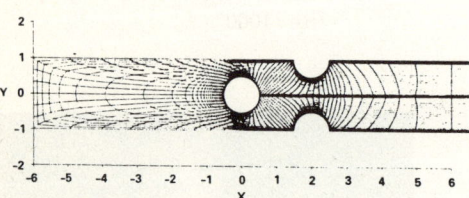

Fig. 7 Top view of the grid for two rows of post.

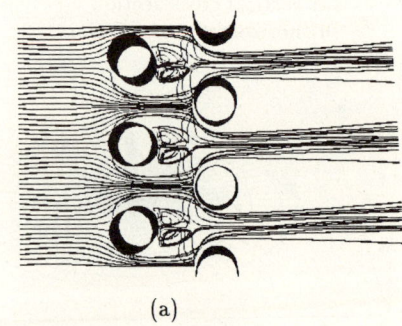

(a)

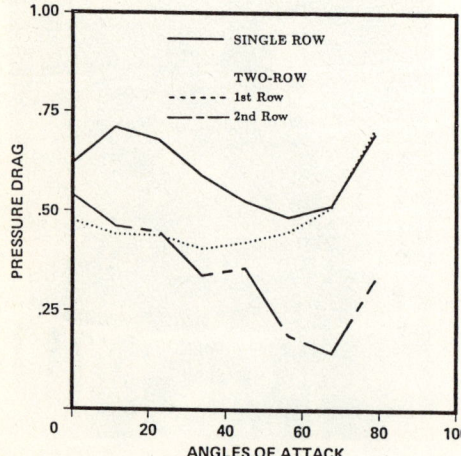

Fig. 8 Total pressure drop per unit of mass flow versus angle of attack (Re=1000).

(b)

Fig. 10 Particle traces for two rows of posts at Re=1000 : (a) 0°, (b) 78° incidence.

On the PIC Method for Elastic-Plastic Flow

Wen Ho Lee
Los Alamos National Laboratory
Los Alamos, New Mexico 87545,
U.S.A.

D. Kwak
Ames Research Center
Moffett Field, California 94036
U.S.A.

Introduction

The present work uses the PIC (particle in cell) numerical approximation with operator splitting to hydro-elastic-plastic flow problems in two-dimensional Eulerian coordinates. The solid materials properties, such as shear modulus and yield strength are dependent on pressure, compression, temperature, and equivalent plastic strain. However, the material melting is dependent on temperature only. Fracture and void or crack propagations are not included although a simple spall model has been implemented currently. The second order PIC method is used for computing hydrodymic flow as well as material behavior. Applying the operator splitting method, the basic set of cylindrical equations is split in radial and axial directions. The calculations, performed in each separate direction, are alternated for each time advancement to maintain the accuracy of one dimensional procedure. A self-forging fragment problem is treated using the present code and the results are compared with the experimental data.

Calculation Procedure

An operator splitting method is applied here. In other words, the governing set of cylindrical equations is split in radial and axial directions. Then calculation is performed in each separate direction. The order of this calculation is alternated for each time advancement to maintian the accuracy of one-dimensional procedure. A brief description of the procedures to solve the pertinent equations is given in Table I and Table II, followed by the notations.

Table I r-Direction Equations

Time Derivative	Remap Phase	Lagrangian Phase		
		Hydro	Stress	
$\rho_t =$	$-\bar{u}\,\rho_r$	$-\dfrac{\rho}{r}(ru)_r$		(1)
$u_t =$	$-\bar{u}\,u_r$	$-\dfrac{1}{\rho}P_r$	$+\dfrac{1}{\rho}\left[S_r^{rr} + \dfrac{1}{r}(2S^{rr}+S^{zz})\right]$	(2)
$v_t =$	$-\bar{u}\,v_r$		$+\dfrac{1}{\rho}\left(S_r^{rz} + \dfrac{1}{r}S^{rz}\right)$	(3)
$e_t =$	$-\bar{u}\,e_r$	$-\dfrac{P}{\rho r}(ru)_r$	$+\dfrac{1}{\rho}\left[(S^{rr}u_r + S^{rz}v_r)\right.$	
			$\left.-\dfrac{u}{r}(S^{rr}+S^{zz})\right]$	(4)
$S_t^{rr} =$	$-\bar{u}\,S_r^{rr}$		$+\dfrac{2}{3}G\left(2u_r - \dfrac{u}{r}\right)$	(5)
$S_t^{zz} =$	$-\bar{u}\,S_r^{zz}$		$+\dfrac{2}{3}G\left(u_r + \dfrac{u}{r}\right)$	(6)
$S_t^{rz} =$	$-\bar{u}\,S_r^{rz}$		$+G\,v_r$	(7)

Table II z-Direction Equations

Time Derivative	Remap Phase	Lagrangian Phase		
		Hydro	Stress	
$\rho_t =$	$-\bar{v}\,\rho_z$	$-\rho\,v_z$		(8)
$u_t =$	$-\bar{v}\,u_z$		$+\dfrac{1}{\rho}S_z^{rz}$	(9)
$v_t =$	$-\bar{v}\,v_z$	$+\dfrac{1}{\rho}P_z$	$+\dfrac{1}{\rho}S_z^{zz}$	(10)
$e_t =$	$-\bar{v}\,e_z$	$-\dfrac{P}{\rho}v_z$	$+\dfrac{1}{\rho}(S^{zz}v_z + S^{rz}u_z)$	(11)
$S_t^{rr} =$	$-\bar{v}\,S_z^{rr}$		$-\dfrac{2}{3}G\,v_z$	(12)
$S_t^{zz} =$	$-\bar{v}\,S_z^{zz}$		$+\dfrac{4}{3}G\,v_z$	(13)
$S_t^{rz} =$	$-\bar{v}\,S_z^{rz}$		$+G\,u_z$	(14)

The hoop stress is given by

$$S^{\theta\theta} = -(S^{rr} + S^{zz}) \quad . \tag{15}$$

For plastic regime, we have the stress deviation tensor S^{ij} defined as

$$(S_t^{ij})_P = (S_t^{ij})_e - \left[\frac{3G}{(Y^\circ)^2}\right] \dot{W} \, S^{ij} \tag{16}$$

where: $(S_t^{ij})_e$ = elastic components, and

$$\dot{W} = \left(u_r - \frac{u}{r}\right) S^{rr} - \frac{u}{r} S^{zz} + v_r S^{rz} \qquad \text{for r-Direction,} \tag{17}$$

and

$$\dot{W} = v_z S^{zz} + u_z S^{rz} \qquad \text{for z-Direction} \tag{18}$$

Equations (15) and (16) are good for both r and z directions.

Notations:

ρ = density

t = time

u = velocity in r direction

$\bar{u}$ = cell edge time average velocity in r direction

r = radial coordinate

P = pressure

S^{rr} = normal stress deviator in r direction

S^{zz} = normal stress deviator in z direction

v = velocity in z direction

S^{rz} = shear stress deviator

e = internal energy

G = shear modulus

$\bar{v}$ = cell edge time average velocity in z direction

Y° = yield strength

Subscripts t, r, and z represent the first derivative, i.e., $\frac{\partial}{\partial t}$, $\frac{\partial}{\partial r}$ and $\frac{\partial}{\partial z}$.

Lagrangian Phase

Step 1: First calculate $P^{n+1/2}$ and $(S^{ij})^{n+1/2}$ as below:

$$f^{n+1/2} = f^n + \frac{\delta t}{2} \frac{Df}{Dt} = f^n + \frac{\delta t}{2} (f_t + u f_r) \qquad (19)$$

$$P^{n+1/2} = g(\rho^{n+1/2}, e^{n+1/2}) \qquad (20)$$

Step 2: Then calculate $\tilde{u}$, $\tilde{e}$, and $\tilde{S}^{ij}$ using the time derivative and Lagrangian phase columns in Eqs. (2) through (7) for the r-direction integration and Eqs. (9) through (14) for the z-direction. For example, in z-direction calculation, we solve the $\tilde{e}$ by

$$\tilde{e} = e^n + (\delta t)[-(P v_z)^{n+1/2} + (S^{zz})^n v_z^{n+1/2} + (S^{rz})^n u_z^{n+1/2}]/\rho^{n+1/2} \qquad (21)$$

Remap Phase and Time Step Control

After Lagrangian phase calculation, we have $\tilde{u}$, $\tilde{v}$, $\tilde{e}$, and $\tilde{S}^{ij}$ at each cell. Consistent with these, temporary values of momentum and internal energy are assigned to cell (i) as below

$$\text{Momentum} = M_i \tilde{u}_i \qquad (22)$$

$$\text{Internal Energy} = M_i \tilde{e}_i \qquad (23)$$

where M_i = Mass of cell (i)
Then particles are transported with average velocities, i.e., in radial direction.

$$\bar{u} = \frac{1}{2}(u + \tilde{u}) \qquad (24)$$

Each particle carries with it a fraction of dynamic and state quantities defined by Eqs. (22) and (23). After the particle transport, total changes in momentum, energy, and stresses are calculated for each cell.

The time step control in the present particle code is done empirically. The following condition has been applied successfully.

$$N_c \triangleq \frac{(|u| + c)\delta t}{\delta x} < 0.4$$

where $c \triangleq (\frac{\delta P}{\delta \rho})_s + \frac{4}{3}\frac{G}{\rho}$ \qquad (25)

Sample Calculation and Conclusion

A self-forging fragment problem with a copper-liner, aluminum and steel case, point initiated Octol 75/25 high explosive, is calculated using the present code and a 2-dimensional Lagrangian code. For equation of state, we use JWL for HE, quadratic form for aluminum, steel and copper. Fig. 1-a shows the initial setup grid for the Lagrangian code, there are 31 zones in z direction and 90 sectors between the coordinates. Fig. 1-b gives the liner grid at time 12 μsec. Fig. 1-c and 1-d give the velocity vectors of the high explosive at time 12 μsec. Fig. 2 shows the particle plots of the fragment at time 20, 40, 60, 80, and 100 μsec, and the last figure presents the comparison of the calculation and the experimental data for the fragment at time 164 μsec. During the calculations, we use several values of yield strength (including elastic-perfect plastic) and find that by using a model of pressure and temperature dependent for yield strength and shear modulus will produce the best results.

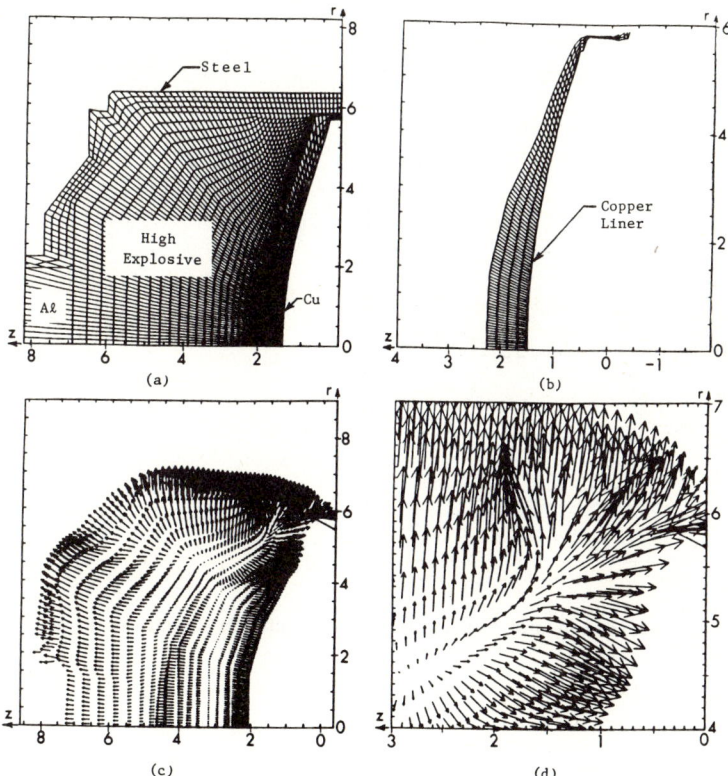

Fig. 1. The Lagrangian meshes and the velocity vectors (The coordinate units are in cm).

(a) The initial Lagrangian meshes with 1 zone in Aluminum, 4 zones in steel, 22 zones in high explosive, and 4 zones in copper liner.
(b) The liner meshes at time = 12 μsec.
(c) The velocity vector plot for high explosive.
(d) A blowup of the velocity vectors near the upper right corner.

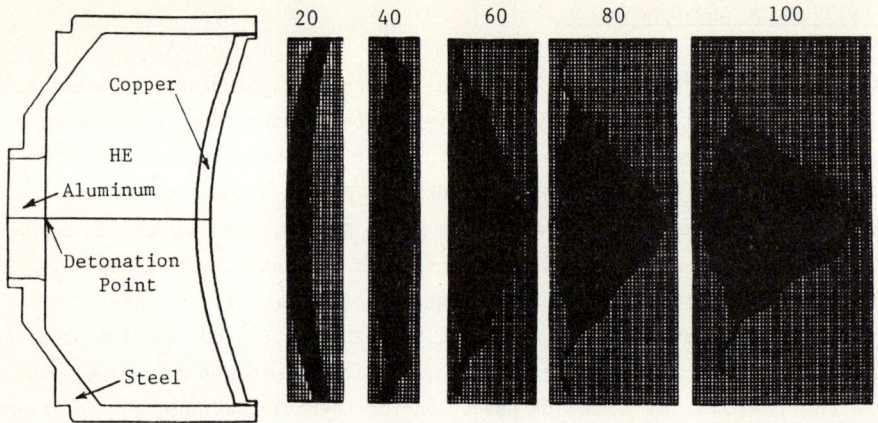

Fig. 2. Computer simulation of the fragment formation at 20, 40, 60, 80, and 100 μsec.

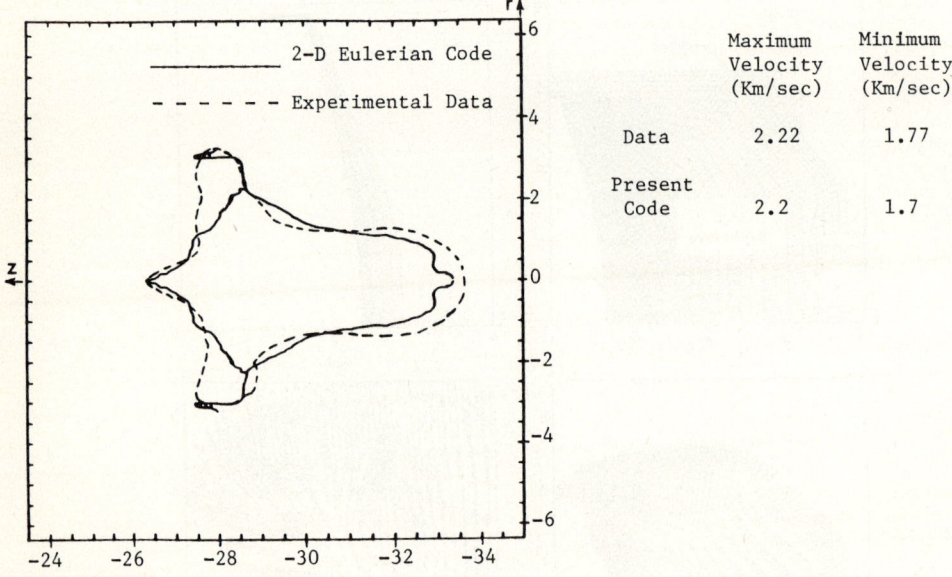

Fig. 3. Computer simulation and the radiograph of the fragment at time = 164 μsec.

IMPLICIT METHODS FOR COMPUTING CHEMICALLY REACTING FLOW

C. P. Li
NASA Johnson Space Center
Houston, Texas 77058

SUMMARY

Modeling the inviscid air flow and its constituents over a hypersonically flying body requires a large system of Euler and chemical rate equations in three spatial coordinates. In most cases, the simplest approach to solve for the variables would be based on explicit integration of the governing equations. But the standard techniques are not suitable for this purpose because the integration step size must be inordinately small in order to maintain numerical stability. The difficulty is due to the stiff character of the difference equations, as there exists a large spectrum of spatial and temporal scales in the approximation of physical phenomena by numerical methods. For instance, in the calculation of gradients caused by shock and by cooled wall on a coarse grid, unchecked numerical errors eventually will lead to violent instability, and in calculations of species near chemical equilibrium, a small error in one species will give rise to a large error in the source term for other species. Despite the different nature of the stiffness in a complex system of equations, the most effective approach is believed to be implicit integration. The step increment is no longer dictated by the stability criteria for explicit methods, but instead is dictated by the degree of linearization introduced to the governing equations and by the order of desired accuracy. The linearization is enacted by means of Jacobian matrices, resulting from the differentiation of the flux as well as the rate production terms with respect to dependent variables. The backward Euler scheme is then applied to discretize the partial differential equations and to convert them into a system of linear difference equations in vector form. As this particular approach has the A-stable property, it is the one recommended by Lomax and Bailey[1] for one-dimensional nonequilibrium flow studies. However, in the practice of solving flow problems in multidimensions, it was not clear then how to deal with the mammoth size of the sparse block matrix equations. The implementation of an implicit method in the solution procedure could be as prohibitively expensive as a modified Runge-Kutta method.[2]

FORMULATION

In view of the drawbacks associated with the implicit methods, other concepts have been evolved to avoid using the fully coupled approach. The most notable concepts probably are the hybrid explicit-implicit techniques that focus on the minimization of the stiffness due to the chemical production[3-4] and the time-split explicit method devised in such a manner that the flow and species equations are integrated separately according to the stability criteria of each.[5] These plausible ideas have provided limited success in two-dimensional problems, for which the flow equations are not stiff. The dilemma in regard to the unrestricted stability conditions and the excessive manipulation of a matrix-vector equation may be resolved now by taking a different path in seeking an efficient method.

The objective of the research reported herein is to evaluate the efficiency and robustness of two variants of an implicit method that has recently been developed to investigate the complete flowfield around an entry vehicle. The most challenging aspect of nonreacting flow computation was the generation of a computational grid so as to allow a single data set structure for both shock layer and trailing wake in one zone. The Euler equations were cast in the domain between the wall and the outer surface consisting of the bow shock and were solved by a modified version of the ADI factorization technique.[6] The methodology can be extended to consider reacting species in several ways; namely, coupled flow and species, decoupled flow from species and simultaneous species solution, and decoupled successive species solution. The differences among the three approaches are mainly in the degree of linearization and the ensuing amount of computation. The coupled approach would be the favorable one to use if the initial conditions are close to the final solution and the temporal accuracy is critical. The last two approaches are more appealing whenever the final solution is obtained after a reasonable, economical number of iterations. In the hypersonic flowfield with large subsonic regions surrounding the body, the initialization is at best very crude. Hence, the issue of efficiency is not as important as the robustness issue. Besides, the third approach has the potential of being the least computationally intensive method as the ionization is considered (11 vs 7 species).

OUTLINE OF THE DECOUPLED METHODS

The ideas can be elucidated by the one-dimensional problem having two species.

$$U_t + F_x + R = 0 \tag{1}$$

where

$$U = \begin{bmatrix} \rho \\ \rho u \\ \rho \varepsilon \\ \rho C_M \\ \rho C_A \end{bmatrix}, \quad F = \begin{bmatrix} \rho u \\ \rho u^2 + p \\ (\rho \varepsilon + p) u \\ \rho C_M u \\ \rho C_A u \end{bmatrix}, \quad R = \frac{A_x}{A} \begin{bmatrix} \rho u \\ \rho u^2 \\ (\rho \varepsilon + p) u \\ \rho \varepsilon_M u - \rho \omega_M \\ \rho C_A u - \rho \omega_A \end{bmatrix}$$

and A is the duct area; the subscripts M and A refer to molecular and atomic species respectively; and w is the production term. Standard notation is used otherwise.

An alternate equation equivalent to Eq. (1) is

$$V_t + AV_x + S = 0 \tag{2}$$

where

$$V = \begin{bmatrix} \rho \\ u \\ e \\ C_M \\ C_A \end{bmatrix}, \quad A = \begin{bmatrix} u & \rho & 0 & 0 & 0 \\ ge/\rho & u & g & 0 & 0 \\ 0 & p/\rho & u & 0 & 0 \\ 0 & 0 & 0 & u & 0 \\ 0 & 0 & 0 & 0 & u \end{bmatrix}, \quad S = \frac{A_x}{A} \begin{bmatrix} \rho u \\ 0 \\ \varepsilon(1-u) + p/\rho \\ -\omega_M \\ -\omega_A \end{bmatrix}$$

$$du = PdV, \quad P = \begin{bmatrix} 1 & 0 & 0 & 0 & 0 \\ u & \rho & 0 & 0 & 0 \\ q & \rho u & \rho & 0 & 0 \\ C_M & 0 & 0 & \rho & 0 \\ C_A & 0 & 0 & 0 & \rho \end{bmatrix}, \quad q = 0.5\, u^2$$

It is readily seen from Eq. (2) that a weak relationship exists between (ρ, u, e) and (C_M, C_A). The loosely coupled property in P and A will be exploited further in the following section. Neither Eq. (1) nor Eq. (2) is sufficient to describe the flowfield. Two equations of T and p

$$T_t + \left[e_t - \sum_\ell (C_\ell)_t e_\ell \right] / \sum_\ell C_\ell e_\ell = 0 \tag{3}$$

$$p = \rho RT \sum_\ell C_\ell / M_\ell$$

also are needed. Relations of the conservation of mass concentration, C_ℓ, and of charges are used for more complicated chemistry.

The production term, $\omega_\ell = \omega_\ell(\rho, T, C_\ell)$, is proportional to $(C_\ell - C_{\ell e})/\tau$, where $C_{\ell e}$ is the equilibrium value and τ is the reaction rate. It can have astronomical value even when normalized by the flow resident time in some portion of the flow. To cope with such stiffness, its relationship with the dependent variable should be analyzed and included in the algorithm by means of a Taylor series expansion as follows:

$$\omega_\ell = \omega_\ell^\circ + \left(\frac{\partial \omega_\ell}{\partial C_M}\right)^\circ \Delta C_\ell + \left(\frac{\partial \omega_\ell}{\partial \rho}\right)^\circ \Delta \rho + \left(\frac{\partial \omega_\ell}{\partial e}\right)^\circ \Delta e \tag{4}$$

The superscript $^\circ$ denotes the known values, and Δ represents the difference between the present and the known values. By incorporating Eq. (4) to Euler's backward scheme, Eq. (2) is converted to difference form in shorthand notation.

$$M_i^k \Delta V_i^{k+1} + \Delta t\, \delta_x \left(A_i^k\, \Delta V_i^{k+1}\right) = RHS \tag{5}$$

where δ_x refers to the centered-difference operator, and i is the index of the grid.

$$RHS = -\Delta t \left(A_i^k \delta_x V_i^k + S_i^k\right)$$

$$M = \begin{bmatrix} 1 & 0 & 0 & 0 & 0 \\ 0 & 1 & 0 & 0 & 0 \\ 0 & 0 & 1 & 0 & 0 \\ (\omega_M)_\rho \Delta t & 0 & (\omega_M)_e \Delta t & 1+(\omega_M)_{C_M}\Delta t & (\omega_M)_{C_A}\Delta t \\ (\omega_A)_\rho \Delta t & 0 & (\omega_A)_e \Delta t & (\omega_A)_{C_M}\Delta t & 1+(\omega_A)_{C_A}\Delta t \end{bmatrix}$$

The appearance of M in Eq. (5) and of the partial derivatives in the elements of M serves the role of tempering the resultant ΔV_i. Simplifications can be made to M by neglecting off-diagonal elements or only those elements associated with $(\ )\rho$ and $(\ )e$. Then, Eq. (5) is effectively decoupled into two groups for flow and species variables.

M has two splitup versions as follows:

$$M_F = \begin{bmatrix} 1 & 0 & 0 \\ 0 & 1 & 0 \\ 0 & 0 & 1 \end{bmatrix}, \quad M_S' = \begin{bmatrix} 1+(\omega_M)_{C_M}\Delta t & (\omega_M)_{C_A}\Delta t \\ (\omega_A)_{C_M}\Delta t & 1+(\omega_A)_{C_A}\Delta t \end{bmatrix}$$

$$M_S = \begin{bmatrix} 1+(\omega_M)_{C_M}\Delta t & 0 \\ 0 & 1+(\omega_A)_{C_A}\Delta t \end{bmatrix}$$

It is interesting to note that, in the simplest version, M_S^{-1} multiplies the RHS and the operator δ_x and results in the reduction of the time increment Δt by a factor of $1 + (w_i)_{C_i}\Delta t$. Since each species equation is then integrated by a different Δt, this approach uses the same idea advocated in Li.[5]

Equation (5) is a tridiagonal block or scalar system of equations depending on the form of M for which the solution procedure is available. It represents one of the three steps in solving three-dimensional problems. The truncation error of the method is $(\Delta t^2, \Delta x^3)$. By contrast, higher order explicit methods can be constructed by using

$$\Delta V_i^{k+1} = M_i^{-1}(RHS) \tag{6}$$

although Δt may be restricted to certain corridors of $\lambda \Delta t$, where λ is the largest eigenvalue of A.

DISCUSSION OF RESULTS

The development of decoupled implicit methods was based on two body configurations of practical interest and under flow conditions that give rise to extreme levels of dissociation and ionization. The fully implicit method has not been implemented to a computer code because it appears to be prohibitively costly for an 11-species model on today's scalar computer. The simultaneous species solver was used primarily to compare its performance with the less expansive successive species solver.

The first case was a sphere of $R_N = 0.328$ ft, at $M = 10$ and $h = 100$ kft. Some results of $M = 10$ reacting flow over the frontal portion of the sphere (table 3.16 in Belotserkoskiy[7]) were used to verify the accuracy of present results. In as much as the body is simple and the dissociation is weak, a coarse grid of 6×20 was found to be adequate.

Figure 1 shows the convergence history of the mass fraction of the oxygen molecule in terms of the maximum incremental and the stagnation values. As shown, both figures are needed to exhibit the realistic rate of convergence. The noticeably slow relaxation process in chemical nonequilibrium is indicative of the mutual interactions between the convection and the production of species allowed by the step size, $CN = 2$. Attempts to increase the courant number to $CN = 4$ failed to control the growth of error; indeed, a wide band of fluctuation was observed in ΔCO_2. After ΔCO_2 had reached and stayed at the level of truncation error, CO_2 began to converge at the forward stagnation point. Using the convergence history obtained from both methods, it was estimated that the successive method is a factor of two slower.

Figure 2 presents the temperature and species distributions across the shock layer in three angular orientations: $\theta = 0$, $\pi/2$, and π. An excellent agreement is found in temperature, but discrepancies are seen in the species distributions. As pointed out in Li,[5] the rate constants usually exert stronger influences on the species than on the temperature.

The second case investigated was an aerobrake of $R_N = 20$ ft at $M = 34.8$ and $h = 250$ kft. The computational grid, 13×36, was similar to that developed in Li[6] for nonreacting flow computations. A comparison of the convergence history between the simultaneous and the successive methods is shown in Fig. 3 for the nitrogen atom. The formal method has a faster rate in the first 100 iterations, but commands as many iterations as the latter method to reach the asymptotic value. The simultaneous method has produced a great deal of fluctuation that is atypical for flow with high levels of dissociation.

The successive method was slightly more efficient (by 15%) than the simultaneous method. Out of the total execute time, about 50% was spent to solve for the flow variables, 43% for the 5×5 ω_C matrix, and the rest of the execute time for the species. During each iteration, the analytical procedure determining w and w_C was called twice for each grid point such that no extra core was assigned to store them.

The temperature contours obtained for the two cases are exhibited in Fig. 4. A complete flowfield can be obtained with 200 to 400 iterations on the coarse grid laying between the wall and the outer surface.

Figure 5 illustrates the convergence history of C_N at the front and the rear stagnation points. This is the same aerobrake considered in case 2, but at a 20° angle of attack. The nonequilibrium flow in the shock layer changed little after 200 iterations, yet, in the near wake, probably more than 400 iterations were required to ensure convergence. Note that the stagnation value of C_N was lower than that predicted for the zero angle of attack case. Figure 6 gives the contour plot of Mach and e^- (no./cm^3) made on the pitch plane. The shear layer was not clearly visible because the grid of $13 \times 36 \times 6$ was quite coarse. The distribution of electrons displays three orders of magnitude developed from the shock layer to the wake.

CONCLUSION

It was found that the conventional implicit techniques can effectively reduce the stiffness associated with the chemical production term in the rate equations. The successive solution for the species was as stable as the simulation solution. The fact that reactive Jacobians were used to scale down the time increment was essentially regarded as a means to separate the rate equations and to perform integrations according to individual reaction time. In summary, the numerical procedure consists of decoupling the flow from species equations, solving them by block and scalar tridiagonal procedures, respectively, and utilizing the factorization ADI technique to tackle two- or three-dimensional problems. The procedure is stable and the computation time of species varies linearly with the number of species. Significant reduction of computation time can be achieved by storing the reactive Jacobians and by updating them periodically in the course of iterations.

REFERENCES

1. Lomax, H. and Bailey, H., NASA TN D-4109, 1967.

2. Treanor, C. E., *Math. Comp.* Vol. 20, 1966, pp. 39-45.

3. Rakich, J. V., and Park, C., Symposium on Application of Computers to Fluid Dynamic Analysis and Design, Polytechnic Institute of Brooklyn Graduate Center, New York, 1973.

4. Widkopf, G. F. and Victoria, K. J., Symposium on Application of Computers to Fluid Dynamic Analysis and Design, Polytechnic Institute of Brooklyn Graduate Center, New York, 1973.

5. Li, C. P., *J. Spacecraft & Rockets*, Vol. 9, 1972, pp. 571-572.

6. Li, C. P., Inter. Symp. Computational Fluid Dynamics - Tokyo, 1985, or NASA TM-58269, 1985.

7. Belotserkoskiy, O. M., NASA TT F-453, 1967.

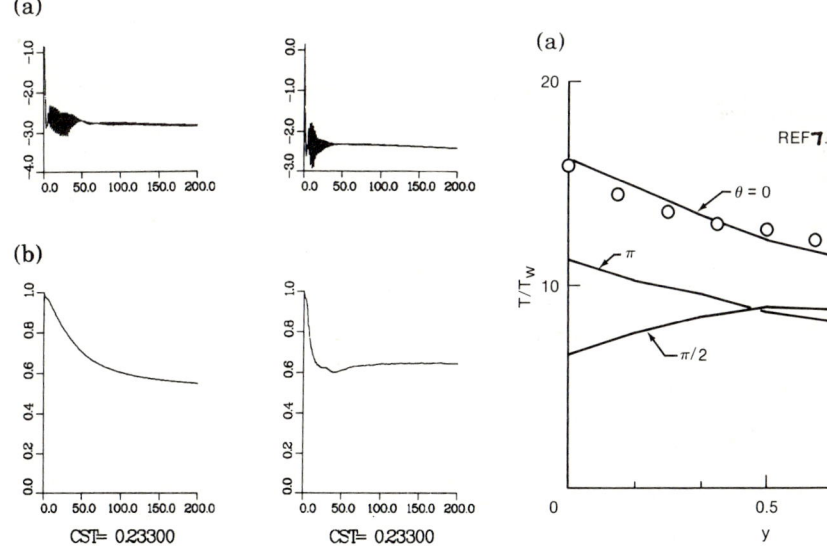

Fig. 1 Comparison of the convergence history between the simultaneous and successive sections for case 1: (a) maximum value of the incremental CO_2; (b) forward stagnation value of CO_2.

Fig. 2 Comparison of temperature and species mass fractions: (a) temperature distributions vs normalized distance from outer surface to wall; (Cont).

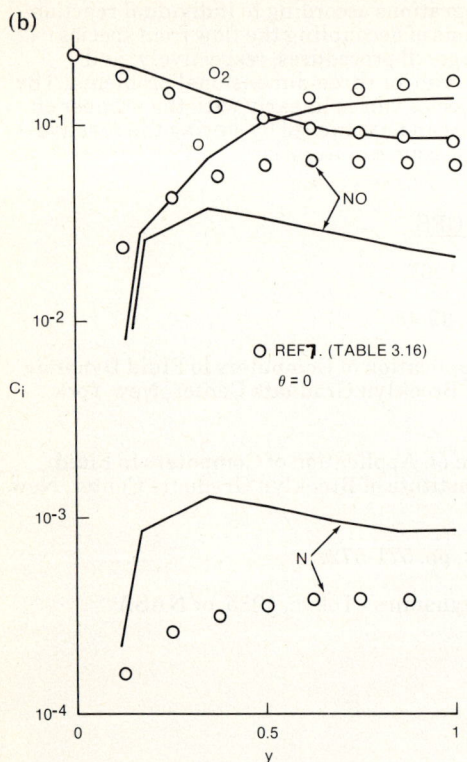

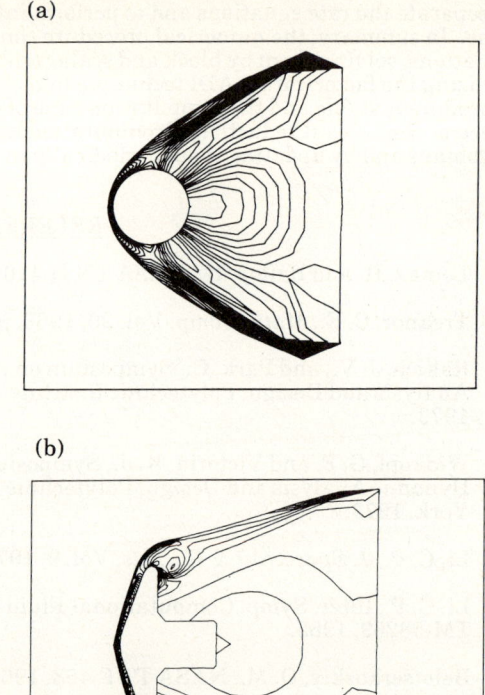

Fig. 2 Comparison of temperature and species mass fractions: (b) species distributions along the forward stagnation line between shock and wall.

Fig. 4 Temperature contours for cases 1 and 2: (a) case 1; (b) case 2.

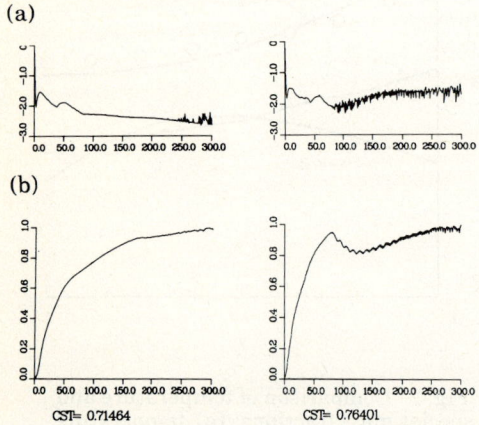

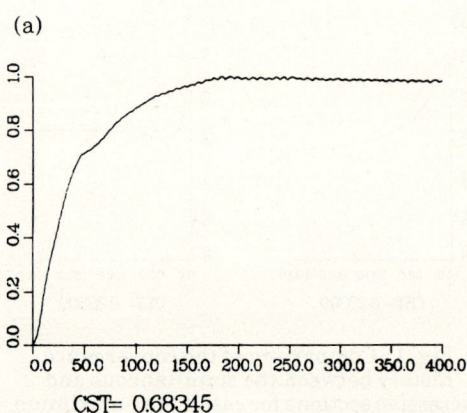

Fig. 3 Comparison of the convergence history between the simultaneous and successive solutions for case 2: (a) maximum value of the incremental C_N; (b) forward stagnation value of C_N.

Fig. 5 Convergence history of case 2 at $\alpha = 20°$: (a) front stagnation value of C_N; (Cont).

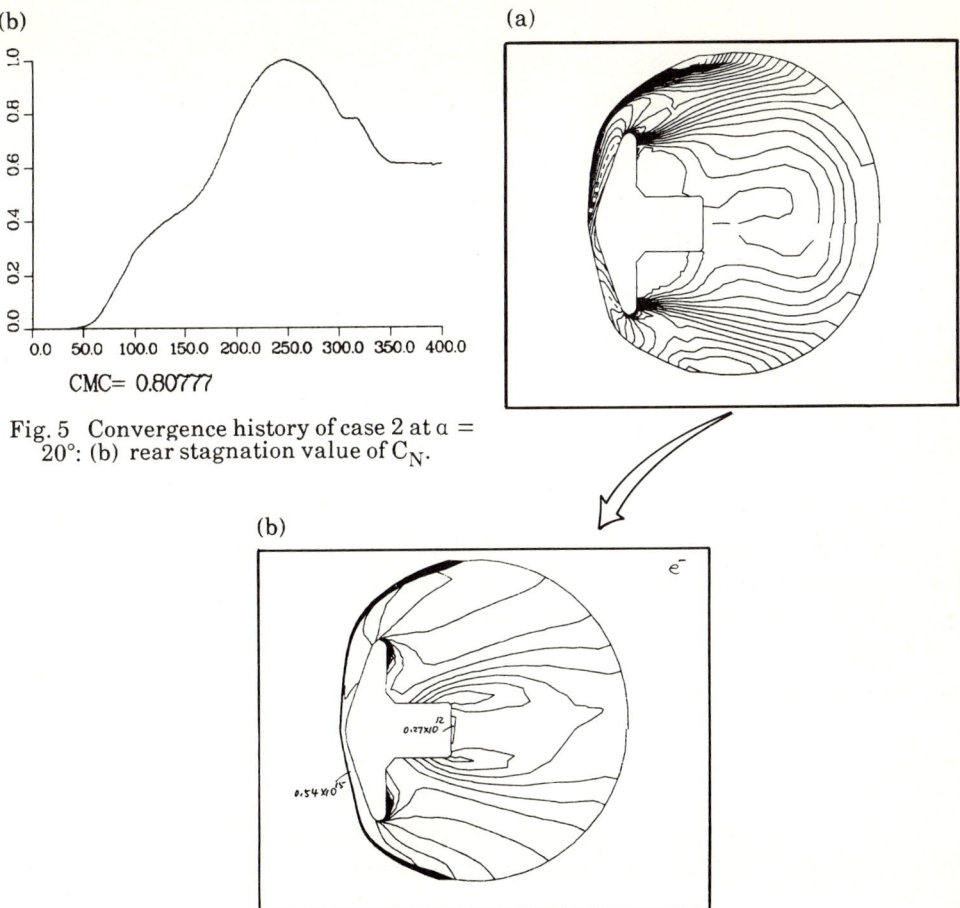

Fig. 5 Convergence history of case 2 at α = 20°: (b) rear stagnation value of C_N.

Fig. 6 Contours of Mach number and e^- (no./cm^3).

A "LARGE-PARTICLE" DIFFERENCE METHOD WITH SECOND ORDER ACCURACY FOR COMPUTATION OF TWO-DIMENSIONAL UNSTEADY FLOWS*

Y.F. Li and E.P. Qian

(Computing Center, Academia Sinica, Beijing)

1. Introduction. In our paper [1], a second-order "Large-Particle" difference method has been posed for computing 1-dimensional unsteady flows governed by Euler equations. The main idea of the method is as follows.

(1) The dependent variables in each cell are approximated by linear functions (as in finite element methods). The slopes of the distributions of variables obey van Leer's monotonicity constraint[3].

(2) The Eulerian time step as a whole is split up into two separate phases. The first phase involves the contributions of the pressure gradients. In this phase, the resolution of the initial discontinuities formed at cell boundaries is the building blocks much similar to Riemann problem in MUSCL method[3]. The second phase evaluates the contributions of the transport of the fluids according to the conservation laws.

In this paper, a second order difference scheme for computing 2-dimensional flows is proposed by using the 1-dimensional second order scheme mentioned above in alternating directions, according to the splitting algorithm of Strong.

2. Description of the Method. Equations of gasdynamics for 2-dimensional unsteady flow of a perfect fluid in Eulerian coordinates are written in conservation form as follows,

$$\frac{\partial U}{\partial t} + \frac{\partial F(U)}{\partial x} + \frac{\partial G(U)}{\partial y} = 0 \qquad (1)$$

where

$$U = \begin{pmatrix} \rho \\ \rho u \\ \rho v \\ \rho E \end{pmatrix} \quad F(U) = \begin{pmatrix} \rho u \\ \rho u^2 + p \\ \rho u v \\ (\rho E + p) u \end{pmatrix} \quad G(U) = \begin{pmatrix} \rho v \\ \rho u v \\ \rho v^2 + p \\ (\rho E + p) v \end{pmatrix}$$

ρ, u, v, p, E are the density, the components of the fluid velocity, the pressure and the total energy, respectively.

The perfect gas state equation is

$$p = (\gamma - 1)\rho e$$

where e is the specific internal energy, γ the ratio of specific heats. In this paper $\gamma = 1.4$.

Let Δt, Δx and Δy denote the time increment, the space increments in x direction and in y direction, respectively.

---

* This work is supported by the science fund of Academia Sinica, Contract R850337.

In each cell, all useful flow variables denoted generally by W, are approximated by a linear function

$$\{W(x,y,t)\}_{approx.} = \bar{W}^0_{i,j} + (x-x_i)\frac{(D_x)_{i,j}\bar{W}^0}{\Delta x} + (y-y_j)\frac{(D_y)_{i,j}\bar{W}^0}{\Delta y} \quad (2)$$

and $x_{i-\frac{1}{2}}$, $x_{i+\frac{1}{2}}$, $y_{j-\frac{1}{2}}$, and $y_{j+\frac{1}{2}}$ are the four sides of the (i,j) cell respectively. $\bar{W}^0_{i,j}$ is the mean of W in the (i,j) cell, and $(D_x)_{i,j}\bar{W}^0/\Delta x$, $(D_y)_{i,j}\bar{W}^0/\Delta y$ are the approximate partial derivatives of W at the centre of the (i, j) cell, (x_i, y_j).

Assuming that the piecewise linear distributions of ρ, u, v, p, etc. are known for certain time level $n\Delta t$, then the new linear distributions of ρ, u, v, p etc. for the next time level $(n+1)\Delta t$ should be evaluated. The procedure is as follows:

1) For each j, one time-step of 1-dimensional second order "Large Particle" method is performed for the equation

$$\frac{\partial U}{\partial t} + \frac{\partial}{\partial x}F(U) = 0 .$$

This procedure is denoted by $(L^x_{\Delta t} U^n)_{i,j} = U^{n+\frac{1}{2}}_{i,j}$

2) For each i, one time-step of 1-dimensional second order "Large Particle" method is performed for the equation

$$\frac{\partial U}{\partial t} + \frac{\partial}{\partial y}G(U) = 0 .$$

This procedure is denoted by $(L^y_{\Delta t} U^{n+\frac{1}{2}})_{i,j} = U^{n+1}_{i,j}$

3) For each i, one time step of $L^y_{\Delta t}$ is performed.

4) For each j, one time step of $L^x_{\Delta t}$ is performed.

Thus combining 1) to 4), we get $U^{n+2}_{i,j} = L^x_{\Delta t} L^y_{\Delta t} L^y_{\Delta t} L^x_{\Delta t} U^n_{i,j}$

The procedure of obtaining $U^{n+2}_{i,j}$ from $U^n_{i,j}$ has second order accuracy.

3. **One Direction Sweep.** In order to explain the method clearly, a brief description of 1-dimensional second order "Large Particle" method as the core of 2-dimensional second order "Large Particle" method is given here. We consider the operator $L^x_{\Delta t}$, which solves numerically the equation

$$\frac{\partial U}{\partial t} + \frac{\partial F(U)}{\partial x} = 0 . \quad (3)$$

It is accomplished in two phases and the first phase considers the contributions of the pressure gradients and the second one the effects of the transport of the fluid.

Phase 1: In this phase actually the following equations are to be solved

$$\begin{cases} \rho \dfrac{\partial u}{\partial t} + \dfrac{\partial p}{\partial x} = 0 \\ \dfrac{\partial \rho E}{\partial t} + \dfrac{\partial pu}{\partial x} = 0 \end{cases} \qquad (4)$$

by following several steps:

1° The resolution of the initial discontinuity

An initial discontinuity is formed at the interface of two adjacent cells, for example at $x = x_{i+\frac{1}{2}}$, the values of variables on the two sides of the interface $x = x_{i+\frac{1}{2}}$ are

$$W^n_{(i+\frac{1}{2},j)_+} = W^n_{i+1,j} - \frac{1}{2}(D_x)_{i+1,j} W^n$$

$$W^n_{(i+\frac{1}{2},j)_-} = W^n_{i,j} + \frac{1}{2}(D_x)_{i,j} W^n.$$

The values at $(x_{i+\frac{1}{2}}, y_j, n\Delta t+0)$ are found by solving the Riemann problem of system (4). Thus

$$p^*_{i+\frac{1}{2},j} = \lim_{t \to n\Delta t + 0} p(x_{i+\frac{1}{2}}, y_j, t),$$

$$u^*_{i+\frac{1}{2},j} = \lim_{t \to n\Delta t + 0} u(x_{i+\frac{1}{2}}, y_j, t).$$

If time increment Δt is choosen properly, each state quantity will be continuous in the forward direction of t after resolving the initial discontinuity. Furthermore, the formulas of $\partial p^*_{i+\frac{1}{2},j}/\partial t$ and $\partial u^*_{i+\frac{1}{2},j}/\partial t$ can be obtained approximately. So, the linear distributions of p^* and u^* can be obtained as follows

$$p^*(x_{i+\frac{1}{2}}, y_j, t) = p^*_{i+\frac{1}{2},j} + \frac{\partial p^*_{i+\frac{1}{2},j}}{\partial t}(t - n\Delta t)$$

$$u^*(x_{i+\frac{1}{2}}, y_j, t) = u^*_{i+\frac{1}{2},j} + \frac{\partial u^*_{i+\frac{1}{2},j}}{\partial t}(t - n\Delta t)$$

$$(n\Delta t \leq t \leq (n+1)\Delta t)$$

The formulas of p^*, u^*, $\partial p^*/\partial t$ and $\partial u^*/\partial t$ have been given in [1] in detail.

2° The difference scheme for the first phase is as follows

$$\tilde{u}_{i,j} = u^n_{i,j} - \frac{\Delta t}{\rho^n_{i,j} \Delta x}[\tilde{p}^{n+\frac{1}{2}}_{i+\frac{1}{2},j} - \tilde{p}^{n+\frac{1}{2}}_{i-\frac{1}{2},j}]$$

$$\tilde{E}_{i,j} = E^n_{i,j} - \frac{t}{\rho^n_{i,j} \Delta x}[(\widetilde{pu})^{n+\frac{1}{2}}_{i+\frac{1}{2},j} - (\widetilde{pu})^{n+\frac{1}{2}}_{i-\frac{1}{2},j}]$$

where $$\tilde{p}^{n+\frac{1}{2}}_{i\pm\frac{1}{2},j} = p^*_{i\pm\frac{1}{2},j} + \frac{\Delta t}{2} \frac{\partial p^*}{\partial t}\bigg|_{i\pm\frac{1}{2},j}$$

$$\tilde{u}^{n+\frac{1}{2}}_{i\pm\frac{1}{2},j} = u^*_{i\pm\frac{1}{2},j} + \frac{\Delta t}{2} \frac{\partial u^*}{\partial t}\bigg|_{i\pm\frac{1}{2},j}.$$

Based on

$$\tilde{p}^{n+1}_{i\pm\frac{1}{2},j} = p^*_{i\pm\frac{1}{2},j} + \Delta t \frac{\partial p^*}{\partial t}\bigg|_{i\pm\frac{1}{2},j}$$

$$\tilde{u}^{n+1}_{i\pm\frac{1}{2},j} = u^{*}_{i\pm\frac{1}{2},j} + \Delta t \left.\frac{\partial u^{*}}{\partial t}\right|_{i\pm\frac{1}{2},j},$$

the slopes of p, u for the phase 2 can be obtained as follows

$$(D_x)_{i,j} \tilde{u}^{n+1} = \tilde{u}^{n+1}_{i+\frac{1}{2},j} - \tilde{u}^{n+1}_{i-\frac{1}{2},j}$$

$$(D_x)_{i,j} \tilde{p}^{n+1} = \tilde{p}^{n+1}_{i+\frac{1}{2},j} - \tilde{p}^{n+1}_{i-\frac{1}{2},j} .$$

Finally, van Leer's monotonicity constraint is used to modify the slopes.

Phase 2. The mean values of the density, the velocity and the total energy in the cell (i,j) at (n+1)Δt are obtained by transporting the intermediate distribution with the velocity $\bar{u} = 0.5(u^n + \tilde{u})$ according to the conservation laws. The least-squares fitting is used to determine the new slopes.

In paper [2], the second order "Large-Particle" difference method for 1 - dimension problem has been improved and simplified. More numerical tests are also given in [2].

4. Numerical Results. The test problem is an unsteady flow in a duct containing a forward-facing step. The duct width is 1, the length 3, and the forward-facing step, 0.2 in height, locates at distance of 0.6 from the entrance. The step and the upper and the lower walls of the duct are specified as reflecting boundaries. On the left inlet there is a uniform inflow with Mach 3. On the right outlet continuation boundary conditions are specified. The initial conditions are specified as a uniform flow throughout the duct:

$$p(x,y,0) = 1, \rho(x,y,0) = 1.4, u(x,y,0) = 3, v(x,y,0) = 0.$$

The step problem, known as the Emery test, has been used by van Leer in [3], Woodward and Colella in [4]. The corner of the step is a singular point of the flow. In our calculation nothing special is done there.

In our computation, Δx=Δy=0.05, Δt=0.01. In Fig.5 and Fig.6, the density and the pressure contours of the solution at t=4 are shown. The contact discontinuity near the upper wall is clearly visible. Also the upper Mach stem is in the correct position directly above the step, and it has the correct length (as proven by [4]). However, there is some confusion near the corner of the step by large numerical errors generated just in the neighborhood of the singular point. For comparison, in Fig.7 and Fig.8 density and pressure contours of the solution at t=4 obtained by the original first order "Large-Particle" method[7] with Δx=Δy=0.05 and Δt=0.01, are also shown.

Obviously, the second order scheme given in this paper greatly improve the accuracy of calculations of the test problem as compared with those given by first order "Large-Particle" method.

Finally, the results of 1-dimensional second order method[1] on Sod's[5] shock tube problem are also given in Figs. 1-4.

References

[1] Li Yin-fan et. al., Scientia Sinica, 28(1985), 10: 1024-1035.
[2] En-ping Qian, Simplified Second-Order "Large-Particle" Difference Method, a M.S. thesis of Computing Center of Academia Sinica.
[3] Van Leer, B., J. Computational physis, 32(1979), 101-136.
[4] Woodward, P., colella, P., ibid., 54(1984), 115-173.
[5] Sod, G.A., ibid., 27 (1978), 1.
[6] Gentry, R.A., Martin, R.E., Daly, B.J., ibid., 1(1966) 87-118.
[7] Belotserkovski O.M., Davydov Yu. M., Z. Vyčisl. Mat. i Mat. Fiz., 11(1971), 1: 182-207 (Russian).

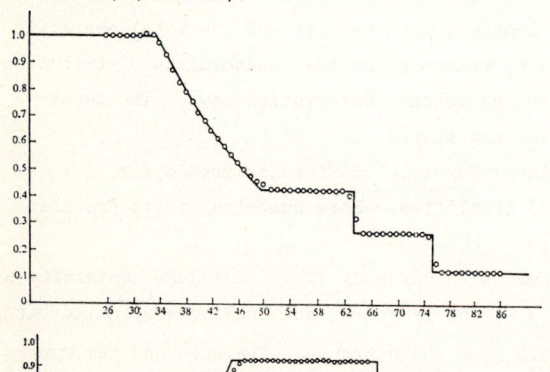

Fig.1. Density at t=0.14154

Δx = 1./100. Δt = t/40.

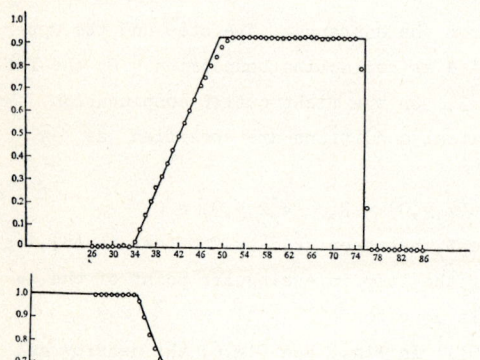

Fig.2. Velocity at t = 0.14154

Δx = 1./100. Δt = t/40.

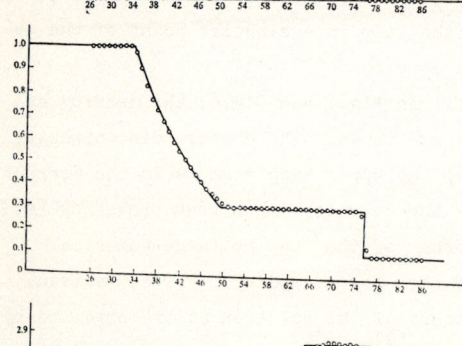

Fig.3. Pressure at t=0.14154

Δx = 1./100. Δt = t/40.

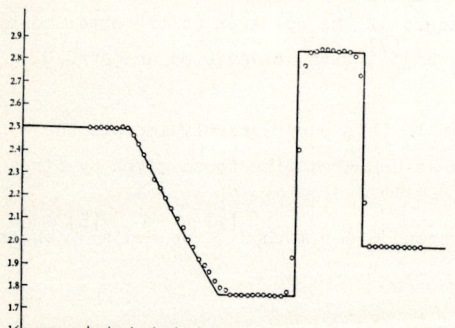

Fig.4. Internal Energy at t=0.14154

Δx=1./100. Δt=t/40.

oooo   Numerical Solution

———   Exact Solution

Fig.5. 30 Density Contours from 1.11 to 6.507
Second Order Method (t=4,$\Delta x=\Delta y=0.05$,$\Delta t=0.01$)

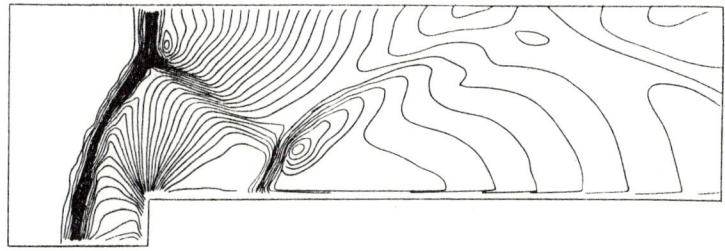

Fig.6. 30 Pressure Contours from 0.8532 to 11.65
Second Order Method (t=4,$\Delta x=\Delta y=0.05$,$\Delta t=0.01$)

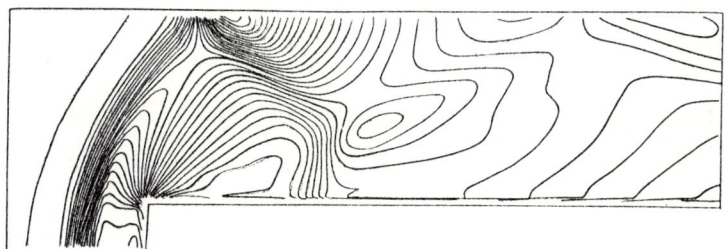

Fig.7. 30 Density Contours from 1.038 to 6.305
First Order Method (t=4,$\Delta x=\Delta y=0.05$,$\Delta t=0.01$)

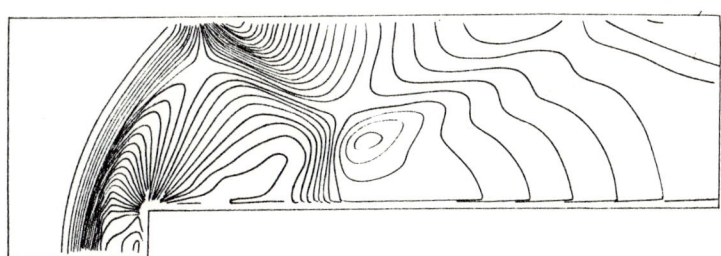

Fig.8. 30 Pressure Contours from 1.178 to 11.52
First Order Method (t=4,$\Delta x=\Delta y=0.05$,$\Delta t=0.01$)

AIRFOIL DESIGN AT SONIC VELOCITY

B.Y. Ling
Nanchang Aircraft Manufacturing Company
Nanchang, China

J.D. Cole
Rensselaer Polytechnic Institute
Troy, NY 12181   U.S.A.

1. Introduction

Airfoil designs by hodograph method have been developed at NLR[1] and CIMS[2]. But their free stream Mach number must be less than 1.0. The pattern of the flow over an airfoil at free stream Mach number 1.0 is quite different from other transonic flows (Fig. 1). We find the sonic line starts on the airfoil and extends in vertical fashion to infinity. Downstream of the sonic line some of Mach waves that come from the airfoil run into the sonic line and are reflected and some of them run into the trailing shock. In between there is one which neither hits the sonic line nor the shock in finite distance. It is named as limiting Mach wave which is asymptotic to the sonic line and the trailing shock at infinity. The shocks start from the tail (such a position is assumed in sonic flow discussion by many and is generally accepted). After the shock the flow is slightly supersonic and is gradually compressed, and becomes sonic at a great distance downstream[3].

In transonic small disturbance theory the far field of the 2-D velocity potential for subsonic free stream ($M_\infty < 1$) becomes singular. Therefore a special far field solution for the sonic case must be developed.

2. Basic Equation on Hodograph Plane

In this paper exact hodograph equation represented with stream function Ψ is obtained by using Chaplygin transformation.

$$q^2 \Psi_{qq} + q(1+M^2) \Psi_q + (1-M^2) \Psi_{\theta\theta} = 0 . \tag{1}$$

where θ is flow angle, q flow velocity.

Let $\tau = (q/q_{max})^2$, q_{max} = Maximum possible speed.

Equation (1) can be written as

$$(1-\tau)\Psi_{\tau\tau} + (1 + \frac{2-r}{r-1}\tau)\Psi_\tau + \frac{1 - \frac{r+1}{r-1}}{\tau} \Psi_{\theta\theta} = 0 . \tag{2}$$

On the sonic line equation (2) becomes

$$\Psi_{\tau\tau} + 9\Psi_\tau = 0 . \tag{3}$$

For convenience characteristic coordinates are used in the hyperbolic region.

Let
$$\xi = -\theta + f(\tau) ,$$

$$\eta = \theta + f(\tau) ,$$

$$f(\tau) = \sqrt{\frac{r+1}{r-1}} \, tg^{-1} \sqrt{\frac{r-1}{r+1} \cdot \frac{1 - \frac{r+1}{r-1}}{\tau - 1}} - tg^{-1} \sqrt{\frac{1 - \frac{r+1}{r-1}}{\tau - 1}} .$$

Equation (2) can be written as

$$(6\tau - 1)^{3/2} \Psi_{\xi\eta} + \frac{7.5\,\tau^2}{\sqrt{1-\tau}} (\Psi_\xi + \Psi_\eta) = 0 \tag{4}$$

Equations (2), (3) and (4) are the basic ones for numerical calculation used in the different velocity regions.

We know that the hodograph method cannot be straightforwardly used to calculate the flow field for a given shape of body. But it can be used to solve an inverse problem. We can compute a smooth transonic flow and then find the body which generates it.

3. Boundary Condition in the Hodograph Plane

(A) Treatment of free stream singularity

The far field in physical plane is mapped to a singular point O in the hodograph plane. On the basis of small disturbance theory the approximate analytical solution of Tricomi equation is employed for establishing the boundary condition near the free stream singular point in hodograph plane. Germain[4] gave the solution in algebraic form which is very handy for computation.

$$\Psi = c\rho^{-3}\left[(\rho+3\vartheta)(\rho-\vartheta)^{1/3} + (3\vartheta-\rho)(\rho+\vartheta)^{1/3}\right],$$

$$\rho^2 = \vartheta^2 - \frac{4}{9}w^3/\vartheta^2, \tag{5}$$

$$\vartheta^2 = \theta^2/(r+1).$$

For the convenience of numerical calculation we use a small box near the point O as the boundary of the free stream singularity. On the box boundary the mesh spacing is very small, the value of Ψ can take on all values from $+\infty$ to $-\infty$.

(B) The curve OA (Fig.2) in the hodograph plane, which is corresponding to a streamline from infinity upstream to the stagnation point, is a zero streamline. It is to be formed naturally during the solution process.

(C) The boundary curves DS_1L_1T and DAS_2L_2T, on which the values of Ψ equal to zero, are arbitrarily determined by using Spline method.

4. Numerical Solution

In this paper a numerical solution for the basic equations (2), (3) and (4) is done using finite difference method. A centered difference scheme is used in the subsonic region. A characteristic grid is used in the supersonic region and a backward difference scheme is employed in the marching direction. A new difference scheme is used for the grids on the sonic line.

The whole flow field is divided into the 'front' and 'rear' parts by limiting lines. In the front part the difference scheme is implicit. It can first be solved independently. After obtaining the results for the front part, we can carry on the calculation to the back part. But the calculation of back part is explicit.

Different coordinate systems are used in the different velocity regions while using successive relaxation. We do a velocity vector(τ) line relaxation procedure in the direction indicated in Fig. 3.

In the paper five examples are calculated. The boundary curves of five examples are shown in Fig. 4. Five airfoil shapes obtained are shown in Fig. 5. The pressure distributions are in Fig. 6.

Reference

(1) Nieuwland, G. F., "Transonic Potential Flow around a Family of Quasi-Elliptical Airfoil Sections", NLR 1967.
(2) Garabedian, B. F., Korn, P. D., and Jameson, A., "Supercritical Wing Section II". Spring-Verlag, New York, 1975.
(3) Tse, E., "Airfoil at Sonic Velocity", Ph.D. Dissertation, University of California, Los Angeles, June 1980.
(4) Germain, P., "Ecoulements Transsoniques Homogenes", Progress in Aeronautical Sciences, Vol. 5, Pergamon Press, 1964.

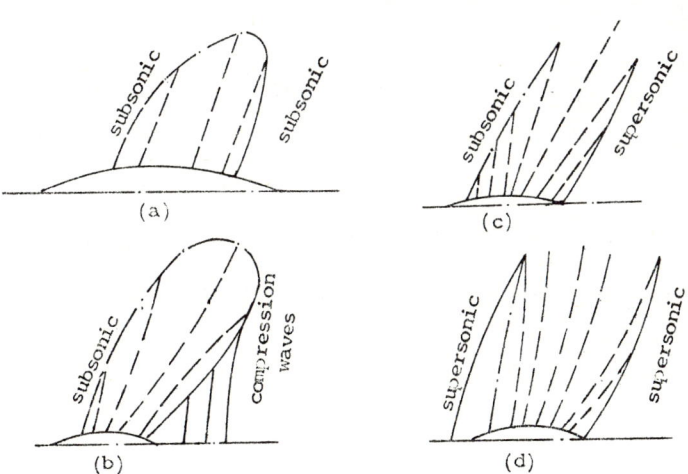

(a) High subsonic (b) Higher subsonic (c) Sonic (d) Slightly supersonic

Fig. 1 Structures of flow past an airfoil at Mach number close to one

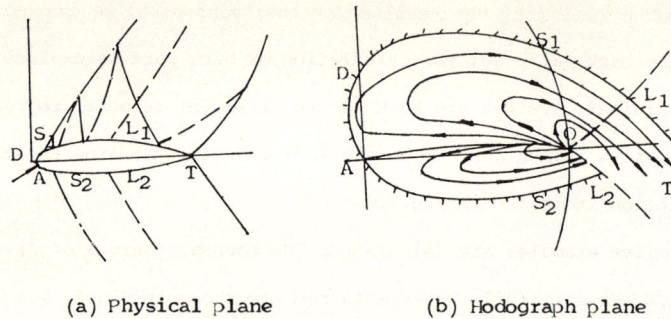

(a) Physical plane   (b) Hodograph plane

Fig.2 Sonic flow pattern over an airfoil

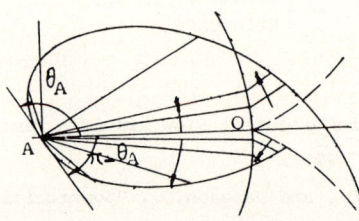

Fig.3 Direction of line relaxation on the front part

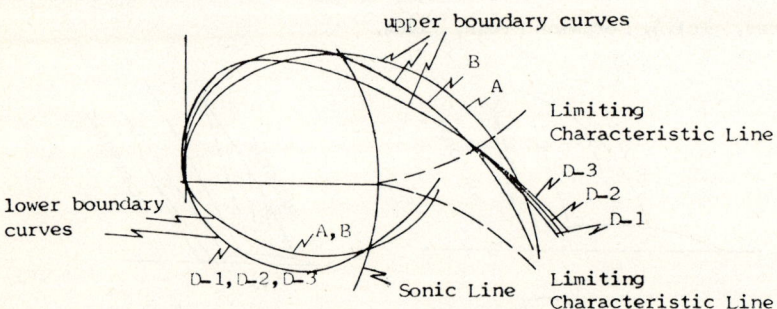

Fig.4 Hodograph boundary curves of five examples

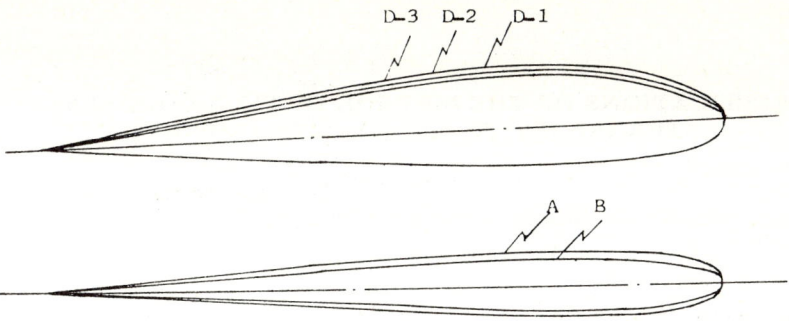

Fig. 5 Airfoil shapes of five examples

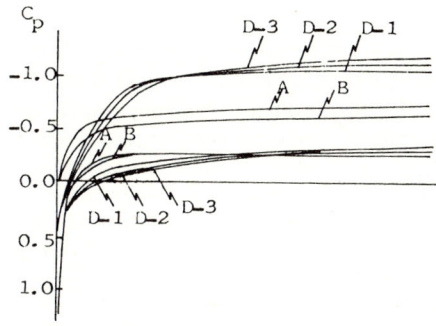

Fig. 6 Pressure distributions of five examples

APPLICATIONS OF THE METHOD OF FLUX-CORRECTED TRANSPORT TO GENERALIZED MESHES

Rainald Löhner and Gopal Patnaik
Berkeley Research Associates
Springfield, VA 22150

Jay P. Boris, Elaine S. Oran, and David L. Book
Laboratory for Computational Physics
Naval Research Laboratory, Washington, DC 20375

FCT on Unstructured Grids

A new technique for numerical solution of fluid equations has been developed by combining the Finite-Element Method with the method of Flux-Corrected Transport. The resulting hybrid method, called FEM-FCT, is useful for problems involving steady and unsteady transonic and supersonic flow in irregular geometries. The main computational advance grows out of the need to find a prescription for limiting fluxes through the sides of a triangular or other nonquadrilateral zone when arbitrarily many zones may meet at a given vertex (Löhner et al., 1986).

The technique described here employs a version of Zalesak's (1979) fully multidimensional FCT on a triangular grid. The underlying transport scheme can be any standard high-order finite-element or finite-volume method, and need not be based on triangular zones. In the present work a Taylor-Galerkin scheme (Löhner et al., 1985) has been used for this purpose. The low-order scheme employed in Zalesak's formulation is obtained by adding a numerical diffusion term to the high-order scheme. Antidiffusion removes this added diffusion except where the flux-limiting process modifies the antidiffusive fluxes to prevent unphysical extrema from forming. For each fluid equation all the high-order fluxes which tend to increase the value in a particular zone, and separately all the fluxes which tend to decrease the value there, are considered together. If in either case a maximum or minimum is formed anywhere which is not produced by the low-order scheme, all the participating fluxes are reduced by the same factor until the extraneous peak or valley is eliminated. The result for the transported quantity is a value which is intermediate between the high- and the low-order results, with just enough of the latter present to guarantee that no extrema form (except those produced physically).

For the test problems used in this paper to illustrate the method, the temporal discretization is a two-step (predictor-corrector) Lax-Wendroff-type method and the spatial discretization is done with triangular finite elements (Löhner et al., 1985). A Lapidus diffusion term of the form

$$h^2 \frac{\partial}{\partial l} \left\{ \max\left[0, \left|\frac{\partial(1 \cdot \mathbf{v})}{\partial l}\right|\right] \frac{\partial \mathbf{u}}{\partial l} \right\},$$

where Δt is the timestep, h is the zone size, $\mathbf{l} = \nabla v / |\nabla v|$ is the unit vector in the direction of the gradient of $v = |\mathbf{v}|$, and $\mathbf{u}$ is the vector of dependent variables, is applied in expansion regions to prevent the formation of "terraces." Running times on a one-pipe Cray X-MP were 42–58 μs per grid point per timestep. These are about a factor of three slower than efficient FCT algorithms on rectilinear grids, e.g., JPBFCT (Boris, 1981). The penalty results from the need to include gather-scatter operations in the algorithm because physically contiguous quantities need not be logically contiguous.

After optimization of the flux-limiting technique through tests on one-dimensional problems, the method was applied to a variety of unsteady flows, including spherical blast waves and shock waves diffracting around hemispheres and half-cylinders. Because the algorithm is not coordinate-timesplit the simulations maintained symmetry perfectly, so that, e.g., the projection of a spherical blast wave on the z-axis is identical to that on the r-axis. Tests were also carried out on the problem of a planar shock impinging on a complex structure composed of two irregular objects (Fig. 1). A strong ($M_s = 10$) shock hits the two structures shown, producing a bow shock and several rarefactions and contact discontinuities, as well as several reflected secondary shocks (e.g., below the structure on the left). Note the high resolution of the shocks and contact discontinuities; the jumps are resolved over 1–2 zones, in contrast with the number (4–6) needed in conventional finite-element schemes. Note also that those contours which are supposed to be straight remain straight, i.e., they essentially ignore the orientation of the underlying grid. Figure 2 shows the results of another calculation, in which a strong ($M_s = 25$) shock interacts with a channel aligned parallel to the shock front. At the time shown ($t = 1.2$ in normalized units) the incident (unreflected) shock has passed beyond the edge of the frame to the right. At the bottom it has reflected off the right wall of the channel and is seen propagating back towards the left. Focusing occurs because of the geometry, resulting in the formation of the "eye" just to the right of the middle of the channel. Note the very strong rarefaction fan attached to the top of the left wall of the channel; features like this in hydrocode calculations are ordinarily very susceptible to dispersive errors or (in the case of FCT algorithms) terracing.

One of the advantages of triangular gridding [Fig. 1(a) and Fig. 2(a)] is that it is easy to refine the mesh in regions where improvements in resolution are needed. The strategy we have followed is one of enrichment, not redistribution. That is, we introduce new triangles in the region of interest, rather than moving triangles from elsewhere. We have implemented an automatic adaptive mesh refinement routine in FEM-FCT (Löhner, 1986). The algorithm can be switched on locally whenever a predefined feature of interest can be identified. The switch is activated by estimating the local error and refining wherever it exceeds a prescribed limit. Since as a rule only small regions require refinement, the overhead involved in mesh refinement is essentially negligible. The timestep, however, is set by requiring that the maximum Courant number be less than unity, so the running time increases inversely with the minimum zone size.

If the physical feature requiring enhanced resolution disappears, the algorithm automatically does away with the extra triangles which had been introduced. Because it "remembers" the original triangulation, it is able to restore the grid to its exact form prior to the refinement. This type of algorithm lends itself readily to vectorization, and a mesh change (performed every 5–10 timesteps) requires only ~ 100 μs on the Cray X-MP-12.

The techniques described here have been extended to axisymmetric (r-z) systems. They have also been used to construct a three-dimensional code (based on tetrahedra), which has been used to solve several supersonic steady-state problems.

A Barely Implicit Correction to FCT for Nearly Incompressible Flow

For explicit finite-difference schemes the timestep Δt is restricted by the Courant condition $\max[(c+v)\Delta t/\Delta x] \leq 1$, where Δx is the zone size and the maximum is taken over the whole mesh. In many nearly incompressible fluid dynamics systems the flow velocity **v** is much less than c, the speed of sound. We are usually interested in phenomena which take place on the slow time scale, so it is of interest to develop implicit algorithms for which the limiting condition becomes $v\Delta t/\Delta x \leq 1$. Since positivity-preserving techniques are needed to maintain sharp concentration or other gradients, we are motivated to construct an implicit FCT algorithm (Patnaik, et al., 1986).

Analysis by Casulli and Greenspan (1984) has shown that the only quantities in the finite-difference form of the fluid equations which have to be differenced implicitly (i.e., defined on the advanced time level) are the pressure in the gradient term of the momentum equation and the velocity in the divergence term of the equation for the energy density E. The other terms of the equations can be differenced explicitly, in the present case by using a form of JPBFCT (Boris, 1981). The essential computational steps include an explicit prediction of density, momentum, and energy, determination of an implicit pressure correction by solving an elliptic equation, and corrections to the momentum and energy obtained from the pressure correction.

This procedure (called Barely Implicit Correction, or BIC) can be readily generalized to work with other explicit schemes, including FEM-FCT. To date, BIC has been implemented in one- and two-dimensional Cartesian explicit FCT routines. Figure 3 illustrates the ability of BIC-FCT to propagate a contact discontinuity without the introduction of additional diffusion. Figure 4 shows the damping of sound waves by BIC-FCT. Damping is seen to be negligible when the method is made semi-implicit, i.e., when the Crank-Nicholson parameter ω satisfies $\omega = 0.5$. Damping increases if the timestep or zone size is increased. In these and other problems to which it has been applied the Mach numbers were as low as 0.01, so that the timestep was up to an order of magnitude longer than would have been possible in an explicit code. The time required for one computational timestep compares very favorably to that required by the explicit two-step JPBFCT module.

Applications have been made to two-dimensional flames ($v \sim 10$ m/s, $c \sim 300$ m/s) and to the transition to turbulence in jets. The decrease in running time permitted by barely implicit differencing makes it possible to include more detailed chemistry models in such simulations.

References

1. J. P. Boris, "ETBFCT: a fully vectorized FCT module," in *Finite-Difference Techniques for Vectorized Fluid Dynamics Calculations*, Ed. D. L. Book, Springer-Verlag, New York (1981).
2. V. Casulli and D. Greenspan, "Pressure gradient method for the numerical solution of transient compressible fluid flows," *Inter. J. Num. Methods in Fluids* **4**, 1001 (1984).
3. R. Löhner, "An Adaptive Finite-Element Scheme for Transient Problems in CFD," *Comp. Meth. Appl. Mech Eng.* (submitted 1986).
4. R. Löhner, K. Morgan, and O. C. Zienkiewicz, "An Adaptive Finite-Element Procedure for High-Speed Flows," *Comp. Meth. Appl. Mech. Eng.* **51**, 441 (1985).
5. R. Löhner, K. Morgan, M. Vahdati, J. P. Boris, and D. L. Book, "FEM-FCT: Combining unstructured grids with high resolution," *J. Comput. Phys.* (submitted 1986).
6. G. Patnaik, J. P. Boris, R. H. Guirguis, E. S.Oran, "A Barely Implicit Correction for Flux-Corrected Transport," *J. Comput. Phys.* (submitted 1986).
7. S. T. Zalesak, "Fully multidimensional Flux-Corrected Transport algorithms for fluids," *J. Comput. Phys.* **31**, 335 (1979).

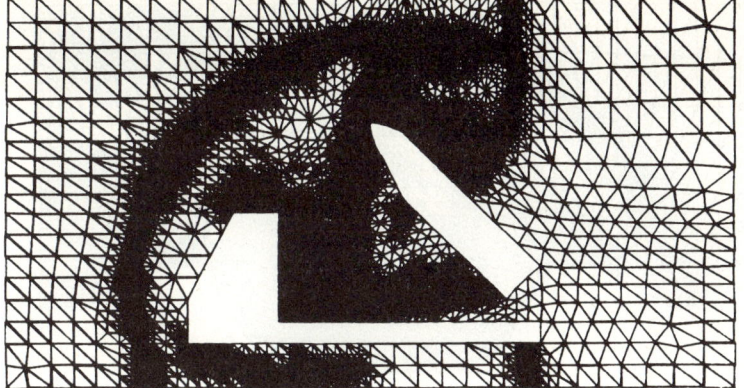

(a) Grid

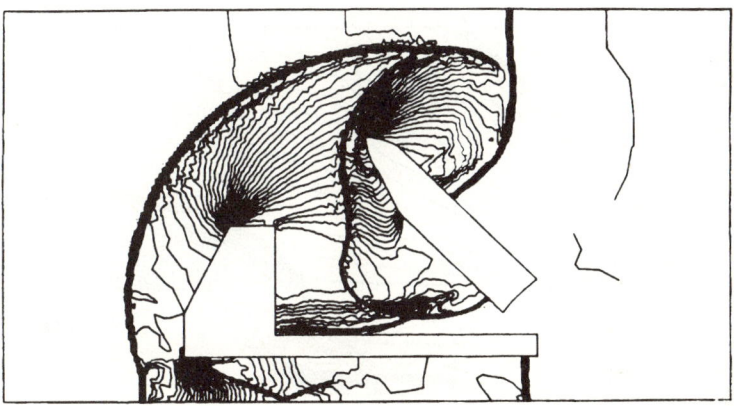

(b) Density

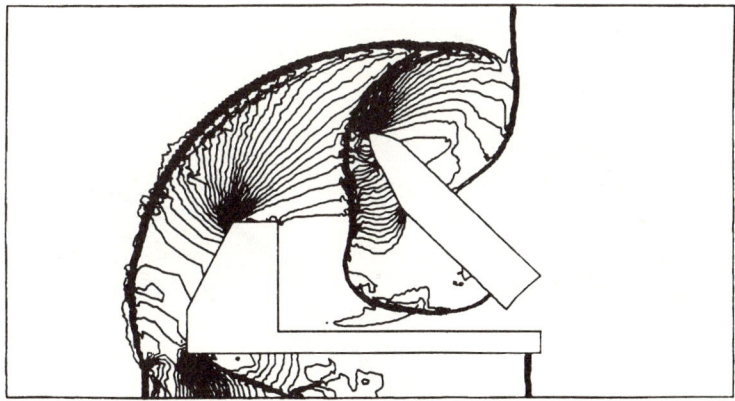

(c) Pressure

Figure 1

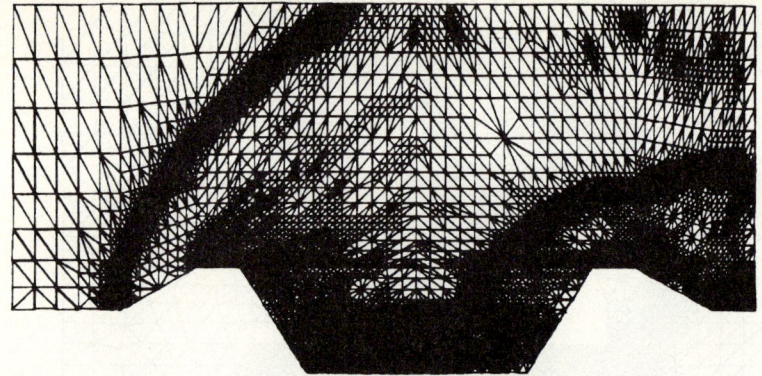

NELEM=13681, NPOIN=6984

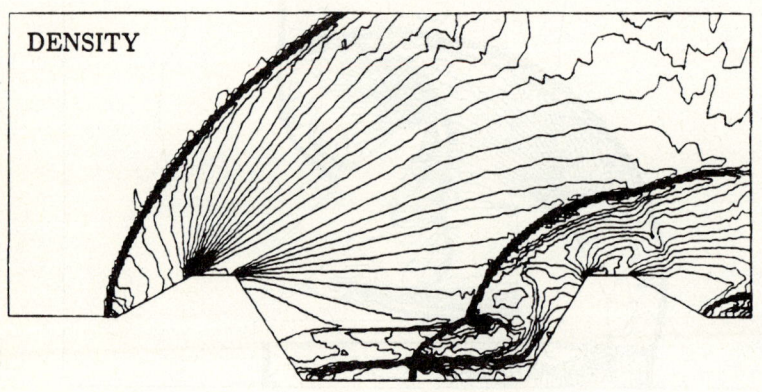

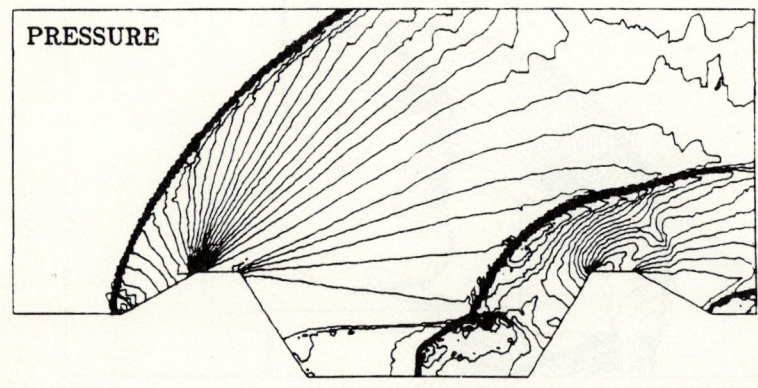

T=0.12

Figure 2

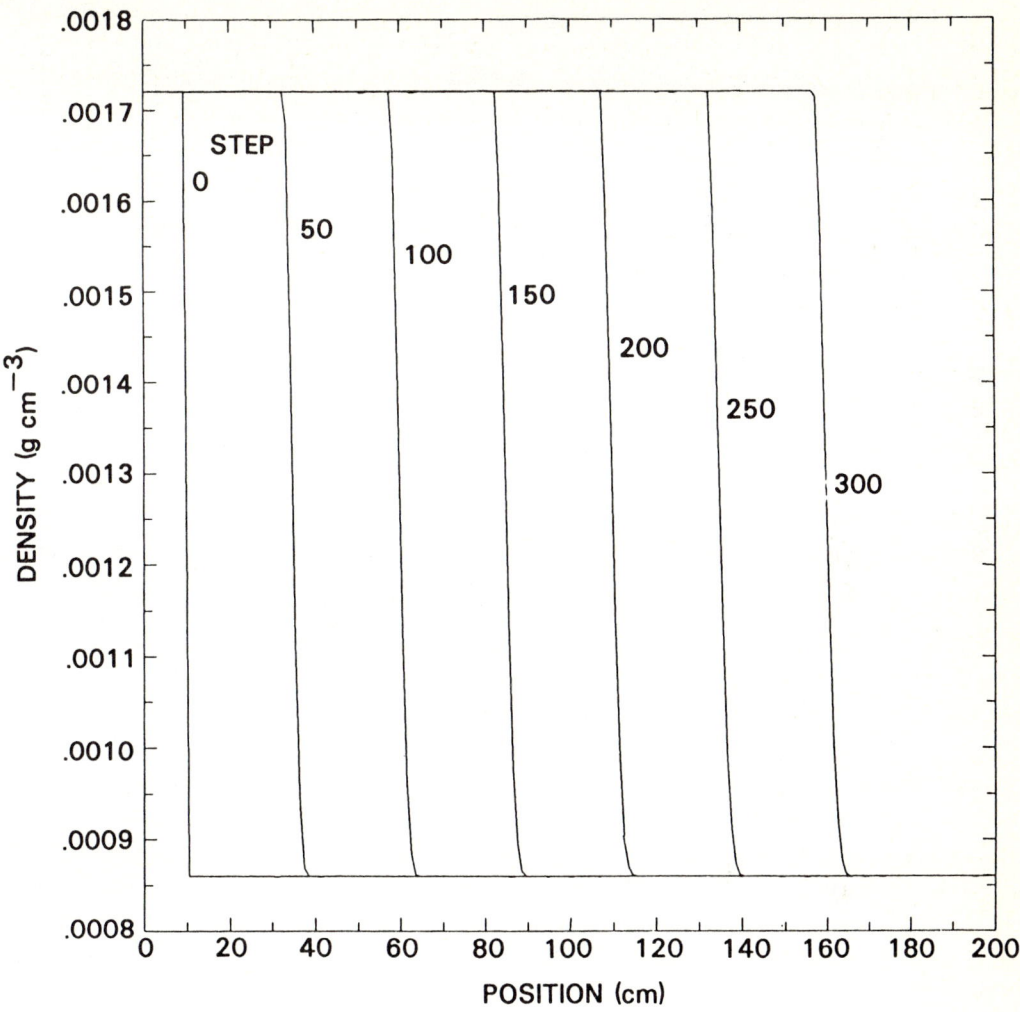

Figure 3

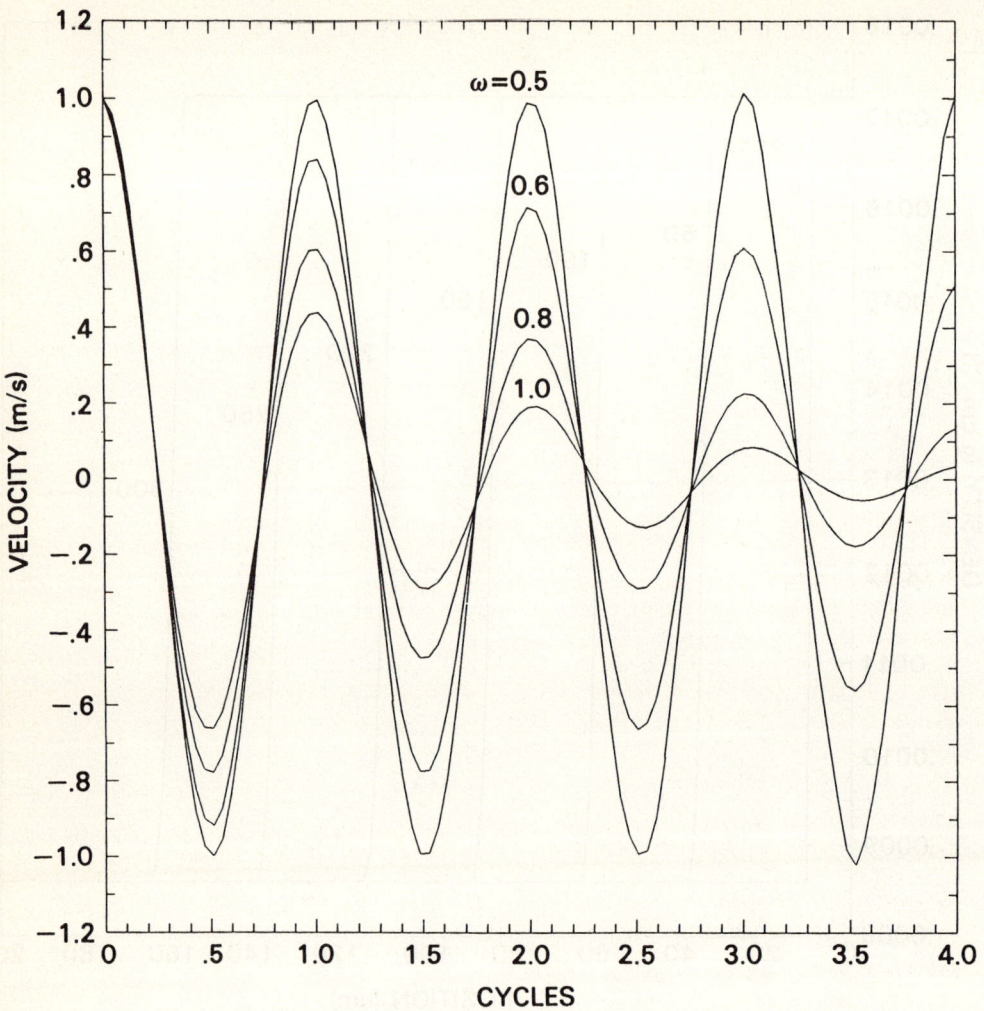

Figure 4

ACCURATE, EFFICIENT AND PRODUCTIVE METHODOLOGY FOR SOLVING TURBULENT VISCOUS FLOWS IN COMPLEX GEOMETRY

C.K. Lombard, J. Bardina, E. Venkatapathy, J.Y. Yang,
R.C.C. Luh, N. Nagaraj and F. Raiszadeh
PEDA Corporation, Palo Alto, CA 94303 USA

Introduction

If one sees the seventies and first half of the eighties as a period of learning how to construct stable, accurate and computationally fast methods to solve compressible flow equations, particularly the Navier-Stokes equations, then it is quite apparent the focus is shifting and for a time the emphasis will be on synthesis, application and enhanced productivity. The present paper introduces improvements in the multi-dimensional formulation, numerical implementation and mode of application of the implicit conservative Supra-Characteristics Method (CSCM). The method for both the compressible Euler and Navier-Stokes equations was first presented for inviscid quasi one dimensional flow in the 8th ICNMF by Lombard, Oliger and Yang[1]. Extension of CSCM to a 2-D and source term free axisymmetric formulation was given by Lombard, Bardina, Venkatapathy and Oliger[2]. Reference 2 introduced solution of the difference equations by a two data level block tridiagonal procedure based on a diagonally dominant approximate factorization DDADI. Subsequent extensions of that approximate factorization were shown by Lombard, Venkatapathy and Bardina[3] to form the basis for a family of single data level operationally explicit implicit relaxation algorithms. One of those algorithms, a symmetric Gauss-Seidel space marching procedure implemented as a method of lines in two space dimensions (or planes[4] in three dimensions) was found to be an order of magnitude more rapidly convergent to steady state than the same equations solved in the two level linearized pseudo time relaxation procedure. In the generalized Roe form[5]

$$\widetilde{A}\,\Delta q \equiv \Delta \widetilde{F} \tag{1}$$

which relation is preserved by the CSCM constructed matrix, the adjacent grid point interval averaged Jacobian matrix $\widetilde{A}$ onto the conservative variable difference is used to represent the flux difference. Thus updating $\widetilde{A}$ and q in the locally iterated pseudo time relaxation procedure at each marching step is equivalent to updating the flux.

Heretofore we wrote[3] the method (here sketched as a first order single level method in one dimension) as

$$D\,\delta q = -\widetilde{A}^+ \nabla q - \widetilde{A}^- \Delta q \tag{2}$$

where $D = J/\Delta t(I + \widetilde{A}^+ - \widetilde{A}^-)$ with J a measure of computational cell volume and $\widetilde{A}^\pm$ are the eigenvector split pieces of the Jacobian matrix according to

$$\widetilde{A}^\pm = (\overline{M}\,\overline{T}\,I^\pm\,\overline{T}^{-1}\,\overline{M}^{-1})\widetilde{A} \equiv \hat{A}^\pm \widetilde{A} \tag{3}$$

with $I^\pm = \frac{1}{2}(I \pm \mathrm{sgn}\overline{\lambda})$. Thus as shown in reference 1,

$$\widetilde{A}^\pm \Delta q \equiv \hat{A}^\pm \Delta \widetilde{F} = \Delta \widetilde{F}^\pm . \tag{4}$$

Flux Divergence Upwind Methods

To begin the present developments, we now write the equivalent

$$D\,\delta q = -\nabla \widetilde{F}^+ - \Delta \widetilde{F}^- = -\hat{A}^+ \nabla \widetilde{F} - \hat{A}^- \Delta \widetilde{F}. \tag{5}$$

In two or more space dimensions flux difference terms such as appear on the right hand side of equation (3) for each coordinate direction can be written either in strong or chain rule (Hindman[6]) conservation law forms according as the metric coefficients of the differential transformation

$$\partial \widetilde{F} = \xi_x \partial F + \xi_y \partial G \tag{6}$$

are represented under or outside of the difference approximations. In the latter case, averaging the metric coefficients on the difference interval is appropriate for Roe form schemes. While in quasi one dimensional flow the upwind method[1] with strongly conservative differencing was found to be very robust and accurate, in 2-D or axisymmetric flow the chain rule approach[2] was found to work far more reliably. We will explain why we believe this is so and present alternative strong conservation law formulations that according to the rationale are both stable and more accurate than chain rule. In the process we will show how to overcome a problem associated with nonorthogonal grids.

Chain Rule Conservation

The chain rule conservation law form for discrete differences is the natural approximation to equation (6). For central difference methods in strong conservation law form, it has long been recognized and practiced[6,7,8] that the numerical divergence of the metric coefficients must be zero under the operations of the finite difference method in order for the method to meet the minimal requirement to preserve a uniform flow. However in upwind methods, particularly those in generalized Roe forms, this fact has been to a greater or lesser extent ignored and stability appealed to based on global conservation of the telescoping difference equations. The problem we will cite also exists for some higher order upwind methods with a numerical flux function in the node centered classical finite volume cell format and when using non cell face local metric coefficients.

The issues associated with local conservation, which we espouse here, can be seen simply geometrically. In Figure 1, we show the computational cell associated with one sided differencing for the ξ direction. For what we may term local geometric conservation in the cell, we consider the flux vector component $\widetilde{F}_\xi$ for the ξ direction The flux tensor has two components $\widetilde{F}_\xi$ and $\widetilde{F}_\eta$. For the moment, the other component of the flux tensor will be regarded as treated in another (η) family of cells aligned with the η coordinate direction. The flux vector must be projected not only through the end faces i and $i+1$ in the ξ direction but also through the lateral side faces $k \pm \frac{1}{2}$. The associated difference relation is symbolized

$$\Delta \bullet \widetilde{F}_\xi \equiv \Delta_\xi \hat{\xi} \bullet \widetilde{F}_\xi + \delta_\eta \hat{\eta} \bullet \widetilde{F}_\xi \qquad (7)$$

Simple globally conservative one sided differencing ignores the second term of the right hand side of relation (7) and, as a consequence, introduces spurious numerical sources and sinks[9] that compromise accuracy and stability.

It is illuminating and easy to show that if the flux components $\widetilde{F}_\xi$ for the lateral η faces of the cell are approximated local to the cell by $\widetilde{F}_\xi \mid_{k \pm \frac{1}{2}} = \frac{1}{2}(F_\xi \mid_i + F_\xi \mid_{i+1})_k \equiv \overline{\widetilde{F}}_\xi$ relation (7) is identically equivalent to chain rule differencing $\overline{\hat{\xi}} \bullet \Delta_\xi \widetilde{F}_\xi \equiv \Delta \bullet \widetilde{F}_\xi - \overline{\widetilde{F}}_\xi \Delta_\xi \bullet \hat{S}$. Here we assume the cell is properly closed and the metrics satisfy $\Delta_\xi \bullet \hat{S} = 0$.

Accordingly, the consequence of applying the chain rule conservation form is that in the laterally adjacent cells the similar flux approximations at $k \pm 1$ through the shared faces $\hat{\eta} \pm \frac{1}{2}$ are inconsistent with the approximation for cell k and the method is in general neither locally nor globally conservative. Two exceptions to the conclusion are found in the cases of uniform flow, which the chain rule method preserves, and no flow through the lateral sides as in the quasi 1-D stream tube approximation for which we had very good success[1] in strongly conservative differencing. Thus in principle the chain rule and simple globally conservative locally one dimensional methods are relatively more effective where the grid is nicely flow aligned, smooth gradients are weak and discontinuities are orthonormal. In practice chain rule conservation, with similar one sided difference terms for the η coordinate direction, works quite well[2,3] for the numerically dissipative first order method. For second order methods with limiters the error in conservation is diminished but the methods also decline in robustness, about which more will be said later.

Locally One Dimensional Strong Conservation

The conclusion from the above discussion is that a method closely akin to the chain rule locally one dimensional one sided upwind differencing procedures of references 2, 3 and 10 but which features consistent local geometric conservation for each flux component in its own family of one sided cells is obtained by replacing the simple one sided flux difference $\Delta_\xi \widetilde{F}_\xi$ for equation (5) by cell flux divergences symbolized by $\Delta_\xi \bullet \widetilde{F}_\xi$ with the lateral $\hat{\eta}$ side fluxes in equation (7) given a symmetric definition, for example averaged between k and $k+1$. Similar generalized one sided upwind difference expressions are given for the η coordinate direction and we can write

$$D\delta q = -\widetilde{A}_\xi^+ \nabla_\xi \bullet \widetilde{F}_\xi - \hat{A}_\xi^- \Delta_\xi \bullet \widetilde{F}_\xi - \hat{A}_\eta^+ \nabla_\eta \bullet \widetilde{F}_\eta - \hat{A}_\eta^+ \Delta_\eta \bullet \widetilde{F}_\eta \qquad (8)$$

This approach is both globally conservative and more consistent with the accuracy aims of local conservation and accordingly is expected to be more stable and robust. The above locally one dimensional approach has a numerical virtue in that the splitting operators apply homogeneously and rationally to their own convective fluxes.

There are, however, drawbacks still to be addressed. The first of these deals with nonorthogonal meshes. While the problem is not observed numerically to be severe for moderate deviations from orthogonality, from equation (8) it is obvious that in the limit as the mesh is skewed and the η and ξ fluxes become identical, other effects aside, the method loses information from one sector of the flow domain and its solution must become inconsistent with the multi-dimensional PDE. A second weakness is that for general curvilinear coordinates the ξ direction in the Cartesian bases changes from cell to cell so there is not a unique and consistent direction on which to effect the flux decomposition and satisfy assumptions. A third weakness is that in the locally one dimensional method neither the partial nor total flux is identically conserved locally in each family of cells. The significance is that residual reduction at convergence may not be as firm as with the last method.

Symmetric Locally One Dimensional Conservation

Objections one and two above may be overcome by constructing methods that are symmetric overall in the two families of cells. By this we mean that all the cells of the procedure are shared among the two coordinate directions; a ξ one sided difference cell for one coordinate is also an η cell for the other coordinate. A last requirement for symmetry is that a unique definition of total flux be given for every cell face. These conditions ensure that regardless of the bases for flux tensor decomposition in adjacent cells, the total flux that numerically flows out of a face of one cell also flows through the same face into its neighbor cell.

Beyond the classical node centered finite volume, two kinds of computational cells can lend themselves to the construction of symmetric strongly conservative biased upwind methods. One of these cell types is the natural mesh cell bounded at the four corners (in 2-D) by its data nodes. The other is the laterally symmetric and upwind (coordinate) biased cell of Figure 1.

Symmetric Methods in Natural Mesh Cells

Of two approaches we have considered, one splits the weighted average of the flux divergences in adjacent mesh cells on either side of the differencing coordinate line with the eigen state determined by adjacent node averages on the difference interval. This scheme has laterally wide support in the difference stencil and thus is not very accurate in high gradient regions and will smear solutions. The other approach splits the side averaged flux difference according to the eigen state at the mesh cell centers. This procedure implies compact differencing schemes and is restrictive. The method also requires a lot of averaging which can be mitigated by placing data nodes at the cell centers. The latter idea is not bad as we shall consider next.

Symmetric Methods on Biased Upwind Cells

Additional data nodes at the (body) centers of the natural mesh cells create a symmetric geometric pattern in which the biased upwind cells (Figure 2) of the base level mesh for one coordinate direction are the similarly biased upwind cells in the other coordinate direction of the body centered mesh. Since the nodes of the body centered mesh are centered on the lateral faces of the base mesh upwind cells and visa versa, there is a unique definition of the total flux at the node on every cell face.

For each face of every cell the flux tensor can be decomposed in orthogonal components according to reference directions that may be determined independently from cell to cell. The orthogonal decomposition can be expressed in the unit base vectors $\hat{\alpha}$ and $\hat{\beta}$ as

$$F = \hat{\alpha}\,\hat{\alpha} \bullet F + \hat{\beta}\,\hat{\beta} \bullet F \equiv F^{\alpha} + F^{\beta} \qquad (9)$$

We assume the base vectors are related to and generally align with the coordinate lines. Then for either component we can construct the locally one dimensional flux divergence $\Delta_{\alpha} \bullet F^{\alpha}$ according to equation (7) with F^{α} replacing $\widetilde{F}_{\xi}$. Of course a similar relation is developed for $\Delta_{\beta} \bullet F^{\beta}$ appropriately interchanging the upwind one sided and lateral central difference symbols among ξ and η.

For either flux divergence component and by analogy with the chain rule CSCM method, generalized flux difference eigenvector splitting may be carried out based on the associated upwind difference interval averaged eigenstate to give

$$\Delta \bullet F^{\pm} = \hat{A}(I^{\pm}) \Delta \bullet F \qquad (10)$$

Finally associating coordinate ξ (j) with the α flux tensor component and η (k) with β, we can write the flux divergence split implicit method as

$$[\hat{A}(I_{\alpha}^{+} + \Lambda_{\alpha}^{+})\nabla_{\xi} + \hat{A}(I_{\alpha}^{-} + \Lambda_{\alpha}^{-})\Delta_{\xi} + \hat{B}(I_{\beta}^{+} + \Lambda_{\beta}^{+})\nabla_{\eta} + \hat{B}(I_{\beta}^{-} + \Lambda_{\beta}^{-})\Delta_{\eta}]\delta q_{j,k}$$
$$= -(\hat{A}(I_{\alpha}^{+})\Delta_{j-1} \bullet F^{\alpha} + \hat{A}(I_{\alpha}^{-})\Delta_{j} \bullet F^{\alpha} + \hat{B}(I_{\beta}^{+})\Delta_{k-1} \bullet F^{\beta} + \hat{B}(I_{k}^{-})\Delta_{k} \bullet F^{\beta}) \qquad (11)$$

Here, for a two level linearized method, we have via Eq. (5) approximated the split Jacobian matrix $\widetilde{A}^{\pm}$ by the similarity transform based on the $\hat{A}$ operator and the associated difference interval estimated eigenvalues

$$\Lambda_{\alpha}^{\pm} \simeq [W^{\xi}, W^{\xi}, W^{\xi} + \hat{\xi}C, W^{\xi} - \hat{\xi}C]^{\pm} \qquad (12)$$

For very skewed meshes, a better estimate of the eigenvalues would be required.

While inherently more accurate, the symmetric staggard grid method described above is seen to closely approximate the chain rule CSCM algorithm. We have given the new derived method the name S-CSCM.

Local Total Flux Strong Conservation

Finally we give a simple approach which provides total flux conservation in each computational cell. In the approach we replace in Eq. (5) the partial fluxes by the total flux to give

$$D\delta q = -\hat{A}_{\xi}^{+}\nabla_{\xi} \bullet \widetilde{F} - \hat{A}_{\xi}^{-}\Delta_{\xi} \bullet \widetilde{F} - \hat{A}_{\eta}^{+}\nabla_{\eta} \bullet \widetilde{F} - \hat{A}_{\eta}^{-}\Delta_{\eta} \bullet \widetilde{F} \qquad (13)$$

With Eq. (13) a sufficient (and we would guess necessary) condition for convergence to steady state is that the flux divergence vanish for every computational cell. Within the total flux divergence framework the subscripts on the symbolic divergence operators $\nabla_{\xi} \bullet \widetilde{F}$ only serve to identify a kind of cell in the grid. Further, the subscripts ξ and η on the projection operators serve to identify characteristic directions in which the velocity vectors will be composed in normal and tangential components for advection. When these directions, which we can free from constraint to either or both the curvilinear coordinate directions, are taken to be orthogonal, it can be shown that within truncation errors similar to those for locally one dimensional conservative eigenvector split upwind differencing the method is consistent with the PDE.

Based in an older and independently conceived notion by Roe and Lombard the local cell total flux divergence $\Delta \bullet \widetilde{F}$ can be identified with a residual ("fluctuation", Roe) that is "sent" along grid lines by the method through the projection operators $\hat{A}^{\pm}_{\xi,\eta}$ to relax the state at appropriate upwind mesh points. With freedom of the characteristic directions from the grid, we achieve greater freedom in how and where to send residual information.

Second Order TVD Schemes

For the new locally one dimensional procedure, higher order spatial difference methods for the right hand side of equation (13) can be constructed of the split partial flux divergence by pieced linear combinations of the standard upwind and central difference correction operators. In reference 10, Lombard proposed a simple limiter (Scheme IIIb) for the conservative flux difference split pieces of such a second order method. That limiter by direct analogy can be used here to enhance nonlinear stability. We note that the natural second order method for a scheme constructed of the total flux divergence involves the "Fromm" weights of half central and half upwind.

Viscous Terms

For both the locally one dimensional and total flux formulations it is feasible following the work of reference 2 to simply add viscous terms in a conventional thin layer central difference formulation. For the locally one dimensional approach this may be appropriate. But from the strong conservation point of view such a practice will interfere with the desirable feature of zeroing the total flux divergence in each computational cell. To preserve that property of the inviscid method, we have only to add the viscous stress terms at the boundaries of the cells such that the terms contribute consistently to the total flux whether or not it is subsequently decomposed into locally one dimensional components.

Numerical Results

We show here results of computations for a topical difficult propulsion base flow problem of a wind tunnel model tactical missile flow involving realistically complex geometry and flow structure including a centered propulsive jet. The freestream Mach number is 1.4, base diameter Re 0.92×10^6 and nozzle exit Mach number 2.7 and static pressure ratio of 2.5. The problem has recently been solved with a geometry flexible multiple patched mesh CSCM code – based on the chain rule conservation formulation.

The problem was solved previously by us with an adapted single block mesh code on a double wraparound grid of Figure 3. The grid generated in patches with a simplified algebraic procedure has nice quality over the external and internal corners. But the constrained topology is seen to waste mesh points through excessive clustering in the wake.

The new, better balanced mesh for the present solutions is shown in Figure 4. A plot of the outline of the multiple patch structure of the mesh in the vicinity of the nozzle exit is shown in Figure 5. In the boundary layer region the mesh has an effective wraparound character on an overset patch local to the corner.

We show in Figure 6 a Mach contour plot for the solution on the new mesh. On the whole the better balanced mesh employing about the same number of points produces a result of crisper flow structural detail than the less flexible topology whose solution is shown in Figure 7. Nicely evident in Figure 6 is a weak lip shock, shear layer, barrel shock, (near) contact discontinuity and, above them all, a recompression shock in the external flow.

Lastly in Figure 8, we show a comparison of computational results and experiment. The wave in the base region is due to two counterrotating persistent vortices in the base bubble. We note at the corners a strong qualitative disagreement between solutions on the wraparound grids and a third nonwraparound step mesh. Evidently there the mesh topology strongly influences the mechanism of separation. This point requires further study into topology and mesh refinement.

References

1. Lombard, C.K., Oliger, J. and Yang, J.Y., *Lecture Notes in Physics, 170*, pp. 364-370, 1982.
2. Lombard, C.K., Bardina, J., Venkatapathy, E. and Oliger, J., AIAA Paper 83-1895, 1983.
3. Lombard, C.K., Venkatapathy, E. and Bardina, J., AIAA Paper 84-1533, 1984.
4. Bardina, Jorge and Lombard, C.K., AIAA Paper 85-1193, 1985.
5. Roe, P.L., *Lecture Notes in Physics, 141*, pp. 354-359, 1981.
6. Hindman, Richard D., *AIAA J.*, Vol. 20, No. 10, pp. 1359-1367, Oct. 1982.
7. Viviand, H. and Ghazzi, W., *Lecture Notes in Physics*, No. 59, Springer-Verlag, 1976.
8. Lombard, C.K., Davy, W.C. and Green, M.J., AIAA Paper 80-0065, 1980.
9. Vinokur, M., *J. of Computational Physics*, Vol. 14, pp. 105-125, 1974.
10. Yang. J.Y., Lombard, C.K. and Bardina, Jorge, Paper presented at the *International Symposium on Computational Fluid Dynamics-Tokyo*, Sept. 1985.
11. Sahu, J. and Nietubicz, C.J., AIAA Paper 84-0527, 1984.
12. Petrie, H.J. and Walker, B.J., AIAA Paper 85-1618, 1985.

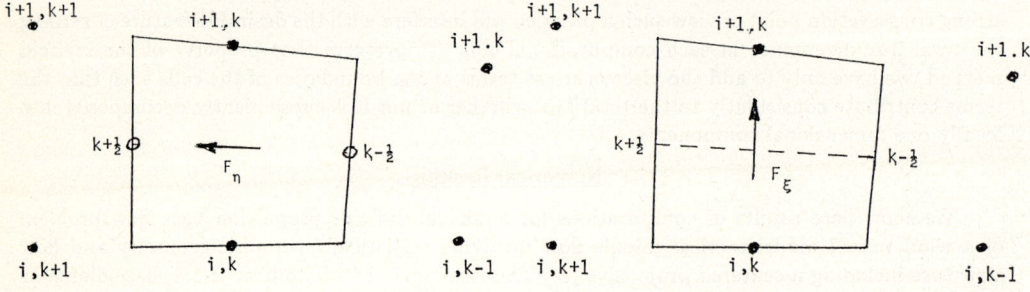

Fig. 2. Coordinate biased cells of base grid are also biased η cells of body centered grid.

Fig. 1. Computational cell for locally one dimensional coordinate-biased one sided difference method.

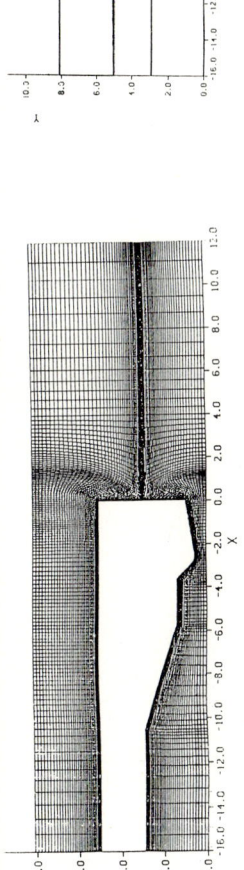

Fig. 3. Double wraparound grid.

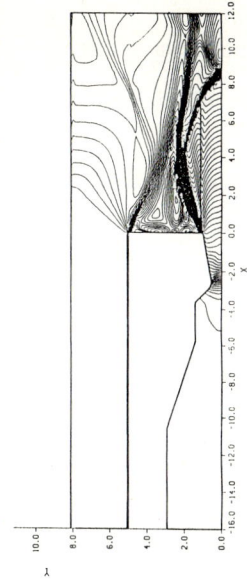

Fig. 7. Mach contour lines for wraparound grid.

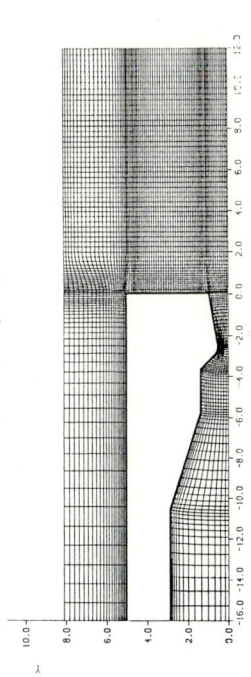

Fig. 4. Multiple patch grid.

Fig. 6. Mach contour lines for wraparound grid.

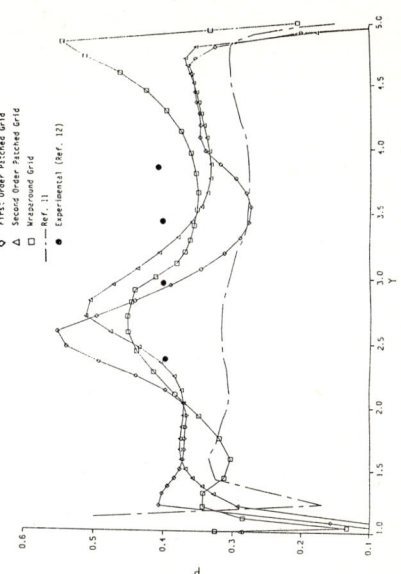

Fig. 8. Base pressure comparison between experiment and numerical solutions.

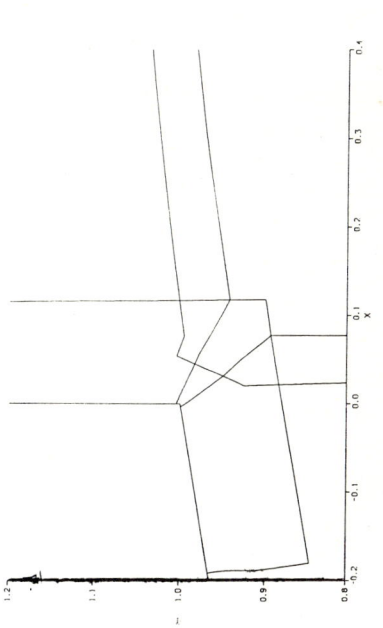

Fig. 5. Overlapping of multiple mesh patches in vicinity of nozzle-base corner.

A SIMPLE AND EFFICIENT IMPLICIT SCHEME
FOR THE COMPRESSIBLE NAVIER-STOKES EQUATIONS

Y. W. Ma   D. X. Fu
Beijing Institute of Aerodynamics
P.O.Box 7215   Beijing China

An efficient implicit scheme with second order accuracy is developed. The scheme is simple and there are no matrix inversions and no matrix operations in computation. The solutions can be expressed explicitly. A model equation is discussed first, then the difference approximation for the system of equations is given. The new developed scheme was used to solve the Couette flow, shock wave-boundary layer interaction problem and supersonic viscous flow past a sphere cone body.

MODEL EQUATION

For simplicity consider the model equation with constant coefficient

$$\frac{\partial u}{\partial t} + c \frac{\partial u}{\partial x} = \gamma \frac{\partial^2 u}{\partial x^2} . \tag{1}$$

The viscous term in computation was treated as it was done in [1]. only case $\gamma = 0$ is discussed here. The equation (1) can be approximated by scheme

$$\delta_t u_m^{n+1} + \frac{\Delta t}{\Delta x}\left(c \frac{\delta_x^- u_m^{n+1}}{2} + c \frac{\delta_x^+ u_m^n}{2} \right) = 0 , \tag{2}$$

$$\delta_t u_m^{n+1} + \frac{\Delta t}{\Delta x}\left(c \frac{\delta_x^+ u_m^{n+1}}{2} + c \frac{\delta_x^- u_m^n}{2} \right) = 0 , \tag{3}$$

where

$$\delta_t u^{n+1} = u^{n+1} - u^n , \qquad \delta_x^+ u_m = u_{m+1} - u_m , \qquad \delta_x^- u_m = u_m - u_{m-1} .$$

The scheme (2) and (3) are stable for any $\Delta t / \Delta x$ and have accuracy of order $O(\Delta t^2, \Delta x^2)$. For the special case $c\Delta t/\Delta x = -2$ for the scheme (2) and $c\Delta t/\Delta x = 2$ for the scheme (3) the solutions u_m^{n+1} can not be obtained because its coefficient is equal to zero. These schemes can be improved. Rewrite the equation (1) as follows

$$\frac{\partial u}{\partial t} + c^+ \frac{\partial u}{\partial x} + c^- \frac{\partial u}{\partial x} = 0 , \tag{4}$$

where

$$c^+ + c^- = c , \qquad c^+ \geq 0 , \qquad c^- \leq 0 .$$

The scheme (2) is used to approximate the term containing c^+ in (4) and (3) is used to approximate the term containing c^-. After rearrangement the scheme obtained can be rewritten into the following form

$$(1+\frac{1}{2}\frac{\Delta t}{\Delta x}c^{+}\delta_{x}^{-}+\frac{1}{2}\frac{\Delta t}{\Delta x}c^{-}\delta_{x}^{+})\delta_{t}u_{m}^{n+1}=-c\frac{\Delta t}{\Delta x}\delta_{x}^{o}u_{m}^{n}, \qquad (5)$$

where

$$\delta_{x}^{o}=\frac{1}{2}(\delta_{x}^{+}+\delta_{x}^{-}).$$

The scheme (5) is stable for any $\Delta t/\Delta x$ and it has second order accuracy. It can be solved easily. Without losing accuracy the scheme (5) can be factored as

$$(1+\frac{1}{2}\frac{\Delta t}{\Delta x}c^{+}\delta_{x}^{-})(1+\frac{1}{2}\frac{\Delta t}{\Delta x}c^{-}\delta_{x}^{+})\delta_{t}u_{m}^{n+1}=-c\frac{\Delta t}{\Delta x}\delta_{x}^{o}u_{m}^{n}. \qquad (6)$$

SYSTEM OF EQUATION

Consider the 1-D Euler equation

$$\frac{\partial U}{\partial t}+\frac{\partial f}{\partial x}=0, \qquad (7)$$

where

$$U=(\rho, \rho u, E)^{T}, \qquad f=(\rho u, \rho u^{2}+p, u(E+p))^{T},$$
$$E=\rho(C_{v}T+u^{2}/2), \qquad p=\rho T/(\gamma M_{\infty}^{2}).$$

As it was done for the scalar equation, using flux splitting for the homogeneous function f of degree one in U and linearization the following stable scheme with second order accuracy can be obtained

$$(I+\frac{1}{2}\frac{\Delta t}{\Delta x}\delta_{x}^{-}A^{+}+\frac{1}{2}\frac{\Delta t}{\Delta x}\delta_{x}^{+}A^{-})\delta_{t}U_{m}^{n+1}=\Delta U_{m}^{n}=-\frac{\Delta t}{\Delta x}\delta_{x}^{o}f_{m}^{n}, \qquad (8)$$

where

$$A=\frac{D(f)}{D(U)}, \qquad A=S^{-1}\Lambda S, \qquad A^{+}=S^{-1}\Lambda^{+}S, \qquad A^{-}=S^{-1}\Lambda^{-}S,$$
$$A^{+}=\frac{D(f^{+})}{D(U)}, \qquad A^{-}=\frac{D(f^{-})}{D(U)}, \qquad f^{+}=A^{+}U, \qquad f^{-}=A^{-}U,$$
$$\lambda_{i}^{+}+\lambda_{i}^{-}=\lambda_{i}, \qquad \lambda_{i}^{+}\geq 0, \qquad \lambda_{i}^{-}\leq 0,$$

and I is a unit matrix, Λ is a diagonal matrix with elements λ_i which are eigenvalues of A, Λ^{+} and Λ^{-} are diagonal matrices with elements λ_{i}^{+} and λ_{i}^{-} respectively. The system of equations(8) with diagonally dominant matrix can be solved easily using standard algorithm. It also can be solved using the following factorization

$$(I+\frac{1}{2}\frac{\Delta t}{\Delta x}\delta_{x}^{-}A^{+})(I+\frac{1}{2}\frac{\Delta t}{\Delta x}\delta_{x}^{+}A^{-})\delta_{t}U_{m}^{n+1}=\Delta U_{m}^{n}, \qquad (9)$$

$$(I+\frac{1}{2}\frac{\Delta t}{\Delta x}A^{+})\delta_{t}\overline{U_{m}^{n+1}}=\Delta U_{m}^{n}+\frac{1}{2}\frac{\Delta t}{\Delta x}A_{m-1}^{+}\delta_{t}U_{m-1}^{\overline{n+1}}=F_{m}^{n}, \qquad (10)$$

$$(I-\frac{1}{2}\frac{\Delta t}{\Delta x}A^{-})\delta_{t}U_{m}^{n+1}=\delta_{t}U_{m}^{\overline{n+1}}-\frac{1}{2}\frac{\Delta t}{\Delta x}A_{m+1}^{-}\delta_{t}U_{m+1}^{n+1}=G_{m}^{n}, \qquad (11)$$

$$U_m^{n+1} = U_m^n + \delta_t U_m^{n+1}.$$

In the equations (9), (10) and (11) $A^{\pm}$ are used instead of $\overset{\pm}{A}$. After careful study it can be shown that scheme (9) has second order accuracy and it is stable for any $\Delta t/\Delta x$. Four kinds of splitting are represented:

a. $\lambda_i^+ = (\lambda_i + |\lambda_i|)/2$, $\qquad\qquad\qquad \lambda_i^- = (\lambda_i - |\lambda_i|)/2$;

b. $u^{\pm} = (u \pm |u|)/2$, $\lambda_1^+ = u^+$, $\lambda_2^+ = u^+$, $\lambda_3^+ = u^+ + c$,

$\qquad\qquad\qquad \lambda_1^- = u^-$, $\lambda_2^- = u^- - c$, $\lambda_3^- = u^-$;

c. $\lambda_i^+ = u^+ + c$, $\lambda_1^- = u^- - c$, $\lambda_2^- = u^- - 2c$, $\lambda_3^- = u^-$;

d. $\lambda_i^- = u^- - c$, $\lambda_1^+ = u^+ + c$, $\lambda_2^+ = u^+$, $\lambda_3^+ = u^+ + 2c$.

There are no matrix operations in "+" direction for splitting c) and in "-" direction for splitting d). If splitting b) is used the solutions of the equations (10) and (11) can be expressed explicitly

$$\delta_t U_m^{n+1} = (I + \alpha A^+)^{-1} F_m^n = \frac{1}{1+\alpha u^+} \left\{ F_m^n - \frac{\alpha c}{1+\alpha(u^+ + c)} \beta_3 \begin{vmatrix} s_{1,3}^{-1} \\ s_{2,3}^{-1} \\ s_{3,3}^{-1} \end{vmatrix} \right\}, \qquad (12)$$

$$\delta_t U_m^{n+1} = (I - \alpha A^-)^{-1} G_m^n = \frac{1}{1-\alpha u^-} \left\{ G_m^n - \frac{\alpha c}{1-\alpha(u^- - c)} \beta_2 \begin{vmatrix} s_{1,2}^{-1} \\ s_{2,2}^{-1} \\ s_{3,2}^{-1} \end{vmatrix} \right\}, \qquad (13)$$

where

$$\alpha = \frac{1}{2}\frac{\Delta t}{\Delta x}, \qquad \beta_i = \sum_k s_{i,k} x_k, \qquad X=F \text{ for (12)}, \qquad X=G \text{ for (13)},$$

and c is the sound speed, x_k are elements of the vector X, $s_{i,j}^{-1}$ and $s_{i,j}$ are elements of matrix S^{-1} and S respectively. The scheme (9) with splitting b) has the following properties

a. it is an implicit and single step scheme and it is stable for any $\Delta t/\Delta x$;
b. it has second order accuracy;
c. there is no singularity in computation;
d. there are no matrix inversions and no matrix operations, the solutions can be expressed explicitly.

APPLICATION

The above described scheme were used to solve the Couette flow, shock wave-boundary layer interaction problem and supersonic viscous flow past a sphere cone body. Two test cases, the interaction problem and viscous flow past the sphere cone problem are presented here.

SHOCK WAVE-BOUNDARY LAYER INTERACTION

2-D shock wave-boundary layer interaction problem sketched in Fig. 1 was solved using scheme (9) with splitting a). The inflow conditions are $M_\infty = 2$, $Re = 2.96 \times 10^5$. The total pressure increase is 1.4. Coordinate transformation is made in normal direction. The computed results are given in Figs. 2 and 3. They are compared with experiment and MacCormack solutions.

SPHERE CONE FLOW

The viscous terms of the Navier-Stokes equations in application to sphere cone flow are represented principally by derivatives normal to the surface only. The axis centerline boundary conditions were that for symmetric flow. Extrapolation was used for down stream exit boundary. The R-H relation were used for boundary conditions on the shock. At the body surface we have no slip and isothermal wall conditions. The boundary conditions for implicit part on the wall were computed using simple iteration with system of equations(9) at the points next to the wall and zero normal gradient of pressure on the wall. Splitting b) was used in computation. The test cases are given in the table. The mesh grid system is given in Fig. 4 and computed results are given in Figs. 5,6 and 7. Agreement with experiment and solutions in [4] are good. The steady state solution can be obtained after 600 steps for group 4 in the table taken from [3]. Because of single step and no matrix operations at least half computing time can be saved comparing with two step scheme in [3].

TABLE

group	1	2	3	4
free Mach number	10.0	13.41	9.82	7.97
Reynolds number	3×10^5	1515	9905	7×10^6
half cone angle	$10°$	$7.5°$	$4°$	$10.5°$
T_w/T_o	.05	.74	.5	.5

REFERENCES

1. Ma Yanwen and Fu Dexun, J. of Computational Physics, No.3, 1985 (in chinese)
2. Ma Yanwen, Computational Mathematics, No.2, 1983 (in chinese)
3. R.W.MacCormack, AIAA Paper No.85-0032, 1985
4. B.N.Srivastava, M.J.Werle and R.T.Davis, AIAA J. Vol.16, No.2, 1978.

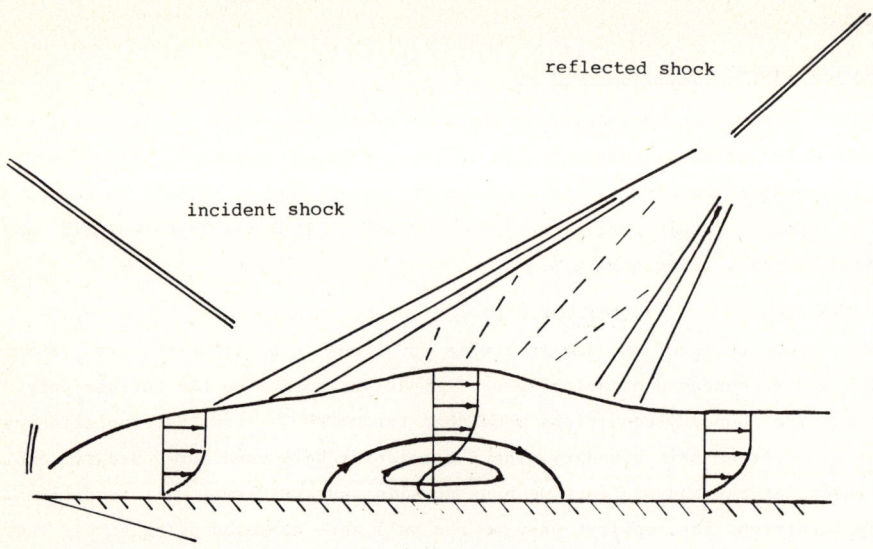

Fig. 1 Shock wave-Boundary layer interaction

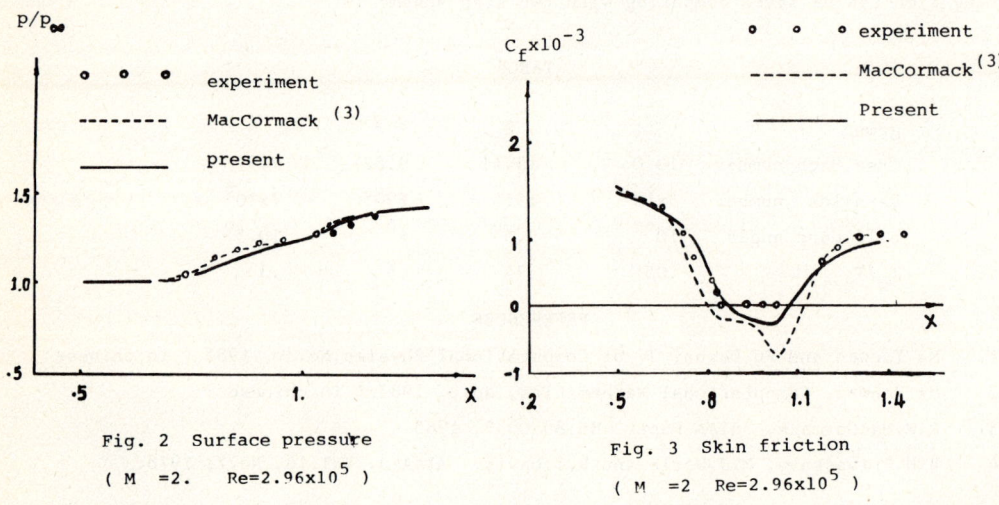

Fig. 2 Surface pressure
($M=2.$ $Re=2.96 \times 10^5$)

Fig. 3 Skin friction
($M=2$ $Re=2.96 \times 10^5$)

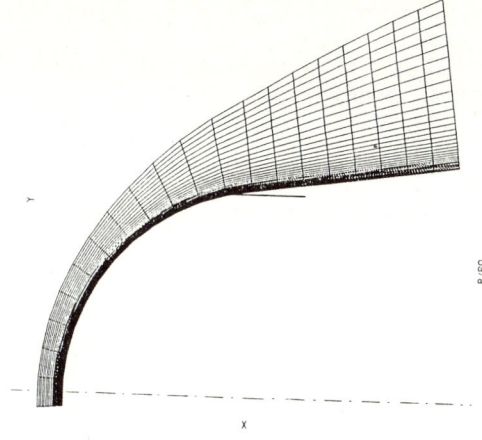

Fig. 4 Mesh grid for the sphere cone

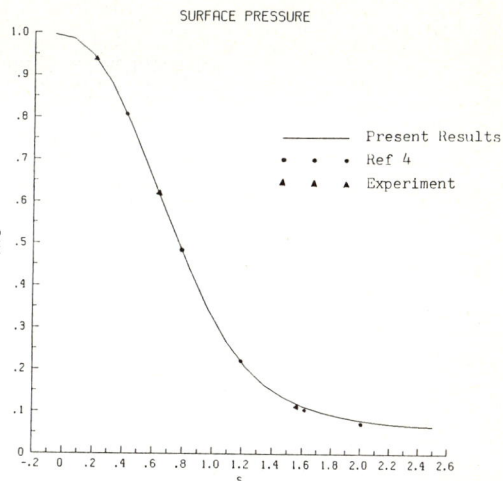

Fig. 5 Pressure on 7.5° sphere cone

(Re=1515, M =13.41, T_w/T_o=.0741, T =200°R)

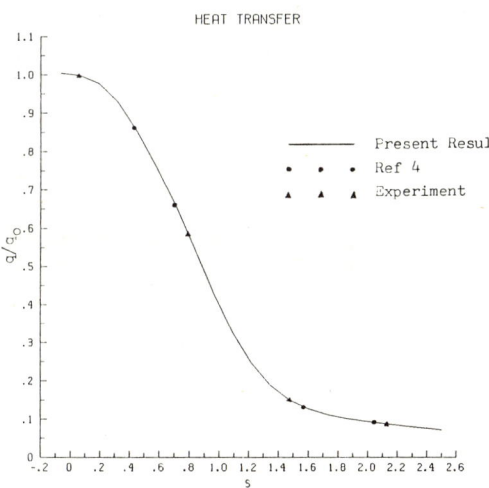

Fig. 6 Heat transfer on 7.5° sphere cone

(Re=1515, M =13.41, T_w/T_o=.0741, T =200°R)

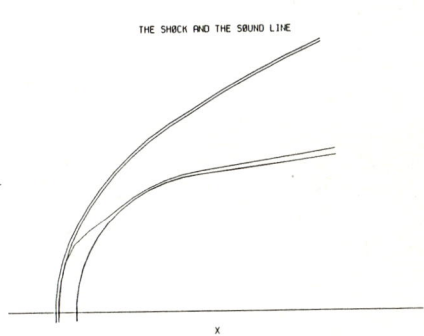

Fig. 7 Shock wave and sound line

(Re=1515, M =13.14, T_w/T_o=.0741, T =200°R)

ON AN IMPLICIT NUMERICAL SCHEME FOR TWO-DIMENSIONAL STEADY NAVIER-STOKES EQUATIONS

N S Madhavan and V Swaminathan
Applied Mathematics Division
Vikram Sarabhai Space Centre
Trivandrum - 695022, India

Introduction

The implicit numerical scheme due to MacCormack for solving viscous compressible flow [1] using Navier-Stokes equations has been extensively applied in the literature for a wide variety of problems [2, 3]. The situation, however, becomes rather tricky and complex when the method is applied to steady Euler [4] and Navier-Stokes [5]. In [5], for instance, two major difficulties are encountered by the authors in applying the implicit MacCormack scheme to the parabolized Navier-Stokes equations: First, the modal analysis, involving the evaluation of the eigenvalues and eigenvectors of the Jacobian matrix; and second, decoding the state vector at each step in order to obtain the corresponding flow variables. To obviate the above two complexities, the authors of [5] had adopted a linearization procedure.

It is the purpose of the present paper to develop an implicit MacCormack scheme for the 2D steady Navier-Stokes equations, analogous to the unsteady case, incorporating therein a full-fledged eigenvalue analysis of the Jacobian matrix and numerical evaluation of the flow variables from the state vector at each marching step.

Governing Equations and Numerical Scheme

The steady Navier-Stokes equations in 2D may be expressed in conservation form as

$$\frac{\partial F}{\partial x} + \frac{\partial G}{\partial y} = 0, \qquad (1)$$

where the symbols have the same meanings as in [1]. If the streamwise viscous derivatives are neglected in (1), we obtain the parabolized Navier-Stokes equations. A detailed analysis reveals that an implicit predictor-corrector marching scheme for the numerical solution of (1) may, following [1], be formulated as below:

Predictor:
$$\Delta F_j^n = -\frac{\Delta x}{\Delta y} D_+ G_j^n$$
$$\left(I - \frac{\Delta x}{\Delta y} D_+ |A|\cdot\right) \delta \overline{F_j^{n+1}} = \Delta F_j^n$$
$$\overline{F_j^{n+1}} = F_j^n + \delta \overline{F_j^{n+1}}$$

(2)

Corrector:
$$\Delta \overline{F_j^{n+1}} = -\frac{\Delta x}{\Delta y} D_- \overline{G_j^{n+1}}$$
$$\left(I + \frac{\Delta x}{\Delta y} D_- |A|\cdot\right) \delta F_j^{n+1} = \Delta \overline{F_j^{n+1}}$$
$$F_j^{n+1} = \frac{1}{2}\left[F_j^n + \overline{F_j^{n+1}} + \delta F_j^{n+1} \right],$$

where $|A|$ has positive eigenvalues and is related to A, the Jacobian matrix, $= \frac{\partial G}{\partial F}$. Detailed computations have shown that

$$A = S^{-1} \Lambda S,$$ (3)

where $S = S_1 \times S_2$,

$$S_1 = \begin{bmatrix} \frac{c^2}{2} & \frac{\rho u}{2} & \frac{\rho v}{2} & 0 \\ 0 & -\frac{\rho v}{\beta} & \frac{\rho u}{\beta} & 1 \\ -\frac{c^2}{2} & \frac{\rho u}{2} & \frac{\rho v}{2} & 1 \\ 0 & \frac{\rho v}{\beta} & -\frac{\rho u}{\beta} & 1 \end{bmatrix}$$ (4)

and

$$S_2 = \begin{bmatrix} 1 & 0 & 0 & 0 \\ -\frac{u}{\rho} & \frac{1}{\rho} & 0 & 0 \\ -\frac{v}{\rho} & 0 & \frac{1}{\rho} & 0 \\ \alpha \beta_1 & -u\beta_1 & -v\beta_1 & \beta_1 \end{bmatrix}$$ (5)

S_1 and S_2 are, respectively, the Jacobians of transformation from nonconservative to characteristic form and conservative to nonconservative variables. Λ is the diagonal matrix given by

$$\Lambda = \begin{bmatrix} v/u & 0 & 0 & 0 \\ 0 & \frac{uv+c^2\beta}{u^2-c^2} & 0 & 0 \\ 0 & 0 & v/u & 0 \\ 0 & 0 & 0 & \frac{uv-c^2\beta}{u^2-c^2} \end{bmatrix} \quad (6)$$

Thus

$$S = \begin{bmatrix} \frac{c^2}{2} - \alpha & \frac{u}{2} & \frac{v}{2} & 0 \\ \alpha\beta_1 & -(\frac{v}{\beta}+u\beta_1) & \frac{u}{\beta}-v\beta_1 & \beta_1 \\ -\frac{c^2}{2}-\alpha+\alpha\beta_1 & \frac{u}{2}-u\beta_1 & \frac{v}{2}-v\beta_1 & \beta_1 \\ \alpha\beta_1 & \frac{v}{\beta}-u\beta_1 & -(\frac{u}{\beta}+v\beta_1) & \beta_1 \end{bmatrix} \quad (7)$$

where

$$\alpha = \tfrac{1}{2}(u^2+v^2),\ \beta = \sqrt{\frac{u^2+v^2}{c^2}-1},\ \beta_1 = \gamma-1,\ c = \sqrt{\frac{\gamma p}{\rho}}. \quad (8)$$

$$S^{-1} = (S_1\,S_2)^{-1} = S_2^{-1}\,S_1^{-1}$$

$$= \begin{bmatrix} 1 & 0 & 0 & 0 \\ u & \rho & 0 & 0 \\ v & 0 & \rho & 0 \\ \alpha & u\rho & v\rho & \frac{1}{\beta_1} \end{bmatrix} \begin{bmatrix} \frac{1}{c^2} & \frac{1}{2c^2} & -\frac{1}{c^2} & \frac{1}{2c^2} \\ \frac{u}{\rho(u^2+v^2)} & -\frac{uv+c^2\beta}{2\rho c^2(v+u\beta)} & \frac{u}{\rho(u^2+v^2)} & -\frac{uv-c^2\beta}{2\rho c^2(v-u\beta)} \\ \frac{v}{\rho(u^2+v^2)} & \frac{u^2-c^2}{2\rho c^2(v+u\beta)} & \frac{v}{\rho(u^2+v^2)} & \frac{u^2-c^2}{2\rho c^2(v-u\beta)} \\ 0 & \frac{1}{2} & 0 & \frac{1}{2} \end{bmatrix} \quad (9)$$

$$= \begin{bmatrix} \frac{1}{c^2} & \frac{1}{2c^2} & -\frac{1}{c^2} & \frac{1}{2c^2} \\ \frac{u(u^2+v^2+c^2)}{c^2(u^2+v^2)} & \frac{\beta(u^2-c^2)}{2c^2(v+u\beta)} & -\frac{u(u^2+v^2-c^2)}{c^2(u^2+v^2)} & -\frac{\beta(u^2-c^2)}{2c^2(v-u\beta)} \\ \frac{v(u^2+v^2+c^2)}{c^2(u^2+v^2)} & \frac{1}{2c^2}\!\left(v+\frac{u^2-c^2}{v+u\beta}\right) & -\frac{v(u^2+v^2-c^2)}{c^2(u^2+v^2)} & \frac{1}{2c^2}\!\left(v+\frac{u^2-c^2}{v-u\beta}\right) \\ \frac{\alpha}{c^2}+1 & \frac{\alpha}{2c^2}-\frac{1}{2}+\frac{1}{2\beta_1} & -\frac{\alpha}{c^2}+1 & \frac{\alpha}{2c^2}-\frac{1}{2}+\frac{1}{2\beta_1} \end{bmatrix} \quad (10)$$

$|A|$ may now be defined as

$$|A| = S^{-1} D_A S, \qquad (11)$$

where

$$D_A = \begin{bmatrix} D_{A_1} & 0 & 0 & 0 \\ 0 & D_{A_2} & 0 & 0 \\ 0 & 0 & D_{A_3} & 0 \\ 0 & 0 & 0 & D_{A_4} \end{bmatrix}, \qquad (12)$$

$$D_{A_1} = D_{A_3} = \max\left\{ \left|\frac{v}{u}\right| + \frac{2\nu}{\rho \Delta y} - \frac{1}{2}\frac{\Delta y}{\Delta x}, \; 0.0 \right\},$$

$$D_{A_2} = \max\left\{ \left|\frac{uv + c^2\beta}{u^2 - c^2}\right| + \frac{2\nu}{\rho \Delta y} - \frac{1}{2}\frac{\Delta y}{\Delta x}, \; 0.0 \right\},$$

$$D_{A_4} = \max\left\{ \left|\frac{uv - c^2\beta}{u^2 - c^2}\right| + \frac{2\nu}{\rho \Delta y} - \frac{1}{2}\frac{\Delta y}{\Delta x}, \; 0.0 \right\}, \qquad (13)$$

$$\nu = \max\left\{ \mu, \; \lambda + 2\mu, \; \frac{\gamma \mu}{P_r} \right\}.$$

For regions of the flow for which Δx satisfies the explicit stability criterion

$$\Delta x \leq \frac{1}{2} \cdot \frac{\Delta y}{\left[\frac{uv + c^2\beta}{u^2 - c^2}\right]}, \qquad (14)$$

all D_A's vanish and the implicit numerical scheme, described above, reduces automatically to the corresponding explicit method.

The flow parameters ρ, u, v, p are determined through an iterative procedure at each marching step. Let us suppose

$$\begin{bmatrix} \rho u \\ \rho u^2 + \sigma_x \\ \rho u v + \tau_{xy} \\ (e + \sigma_x) u + \tau_{yx} v - k \frac{\partial T}{\partial x} \end{bmatrix} = \begin{bmatrix} f_1 \\ f_2 \\ f_3 \\ f_4 \end{bmatrix} \qquad (15)$$

and assuming as a first approximation

$$\rho u = f_1, \quad \rho u^2 + p = f_2, \quad \rho u v = f_3,$$

$$(e+p)u = f_4 = \rho u \left[\frac{\gamma}{\gamma-1} \cdot \frac{p}{\rho} + \frac{1}{2}(u^2+v^2) \right],$$

calculations yield

$$u = \frac{\gamma}{f_1(\gamma+1)} \left\{ f_2 \pm \left[f_2^2 - \frac{\gamma^2-1}{\gamma^2}(2 f_4 f_1 - f_3^2) \right]^{1/2} \right\},$$

$$\rho = f_1/u, \quad v = f_3/f_1, \quad p = f_2 - f_1 u. \qquad (16)$$

f_1, f_2, f_3, f_4 are now modified in Eqn. (15) and the procedure applied iteratively to determine the flow parameters.

Details of the Computer Program and Results

A computer program was developed and made operational on the CDC CYBER 170/730 computer of Vikram Sarabhai Space Centre, Trivandrum and applied for the interaction of an oblique shock wave with a laminar boundary layer on a flat plate with attached flow by solving steady two-dimensional Navier-Stokes equations using the above implicit numerical scheme. The features of the flow field under investigation are depicted in Fig.1. The initial condition at $x=0$ for the flow field was taken to be uniform flow ($M=2$, $Re_L = 0.284 \times 10^6$), where the characteristic length L is the distance of the shock impingement point from the leading edge. The conditions at the top mesh boundary were such that a shock wave of given strength would be generated and impinge upon the flat plate at a given point (shock angle = $31°.347$), while those of the lower boundary were got by reflection. A two mesh system of a fine stretched mesh of 16 points close to the flat plate and a coarse uniform mesh of 16 points away from the plate were applied in the $y-$ direction. The solution was found to be more difficult in the subsonic region of the boundary layer, where measures analogous to [5] were introduced to minimize the destabilizing influence.

About 1000 steps were made use of in the $x-$ direction which covered approximately twice the characteristic length. The computation time was around 150 secs. corresponding to a Courant number of 5, as against 540 secs. in the corresponding unsteady case with a Courant number of 20. The implementation of the numerical scheme on the computer required only about 8×10^3 60-bit words of core memory as against a requirement of 3×10^4 words of storage in the case of the time asymptotic steady state solution of the same problem.

The results for surface pressure distribution, obtained through the program, are compared in Fig.2 with known experimental data [6] as well as previously calculated values available in the literature [1]..

Conclusion

The present paper attempts to develop an implicit MacCormack scheme for the two-dimensional steady Navier-Stokes equations, analogous to the unsteady case, incorporating a full-fledged eigenvalue analysis of the Jacobian matrix and numerical evaluation of the flow variables from the state vector at each marching step. A software is developed employing the above numerical scheme for the interaction of a shock wave incident upon a boundary layer with attached flow. The results obtained through the program for surface pressure distribution exhibit reasonable agreement with existing computations and available experimental data.

REFERENCES

1. MacCormack, R.W., 'A numerical method for solving the equations of compressible viscous flow', AIAA Journal, 20, 1982, 1275-1281.

2. Swaminathan, V. and Madhavan, N.S., 'Computer solutions of Navier-Stokes equations for shock wave-boundary layer interaction far away from the leading edge', Proc. Int. Conf. Num. Methods for Laminar and Turbulent Flow, Univ. of Washington, Seattle, Aug 8-11, 1983, Pineridge Press, Swansea, U.K., C. Taylor, J.A. Johnson and W.R. Smith (Eds.), Paper No.21, 221-231, 1983.

3. Madhavan, N.S. and Swaminathan, V., 'Numerical solution of the supersonic laminar flow over a two-dimensional compression corner using an implicit approach', Int. J. for Num. Methods in Fluids (in print).

4. Madhavan, N.S. and Swaminathan, V.: 'An implicit algorithm for solving steady Euler equations in two dimensions', Proc. 29th ISTAM Congress, Surathkal, India, Dec 27-30, 1984.

5. Lawrence, S.L., Tannehill, J.C. and Chausee, D.S., 'Application of the implicit MacCormack scheme to the parabolized Navier-Stokes equations', AIAA Journal 22, 1984, 1755-1763.

6. Hakkinen, R.J., Greber, I., Trilling, L., and Abarbanel, S.S., The interaction of an oblique shock wave with a laminar boundary layer, NASA Memorandum 2-18-59 W, 1959.

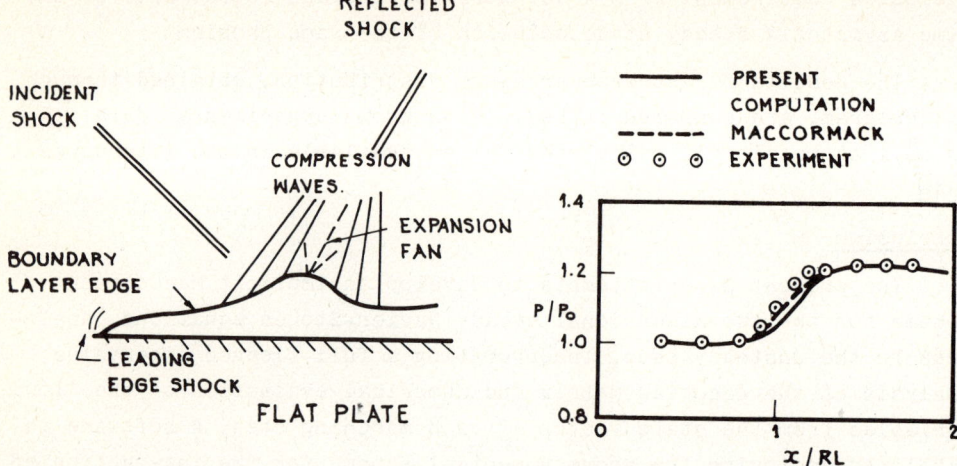

Figure 1. Sketch of Shock Boundary Layer Interaction

Figure 2. Comparison of Surface Pressure Distribution

NUMERICAL SIMULATION OF TRANSITION IN A THREE-DIMENSIONAL BOUNDARY LAYER

Mujeeb R. Malik
High Technology Corporation
P. O. Box 7262
Hampton, Virginia 23666-0262

Abstract

A Fourier-Chebyshev spectral method is used to study the nonlinear stability and the breakdown of crossflow vortices in a three-dimensional boundary layer present on a disk rotating in an otherwise quiescent ambient. The results obtained to date in a continuing research effort are presented.

Introduction

The transition mechanism in a swept-wing boundary layer is a subject of both fundamental and practical importance in fluid mechanics. This boundary layer is amenable to inflectional instability which manifests itself in the form of stationary co-rotating vortices known as crossflow vortices since they are caused by the instability of the crossflow velocity profile. These vortices align themselves approximately with the external streamlines. While a great deal is now known about the breakdown of Tollmien-Schlichting waves to turbulence in a two-dimensional boundary layer, no theoretical or experimental study has yet fully explained the breakdown mechanism of the steady crossflow vortices to chaotic motion.

Here, this problem is studied numerically by solving three-dimensional Navier-Stokes equations using a spectral collocation method. The particular three-dimensional boundary considered is that formed on a rotating disk. This flow has exact steady solution [1] to the Navier-Stokes equations and is also subjected to the inflectional instability present near the leading edge of a swept wing. The rotating disk flow has long been used as a model for studying crossflow instability phenomenon on swept wings.

Stationary disturbances in the rotating disk flow originate at discrete roughness sites and spread radially outwards in the form of wave packets [2]. When these wave packets fully cover the disk circumference, the number of vortices turns out to be approximately 30 and the normals to these vortices are oriented at angles between 11 and 13 degrees with respect to the radial direction. As shown by Mack [3], the initial evolution of each wave packet can be well described by the linear theory. It is the breakdown of the individual vortices which is the subject of the present investigation.

The Governing Equations and Numerical Method

The governing equations for three-dimensional incompressible flow are written in a rotating frame of reference:

$$\mathbf{u}_t + \mathbf{u} \cdot \nabla \mathbf{u} = -2\mathbf{\Omega} \times \mathbf{u} - \frac{1}{\rho} \nabla p + \nu \nabla^2 \mathbf{u}, \quad (1)$$

$$\nabla \cdot \mathbf{u} = 0, \quad (2)$$

where $\mathbf{\Omega}$ ($\mathbf{\Omega} = (0,0,\Omega)$) is the angular speed of rotation of the disk. Cylindrical coordinates (r,θ,z) are chosen so that $\mathbf{u} = (v_r, v_\theta, v_z)$. Equations (1) and (2) are solved by the Fourier-Chebyshev spectral method described by Malik, Zang, and Hussaini [4]. Periodicity is assumed in r and θ directions. The dependent variables have Fourier-Chebyshev series of the form

$$v_r(r,\theta,z,t) = \sum_{k_r = -K_r/2}^{K_r/2 - 1} \sum_{k_\theta = -K_\theta/2}^{K_\theta/2 - 1} \sum_{n=0}^{N} \hat{v}_r(t) e^{2\pi i (k_r r/L_r)} e^{2\pi i (k_\theta \theta / L_\theta)} T_n(\eta), \quad (3)$$

where L_r and L_θ are the periodicity lengths in r and θ directions, respectively, and T_n is the Chebyshev polynomial of degree n. The normal computational coordinate η is related to z through the algebraic transformation

$$z = a \frac{1 + \eta}{1 + \frac{2a}{z_\infty} - \eta}, \quad (4)$$

where a is a scaling constant used for proper distribution of points and z_∞ is the location where free stream boundary conditions are imposed.

The spatial discretization employs spectral collocation. The collocation points for the periodic directions are

$$r_j = j L_r/K_r, \quad j = 0, 1, \cdots, K_r - 1, \quad (5)$$

$$\theta_m = m\, L_\theta/K_\theta, \qquad m = 0,1,\cdots,K_\theta - 1. \tag{6}$$

A staggered grid is employed in the normal direction. Velocities are defined at the points

$$\eta_q = \cos\left(\tfrac{\pi q}{N}\right), \quad q = 0,1,\cdots,N \tag{7}$$

and the pressures at

$$\eta_{q+1/2} = \cos\left(\pi(q+\tfrac{1}{2})/N\right), \quad q = 0,1,\cdots,N-1. \tag{8}$$

No artificial pressure boundary conditions are therefore needed. The momentum equations are imposed at the points given by Eq. (7) and the continuity at those given by Eq. (8).

In the spectral collocation method, spatial derivatives of v_r are obtained by differentiating the series expansion coefficients $\hat{v}_r(t)$ determined by discrete Fourier and Chebyshev transforms of the grid-point values of v_r. The temporal discretization involves using Crank-Nicolson on the pressure gradient and vertical diffusion terms. The remaining terms in the momentum equations are handled explicitly by using second-order Adams-Bashforth method. The incompressibility constraint is imposed implicitly. The resulting implicit equations have the form:

$$-\beta \hat{v}_r'' + \hat{v}_r + ik_r \gamma \hat{Q} = H_1, \tag{9}$$

$$-\beta \hat{v}_\theta'' + \hat{v}_\theta + ik_\theta \gamma \hat{Q} = H_2, \tag{10}$$

$$-\beta \hat{v}_z'' + \hat{v}_z + \gamma \hat{Q}' = H_3, \tag{11}$$

$$-ik_r \gamma \hat{v}_r - \gamma \frac{\hat{v}_r}{R} - ik_\theta \gamma \hat{v}_\theta - \gamma \hat{v}_z' = H_4, \tag{12}$$

where the primes denote derivative with respect to z, $\beta = \mu\Delta t/2$, $Q = (\Delta t/2\gamma)\hat{P}$, $R = r\sqrt{\Omega/\nu}$, and $i = \sqrt{-1}$. The right-hand sides of these equations contain the explicit terms. In the above equations, all lengths have been scaled with $\ell = \sqrt{\nu/\Omega}$ and all velocities with $U_\infty = r\Omega$. An iterative solution of these equations is obtained using minimum residual method and finite-difference preconditioning [4]. The parameter γ is introduced to help the convergence of higher Fourier modes. It has been assigned a value of $\gamma = 1/\sqrt{k_r^2 + k_\theta^2}$.

No-slip boundary conditions are imposed at the solid wall. In the freestream, the first-order boundary conditions are imposed at $z_\infty = 20$. The mean flow is obtained by solving the following ordinary differential equations:

$$F^2 - (G+1)^2 + F' H - F'' = 0, \qquad (13)$$

$$2F(G+1) + G' H - G'' = 0, \qquad (14)$$

$$\Pi' + HH' - H'' = 0, \qquad (15)$$

$$2F + H' = 0, \qquad (16)$$

where the primes denote differentiation with respect to z. The boundary conditions for Eqns. (13) - (16) are

$$F = 0, \quad G = 0, \quad H = 0, \quad (z = 0); \quad F = 0, \quad G = -1, \quad (z \to \infty).$$

In these equations F, G, H, and Π are related to the velocity components and pressure according to

$$V_r = r \Omega F(z), \qquad V_\theta = r \Omega G(z),$$

$$V_z = \sqrt{\nu\Omega} \; H(z), \qquad P = \rho \Omega \nu \Pi(z).$$

The initial conditions consists of the mean flow with superimposed disturbance eigenfunctions obtained by solving the linear eigenvalue problem described by Malik [5].

Results and Discussion

The neutral curves (in α-R and ε-R planes) for stationary disturbances are plotted in Fig. 1(a) and 1(b). Here, α is the wavenumber in the radial direction and ε is the wave angle. The theoretical critical Reynolds number is about 285.4. A second mode becomes unstable at Reynolds numbers in excess of about 440.9. However, the associated wave angles are around 20°. It is the upper branch solution with wave angles around 11°-13° that is commonly observed in experiments. Transition to turbulence takes place at Reynolds numbers in excess of about 530.

The particular problem chosen for study using the Navier-Stokes code has R = 500, α = .375148, β = .077689 (here, β is the azimuthal wavenumber). This is the most amplified stationary mode associated with the upper branch solution. It has a temporal growth rate from linear theory of ω_i = .005159. The amplitude of the perturbation is taken to be .1% with respect to the mean flow and the computation is performed by using 8 collocation points in both r and θ directions and 33 Chebyshev polynomials in z direction. The evolution of the disturbance energy in the (1,1) mode (i.e., the fundamental mode) is plotted in Fig. 2. The disturbance energy has been normalized with the mean flow energy. The solid line in the figure represents the linear theory result. The computed growth rates for this relatively low amplitude calculation are in excellent agreement with the linear theory.

Next, a calculation is made in which the initial conditions contain linear eigenfunction but the amplitude assigned to the maximum of the eigenfunction is 10% with respect to the maximum of the mean flow. This calculation is made by using 32×32×33 grid. The results for the energies of (1,1), (2,2), (3,3), (4,4), and (5,5) modes are presented in Fig. 3. The solid curve represents the linear theory result for the (1,1) mode. Spectral broadening takes place and the higher harmonics gain energy as the growth of the fundamental slows down. The energy cascading seems to take place through (2,2), (3,3), (4,4), etc. modes.

Hot-wire traces at various radii in a wave packet from the experiment of Wilkinson and Malik [2] are presented in Fig. 4(a). The hot wire is located at $z \simeq 1.8$ and is oriented such that it essentially senses the perturbations in the azimuthal velocity component. The traces develop secondary instability prior to transition. Corresponding computational traces at various radii are presented in Fig. 4(b). The time of the simulation is t = 170. The computational trace appears to develop the same type of ´kink´ as in the experiment.

The azimuthal velocity profiles at a fixed radius but at various azimuthal locations within the computational domain are presented in Fig. 5. The time of the simulation is again t = 170. While the von Karman profile is full with no inflection point, the computed profiles develop shear layer type profiles at z around 2.5. These profiles are amenable to shear layer instability which would manifest in the form of vortices normal to these velocity profiles. In the case of the rotating disk, these vortices may appear wrapped around the primary crossflow vortices. In the present simulation, this shear layer instability has not been observed perhaps because the round-off errors are not sufficient to provide the seed for such disturbances. A simulation with small random initial distribution of modal energy is being carried out.

This work was sponsored by NASA Langley Research Center under Contracts NAS1-16916 and NAS1-18240. The author would like to thank Mr. Ivan Beckwith of NASA Langley for his encouragement for this work and Dr. M. Y. Hussaini of ICASE for very useful discussions on the subject.

References

[1]   von Karman, T., "Uber laminare und turbulente Reibung," <u>Z. Angew. Math. Mech.</u>, Vol. 20, 1940, pp. 241-253.

[2] Wilkinson, S. P. and Malik, M. R., "Stability Experiments in the Flow over a Rotating Disk," AIAA J., Vol. 23, 1985, p. 588.

[3] Mack, L. M., "The Stationary Wave Pattern Produced by Point Source on a Rotating Disk," AIAA Paper No. 85-0490.

[4] Malik, M. R., Zang, T. A., and Hussaini, M. Y., "A Spectral Collocation Method for the Navier-Stokes Equations," J. Comput. Phys., Vol. 61, 1985, pp. 64-88.

[5] Malik, M. R., "The Neutral Curve for Stationary Disturbances in Rotating Disk Flow," J. Fluid Mech., Vol. 164, 1986, pp. 275-288.

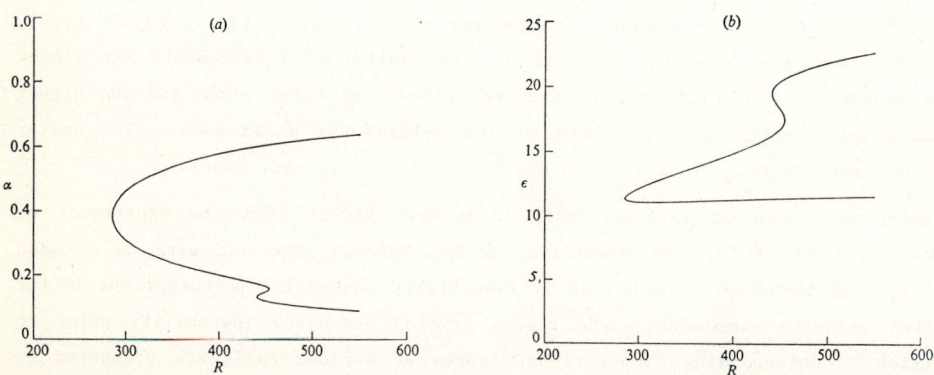

Fig. 1. Neutral curves for stationary disturbances in rotating disk flow. (a) wavenumber; (b) wave angle.

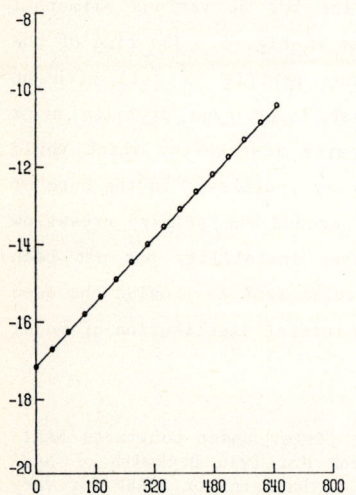

Fig. 2. Computed evolution of the energy of a small amplitude stationary disturbance with $\alpha = .375148$, $\beta = .077689$, and $R = 500$. The solid line represents the linear theory result.

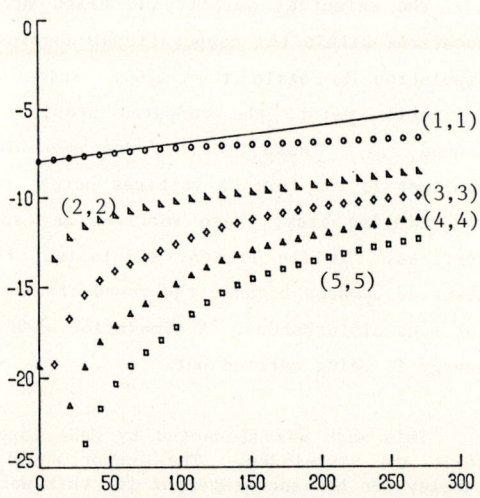

Fig. 3. Computed evolution of energies of various modes in the presence of a finite-amplitude crossflow vortex ((1,1) mode). In this calculation $R = 500$, $\alpha = .375148$, $\beta = .077689$. Initial amplitude of the (1,1) mode is 10% of the mean flow.

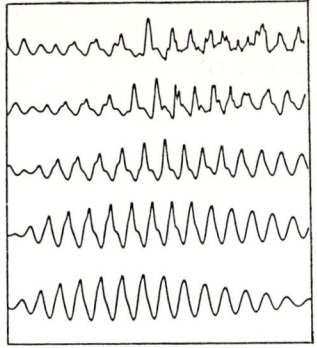 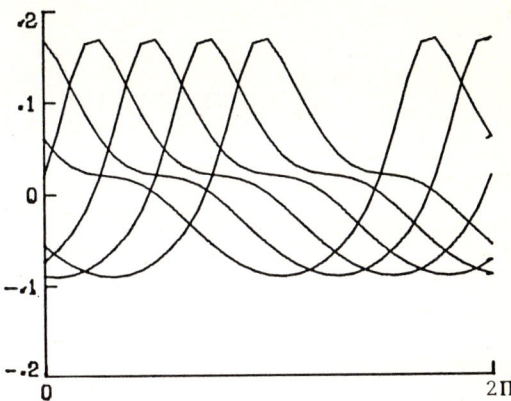

Fig. 4(a). Hot-wire traces at various radii near transition in rotating disk boundary layer. Hot-wire is located at $z \simeq 1.8$ and it is oriented such that it senses the perturbation in azimuthal velocity component.

Fig. 4(b). Computed traces of azimuthal velocity perturbations at $z = 1.8$. Only one fundamental wavelength is shown. The The calculations are for the conditions of Fig. 3. Time $t = 170$.

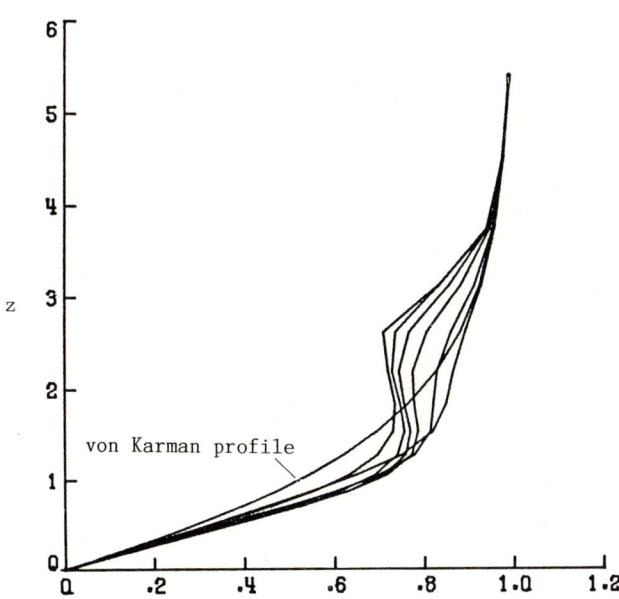

Fig. 5. Computed azimuthal velocity profiles at a fixed radial location and within 1/2 fundamental wavelength. Time $t = 170$.

A DISCRETE VECTOR POTENTIAL METHOD FOR
UNSTEADY 3-D NAVIER-STOKES EQUATIONS

D. Mansutti, U. Bulgarelli
Istituto Applicazioni Calcolo, C.N.R., Roma

R. Piva, G. Graziani
Dipartimento di Meccanica e Aeronautica
Università di Roma

1. Introduction

The numerical simulation of unsteady incompressible flow in two dimensions has been performed, mostly in the framework of finite difference methods, by following three different approaches. The first one assumes the stream function (ψ) as a variable satisfying a fourth order equation. The second approach splits this equation in two second order equations by introducing the vorticity (ω). In both cases the boundary conditions should be given on the two velocity components that is on ψ and $\partial\psi/\partial n$. From these ones we can obtain a boundary conditions on the vorticity (ω) to be used for the split system, that is an integral boundary condition [10], or an approximated form to be evaluated by means of previous ψ values, when iterative schemes are used [11]. Finally in the third approach the primitive variables (u,v,p) are adopted and the boundary conditions are given on a subset of these variables.

Advantages and disadvantages of the above methods are well known and have been largely discussed. The most appealing feature of both the streamfunction and the vorticity-streamfunction formulation is the capability to generate automatically a zero divergence velocity field. As a drawback the boundary conditions, particularly when the pressure is involved, (e.g. flows with free surface) are not straightforward to be handled. The alternative choice of numerical schemes in primitive variables is more convenient from the point of view of boundary conditions which are easily assigned for all the physically relevant cases. The satisfaction of the mass conservation at each time step, indeed, is a more difficult task and it usually requires cumbersome iterative procedures.

In a previous paper [6] we proposed a 2-D computational model, based on the generation of a discrete streamfunction, which was presenting some interesting features with respect to the more classical methods above briefly discussed. In fact the proposed model allows for the boundary conditions to be given in primitive variables and, at the same time, by eliminating the pressure as a variable to be computed, produces a velocity field which exactly satisfies the mass conservation in discretized form.

Three-dimensional flow fields can be represented by following completely analogous schemes. In fact the equations can be formulated either by adopting the scalar and vector potential or the primitive variables. In particular, in the first case, by taking the curl of the Navier Stokes equations, and defining the variables in terms of the scalar and vector potential, a system of fourth order equations (one for each component of the vector potential) plus a Laplace equation for the scalar potential, are obtained. By introducing the vorticity vector as a variable, a system of second order equations is obtained. These approaches show the same features of the correspondent ones in the 2-D case. Moreover the first formulation is not usually considered, and the boundary conditions on the vector potential approach are not as simply formulated as those for ψ in 2-D. In particular for multiply connected fields the correct assignment of the boundary conditions is still a matter of debate.

The 2-D model previously presented by the Authors [6] is here extended to

3-D flows by introducing a discrete vector potential. In particular, we determine a particular base of the null space of the discrete divergence operator and, by expressing the velocity in this base, we introduce a reduced set of unknowns which can be interpreted as the components of a discrete vector potential, Hence a discrete curl operator is applied to the Navier-Stokes equations to eliminate the pressure as a variable to be computed. The resulting system of equations is finally solved by a fully implicit integration in time. The proposed numerical procedure provides a discrete divergence-free velocity field and, at the same time, allows for the boundary conditions to be given in primitive variables. These features imply a straightforward solution to be easily obtained even by for multiply connected domains and for free surface flows. In section 2 the procedure and the general features of this method are described. Several numerical results of 3-D flow problems in enclosures are presented and discussed in section 3.

2. Mathematical Formulation and Computational Model

The mathematical model of the flow problem in a bounded domain D, with boundary ∂D, is given by the Navier-Stokes equations for incompressible flows

$$\frac{\partial u_i}{\partial t} + (u_i u_j - T_{ij})_{,j} = 0 \tag{1}$$

$$u_{i,i} = 0 \tag{2}$$

and the boundary and initial conditions

$$u_i(x,y,z,t) = \bar{u}_i \quad \text{for } x,y,z \in \partial D, \ t > 0 \tag{3}$$

$$u_i(x,y,z,0) = u_{oi} \quad \text{for } x,y,z \in D \tag{4}$$

where u_{oi} is a divergence-free initial velocity field and the other symbols have their usual meaning. The presentation is confined to Cartesian coordinates to simplify the description of the computational procedure that follows the one given in [7].

The finite difference approximation of the governing equations is generated, by using a centered space-forward time scheme, on a staggered grid consisting of cubic cells. We obtain the following system of algebraic equations

$$\frac{1}{\Delta t}(u^m - u^{m-1}) + B_{m-1} u^m = A^T p^m + f^m \tag{5}$$

$$A u^m = 0 \tag{6}$$

where m indicates the time level, u and p are vectors containing respectively all discrete velocity components and discrete pressures, A is a rectangular matrix resulting from the discretization of the divergence operator, B_{m-1} is a square matrix resulting from the discretization of the convective and viscous terms and f^m is the vector of the mass force components. If N is the number of cells and L is the number of unknown velocity components, B_{m-1} is a L×L matrix, A is a N×L matrix. Moreover A, that represents the discrete divergence operator, has rank N-1 [9]. Hence its null space has dimension L-N+1 and it is possible to find L-N+1 vectors of R^L, linearly independent, forming a matrix C with dimension L×(L-N+1), such that:

$$A C = 0 \tag{7}$$

From (6) it follows

$$u^m = C \gamma^m \tag{8}$$

for some $\gamma^m \in R^{L-N+1}$. If we write eq.(5) as

$$Q_{m-1} u^m = \Delta t\, A^T p^m + b^m \tag{9}$$

where

$$Q_{m-1} = I + \Delta t\, B_{m-1}$$

$$b^m = u^{m-1} + \Delta t\, f^m$$

and multiply (9) by C^T, using (7) we obtain

$$C^T Q_{m-1} u^m = C^T b^m \tag{10}$$

that combined with (8) gives

$$C^T Q_{m-1} C \gamma^m = C^T b^m \tag{11}$$

The original system, which is of order 4N, has been reduced to the system (11) of L-N+1 equations in L-N+1 unknowns. The matrix A can be interpreted as a node-arc incidence matrix of the nonplanar network, whose nodes are the points where the discrete pressure is localized and whose arcs are the unknown velocity components. A base in the null space of the operator A is then determined by selecting L-N+1 elementary cycles of the network [8]. For a particular choice of the cycles the matrix C represents the discrete curl operator [9]. Being L of order 3N, the dimension of the system (11) is of order 2N. The coefficient matrix, $C^T Q_{m-1} C$, has a banded structure and efficient solution algorithms may be used.

In presence of inflow-outflow boundary conditions the mass conservation gives an inhomogeneous equation and a particular solution has to be determined.

Moreover, if a boundary condition on pressure is given, the matrix A is not singular any more and one unknown velocity component appears in each cell where the pressure is assigned. If L_o is the number of such cells, the matrix A is a linear operator from $\mathbb{R}^{L+L_o}$ to $\mathbb{R}^N$ and its null space has dimension $L+L_o-N$. Hence in this case L_o-1 linearly independent vectors have to be determined in addition to the previously considered L-N+1. Either the particular solution required for inflow-outflow boundary conditions, and the additional L_o-1 linearly independent vectors for the case of pressure boundary conditions may be determined by using the network theory [8].

3. Numerical Results and Comments

The flow field equations (1),(2) have been solved by means of the numerical procedure described in section 2. First the method has been applied to a two-dimensional problem with analytical solution, which can be assumed as a preliminary test, also in the 3D case. The solution of the unsteady Navier Stokes equations is given by

$$u_1(x_1,x_2,x_3,t) = \sin x_1\ \cos x_2\ e^{-2\nu t}$$

$$u_2(x_1,x_2,x_3,t) = -\cos x_1\ \sin x_2\ e^{-2\nu t}$$

$$u_3(x_1,x_2,x_3,t) = 0 \tag{12}$$

$$p(x_1,x_2,x_3,t) = \tfrac{1}{4} [\cos 2 x_1 + \cos 2 x_2]\, e^{-4\nu t}$$

in $D \equiv [0,\Pi]\times[0,\Pi]\times[0,\Pi]$, with boundary and initial conditions consistent with (12). The flow field has been determined for each of the three cases of flow parallel to the faces of D. The numerical solution for

$t = 0.01$ and $t = 0.1$, with $\Delta t = 0.1$, perfectly agree with the analytical solution, also in the case of boundary condition given on pressure (fig. 1).

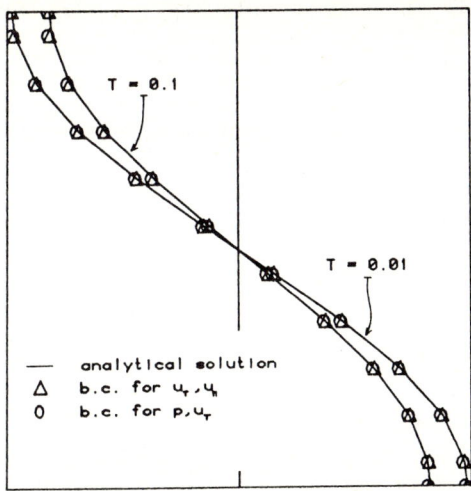

Fig. 1 - Numerical results versus analytical solution.

Then the method has been applied to a more significant and really three dimensional test case, the driven cavity flow, for which recent numerical results [3,4,5] are available in the literature. Exploiting the simmetry of the flow only one half of the domain has been considered, the computation has been performed for Re = 400. The vertical profile, at the centerline of the cavity for the velocity component parallel to the driving direction is shown in fig. 2. The computation has been made both with free slip condition and with no slip condition at the lateral walls.

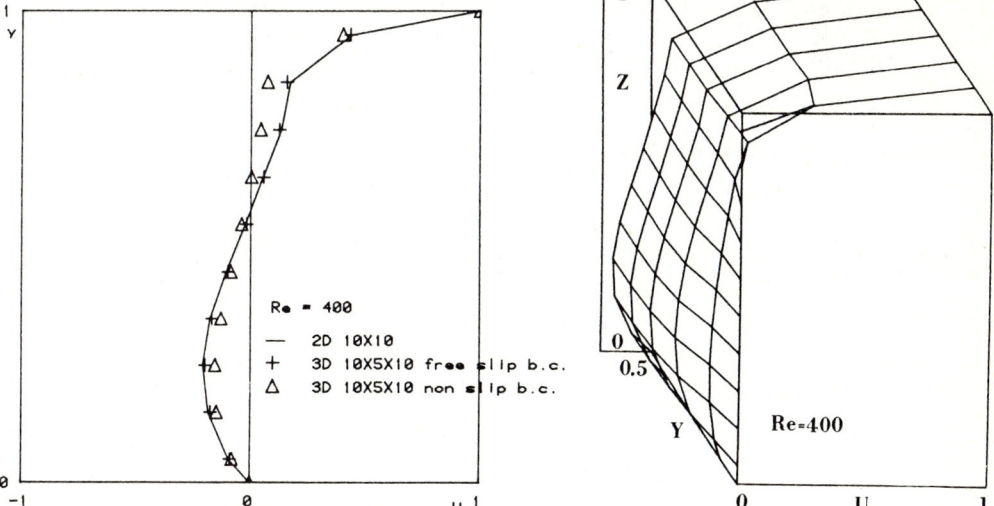

Fig. 2 - 3D and 2D driven cavity: profile of the driven component velocity u_1 along the central vertical line.

Fig. 3 - 3D driven cavity: profile of the driven component velocity u_1 in the plane $x_1 = 0.5$.

A comparison between the profiles obtained for these two boundary conditions, and the correspondent one for the 2D case is shown in fig. 2. As expected the two profiles are identical when the slip condition is used, while a less pronounced profile is obtained with the no slip condition due to the energy dissipation at the lateral walls.

The present solution agrees reasonably well (fig. 3) with that one plotted by Goda in [4], even if our calculation has been performed with a coarser mesh. Computations at larger Reynolds number presently under investigation should make possible a more complete comparison with several other solutions [3,5] recently appeared in the literature.

References

1. Richardson, S.M., Cornish, A.R., "Solution of three-dimensional incompressible flow problems", J. Fluid Mech., 82, 2, 309-319, 1977.
2. Reizes, J.A., Leonardi, E., de Vahl Davis, G., "Problems with derived variable methods for the numerical solution of three-dimensional flows", Computational Techniques and Application Conference, Australia, 1983.
3. Koseff, J.R., Street, R.L., "The lid driven cavity flow: a synthesis of qualitative and quantitative observations", J. Fluids Eng., 106, 390-398, 1984.
4. Goda, K., "A multistep technique with implicit difference schemes for calculating two or three dimensional cavity flows", J. of Comp. Phys., 30, 76-95, 1979.
5. Kim, J., Moin, P., "Application of a fractional-step method to incompressible Navier-Stokes equation", NASA Tech. Mem. 85898, March 1984.
6. Bulgarelli, U., Graziani, G., Mansutti, D., Piva, R.,"A reduced implicit scheme via discrete stream function generation for unsteady Navier Stokes equations in general curvilinear coordinates", VI GAMM Conference, Gottingen, 1985.
7. Amit, R., Hall, C.A., Porsching, T.A., "An application of network theory to the solution of implicit Navier Stokes difference equations", J. Comp. Phys., 40, 183-201, 1981.
8. Kennington, J.L., Helgason, R.C., "Algorithms for network programming", John Wiley & Sons, New York.
9. Mansutti, D., Bulgarelli, U., Piva, R., Graziani, G., "Generazione di un potenziale vettore discreto per flussi incomprimibili viscosi non stazionari", AIMETA, Torino, 1986.
10. Quartapelle, L., Valz Gris, F., "Projection conditions on the vorticity in viscous incompressible flows", Int. J. Numer. Meth. Fluids, 1, 129-144, 1981.
11. Ehrlich, L.W., Gupta, M.M., "Some difference schemes for the biharmonic equation", SIAM J. Numer. Anal., 12, 5, 773-790, October 1975.

INTERACTION BETWEEN STRUCTURE AND FREE SURFACE FLUID
WITH LARGE DISPLACEMENTS BY FINITE ELEMENTS

Jean MATHIEU[**], Philippe RAVIER[+], Jacqueline BOUJOT[**],
Patrick GENDRE[+], Marc HITTINGER[+]

(+) INFORMATIQUE INTERNATIONALE, Agence de SACLAY
BP.24 - 91190 GIF-s/-YVETTE.(FRANCE)

(**) UNIVERSITE d'ORLEANS
45046 ORLEANS-CEDEX.(FRANCE)

Key-words : Fluid-Structure Interaction
Euler-Lagrange Description
Transient Calculation
Large Displacements
Finite Elements

In many industrial fields of interest, engineers are confronted with fluids (water, oil, kerosene, gases, ...) interacting with structures (tanks, containers, obstacles, ...), along a more or less extended area.

In cases deformations of the structure may be fairly important and even affect the movement of the fluid, engineer has then to solve problems including a coupling between fluid displacements and elastic deformations of the structure.

Depending on the time scale for the movement, several approaches are at hand : steady-state calculations corresponding to asymptotic values, slow transient coupling, coupled vibration analysis, fast movements (explosion), etc. ...

Our contribution will be dealing with slow transient coupling or evolution towards a steady-state equilibrium, supposed to be the asymptotic value of the transient evolution.

We are interested in the interaction between a free surface fluid and the structure, assumed to be elastic, which contains it, in an external forces field. We intend to present the transient simulation of such a process with our code BACCHUS.

1. - ARBITRARY LAGRANGE-EULER (ALE) DESCRIPTION.

1.a The differential problem :

Consider a moving domain $D^F(t)$, with deformation velocity $\underline{w}(\underline{x},t)$, filled with an incompressible fluid of density ρ^F and obeying the Navier equations.

Consider a second domain $D^S(t)$, made of elastic material, and limiting D^F.

The problem to solve can be stated as follows :

$$\begin{cases} \rho^F \{\frac{\delta}{\delta t} \underline{v}^F (\underline{x},t) + ((\underline{v}^F - \underline{w}) \nabla) \underline{v}^F (\underline{x},t)\} = \text{Div } \underline{\sigma}^F + \rho^F \underline{F}^F \\ \text{Div } \underline{v}^F (\underline{x},t) = 0 \end{cases} \quad \text{in } D^F$$

$$\begin{cases} \rho^S \frac{\partial}{\partial t} \underline{v}^S (\underline{x},t) = \text{Div } \underline{\sigma}^S + \rho^S \underline{F}^S \\ \frac{d}{dt} (\rho^S |J|) = 0 \end{cases} \quad \text{in } D^S$$

where :

$\underline{v}^i(\underline{x},t)$ stands for the velocity field in $D^i(t)$;
$\underset{\sim}{\sigma}$ stands for the stress tensor in D^i ;
$\underline{F}$ stands for external forces ;
ρ^i stands for density in D^i (assumed constant in D^F) ;
J   stands for the Jacobian of the mapping $D^S(0) \to D^S(t)$.

The actual ALE description is determined by the actual $\underline{w}(\underline{x},t)$ velocity field, which we define as follows :

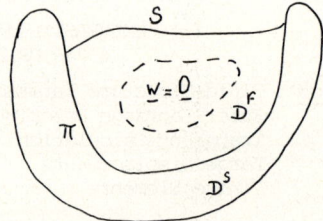

$\underline{w} = \underline{0}$ in an internal eulerian region ;
$\underline{w} = \underline{v}^F$ on the free-surface $S(t)$;
$\underline{w} = \underline{v}^S$ on the wet part of the wall $\pi(t)$.

1.b Constitutive laws :

The fluid is assumed to be stokesian, which is stated, for an incompressible fluid, as :

$$\underset{\sim}{\sigma}^F = \mu \, [\underline{\nabla} \, \underline{v}^F + (\underline{\nabla} \, \underline{v}^F)^t] + p \, \underline{1}$$

where :

μ stands for the dynamic viscosity, and
p   for the hydrostatic pressure.

The Cauchy tensor $\underset{\sim}{\sigma}^F$ can be related to the standard second Piola Kirchhoff tensor $\underset{\sim}{\tau}$ as follows :

$$\underset{\sim}{\sigma}^S(t) = \frac{\rho^S(t)}{\rho^S(o)} \, F(t) \, \underset{\sim}{\tau}(t) \, F^t(t)$$

where $F(t)$ is the gradient of the mapping $D^S(o) \to D^S(t)$, with $\underset{\sim}{\tau}(o) = \underset{\sim}{\sigma}^S(o) = \underset{\sim}{0}$.

1.c Boundary conditions :

1.c.1 On the structure : displacements are given on some part $\Gamma_u(t)$ of $\partial D^S(t) \backslash \pi(t)$, and stresses are given along the remaining part of this boundary.

1.c.2 On the wet part of the wall $\pi(t)$:

• We impose a partial slip-condition of the fluid along the wall :

$$\underline{\sigma}^F_{tg} = - \beta (\underline{v}^F_{tg} - \underline{v}^S_{tg})$$

This condition is intended to authorize displacements of the free-surface along ∂D^S in spite of the viscous asumption, without dealing with nasty boundary-layer difficulties.

• We suppose the lack of any cavitation :

$$\underline{v}^F \cdot \underline{n} = \underline{v}^S \cdot \underline{n}$$

- We impose the continuity of normal stresses :

$$(\sigma^F - \sigma^S) \cdot \underline{n} = \underline{0}$$

1.d Initial conditions :

At the initial time $t = 0$, the domain $D^S(o)$ and the boundary $S(o)$ are known, and the initial stress $\sigma^S(o)$ is null.

2. - DISCRETIZATION OF THE PROBLEM.

2.1 Space and time discretization :

We use a finite-element representation in space, in order to provide easy handling of complex geometric configurations which may be subject to large displacements. The elements are of degree one for the velocities, and piecewize constants for the hydrostatic pressure.

A one-step explicit finite-difference scheme is used in time. This scheme is subject to some stability conditions, limiting the time step. The criteria taken into account are :

- free-surface wave and viscous wave stability in the fluid region ;
- acoustic wave stability in the structure.

Because of the more drastic limitation due to the stability criterion in the structure, a subcycling process is used for the structure calculation.

2.2 Numerical methods :

In the structure, the constitutive law is used to deduce from the dependancy of $\tau(t)$ on t, a linearization of $\sigma^S(t)$, leading to an incremental method relating $\sigma^S(t^n)$ to $\sigma^S(t^{n+1})$.

The incompressibility condition for the fluid is imposed using a SOLA-type algorithm. The corrective pressure-gradient to be applied to the predicted velocity field is evaluated using SOR method.

The mass matrices are lumped, in order to obtain a truely explicit algorithm.

3. - MESH ADAPTATION.

At each fluid-calculation step, the free surface has to be repositioned, in order to make the incompressibility condition ascertained in the mixed Lagrange-Euler region.

The displacements of the free-surface and the interface between fluid and structure induces a modification of the mesh in the mixed Lagrange-Euler region, and possibly a modification of the Euler region (if its boundary happens to intersect the free-surface or the wall), or a degeneracy of some element.

Thus a rezoning is automatically performed during the calculation. This rezoning is subject to some admissibility criteria and, for the seek of lower cost, is performed only when some "necessarity-criterion" is verified.

At the intersection between the free-surface $S(t)$ and the structure $\partial D^S(t)$, the admissibility of the mesh makes necessary that any node belonging to $S(t)$ $\partial D^S(t)$ should be a vertex of some element of $D^S(t)$ <u>and</u> of some element of $D^F(t)$. Forbidden are situations like the following :

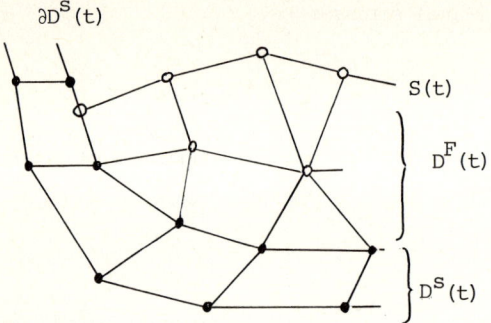

To be avoided, this situation is likely to lead to a rezoning at each calculation step. For the seek of lower cost, we always identify a "node" of $\partial D^S(t) \cap S(t)$ with its nearest neighbour in $D^S(t)$, thus limiting the need for rezoning. The influence of such an approximation has not been yet explored, but is likely to be weak if the stability criteria of the explicit scheme are largely enough respected.

4. – A TWO-DIMENSIONAL SAMPLE CALCULATION.

A test calculation has been performed in plane geometry, whose data are the following (in SI units) :

4.1 Physical parameters (plane strain hypothesis).

external force	-10^{-2}	
Young modulus	100	(very soft material !)
Poisson coefficient	0,25	
ρ^F	10^3	
ρ^S	2.10^3	
μ	$1,08.10^{-5}$	
β	0,01	

4.2 Time step.

0,01   for a 0,12 x 0,1 structure

4.3 Boundary and initial conditions.

$\underline{v}^F(o) = \underline{v}^S(o) = \underline{0}$

$S(o)$ at $\sim 15°$ from the horizontal

no displacement at the bottom of the tank.

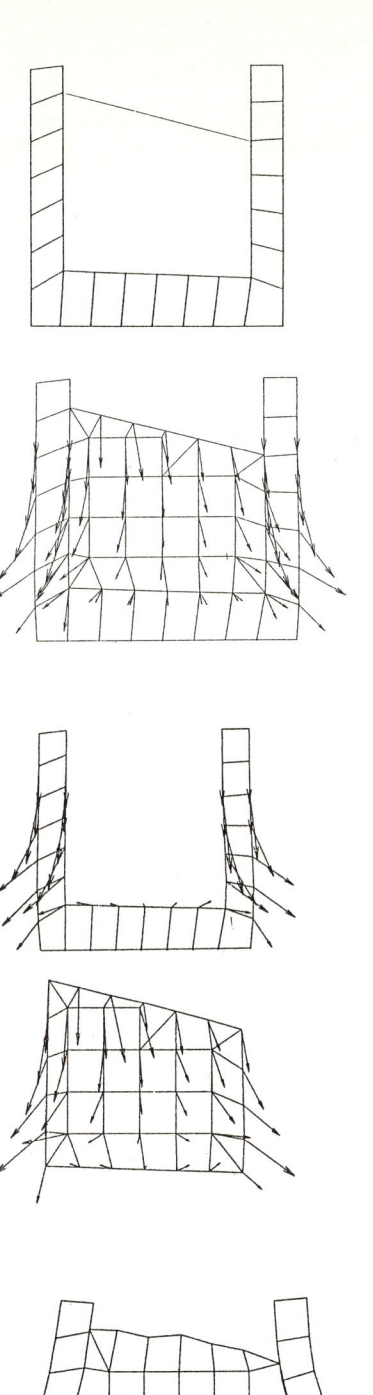

$\underline{t = 0}$ D^S and S are given.

The internal mesh will be automatically designed.

$\underline{t = 0,2}$ The bottom of the tank is in a compressive phasis.

Notice the behaviour of the velocity fields on π : double-valued with normal component continuity.

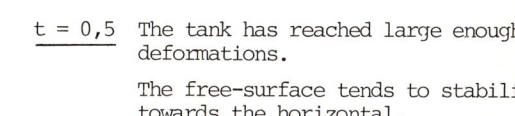

$\underline{t = 0,3}$ The bottom of the tank has entered an expansive phasis.

$\underline{t = 0,5}$ The tank has reached large enough deformations.

The free-surface tends to stabilize towards the horizontal.

Notice the modification of the mesh in D^F.

5. - CONCLUSION.

The ALE technic has proved to be very convenient for the handling of great displacements for free-surface and fluid-structure interaction problems.

Extensions of this application will currently be performed, using our code BACCHUS, and three-dimensional tests will be experimented later, because of their much greater cost, mainly due to the price of automatic rezoning in 3D.

REFERENCES.

BOUJOT, J. - "Interaction fluide-structure en régime transitoire" - La Recherche Aérospatiale, n°3 (1984) 203-209

BOUJOT, J. - "Problèmes mathématiques de la mécanique" - C.R. Acad.Sc.Paris, t.301, série I, n°7 (1985)

CIARLET, PG., GEYMONAT, G. - "Mécanique des solides élastiques. Sur les lois de comportement en élasticité non linéaire compressible" - C.R. Acad.Sc.Paris, t.295 (4 Octobre 1982)

GENDRE, P., HITTINGER, M., RAVIER, Ph. - "Free-surface flows with large displacements by finite elements" - Submitted to HYDROSOFT86 (Southampton)

Van GOETHEM, G. - "Description mixte d'Euler-Lagrange et modèle d'éléments finis à domaine variable" - Thèse, Université Catholique de Louvain (1980)

MATHIEU, J. - "Simulation des interactions fluide-structure en théorie des grands déplacements" - Thèse, Université de Paris-Sud, Orsay (1985)

RAVIART, P.A. - "Les méthodes d'éléments finis en mécanique des fluides" - Eyrolles, Paris

TEMAM, R. - "Navier-Stokes equations" - North-Holland, Amsterdam (1977)

COMPUTATION OF TURBULENT SEPARATED FLOW WITH AN
INTEGRAL BOUNDARY LAYER METHOD

R.E. Melnik, J.W. Brook and P. DelGuidice
Grumman Corporate Research Center
Bethpage, New York 11714-3580

INTRODUCTION

It is well known that flow separation is an important fluid dynamic phenomenon that is often a performance limiting factor in many practical aerodynamic problems. As a result, the development of numerical methods and turbulence models for separated flow continues to be an active area of fluid mechanics research. In the present paper, we report on progress that has been achieved with an integral boundary layer method for computing two-dimensional separated flows. It is implemented in a computer code, GRUMFOIL, that uses a zonal approach to solve the coupled inviscid and integral boundary layer equations iteratively with Carter's[1] semi inverse technique. The formulation, described in Ref. 2-4, incorporates a complete set of viscous matching conditions which allows for viscous effects of the wake in addition to the usual displacement thickness terms on solid surfaces.

The GRUMFOIL code employed in the present study is based on a zonal viscid/inviscid formulation which uses a conservative full potential flow representation for the inviscid flow and integral methods for the laminar and turbulent boundary layers and wake. The laminar boundary layer is solved using Thwaites method while the attached parts of the turbulent boundary layer and wake are solved using Green's lag entrainment method. The separated turbulent boundary layer regions are solved with an extension of Green's method previously developed by the first two authors[4].

In the present paper, we report on extensions we have made to the GRUMFOIL code to include 1) the Reynolds normal stress terms in the boundary layer equations and 2) a more accurate implementation of the viscous matching conditions in the wake. The question of whether or not it is important to retain the normal stress terms in theoretical formulations arose from the experimental observations made by Simpson, et.al.[5] of a separated flow in a diffuser. The issue of the numerical implementation of the wake coupling condition arises because, in the original GRUMFOIL code, the wake conditions are set along a fixed reference curve from the trailing edge which does not follow the large wake deflections that can arise at high angles of attack. In order to reduce this source of error, we have incorporated a solution adaptive mesh that allows us to set the boundary conditions along the actual wake position.

BOUNDARY LAYER EQUATIONS WITH REYNOLDS NORMAL STRESSES

The effects of Reynolds normal stresses occur in two places in the integral boundary layer equations--one as an additional apparent skin friction and the other as an additional production term in the turbulent kinetic energy (lag entrainment) equation. With these terms included, the three lag entrainment equations for the edge velocity, U_e, incompressible shape factor, $\bar{H}$, and entrainment coefficient, C_E, can be written in the form,

$$\frac{dU_e}{dx} = \frac{\bar{C}_1 \rho_e U_e - KdQ/dx}{C_2 \rho_e \theta} \tag{1}$$

$$\frac{d\bar{H}}{dx} = \frac{[\bar{C}_3 \rho_e U_e - H_1 dQ/dx][H+1]}{C_2 \rho_e U_e \theta} \tag{2}$$

$$\frac{dC_E}{dx} = (\frac{C_\tau}{\theta \partial C_\tau / \partial C_E}) \{ \frac{2a_1(U_e/U_m)(\delta/L)}{H+H_1} [FC_{\tau_{EQ_0}}^{\frac{1}{2}} - \lambda C_\tau^{\frac{1}{2}}]$$

$$+ 2[\frac{\theta}{U_e} \frac{dU_e}{dx}]_{EQ} - 2[\frac{\theta}{U_e} \frac{dU_e}{dx}] [1 + 0.075 M_e^2 (\frac{1+0.2M_e^2}{1+0.1M_e^2})] \} \tag{3}$$

These equations are identical to the original lag entrainment equations appearing in Ref. 4 except for the appearance of modified coefficients $\bar{C}_1$, $\bar{C}_3$ in place of C_1, C_3 and the presence of an extra production term, F, due to normal stresses, appearing in Eq. (3). The modified coefficients agree with the definition of the original coefficients except for the replacement of the usual skin friction coefficient, C_f, that appears in the latter, with a "total" skin friction coefficient C_{ft} defined by,

$$C_{ft} = C_f + C_{fa} \tag{4}$$

where,

$$C_{fa} = \frac{2}{\rho_e U_e^2} \frac{d}{ds} (\rho_e U_e^2 \theta I_\sigma) \tag{5}$$

and

$$I_\sigma = \frac{1}{\theta} \int_0^\infty \overline{(\frac{u'^2 - v'^2}{u_e^2})} \frac{\rho}{\rho_e} dy \tag{6}$$

The extra production term, F, is defined as the ratio of total turbulent kinetic energy production to that due to the shear stresses alone, with both evaluated at the position of maximum shear stress across the boundary layer. Thus,

$$F = 1 + \frac{\overline{(u'^2 - v'^2)}}{\overline{u'v'}} \frac{\partial u / \partial x}{\partial u / \partial y} \Big|_{y=y_m} \tag{7}$$

For further discussion and definitions of the quantities employed in the above equations the reader is referred to the original references. The above system of equations does not form a complete system for the three primary unknowns U_e, $\bar{H}$, and C_E. Relations for the other unknown functions must be provided to close the system.

For attached flows, we employed the closure relations used in Green's original method. There, the structure function was set to $a_1=0.15$, the dissipation length scale was taken as $L=0.08\delta$, and the velocity ratio, as determined from Mellor and

Gibson's equilibrium solution, was set to the value $U_m/U_e = 2/3$.

In Ref. 4 we developed corresponding closure relations for separated flow using Le Balleur's velocity profile and mixing layer concepts. We found that good predictions of several separated flows were obtained if the value of a_1 was increased to $a_1 = 0.225$ in regions of separated flow.

In order to include the normal stress terms in the method, additional closure relations are needed for I_σ and F. In this paper I_σ is simply taken as a function of the local value of $\bar{H}$,

$$I_\sigma = R_1(\bar{H}) \quad \text{Equilibrium Model} \quad (8)$$

The data set and the chosen correlation are shown in Fig.1. Two linear fits have been used for simplicity: the one for $J \leq 0.6$ is that proposed by East Sawyer and Nash for their attached flow data, while that for $0.6 \leq J \leq 1$ was chosen to join the East, Sawyer, Nash correlation with the separated flow data point of Hastings and Moreton. Note the large increase in the growth of I_σ in separated flow, $J \geq 0.5$. Also shown for reference are the data of Shiloh, Shivaprasad, and Simpson[8] and Hastings and Williams[9], the latter for a nonequilibrium flow situation. These data show qualitatively the same behavior as the equilibrium data but were not used in the curve fits.

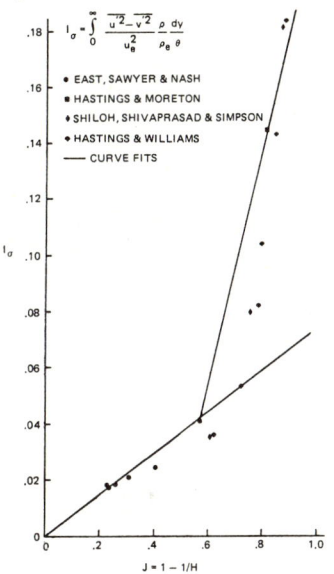

Fig. 1 Correlation for the Normal Stress Integral

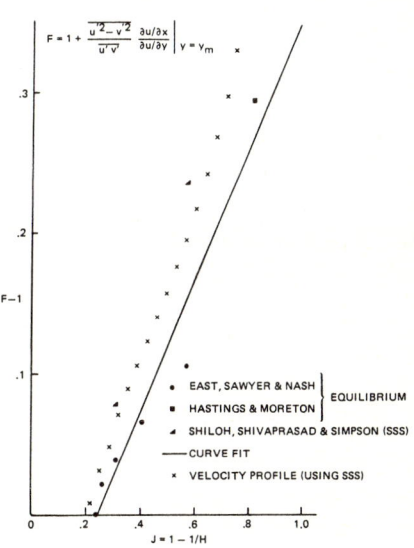

Fig. 2 Correlation for the Extra Production Term

The same equilibrium data sets are used to model the extra production term, F. The data indicate that F-1 is essentially zero up to separation and then increases rapidly in a monotonic fashion. Lacking any further information, we have chosen to use the linear fit shown in Fig. 2. Also shown are the data of Shiloh, Shivaprasad, and Simpson[8] for F as well as a computation using their data for the velocity fluctuations, and Le Balleur's velocity profile for the mean flow gradients appearing in the definition of F. Although these data sets are not in equilibrium, they show the same trend.

Equations (1) through (8) and the correlations given in Fig. (1)-(2) represent a further generalization of the lag entrainment method to include the effects of normal stresses. In the following section, we will present results of computations of several flows involving separation that illustrate the effect of these terms on the solution.

RESULTS AND DISCUSSION

In this section, we present several sets of results for three applications to flows involving extensive regions of turbulent separated flow. All the solutions were obtained with the upgraded GRUMFOIL code described in this paper. Results for the same cases are presented with and without the normal stress terms included and, where applicable, with and without the floating wake option.

The first application is to the low-speed diffuser flow experiment of Simpson, Chew, and Shivaprasad[5] carried out at a Reynolds number of Re = 4.19×10^6. In Fig. 3, we compare the present solutions with the data for the inviscid core velocity and boundary layer shape factor. Two sets of solutions, obtained with the present method, are included in the figure, one including the normal stress terms and the other not. Each of these computations employed the standard value of a_1 = 0.15 for the turbulent structure constant. The inclusion of the normal stress terms is seen to significantly improve the agreement of the present method with data. We have previously shown[4] that this flow could be adequately predicted without the normal stress terms if the value of the structure constant were increased by 50%, to a value, a_1 = 0.225. Since these results plot very close to the solid curve shown in Fig.3, they are not presented in the interest of clarity.

For reference, we have also included in Fig. 3 the inviscid solution and the computations of Johnson and King[10]. The latter use a new turbulence model developed by them specifically to improve the predictions of the Cebeci-Smith (zero equation) model in adverse pressure gradients and separated flow. Their model bears many similarities to Green's lag entrainment method. The Johnson-King results included in Fig. 3 clearly demonstrate the inadequacy of the basic Cebeci-Smith

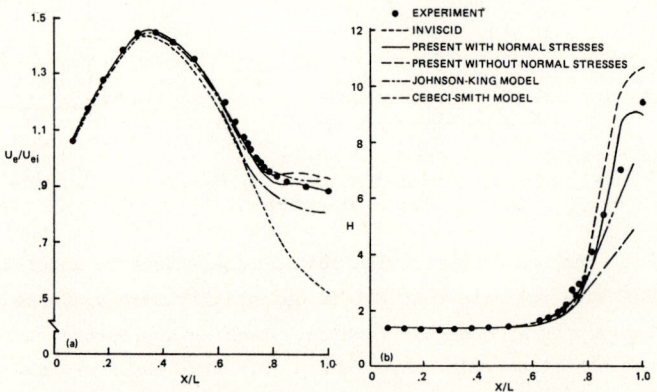

Fig.3 Low-Speed Diffuser Flow: (a) Core Velocity Distribution; (b) Shape Factor Distribution

model for separated flow, a not surprising finding. These results also illustrate the significant improvement in the prediction of separated flow afforded by their new turbulence model.

The second application considered in this paper is the transonic flow over the 12% thick RAE 2822 airfoil case 10. This is a moderately rearloaded airfoil that has been tested by Cook, McDonald, and Firmin[11]. Case 10 is at a nominal free stream Mach number M_∞ = 0.730, Reynolds Number Re = 6.5×10^6, and geometric incidence of α = 3.19°. These conditions produce a flow that separates at the shock wave and reattaches somewhat upstream of the trailing edge. Computations indicate that this flow is near the point of maximum lift. Estimated angle of attack corrections due to wall interference are provided with the data[11]. The airfoil was tested with transition fixed using roughness strips at 3% chord on the upper and lower surfaces of the airfoil. Accordingly, the computations were also run with the transition point location set at 3% chord and a corrected incidence of α_c = 2.81°. In addition, except where otherwise noted, the computations for the RAE 2822 airfoil employed a jump in momentum thickness at the upper surface transition point of $\Delta\theta_{tr}$ = 0.00010. This value was chosen in order to match the computations to the experimental value of the momentum thickness at the first measurement location downstream of the roughness strip.

Five sets of computations obtained with the present code exercising various options are included. The baseline result includes the normal stress terms, the wake curvature terms evaluated along a fixed line as explained earlier, and the momentum thickness jump at transition. The four other results illustrate the effect of selectively dropping these terms from the baseline method.

The results for the RAE 2822 airfoil are given in Table 1 and Fig. 4.

Table 1 Forces for RAE 2822 Case 10-M_∞ =0.750, a=3.19°

	α_c°	C_L	C_D	$-C_M$	Option Exercised
Experiment[24]	2.81	0.743	0.0242	0.1060	
Pulliam[27]	2.81	0.830	0.0303	0.1132	Navier Stokes
Present 1	2.81	0.773	0.0294	0.1028	Baseline
Present 2	2.81	0.755	0.0284	0.0991	Normal Stress Terms Out
Present 3	2.81	0.774	0.0284	0.1017	Wake Curvature Out
Present 4	2.81	0.763	0.0286	0.0983	Floating Wake In
Present 5	2.81	0.777	0.0299	0.1045	$\Delta\theta_{tr}$ = 0

The solution with normal stress terms deleted produces nearly identical results in these plots and is therefore not shown. We have, however, included in the figure for the momentum thickness, the solution without the normal stress terms, as the effect is more significant here. Basically, the results in the table show that the

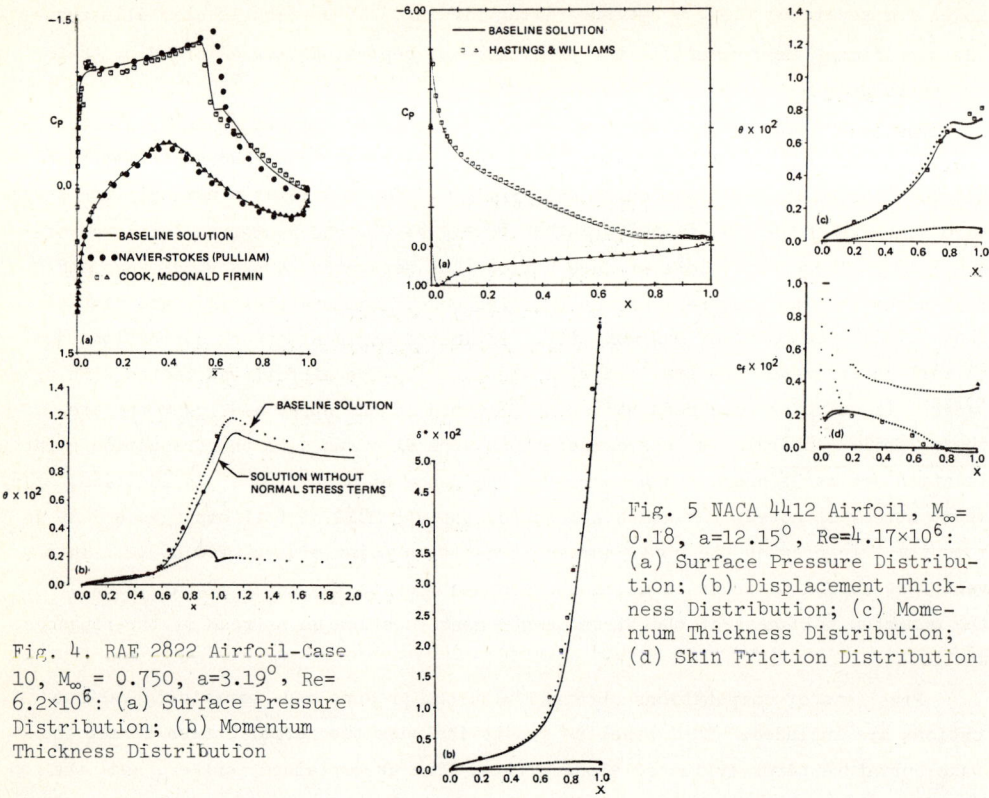

Fig. 4. RAE 2822 Airfoil-Case 10, $M_\infty = 0.750$, $a=3.19°$, $Re=6.2\times10^6$: (a) Surface Pressure Distribution; (b) Momentum Thickness Distribution

Fig. 5 NACA 4412 Airfoil, $M_\infty = 0.18$, $a=12.15°$, $Re=4.17\times10^6$: (a) Surface Pressure Distribution; (b) Displacement Thickness Distribution; (c) Momentum Thickness Distribution; (d) Skin Friction Distribution

present solutions for this case are insensitive to the effects of the normal stress terms, the various forms of the wake curvature terms and the momentum thickness jump at transition. The lift is predicted to within 4% by the baseline method, a relatively good result, considering the uncertainty due to wind tunnel wall interference. However, the result for the drag is not nearly as good, with an overprediction by about 30% for this quantity. Also included in the comparisons of the pressure distributions are Pulliam's[12] Navier-Stokes solution. The reason for the poor performance of the ARC2D Navier-Stokes code, in this case, is likely the inadequacies of the zero equation turbulence model for separated flow.

The last application we consider is that for the low-speed flow over the NACA 4412 airfoil. The data[9] include measured lift coefficients over a range of angles of attack up to near maximum lift, as well as detailed boundary layer measurements at an angle of attack, $a = 12.15°$, which is close to the maximum lift condition. The transition point was fixed to $x/c=0.14$ and 0.11 on the upper and lower surfaces, respectively, which correspond to the location of roughness strips used in the experiment. As discussed previously, we imposed a momentum thickness jump at transition to simulate the effect of the roughness strip on the boundary layer. From numerical experimentation, we have found that the value $\Delta\theta_{tr} = 0.00018$ results in the best matching to the experimental value of the momentum thickness just downstream

Table 2 Forces for the NACA 4412 Airfoil—$M_\infty=0.18$, $\alpha=12.15°$

	$\alpha_c°$	C_L	C_D	$-C_M$	Option Exercised
Experiment[21]	12.15	1.460	---	---	
Present 1	12.15	1.448	0.0327	0.0359	Baseline
Present 2	12.15	1.453	0.0298	0.0336	Normal Stress Terms Out
Present 3	12.15	1.367	0.0315	0.0313	Wake Curvature Out
Present 4	12.15	1.442	0.0313	0.0304	Floating Wake In
Present 5	12.15	1.608	0.0214	0.0469	$\Delta\theta_{tr} = 0$

of transition. It is shown in Table 2 that the neglect of the momentum thickness jump leads to a significant overprediction of the lift and underprediction of the drag. The tabulated results indicate that the normal stress terms and the floating wake option have only a very small effect on the forces. However, the effect of the wake curvature term has a much more significant effect on both the lift and the drag.

In Fig.5, we present the $\alpha = 12.15°$ results for the pressure, displacement, and momentum thicknesses, and shape factor and skin friction distributions. Included in the figure are results obtained with the baseline code and results computed with the normal stress terms deleted. We see from these results that the effect of the normal stress terms show up primarily in the solutions for the momentum thickness and shape factor. The effects are relatively small but in a direction to improve agreement with data.

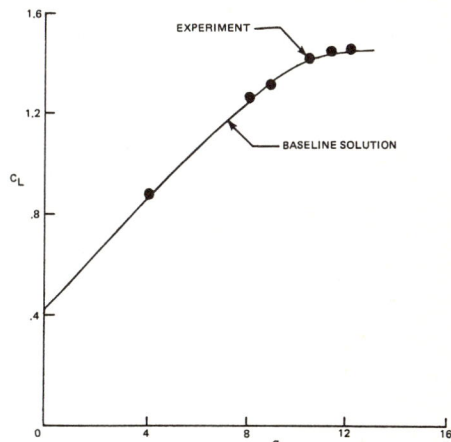

Fig.6 NACA 4412 Airfoil: Variation of Lift Coefficient with Incidence $M_\infty=0.18$, $Re = 4.17\times10^6$

From the skin friction results, we see that separation occurs at about 80% chord and that this is well predicted by the present method. In Fig.6, we present a comparison of the experimental lift curve $C_L(\alpha)$ with the results of the computation using the baseline method. We see that the agreement with data is good over the entire range of incidence which extends nearly up to maximum lift. The present method could not be converged beyond the incidence shown in the figure, because of inadequacies of the semi-inverse method for large separation zones.

By carrying out repeated computations with selected terms deleted from the model, we have been able to determine the sensitivity of the diffuser solutions to the various terms contributing to the normal stresses. This study indicated that the effect of the extra term in momentum integral equation is negligible and that the primary effect of the normal stresses is due almost entirely to the extra production term in the entrainment equation.

CONCLUSIONS

In the present paper, we have described modifications we made to the GRUMFOIL code to include a model for Reynolds normal stresses and to allow for a more exact numerical treatment of the viscous matching condition in the wake. The principal conclusion of the present study is that an integral boundary layer method can give good predictions of separated two-dimensional flow when included in an interactive viscid/inviscid procedure. The inclusion of Reynolds normal stresses, principally those due to extra turbulence energy production, was shown to lead to significant effects in the diffuser flow but to relatively small effects in the airfoil flows. The effect of using a more exact floating wake implementation of the wake curvature conditions turned out to be negligable.

REFERENCES

[1] Carter, J.E., AIAA Paper No. 79-1450 (1979).
[2] Melnik, R.E., Chow, R.R., Mead, H.R., and Jameson, A., NASA CR 3805, Oct. 1985.
[3] Melnik, R.E., Mead, H.R., and Jameson, A., AIAA Paper 83-0234 (1983).
[4] Melnik, R.E. and Brook, J.W., Numerical and Physical Aspects of Aerodynamic Flows III, Cebeci, T., ed., Springer Verlag (to be published 1987), (see also Grumman Aerospace Corp R&D Report RE-697, April 1985).
[5] Simpson, R.L., Chew Y.J., and Shivaprasad, B.G., J. Fluid Mech., Vol 113 (1981).
[6] East, L.F., Sawyer, W.G., and Nash, C.R., RAE Tech. Report 79040, April 1979.
[7] Hastings, R.C. and Moreton, K.G., in Laser Anemometry in Fluid Mechanics, Published by Ladoan-Instituto Superior Technico, Portugal (1984).
[8] Shiloh, K., Shivaprasad, B.G., and Simpson, R.L., J. Fluid Mech., Vol 113(1981).
[9] Hastings, R.C. and Williams, B.R., Numerical and Physical Aspects of Aerodynamic Flows III, Cebeci, T., ed. (Springer-Verlag, to be published 1987).
[10] Johnson, D.A. and King, L.S., AIAA Paper No. 84-0175 (1984).
[11] Cook, D.H., McDonald, M.A. and Firmin, M.C.P., in AGARD AR-138 (1979).
[12] Pulliam, T.H., Private communication.

ACKNOWLEDGEMENT

This work was supported by the Office of Naval Research and the NASA Lewis Research Center.

NUMERICAL INVESTIGATIONS OF THE STRUCTURE OF
THREE-DIMENSIONAL CONFINED WAKES BEHIND A
CIRCULAR CYLINDER

N. K. Mitra, P. Kiehm and M. Fiebig
Institut für Thermo- und Fluiddynamik
Ruhr Universität Bochum, Fed. Rep. Germany

INTRODUCTION

Although flows in a channel around built-in circular cylinders appear in many engineering applications, e.g. in plate-fin heat exchangers, numerical investigations of such fows are rare (1). The most interesting feature of these flows are the structure of the three-dimensional confined wakes. Experimental and numerical studies of three-dimensional wakes behind a circular cylinder with end plates have been reported by Baker (2) and Kaul et. al (3) respectively. The main thrust of Baker's (2) work lies in the flow visualization and pressure measurement of the steady and unsteady horse shoe vortex system on the end plate. The numerical work of Kaul et. al.(3) shows a nonuniform pressure distribution along the length of the cylinder and the existence of spiraling vortex filaments behind the cylinder. In these studies the Reynolds numbers based on the cylinder diameter lie between 10^3 and 10^4. The present work addresses to the numerical investigations of the structure of three-dimensional confined wakes behind a circular cylinder in rectangular channels in a smaller Reynolds number regime (Re < 500).

BASIC EQUATIONS AND METHOD OF SOLUTION

Figure 1 shows the computational domain of a rectangular channel of width B and height 2H with a built-in cylinder of diameter D. The three-dimensional flow field in the channel is described by incompressible unsteady continuity and three momentum (Navier-Stokes) equations. These equations are solved by a modified Marker-and-Cell (MAC) technique that uses second order upwind differences. Fully developed laminar flow profile have been used at the channel inlet. The velocity components at the channel exit are extrapolated from two interior points by assuming that the second derivatives of the velocity components in the main flow direction are equal to zero. On the walls and on the cylinder surface no-slip boundary conditions have been used. In MAC-scheme, boundary conditions for pressure on solid surface are not required. The geometrical complexity of the problem - a circular cylinder in a rectangular channel - can be overcome by generating boundary fitted coordinates. The basic equations have to be then transformed in this coordinate system for numerical solution. Alternatively one can use two overlapping rectilinear and polar coordinate system (4). This technique of overlapping coordinates have been used in the present work (Fig. 2). In this technique, first the flowfield in the channel is calculated by using a cartesian mesh. The cylinder surface in this phase of computation is simulated by the cartesian mesh. For better accuracy, large mesh densities are used near the cylinder surface and the channels walls. Once steady or periodic flows are obtained as numerical solutions, the flow in polar coordinates near cylinder are computed. The flowfield on the outer boundary of the computational domain in polar coordinates are obtained by inerpolation from the flowfield in cartesian mesh. The details of the computational scheme can be found in Ref. (5).

RESULTS AND DISCUSSION

Figure 3 shows computed particle traces in a steady three-dimensional flow around a cylinder in a channel of D/B = 0.33, H/B = 0.36 and Re = 67. This Re is based on the average incoming velocity and the cylinder diameter. The structure of the wake in the vicinity of the symmetry plane with two symmetrical separation bubbles is similar to that of a steady two-dimensional flow. Behind the cylinder spiral vortex filaments (shown in Fig. 3 for the left side wall only) are noticed. Such vortex filaments have also been obtained by Kaul et. al. (3). The normal velocity component becomes large at some part between the bottom plate and the symmetry plane in front and back of the cylinder. In the symmetry plane it becomes zero resulting in weakening of the vortex intensity and spreading the vortex itself. Figure 4 shows the cross sectional view at the centerplane for flow with D/B = 0.2, H/B = 0.22 and Re = 40. The vortex on the symmetry plane has not been plotted. The fluid coming on this plane at the forward stagnation side moves in a screwing motion to the rear side of the cylinder and is concentrated in the upper region only. The unplotted white area between the concentrated fluid stream and the bottom plate is filled with fluid from other regions of the channel.

The structure of the limiting streamlines on the cylinder surface and the bottom plate have been numerically investigated and are shown in figures 5,6,7 and 8. In fig. 5 the positive stream surface bifurcation can be seen at $\theta = 0$ and 6.28 (forward stagnation line) and at $\theta = 3.14$ (rearward stagnation line) and negative stream surface bifurcation at $0.7\pi < \theta < 0.8\pi$ and $-0.7\pi > \theta > -0.8\pi$. On the negative stream surface bifurcation, two streamlines combine to form one streamline which represents the separation line on the cylinder surface. Figure 5b and 5c represent periodic flows at Re = 440. Although the separation lines on the cylinder remain almost symmetrical at this Re, the rear stagnation line moves slightly away from $\theta = \pi$. The asymmetry in the rear stagnation area charaterizes the periodic flow. Figure 6 shows the limiting streamlines for a larger blockage ratio (D/B = 0.5) and H/B = 0.5. Here the separation lines (negative stream bifurcation) for higher Reynolds numbers (220 and 440) contain spiral focus points (shown in detail beside each picture) where both components of the surface shear stress vanish. With increasing Re the spiral point moves towards the bottom plate. The structure of limiting streamlines are determined by the geometrical parameters D/B, H/B and the Re. Our computations show that the geometral parameters can be combined into a single similarity parameter H/D and the structure of the limiting streamlines will be largely determined by H/D and Re. Only for small H/D (H/D $\leq$ 1) a spiral point appears in the Reynoldsnumber regime of our investigation. No spiral point for H/D = 2.5 appears (Fig.5). Figure 7 shows limiting streamlines for D/B = 0.5 and H/B = 0.22 (H/D = 0.44). The spiral points here are even more remarkable than in fig.6 (H/D = 1). Figure 8 shows the limiting streamlines on the bottom plate. The flow structure here is somewhat similar to that of Kaul (3). A separation point in front of the forward stagnation point is clear. With D/B = 0.2, the wake is steady. With D/B = 0.5 the wake is unsteady. Unfortunately, the resolution here is not fine enough to clearly show the existence of a horse shoe vortex. It should be noted that the Reynoldsnumber range of our investigation is also too small for the formation of a horse shoe vortex (2).

CONCLUSION

The three-dimensional flow in a channel behind a circular cylinder show the existence of spiraling vortex filaments. On the cylinder itself spiral points of separation are noted for flows with small ratio of channel height to cylinder diameter.

ACKNOWLEDGEMENT

This work has been supported by the Deutsche Forschungsgemeinschaft.

REFERENCES

(1) Kiehm, P., Mitra, N. K. and Fiebig, M., AIAA Paper 86-0035 (1986)
(2) Baker, C. J. Jour. Fluid Mechanics, Vol. 95, part 2, (1979), pp. 347 - 367
(3) Kaul, U. K., Kwak, D. and Wagner, C., AIAA Paper 85-0182, (1985)

(4) Launder, B. E. and Massey, T. H., Jour. Heat Transfer, Vol. 100, (1978), pp. 565 - 571
(5) Kiehm, P., Numerische Untersuchung der laminaren Strömungsfelder und des Wärmeübergangs in einem Kanal mit quereingebautem Kreiszylinder, Dr.-Ing.-Dissertation, Ruhr-Universität Bochum, (1986).
(6) Hornug, H. and Perry, A. E. ZfW. 8 - 2, (1984), pp. 77 - 87

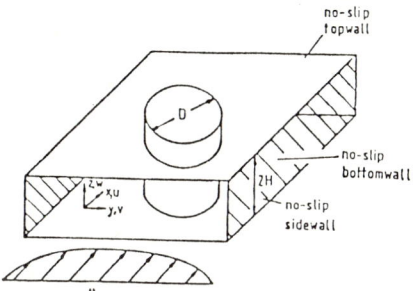

Fig. 1: Configuration definition for 3-d channel flow

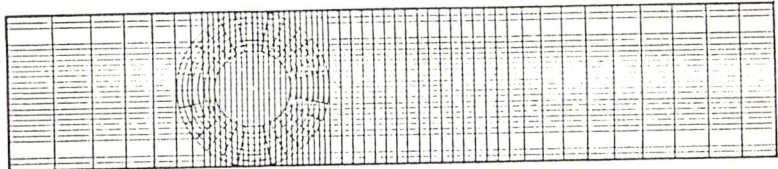

Fig. 2: Meshgrid in the computational domain

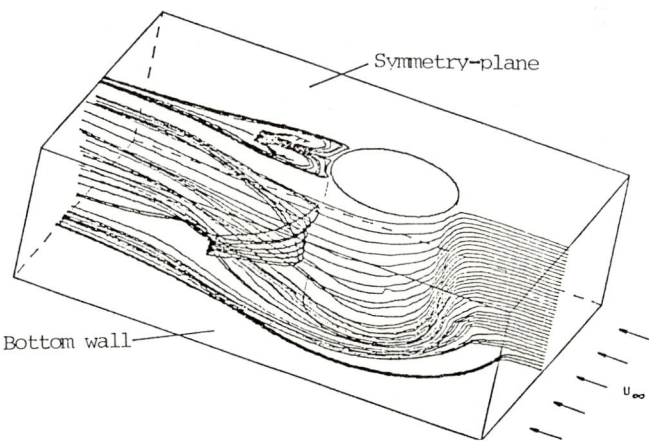

Fig. 3: Structure of 3-d flow, Re = 67.
D/B = 0.33, H/B = 0.36

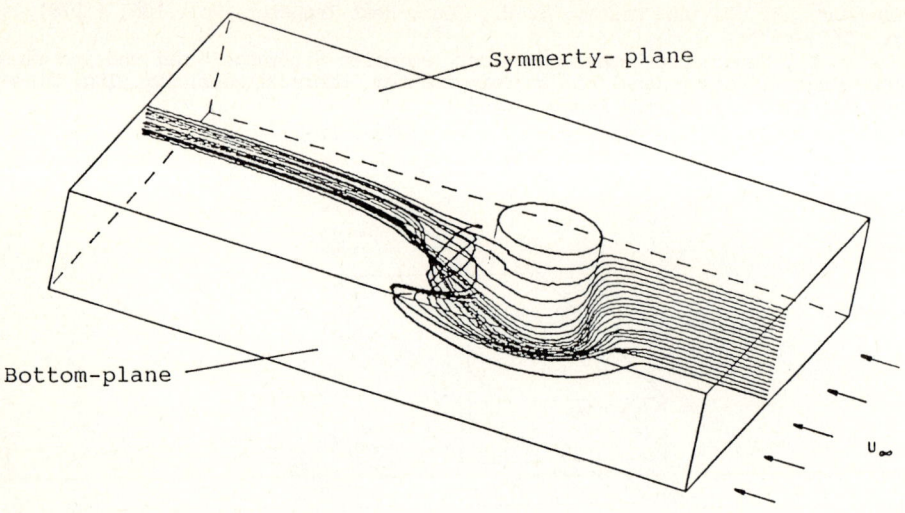

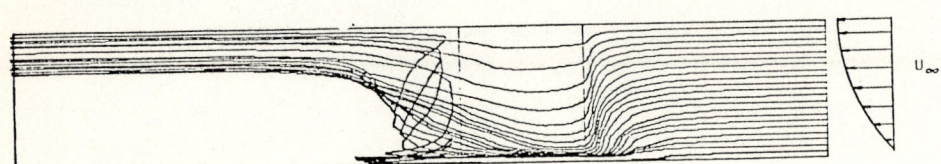

Fig.4: Structure of the spiral vortex filament, D/B = 0.2, H/B = 0.22, Re = 40, (view from the left side wall).

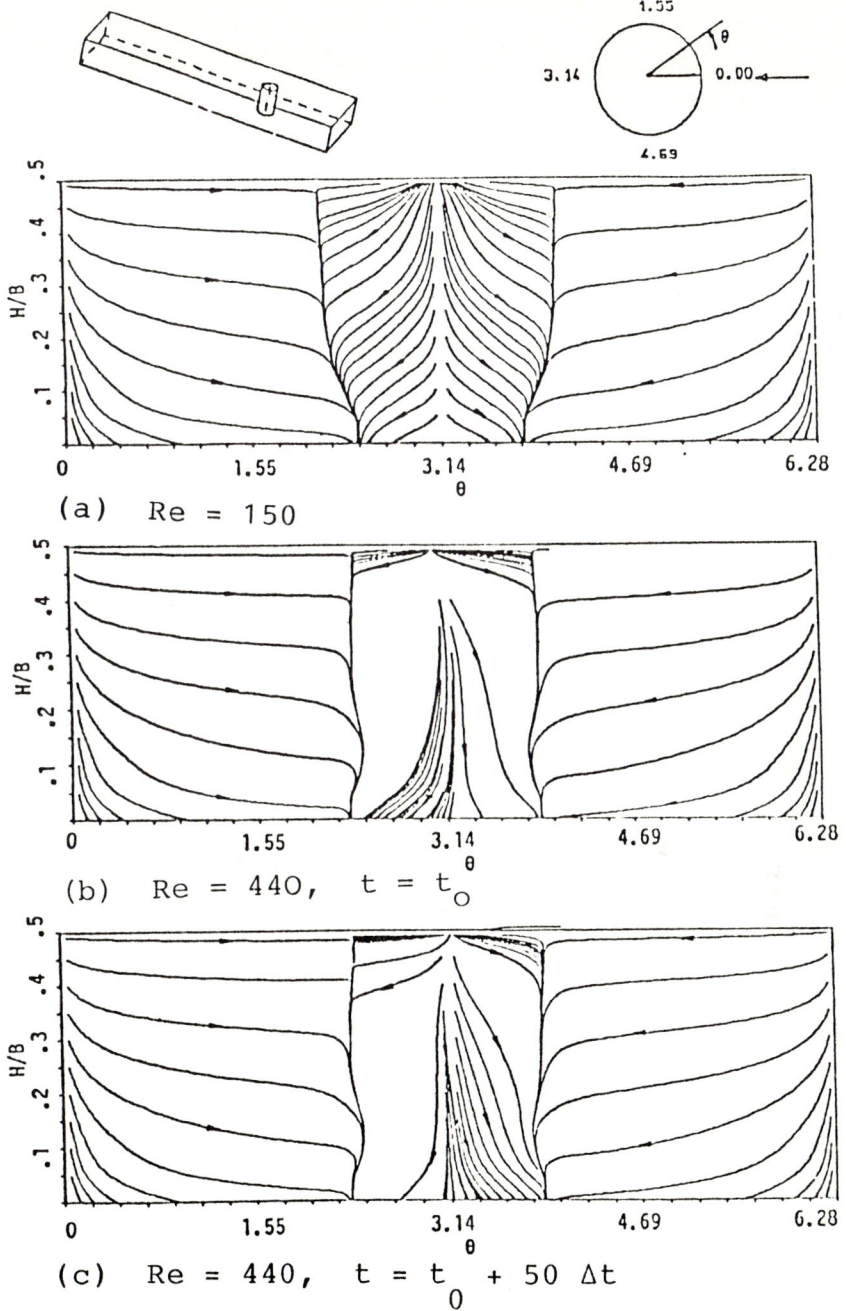

Fig.5: Limiting streamlines on the cylinder surface, $D/B = 0.2$, $H/B = 0.5$

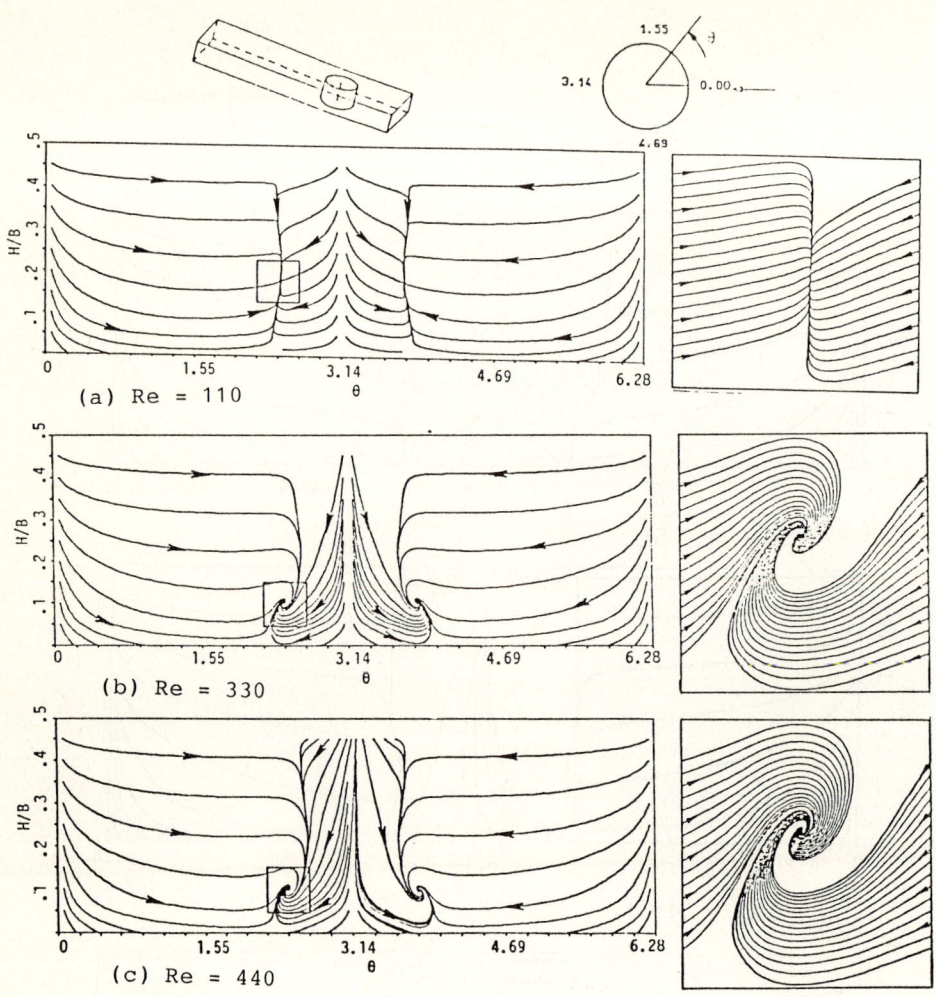

Fig.6: Limiting streamlines on the cylinder surface, D/B = 0.5, H/B = 0.5.

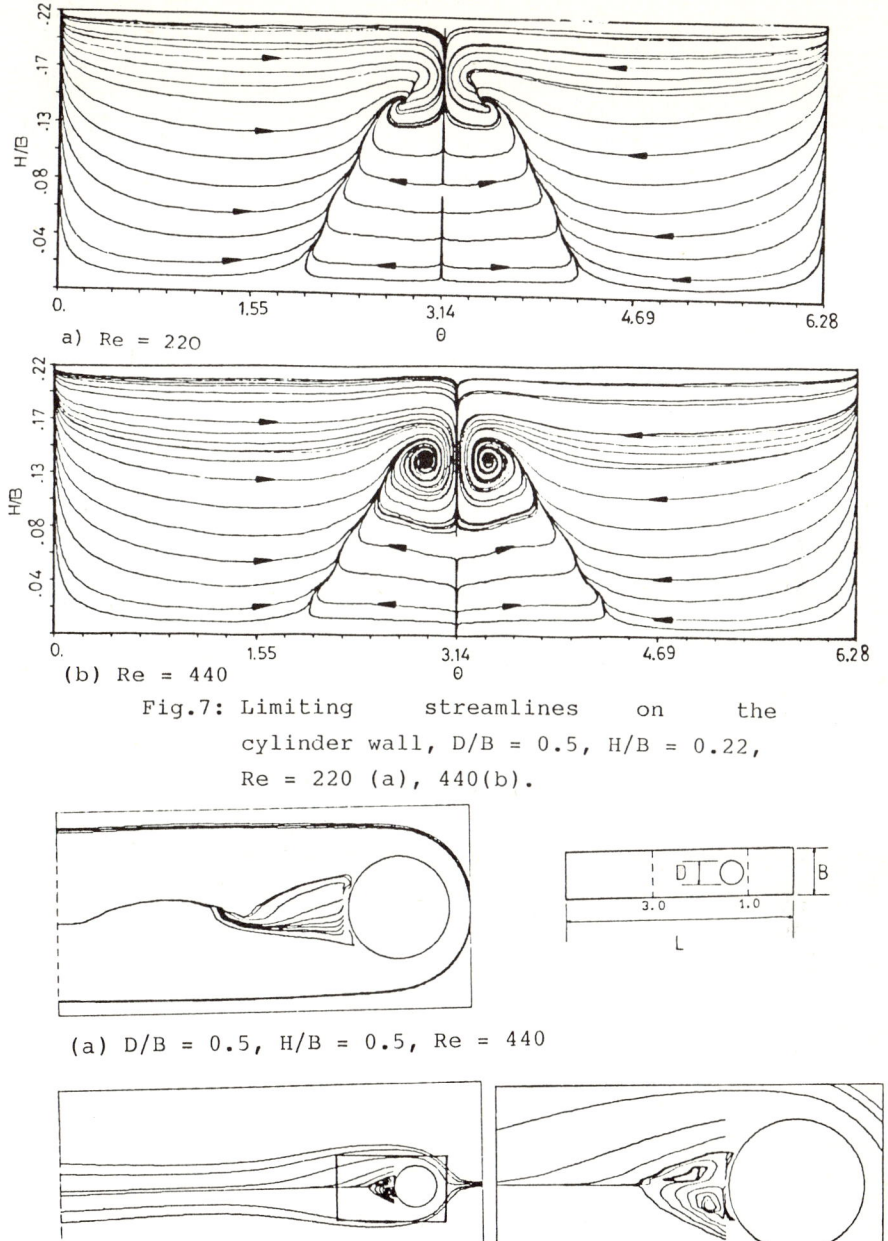

a) Re = 220

(b) Re = 440

Fig.7: Limiting streamlines on the cylinder wall, D/B = 0.5, H/B = 0.22, Re = 220 (a), 440(b).

(a) D/B = 0.5, H/B = 0.5, Re = 440

(b) D/B = 0.2, H/B = 0.5, Re = 440

Fig.8: Limiting streamlines on the bottom wall, for Re = 440

ON THE CELL-CENTRE AND CELL-VERTEX
APPROACHES TO THE STEADY EULER
EQUATIONS AND THE USE OF SHOCK FITTING

K.W. Morton and M.F. Paisley
Institute for Computational Fluid Dynamics
Oxford University Computing Laboratory
8-11 Keble Road, Oxford OX1 3QD, England

1. Introduction

Finite volume calculations of the two-dimensional Euler equations

$$f(u)_x + g(u)_y = 0 \qquad (1)$$

are commonly undertaken on a general convex quadrilateral mesh and lead to equations for the approximation U of the form

$$R_P(U) \equiv (1/V_P) \sum_1^4 [\bar{f}(U)_i \Delta y_i - \bar{g}(U)_i \Delta x_i] = 0, \qquad (2)$$

where $\bar{f}_i = \bar{f}(U)_i$ and $\bar{g}_i$ are averages of the fluxes over side i, and V_P is the area of the cell. There are two natural ways of computing these averages: U can be carried at the cell-centres and, for example referring to Fig.1, the fluxes through side 1 averaged from those at points P and W so that the five points P, E, W, N, S are involved for the cell centred at P; or U can be carried at cell-vertices so that fluxes through side 1 are averaged from those at A and B and, in all, the four points A, B, C, D are involved for the cell at P. The former is associated with the name of Jameson [1] and the latter with those of Ni [2] and Hall [3].

The cell-vertex approach has important advantages in terms of accuracy on non-uniform meshes, imposition of boundary conditions, fewer spurious solution modes and ensuring convergence with minimal added dissipation. However, the objective of setting to zero the residual obtained from the vertices of each cell is appropriate only for smooth flows or simple geometries. If a shock crosses a cell, in general the residual should not be set to zero and trying to do so leads to the use of dissipative terms that need careful tuning. In the work described here, the shock fitting has been used to give highly accurate, sharply defined results in a straight-forward robust manner.

2. Accuracy and spurious modes

With the scaling of R_P given in (2), $R_P(u)$ represents the truncation error of the scheme. Since the cell-vertex scheme is equivalent to using the trapezoidal rule to approximate flux integrals along the sides of the cell, it is easy to see that its truncation error is dominated by such terms as $\frac{1}{12}f''[(y_D-y_C)^3-(y_A-y_B)^3]/V_P$: thus it is clearly second order accurate on a non-uniform mesh if opposite sides of a cell are in ratio $1+O(h)$. On the other hand, the cell-centre scheme requires averages between the cell-centre P and its four neighbours N,S,E,W which are not easy to choose when the mesh is distorted. On a uniform mesh, simple arithmetic averages give second order accuracy: but it is easy to show that an order of accuracy is lost if opposite cell sides are in ratio $1+O(h)$ and they need to be in ratio $1+O(h^2)$ to remain second order. The averages can be modified to rescue this situation, but then the scheme gives an iterative method with much slower convergence.

More seriously, the cell-centre scheme suffers from many more spurious oscillatory solution modes. For example, on a uniform mesh the terms in f(U) can be rewritten

$$\tfrac{1}{2}(f_P+f_E) - \tfrac{1}{2}(f_P+f_W) = \tfrac{1}{2}(f_E-f_W), \qquad (3)$$

yielding the notorious two mesh-length central difference with its $(-1)^j$ oscillatory mode: and in two dimensions there are three such modes. In one dimension the compact cell-vertex scheme has no spurious modes: while in two dimensions it has only the one in which diagonally opposite vertices carry equal values. Absence of oscillatory modes is important both in designing rapidly convergent iterations and in ensuring good final accuracy.

3. Iteration and boundary conditions

The advantage and initial attraction of the cell-centre scheme is that the residual $R_P(U)$ is naturally centred for updating U_P. Jameson pioneered the use of multi-stage algorithms for this, based on Runge-Kutta methods. However, there is a conflict between using the averaging necessary to get good accuracy in R_P and obtaining a large stability region for the resulting multistage algorithm. It is for this reason that some fourth order smoothing is usually found necessary to achieve convergence: a danger then is that individual residuals are not as small as they should be on termination of the iteration.

For the cell-vertex approach both Ni [2] and Hall [3] used a Lax-Wendroff algorithm for updating U: in the interior, four neighbouring residuals are averaged to obtain an update for the common vertex. Hall found that the first order residuals had to be weighted by the areas of the individual cells to achieve convergence. The whole process results naturally from a Taylor-Galerkin formulation which can be regarded as a finite element version of the Lax-Wendroff technique. Thus the second order scheme for approximation using basis functions $\{\phi_i\}$ takes the form

$$<U^{n+1}-U^n,\phi_i> + \Delta t<\partial_x f^n+\partial_y g^n,\phi_i>$$
$$+ \frac{1}{2}(\Delta t)^2 \{<A(\partial_x f^n+\partial_y g^n),\partial_x \phi_i> + <B(\partial_x f^n+\partial_y g^n),\partial_y \phi_i>\} = 0 \quad (4)$$

where A and B are the Jacobians for f and g respectively. This gives a mass matrix extending in both the the x- and y-directions for the time difference term, in the y-direction for $\partial_x f$ and in the x-direction for $\partial_y g$: but if these are modified in simple ways, an explicit iteration of the correct form is obtained.

A key advantage of the cell-vertex scheme is its treatment of boundary conditions. Thus a body surface lies along a mesh line and one can impose directly the condition $\underline{v}\cdot\underline{n} = 0$: at outer (artificial) boundaries the number and the type of conditions imposed corresponds to the ingoing characteristics - see Hall [4] for details. For the remaining quantities the updating process is modified so that only residuals from interior cells are used. The result is that by working away from the corners and then the sides of the computing domain one can show that at convergence all the individual residuals will be zero, rather than any averages.

4. Treatment of shocks

Suppose a shock crosses the cell of Fig.1, dividing AD in proportions $\theta_1:1-\theta_1$ and BC in proportions $\theta_2:1-\theta_2$; and suppose there is a constant state u_L on the left and u_R on the right. Then one finds for the residual

$$V_P R_P(u) = [(\tfrac{1}{2}-\theta_1)(y_A-y_D)+(\tfrac{1}{2}-\theta_2)(y_C-y_B)](f_R-f_L)$$
$$-[(\tfrac{1}{2}-\theta_1)(x_A-x_D)+(\tfrac{1}{2}-\theta_2)(x_C-x_B)](g_R-g_L). \quad (5)$$

Clearly this will be zero only under special circumstances: for example, if $\theta_1 = \theta_2 = \frac{1}{2}$; or if the cell is a parallelogram and the shock is parallel to a pair of sides. Truncation errors which are $O(1/h)$ will be committed if $R_p(U)$ is set to zero: and if shocks are finitely spaced the global error is likely to be $O(1)$. Numerical experiments and analysis of simple cases confirm this and show that at best oscillatory errors spread out from the shock, but often convergence is impossible to achieve. In practice, additional dissipative terms are often used so as to spread the shock, but this makes its position rather arbitrary unless an extremely fine mesh is used - see Pulliam and Barton [5].

We have therefore used the shock capturing capability of the Lax-Wendroff iteration when applied to conservation laws to locate shocks approximately and then adjusted the mesh in their neighbourhood. Shocks are fitted by curves as in [6] and treated as internal boundaries with double-valued variables: the same account of characteristics is taken as for the outer boundaries. The Rankine-Hugoniot conditions are used to calculate a shock "speed" in order to adjust the position of the shock. Once the fitting procedure has been started, the dissipative terms used for the shock capturing phase can be removed and all the residuals driven close to zero. For further details see [7].

5. Computational details and results

Results are shown for Ni's 10% circular arc in a channel on a 64 x 16 mesh and the NACA 0012 aerofoil on a 128 x 16 mesh. Fig.2 shows the comparative Mach Number distributions for shock capturing and shock fitting for the channel problem at $M_\infty = 0.675$. Note the good resolution of the Zierep singularity behind the shock - especially marked for this problem where surface curvature is high. The shock and surrounding mesh are shown in Fig.3. Fig.4 is for the NACA 0012 aerofoil at $M_\infty = 0.8$, $\alpha = 1.25$ where only the strong upper shock has been fitted. The more testing case at $M_\infty = 0.85$, $\alpha = 1°$ is shown in Fig.5 where now both shocks have been fitted. Part of the mesh for this case is shown in Fig.6.

A multigrid scheme has been implemented in the channel calculation with equal success for shock fitting as previously obtained by Ni and Hall for shock capturing.

Approximate computing times for these calculations were as follows: 150s on a CYBER 205 for channel flow (multigrid but without vectorisation) and 40-50s on a CRAY 1S for the aerofoil cases (no multigrid but vectorised). The level of convergence achieved corresponded to the average relative change $(\rho^{n+1}-\rho^n)/\Delta t$ reaching about 10^{-5}.

Acknowledgements

The authors would like to thank RAE Farnborough for partially supporting the work described here, particularly Drs M.G. Hall and C.M. Albone for advice and encouragement in discussions, and Dr M.G. Hall for helpfully providing access to his aerofoil code on the CRAY 1S.

The second author also gratefully acknowledges financial support from the Science and Engineering Research Council.

References

[1] A. Jameson, Proc. I.M.A. Conf. on Num. Meth. in Aero Fluid Dyn., Univ. of Reading 1981, (ed. P.L. Roe), Academic Press, 1982.

[2] R.-H. Ni, A.I.A.A. Jnl. Vol 20, No 11, pp 1565-1571 (1982).

[3] M.G. Hall, Proc. Conf. on Num. Meth. for Fluid Dyn. Univ. of Reading, 1985, (eds. K.W. Morton and M.J. Baines), Oxford University Press, 1986, 303-345.

[4] M.G. Hall, R.A.E. Tech. Rept. 84116, 1984.

[5] T.H. Pulliam and J.T. Barton, A.I.A.A. Paper 85-0018, A.I.A.A. 23rd Aerosp. Sci. Mtg. Jan 1985.

[6] C.M. Albone, Proc. Conf. on Num. Meth. for Fluid Dyn. Univ. of Reading, 1985, (eds. K.W. Morton and M.J. Baines), Oxford University Press, 1986, 427-437.

[7] K.W. Morton and M.F. Paisley, Finite Volume Methods and Shock Fitting for the Euler Equations (to appear).

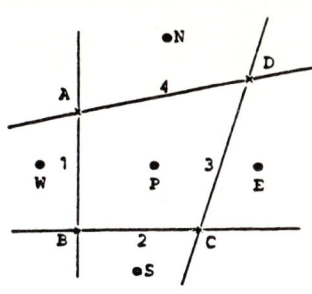

Fig.1 Cell geometry

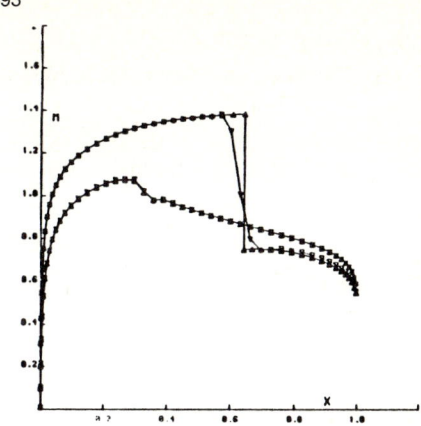

Fig.4 Mach No. distribution for NACA0012 aerofoil, $M_\infty = 0.8$, $\alpha = 1.25°$

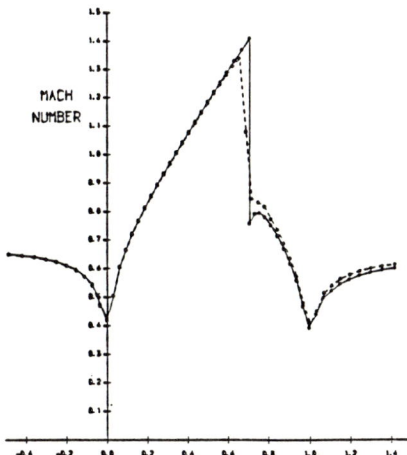

Fig.2 Mach No. distribution for channel flow, $M_\infty = 0.675$

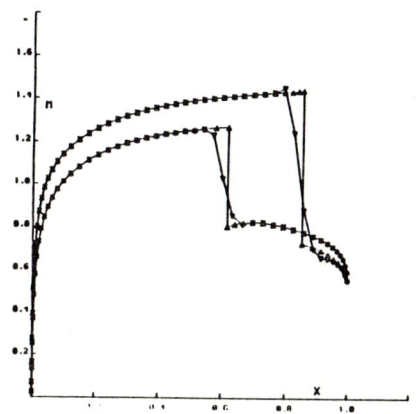

Fig.5 Mach No. distribution for NACA0012 aerofoil, $M_\infty = 0.85$, $\alpha = 1$

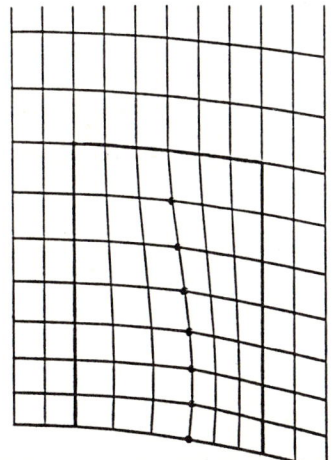

Fig.3 Detail of mesh around shock for channel flow

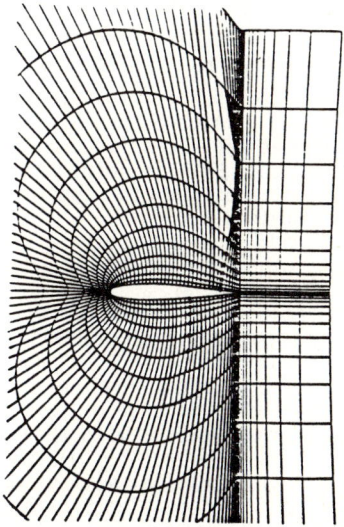

Fig.6 Detail of mesh around aerofoil. $M_\infty = 0.85$, $\alpha = 1°$

FDM-FEM Zonal Approach for Computations of Compressible Viscous Flows

Kazuhiro Nakahashi
National Aerospace Laboratory
7-44-1, Jindaiji-Higashi, Chofu, Tokyo 182, JAPAN

Abstract

A new technique is developed for computations of compressible, viscous flow over multiple-bodies. In this scheme, an implicit finite-difference method(FDM) is applied to regions of near-bodies with body-fitted grids, and those FDM-regions are patched by solving the remaining regions using an explicit finite-element method(FEM). With this zonal approach, the computational efficiency and the solution quality of FDM are retained, while the geometric flexibility is given by the FEM. The most important advantage the present technique yields is that the flow solver with this FDM-FEM zonal approach can be applied to various kinds of multiple-body flow fields without modifying the basic solution procedure. Several test problems are computed by the method in order to validate the capabilities of the FDM-FEM zonal approach for compressible viscous flows over multiple-bodies.

Introduction

In recent years, there has been considerable improvement in our ability to simulate compressible viscous flow fields. The finite-difference method(FDM) is currently very popular for this purpose because of its computational efficiency. However, the complexity of the geometric configurations which can be modeled is severely restricted by the difficulty of generating appropriate computational grid. One of the ways to overcome this difficulty is to use a zonal-grid methodology[1-3]. However these FDM-zonal grid approaches usually require some type of internal boundary interpolations, which degrade the accuracy, or boundary patching procedure, which complicate coding of the algorithm.

The finite-element method(FEM) is well-suited for the analysis of flow problems with complex or multiple-body geometry, since the solution domain of any shapes can be accurately represented by unstructured meshes. However a critical drawback of the FEM is the computational efficiency compared to the FDM for computations of the Navier-Stokes equations.

In this paper, a method which retains both FDM's computational efficiency and FEM's geometrical flexibility is developed. Here a computational flow field over multiple-bodies is divided into several zones as shown in Fig.1. FDM-zones (FDM-ZONE$_1$ to FDM-ZONE$_5$ in Fig.1) mainly cover viscous flow fields near bodies and are computed by an implicit finite-difference method with body-fitted coordinates. FEM-zones (FEM-ZONE$_1$ and FEM-ZONE$_2$) fill up the connecting zones between the FDM-zones and are computed by an explicit finite-element method. The FEM is used here as a paste for the FDM-zones so that the excessive computational time required for the FEM is minimized, while the flexibility of the FEM to complex geometry is still fully retained.

Each FDM-zone and FEM-zone are slightly overlapped each other on their boundaries. In the overlapped region, FDM and FEM grid points are distributed so as to coincide with each other utilizing the flexibility of FEM's unstructured grid. Thus the zonal patching procedure of the present approach does not require internal boundary interpolations. Because of this simple zonal patching, a flow solver with the FDM-FEM zonal approach can be applied to various kinds of multiple-body flow fields without modifying the basic solution procedure. The flexibility in grid

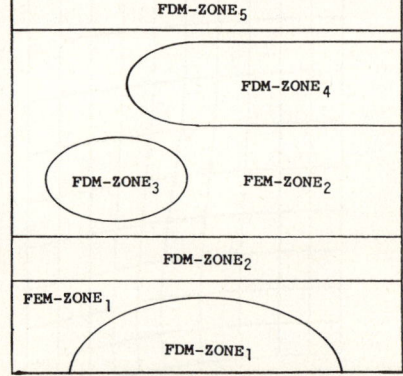

Fig.1 Domain partition of a multiple-body flow field.

generation would not be useful unless all the possible types of grids could be utilized by the numerical scheme without coding changes. In this sense, the present technique make the best use of both body-fitted FDM-grid and unstructured FEM-grid.

FDM-FEM Zonal Approach

Governing Equations : The two-dimensional, time-dependent, compressible, Reynolds-averaged Navier-Stokes equations in Cartesian coordinates can be written as;

$$\frac{\partial Q}{\partial t} + \frac{\partial E}{\partial x} + \frac{\partial F}{\partial y} = \frac{1}{Re} \{ \frac{\partial G}{\partial x} + \frac{\partial H}{\partial y} \} \tag{1}$$

where Q, E, F, G and H are vectors.

Finite-Difference Computation : The Navier-Stokes equations, Eq.(1), are transformed from Cartesian coordinates, (x, y, t), to generalized curvilinear coordinates, (ξ, η, τ). The transformed equations are solved by an implicit approximate factorization scheme[4]. The Baldwin-Lomax turbulence model is used for turbulent shear layers.

Finite-Element Computation : As mentioned in the introductory section, the finite-element solver is applied to connecting regions for the FDM-zones. Since those regions are essentially inviscid, we consider here a finite-element method for the Euler equations which are recovered from Eq.(1) by dropping the viscous terms.
The computational domain is discretized by triangular elements with nodes placed at the vertices of the triangles. A variant of a two-step, Taylor-Galerkin algorithm[5] is used in this study.
At time level $t^{n+1/2}$, values for $Q_e^{n+1/2}$ which are constant within each element are computed explicitly by an equation;

$$Q_e^{n+1/2} A = \Sigma \hat{Q}_j^n \int N_j dA - \frac{\Delta t}{2} \Sigma \hat{E}_j^n \int \frac{\partial N_j}{\partial x} dA - \frac{\Delta t}{2} \Sigma \hat{F}_j^n \int \frac{\partial N_j}{\partial y} dA \tag{2}$$

where $\hat{Q}_j^n$, $\hat{E}_j^n$, and $\hat{F}_j^n$ are nodal values at time level t^n, N_j denotes element interpolation functions, and A is an area of the element.
At time level t^{n+1}, a equation for $\hat{Q}_j^{n+1}$ can be derived by the method of weighted residuals using the interpolation functions N_i as weighting functions;

$$\Sigma \Delta \hat{Q}_j^{n+1} \int N_j N_i dA = \Delta t \int \hat{E}_e^{n+1/2} \frac{\partial N_i}{\partial x} dA + \Delta t \int \hat{F}_e^{n+1/2} \frac{\partial N_i}{\partial y} dA$$

$$- \Delta t \int \{ l \hat{E}_s^{n+1/2} + m \hat{F}_s^{n+1/2} \} N_i ds + \Delta t \int D N_i dA \tag{3}$$

where ΔQ is the vector of changes in the nodal values of Q over the time step, and D is an artificial viscosity due to Lapidus[6]. Eq.(3) can be rewritten as, $M \Delta Q = [RHS]$, where M is a consistent mass matrix. To yield an explicit algorithm, it is necessary to replace the consistent mass matrix by some approximated mass matrix, M_d which is the lumped diagonal matrix.

FDM-FEM Patching : A patching procedure between two FDM-zones and one FEM-zone is schematically depicted in Fig.2. Here the FDM-

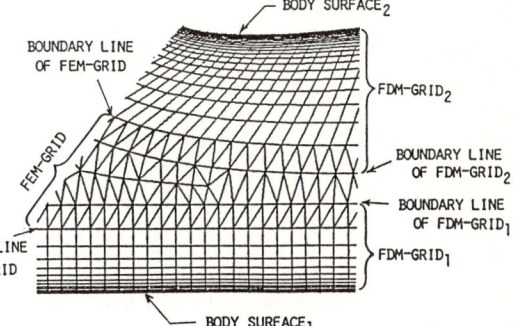

Fig.2 Patching of two FDM-grids and a FEM-grid.

grids and the FEM-grid are overlapped each other at their boundaries. The computation is performed alternately in FDM and FEM. The boundary conditions for the FEM are taken from the FDM-results on one-inside lines of FDM-grids, and same for the FDM boundary conditions. Since grid points of FDM and FEM in the overlapped region coincide with each other, interpolation of solutions between grids on the internal boundaries is not required so that the patching of the FDM and FEM is simple and straightforward.

Computational Results

Airfoil and Flat Plate : The usefulness of the FDM-FEM zonal approach will be demonstrated especially for flow over multiple-bodies. A typical example of multiple-body problems is shown in Fig.3 where an airfoil is located parallel to a flat plate. This kind of configuration is difficult to be discretized by a single FDM-grid because of two viscous regions in a flow field. With the present scheme, the grid generation becomes quite simple. At first, FDM-grid for each body is generated independently, then the remainning region between the FDM zones is filled up by triangular elements of the FEM-grid.

Shown in Fig.3(a) are three FDM-grids. The bottom FDM-grid in Fig.3(a) is for a viscous flow near the flat plate and the upper one is for the upper boundary zone which is solved using the Euler equations. A NACA0012 airfoil is wrapped by a FDM-C-grid. The FEM-grid in Fig.3(b) connects these FDM-zones and a patched FDM-FEM zonal grid is shown in Fig.3(c).

Computed density contours for a free stream Mach number 2 is shown in Fig.3(d). A shock wave generated by the airfoil goes through the FDM-FEM zone boundaries without being disturbed, and impimges on a laminer boundary layer on a flat plate causing a separation.

Turbine Cascade Flow : A cascade flow is a typical example of multiple-body problems. Although this flow field can be computed by a single FDM-grid because of periodic boundary conditions, to generate a FDM-grid for a cascade problem is somewhat time-consuming. Moreover, the periodic boundary condition often causes a severe grid skewness especially for cascades with high-turning angle and short pitch. By using the FDM-FEM zonal approach, the effort for the grid generation is considerably reduced without any problems about the grid skewness.

Shown in Fig.4 is a FDM-FEM zonal grid for a typical turbine cascade with a high-turning angle. At first, each turbine blade is wrapped by a FDM-C-grid(Fig.4(a)), then a gap between blades as well as the upstream region are covered by a FEM-grid(Fig.4(b)).

Computed Mach contours are shown in Fig.5 with a Schlieren photograph of an experiment[7]. A fish-tail shock generated at the trailing edge impinges on the suction side of the next blade where the shock causes a local boundary layer separation. Computed result reproduces this flow field quite well. Surface pressure distributions in Fig.7 show a good agreement between the computed and the experimental results over most of the airfoil surface.

Conclusion

Test problems discussed in the above section have validated the capabilities of the FDM-FEM zonal technique. The scheme has several important advantages and unique features to produce a flexible and accurate solution procedure for computations of compressible viscous flows over multiple-bodies.
(1) The scheme has both FDM's computational efficiency and FEM's geometric flexibility. Since the major viscous flow fields are computed by the FDM, existing well-developed various computational techniques for the FDM can be easily incorporated.
(2) The FDM-FEM patching procedure is quite simple and does not require interpolations between grids even in three-dimensional flow fields. Therefore the scheme allows easy tailoring of the grids to the flow field being studied without modifying the basic solution procedure. This feature is very important for practical applications of the scheme to engineering flow problems.

These advantages of the FDM-FEM zonal approach will be more realized for computations of three-dimensional viscous flow problems.

References

[1] Benek, J. A., Buning, P. G. and Steger, J. L., "A 3-D Chimera Grid Embedding Technique," AIAA Paper 85-1523-CP, 1985.
[2] Holst, T. L., Kaynak, U., Gundy, K. L., Thomas, S. D., and Flores, J., "Numerical Solution of Transonic Wing Flows Using An Euler/Navier-Stokes Zonal Approach," AIAA Paper 85-1640, 1985.
[3] Rai, M. M., "Navier-Stokes Simulations of Rotor-Stator Interaction Using Patched and Overlaid Grids," AIAA Paper 85-1519-CP, 1985.
[4] Obayashi, S. and Kuwahara, K., "LU Factorization of an Implicit Scheme for the Compressible Navier-Stokes Equations," AIAA Paper 84-1670, 1984.
[5] Lohner, R., Morgan, K., Peraire, J., and Zienkiewicz, O. C., "Finite Element Methods for High Speed Flows." AIAA Paper 85-1531-CP, 1985.
[6] Lapidus, A., "A Detached Shock Calculation by Second-Order Finite Differences," J. Comp. Phys. Vol.2, pp.154-177, 1967.
[7] Graham, C. G. and Kost, F. H., "Shock Boundary Layer Interaction on High Turning Transonic Turbine Cascades," ASME Paper No.79-GT-37, 1979.

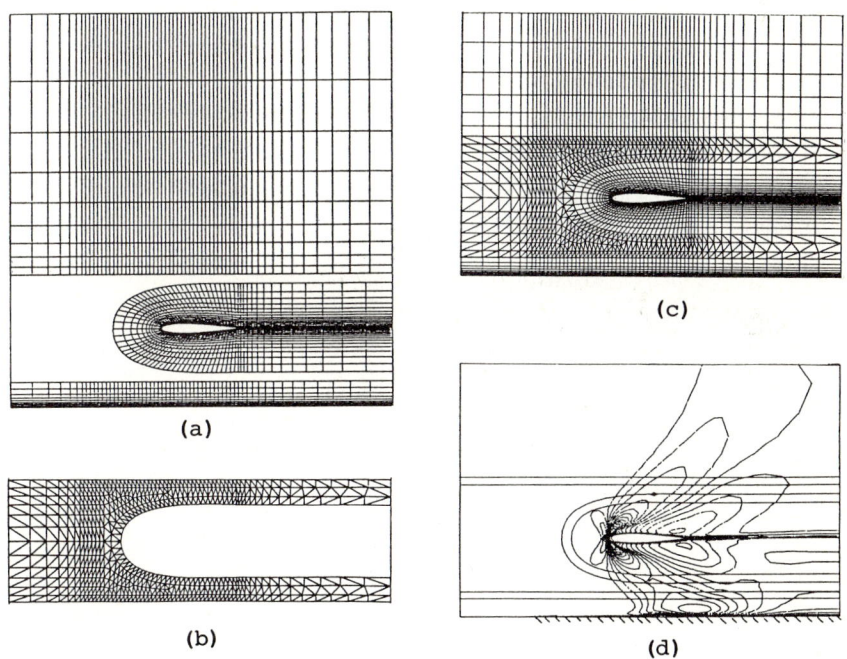

Fig.3 Computations of an interactive flow field between an airfoil and a flat plate ; (a) FDM-grids, (b) FEM-grid, (c) patched FDM-FEM grid, (d) computed density contours.

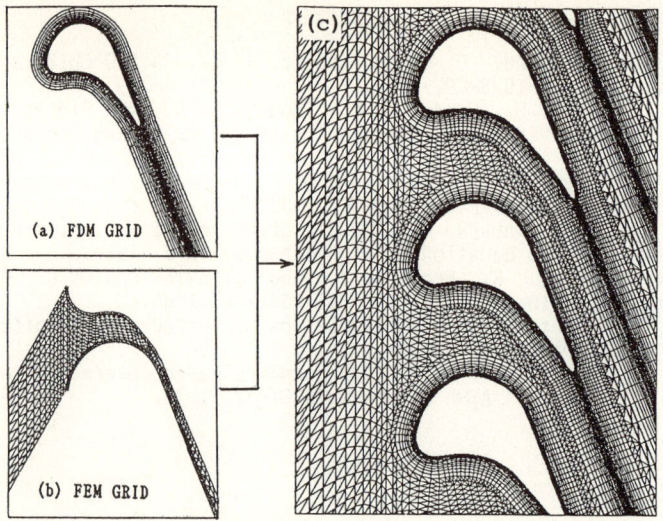

Fig.4 FDM-FEM zonal grid for a turbine cascade flow;
(a) FDM-grid, (b) FEM-grid, (c) patched FDM-FEM grid.

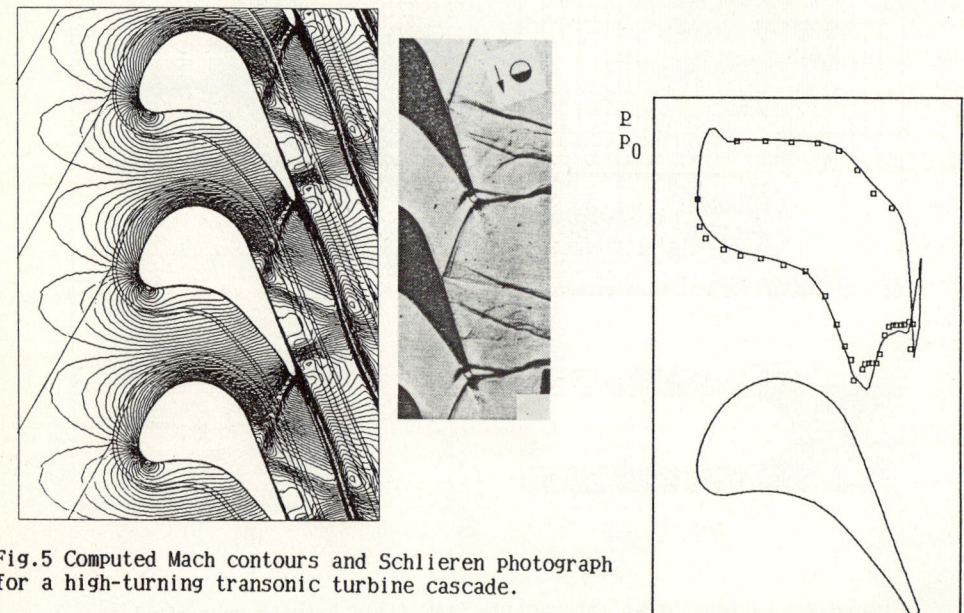

Fig.5 Computed Mach contours and Schlieren photograph for a high-turning transonic turbine cascade.

Fig.6 Comparison of computed surface pressure distribution with expetiment.

Numerical Simulation of Splash of Droplet

Nobu. Nishikawa, Taka. Suzuki, Akira Suzuki

Faculty of Engineering, Chiba University
Yayoi 1-33 Chiba, Japan

1. Introduction

As known widely, the numerical method for incompressible flows, Marker and Cell method was devised specifically for problems involving free surface flows. A great simplification in solving incompressible flow appeared in the SOLA computer program by Hirt et al[1] where the necessity of solving Poisson equation is completely eliminated by their Highly Simplified MAC scheme. One of the cause of inaccuracy in MAC Schemes due to the treatment of free boundaries were overcome by two investigators[2,3] in the papers in 7th ICNMFD Stanford. Nichols et al[2] introduced a volume of fluid function in SOLA-VOF scheme and accurately treated the free boundaries intersecting grid lines, which has been applied several problems including low gravity flow[4]. While Fromm[3] applied surface coordinates for liquid column as for dynamics of liquid jet into air. He investigated the formation of droplets as for ink jet by solving Poisson equation for the pressure together with vorticity-streamfunction formulation.

In this paper descriptions are given for a series of calculations carried out as an effort aimed at studying fluid motion of liquid after splash of droplet into a liquid surface. A few improvements are attempted to a Highly Simplified MAC scheme in eliminating pressure terms in the iteration procedure. The typical results are shown in the images of three dimensional Computer Graphics for understanding of dynamics of fluid.

2. Formulation

In the present problem the physical process can be described as follows. After impingement a crater of liquid appears around upper part of droplet and the crater becomes lower with simultaneous growth of a collumn at the impinged point. Then, the collumn so called Rayleigh Jet elongates and a droplet separates from the collumn.

Here we assume the axisymmetry of phenomena to neglect the break up of the crater. Incompressible Navier Stokes equations in a divergence form are treated in terms of primitive variable form in a cylindrcal coordinate. A droplet impinges along Z coordinate normal to an initially still liquid surface parallel to r coordinate.

The basic equations are written as follows.

$$\frac{\partial u}{\partial r} + \frac{\partial v}{\partial z} + \frac{u}{r} = 0 \qquad (1)$$

$$\frac{\partial u}{\partial t} + \frac{\partial u^2}{\partial r} + \frac{\partial uv}{\partial z} + \frac{u^2}{r} = -\frac{1}{\rho}\frac{\partial p}{\partial r} + g_r + \nu\frac{\partial \zeta}{\partial z} \qquad (2)$$

$$\frac{\partial v}{\partial t} + \frac{\partial uv}{\partial r} + \frac{\partial v^2}{\partial z} + \frac{uv}{r} = -\frac{1}{\rho}\frac{\partial p}{\partial z} + g_z + \frac{\nu}{r}\frac{\partial (r\zeta)}{\partial r} \qquad (3)$$

where $\quad \zeta = \frac{\partial u}{\partial z} - \frac{\partial v}{\partial r}$

3. Numerical scheme

Preserving the simple structure of Highly Simplified MAC scheme [1] the convergence can be very rapid in the following technique. The improved value of velocity through iteration are obtained from simultaneous equations for corrections of velocity components. These equations are deduced from elimination of pressure term in a iteration procedure of original HSMAC scheme.

1. Integration with respect to time give advanced values of velocity components.
2. Renewed velocity components are substituted to the continuity equation and residual appears.
3. Simultaneous equations for velocity variation on upper and left

grids are solved and components of velocity are renewed.
4. Pressure is calculated from sum of residual of continuity equation in each iteration cycle.

The steps 2,3 are iterated until negligible residual appears.

One of the new attempts of the present study is the multiplication of velocity component ratio to give modified value of velocity, to avoid the equal modifications to each velocity components. Another is weighted modification in the iteration steps as follows

$$\delta u_{I,J}^{k+1} = \frac{\omega \Delta R^2 \Delta z^2}{2\Delta R(\Delta R^2 + \Delta z^2)} (u_{I,J}^k + \delta u_{I,J}^{k+1} - u_{I-1,J}^{k+1}) + \ldots$$

where k is the iteration cycle. A corresponding form can be found as a weighted sum of two terms for variation per one time step in the three time level scheme in p207 of the textbook by Richtmeyer & Morton[5].

4. Boundary Condition

At the walls the rigid no-slip is assumed. For free surface the liquid phase pressure measured from gas phase is expressed by the radius of curvatures along a azimuthal plane and a horizontal one.

5. Numerical Results

The liquid was chosen as water. The velocity of droplet of d=2.3mm normal to the still surface depth at the instance of contact each other is in the range smaller than 3 m/s. The depth of still liquid was chosen as 3mm,6mm,and 8mm. Forty grid points are distributed for each direction with the uniform mesh size of 0.1cm. In each grid cell 4 markers were distributed and redistribution was employed properly to maintain accuracy of surface position which seems to have larger contribution than applying a smaller mesh size. The CPU time to attain the vanishing of Rayleigh jet in a typical case was 44 seconds by 7 MIPS machine for 240 time cycles. As shown in Fig.1

comparison with experiment shows a satisfactory agreement with experimental results as for maximum height of Rayleigh jet and the time elapsed to reach the height. The surface shape given by computer graphics is compared with the experimental photograph by MacLin & Hobbs[7] in Fig 2.3. Our numerical solution show a smeared ridge of crown,while the inner-lower region of crown has similar shape to the experimental photograph. Here,we show solutions for surface shape in surface model computer graphic in three dimensional image

In Fig.4 the pressure contour in the liquid is shown for time t=0.0011 4sec,where the disturbance-propagated region is limited and the stagnant region is predicetd by concentrated contours.

The cross sections of splashed area are shown in Figs.5,10 In these time steps the up-down motion develops and the fluid outside of such fluid a centrifugal wave is induced.

Thus,the 'simulation' is successfully completed as if the experimental photographs appear in the CRT of our graphic apparatus and in Figs 2,3. In the future study it is necessary to employ more meshes,because in this study a droplet is located over only 5x5 grids.

REFERENCES

[1] Hirt,C.W.,Nichols,B.D.,and Romero,N.C. Los Alamos Sci.Labo.Rep. LA-5852(1975).

[2] Nichols,B.D.,Hirt,C.W.-and-Hotchikiss,R.S.-Proc.-7th-ICNMFD -.... pp304-309(1981) Springer Ver.

[3] Fromm,J.,Proc. 7th ICNMFD pp188-193(1981) Springer Ver.

[4] Hochstein,J.I.,Gerhart,P.M.,and-Aydelott,J.C.,AIAA-paper-84-1344

[5] Richtmeyer,R.D.,Morton,K.W.,Difference methods for Initial value Problems,p207,InterSci.Pub.,John Wiley &Sons(1967).

[6] Hobbs,P.V.,and Osheroff,T.,Science Vol.158,p1184(1967)

[7] Macklin,W.C.,and Hobbs,P.V.,Science Vol.66,pp107-108(1969).

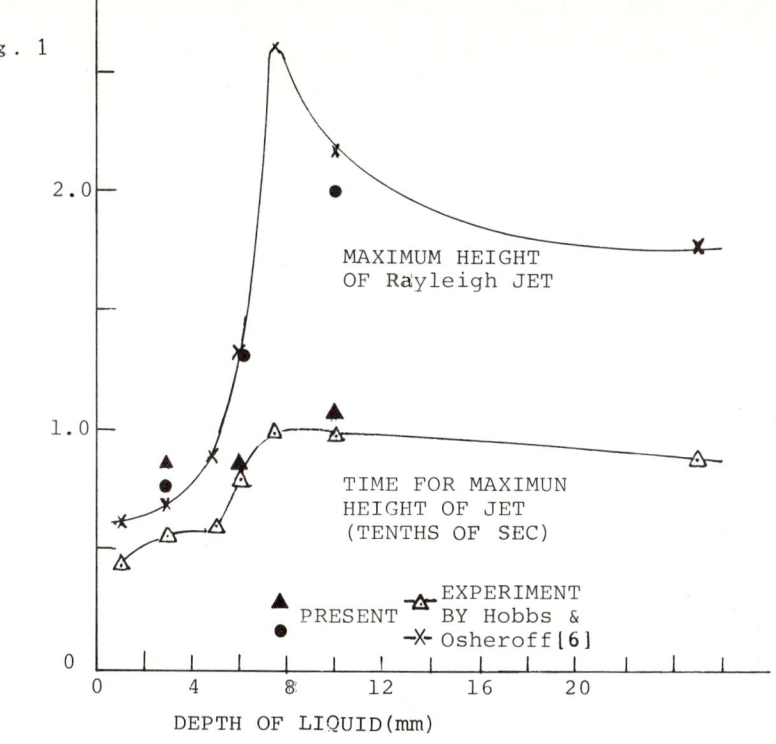

Comparison with Experiment

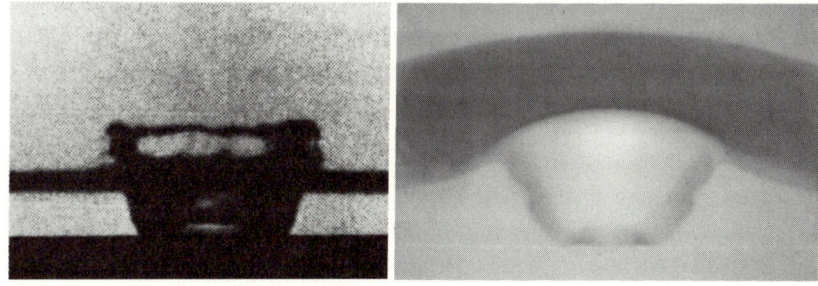

Fig. 2 MacKlin & Hobbs

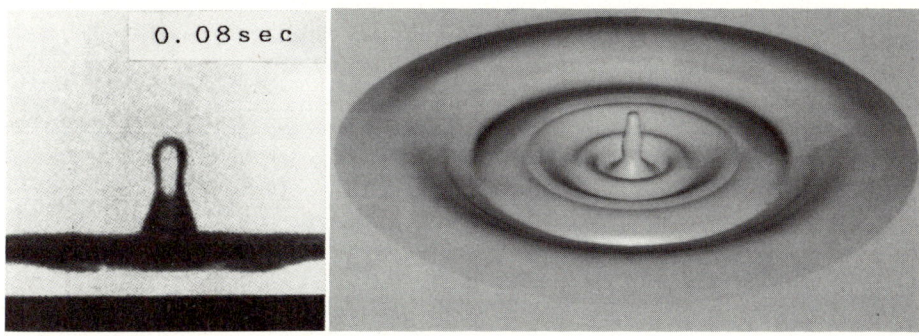

Fig. 3 MacKlin & Hobbs

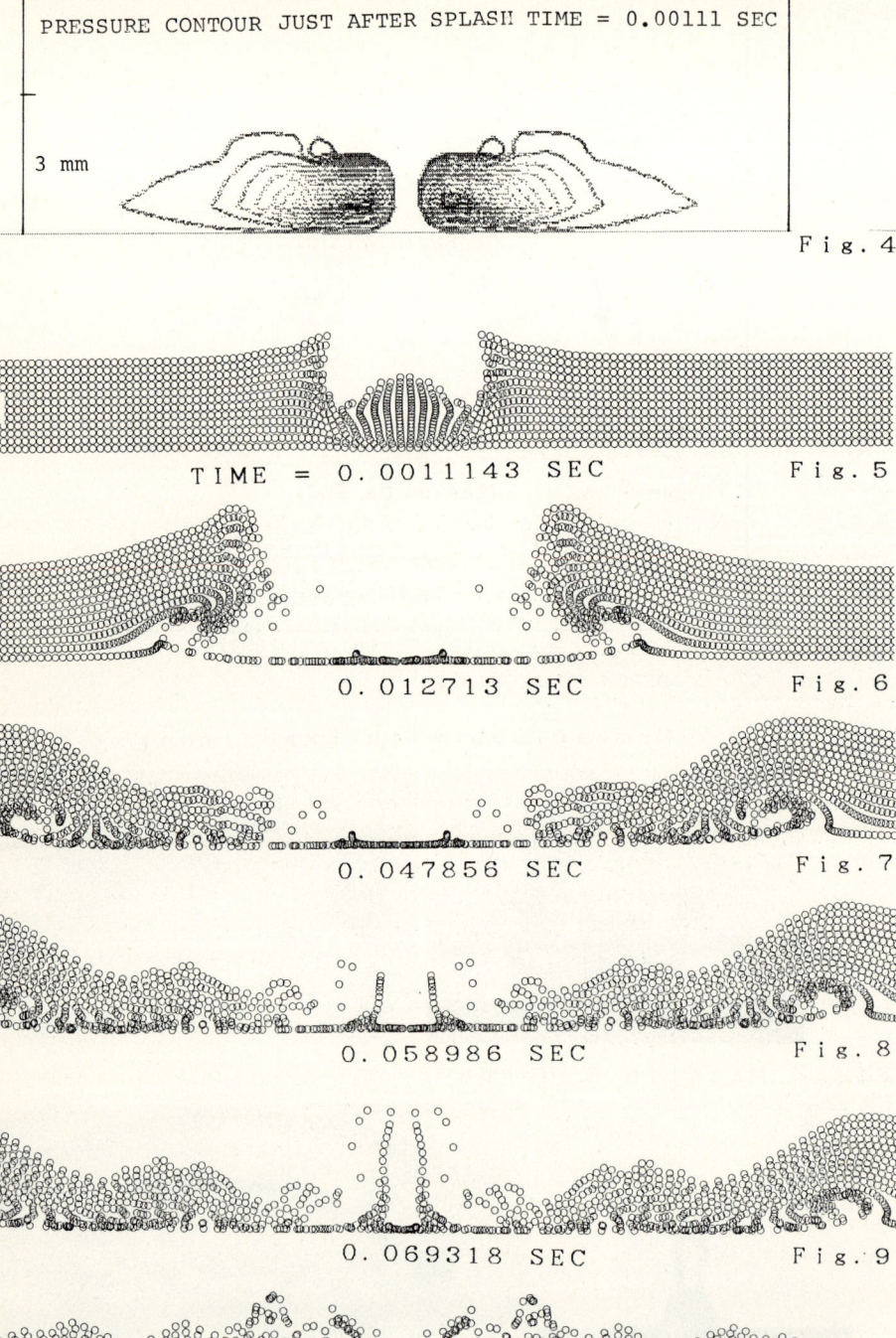

Fig. 4 PRESSURE CONTOUR JUST AFTER SPLASH TIME = 0.00111 SEC

Fig. 5 TIME = 0.0011143 SEC

Fig. 6 0.012713 SEC

Fig. 7 0.047856 SEC

Fig. 8 0.058986 SEC

Fig. 9 0.069318 SEC

Fig. 10 0.11947 SEC

ENERGY ABSORBING BOUNDARY CONDITIONS
FOR THE NAVIER-STOKES EQUATION

Jan Nordström
FFA, The Aeronautical Research Institute of Sweden
S-116 11 BROMMA, Sweden

Summary

In this paper well-posed boundary conditions for the Navier-Stokes equation are derived using the energy method. A new set of boundary conditions are compared with well-posed boundary conditions for the Euler equations when used to compute the subsonic flow over a flat plate. It is shown that the new set of boundary conditions are superior with regard to rate of convergence except for the highest Reynolds number tested. By combining the new boundary conditions with the Euler conditions the best overall performance was achieved.

Introduction

When numerically integrating the Navier-Stokes equation the most frequently used approach is to use the boundary conditions given by the Euler equation and then adding numerical boundary conditions of extrapolation type. Theoretically this means that there is not a clear separation between the boundary conditions required by the equation and the additional conditions. In practice it means that the choice of additional conditions is made very much on a cut and trial basis. In an attempt to clearify the situation Gustafsson and Sundström [1] derived well-posed boundary conditions for the Navier-Stokes equation such that the energy growth is bounded and the boundary conditions transform to well-posed boundary conditions for the Euler equations when the Reynolds number goes to infinity. Recently Dutt [2] considered the full nonlinear equation. An entropy function is used to define a norm, and the energy method is used to prove the boundness of the solution. In this paper a similar derivation as the one made in Ref. 1 is done and the result is the same for inflow boundaries and supersonic outflow boundaries. For a subsonic outflow boundary which has a great influence on the convergence rate [3], a new set of boundary conditions is proposed.

Analysis

Consider the Navier-Stokes equation in one space-dimension:

$$f_t = -(F_i + \varepsilon F_v)_x \quad ; \quad f = (\rho, \rho u, \rho v, e)^T \tag{1}$$

$$F_i = (\rho u, \rho u^2 + P, \rho uv, u(e+P))^T \quad ; \quad F_v = \varepsilon(0, \tau_{xx}, \tau_{xy}, u\tau_{xx} + v\tau_{xy} + Q) \tag{2}$$

$$\tau_{xx} = -\theta u_x, \tau_{xy} = -\mu v_x, Q = -\phi T_x, \varepsilon = Re^{-1}, \theta = \lambda + 2\mu, \phi = \gamma k/Pr, \gamma = C_p/C_v$$

In the formulas above the various quantities $\rho, u, v, \mu, \lambda, e, k, \tau_{xx}, \tau_{xy}, Q$ and P are respectively the density, x and y components of velocity, the shear and second viscosity, the total energy per unit volume, the coefficient of heat conduction, the x and y components of the viscous stress tensor, the heat flux and the pressure. Now we linearize (1) around a state $\overline{f}$ and symmetrize the equation using a matrix $S(\overline{f})$ [4]. We get a system of equations with symmetric constant matrixces. Using the energy method we can derive an equation for the L_2-norm of f:

$$\|f\|_t^2 = -(f^T A f + 2\varepsilon f^T B f_x)\Big|_a^b + 2\varepsilon \int_a^b f_x^T B f_x \, dx = -\Phi\Big|_a^b + 2\varepsilon \int_a^b f_x^T B f_x \, dx \tag{3}$$

The matrix B is negativ semi-definit, and therefore the sufficient condition for a well-posed problem can be written:

$$\Phi = (\overline{u}-\overline{c})[\sqrt{\frac{(\gamma-1)\overline{T}}{2}}[\rho/\overline{\rho}+T/\overline{T}] - \sqrt{\frac{\gamma}{2}}\, u]^2 + (\overline{u})[\sqrt{\overline{T}}\,[T/\overline{T} - (\gamma-1)\rho/\overline{\rho}]]^2 + (\overline{u})[\sqrt{\gamma}\,v]^2 +$$
$$+ (\overline{u}+\overline{c})[\sqrt{\frac{(\gamma-1)\overline{T}}{2}}[\rho/\overline{\rho}+T/\overline{T}] + \sqrt{\frac{\gamma}{2}}\, u]^2 - (2\gamma\varepsilon/\overline{\rho})[\theta u u_x + \mu v v_x + \phi (T/\sqrt{\overline{T}})(T/\sqrt{\overline{T}})_x] \geqslant 0 \tag{4}$$

The number of boundary conditions to the Euler equation depends on the sign of $\overline{u}-\overline{c}$, $\overline{u}$, $\overline{u}+\overline{c}$. For the Navier-Stokes equation we have to give four boundary conditions on an inflow boundary and three on an outflow boundary [5]. When the Reynolds number goes to infinity the system (1) has the character of a singular perturbation problem. In the limit one expects to obtain solutions to the Euler equations. When constructing boundary conditions to the Navier-Stokes equation it is therefore suitable to begin with a set of well-posed boundary conditions to the Euler equation and then augment these with derivative boundary conditions. In this way non-physical boundary layers can be avoided [6],[7].

Supersonic inflow: We have to specify all variables on the boundary for the inviscid problem and also for the viscous one, we get:

$$\rho = G_1(t) \quad ; \quad u = G_2(t) \quad ; \quad v = G_3(t) \quad ; \quad T = G_4(t) \tag{5}$$

Subsonic inflow: The inviscid problem requires three boundary conditions while the Navier-Stokes equation need one additional condition:

$$u - 2\sqrt{\frac{(\gamma-1)T}{\gamma}} = G_1(t); \quad T/\rho^{(\gamma-1)} = G_2(t); \quad v = G_3(t) \tag{6}$$

$$\varepsilon[\theta u_x - 2\phi\sqrt{\frac{\gamma}{\gamma-1}}(\sqrt{T})_x] = \varepsilon G_4(t)$$

Supersonic outflow: No boundary condition is neccessary for the inviscid problem, the Navier-Stokes equation is well-posed if we add:

$$\varepsilon\tau_{xx} \equiv -\varepsilon\theta u_x = \varepsilon G_1(t); \quad \varepsilon\tau_{xy} \equiv -\varepsilon\mu v_x = \varepsilon G_2(t); \quad \varepsilon Q \equiv -\varepsilon\phi T_x = \varepsilon G_3(t) \tag{7}$$

Subsonic outflow: By specifying the pressure or the normal velocity we get a well-posed problem for the Euler equation. If we specify the normal velocity and add two conditions on the gradient of v and T we get:

$$u = G_1(t); \quad \varepsilon\tau_{xy} \equiv -\varepsilon\mu v_x = \varepsilon G_2(t); \quad \varepsilon Q \equiv -\varepsilon\phi T_x = \varepsilon G_3(t) \tag{8a}$$

By specifying the pressure it is not possible to add 2 more conditions and get the right sign on Φ, however, by specifying a linear combination of pressure and the gradient of the normal velocity one can find the following well-posed boundary conditions for the Navier-Stokes equation:

$$P + \varepsilon\tau_{xx} \equiv P - \varepsilon\theta u_x = G_1(t,\varepsilon); \quad \varepsilon\tau_{xy} \equiv -\varepsilon\mu v_x = \varepsilon G_2(t); \quad \varepsilon Q \equiv -\varepsilon\phi T_x = \varepsilon G_3(t) \tag{8b}$$

The conditions (5), (6), (7) and (8a) are given in [1], while (8b) is a new boundary condition for subsonic outflow.

Computations

The numerical method used to integrate the Navier-Stokes equation is a finite-volume method with a fourth order Runge-Kutta time integrator [8]. Preliminary numerical experiments showed that computations using condition (8a) did not converge so the conditions used was (6) and (8b). A reference solution was computed using a common set of boundary conditions for the Navier-Stokes equation [9],[10],[11]:

Subsonic inflow: $\quad \rho = G_1(t); \quad u = G_2(t); \quad v = G_3(t)$ (9)

Subsonic outflow: $\quad P = G_1(t)$ (10)

For the Euler equations (9) and (10) are well-posed conditions [12]. The boundary conditions on the solid surface and symmetry line was:

Solid surface: $u = 0$; $v = 0$; $T_x = 0$ (11)

Symmetry line: $u = 0$; $v_x = 0$; $T_x = 0$ (12)

Note that (11) and (12) are well-posed in the sense that they satisfy (4). The additional numerical boundary conditions needed were taken as linear extrapolation. The test case was laminar flow over a flat plate where the boundary conditions (6) and (9) were applied at boundary (A), (8b) and (10) at boundary (B-C-D), (11) at boundary (E) and (12) at boundary (F), see Fig. 1. The computations were made at Mach number 0.5 for four different Reynolds numbers ranging from 2×10^3 to 10^5. As initial condition we used freestream values everywhere in the field. A computation was considered converged if the RMS value of $f^{n+1}-f^n$ was less than 10^{-6}.

Results

The convergence rate for the different cases are shown in Fig. 2, the RMS-value of $f^{n+1}-f^n$ is plotted against the nondimensional time. The convergence for the boundary conditions (6),(8b) are much better than (9),(10) for the three lowest Reynolds numbers. For the highest one, however, the boundary conditions (9),(10) are superior. The fact that the Euler conditions performed better at the highest Reynolds number led to the idea of a combination of (6),(8b) and (10) where (8b) were applied on the "viscous" outflow boundary (D) and (10) on boundary (B-C). This combination of outflow boundary conditions gave the best overall convergence. Fig. 3 shows the accuracy at steady-state for the case $Re=40 \times 10^3$, the difference in accuracy is small.

Conclusions

The boundary conditions derived in this paper was shown to give faster convergence to steady-state than the conventional Euler conditions except for the highest Re. By combining the Euler conditions with the new set of boundary conditions the best overall convergence to steady-state was achieved. The reason for the "failure" by the new set of boundary conditions at the highest Re is not known at the present time. However, one can speculate that the correct boundary conditions are of importance if the boundary is "viscous" enough (gradients different from zero), and that numerical considerations are more important otherwise.

References

[1] Gustafsson, B., Sundström, A.: "Incompletely Parabolic Systems in Fluid Dynamics". SIAM J. Appl. Math. Vol. 35, No. 2, 1978.

[2] Dutt, P.: "Stable Boundary Conditions and Difference Schemes for Navier-Stokes Equations". To appear in SIAM J. Numer. Anal.

[3] Ruby, D., Strickwerda, J.C.: "Boundary Conditions for Subsonic Compressible Navier-Stokes Calculations. Computers and Fluids Vol. 9, 1981.

[4] Nordström, J.: "Stability Criteria for a Second-Order Accurate, Time-Split Finite Volume Scheme to Solve the Navier-Stokes Equation". FFA Report TN 1985-08, The Aeronautical Research Institute of Sweden, 1985.

[5] Strickwerda, J.C.: "Initial-Boundary Value Problems for Incompletely Parabolic Systems". Comm. Pure Appl. Math, Vol. X, 1977.

[6] Michelson, D.: "Initial-Boundary Value Problems for Incomplete Singular Perturbations of Hyperbolic Systems". Lectures in Appl. Math. Vol. 22, 1985.

[7] Gustafsson, B.: "Numeriska metoder i strömningsmekanik". Course at Uppsala University, 1986.

[8] Nordström, J.: "The Use of "Viscous Splitting" when Solving the Navier-Stokes Equation for High Reynolds Numbers". Conference Proceedings, ISCFD Conference, Tokyo 1985.

[9] Swanson, R.C., Turkel, E.: "A Multistage Time-Stepping Scheme for the Navier-Stokes Equation". AIAA Paper No. 85-0035, Nevada 1985.

[10] Müller, B.: "Implicit Central Difference Simulation of Compressible Navier-Stokes Flow Over a NACA0012 Airfoil". Proceedings of GAMM-workshop on "Numerical Simulation of Compressible Navier-Stokes Flows". INRIA-Sophiaantipolis-France 1985.

[11] Steger, J.L.: "Implicit Finite-Difference Simulation of Flow About Arbitrary Two-Dimensional Geometries". AIAA J. Vol. 16, 1978.

[12] Oliger, J., Sundström, A.: "Theoretical and Practical Aspects of some Initial Boundary Value Problems in Fluid Dynamics". SIAM J. Appl. Math. Vol. 35, 1978.

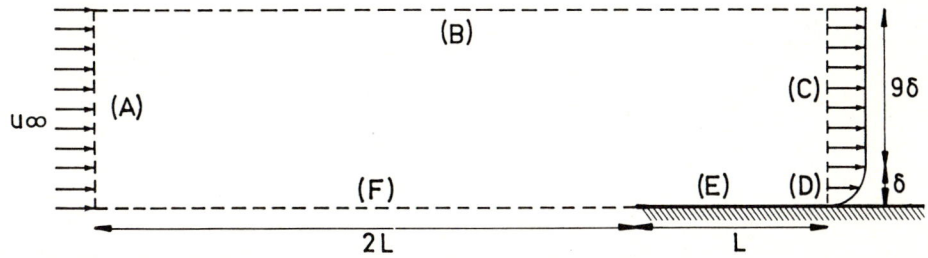

Fig. 1 Geometry and location of boundary conditions.

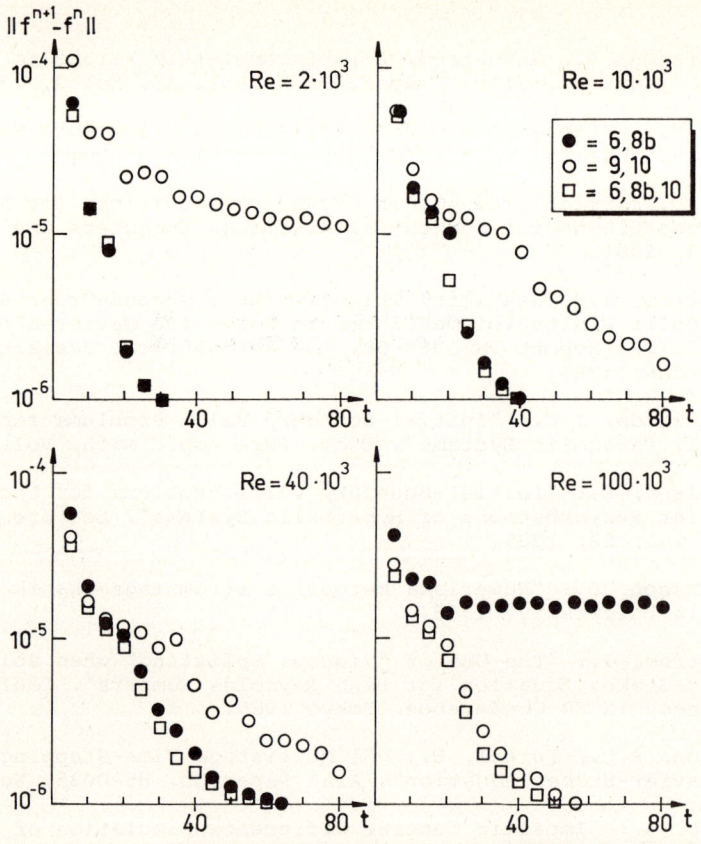

Fig. 2 Convergence history

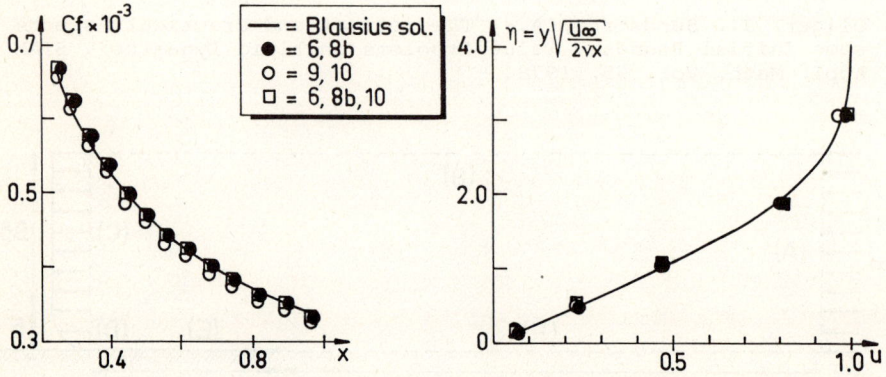

Fig. 3a Skin friction distribution for Re=40×10³.

Fig. 3b Velocity profiles at x=0.55 for Re=40×10³.

INTERACTION OF VORTICAL FLOW REGIONS

Koichi OSHIMA*, Yuko OSHIMA**, Naoki IZUTSU*,
Yoshio ISHII* and Toshihiko NOGUCHI*

*Inst. Space & Astronautical Science
Komaba, Meguro-ku, Tokyo 153 JAPAN
** Dept. Physics, Ochanomizu University
Ohtsuka, Bunkyo-ku, Tokyo 112 JAPAN

Abstract

Discrete vortex method, which simulates vortical flow region by many, small vortex filaments and has been proved to be successful for two-dimensional flow field, is extended for the three-dimensional vortical flow, based on Rosenhead-Moore's approximation. Cross-linking of two vortex rings travelling in parallel were successfully simulated by this method. Full Navier-Stokes code was written for incompressible viscous vortical flow field surrounded by uniform, steady flow, utilizing upwind ADI scheme. The boundary condition along the outer boundary of the computing region was obtained using the asymptotic solution by Lu Ting. These computed results are compared with the experimental observation carried out at ISAS Tokyo and the applicability of this method was confirmed. During these comparative study, three-dimensional, color graphics are extensively utilized.

1. INTRODUCTION

Vortical flow which contains regions with vorticity immersed in mostly irrotational fluid has been attracting much attension of fluid-dynamicists due to its analytical difficulty to treat as well as to its practical importance for engineering and nature. Many good reviews are available for these problems, for example (1)-(2). Behaviors of the vortical region which migrate through the flow field under mutual and self induced velocity are considered, using extended discrete vortex methods as well as numerical solutions of the Navier-Stokes equations. Movement of isolated vortex rings and interaction between two of them will be discussed. These are extensions from the previous work (3)-(6).

2. THREE-DIMENSIONAL DISCRETE VORTEX METHOD

As an idealization of the vortical region confined within an isolated tubular region, we consider a space curve C with the finite circulation Γ which is a vortex filament with the zero cross sectional area. The velocity field induced by this vortex filament is expressed by the Biot-Savart law

$$u(x) = -\frac{\Gamma}{4\pi} \int_C \frac{[x-r(\xi')] \times \partial r/\partial \xi'}{|x-r(\xi')|^3} d\xi' , \qquad (1)$$

where $r(\xi)$ is the space coordinates of the filament C parametrized by the arc length ξ.

A difficulty arises here because the Biot-Savart integral (1) diverges logarithmically as the field point x approaches to the space curve $r(\xi)$. If the filament has no curvature, the divergence disappears. This is the reason of success of discrete vortex method in two-dimensional case. Rosenhead modified the Biot-Savart integral to the expression

$$u(x) = -\frac{\Gamma}{4\pi} \int_C \frac{[x-r(\xi')] \times \partial r/\partial \xi'}{[|x-r(\xi')|^2 + \mu^2]^{3/2}} d\xi' , \qquad (2)$$

where the singularity is excluded by the constant factor. Comparing this asymptotic expression with the analytical solution of axisymmetric vortex ring with small core, Moore chose the factor as;

$$\mu = \sigma \exp\left[-\frac{1}{2} - \int_0^\sigma \frac{\chi^2(s)}{s} ds\right] , \qquad (3)$$

where σ is the radius of the vortex core and $\chi(s)$ is the fraction of the circulation within the radius s. By this procedure, the modified Biot-Savart integral (3) bears the effect of the structure of the vortex core. It is known that this matching procedure produces asymptotically correct equation for thin vortex filament having an arbitrary space curve, provided that $\kappa\sigma \ll 1$ where κ is the local curvature of the filament, and that the structure of the vortex core remains relatively undisturbed.

Based on this Rosenhead-Moore's approximation, an approximate, analytical expression of the induced flow field due to three-dimensional vortex filaments with finite core region is derived, which is applicable except the very close region to the vortex core (7). This is a straight extension of two-dimensional vortex blob method to three-dimensional case, and relatively short computing time is required to follow the interactions of vortices.

3. INTERACTION OF TWO VORTEX RINGS

Interaction of two identical circular vortex rings placed initially on the same plane is simulated. The radius of the vortex rings R is chosen as the characteristics length, and the separation distance of the centers of the rings is 3.0*R. Each vortex ring is divided into 32 vortex elements and the element is composed of 7 vortex filaments. That is, mutual interactions of 2X32X7 vortex segments in total are accounted. Fig.1 shows the interaction patterns in three direction projections and bird's eye view. Fig.2 and 3 are a series of the front view photographs which were visualized by smoke and which correspond to the view of the X-Z and Y-Z planes, respectively. At the initial stage, two rings proceed along their ways in the Z-direction, and get closer to each other. When the separation distance between the nearest portions of the two rings becomes small, the core deforms severly, and they touch each other. Continuing further calculation results in catastrophic instability with sudden increase of the fluid impulse, which is supposed to conserve. According to the experimental results, the crosslinking of the vortices occures to form one distorted vortex ring at this stage. Since the crosslinking of the vortices is a process which cannot be followed by the present method, some artificial operation is needed. When the two vortex cores having the equal vorticity with the opposit sign touch, viscous diffusion would annihilate the vorticity locally in this overlapped portion, and the vortex elements of this region cancel each other. After this artificial crosslinking process is taken, the fluid impulse remains almost unchanged. This justifies this operation.

4. FULL NAVIER-STOKES SOLUTION AND EXPERIMENTS

Utilizing upwind ADI scheme (9), numerical simulation code for three-dimensional, incompressible viscous flow field around free vortices was written, in which far field boundary condition is derived by the asymptotic relation given by Lu Ting (10). The above-discussed problems were successfully simulated numerically. Because the capability of present super-computers is still limited, only limited number of cases were solved and the resolution of the computed results is still marginal.

Computer graphics combined with the high speed data processing system have certainly introduced new world of flow visualization. Creation, growing, migration, decaying and diminishing of vortices are clearly made visible. Experimental data obtained by image-processing of visualized flow field or by digital data-processing of scanning hot-wire measurements have provided complete time history of the flow field. Because these experiments have far higher spacial as well as time resolution (less than 0.1 mm spacially and about 100 microsecond in time for typical hot-wire scanning) than by numerical simulations, subtle mechanisms such as vortex pairing or cross-linking were able to be recognized on the graphics only by those experimental data.

5. CONCLUSIONS

Three-dimensional, inviscid vortical flow can be simulated by discrete vortex method modified with finite potential core, bundle of filament and Rosenhead-Moore's approximation.

REFERENCES

1. Oshima, Y. & Oshima, K.: Vortical Flow behind an Oscillating Airfoil, Proc. 15 ICTAM, North Holland pp.357-368 (1980)
2. Lugt, H.J.: Vortex Flow in Nature and Technology, Wiley Interscience (1983)
3. Oshima, K., Oshima, Y. & Kuriyama, Y.: Finite Element Analysis of Viscous Incompressible Flow Around an Oscillating Airfoil, Proc. 6ICNMFD pp.433-438 (1978)
4. Ono, K., Kuwahara, K. & Oshima, K.: Numerical Analysis of Dynamic Stall Phenomena of an Oscillating Airfoil by the Discrete Vortex Approximation, Proc. 7ICNMFD pp.310-315 (1980)
5. Oshima, K. & Oshima, Y.: Flow Simulation by Discrete Vortex Method, Proc. 8ICNMFD pp.94-106 (1982)
6. Oshima, K., Oshima, Y. & Izutsu, N.: Numerical Simulation of Dynamics of an Autorotating Airfoil, 9ICNMFD pp.442-446 (1984)
7. Moore, D.W.: Finite Amplitude Waves on Aircraft Trailing Vortices, Aeron. Quart. vol.23 pp.307-314 (1972)
8. Leonard, A.: Computing Three-Dimensional Incompressible Flows with Vortex Elements, Ann. Rev. Fluid Mech. vol.17 pp.523-559 (1985)
9. Peyret, R. & Taylor, T.D.: Computational Method for Fluid Flow, Springer-Verlag (1985)
10. Lu Ting: Theoretical and Numerical Studies of Vortex Interaction and Merging, Preprints of ISCFD-Tokyo pp.582-595 (1986)

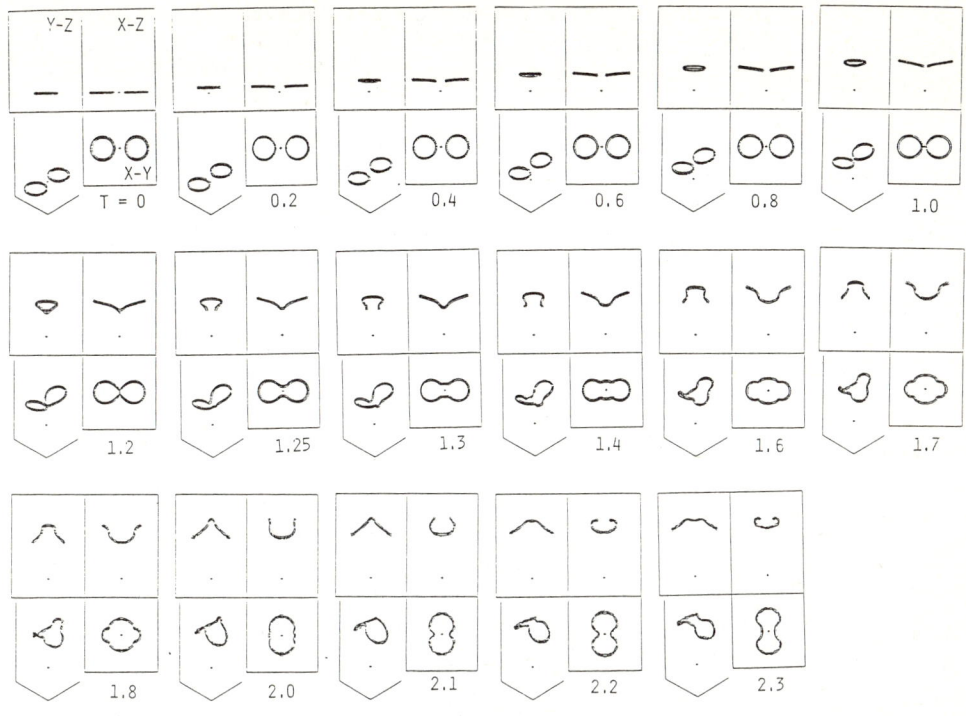

Fig.1 Interactions of two vortex rings

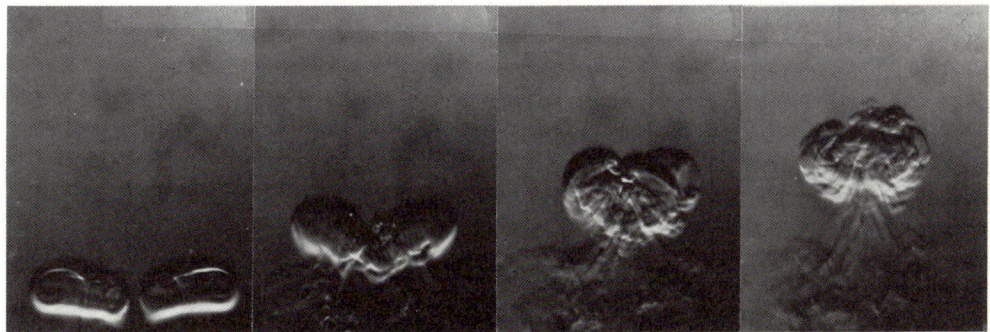

Fig.2 X-Z view of two vortex interaction

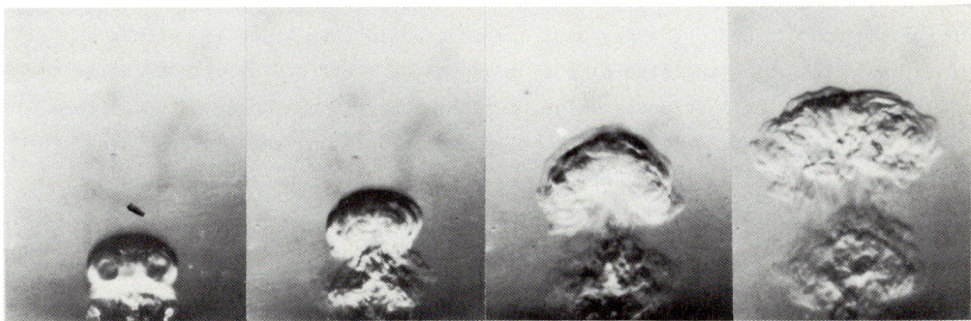

Fig.3 Y-Z view of two vortex interaction

SIMULATION OF BUFFETTING STALL FOR A CAMBERED JOUKOWSKI AIRFOIL USING A FULLY IMPLICIT METHOD[†]

G.A. Osswald, K.N. Ghia and U. Ghia*

Department of Aerospace Engineering and Engineering Mechanics
*Department of Mechanical Engineering
University of Cincinnati, Cincinnati, Ohio 45221 USA

Introduction

Persistent unsteadiness, resulting from natural instabilities within the flow itself, plays a critical role in many important aerodynamic phenomena. Under certain circumstances, this unsteadiness results in a limit-cycle solution of the Navier-Stokes equations. Such time-asymptotic limit-cycle solutions correspond to attractors in the phase or state space of the governing equations. Other examples of attractor solutions are fixed points, i.e., time-asymptotic steady-state solutions, and the so-called "strange attractors" which represent time-asymptotic persistently chaotic solutions of the Navier-Stokes equations. The complexity, or frequency content, of the attractor solution will vary with the critical parameters of the flow. Typically, for small values of the critical parameter, time-asymptotic steady-state flow results. As the critical parameter is increased, the fixed-point solution will become unstable to disturbances, leading to the evolution of a time-asymptotic limit-cycle solution dominated by the most unstable mode. Further increases in the critical parameter can lead to a strange-attractor solution, as additional unstable modes emerge in the persistently unsteady solution. This approach to chaos is examined in the present study while pursuing the primary objective of analyzing the buffeting stall phenomenon for a cambered Joukowski airfoil, through direct unsteady simulation. It is also intended to provide the detailed flow structure for unsteady flow configurations at low Re and a range of angle of attack α_f.

Comments on Analysis, Grid Generation Technique and Numerical Method

The conservation form of the incompressible Navier-Stokes equations are formulated using vorticity and stream function (ω, ψ) in generalized orthogonal curvilinear coordinates. The Reynolds number used in this study is defined as $Re = U_\infty c/\nu$. The characteristic time is given as $t = t^*/(c/U_\infty)$, with U_∞ being the undisturbed free-stream velocity. The boundary conditions correspond to uniform flow at infinity, together with the no-slip condition along the body surface. To facilitate the numerical implementation of the far-field boundary conditions, the

---

This research was supported, in part, by AFOSR Grant No. 85-0231 and, in part, by NASA Grant no. NAG-1-465.

stream function ψ is decomposed into two parts such that $\psi = \psi_{in} + \psi_v$, with ψ_{in} being the inviscid stream function. The initial conditions for the viscous flow are taken as the corresponding steady-state inviscid solution. A clustered conformal grid is generated; the clustering is controlled by appropriate one-dimensional (1-D) stretching transformations. An attempt is made to resolve many of the multiple scales of an unsteady flow with massive separation, while maintaining the transformation metrics to be smooth and continuous in the entire flow field. A typical clustered conformal C-grid, with (230, 46) points, is shown in Fig. 1 for the Göttingen 580 airfoil (i.e., an 11.8 percent thick cambered Joukowski airfoil with a zero-lift angle of attack $\alpha_0 = -5.711°$), at effective flow angle of attack $\alpha_e = 30°$. A fully implicit time-marching method of Osswald, K. Ghia and U. Ghia (1985) is used, in which all spatial derivatives are approximated using central differences and no use is made of any artificial dissipation. The numerical method solves the discretized equations using the alternating-direction implicit (ADI)-block Gaussian elimination (BGE) method and has overall $O[\Delta t, (\Delta \xi^1)^2, (\Delta \xi^2)^2]$ accuracy.

Results and discussion

Results are obtained for three flow configurations with Re = 1000 and α_e = 5.711°, 15° and 30°, the last two being in the stall and post-stall flow regimes, respectively. Here, the effective angle of attack $\alpha_e = \alpha_f - \alpha_0$, with α_f being the geometric angle of attack between the chord and the free-stream direction.

Steady Flow: Re = 1000, α_e = 5.711°

Figure 2 shows the inviscid starting solution, the grid distribution in the near field and the steady-state stream-function and vorticity contours. The stream-function contours show a mildly separated region in the vicinity of the trailing edge (TE). Laminar boundary layers prevail on both the suction and pressure surfaces, as observed in the vorticity contours, which also show a tongue-like behavior. The streamwise extent of the separated flow near the TE is 0.15c, as compared to approximately 0.4c for the symmetric Joukowski airfoil with α_e = 5°, as given by K. Ghia, Osswald and U. Ghia (1985a).

Unsteady Flow Near Stall: Re = 1000, α_e = 15°

The time history of the lift and drag coefficients C_L and C_D, respectively, is depicted in Fig. 3. A time asymptotic limit-cycle solution evolves by approximately t = 12, as compared to t = 31 for the of the corresponding symmetric airfoil considered by K. Ghia, Osswald and U. Ghia (1985a). They had defined the period as the time interval required for the L_2-norm of the deviation of the instantaneous state of the flow from a reference initial state to become smaller than a specified tolerance. For the present configuration, the period is established by examining the successive maxima of C_L in Fig. 3. This limit-cycle solution is an "ordinary attractor", the attractor being a 1-D object to which the phase-space trajectories are asymptotically attracted at large times. Its period is 1.046 characteristic time units and the corresponding Strouhal number $S_{\alpha_f} = fc\ (\sin \alpha_f)/U_\infty = 0.154$; based on α_e, its value is $S_{\alpha_e} = 0.247$. The corresponding period for the symmetric airfoil is 1.416 and $S_{\alpha_e} = 0.18$. Hence, the effect of camber is to increase the frequency of the periodic shedding of large-scale vortices from the TE region and from the separation zone on the suction surface. Network communication difficulties are being experienced in retrieving large graphics data files and, hence, the near-wake coherent structure depicted in Fig. 4 is for t between 21 and 22, rather than for a time interval between two consecutive TE sheddings. The time instants in Fig. 4 correspond to points 1 through 5 in Fig. 3. The instantaneous stream-function contours in Fig. 4 a,c,e,g and i show separated flow over the suction surface. Comparison of these with the results for the corresponding symmetric airfoil shows that the effect of camber appears to make this flow field be dominated by the TE region characterestics. The corresponding instantaneous vorticity contours show well developed eddies, some of which are merging due to the interaction between them. In Fig. 4b, at t = 21, a counterclockwise eddy B_T is evident at the TE. The interaction between the shear layer from the leading edge (LE) and the TE eddies is also clearly seen. The eddy A_T is being pushed downstream and will soon separate from the shear layer emanating from the LE. At t = 21.2, the airfoil develops nearly maximum lift; this time instant corresponds to the onset of the shedding of eddy B_T from the TE. In Fig. 4f, the eddy is being slowly pushed by the LE shear layer into the wake. A new eddy evolves at the TE, as seen in Fig. 4h; this state is associated with nearly minimum lift. The state of the flow at t = 22.0 shows resemblance to that at t = 21.0, thereby suggesting the completion of one cycle. K. Ghia, Osswald and U. Ghia (1985a) have provided the flow structure for the corresponding symmetric airfoil; there, the interactions between the LE and TE vortices are stronger.

Unsteady Flow in Post-Stall Regime: Re = 1000, $\alpha_e = 30°$

The coefficients of lift C_L and drag C_D for this case are shown in Fig. 5. These curves show that, even at t = 52, the flow has not yet reached an asymptotic state. Also, these curves show that one cycle consists of two TE sheddings. Thus, there are two shedding frequencies associated with this attactor; these correspond to the sheddings associated with points 1 and 2 in Fig. 5. The LE shear layer associated with the first frequency is thinner and more intense as compared to that associated with the second frequency. This flow field, with its two natural incommensurate frequencies, is referred to as a quasiperiodic flow. The phase-space portrait (not shown here), is complex, and is tending towards being a "strange attractor". The corresponding Poincaré section indicates that the attractor may be a thin torus. These results are similar to those of the symmetric airfoil given by

K. Ghia, Osswald and U. Ghia (1985b), except that the C_L history appears more chaotic and does not seem to be tending towards a limit cycle. Also, there is a peak due to the shedding from the LE at point 3; this was not present in the results for the symmetric airfoil. The time instants at which the detailed flow results are presented correspond to the TE-LE-TE sheddings.

The instantaneous stream-function contours, presented in Figs. 6a,c,e,g,i, show massively separated flow, with large eddies present over the suction surface as well as in the wake. Figures 6a,g,i show the presence of multiple separations, whereas Fig. 6c shows the presence of two counterclockwise co-rotating bubbles, aft of the shoulder, towards the TE and these bubbles are in the process of coalescing. The corresponding vorticity contours are shown in Fig. 6b,d,f,h,j and various vortex interactions can be observed from this figure. In Fig. 6b, the TE vortex has just been shed, a new TE vortex intensifies as shown in Fig. 6d and is being just separated from the TE by the growing LE vortex. Figure 6f corresponds to shedding of the LE vortex. The state shown in Fig. 6j corresponds to TE shedding and has features similar to those in Fig. 6b.

Conclusion

The unsteady Navier-Stokes analysis is developed for analyzing the massively separated, persistently unsteady flow in near- and post-stall regimes of a cambered Joukowski airfoil. The aerodynamic coefficients C_L and C_D, as well as the detailed spatial structure, for three separate flow configurations have been provided. The results are examined carefully to study the effect of airfoil camber on the flow field. Up to the stall regime, camber causes the flow fields to be dominated by the TE geometry. Also, the extent of the separated flow is diminished both in the streamwise and the lateral dimensions and the shedding frequency is increased. For flow fields in the post-stall regime, the C_L time history becomes more chaotic and shows a peak associated with LE shedding. The computation of the flow field needs to be continued to larger t, to predict its further behaviour with definite assurance.

References

Ghia, K.N., Osswald, G.A., and Ghia, U., (1985a), Proceedings of Third Symposium on Numerical and Physical Aspects of Aerodynamic Flows, Long Beach, California; also, Numerical and Physical Aspects of Aerodymanic Flows, Editor: T. Cebeci, Springer Verlag, New York.

Ghia, K.N., Osswald, G.A. and Ghia, U., (1985b), High Reynolds Number Flow Computations. Editor: K. Kuwahara, Springer Verlag, New York.

Osswald, G.A., Ghia, K.N. and Ghia, U. (1985), AIAA CP 854, pp. 25-37.

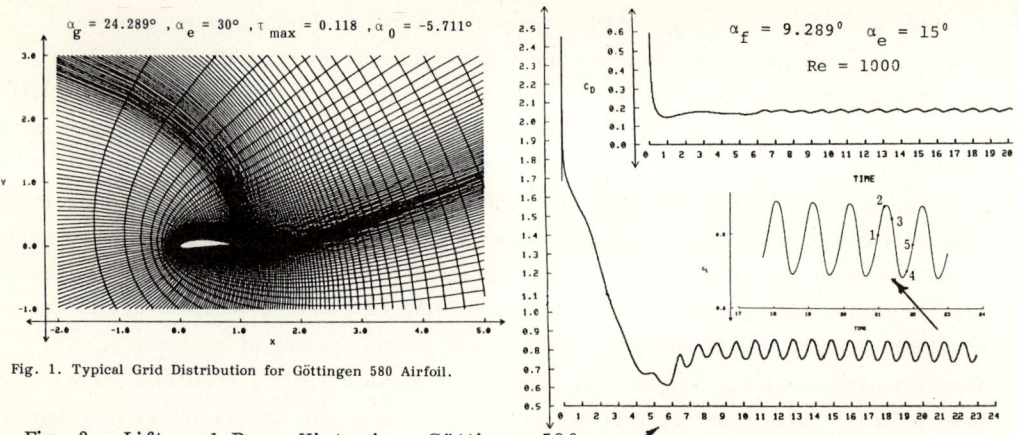

Fig. 1. Typical Grid Distribution for Göttingen 580 Airfoil.

Fig. 3. Lift and Drag Histories; Göttingen 580.

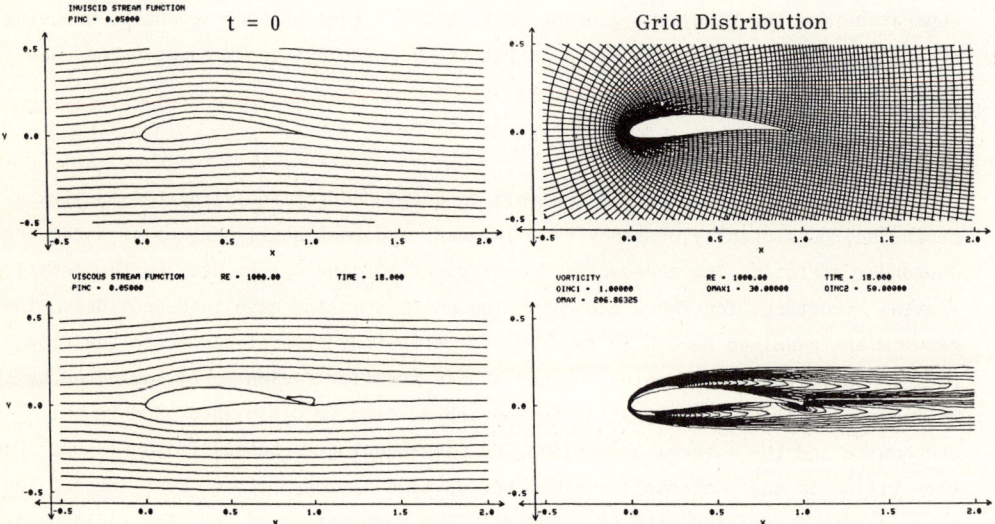

Fig. 2. Göttingen 580 at Re = 1000, α_e = 5.711°; at t = 0: Inviscid Stream Function and Grid Distribution; at t = 18.0 Steady-State Solution.

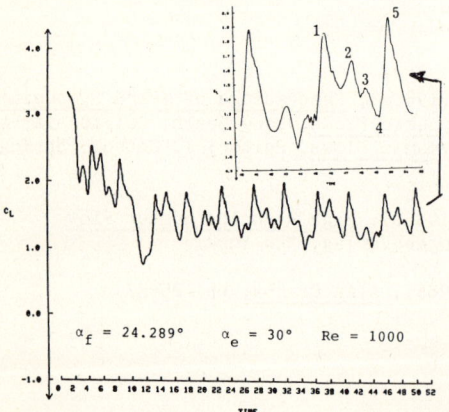

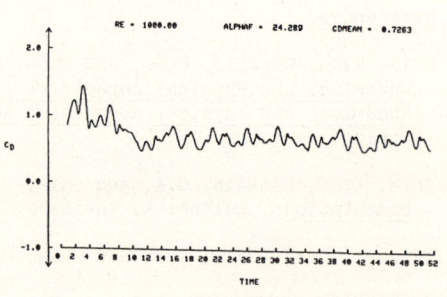

Fig. 5. Lift and Drag Histories; Göttingen 580.

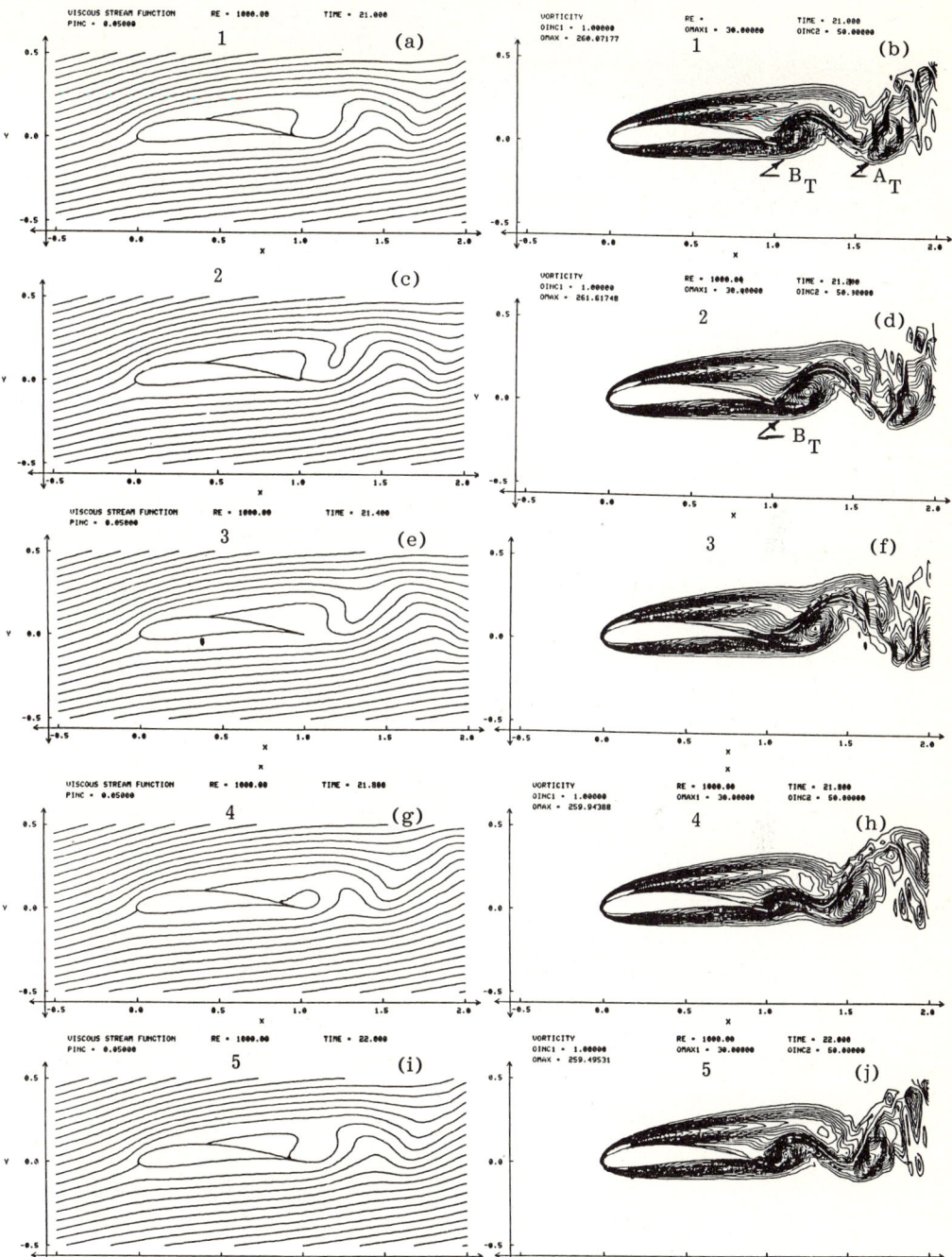

INSTANTANEOUS STREAM-FUNCTION CONTOURS   VORTICITY CONTOURS
Fig. 4.   Limit-Cycle for Göttingen 580 Airfoil at Re = 1000, α_e = 15°.

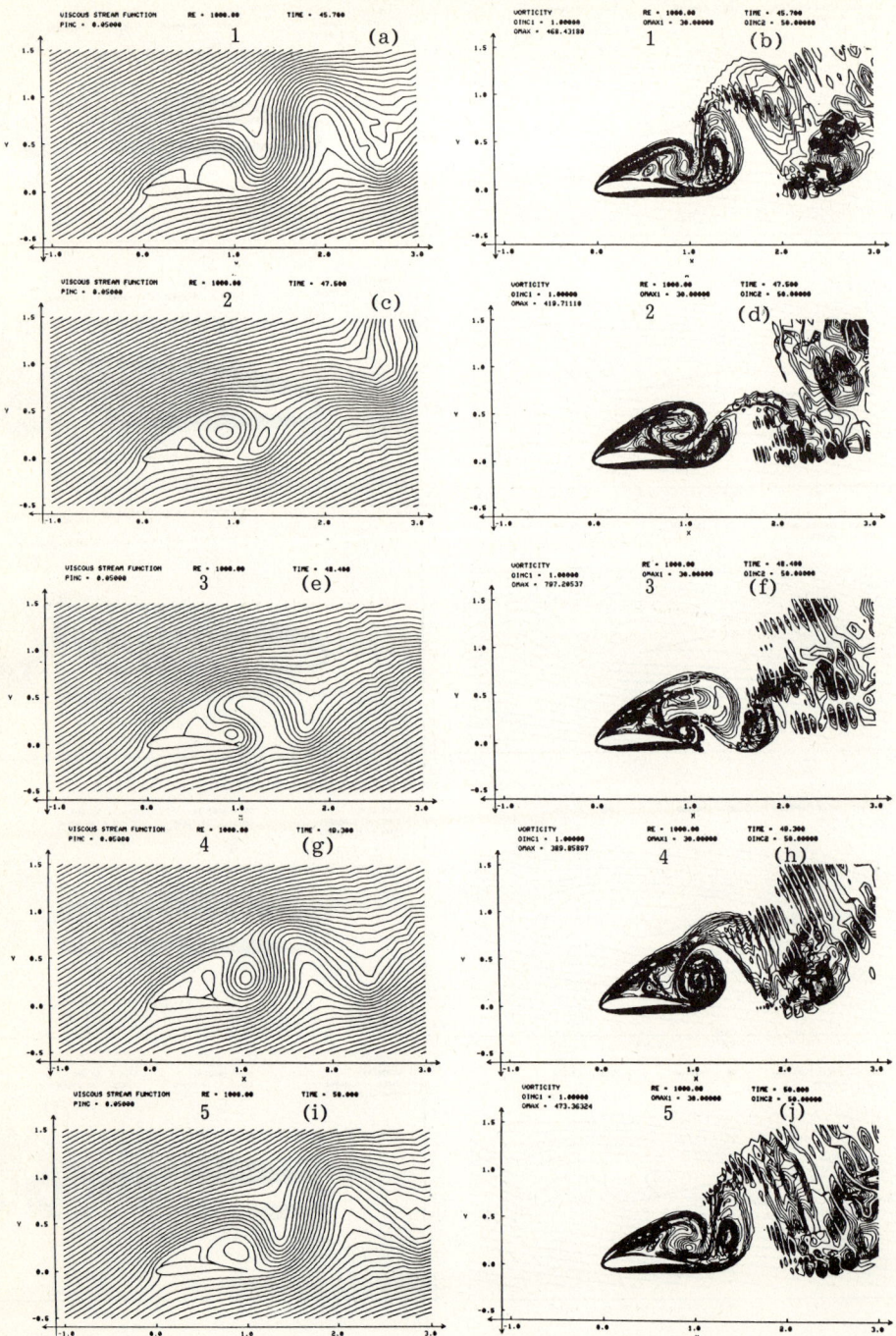

INSTANTANEOUS STREAM-FUNCTION CONTOURS   VORTICITY CONTOURS

Fig. 6. TE-LE-TE Vortex-Shedding Cycle for Göttingen 580 Airfoil at Re = 1000, α_e = 30°.

ADAPTIVE FULL-MULTIGRID FINITE ELEMENT METHODS

FOR SOLVING THE TWO-DIMENSIONAL EULER EQUATIONS

E. PEREZ, J. PERIAUX, J.P. ROSENBLUM, B STOUFFLET AMD-BA, BP 300, 78 Quai Carnot, 92214 SAINT-CLOUD (France)

A. DERVIEUX, M.H. LALLEMAND INRIA Sophia-Antipolis, Avenue Emile Hugues, 06560 VALBONNE (France)

1. INTRODUCTION

Finite Element schemes applying on unstructured triangulations (2-D) or tetrahedrizations (3-D) have proved to be a convenient tool for Euler flow calculations around complete geometries. This option has permitted calculations around complete civil aircraft configurations [1]. It has also permitted hypersonic flow predictions around a space vehicle [2].

For such calculations, **efficiency** is an important question (among others such as accuracy, not discussed here).

An efficient and robust **implicit algorithm** has been introduced for calculations with unstructured meshes [3] and extended to 2-D and 3-D upwind approximations [4]. Important gains in efficiency are obtained but the approach is core memory costly.

The purpose of this paper is to study the **explicit multigrid** approach applied to **unstructured meshes.** This option is felt as an important way to reach higher efficiency and have been previously advocated by LOHNER and MORGAN [5].

The first keyword of the study is thus **"unstructured meshes".**

The second main keyword will be **"nested grids":**

The "nested" option implies the following properties :

(a) inter-grid transfers will be straightforward derived (cheap and simple)

(b) the different levels or grids are easily derived, saving computer and user time. This point is essential when 3-D extension is the ultimate goal.

Now the method used to generate the different grids is an important point for the distinction between the various approaches. Three families of them will be considered :

(1) Uniform refinement methods

Starting from an unstructured coarse grid, finer grids are generated by element division over the whole domain of computation.

(2) Adaptive local refinement methods

The initial "coarse" grid is refined only locally, after a posteriori error estimates.

(3) Agglomerated Multigrid Methods

Starting from an arbitrary unstructured fine mesh, coarser meshes are generated by "agglomerating" neighboring nodes. This approach is related to both usual MG and Algebraic MG (see [7]).

2. UNIFORM MESH REFINEMENT MG METHOD

This methods corresponds to the basic conditions for the validation of new MG algorithms. In the first phase of the work, the extension of two usual families of schemes to F.E.M multigrids has been studied. **Sec. 2.1 and 2.2 are reported in details in [6].**

2.1. The multigrid algorithm

Jameson's multigrid time stepping [11] is adapted to triangulations : the resulting algorithm is based on :

A Runge-Kutta time stepping :

The four-step linearized version is used with various coefficients depending on the spatial approximation.

The R-K scheme is applied on each grid with the corresponding spatial approximation and using the residual of finer grids as forcing functions.

Inter-grid transfers :

Coarse to fine residual collection is performed using an adhoc operator derived from the Galerkin F.E.M. basis functions.

2.2. Galerkin with artificial viscosity

The usual Galerkin method is applied ; two stages of second order artificial viscosity are added ; they result in a blend of second order and fourth order viscosities. The Full-MG convergence is observed : we present as an example the computation of a flow around a bi-plane NACA0012 (Fig.1 and 2).

2.3. Upwind schemes

This section will be developped in [10]. In short the spatial scheme is the upwind Finite Element scheme introduced in [4] ; the R-K time stepping is adapted in a way close to that of [9].

3. ADAPTIVE MESH REFINEMENT MG METHOD

While the accuracy of the global mesh refinement method is rather limited by the uncontrolled quality of the finest mesh, the use of local adaptive mesh refinements allows much more flexibility via full-multigrid strategies as described in [8].

Numerical experiments

The abilities of this approach is demonstrated by its application to flow calculations (using the Galerkin approximation) around a NACA0012 airfoil (Figs.3 to 6).

4. AGGLOMERATED UNSTRUCTURED MULTIGRID

The above method is rather efficient but does not directly apply to the solution of the Euler equation with an arbitrary unstructured fine mesh. To reach such a reliability level, we explorate the possibility of generating coarser meshes by sticking cells together ; in such a research program, an algorithm to do this is first to be found :

4.1. Agglomerating algorithms

The purpose is (1) to decrease mesh size while increasing maximum timesteps (explicit iterations), (2) keep a sufficiently accurate representation of the solution on coarser grids so that we get a good initialization (Full-MG) and good preconditionners and (3) to have a system of nested grids which allows the damping of a dense enough collection of frequencies.

One approach consists in using some auxiliary regular coarser mesh which divides the domain in regions, in order to gather the cells the center of which belongs to the same region. Such an approach may not take enough into account the density of the initial mesh ; some more sophisticated method can be required : we can derive them from the works motivated by multi-tasking in super-computer ; the problem is to divide the domain in regions which are (1) of comparable size (number of nodes) and (2) with as few connections between each others as possible (therefore with as much connection in each region). Finally the sophistication can be increased up to the examination of the discrete equation as in Algebraic MG methods.

The study of these possibilities is in progress ; in this paper, some experiments will be presented with coarser meshes generated with a trivial (and very cheap) algorithm : in only one double do loop, we consider successively each cell ; if it has not yet been included in a coarser zone, we create a new zone containing this cell and each of its direct neighbour (common boundary) which is not yet included in a previous zone.

An example is presented in Figs.7 to 9 : from an initial triangulation (800 vertices) we derive the FVM dual mesh by constructing cells around vertices with medians as boundaries (obtaining 800 cells); then agglomeration is performed, yielding 262 zones.

4.2. Spatial approximation

We restrict ourselves to first-order accurate upwind approximations. The spatial scheme applied to coarser grids can be defined in short as follows : the main difference between the fine grid scheme and a coarser one is that only fluxes between nodes belonging to two different zones are applied.

4.3. Time stepping

The time-stepping applied is essentially the same as in the two above approaches : a four-stage RK4 but also a RK1 scheme have been tried. One simplification is that a same constant time-step is used in this preliminary study for all the nodes and grids. A theoretical study of the possible time-steps on the coarse grids is in progress.

4.4. A first numerical experiment

A transonic case (M=.85) is computed first with the coarse grid, then used as initial data with the fine grid : the two-grid scheme is then applied and compared with the one-grid scheme : a ratio of 2 is obtained (Fig.10).

5. CONCLUSION

In these few pages, we have attempt to describe the main trenbds of our multigrid research program. The adaptative MG method is already an efficient 2-D solver, which in several cases is twice as faster as an implicit time-stepping previously derived [12]. The agglomerating MG method is still in an experimental phase but seems very promising.

6. REFERENCES

[1] JAMESON A., BAKER T.J., WEATHERILL N.P., Calculation of inviscid transonic flow over a complete aircraft, AIAA Paper 86-0103 (1986)
[2] BILLEY V., PERIAUX J., PERRIER P., STOUFFLET B., 2-D and 3-D Euler computations with Finite Elements Methods in Aerodynamics, International Conference on Hyperbolic Problems, Saint-Etienne, Jan. 13-17 (1986)
[3] STOUFFLET B., Implicit Finite Element methods for the Euler equations, in **Numerical Methods for the Euler Equations of Fluid Dynamics**, F.Angrand et al eds., SIAM Philadelphia (1985)
[4] FEZOUI F., STOUFFLET B., PERIAUX J., DERVIEUX A., Implicit high-order upwind Finite-Element schemes for the Euler equations, Fourth Int. Symposium on Numerical Methods in Engineering (Atlanta, USA, March 24-28 1986), to be published by computational Mechanics Pub., Southampton(UK)
[5] LOHNER R., MORGAN K., Unstructured Multigrid methods, Second European Conference on Multigrid Methods, Koln (RFA), Octobe 1-4, 1985
[6] PEREZ E., Finite Element and Multigrid solution of the two-dimensional Euler equations on a non-structured mesh, INRIA Report 442 (1985)
[7] BRANDT A., MACCORMICK S.F., RUGE J., Algebraic Multi-Grid (AMG) for sparse matrix equations, in **Sparsity and its applications**, (D.J. Evans Ed.), Cambridge University Press (1984)
[8] BANK R.E., SHERMAN A.H., An adaptive multi-level method for elliptic boundary value problems, Computing, 26, 91-105 (1981)
[9] TURKEL E., VAN LEER B., Flux vector splittinng and Runge-Kutta methods for the Euler equations, ICASE Report 84-27, june 1984
[10] LALLEMAND M.H., INRIA Research Report (in preparation)
[11] JAMESON A., Numerical solution of the Euler equations for compressible inviscid fluids, **Numerical methods for the Euler equations of Fluid Dynamics**, F.Angrand et al. Eds., SIAM Philadelphia (1985)
[12] ANGRAND F., BILLEY V., PERIAUX J., POULETTY C., ROSENBLUM J.P., 2-D and 3-D Euler computations around lifting bodies on self adapted finite element meshes, Sixth Int. Symp. Finite Element in Flow Problems, Antibes (France),June 16-20, 1986

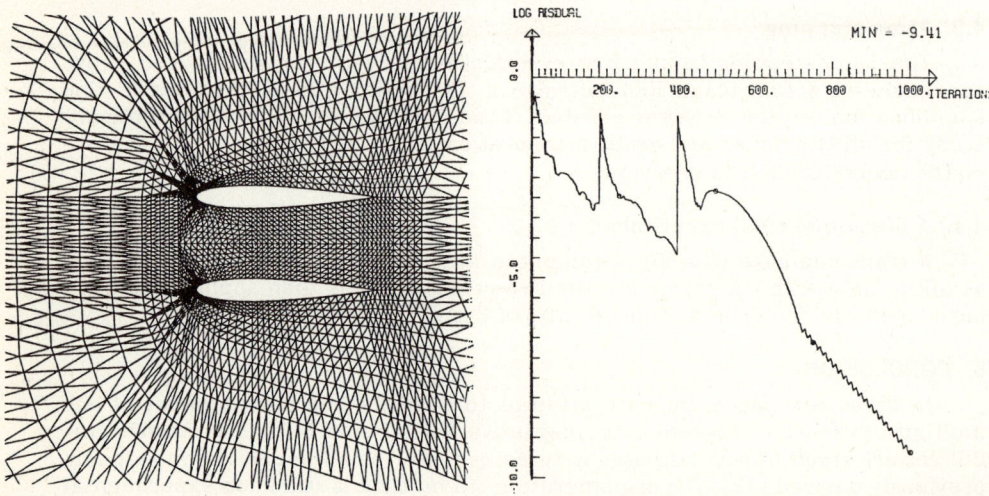

FIGURES 1 AND 2 :

FULL MULTI-GRID METHOD : Flow around a bi-NACA0012 airfoil.

Mach number = .55 , angle of attack = 6 degrees.

Partial view of the mesh and convergence history.

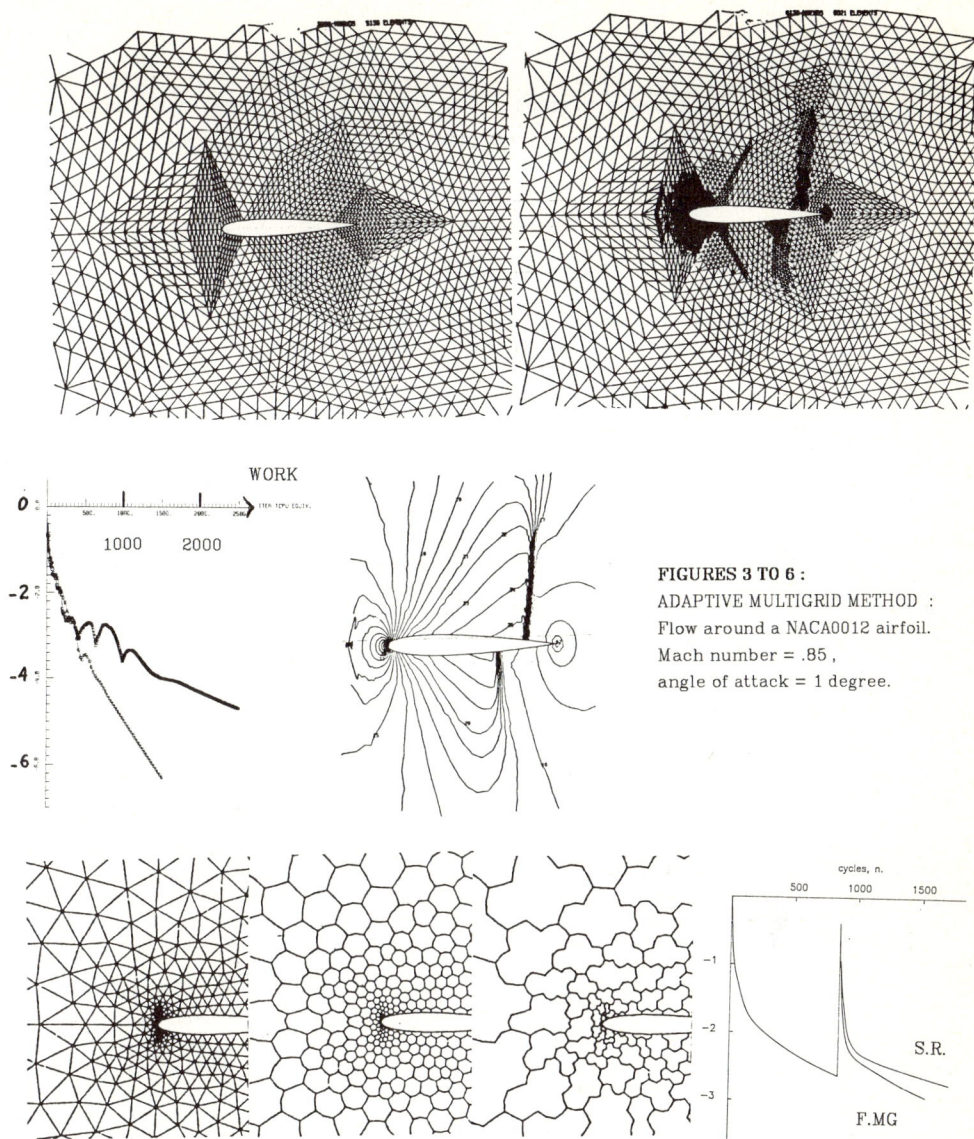

FIGURES 3 TO 6 :
ADAPTIVE MULTIGRID METHOD :
Flow around a NACA0012 airfoil.
Mach number = .85 ,
angle of attack = 1 degree.

FIGURES 7 to 10 : AGGLOMERATED MULTI-GRID METHOD :
Initial fine grid, dual grid and coarse grid, comparison between
successive refinement method (S.R.) and full multigrid mathod (F.MG).
Mach number = .85 , angle of attack = 0 degree.

SIMULATION OF HYPERSONIC VISCOUS FLOWS AROUND A CONE-DELTA-WING COMBINATION
BY AN IMPLICIT METHOD WITH MULTIGRID ACCELERATION

Ning Qin[+] and B. E. Richards

Department of Aeronautics and Fluid Mechanics
Glasgow University, Glasgow G12 8QQ, UK

[+]also Aerodynamics Department
Nanjing Aeronautical Institute, Nanjing, China

Introduction

There is a general tendency to use a time marching approach for steady state solutions in fluid dynamics because of its robustness in following true physical processes. Implicit methods can generally yield faster convergence to steady state than explicit ones by taking larger time steps, which are not subjected to the severe CFL stability criterion(see, e.g.,[1]). Even quadratic convergence will be achieved if the implicit time marching can reduce to Newton iteration for the corresponding steady state problem when $\Delta t \to \infty$. One of these "perfect" implicit operators has recently been constructed by Mulder and van Leer[2] for one-dimensional problems in gas dynamics. Unfortunately, this perfect implicit operator is usually very difficult and sometimes even impossible to construct. Furthermore the way forward into multidimensional problems is blocked by the fact that no efficient direct inversion of the unfactored multidimensional implicit operator exists and approximate factorization is often introduced. Therefore we should make the errors from the approximate factorization balance those from the discretization of the continuous model. This imposes a limitation on time steps and prevents the convergence of the implicit method from achieving the fast rate achieved in one-dimension. Too large time steps can only create severe errors instead of yielding quadratic convergence.

The purpose of this research is to simulate in an efficient way the interaction flow structure involving shock waves as well as strong viscous effects, such as separation, reattachment and vortex flow. For supersonic and hypersonic viscous flow around conical bodies, the unsteady "conical" Navier-Stokes equations are solved by a two step implicit MacCormack scheme. Very fine grids in both θ and ϕ directions are often involved near the wall for sufficient resolution of the viscous effects. The time step, though it can be an order higher than that of the corresponding explicit scheme, is limited by the error from approximate factorization and the convergence is still not satisfactory. How can we further accelerate the convergence of an implicit method for a multidimensional problem? We note that while multidimension presents an obstacle for improving the convergence of an implict method, it provides an opportunity for exploiting efficiency by means of a multigrid procedure. Inspired by the recent success of Ni[3] and Johnson[4], we propose a multigrid procedure to further accelerate the convergence of the implicit scheme to steady state without sacrificing the advantages of the original method. The numerical method is then used as an experimental tool to simulate the hypersonic flow beneath a cone-delta-wing combination. This first Navier-Stokes solution of this problem reveals a far more detailed interaction structure in the flow field than had been expected.

Governing Equations

The three-dimensional unsteady, compressible Navier-Stokes equations are first written in weak conservation form for a spherical polar coordinate system (r,θ,ϕ). For supersonic and hypersonic viscous flows around conical bodies, it is assumed that the gradients in the radial direction are much smaller than those in the crossflow direction. The validity of this "locally conical" approximation downstream from the nose region has been well established through experiment and computation even though a relatively large viscous region exists[5]. After this approximation, i.e. $\partial/\partial r = 0$, is applied to all fluid quantities in the above equations, we may write the resulting "conical" Navier-Stokes equations as follows,

$$U_t + F_\theta + G_\phi + H = 0 \qquad (1)$$

where

$U = \sin\theta(B)$, $F = \sin\theta(vB-S_2)$, $G = (wB-S_3)$
$B^T = (\rho, \rho u, \rho v, \rho w, \rho e)$
$S_i^T = (0, \tau_{i1}, \tau_{i2}, \tau_{i3}, -q_i + u\tau_{i1} + v\tau_{i2} + w\tau_{i3})$

$$H = \sin\theta \begin{bmatrix} 2\rho u \\ 2\rho u^2 - \rho v^2 - \rho w^2 - (\tau_{11}+p) + (\tau_{22}+p) + (\tau_{33}+p) \\ 3\rho uv - \cot\theta(\rho w^2+p) + \cot\theta(\tau_{33}+p) - 2\tau_{12} \\ 3\rho uw + \cot\theta\rho vw - \cot\theta\tau_{23} - 2\tau_{13} \\ 2u(\rho e+p) - u(\tau_{11}+p) - v\tau_{12} - w\tau_{13} \end{bmatrix}$$

$\tau_{11} = -p - (2/3)(\mu/Re_{\infty,r})(2u + v_\theta + v\cot\theta + w_\phi/\sin\theta)$
$\tau_{22} = 2(\mu/Re_{\infty,r})(v_\theta + u) + \tau_{11}$
$\tau_{33} = 2(\mu/Re_{\infty,r})(w_\phi/\sin\theta + u + v\cot\theta) + \tau_{11}$
$\tau_{12} = \tau_{21} = (\mu/Re_{\infty,r})(-v + u_\theta)$
$\tau_{23} = \tau_{32} = (\mu/Re_{\infty,r})(w_\theta - w\cot\theta + v_\phi/\sin\theta)$
$\tau_{31} = \tau_{13} = (\mu/Re_{\infty,r})(u_\phi/\sin\theta - w)$
$q_2 = -(1/2)(\mu/Re_{\infty,r}Pr)T_\theta$, $q_3 = -(1/2\sin\theta)(\mu/Re_{\infty,r}Pr)T_\phi$

Note that, for viscous flow, a length scale dependence remains and is contained in the Reynolds number, $Re_{\infty,r} = \rho_\infty V_\infty r/\mu_\infty$, which determines the location of a crossflow plane in which a solution is computed.

The pressure is given by

$$p = (\gamma-1)\rho[e - (u^2+v^2+w^2)/2]$$

and viscosity is accounted for by the Sutherland formula.

Numerical method

Basic implicit method as a fine grid scheme

The basic integration scheme employed in this work is the two-step implicit method due to MacCormack[1]. We apply it to Eqs.(1) in the following form.

Predictor:

$$\Delta U_{i,j}^n = -\Delta t(\Delta_+ F_{i,j}^n/\Delta\theta + \Delta_+ G_{i,j}^n/\Delta\phi + H_{i,j}^n)$$
$$[I - (\Delta t/\Delta\theta)\Delta_+|A^n|][I - (\Delta t/\Delta\phi)\Delta_+|B^n|]\delta U_{i,j}^{\overline{n+1}} = \Delta U_{i,j}^n$$
$$U_{i,j}^{\overline{n+1}} = U_{i,j}^n + \delta U_{i,j}^{\overline{n+1}}$$

Corrector: $\qquad\qquad\qquad\qquad\qquad\qquad\qquad\qquad\qquad\qquad\qquad\qquad\qquad (2)$

$$\Delta U_{i,j}^{\overline{n+1}} = -\Delta t(\Delta_- F_{i,j}^{\overline{n+1}}/\Delta\theta + \Delta_- G_{i,j}^{\overline{n+1}}/\Delta\phi + H_{i,j}^{\overline{n+1}})$$
$$[I + (\Delta t/\Delta\theta)\Delta_-|A^{\overline{n+1}}|][I + (\Delta t/\Delta\phi)\Delta_-|B^{\overline{n+1}}|]\delta U_{i,j}^{n+1} = \Delta U_{i,j}^{\overline{n+1}}$$
$$U_{i,j}^{n+1} = (1/2)(U_{i,j}^n + U_{i,j}^{\overline{n+1}} + \delta U_{i,j}^{n+1})$$

Although the above implicit scheme has some advantages over other implicit schemes[1], one major problem with it is the difficulty in appropriately treating the implicit boundary conditions on the wall. To sweep from the wall, the implicity is often degraded or destroyed at the wall boundary by simply setting the wall flux to zero or reflecting it from the previous sweep. These operations are generally inconsistent with the physical explicit wall conditions. We propose a new technique to start the implicit sweep from the wall boundary. According to the physical explicit boundary conditions we can express the changes on the wall as functions of the changes of the neighbouring interior points. After linearization, the implicit wall boundary conditions are wholly embedded into the inversion of the implicit operator. A detailed comparison of the different treatment of the implicit boundary conditions and their influence on robustness, accuracy and convergence will be reported elsewhere.

To cope with the oscillations near strong shock waves, we use an adaptive artificial viscosity in consevertive form similar to Jameson[6].

Multigrid procedure and coarse grid scheme

We define successively coarser grids by successive deletion of every other line in each coordinate direction. The multigrid cycling strategy used here is the sawtooth cycling, which is illustrated in Fig. 1 for a three-level multigrid procedure. The coarse grid scheme is a one-step Lax-Wendroff explicit scheme expressed in finite volume integration form as distribution formulae by Ni[3]. For our problems, the distribution formulae can be written as

$$\begin{aligned}\delta U_{i,j} = &(1/4)[\Delta U + \Delta t(\ \Delta F/\Delta\theta + \Delta G/\Delta\phi + \Delta H)]_a \\ &+(1/4)[\Delta U + \Delta t(-\Delta F/\Delta\theta + \Delta G/\Delta\phi + \Delta H)]_b \\ &+(1/4)[\Delta U + \Delta t(\ \Delta F/\Delta\theta - \Delta G/\Delta\phi + \Delta H)]_c \\ &+(1/4)[\Delta U + \Delta t(-\Delta F/\Delta\theta - \Delta G/\Delta\phi + \Delta H)]_d\end{aligned} \quad (3)$$

where a,b,c and d represent the four control volumes surrounding the grid point (i,j).

Based on physical argument[4], the coarse grid scheme only involves inviscid Jacobians, which makes the procedure more efficient. That is, in Eq.(3),

$$\Delta F = (\partial F_I/\partial U)\Delta U, \quad \Delta G = (\partial G_I/\partial U)\Delta U, \quad \Delta H = (\partial H_I/\partial U)\Delta U$$

For stability a local variable time step, which is determined by the local inviscid stability condition, is used in the coarse grid scheme.

$$\Delta t = CFL \min[(\Delta\theta/(|v|+c)), \sin\theta\Delta\phi/(|w|+c)], \quad CFL \leq 1$$

The restriction and prolongation operators are respectively full-weighting average and bilinear interpolation.

Results

We examine the efficiency of different methods by studying a simpler case. Fig.2 compares the convergence histories of the explicit and implicit methods with and without multigrid acceleration by plotting the maximum residuals against work units. One work unit is defined as the CPU time needed for the explicit MacCormack scheme on the finest grid per time step. In this case we use a 65×9 mesh which is stretched to have a fine grid near the wall. For the implicit method a maximum CFL number of about 30 is used. The efficiency of the implicit operator is quite clear as compared to the performance of the corresponding explicit method. Also shown in the picture is the further acceleration achieved by the multigrid methods, which are found to give more benefit when combined with the implicit method than that with the explicit method in this case.

We have numerically simulated a variety of flow problems around conical shapes, where viscous effects and their interaction with shock waves are significant, including supersonic and hypersonic flows around cones at high angle of attack, delta-wings with subsonic and supersonic leading edges and cone-delta-wing combinations at various angles of attack. The numerical results compare favourably with published experimental information. Further insight into the hypersonic flow around a cone-delta-wing combination has been gained.

It had been expected that lifting effectiveness would be derived from a favourable interference of the cone pressure field with the delta wing lower surface. However, attempts to make use of it had been complicated by the boundary layer interaction with strong shock waves and the difficulty in predicting the resulting separation, reattachment and vortex flow[7]. The shapes similar to it therefore became unattractive to the design engineer. Numerical experiments using the method outlined provide an opportunity to reexamine the idea.

The cone-delta-wing configuration studied experimentally by Meyer and Vail[7] is chosen. Fig.3 shows the sketch of the geometry and the solution plane. The computation is made at the spherical plane of r = 3.7in corresponding to the experimental measurement. The mesh of 65x65, illustrated in Fig.4, is stretched in both θ and ϕ directions to ensure sufficient resolution of the viscous effects. A three-level multigrid procedure is used. The initial flow field is set to the flow properties of infinity. In Fig.5, 6 and 7 a case at 5° angle of attack is selected for demonstration. At this small angle of attack, the flowfield is characterised by the interaction of the cone shock wave and the wing shock wave and the further interaction of the resulting flow with the boundary layer beneath the wing. In Fig.5(a) the cone shock and the wing shock and their interaction can be seen as a sharp transition in the velocity direction and/or magnitude under the surface. The computation simulates the significant primary separation beneath the wing given by the experiment. A striking result however is that the flow field is richer in detail than Meyer and Vail had expected based on their wall measurement. A secondary separation beneath the wing and a vortex rolling up from the cone wall are predicted as illustrated in Fig.5(b). In the cross-flow Mach contour, Fig.6, the "three-shock configuration" is well represented, while the contact discontinuity in inviscid theory is substituted by a viscous shear layer. And an embedded shock standing on the primary vortex due to the reaccelerated supersonic cross-flow can be clearly seen. Fig.7 compares the computed wall pressure distribution with the wall measurement.

Concluding Remarks

A multigrid procedure has been used to speed up an implicit scheme. From numerical experiments on Navier-Stokes equations, we find the coarse grid scheme can further accelerate the convergence of the implicit method even though a fairly large CFL number had already been used in the basic implicit method.

Numerical simulation of hypersonic flow beneath a conical wing-body combination provides a further insight into the complex flow field resulting from interaction among shock waves and strong viscous effects.

Acknowledgment

We wish to thank Mr. D.C. Jiang for helpful discussion on the implicit method and Prof. C.P. Cao of NAI for instructive suggestions. This work was supported by a Postgraduate Scholarship of Glasgow University to the first author.

References

[1] MacCormack, R.W., *AIAA J*, v 20, n 9, 1275-1281 (1982)
[2] Mulder, W.A. and van Leer, B., *J Comp Phys* v 59, 232-246 (1985)
[3] Ni, R.H., *AIAA J*, v 20, n 11, 1565-1571 (1982)
[4] Johnson, G. M., *Appl Math Comp*, v 13, 375-398 (1983)
[5] MacRae, D.S. and Hussaini, M.Y., *AGARD-CP-*247 paper 23 (1979)
[6] Jameson, A., *Lecture Notes in Math*, v 1127 156-242 (1985)
[7] Meyer, R.F. and Vail, C.F., *NAC, NAE Aero Report LR-*475 (1967)

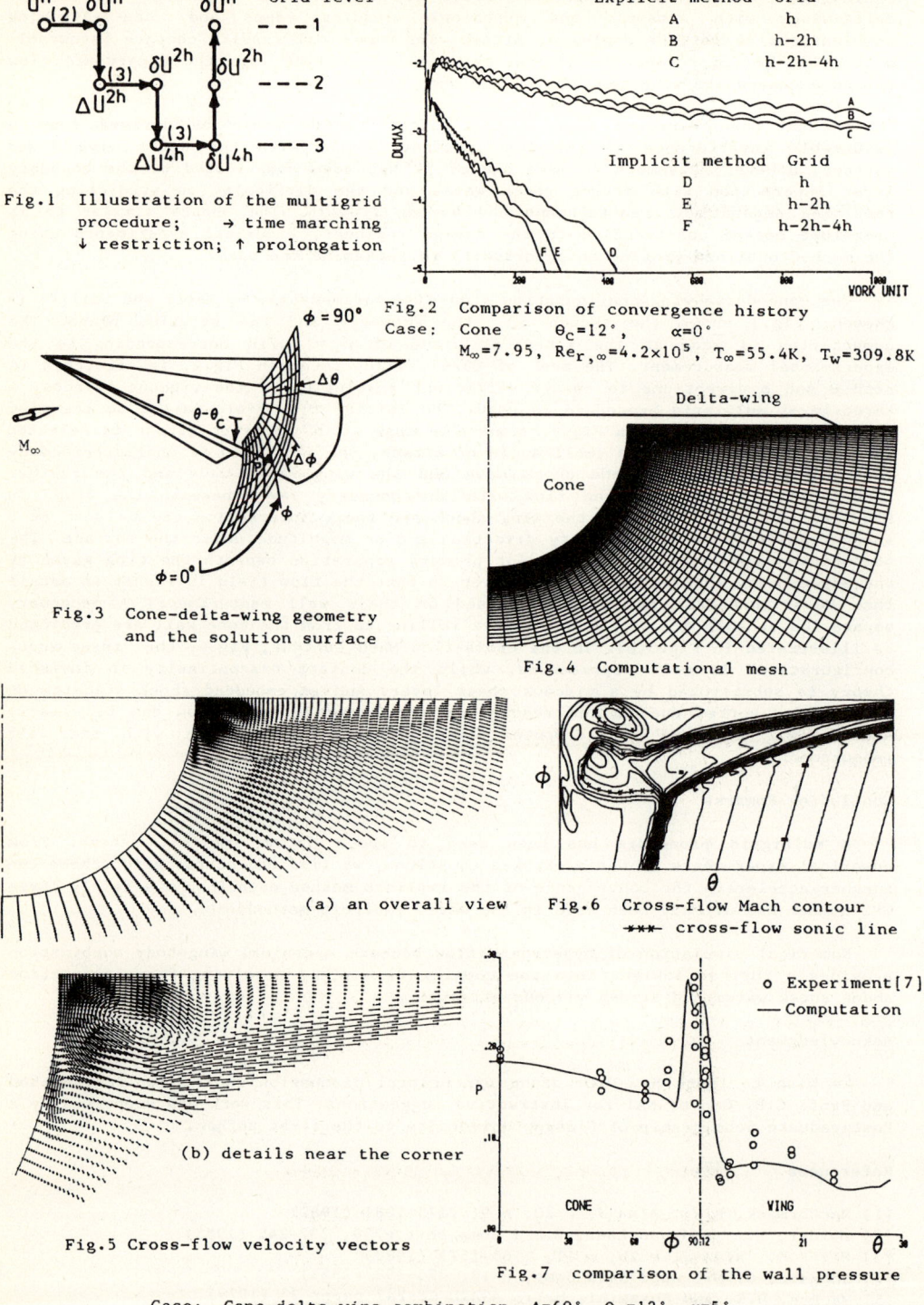

Fig.1 Illustration of the multigrid procedure; → time marching ↓ restriction; ↑ prolongation

Fig.2 Comparison of convergence history
Case: Cone $\theta_c=12°$, $\alpha=0°$
$M_\infty=7.95$, $Re_{r,\infty}=4.2\times10^5$, $T_\infty=55.4K$, $T_w=309.8K$

Fig.3 Cone-delta-wing geometry and the solution surface

Fig.4 Computational mesh

(a) an overall view

Fig.6 Cross-flow Mach contour
-✱✱✱- cross-flow sonic line

(b) details near the corner

Fig.5 Cross-flow velocity vectors

Fig.7 comparison of the wall pressure

Case: Cone-delta-wing combination $\Lambda=60°$, $\theta_c=12°$, $\alpha=5°$
$M_\infty=12.65$, $Re_{r,\infty}=3.78\times10^5$, $T_\infty=30K$, $T_w=160K$

VISCOUS PRESSURE WAVE BOUNDARY LAYER INTERACTION

H. Reister and D. Schwamborn
Institute for Theoretical
Fluid Mechanics
DFVLR-AVA Göttingen

The trailing edge boundary layer separation on airfoils induces upstream propagating pressure waves. Of particular interest is the unsteady viscous interaction with the boundary layer, which may cause unsteady separation on the airfoil. A fourth-order Runge-Kutta finite volume method was developed, solving the compressible Navier-Stokes equations in integral form. The laminar flat plate boundary layer was used as a test case for developing the method as well as for interpreting the fundamental physical mechanisms of the acoustic and viscous interaction.

Numerical method

A finite volume method is used to calculate the pressure wave boundary layer interaction. We start from the integral form of the two-dimensional compressible Navier-Stokes equations.

$$\frac{\partial}{\partial t} \int_V \vec{U} dV + \int_O \bar{\bar{H}} \vec{n} dO = 0$$

$\vec{U}$ is the solution vector and $\bar{\bar{H}}$ describes the flux of the respective fundamental quantities through the surface O of the volume V under consideration.

$$\vec{U} = \begin{pmatrix} \rho \\ \rho u \\ \rho v \\ e \end{pmatrix} \quad \bar{\bar{H}} = \begin{pmatrix} \rho \vec{v} \\ \rho u \vec{v} + (p+\tau_{xx})\vec{i}_x + \tau_{xy}\vec{i}_y \\ \rho v \vec{v} + \tau_{xy}\vec{i}_x + (p+\tau_{yy})\vec{i}_y \\ (e+p)\vec{v} + (\tau_{xx}u + \tau_{xy}v)\vec{i}_x + (\tau_{xy}u + \tau_{yy}v)\vec{i}_y + \vec{q} \end{pmatrix}$$

The principle of the finite volume method consists in the division of the entire flowfield into volume elements, for the two-dimensional case in surface elements, and the discretisation of the integral conservation laws for each volume element V_{ij}.

$$\frac{d}{dt}(\vec{U}_{ij}) V_{ij} + \sum_{k=1}^{4} (\bar{\bar{H}}^{(k)} \vec{O}_k)_{ij} = 0$$

Various approximations of the fluxes $\bar{\bar{H}}^{(k)}$ through the element sides $\vec{O}_k$ as well as various methods for the integration of the resulting systems of ordinary differential equations in time direction result in a number of different finite volume methods.

It is of crucial importance that the temporal development of the flow is adequately numerically simulated. We therefore used the Runge-Kutta time-stepping method introduced by JAMESON et al. [1], which is fourth-order accurate in time and where the fluxes are averaged from neighbouring cells.

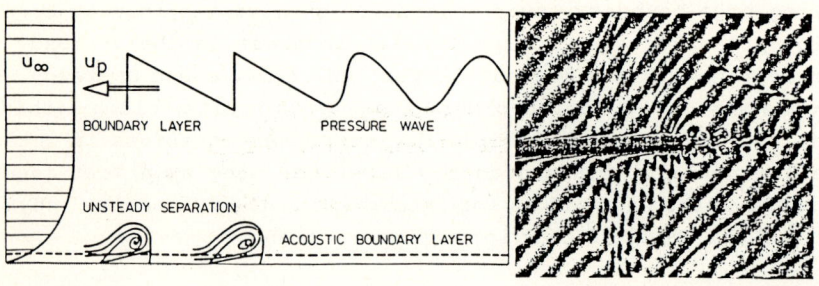

Principle Sketch of the Interaction Domain

Unsteady Pressure Wave Boundary Layer Interaction

Fig. 1 Experimental Trailing Edge induced Pressure Waves moving Upstream.

Pressure wave boundary layer interaction

Pressure waves moving upstream are the result of periodic separation at the trailing edge and vortex shedding. The experiment in fig. 1 (OERTEL [2]) shows vortex shedding behind a plate and pressure waves propagating upstream with a relative velocity $U_\infty - a_\infty$. The amplitude of the pressure waves is only a small percentage of the undisturbed velocity. The principle sketch of fig. 1 demonstrates the model for the numerical simulation performed to investigate the viscous interaction of the periodic pressure waves with the laminar boundary layer on a flat plate. The pressure waves are imposed at the right-hand boundary of the integration domain. Depending on their amplitude, frequency, and the free-stream Mach Number M_∞, they can steepen to a sawtooth pattern in the inviscid outer field. At the wall, these waves are damped in the acoustic boundary layer. A periodic unsteady separation takes place due to the interaction with the viscous boundary layer, which is larger by one order of magnitude compared with the acoustic boundary layer.

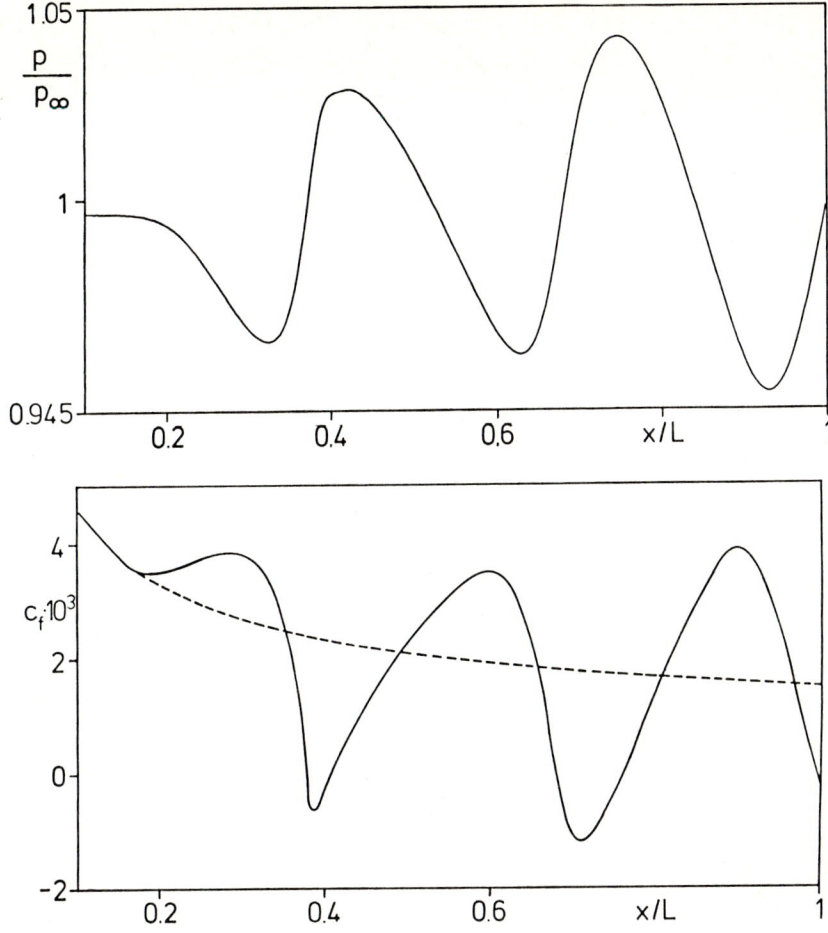

Fig. 2 Unsteady Separation.
--- undisturbed flat plate solution
$M_\infty = 0.5$, $Re = 2 \cdot 10^5$, $(p_2-p_\infty)/p_\infty = 0.056$

For the numerical simulation of the pressure wave boundary layer interaction, the steady laminar flat plate boundary layer solution of CHAPMAN and RUBESIN [3] was taken as an initial distribution. This solution was maintained at the inflow boundary. At the right-hand boundary we impose the pressure waves by sinusoidal variation of the pressure. For the remaining variables we studied various extrapolations. We found that the numerical disturbances are very small compared with the amplitudes of the physical waves if we set the third derivatives to zero. First or second derivatives being zero did not allow the waves to travel into the integration domain without strong numerical disturbances. At the upper boundary we set the first derivatives of all variables to zero. Once the waves had reached the inflow boundary, the calculation was stopped. No numerical dissipation was added for this calculation. Fig. 2 shows the

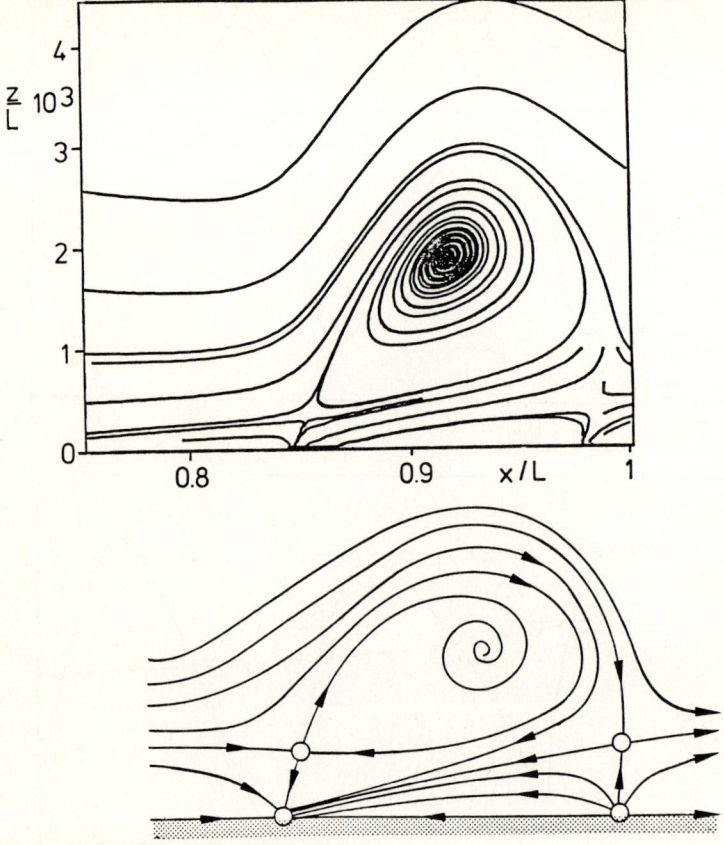

Fig. 3 Streamlines and Topological Structure.

results of the viscous interaction with the boundary layer. The wall pressure distribution and the dimensionless wall shear stress are plotted. The wall shear stress periodically becomes negative, which indicates unsteady separation. If we integrate the instantaneous streamlines in the interaction region for a free-stream Mach number of 0.5 and a pressure amplitude of 0.112 (fig. 3) we find that the structure is determined by two nodes, a source and a sink, which show a back flow area at the wall, two saddle points, and a focus. This unsteady separation structure remains unchanged with increasing distance from the trailing edge. In fig. 4 the critical amplitudes of the pressure waves, at which the unsteady separation sets in, are shown. The results of the Navier-Stokes calculations are compared with a linear superposition of the initial distribution of CHAPMAN and RUBESIN [3] and the acoustic boundary layer solution of CREMER [4]. In the Navier-Stokes solution the velocity-amplitudes are increasing in the boundary layer due to the interaction and thus the critical amplitudes are smaller compared with those resulting from the linear superposition.

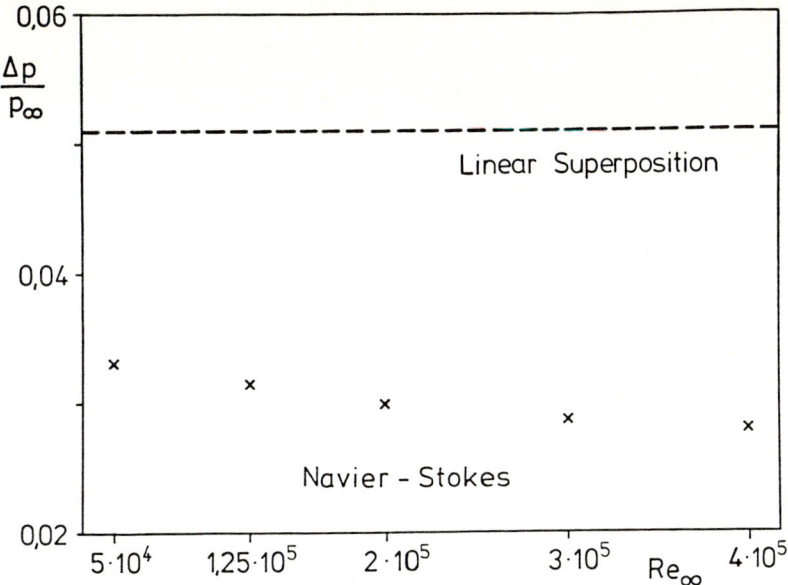

Fig. 4 Critical Pressure Wave Amplitude.

References

[1] JAMESON, A., SCHMIDT, W., TURKEL, E., Numerical Solution of the Euler Equations by Finite Volume Methods Using Runge-Kutta-Time-Stepping Schemes, AIAA Paper 81-1259 (1981).

[2] OERTEL, H., Vortices in Wakes Induced by Shock Waves, Proc. of the 14th Int. Symp. on Shock Tubes and Waves, pp. 293-300 (1983).

[3] CHAPMAN, D.R., RUBESIN, M.W., Temperature and Velocity Profiles in the Compressible Laminar Boundary Layer with Arbitrary Distribution of Surface Temperature, J. Aeronautical Sciences, Vol. 16, pp. 547-565 (1946).

[4] CREMER, L., Über die akustische Grenzschicht vor starren Wänden, A.E.Ü, 2, S. 136-139 (1948).

SOME NONSTANDARD FINITE ELEMENT METHODS FOR THE
NUMERICAL SOLUTION OF VISCOUS FLOW PROBLEMS

V. Ruas
Departamento de Informática
Pontifícia Universidade Católica
22.453 Rio de Janeiro, Brazil

1 - PARAMETRIZED DEGREES OF FREEDOM

Let us first consider the variational formulation of the incompressible stationary Navier-Stokes equations defined in a three-dimension flow region Ω, in terms of primitive variables $(\vec{u},p)$, with admissible non square integrable pressure gradients.

As pointed out at the 9th ICNMFD[1] if one wishes to construct optimally convergent tetrahedral finite element methods of solution, one must use nonstandard degrees of freedom called parametrized, in conjunction with a nonconforming velocity. Details can be found in [2].

In this work we show that a similar technique enables us to construct a particularly simple quadratic element for fourth order problems, such as the potential vector formulation of the Navier-Stokes equations studied in [3], related to the biharmonic equation modified by the convective term. This element can be viewed as the three-dimensional version of Morley's element, introduced for solving biharmonic problems in 2-D space [4]. Notice that this element is particularly suitable for the stream function formulation of the Stokes problem or of its natural extension to the Navier-Stokes equations. Such a formulation is precisely the 2D analogue of the potential vector formulation that we consider in this paper.

As we should recall, a <u>parametrized degree of freedom</u> is any linear functional defined on a space polynomials, that is a fixed linear combination of usual degrees of freedom, that is to say, functional values of derivatives at given points, or yet mean values of functions or derivatives over faces, edges, etc. The sum of the parameters defining such a linear combination is one, and there are at most $int(k/2)$ (resp,k-1) of them, k being the degree of the restriction of the polynomial over the edge (resp. face) to which the parametrized degree of freedom is attached. In such a way one increases the number of coefficients to be determined in order to define the associated basis func tions, and to satisfy minimum continuity requirements at interelement boundaries, due to nonconformity, as specified below.

2. MINIMUM CONTINUITY REQUIREMENTS

Let $H^k(D)$ be the Sololev space of functions defined in an open set $D \subset \mathbb{R}^3$ with the usual norm $\|.\|_{H^k(D)}$ (see e.g. [5]), for a certain integer k, and m be the index corresponding to half an order of the problem under study. Here we confine ourselves to the case where the problem is a <u>linear elliptic equation in Ω, involving no differentiation of order lower than 2m, with Dirichlet boundary conditions, for a polyhedric Ω</u>, in order to avoid non essential difficulties. Other cases can be treated in a similar way.

Let now V_h be the finite dimensional space corresponding to the finite element method being used to solve the problem. V_h consists of piecewise polynomials of degree $\ell-1$, over each element of a partition τ_h of Ω, belonging to a regular family of partitions $(\tau_h)_h$ into tetrahedrons with maximum edge length h.

The approximation properties of space V_h are such that, for every function $u \in H^\ell(\Omega)$, $\ell \geq 2m-1$, we have:

$$\inf_{v_h \in V_h} \|u - v_h\|_{m,h} \leq Ch^{\ell-m} \|u\|_{H^\ell(\Omega)}$$

where C is a constant independent of h and

$$\|v\|_{m,h} = \left[\sum_{K \in \tau_h} |v|^2_{H^m(K)} \right]^{1/2} \qquad (1)$$

Let now v' and v'' be either the respective restrictions of a function $v \in V_h$ to two tetrahedrons K' and K'' having a common face F, or the restriction of $v \in V_h$ to a tetrahedron having a face F on the boundary of Ω and the zero function respectively. Then if the following conditions are satisfied:

(i) $\|.\|_{m,h}$ defined by (1) is a norm for V_h;

(ii) $\int_F (\partial^\beta v' - \partial^\beta v'') q = 0 \qquad \forall q \in P_{\ell-|\alpha|-1}(F) \qquad$ provided $\ell \geq |\alpha|+1$;

(iii) $\partial^\beta v' = \partial^\beta v''$ if $\partial^\beta v''$ and $\partial^\beta v''$ are polynomials of degree $m-1-|\beta|$

where α and β are tripleindices $\alpha = (\alpha_1, \alpha_2, \alpha_3)$, $\beta = (\beta_1, \beta_2, \beta_3)$ with $\alpha_i, \beta_i \geq 0$, $|\alpha| = \alpha_1+\alpha_2+\alpha_3$, $|\beta| = \beta_1+\beta_2+\beta_3$ and $|\alpha|+|\beta| = 2m-1$, $m \leq |\alpha| \leq 2m-1$, $0 \leq \beta \leq m-1$, $\partial^\alpha v$ denoting the partial derivative $\partial^{|\alpha|} v / \partial x_1^{\alpha_1} \partial x_2^{\alpha_2} \partial x_3^{\alpha_3}$, a global error estimate of the same order $\ell-m$ in the $\|.\|_{m,h}$ norm, holds for the approximate solution in V_h, if the exact solution belongs to $H^{\max(\ell, 2m)}(\Omega)$.

<u>Remark 1</u>: Some partial derivatives can be eliminated from conditions (ii) and (iii) according to the used variational formulation. □

Conditions (i) and (iii) are easily verified for most of noncon forming methods, while (ii) is usually the problematic one. The use of free parameters in the degrees of freedom themselves, introduces the ad ditional coefficients needed to satisfy this condition in some cases.

For example, we recall here the essential features of the second order method studied in [2] for the case of the $(\vec{u},p)$ formulation: The velocity was taken quadratic over each tetrahedron with vertices S_1, S_2, S_3, S_4, with degrees of freedom D_n associated with face F_n opposite to S_n, and D_{ij} associated with edge $\overline{S_i S_j}$. The respective basis functions are of the forms:

$$\phi_n = \gamma[a(\lambda_i + \lambda_j + \lambda_k) + b\lambda_n + c\lambda_n^2 + d(\lambda_i\lambda_j + \lambda_i\lambda_k + \lambda_j\lambda_k)] \quad (2)$$

$$\phi_{ij} = \delta[e_k\lambda_k + e_n\lambda_n + f(\lambda_i + \lambda_j) + g_k\lambda_k^2 + g_n\lambda_n^2 + p(\lambda_i^2 + \lambda_j^2) + q\lambda_i\lambda_j] \quad (3)$$

where i,j,k and n are distinct integers between one and four, and λ_i de notes the barycentric coordinate with respect to vertex S_i.

Here γ and δ must be adjusted so that $D_n(\phi_n) = D_{ij}(\phi_{ij}) = 1$, and due to symmetry $e_k = e_n$ and $g_k = g_n$. Thus we had only nine coefficients to be determined, but using one additional parameter in each type of degree of freedom, we could satisfy (ii) and (iii) with $\ell=3$ and $m=1$ as required.

3. THE POTENTIAL VECTOR FORMULATION OF THE NAVIER-STOKES EQUATIONS

For the sake of conciseness we consider the formulation corres ponding to potential vectors with zero trace of its tangential com ponent over the boundary Γ of Ω, and only the case where Ω is simply connected.

We wish to solve the following problem.
Find a potential vector field $\vec{\phi}$ such that,

$$(E_1) \begin{cases} -\nu\Delta^2 \vec{\phi} + \vec{rot}((\vec{rot}\vec{\phi} \cdot \vec{\nabla}) \vec{rot}\vec{\phi}) = \vec{rot} \vec{f} \\ \text{div } \vec{\phi} = 0 \end{cases} \text{ in } \Omega$$

$$\begin{cases} \vec{\phi} \wedge \vec{\nu} = \vec{0} \quad \vec{\nu} \text{ being the unit outer normal vector} \\ \vec{rot}\vec{\phi} \wedge \vec{\nu} = \vec{0} \end{cases} \text{ on } \Gamma$$

$\vec{f}$ being a given force field and ν the viscosity.
The velocity of the fluid is given by $\vec{u} = \vec{rot}\vec{\phi}$

By introducing the space,
$$V = \{\vec{\psi}/\vec{\psi}\epsilon[H^2(\Omega)]^3, \vec{\psi} \wedge \vec{\nu} = \vec{rot}\vec{\psi} \wedge \vec{\nu} = \vec{0} \text{ and div } \vec{\psi} = 0 \text{ on } \Gamma\}$$
the above fourth order equation in variational form writes:

$$(P_1) \begin{cases} \text{Find } \vec{\phi}\epsilon V \text{ such that} \\ a(\vec{\phi}, \vec{\phi}, \vec{\psi}) = \int_\Omega \vec{f} \cdot \vec{rot} \psi \quad \forall \vec{\psi} \epsilon V \end{cases}$$

where

$$a(\vec{\chi},\vec{\phi},\vec{\psi}) = \nu \int_\Omega \Delta\vec{\phi}\cdot\Delta\vec{\psi} + \int_\Omega (\vec{\text{rot}}\,\vec{\chi}\cdot\vec{\nabla})\,\vec{\text{rot}}\,\vec{\phi}\cdot\vec{\text{rot}}\,\vec{\psi}$$

A finite element method that works for problem (P_1) should necessarily work for the scalar biharmonic equation in a polyhedron Ω with Dirichlet boundary conditions, namely:

$$(E_2) \begin{cases} \Delta^2 u = f \text{ in } \Omega \text{ for a given } f \in L^2(\Omega) \\ u = \dfrac{\partial u}{\partial \nu} = 0 \text{ on } \Gamma \end{cases}$$

For the sake of simplicity we consider at first only the essential aspects of the scalar version of the element. The vectorial one can be treated as a rather natural, though non trivial extension of it, as pointed out later on. A detailed analysis of the scalar case can be found in [6].

4. A 3D VERSION OF MORLEY'S ELEMENT

Let us first pose an academic problem, which is a variational form for equation (E_2). This variational form is analogous to the one equivalent to the two dimensional biharmonic equation, to be used in connection with certain non conforming finite element methods such as Morley's element (see e.g. [7]). This is also the case of our new element.

$$(P_2) \begin{cases} \text{Find } u \in H_0^2(\Omega) \text{ such that} \\ a(u,v) = \int_\Omega fv \quad \forall v \in H_0^2(\Omega), \text{ where} \end{cases}$$

$$a(u,v) = \nu \left[\sigma \int_\Omega \Delta u \Delta v + (1-\sigma) \int_\Omega \left(\sum_{i,j=1}^{3} \frac{\partial^2 u}{\partial x_i \partial x_j} \frac{\partial^2 v}{\partial x_i \partial x_j} \right) \right] \quad (4)$$

for any $0 \leq \sigma < 1$, and $H_0^2(\Omega) = \{v/v \in H^2(\Omega), v = \dfrac{\partial v}{\partial \nu} = 0 \text{ on } \Gamma\}$

In our case $m=2$, and approximation result (1) holds with $\ell=3$, if V_h consists of piecewise complete quadratic polynomials.

Hence, according to the a priori arguments given in Section 2, if one wishes to approximate (P_2) by a first order convergent nonconforming finite element method, it is necessary to take a space V_h that, besides being normed by $\|\cdot\|_{2.h}$, satisfies (ii) and (iii). This means that for every face F of the partition τ_h (refer to Section 2),

(ii) $\int_F \vec{\nabla} v' = \int_F \vec{\nabla} v''$

(iii) $\begin{cases} v'/_F = v''/_F \text{ whenever } v' \text{ and } v'' \text{ are linear functions over F} \\ \vec{\nabla} v'/_F = \vec{\nabla} v''/_F \text{ whenever } \vec{\nabla} v' \text{ and } \vec{\nabla} v'' \text{ are constant over F} \end{cases}$

Notice that conditions (iii) are automatically fulfilled if the finite element method has three degrees of freedom associated with functional values attached to each face F and one degree of freedom associated with the first order normal derivative over each face. Moreover, this allows the immediate fulfillment of (i).

From these remarks it is readily seen that the natural choice for the degrees of freedom of the quadratic element is the following:

- $D_n(v) = \frac{\partial v}{\partial \nu}(G_n)$: Normal derivative at the barycenter G_n of face F_n
- $D_{ij}(v) = \lambda v(M_{ij}) + (1-\lambda) \int_{S_i}^{S_j} v / \text{lenght}(\overline{S_i S_j})$, M_{ij} being the mid-point

of $\overline{S_i S_j}$, i,j,k and n being four distinct integers between one and four.

Here λ is the only parameter of degree of freedom D_{ij}, to be determined in such a way that (ii) also holds. Recalling (2) and (3) we have:

$\frac{1}{\gamma} \vec{\nabla} \phi_n = a(\vec{\nabla}\lambda_i + \vec{\nabla}\lambda_j + \vec{\nabla}\lambda_k) + b \vec{\nabla}\lambda_n + 2c\lambda_n \vec{\nabla}\lambda_n + 2d(\lambda_i \vec{\nabla}\lambda_i + \lambda_j \vec{\nabla}\lambda_j + \lambda_k \vec{\nabla}\lambda_k)$

$\frac{1}{\delta} \vec{\nabla} \phi_{ij} = e_k \vec{\nabla}\lambda_k + e_n \vec{\nabla}\lambda_n + f(\vec{\nabla}\lambda_i + \vec{\nabla}\lambda_j) + 2g_k \lambda_k \vec{\nabla}\lambda_k + 2g_n \lambda_n \vec{\nabla}\lambda_n + 2p(\lambda_i \vec{\nabla}\lambda_i + \lambda_j \vec{\nabla}\lambda_j) + q(\lambda_i \vec{\nabla}\lambda_j + \lambda_j \vec{\nabla}\lambda_i)$.

Thus the equations to be satisfied by the twelve coefficients, $a,b,c,d,e_n,e_k,f,g_n,g_k,p,q$ and λ are:

$\left. \begin{array}{l} D_{in}(\phi_n) = 0 \\ D_i(\phi_n) = 0 \end{array} \right\}$ Unisolvence $\quad \left\{ \begin{array}{l} D_k(\phi_{ij}) = D_n(\phi_{ij}) = 0 \\ D_{ik}(\phi_{ij}) = D_{in}(\phi_{ij}) = 0 \\ D_{kn}(\phi_{ij}) = 0 \end{array} \right.$

$\left. \begin{array}{l} \vec{\nabla}\phi_n(G_i) = \vec{0} \\ \\ \text{(two equations)} \end{array} \right\}$ Unisolvence and condition (ii) $\quad \left\{ \begin{array}{l} \vec{\nabla}\phi_{ij}(G_i) = \vec{0} \\ \\ \text{(three equations)} \end{array} \right.$

The resulting system of twelve homogeneous equations admits a solution for which $\lambda=0$. Hence, adjusting γ and δ in order to have

$D_n(\phi_n) = 1$ and $D_{ij}(\phi_{ij}) = 1$, we obtain:

$$\boxed{\phi_n = (\lambda_n - \frac{3}{2}\lambda_n^2)/\alpha_{nn}}$$

$$\boxed{\phi_{ij} = -2(\lambda_i + \lambda_j) + 3(\lambda_i + \lambda_j)^2 + 2\phi_k(\alpha_{kn} + \alpha_{kk}) + 2\phi_n(\alpha_{nk} + \alpha_{nn})}$$

where $\alpha_{\ell m} = D_\ell(\lambda_m)$, $1 \le \ell, m \le 4$.

As for the vector version V_h of the space constructed upon this element, let us just make a remark on the two crucial issues:

1) $\|\cdot\|_{2,h}$ is a norm for V_h if we adapt the boundary conditions of V in the natural way, namely, $D_{ij}(\vec{v}) \wedge \vec{v} = (\vec{\mathrm{rot}}\,\vec{v} \wedge \vec{v})(G_n) = \vec{0}$ and $\mathrm{div}\,\vec{v}(G_n) = 0$, $\forall \vec{v} \in V_h$, whenever S_i, S_j and G_n lie on Γ.

2) In the case of (E_2), the modified variational formulation (P_2) involving a non physical coefficient σ had to be introduced, in order to guarantee that the corresponding form $\underline{a}$ is coercive over V_h equiped with the $\|\cdot\|_{2,h}$ norm. For equation (E_1), an analogous modified variational formulation has to be used, but with a mesh dependent coefficient σ, and coerciveness applies to vectorial V_h with a constant of order $\sqrt{h}$. This leads to convergence results of the same order in the vectorial $\|\ \|_{2,h}$ norm as described in a forthcoming paper [8].

Nevertheless the author conjectures that a completely equivalent numerical analysis to the one sketched in this paper, leading to first order convergence for equation (E_2), also, applies to the potential vector equation (E_1). However, such a finer and complete analysis of the latter case is still a subject for further investigation.

REFERENCES

[1] - V.RUAS, Nonconforming 3D analogues of conforming triangular finite element methods in viscous flow, in Ninth Int. Conf. on Num. Meth. in Fluid Dynamics, Ed. Soubbaramayer, Boujot, Springer Verlag, Berlin-Heidelberg-New-York, 465-469, 1985.

[2] - V.RUAS, Finite element solution of 3D viscous flow problems using nonstandard degrees of freedom, Japan Journal of Applied Mathematics, 2-2 (1985), 415-431.

[3] - S.GALLIC, Système de Stokes stationnaire em dimension 3; Formulation on ψ et formulation en u,p, dans le cas axisymétrique, Thè-

se de Doctorat 3ème cycle, Univ. Pierre et Maire Curie, Paris, 1982.

[4] - L.S.D. MORLEY, The triangular equilibrium element in the solution of plate bending problems, Acro-Quart. 19, (1968), 149-169.

[5] - R.A.ADAMS, Sobolev Spaces, Academic Press, N.Y., 1975

[6] - V. RUAS, A quadratic finite element method for solving biharmonic problems in $\mathbb{R}^n$, Monografia em Ciência da Computação PUC/RJ, 1/1986.

[7] - P.LASCAUX & P.LESAINT, Some nonconforming finite elements for the plate bending problem, RAIRO Analyse Numérique R1, (1975), 9-53.

[8] - V.RUAS, Solution of the 3D Navier-Stokes equations in potential vector formulation via the constant stress finite element method, submitted to the first International Conference on Computational Methods in Flow Analysis, to be held in Japan (1988).

EXACT SOLUTION OF NONLINEAR DIFFERENCE EQUATIONS FOR DISCRETE SHOCK WAVES

V.V. Rusanov
Keldysh Institute of Applied Mathematics
Moscow, USSR

Consider the equation

(1) $\quad u_t + [F(u)]_x = 0$

where $F(u)$ is continuous, differentiable function which satisfies the condition of convexity with respect to some interval (u_1, u_2):

(2) $\quad \{F(u) - F(u_1)\}/(u - u_1) > \{F(u_2) - F(u_1)\}/(u_2 - u_1)$

For (1) we consider the difference scheme on a floating grid with steps $\Delta x = h$ and $\Delta t = \tau$, $\varkappa = \tau/h$.

(3) $\quad u_x^{n+1} = u_x^n - (\varkappa/2)[F(u_{x+h}^n) - F(u_{x-h}^n)] + \Phi_{x+h/2}^n - \Phi_{x-h/2}^n$

$\quad u_x^h = u^n(x), \quad \Phi_{x+h/2}^n = (1/2)\,\Omega(\sigma_{x+h/2})(u_{x+h}^n - u_x^n)$

$\quad \sigma_{x+h/2} = \varkappa F'(u_{x+h/2}), \quad u_{x+h/2}^n = (u_{x+h}^n + u_x^n)/2, \quad \sigma^2 \leq \Omega(\sigma) \leq 1$

Let u^-, u^+ and D satisfy the Hugoniot conditions
$\quad F(u^+) - F(u^-) = D(u^+ - u^-), \quad F'(u^-) > D > F'(u^+), \quad \delta = \varkappa D$.

A discrete stationary shock wave is the function that satisfies the equation [1]:

(4) $\quad V_{x-\delta h} = V_x - (\varkappa/2)[F(V_{x+h}) - F(V_{x-h})] + \Phi_{x+h/2} - \Phi_{x-h/2}$

$\quad \Phi_{x+h/2} = (1/2)\,\Omega(\sigma_{x+h/2})(V_{x+h} - V_x), \quad \sigma_{x+h/2} = \varkappa F'(V_{x+h/2})$

and the boundary conditions

(5) $\quad \lim_{x \to \pm\infty} V_x = u^\pm$

The problem is to find the function V_x and study its properties. In this communication the solution of this problem is considered for a rational value $\delta = p/q$ and some special function $F(u)$. Assuming without loss of generality $h = q$ we fix a certain value of $x = x_0$. Then the points $x = x_0 + \ell$ where ℓ is any integer, form a grid, and $V_{x_0 + \ell}$ are values of a grid function $w_\ell(x_0)$ on this grid. We obtain for $w_\ell(x_0)$ the system of an infinite number of nonlinear difference equations:

(6) $\quad w_{\ell-p} = w_\ell - (\varkappa/2)[F(w_{\ell+q}) - F(w_{\ell-q})] + \Phi_{\ell+q/2} - \Phi_{\ell-q/2}$

$\quad \Phi_{\ell+q/2} = (1/2)\,\Omega(\sigma_{\ell+q/2})(w_{\ell+q} - w_\ell)$

If (4) has a continuous solution on the interval $(-\infty, \infty)$ then (6) should have an one-parameter family of solutions $w_\ell(x_o)$, such that $w_\ell(x_o+1) = w_{\ell+1}(x_o)$

The existence of the solution (6) is proved for the case when the scheme (3) is monotonic [1]. Numerical experiment shows that system (6) has a solution for a nonmonotonic scheme too. In this case the grid function w_ℓ is also nonmonotonic. It proves that for a special function $F(w)$ system (6) is equivalent to a nonlinear system with a finite number of unknowns and for that system the direct method of solution may be shown applicable to both monotonic and nonmonotonic schemes.

We determine $F(w)$ as

$$F(w) = \begin{cases} g^- w + f^-, & |w-w^-| \le \varepsilon_L \\ G(w), & \{|w-w^-| > \varepsilon_L \wedge |w-w^+| > \varepsilon_R\} \\ g^+ w + f^+, & |w-w^+| \le \varepsilon_R \end{cases}$$

where $\varepsilon_L, \varepsilon_R$ some small numbers

The function $G(w)$ is chosen so that the function $F(w)$ is continuous, differentiable and convex in the interval (w^-, w^+) and in any interval containing the interval (w^-, w^+).

By assuming the existence of w_ℓ satisfying (6) and the boundary conditions at $\pm\infty$, we obtain that there exist ℓ_L and ℓ_R such that $|w_\ell - w^-| \le \varepsilon_L$ for $\ell \le \ell_L$, $|w_\ell - w^+| \le \varepsilon_R$ for $\ell \ge \ell_R$. Then for $\ell \le \ell_L - q$ and $\ell \ge \ell_R + q$ we have, respectively,

(7) $w_{\ell-p} = a^\pm w_{\ell+q} + b^\pm w_\ell + c^\pm w_{\ell-q}$

where $a^\pm = (\omega^\pm - \sigma^\pm)/2$, $b^\pm = 1 - \omega^\pm$, $c^\pm = (\omega^\pm + \sigma^\pm)/2$

$\omega^\pm = \Omega(\sigma^\pm)$, $\sigma^\pm = æg^\pm$ the sign "$-$" stands for $\ell \le \ell_L - q$ and the sign "$+$" for $\ell \ge \ell_R + q$

A general solution of (7) may be written in the form

(8) $w_\ell = \sum_{k=1}^{k^\pm} A_k^\pm (\xi_k^\pm)^\ell + w^\pm$

where $\xi_k^\pm$ are roots of the equations

(9) $\xi^{q-p} = a^\pm \xi^{2q} + b^\pm \xi^q + c^\pm$

For the simlicity of presentation only we suppose that $a^\pm \ne 0$ and $c^\pm \ne 0$. Only the $\xi_k^\pm$ satisfying the conditions $|\xi_k^+| < 1$, $|\xi_k^-| > 1$ should be included in (8).

Analysing equation (9) and applying the Kreiss theorem [2] enable us to found that there are exactly $q - p = k^+$ roots of $|\xi_k^+| < 1$ and exactly $q + p = k^-$ roots of $|\xi_k^-| > 1$.

Now we consider the equations of system (6) for ℓ from $\ell_L - q + 1$ to $\ell_R + q - 1$, and call this set of equations the system S_o. We add to S_o equations (9) namely the S_- with the sign "$-$" for $\ell_L - 2q + 1 \le \ell \le \ell_L$

and the S_+ with "+" for $l_R \leq l \leq l_R + 2q - 1$. We obtain a system S of $l_R - l_L + 2q - 1$ equations for the same number of unknowns $w_{l_L+1}, \ldots, w_{l_R-1}, A_1^-, \ldots, A_{q+p}^-, A_1^+, \ldots, A_{q-p}^-$. A direct check shows that these equations are dependent and have an one-parametric set of solutions, as it should be. A specific character of the system S allows rather simple algorithms for its solution providing high accuracy. A particularly simple algorithm may be obtained if $q - p = 1$. In this case there is only one root ξ_1^+ and the coefficient A_1^+ may be taken as a parameter. By giving it we may successively calculate from right to left all w_l by using equations (6), and then calculate A_k^- from (8).

In a general case the quantity α_w determining "the middle point" of the profile may be taken as a parameter [1,3].
Let $\alpha^\lambda(l_1, l_2) = \sum_{l=l_1}^{l_2}(\lambda + lq + q/2)(w_{\lambda+lq+q} - w_{\lambda+lq})$
Then $\alpha_w^\lambda = \alpha^\lambda(-\infty, \infty)$, $\lambda = 1, 2, \ldots q$. It had been proved [3] that for the stationary solution the α_w^λ does not depend on λ. It is easy to found that

(10) $\alpha_w^\lambda = \sum_k A_k^- \Psi_k^-(\xi_k^-, q, \lambda) + \alpha^\lambda(l_L, l_R) + \sum_k A_k^+ \Psi_k^+(\xi_k^+, q, \lambda)$

The solution of system S for given α_w could be found using the following iterative process.

1) Let the value of w_l at iteration j be w_l^j, $l_L - 2q + 1 \leq l \leq l_R + 2q - 1$; 2) w_l^{j+1} for these l is computed using the difference scheme (3); 3) the predicted $\tilde{A}_k^\pm$ are determined solving equations (8); 4) the value of coefficients $A_k^\pm$ are corrected using (10) for the given value of α_w and some fixed λ.

The numerical experiment shows that the convergence of the iterative process is very good for monotonic solution and satisfactory for nonmonotonic. In the figures [1-4] several examples of discrete shock waves are shown for the function $G(w) = aw^2 + bw$, $w^- = 0$, $w^+ = 1$

Figure n	δ	σ^-	σ^+	ω
1	0,5	0,6	0,4	0,8
2	1/3	0,5	1/6	σ^2
3	1/3	0,5	1/6	0,6
4	2/3	5/6	0,5	σ^2

The method proposed for finding the exact solution of system [6] allows a qualitative as well as quantitative analysis of difference solutions for any function $F(w)$. It is due to the fact that $w \to w^\pm$ exponentially. Therefore even at ε_L and ε_R rather small, a number of nonlinear terms in the system, that depends on $l_R - l_L$ only, is not large. A number of linear terms depends on q only and always may be chosen not to large.

The method could be also applied to hyperbolic quasilinear systems and such general schemes as [4]:

(11) $w_m^{n+1} = w_m^n - æ\{H_{m+1/2} - H_{m-1/2}\}, \quad H_{m+1/2} = H(w_{m-\jmath_1+1}, \ldots, w_{m+\jmath_2})$

REFERENCES

1. V.V. Rusanov, and I.V. Bezmenov. 1981, Proceeding of the Mathematical Steklov Institute Ac.Sci.USSR, v.157, 178-190 (in Russian).
2. H.O. Kreiss. 1968, Math. Comp., v.22, 703-714.
3. V.V. Rusanov. 1975, Lecture Notes in Physics, v.35, 270-278.
4. A.Harten, J.M. Hyman and P.D.Lax. 1976, Comm. Pure Appl. Math., v.29, 297-322.

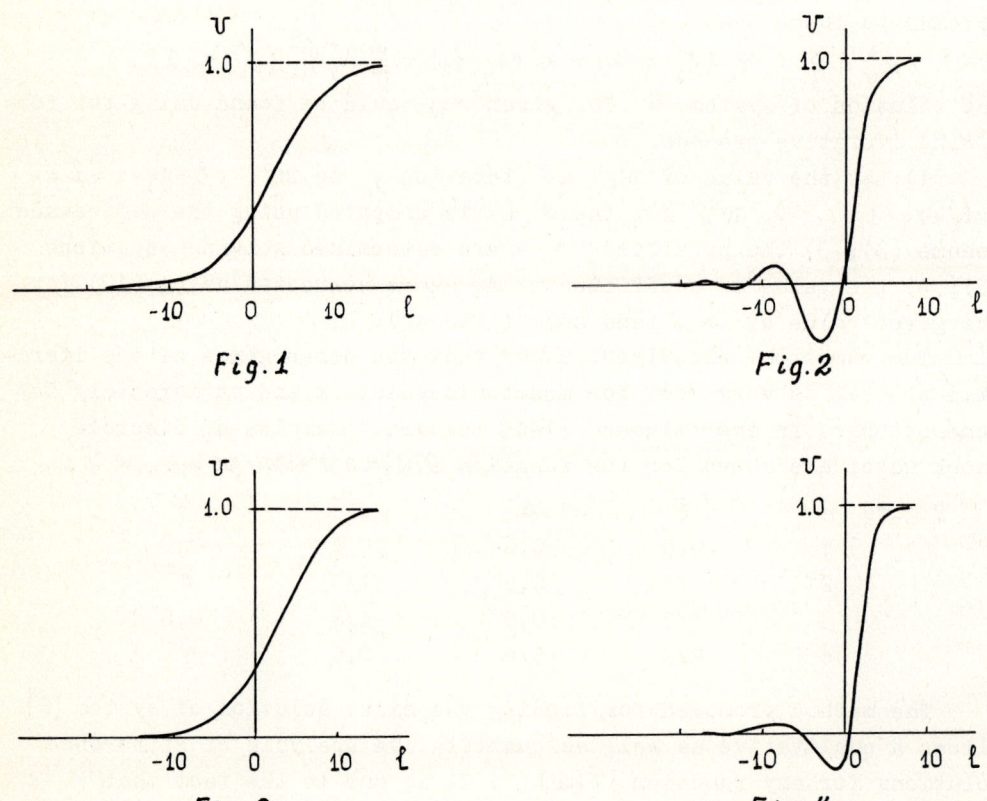

Fig.1   Fig.2   Fig.3   Fig.4

A CELL-VERTEX MULTIGRID SCHEME FOR SOLUTION OF THE
EULER EQUATIONS FOR TRANSONIC FLOW PAST A WING

* Deborah J Salmond
Aerodynamics Department
Royal Aircraft Establishment
Farnborough, Hants, U.K.

1 INTRODUCTION

A multigrid method is presented for solution of the three-dimensional Euler equations for the steady transonic flow past a wing. It is an extension to three-dimensional flow of Hall's[1] two-dimensional method for aerofoils.

The distinguishing features of Hall's[1] aerofoil method are its exceptional accuracy and speed. An important aim in the work described in this paper is to investigate the extent to which these features may be retained when the method is extended to three dimensions. This would provide a basis for deciding whether the method is suitable for further extension to provide the aircraft designer with a fast, accurate method for predicting aerodynamic characteristics of more complex configurations than a simple wing.

The method adopts a Lax-Wendroff time-marching algorithm in which the flux balance for each computational cell is approximated using values of the dependent variables specified at the cell vertices. Convergence is obtained to a solution which is second-order accurate, even at the boundaries and where the grid is highly stretched. This 'cell-vertex' scheme is different from the finite-volume scheme of Jameson[2] which has the dependent variables specified as cell averages, thus losing second-order accuracy at boundaries and for highly stretched grids. The method described in the present paper is explicit and convergence to the steady state is accelerated by proceeding in multigrid cycles. In each cycle the solution is updated on a sequence of successively coarser grids. On each grid the maximum time-step which still maintains stability is taken at each grid point. Convergence to the solution on the finest grid is ensured by formulating the problem so that coarse-grid changes to the solution vanish when changes on the previous grid vanish, and also by enforcing the boundary conditions on each grid.

Some new problems not encountered in Hall's two-dimensional method have had to be overcome. These are associated mainly with the development of a second-order accurate method for calculating the fluxes across a cell face and with the need to store and transfer large amounts of data in the course of a calculation.

---

*Present address Cray Research (UK) Ltd, Cray House, London Road, Bracknell, Berks, UK

2 BRIEF DESCRIPTION OF THE METHOD

2.1 Flow equations

Since we are seeking steady-state solutions of the Euler equations it is sufficient to consider the equations

$$\frac{\partial \underline{U}}{\partial t} + \frac{\partial \underline{F}}{\partial x} + \frac{\partial \underline{G}}{\partial y} + \frac{\partial \underline{H}}{\partial z} = 0 \quad , \tag{1}$$

where the quantities $\underline{U}, \underline{F}, \underline{G}$ and $\underline{H}$ are given by

$$\underline{U} = \begin{bmatrix} \rho \\ \rho u \\ \rho v \\ \rho w \end{bmatrix}, \quad \underline{F} = \begin{bmatrix} \rho u \\ \rho u^2 + p \\ \rho uv \\ \rho uw \end{bmatrix}, \quad \underline{G} = \begin{bmatrix} \rho v \\ \rho vu \\ \rho v^2 + p \\ \rho vw \end{bmatrix}, \quad \underline{H} = \begin{bmatrix} \rho w \\ \rho wu \\ \rho wv \\ \rho w^2 + p \end{bmatrix} \tag{2}$$

and the pressure, p is given by the Bernoulli equation

$$p = \frac{\rho}{\gamma} \left(1 - \frac{(\gamma-1)}{2} (u^2 + v^2 + w^2) \right) \quad . \tag{3}$$

Equation (1) describes a flow of constant enthalpy, the energy equation having been replaced by the Bernoulli equation. The quantities u, v and w are the velocity components in the cartesian space directions x, y and z respectively, ρ is the density and t is the time.

2.2 Cell-vertex scheme for the finest grid of each multigrid cycle

To derive the cell-vertex scheme we take an arbitrary grid of points in cartesian space. Firstly we expand $\underline{U}$ in a Taylor series

$$\delta \underline{U}^{n+1} = \left(\frac{\partial \underline{U}}{\partial t}\right)^n \delta t^{n+1} + \left(\frac{\partial}{\partial t}\left(\frac{\partial \underline{U}}{\partial t}\right)\right)^n \frac{(\delta t^{n+1})^2}{2} + O\left((\delta t^{n+1})^3\right) \quad , \tag{4}$$

where the superscript 'n' denotes the quantity evaluated at time t^n and

$$\left. \begin{array}{l} \delta \underline{U}^{n+1} = \underline{U}^{n+1} - \underline{U}^n \\ \delta t^{n+1} = t^{n+1} - t^n \end{array} \right\} \quad . \tag{5}$$

Substitution of equation (1) into equation (4) and changing the order of differentiation gives

$$\delta\underline{u}^{n+1} = \Delta_1\underline{u}^n \delta t^{n+1} + \Delta_2\underline{u}^n \frac{(\delta t^{n+1})^2}{2} \qquad (6)$$

where

$$\Delta_1\underline{u}^n = -\left\{\frac{\partial F}{\partial x} + \frac{\partial G}{\partial y} + \frac{\partial H}{\partial z}\right\}^n \qquad (7)$$

and

$$\Delta_2\underline{u}^n = -\left\{\frac{\partial}{\partial x}\left(\frac{\partial F}{\partial \underline{U}}\frac{\partial \underline{U}}{\partial t}\right) + \frac{\partial}{\partial y}\left(\frac{\partial G}{\partial \underline{U}}\frac{\partial \underline{U}}{\partial t}\right) + \frac{\partial}{\partial z}\left(\frac{\partial H}{\partial \underline{U}}\frac{\partial \underline{U}}{\partial t}\right)\right\}^n . \qquad (8)$$

The average first-order change, $\Delta_1\underline{u}^n \delta t^{n+1}$ for each cell is calculated by integrating over the cell to give

$$(\Delta_1\underline{u}^n)_{cell} \delta t^{n+1} = -\frac{\delta t^{n+1}}{V_{cell}} \int_{\substack{\text{over}\\\text{cell}}} \left\{\frac{\partial F}{\partial x} + \frac{\partial G}{\partial y} + \frac{\partial H}{\partial z}\right\}^n dV , \qquad (9)$$

where V_{cell} is the volume of the cell. The integral on the right-hand side of equation (9) is evaluated by reducing it (using the divergence theorem) to a sum of surface integrals which are evaluated numerically from values at the cell vertices. The second-order part $\Delta_2\underline{u}^n \frac{(\delta t^{n+1})^2}{2}$ is evaluated similarly. Smoothing terms are also added for stability, since the above discretisation does not have a directional bias.

2.3 Multigrid scheme applied to each of the coarser grids

For each of the coarser grids in the multigrid cycle the solution was advanced using the following formula, derived using Taylor series expansions

$$\delta\underline{u}^{n+1} = \Delta_1^M\underline{u}^n \left(\frac{\delta t^{n+1}}{\delta t^n}\right) + \Delta_2^M\underline{u}^n \left(\frac{(\delta t^{n+1})^2 + \delta t^{n+1}\delta t^n}{2}\right) , \qquad (10)$$

where

$$\Delta_1^M\underline{u}^n = \delta\underline{u}^n \qquad (11)$$

and

$$\Delta_2^M\underline{u}^n = -\left\{\frac{\partial}{\partial x}\left(\frac{\partial F}{\partial \underline{U}}\frac{\partial \underline{U}}{\partial t}\right) + \frac{\partial}{\partial y}\left(\frac{\partial G}{\partial \underline{U}}\frac{\partial \underline{U}}{\partial t}\right) + \frac{\partial}{\partial z}\left(\frac{\partial H}{\partial \underline{U}}\frac{\partial \underline{U}}{\partial t}\right)\right\}^n . \qquad (12)$$

The first-order and second-order parts of expression (10) are then calculated in a way similar to that of the fine grid.

3   NUMERICAL RESULTS

Numerical results for the ONERA M6 wing are presented and show that for a subcritical flow very good agreement is obtained with a well established full potential method[3]. Figure 1 shows a chordwise pressure distribution at one spanwise station from the present method run with 129 x 17 x 41 grid points and with 65 x 9 x 21 grid points, compared with the full potential method run with 162 x 22 x 24 grid points. For a supercritical flow good agreement is obtained with results from an Euler method of Jameson type[4]. In Figure 2 calculated distributions of the entropy function, $S = \frac{p}{p_\infty}\left(\frac{\rho_\infty}{\rho}\right)^\gamma - 1$, for the supercritical case run with 55,000 grid points are compared with the corresponding distributions obtained by the Jameson-type method[4] run with 130,000 grid points. The comparison shows that improved entropy distributions are obtained from the present method; in particular there are no oscillations extending forward from the trailing-edge as are seen in the results from the Jameson-type method.

The use of multigrid has been shown to give a large reduction in the CPU time required for a well converged solution when compared with the algorithm without multigrid. A converged result for the steady flow past a swept wing can be obtained in about 6-7 minutes of CPU time on the CRAY 1S.

However, to maintain the speed advantages of this three-dimensional algorithm, when run on a computer without a very large main memory, careful consideration of any additional disk backing store used is important. For example, the algorithm was developed using a CRAY 1S with a 1 M word main memory which was insufficient for a fine grid calculation. Even with careful use of asynchronous Input/Output instructions a large overhead due to the time waiting for I/O was incurred with conventional disks. However, when a Solid State Disk was used the I/O wait time was reduced by about two orders of magnitude, from about 6 hours to about 4 minutes for a test case taking 1 hour of CPU time.

REFERENCES

1. M.G. Hall, Cell-vertex multigrid schemes for solution of the Euler equations. Invited paper at conference on Numerical Methods for Fluid Dynamics 1-4 April 1985. University of Reading, UK. To be published in proceedings in IMA conference series, Oxford University Press.
2. A. Jameson, Solution of the Euler equations by a multigrid method, Applied mathematics and computation 13, 327(1983).
3. M. T. Arthur, A method for calculating subsonic and transonic flow over wings or wing-fuselage combinations with an allowance for viscous effects, AIAA paper 84-0428(1984).
4. C.A. Nelson, Private communication.

Fig 1 Comparison of Pressure Distributions

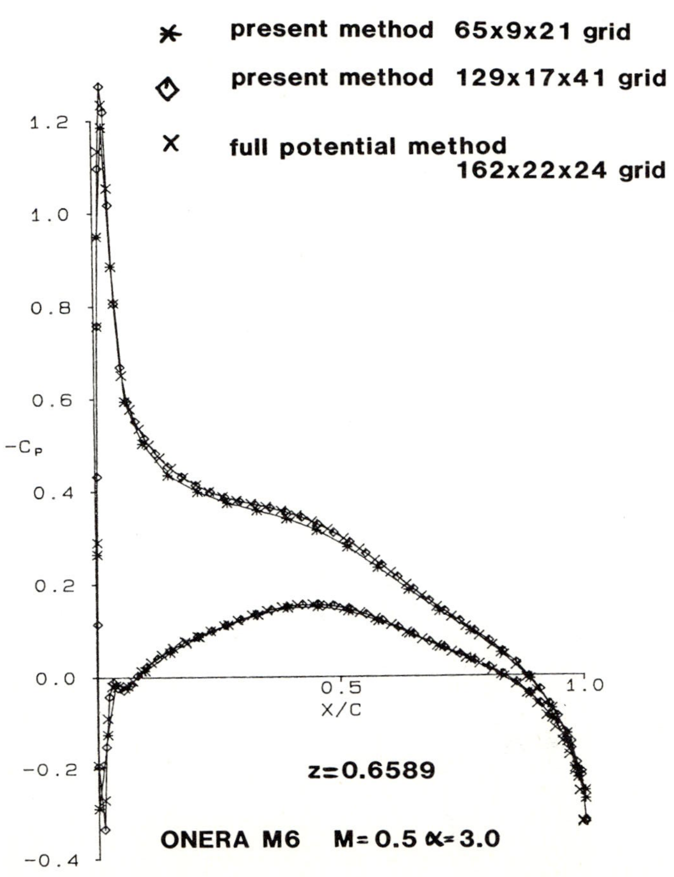

* present method 65x9x21 grid
◇ present method 129x17x41 grid
× full potential method 162x22x24 grid

ONERA M6   M=0.5 α=3.0

Fig 2 Comparison of Entropy Distributions

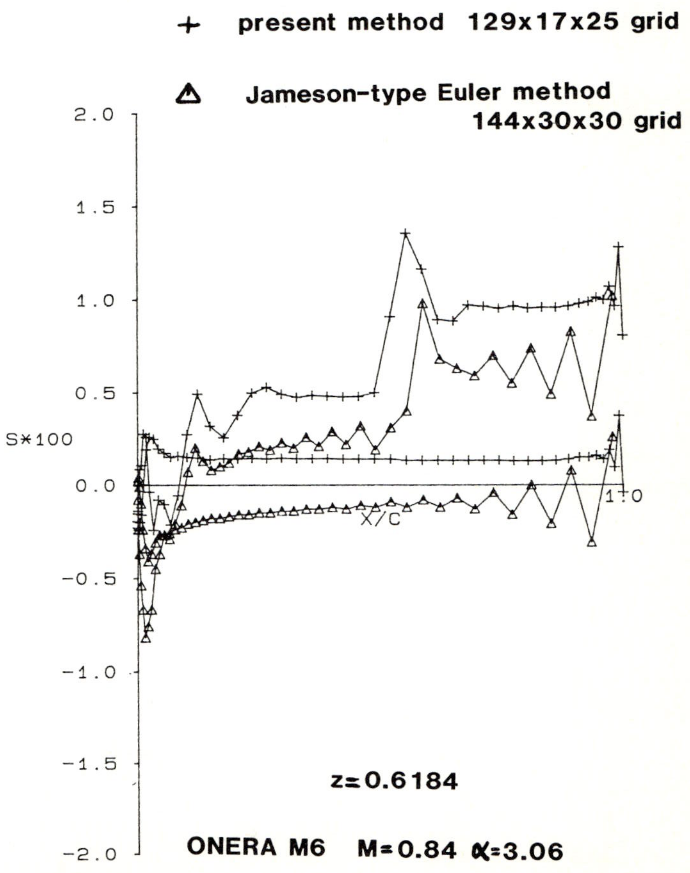

+ present method 129x17x25 grid
△ Jameson-type Euler method 144x30x30 grid

ONERA M6   M=0.84 α=3.06

NUMERICAL PREDICTION OF THE AERODYNAMIC BEHAVIOUR OF POROUS AIRFOILS

G. Savu, O. Trifu
The Aviation Institute - INCREST
77538 Bucharest, ROMANIA

Introduction

Since 1983 numerical experiments were made on thick porous airfoils at transonic speeds by a number of authors[1...5].

The use of a permeable surface was suggested over the region where usually the shock waves occur, the external transonic flow around the airfoil beeing, through the permeable wall, in a dynamic mass exchange with a flow established in the inner airfoil's cavity (Fig. 1.). The higher static pressure behind the shock wave determines the air to penetrate to the inner cavity, and from here, to emerge in front of the shock wave in such a manner that the abrupt pressure jump is smoothed.

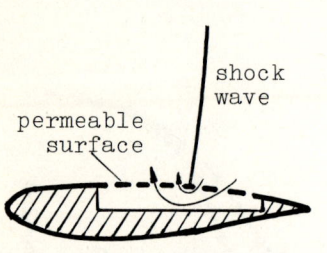

Fig.1. The basic concept

Initially, in the papers[1,2,3], the secondary flow in the airfoil's inner cavity, was **substituted** by a fictitious constant velocity flow, and the flow through the permeable wall was modeled using a linear Darcy type law. In the present paper a more adequate model for the flow in the inner cavity of the airfoil was adopted, which takes into account the influence of some geometrical parameters: inner flow cavity length, height and porosity magnitude and distribution along the permeable wall or viscous effects which may become important in certain circumstances. Also, for the flow through the porous region, a nonlinear Darcy type law was used, based on a turbulent flow model. Consequently, a more realistic image about the way in which "the porous airfoil" concept can really work, was obtained.

The external and internal flows equations and boundary conditions

The flow around the airfoil (external flow) was considered to be governed by the nonlinear, small disturbance, transonic 2-D, von

Kármán equation (1):

$$(1 - M_\infty^2)\varphi_{xx} + (\kappa+1) M_\infty^2 \varphi_x \varphi_{xx} + \varphi_{yy} + \varphi_{zz} = 0 \qquad (1)$$

with the boundary conditions:

$$\bar{v}_n = 0 \quad \text{for the solid region} \qquad (2)$$

$$\bar{v}_n = \bar{\sigma} \cdot \text{sign}(\bar{u}_e - \bar{u}_p) |\bar{u}_e - \bar{u}_p|^{4/7} / \bar{\rho}^{0.474} \quad \text{for the} \qquad (3)$$

porous regions (nonlinear Darcy type law), where:
$\bar{v}_n$ - normal velocity to the airfoil's surface, $\bar{u}_e, \bar{u}_p$ - nondimensional external/internal (plenum) flows perturbations velocities,
$\bar{\sigma}$ - nondimensional porosity factor, $\bar{\rho}$ - density (all the quantities, except $\bar{\sigma}$ were nondimensionalized with respect to the freestream quantities ρ_∞, U_∞, p_∞, T_∞).

Relation (3) was deduced assuming the turbulent flow through the holes of the permeable wall[6].

In order to determine the values of $\bar{u}_p$, the momentum theorem and the continuity condition, on the inner cavity domain, were written. Considering that in the inner cavity the flow is turbulent, after some transformations, the above conditions were combined in the final equations for the inner flow (4),(5).

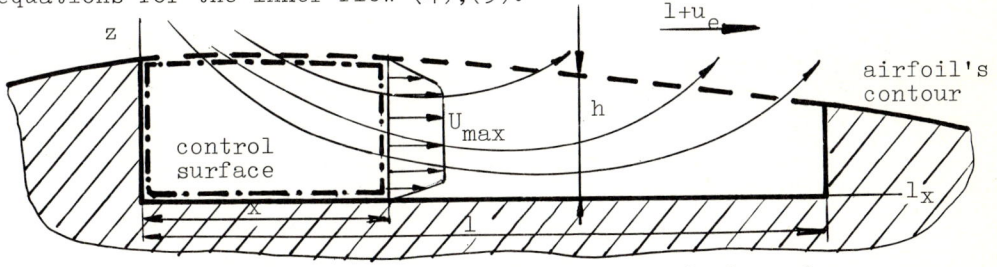

Fig.2. The inner cavity with the control surface

$$(\bar{q} \cdot \bar{U})_x = 1.2 \, h \cdot (\bar{u}_p)_x - 19.2 \, \bar{U}/Re \qquad (4)$$

$$\bar{u}_p = \bar{u}_e + \text{sign}((q)_x) \cdot (7/8 \, (q)_x/s)^{7/4} \qquad (5)$$

with: $\quad q = \bar{\rho} \, U_{max} \cdot h$

$\quad\quad\quad s = \bar{\sigma} \cdot \bar{\rho}^{0.526} \quad$ and

$\quad\quad\quad Re = U \cdot h / \nu_\infty \quad\quad$ subscript $x = d/dx$

One may say that the inner flow (eqs.(4),(5)) is a boundary condition for the external flow (eq.(1)) and viceversa[7]. Both suffer the influence of the mutual coupling (eq.(3)) which establishes between them.

The numerical solving of the external and inner flow

The external transonic flow is solved by writing in finite differences, for a 2-D domain (Fig.3.), the differential operator (1). The domain is divided using a mesh with expanded unequal steps: near the airfoil the steps are smaller than farther from it. For the subsonic zones, where the eq. (1) has elliptic character, the centered[8] differences were adopted, while in the supersonic inclusions (hyperbolic character), upwind differences are used, for the simulation of perturbations propagation along the characteristics and for the shock capturing and fitting. To ensure the stability and convergence of the solution (CFL condition) the Courant number C must be less than one, in other words, the step on Oz axis must be included in the cone of the physical characteristics from point (i,j).

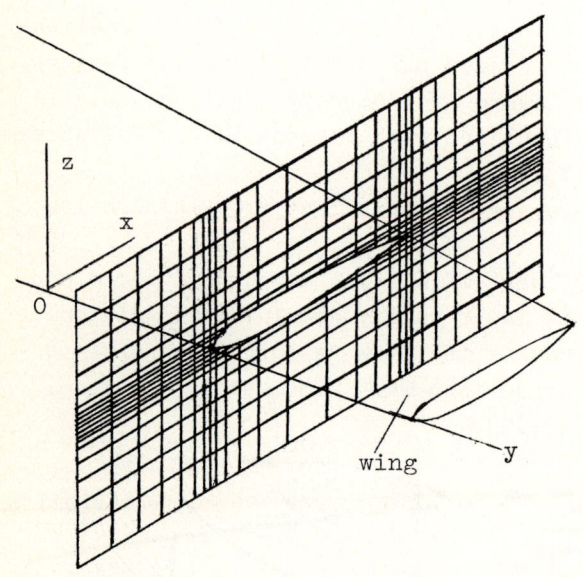

Fig.3. 2-D mesh domain

When writing in finite differences the operator (1), we obtain a nonlinear equations system in $\varphi(i,j)$ unknowns. For it's effective solving, a second order Newton iterative procedure was adopted:

$$G(\varphi(i,j)) = 0 \qquad (6)$$

$$\varphi_{i,j}^{(n)} = \varphi_{i,j}^{(n-1)} - G/G' - \tfrac{1}{2}G''\cdot G^2/{G'}^2 \qquad (7)$$

The same method is used to solve the nonlinear equations system obtained as the finite differences equivalent of eq.(4) and hence to get the solution for the inner flow.

To increase the speed of computations, meaning to obtain the solution after the smallest number of iterations, a special algorithm was applied to the computer FORTRAN code: the unknowns satisfying the relative ε error condition (8), were "frozen", and the next iterations were done only for the equations containing the $\varphi(i,j)$ unknowns which did not satisfy yet the condition (8):

$$\varepsilon \geqslant ABS(1 - \varphi_{i,j}^{(n)} / \varphi_{i,j}^{(n-1)}) \qquad (8)$$

Numerical results did show that in some conditions this procedure provides a fast solution (ex. 6 iterations for a 384x384 equations system) but in certain cases, it is divergent.

Results

Many cases of flow were tested at several Mach numbers, incidences, porosities, distributions of porosity along the airfoil chord and viscosity coefficients. The tested airfoil was NACA 0012.

The pressure distribution diagrams show that when using airfoils with permeable surfaces over the zones where on a similar solid airfoil shock waves occur, the recompression becomes a more gradual one, without shocks. While a solid supercritical airfoil works shockless in a very narrow range of Mach numbers and incidences, the permeable airfoil is "self-adapting shockless" for a broader interval of speeds and attitudes[1,2].

We give here only two examples of a much broader serial that were numerically tested. In Figs. 4,5 are comparatively plotted the pressure distributions for the solid and the porous airfoils at transonic speed and two different incidences. The porosity magnitude and distribution, represented in Fig.4., was the same for both cases. One may observe that the jumps in the pressure distribution associated to the solid airfoil were transformed into a gradual recompression when the upper side of the airfoil was permeable. On Fig.4. the solid equivalent airfoil[1,2,7] corresponding to the pressure distribution of the porous NACA 0012

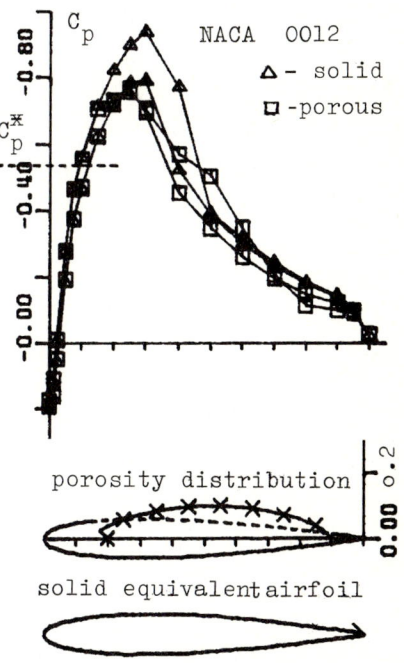

Fig.4. Pressure distributions at M=o.78, incidence 0.5

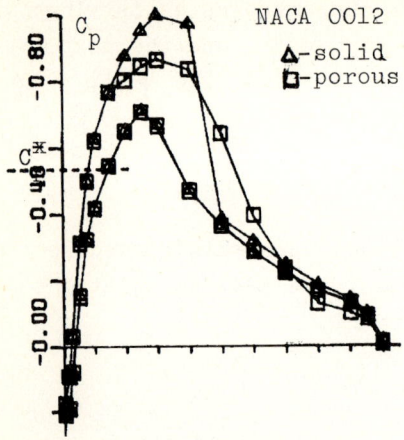

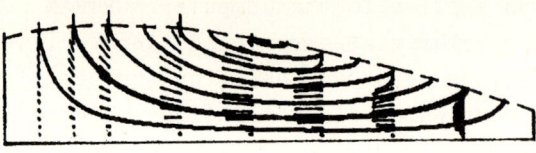

Fig.6. Inner flow velocities and streamlines at M=o.78,incid.=o.5 deg.

Fig.5. Pressure distributions at M=o.78,inc.=1 deg.

airfoil at M=o.78 and o.5 degs. incidence is also represented. Corresponding to the same situation, in Fig.6. some features of the inner airfoil's cavity flow may be observed from the velocities and streamlines display.

Even if the present study is restricted only to several examples, in order to demonstrate the influence exerted on the external flow around a porous airfoil by a better modelling of the inner flow inside its cavity, one may say that the porosity distribution and magnitude seem to be of maximum importance if beneficial gains are desired.

References

[1] G.Savu, O.Trifu, L.Z.Dumitrescu "Suppression of Shocks on Transonic Airfoils", Proceedings of the 14-th Intnl. Symp. on Shock Tubes & Waves, Sydney, Australia, 1983, N.S.W. University Press, 1984, pp. 92-lol.

[2] G.Savu, O.Trifu "Porous Airfoils in Transonic Flow", AIAA Journal, vol. 22, no.7, July 1984, pp. 989-991.

[3] E. Serbănescu, G.Savu "Drag Reduction of Perforated Axisymmetric Bodies in Supersonic Flow", J. of Spacecraft and Rockets, vol.22, no.6, Nov.-Dec. 1985, pp.663-665.

[4] C.F.Olling, G.Dulikravich "Viscous/Inviscid Computations on Transonic Separated Flows, over Solid and Porous Cascades", Texas University Report, 1985, Austin, Texas.

[5] C.-L.Chen, C.-Y.Chow, T.L.Holst, W.R.van Dalsem "Numerical Study of Porous Airfoils in Transonic Flow", NASA TM 86713, 1985.

[6] G.Savu, O.Trifu "The porosity distribution for 3-D supercritical wings", INCREST Report N-6576, 1986.

[7] G.Savu, O.Trifu "The inner flow influence on the aerodynamics of porous airfoils", presented as paper at the IV-th International Conference on Boundary and Internal Layers: Computational and Asymptotic Methods - BAIL IV, 7-11 July 1986, Novosibirsk, U.S.S.R.

[8] E.Murman, J.D.Cole "Calculation of Plane Steady Transonic Flow", AIAA Journal, vol.9, no. 1, Jan. 1971, pp. 114-121.

FINITE ELEMENT SOLUTION TO THE EULER EQUATIONS

V. Selmin and L. Quartapelle

Istituto di Fisica, Politecnico di Milano
Piazza Leonardo da Vinci, 32
20133 Milano, Italy

SUMMARY

A Taylor-Galerkin finite element method for solving the time dependent Euler equations for a perfect gas is presented. The equations in conservation-law form are first discretized in time according to the second-order-accurate, single-step, Lax-Wendroff method. The problem is then expressed in a weak variational form in order to employ the standard Galerkin finite element method for the spatial discretization. A continuous interpolation of the conservation variables is considered using bilinear quadrilateral elements. The resulting scheme represents an extension to finite elements of the classical Lax-Wendroff finite difference method. An explicit artificial dissipation is finally introduced to improve the shock capturing properties of the method and to prevent nonlinear instabilities in the presence of strong shocks. Numerical results are presented for supersonic flows in two dimensions to illustrate the properties of the proposed method.

EULER EQUATIONS

For computational purposes, it is convenient to express the Euler equations /1/ in conservation-law form using as dependent variables the density of mass ρ, of momentum $\underline{u}$ and of total energy e. The single unknown $u = (\rho, \underline{u}, e) = (u_0, u_1, u_2, u_3, u_4)^T$ is then introduced, the underline denoting vectors in the spatial coordinate system (x, y, z), so that $\underline{u} = (u_1, u_2, u_3)$. The time dependent Euler equations governing compressible inviscid flows can be written in the divergence form

$$\underline{u}_t + \underline{\nabla} \cdot \underline{f}(u) = 0 , \qquad (1)$$

where $\underline{f}(u) = (\underline{u}, \underline{u}\, u_\alpha/\rho + \underline{\hat{e}}_\alpha P, \underline{u}(e+P)/\rho)^T$ is the flux matrix. Here, P is the pressure whereas $\underline{\hat{e}}_1 = \underline{\hat{x}}$, $\underline{\hat{e}}_2 = \underline{\hat{y}}$ and $\underline{\hat{e}}_3 = \underline{\hat{z}}$ are the unit vectors directed as the coordinate axes. Notice that the matrix $\underline{f}(u)$ can also be regarded as a five-component vector whose elements are three-component vectors. The statement of the problem is made complete by providing an equation of state which, for the case of a perfect gas with constant specific heats, assumes the form

$$P = (\gamma-1)(e - \tfrac{1}{2}\rho|\underline{v}|^2) = (\gamma-1)(e - \tfrac{1}{2}|\underline{u}|^2/\rho) , \qquad (2)$$

where $\gamma = c_p/c_v$ is the specific heat ratio and $\underline{v} = \underline{u}/\rho$ is the fluid velocity. In smooth regions of the flow, the Euler equations (1) could also be expressed in the quasi-linear form

$$\underline{u}_t + \underline{A}(u) \cdot \underline{\nabla} u = 0 , \qquad (3)$$

where $\underline{A}(u) = [\underline{A}_{ij}(u)] = [\partial \underline{f}_i(u)/\partial u_j]$, $0 \leq i,j \leq 4$, the entries of the Jacobian matrix being vectors with three components. Starting from the definition for flux and using the equation of state (2), the explicit expressions of the vector elements of $\underline{A}(u)$ in the case of a perfect gas are easily obtained:

$$\underline{A}_{oo} = \underline{0} , \qquad \underline{A}_{o4} = \underline{0} , \qquad (4a)$$

$$\underline{A}_{o\alpha} = \underline{\hat{e}}_\alpha , \qquad \underline{A}_{\alpha 4} = (\gamma-1)\underline{\hat{e}}_\alpha , \qquad (4b)$$

$$\underline{A}_{\alpha o} = -\underline{u}\, u_\alpha/\rho^2 + \tfrac{1}{2}(\gamma-1)|\underline{u}|^2/\rho^2\, \underline{\hat{e}}_\alpha , \qquad (4c)$$

$$\underline{A}_{\alpha\beta} = \{\underline{u}\,\delta_{\alpha\beta} + u_\alpha \underline{\hat{e}}_\beta - (\gamma-1) u_\beta \underline{\hat{e}}_\alpha\}/\rho , \qquad (4d)$$

$$\underline{A}_{4\alpha} = -(\gamma-1)\underline{u}\, u_\alpha/\rho^2 - \{\tfrac{1}{2}(\gamma-1)|\underline{u}|^2/\rho^2 - \gamma e/\rho\}\underline{\hat{e}}_\alpha , \qquad (4e)$$

$$\underline{A}_{4o} = -\gamma \underline{u}\, e/\rho^2 , \qquad \underline{A}_{44} = \gamma \underline{u}/\rho , \qquad (4f)$$

where $\delta_{\alpha\beta}$ is the Kronecker delta.

TIME DISCRETIZATION

According to the Taylor-Galerkin method for initial-boundary value problems /2/, the system of hyperbolic equations (1) is discretized in time <u>before</u> its spatial approximation. Consider the following Taylor series expansion in the time step Δt

$$u^{n+1} = u^n + \Delta t \, u_t^n + \tfrac{1}{2}(\Delta t)^2 \, u_{tt}^n + O[(\Delta t)^3] \,. \tag{5}$$

In the classical Lax-Wendroff method /3-4/ or in the Taylor-Galerkin scheme of second-order accuracy in time /5-6/, u_t^n is provided directly by the governing equation (1) whereas u_{tt}^n is obtained by taking the time derivative of Eq. (1), which gives

$$u_{tt} = -[\underline{\nabla} \cdot \underline{f}(u)]_t = -\underline{\nabla} \cdot [\underline{f}(u)_t] =$$
$$= -\underline{\nabla} \cdot [\underline{A}(u) u_t] = \underline{\nabla} \cdot [\underline{A}(u) \, \underline{\nabla} \cdot \underline{f}(u)] \,. \tag{6}$$

It should be remarked that, only if the governing equations can be expressed in conservation-law (divergence) form, the second-order term in Eq. (6) does not involve the derivative of the Jacobian matrix $\underline{A}(u)$. By substituting Eqs. (1) and (6) in the Taylor series (5), one obtains the time discretized equation

$$(u^{n+1} - u^n)/\Delta t = -\underline{\nabla} \cdot \underline{f}^n + \tfrac{1}{2} \Delta t \, \underline{\nabla} \cdot [\underline{A}^n \, \underline{\nabla} \cdot \underline{f}^n] \,, \tag{7}$$

where $\underline{f}^n = \underline{f}(u^n)$ and $\underline{A}^n = \underline{A}(u^n)$.

SPATIAL DISCRETIZATION

In order to introduce a finite-element-based spatial approximation, Eq. (7) is recast in a weak form using the weighted residual formulation /7/. One takes the scalar product of the time discretized equation (7) with suitable weighting functions $w = (w_o, \underline{w}, w_4)$ and integrates by parts the term containing the second-order spatial derivatives, to obtain

$$\langle w, u^{n+1} - u^n \rangle / \Delta t = -\langle w, \underline{\nabla} \cdot \underline{f}^n \rangle$$
$$- \tfrac{1}{2} \Delta t \, \{\langle \underline{A}^{nT} \cdot \underline{\nabla} w, \underline{\nabla} \cdot \underline{f}^n \rangle - \int \underline{n} \cdot \underline{A}^{nT} w \, \underline{\nabla} \cdot \underline{f}^n \, d\Gamma\} \tag{8}$$

where $\langle , \rangle$ denotes the L^2 scalar product, Γ is the boundary of the integration domain and $\underline{n}$ is the unit vector normal to the boundary. The fully discretized equations are subsequently obtained by means of the standard Galerkin method. It consists in approximating the unknown u^n and the weighting functions w in the same finite dimensional space generated by piecewise polynomial interpolations over simple regions. By denoting with u_h^n and w_h the finite element approximation to u^n and w, respectively, the spatially discretized equivalent of Eq. (8) is

$$\langle w_h, u_h^{n+1} - u_h^n \rangle_h / \Delta t = - \langle w_h, \underline{\nabla} \cdot \underline{f}_h^n \rangle_h$$

$$- \tfrac{1}{2} \Delta t \{ \langle \underline{A}_h^{nT} \cdot \underline{\nabla} w_h, \underline{\nabla} \cdot \underline{f}_h^n \rangle_h - \int \underline{n} \cdot \underline{A}_h^{nT} w_h \underline{\nabla} \cdot \underline{f}_h^n \, d\Gamma_h \}$$

(9)

where $\underline{f}_h^n = \underline{f}(u_h^n)$, $\underline{A}_h^n = \underline{A}(u_h^n)$ and Γ_h is the finite element discretization of Γ. We consider a continuous approximation and use piecewise bilinear finite elements for computing flows in two dimensions. By virtue of the integral character of the weak equation (9), the conservative (i.e., divergence) form $\langle w, \underline{\nabla} \cdot \underline{f} \rangle$ and the quasi-linear (i.e., advective) form $\langle w, \underline{A} \cdot \underline{\nabla} u \rangle$ are equivalent provided that the interpolation is continuous at inter-element boundaries and the spatial integrals are determined exactly. However, the integrals are computed only approximately by means of numerical quadrature so that the above two forms are not equivalent in the actual computations. On the other hand, the conservative form has the property that the Rankine-Hugoniot jump condition is satisfied exactly and the shock speed is computed correctly, regardless of any approximation in evaluating the integrals. For this reason, the conservative form has to be used when shock waves are present. On the contrary, the second-order term can be evaluated using alternatively the form $\langle \underline{A}^T \cdot \underline{\nabla} w, \underline{\nabla} \cdot \underline{f} \rangle$ or $\langle \underline{A}^T \cdot \underline{\nabla} w, \underline{A} \cdot \underline{\nabla} u \rangle$. All integrals in Eq. (9) are evaluated using the isoparametric transformation technique /7/ and employing full integration with the 2x2 Gaussian product formula.

A characteristic feature of the Taylor-Galerkin scheme is the presence of the consistent mass matrix. It assures a fourth-order spatial accuracy when the time variable is left continuous. In the time discretized case, however, the scheme has only a second-order global accuracy. The condition of numerical stability for the linear equation in one dimension is found to be $c^2 \leq 1/3$, which is more restrictive than the optimal stability limit $c^2 \leq 1$ of the finite difference Lax-Wendroff method.

NUMERICAL EXAMPLES

The proposed Taylor-Galerkin method has been applied to compute the propagation /8/ and interaction /9/ of shock waves in one dimension. The comparison with results provided by finite difference methods shows that the finite element solutions are characterized by a more accurate resolution of all the waves and exhibit a tolerable amount of spurious spatial oscillations /6/.

As a first example in two dimensions, the problem of a Mach 3 flow in a wind tunnel with a step /9/ has been considered. The computational domain and its discretization by means of $\sim$ 1000 four-noded quadrilateral elements are shown in Fig. 1. The pressure and density fields calculated at t = 0.5 and t = 4 using the artificial dissipation described in /6/ are also given in Fig. 1. In the present calculations no special treatment has been adopted to dominate the singularity of the flow at the corner of the step which is the centre of a

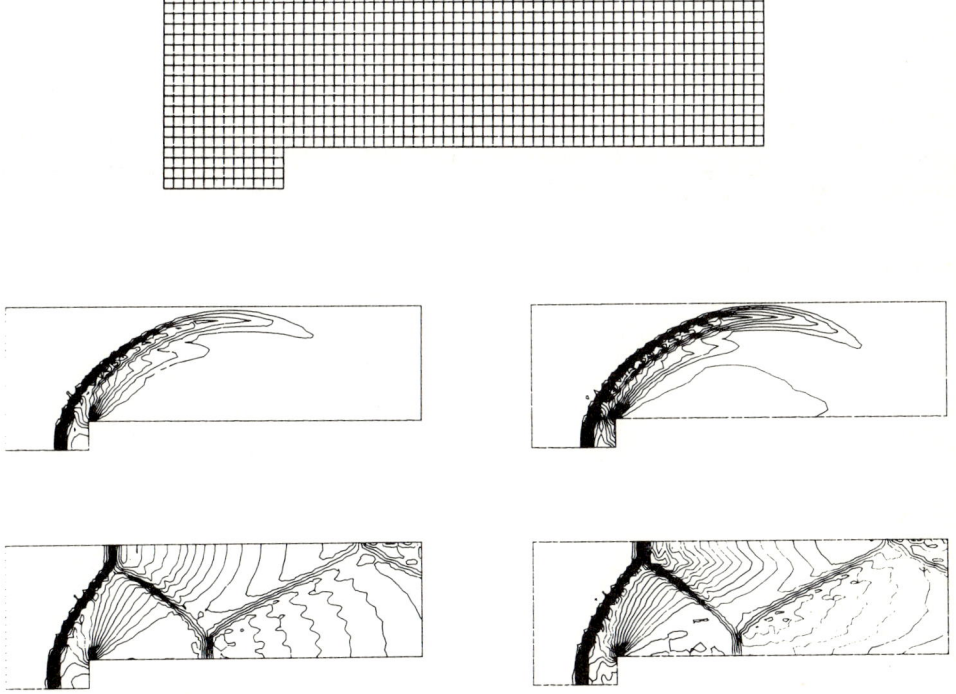

Figure 1 Finite element mesh, pressure (left) and density (right) contours for the Mach 3 wind tunnel problem with a step at t = 0.5 and t = 4.

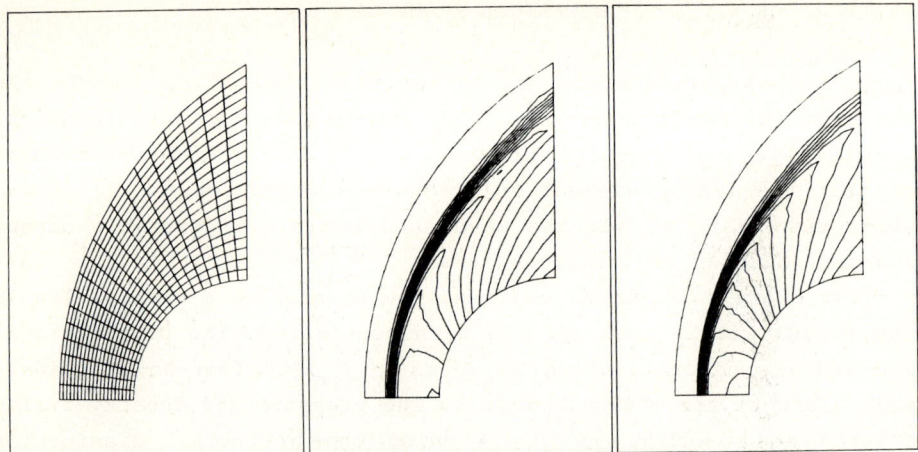

Figure 2 Finite element mesh, density and pressure contours for the Mach 8 flow past a circular cylinder.

rarefaction wave. As a consequence, the solution using the rather coarse mesh contains a numerical boundary layer above the step with an associated spurious Mach reflection. Furthermore, the position of the shock at the top boundary displaced downstream with respect to the reference solution calculated by Woodward and Colella /9/.
The last example is the flow past a circular cylinder with a freestream Mach value of 8. In this problem, the shock capturing properties of the scheme (9) are improved and the occurrence of nonlinear numerical instability is prevented by introducing the following artificial dissipation. In the case of a single hyperbolic equation for a scalar unknown s, the term

$$\nu < \underline{g}_{hc} \cdot \underline{\nabla} w_h, \underline{g}_{hc} \cdot \underline{\nabla} s_h >_h$$

is added to the right hand side of the weak equation, where $\underline{g}$ and ν are defined by the following relationships

$$\underline{g} = \underline{\nabla} \rho \quad , \qquad \nu = \tfrac{1}{2} \left(\left| \frac{\partial \rho}{\partial x} \right| + \left| \frac{\partial \rho}{\partial y} \right| \right) .$$

Due to its tensorial character, this artificial dissipation is anisotropic and acts only in the direction of the density gradient but not transversely. The computational mesh of 20x20 quadrilateral elements is shown in Fig. 2 together with density and pressure fields at the steady state. Some spatial oscillations are present in the

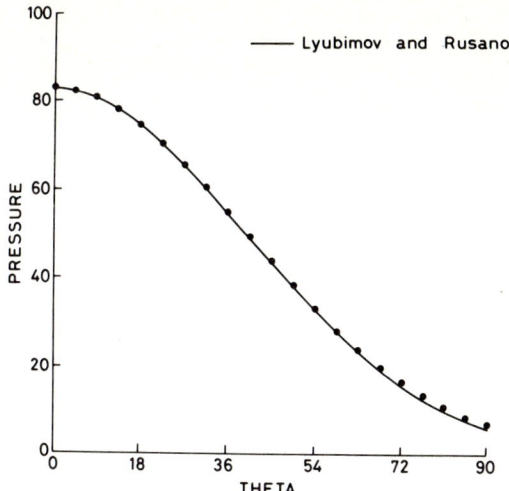

Figure 3   Surface pressure distribution for the Mach 8 flow past a circular cylinder. (---) Lyubimov and Rusanov, (O) present finite element method.

after-shock region, but they do not seem to affect the global accuracy of the solution. In fact, the pressure profile over the cylinder compares very well with the results of Lyubimov and Rusanov /10/ and is found to be more accurate than results obtained by means of finite differences /11/ and other finite element schemes.

In conclusion, the present finite element method seems attractive and worthy of further investigations and developments especially if one considers its rather unsophisticated character.

REFERENCES

1. A. Majda, Compressible Fluid Flow and Systems of Conservation Laws in Several Space Variables, Springer-Verlag, New York, 1984.
2. J. Donea, Int. J. Numer. Meths. Engrg. 20,(1984),101-120.
3. P. D. Lax and B. Wendfoff, Comm. Pure Appl. Math. 13(1960), 217-237.
4. R. D. Richtmyer and K. W. Morton, Difference Methods for Initial-Value Problems, 2nd ed, Interscience, New York, 1967.
5. V. Selmin, J. Donea and L. Quartapelle, Comput. Meths. Appl. Mech. Engrg. 52(1985), 817-845.
6. V. Selmin, Simulation par une methode d'elements finis de problemes hyperboliques avec discontinuites, Ph.D. Thesis, 1986.
7. G. Strang and G.J. Fix, An Analysis of the Finite Element Method, Prentice-Hall Englewood Cliffs, New Jersey, 1973.
8. G. Sod, J. Comput. Phys. 29(1978), 1-31.
9. P. Woodward and P. Collela, J. Comput. Phys. 54(1984), 115-173.
10. A. N. Lyubimov and V. V. Rusanov, Gas Flows past Blunt Bodies, NASA-TT-F 715, 1973.
11. S. Osher and S. Chakravarty, J. Comput. Phys. 50(1983), 447-481.

MULTIGRID SOLUTION OF THE COMPRESSIBLE NAVIER-STOKES
EQUATIONS ON A VECTOR COMPUTER

G. Shaw   and   P. Wesseling
Oxford Univ. Comp. Laboratory   University of Technology
8-11 Keble Road   Dept. of Math. and Informatics
Oxford OX1 3QD   P.O. Box 356, 2600 AJ Delft
U.K.   The Netherlands

1. FINITE VOLUME DISCRETIZATION OF THE COMPRESSIBLE NAVIER-STOKES
   EQUATIONS ON IRREGULAR GRIDS

The compressible Navier-Stokes equations can be written as

$$\frac{\partial q}{\partial t} + \frac{\partial F(q)}{\partial x} + \frac{\partial G(q)}{\partial y} = 0 \tag{1.1}$$

with $q = (\rho, \rho u, \rho v, e)^T$, and F, G the well-known flux functions. Two-dimensional flow is assumed. Eq. (1.1) is discretized on an irregular grid, assumed to be topologically equivalent to a uniform rectangular grid. Fig. 1.1. represents part of a suitable grid. Finite volume discretization is carried out as follows. The shaded cell of Fig.1.1, its boundary and its area are called Ω_{ij}, Γ_{ij} and a_{ij}, respectively. Let q_{ij} be the average of q over Ω_{ij}. Integration of (1.1) over Ω_{ij} gives:

$$a_{ij} \frac{d}{dt} q_{ij} + \int_{\Gamma_{ij}} H \cdot n d\gamma = 0 \tag{1.2}$$

where $H = (F,G)^T$ and n is the outward unit normal on Γ_{ij}. We split H into its viscous and inviscid parts:

$$H = H^V + H^I. \tag{1.3}$$

The inviscid flux is treated as follows. There exist numbers c and s such that the rotation matrix

$$R = \begin{pmatrix} 1 & 0 & 0 & 0 \\ 0 & c & -s & 0 \\ 0 & s & c & 0 \\ 0 & 0 & 0 & 1 \end{pmatrix} \tag{1.4}$$

rotates q into $\bar{q}$ in $(\bar{x}, \bar{y})$ coordinates, with the $\bar{x}$-axis parallel to the outward normal on the straight line segment AB (Fig. 1.1). The function F^I is split as follows:

$$F^I = F^+ + F^- \tag{1.5}$$

such that $\partial F^{\pm}/\partial q$ is $\pm$ definite. This is done with the method proposed by Van Leer (1982). We then approximate the inviscid flux over AB by:

$$\int_A^B H^I \cdot n d\gamma \simeq R^{-1}[F^+(Rq_{i,j}) + F^-(Rq_{i+1,j})]L_{AB} . \qquad (1.6)$$

The other sides of ABCD are handled analogously. L_{AB} is the length of AB.

The viscous flux is treated as follows. We introduce the following approximation:

$$\int_A^B H^V \cdot n d\gamma \simeq F^V_{AB}\Delta y_{AB} - G^V_{AB}\Delta x_{AB} \qquad (1.7)$$

where $\Delta x_{AB} = x_B - x_A$, $\Delta y_{AB} = y_B - y_A$, and F^V_{AB}, G^V_{AB} mean values on AB. In order to compute these mean values one requires approximations to the derivatives of the velocity components and the temperature on AB. These are obtained with a variation of a commonly used technique in finite volume discretization, which is described by Peyret and Taylor (1983). Consider for example $\partial u/\partial x$ on AB (Fig. 1.2). This quantity is approximated as a mean value over the intermediate cell $\Omega_{i+\frac{1}{2},j}$ (the dashed region in Fig. 1.2) between $\Omega_{i,j}$ and $\Omega_{i+1,j}$. Then

$$\left.\frac{\partial u}{\partial x}\right|_{AB} \simeq a^{-1}_{i+\frac{1}{2},j} \int_{\Omega_{i+\frac{1}{2},j}} \frac{\partial u}{\partial x} dV = a^{-1}_{i+\frac{1}{2},j} \int_{\Gamma_{i+\frac{1}{2},j}} u dy \qquad (1.8)$$

and so we obtain

$$\left.\frac{\partial u}{\partial x}\right|_{AB} \simeq a^{-1}_{i+\frac{1}{2},j} [u_{EF}\Delta y_{EF} + u_{FG}\Delta y_{FG} + u_{GH}\Delta y_{GH} + u_{HE}\Delta y_{HE}] \qquad (1.9)$$

where u_{EF} is a mean value of u on EF, etcetera. The following approximations complete the specification of this approximation:

$$u_{EF} = u_{i+1,j}, \quad u_{FG} = \frac{1}{2}(u_{i+1,j} + u_{i,j+1})$$
$$u_{GH} = u_{i,j}, \quad u_{HE} = \frac{1}{2}(u_{i,j} + u_{i+1,j-1}).$$

The other sides of ABCD and other derivatives are handled analogously. We take care to use only q-values from cells marked with a cross in Fig. 1.2, so that a 7-point stencil is obtained, as opposed to a 9-point stencil as obtained with the method described by Peyret and Taylor (1983). This allows vectorized computation with a diagonally ordered Gauss-Seidel relaxation scheme.

Eq. (1.1) is discretized in time with the backward Euler scheme:

$$q^n_{ij} + \frac{\Delta t}{a_{ij}} S^n_{ij} = q^{n-1}_{ij} \qquad (1.10)$$

where S^n_{ij} is a nonlinear vector function of q^n_{ij} and its six neighbours in the 7-point stencil indicated with crosses in Figs. 1.1 and 1.2. The time-step in (1.10) may be taken arbitrarily large, so that we may find steady solutions directly.

2. MULTIGRID METHOD

Eq. (1.10) gives us a nonlinear algebraic system, which is solved with a multigrid method, starting with nested iteration (full multigrid), followed with a nonlinear multigrid (full approximation storage) cycling algorithm. The same multigrid approach has been used for the Osher discretization of the Euler equations by Hemker and Spekreijse (1985a, b).

Since the grid is assumed to be topologically equivalent to a uniform rectangular grid, it is easy to define a nested sequence of grids Ω_ℓ: $\ell = 1(1)m$. Here Ω_m is the finest grid. Each cell of Ω_{m-1} consists of 4 cells of Ω_m, and so on to the coarsest grid Ω_1. The interpolation (prolongation) operator consists of bilinear interpolation over four coarse cells; the restriction operator consists of taking the areaweighted average. The V-cycle is used with 2 pre- and 1 post-smoothing operation.

The main and also the most expensive component of the algorithm is the nonlinear Newton-Gauss-Seidel relaxation used for smoothing. This is defined by a single Newton iteration applied to (1.10):

$$q_{ij}^n \leftarrow q_{ij}^n - J^{-1} (q_{ij}^n + \frac{\Delta t}{a_{ij}} S_{ij}^n - q_{ij}^{n-1}) \tag{2.1}$$

with $J = I + \frac{\Delta t}{a_{ij}} \partial S_{ij}/\partial q_{ij}^n$.

Here J, I and $\partial S_{ij}/\partial q_{ij}^n$ are 4×4 matrices. Note that all four components of q are updated simultaneously. The order in which the cells are visited is along diagonal lines (Fig. 2.1), taking the diagonals first from left to right and then right to left. Because of the 7-point structure of the stencil this is completely equivalent to symmetric point Gauss-Seidel ordering, and the cells in a diagonal are independent from each other. This makes vectorized evaluation of J in (2.1) along diagonals possible. Furthermore, eq. (2.1) is solved with QR decomposition, which is identical for all systems to be solved, so that this also vectorizes along diagonals. Gaussian elimination would involve different pivot orderings in different cells, destroying vectorization. QR is stable with respect to rounding errors.

3. RESULTS

The method has been tested on two bicircular arc cascade problems. The geometry of these problems is illustrated in Fig. 3.1. The number of cells is given by: Ω_1: 6×2, Ω_5: 96×32. Fig. 3.1 shows Ω_5. By severe stretching cells are concentrated in boundary layer and wake. For each case four computations were made with $\Omega_2, \cdots, \Omega_5$ as finest grid, respectively. The multigrid method uses all available coarser grids.

At inflow and outflow we (over-) specify the full state q in virtual columns of cells by means of:

$$\rho = 1, \quad u = M_p, \quad v = 0, \quad T = 1 \tag{3.1}$$

so that M_p is the prescribed Mach number. The state q selected by the algorithm in the cells adjacent to the virtual cells defines the physical problem that is in fact being solved. For the two problems that were solved, table 3.1 lists M_p; M_1 and M_2 are the observed Mach numbers in the bottom and top cells of the first (non-virtual) column of cells at inflow; Mach is found to vary monotonically between bottom and top.

Problem	M_p	Ω_5		Ω_4		Ω_3		Ω_2	
		M_1	M_2	M_1	M_2	M_1	M_2	M_1	M_2
1	.92	.813	.818	.787	.795	.738	.761	.648	.723
2	.90	.813	.817	.783	.791	.734	.757	.645	.720

Table 3.1: Prescribed and effective inflow Mach numbers

Figures 3.2 and 3.3 present Mach contours on Ω_5 and convergence histories. Fig.3.2 shows a supersonic diffusor type of flow. On Ω_4 the solution initially converges (slowly) to a choked flow, with a shock extending to the upper boundary, but it is found that convergence (to the type of flow of Fig.3.2) with about the same rate as on the other grids sets in after 13 iterations. Although the inflow conditions differ only little, the type of flow of Fig.3.3 is appreciably different from that of Fig. 3.2; this type of transonic flow is very sensitive to inflow conditions. Fig.3.3 shows a transonic airfoil type flow, with a shock that does not extend to the upper boundary. The differences in rate of convergence arise mainly from the fact, that different inflow states are selected, so that different problems are in fact being solved on the various grids.

An important feature of the method, not exhibited in the figures, is the rapid selection of a good initial approximation by nested iteration; the figures only show the subsequent convergence of the V-cycles.

In order to evaluate the speed-up provided by the multigrid method the problems were also run in single-grid mode. Forty smoothing iterations were performed on Ω_5. Defining one work unit (WU) to be the cost of one smoothing iteration on Ω_5, the cost of obtaining a first guess of the solution with nested iteration is 1.3 WU, whereas the cost of one V-cycle iteration is 4 WU. Table 3.2 gives the cost in WU required by multigrid to surpass the accuracy obtained by the single grid computation. The error reduction involved is roughly a factor 100.

Problem	1	2
WU	14.3	9.3

Table 3.2: Number of Work Units equivalent to 40 single grid Work Units

The advantage of MG increases as the grid is refined.

Problem 2 has been run on a CRAY-1S. No special intrinsic functions are used to enhance vectorization; the code consists entirely of standard Fortran-77. Most of the work is done in subroutines EULER and QR. EULER calculates inviscid fluxes and their Jacobians; QR solves (2.1). Table 3.3 gives computing times.

	Scalar	Vector	Ratio
EULER	32.9	15.8	2.1
QR	20.3	2.0	10.1
TOTAL	63.7	19.0	3.3

Table 3.3: Computing times (seconds) on CRAY-1S for problem 2

The speed-up ratio may be improved by further code optimization: reducing scalar temporary references and increasing chaining potential; but this has not been attempted. Note that engineering accuracy is already reached within about a third of the number of iterations carried out, so that the time to reach engineering accuracy is substantially less than that quoted in table 3.3.

ACKNOWLEDGEMENT

This work was partly supported by the Netherlands Organization for the Advancement of Pure Research.

REFERENCES

Hemker, P.W. and S.P. Spekreijse: Multiple grid and Osher's scheme for the efficient solution of the steady Euler equations. Report NMR8507, Centre for Math. and Computer Science, P.O. Box 4079, 1009 AB Amsterdam, The Netherlands, 1985a.

Hemker, P.W. and S.P. Spekreijse: Multigrid solution of the steady Euler equations. In: D. Braess, W.Hackbusch and U. Trottenberg (eds.): Multigrid methods. Proc., Oberwolfach 1984. Notes on Numerical Fluid Mech. 11. Vieweg, Braunschweig. Wiesbaden, 1985b.

Peyret, R. and T.D.Taylor: Computational Methods for Fluid Flow. Springer-Verlag, Berlin, 1983.

Van Leer, B.: Flux-Vector Splitting for the Euler Equations. In: E. Krause (ed.): Eighth Int. Conf. on Num. Methods in Fluid Dyn. Proceedings, Aachen 1982. Lecture Notes in Physics 170, 507-512, Springer-Verlag, Berlin, 1982.

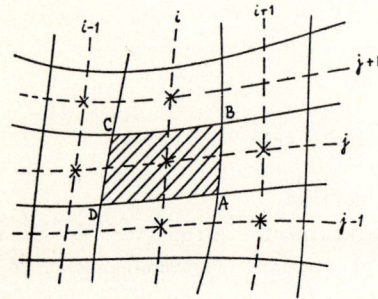

Figure 1.1 Finite volume grid

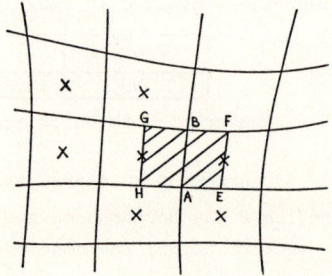

Figure 1.2 Seven-point discretization of viscous terms

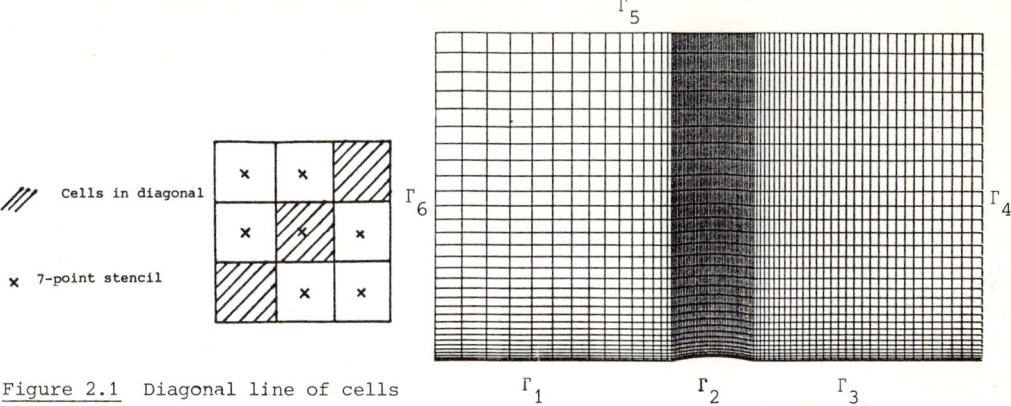

Figure 2.1 Diagonal line of cells

Figure 3.1 Geometry and finest grid. Γ_6 and Γ_4 are inflow and outflow boundaries; Γ_1, Γ_3 and Γ_5 are lines of symmetry; Γ_2 is a circular arc airfoil with thickness 5%.

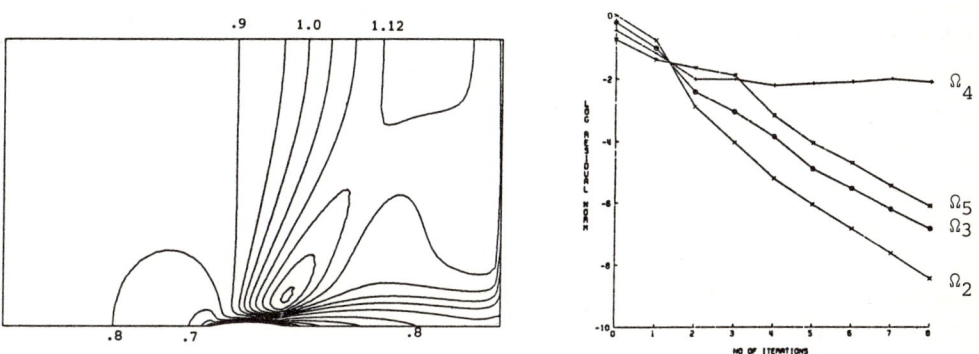

Figure 3.2 Mach contours and convergence history for problem 1. Mach=0.1(0.1)0.9, 0.96(0.04)1.20

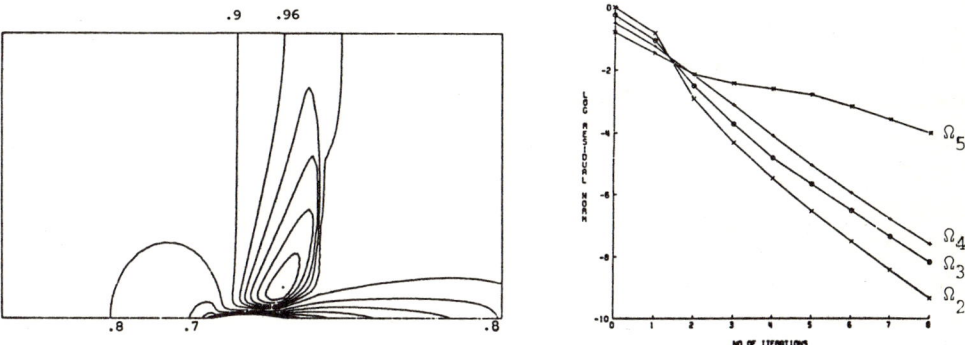

Figure 3.3 Mach contours and convergence history for problem 2. Mach=0.1(0.1)0.9, 0.96(0.04)1.16.

USING OF AN ARBITRARY COORDINATE FOR THREE-DIMENSIONAL FLUID DYNAMIC PROBLEMS

Yu.D.Shevelev
The Institute for Problems in Mechanics
of Academy of Sciences of USSR,
Moscow

In the past decade the computational fluid dynamics had studied the problems near the bodies moving in a stream of relatively simplest forms. The problems of numerical modelling flow field near body of real geometry are more complicated. In this case the ability to model a geometry is an essential moment of the solution of the problem. The generation of a curvilinear coordinate had provided the key to the development of a numerical solution of partial differential equation on regions with arbitrarely shaped boundaries. The conservative form of the governing equations are desirable for its analytical and numerical properties.

Problems associated with 1) geometry description, 2) generation of curvilinear coordinate systems, 3) conservative form, 4) discretization of curvilinear coordinate systems on arbitrary mesh are considered.

1. The initial information for a modelling of geometry used a point geometry data structure that are defined the configuration by tables of the surface coordinates.

Depending to a concrete problem it may be the different requments to a surface smoothness.

In framework of boundary layer and Navier-Stokes theory the region adjacent to body surface generally dominate the character of the flow field. The viscous shear stress depend directly on the large gradients in the direction normal to the body surface. In this case geometry must be specified by continuous derivatives of the second order.

For solving a supersonic fluid flow it is necessary to demand a continiaty of first derivatives with respect to surface coordinate.

A standard technique of subsonic aerodynamic used a panel method or method of discrete **vorticities.** In this case surface geometry is approximated by a large number of flat panel that are generated by connecting surface points from adjacent cross section.

The different methods of solution demand the different restrictions to a surface smoothness.

A description of the geometry are available by different means. The methods of a modelling of geometry are well known. The examples are algebraic, differential and conformal mapping techniques[3].

a) **If the information are given by tables of coordinates then it is problem of interpolation. For this purpose is used an interpolation by the polynomial.** Unfortunately it's difficult to respond to changes of part geometry. All changes require a careful review of all data structure. The process of measuring, checking surface coordinates become a very tedious and time consuming task.

The application of approximation by spline are connected with the facts that the interpolation is nonsensentive to the disturbances of an initial data. The approximation by splines give a minimum error for a special class of function and good interpolation of the derivatives.

The approximation by spline technique are universal, but for a real geometry using a spline-approximation become a time consuming task. For purpose a fluid dynamic is available to represent the surface by analytical technique. In this case it is used a **description of the** geometry of cross-section cuts at select stations and then a description of the surface geometry between cross-section cuts. The geometry between cross-section cuts described by the elements of simplest surface: second order, linear and minimum surfaces. A description of the geometry of cross-section cuts are available by different means: approximation by splines and analytical curves, conformal mapping.

An advantage of this approach is given by fact that it's easy to use for a different level of designing. The mathematical model are taken to provide a continuous analytical model of geometry. All derivatives, the normals, the symbols Christofell are developed analytically or numerically with a sufficient accuracy (fig. 1).

b) Let's to consider of a modelling of geometry by methods of differential geometry. Along a body surface the next conditions are satisfied

$$\bar{\tau}_{\alpha\beta} = G_{\alpha\beta}^{\sigma} \bar{\tau}_{\sigma} + b_{\alpha\beta} \bar{n}$$
$$\bar{n}_{\alpha} = - b_{\alpha}^{\sigma} \bar{\tau}_{\sigma}, \quad (\alpha,\beta,\sigma = 1,2)$$

These formulas are known as first and second groups of derivative formulas. Here $-\bar{\tau}_\alpha = \partial\bar{\tau}/\partial\xi^\alpha$. The coordinates ξ^1 and ξ^2 are choosen on body surface. The values $G^\sigma_{\alpha\beta}$ are symbols of Christoffel on body surface, $b_{\alpha\beta}$ - second characteristic form of surface.

The formulas are presented the systems of partial differential equation of first order for unknown vectors $\bar{\tau}_1$, $\bar{\tau}_2$, $\bar{n}$. This system of equation are overdetermined and thus its solvability should depend on certain compatability conditions. The conditions $\bar{\tau}_{\alpha\beta\gamma} = \bar{\tau}_{\alpha\gamma\beta}$, $\bar{n}_{\alpha\beta} = \bar{n}_{\beta\alpha}$ ($\alpha, \beta, \gamma = 1,2$) are the compatibility conditions. The compatibility conditions give a three vector equations or nine scalar equations. Among them just only three is independent: Gauss formula and Petterson-Codazzi formulas. If the functions $a_{\alpha\beta}$ ($a > 0$, $a_{11} > 0$) and $b_{\alpha\beta}$ ($\alpha, \beta = 1,2$) are satisfied the compatibility conditions then it existed a surface $-\bar{\tau} = \bar{\tau}(\xi^1, \xi^2)$. The coefficients of first and second form are coincided with given functions. If we suppose that the functions $a_{\alpha\beta}$ has the second continuous derivatives and $b_{\alpha\beta}$ has the first continuous derivatives then a surface $\bar{\tau} = \bar{\tau}(\xi^1, \xi^2)$ have the third continuous derivatives. A solution of equations is exist and unique.

This surface may be constructed by next method. The vectors $\bar{\tau}_1$, $\bar{\tau}_2$, $\bar{n}$ are functions of values ξ^1, ξ^2. Let us consider the curve on body surface: $\xi^1 = \xi^1(t)$, $\xi^2 = \xi^2(t)$. The vectors $\bar{\tau}_1$, $\bar{\tau}_2$, $\bar{n}$ are functions of parameter t. So we receive the normal form of the systems of the ordinary differential equations.

c) Contemporary mapping methods are used in cross-sections cuts. Let us consider next conformal mapping

$$z = \sum_{n=1}^{N} a_n \zeta^{B_n}$$

Coefficients a_n must be taken to minimize next equation

$$J = \sum_{p=1}^{P} \left(\left(\sum_{n=1}^{N} a_n \cos\beta_n \theta_p - x_p \right)^2 + \left(\sum_{n=1}^{N} a_n \sin\beta_n \theta_p - y_p \right)^2 \right)$$

2. One of the most important moment of a numerical modelling of fluid dynamics problems is the generation of a curvilinear coordinate system near body surface. A convenient coordinate systems of external fluid dynamic problems have the several particular features: a dominant direction, plane of symmetry, one coordinate should be constant on body surface and others. The accuracy of solving the problem is

defined by a mesh size a physical grid spacing and an ability to control a physical grid shape. A choosen coordinate system must be more close to follow the physical reality of the computational region.

The methods of a generation of three-dimensional coordinate are similar to the methods of a generation of a surface body: algebraic, differential and conformal mapping techniques. Sometimes the algebraic methods are preferable. The physical boundary topology and grid derivative smoothness requirements are easy to satisfy. It's usefull the algebraic coordinate systems which are qualities of a spherical coordinate $\bar{R} = \bar{R}_1$ (near blunt part of body), a cylindrical coordinate $\bar{R} = \bar{R}_2$ (near most part of geometry) and natural coordinate ($\bar{R} = \bar{\tau} + \zeta \bar{n}$, $\bar{\tau}$ - body, $\bar{n}$ - surface normal) near body surface. The different types of streching of coordinate are available. The advantages of the methods are explicit control of physical grid shape and the physical grid spacing. Disadvantage this approach is connected with fact that the coordinate are nonortogonal. Using the governing equation for an arbitrary nonortogonal coordinate does not change a character of equation, but in this case there are the additional terms.

Let us to consider a differential method of a construction of the curvilinear coordinate systems. Let $\bar{R}_1$, $\bar{R}_2$, $\bar{R}_3$ are covariant base vectors. The second derivatives are defined by

$$\bar{R}_{ij} = \Gamma_{ij}^k \bar{R}_k, \quad (i,j,k = 1,2,3)$$

Here Γ_{ij}^k - symbols of Christoffel. This systems can be to consider as the partial differential systems. The compatibility condition is given by condition that Riemann curvature tensor (R_{ijkl}) must be equal zero. In order the functions g_{ij} are defined the coordinate system it's necessary that Riemann tensor is equal a zero. The metric coefficients of 3-D ortogonal coordinate must be satisfied the partial differential equation as follow

$$R_{pip} = h_p h_i \left\{ \partial_p \left(\frac{1}{h_p} \partial_p h_i \right) + \partial_i \left(\frac{1}{h_i} \partial_i h_p \right) + \frac{1}{h_k^2} \partial_k h_p \partial_k h_i \right\} = 0, \quad R_{pik} = h_i \left\{ \partial_{pk}^2 h_i - \frac{1}{h_p} \partial_p h_i \partial_k h_p - \frac{1}{h_k} \partial_k h_i \partial_p h_k \right\} = 0$$

$(p \neq k \neq i)$

Here $\partial_p = \frac{\partial}{\partial x^p}$, $h_1 = \sqrt{g_{11}}$, $h_2 = \sqrt{g_{22}}$, $h_3 = \sqrt{g_{33}}$. It's possible to find out a solution in a closed form. 1. $h_1 = h_2 = F(x^1, x^2)$, $h_3 = 1$. In this case the above equation reduced to a next form $\frac{\partial^2 \ln F}{\partial (x^1)^2} + \frac{\partial^2 \ln F}{\partial (x^2)^2} = 0$. A solution $- \ln F = \phi(z) + \bar{\phi}(\bar{z})$, $z = x^1 + i x^2$.

2. $h_1 = h_2 = (C_0 + C_1 f(x^3)) F(x^1, x^2)$, $h_3 = f'(x^3)$. In this case we obtain the basic equation[['1]] $\frac{\partial^2 \ln F}{\partial (x^1)^2} + \frac{\partial^2 \ln F}{\partial (x^2)^2} = - k F^2$, $(k = C_1^2 > 0)$, $u = F^2 = 4|\phi'(\zeta)|/(1 + |\phi(\zeta)|^2)^2$,

$\zeta = \xi + i\eta$, $\xi = \sqrt{k} x^1$, $\eta = \sqrt{k} x^2$.

3. It's convinient to use the conservative form of the governing equations. This form is desirable for its analytical and numerical properties. Usually the governing equation used a scalar and vector form. Under the transformation coordinate the scalar equation are preserved the conservative form in general case. For vector equation there are the appearence the additional terms. But it's possible to reduce a vector equation to conservative form by different means. The differential equation are approximated by the sets of difference equations. Then this set of algebraic equations for the discrete values of the functions is solved. The nonuniform difference meshs which will concentrate the points in the large gradient domains are usefull.
For the case of nonequidistant modes the difference approximation of first and second derivatives yields the next form ($[^3]$)

$$\frac{\partial u}{\partial x} = \delta_{1,\ell} u_{\ell+1} + \delta_{2,\ell} u_\ell + \delta_{3,\ell} u_{\ell-1} + O(\ )$$

$$\frac{\partial^2 u}{\partial x^2} = \gamma_{1,\ell} u_{\ell+1} + \gamma_{2,\ell} u_\ell + \gamma_{3,\ell} u_{\ell-1} + O(\ )$$

The second order accuracy we obtain for an equidistant nodes. The discretization of governing equation on arbitrary mesh give an opportunity to concentrate points in near the large gradient values region and to control a spacing.

REFERENCES

1. Векуа И.Н. Основы тензорного анализа и теории ковариантов, "Наука", 1978.
2. Седов Л.И. Механика сплошной среды. "Наука", 1970, т. I.
3. Шевелев Ю.Д. Пространственные задачи вычислительной аэрогидродинамики. "Наука", 1986.
4. Shevelev Yu.D. Numerical calculation of the 3-D boundary layer in an incompressible fluid. Fluid Dynamics, vol. 1, No. 5, 1966.

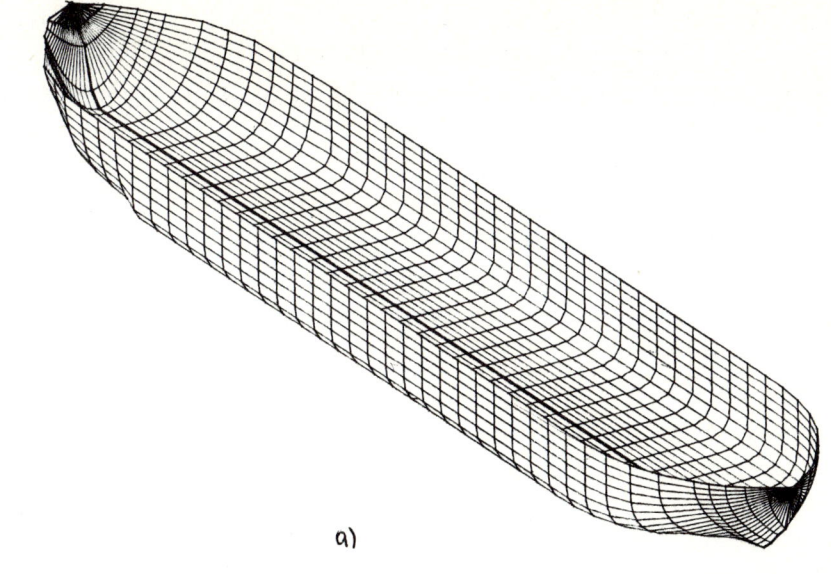

a)

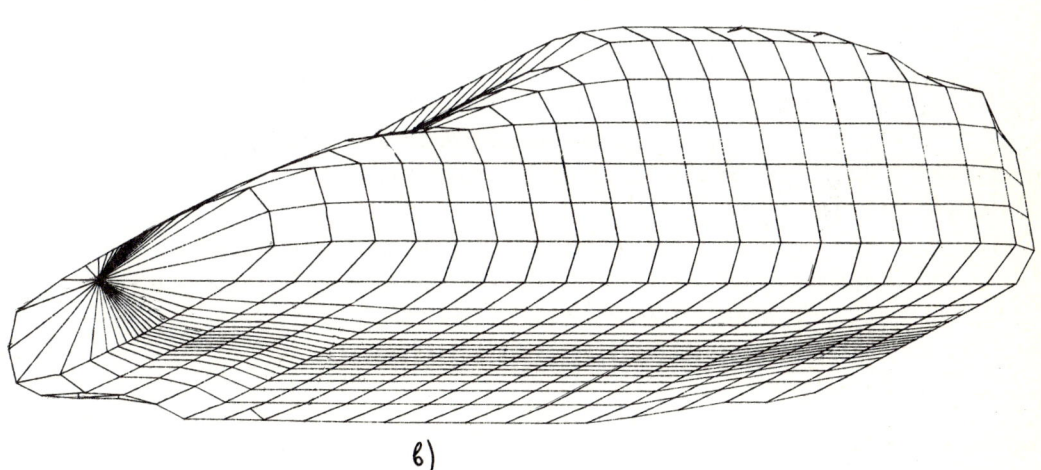

в)

fig.1

ON CONSERVATISM OF DIFFERENCE SCHEMES OF GAS DYNAMICS

Yu.I.Shokin
Computing Center
Siberian Branch Academy of Sciences of the USSR
Akademgorodok, 660036, Krasnoyarsk-36, USSR

Numerical modelling of gasdynamics flows indicates that one of the important properties of difference schemes is their complete conservatism [1].
Difference scheme for equations of gas dynamics are completely conservative if with the help of equivalent transformations, corresponding to the continuous case, one can obtain from it difference approximations of complete linear independent system of conservation laws and balance relations of gas dynamics equations, divergent difference schemes.
In this paper an approach to the study of this property by the differenyial approximation method is described [2]. New families of completely conservative difference schemes for gas dynamics equations in Eulerian coordinates are constructed. The results of numerical calculations are given.
1. Show how complete conservatism of difference schemes is related to the properties of their first differential approximations (FDA). Approximate a system of gasdynamics equations in Lagrangian coordinates

$$\widetilde{W}_t + A\widetilde{W}_m = 0,$$

$$\widetilde{W} = \begin{pmatrix} u \\ v \\ \varepsilon \end{pmatrix}, \quad A = \begin{pmatrix} 0 & P_v & P_\varepsilon \\ -1 & 0 & 0 \\ P & 0 & 0 \end{pmatrix}, \quad p = p(\varepsilon, v),$$

by the following difference scheme

$$\frac{\Delta_0 u^n(m)}{\tau} + \frac{\Delta_1}{h} p^{\zeta_1}(m - \tfrac{h}{2}) = 0,$$

$$\frac{\Delta_0 v^n(m + \frac{h}{2})}{\tau} - \frac{\Delta_1}{h} u^{\sigma_2}(m) = 0, \qquad (1)$$

$$\frac{\Delta_0 \varepsilon^n(m + \frac{h}{2})}{\tau} + p^{\sigma_3}(m + \frac{h}{2}) \frac{\Delta_1}{h} u^{\sigma_4}(m) = 0.$$

Here m is the Lagrangean variable, $\Delta_0 = T_0 - E$, $\Delta_1 = T_1 - E$, $T_0 \varphi(m,t) = \varphi(m,t+\tau)$, $E\varphi(m,t) = \varphi(m,t)$, $T_1 \varphi(m,t) = \varphi(m+h,t)$, $\varphi^{\sigma_i} = \sigma_i \varphi^{n+1} + (1 - \sigma_i)\varphi^n$ $(i = 1,2,3,4)$;

the rest of notations are in general use. The FDA of difference scheme (1) has the form

$$\widetilde{W}_t + A\widetilde{W}_m = \widetilde{Q}, \qquad (2)$$

where

$$\widetilde{Q} = \begin{pmatrix} -(0.5 - \sigma_1)(a^2 u_m)_m \\ (0.5 - \sigma_2) p_{mm} \\ -(0.5 - \sigma_3) a^2 u_m^2 - \tau(0.5 - \sigma_4) pp_{mm} \end{pmatrix},$$

Having multiplied system of equations (2) by the matrix

$$R = \begin{pmatrix} 1 & 0 & 0 \\ 0 & 1 & 0 \\ u & 0 & 1 \end{pmatrix},$$

obtain the system of equations

$$W_t + F_m = Q,$$

$$W = \begin{pmatrix} u \\ v \\ E \end{pmatrix}, \quad F = \begin{pmatrix} p \\ -u \\ up \end{pmatrix}, \quad E = \varepsilon + u^2/2,$$

$$Q = \begin{pmatrix} -(0.5 - \sigma_1)(a^2 u_m)_m \\ (0.5 - \sigma_2) p_{mm} \\ -(0.5 - \sigma_1) u (a^2 u_m)_m - \tau(0.5 - \sigma_3) a^2 u_m^2 - \tau(0.5 - \sigma_1) pp_{mm} \end{pmatrix}, (3)$$

which is the FDA of the difference scheme, approximating the system of equations of gas dynamics in a divergent form $W_t + F_m = 0$, obtained by multiplication by the matrix

$$R = \begin{pmatrix} 1 & 0 & 0 \\ 0 & 1 & 0 \\ u^{0.5}(m) & 0 & 1 \end{pmatrix}$$

of difference scheme (1). The following theorem takes place:

THEOREM. For difference scheme (1) to be completely conservative it is necessary and sufficient that it should possess the property K [2] and system of equations (3) be divergent.

CONSEQUENCE. If the theorem holds difference scheme (1) belongs to the class M [2].

2. Pass on to considering Eulerian coordinates. In this case the differential approximation method allowed us to construct a new class of completely conservative difference schemes. There was considered a family of difference schemes with sixteen parameters

$$\frac{\Delta_0}{\tau} \rho^{n-\frac{1}{2}}(x+\frac{h}{2}) + \frac{\Delta_1}{h} f(x) \Big|_{\alpha_1}^{\sigma_1, \sigma_2} = 0,$$

$$\frac{1}{2}\left\{\rho^{n+\frac{1}{2}}(x+\frac{h}{2})\frac{\Delta_0}{\tau} u^n(x) + \rho^{n-\frac{1}{2}}(x+\frac{h}{2})\frac{\Delta_0}{\tau} u^{n-1}(x)\right\} +$$

$$+ \frac{1}{2}\left\{ f(x+h) \Big|_{\alpha_2}^{\sigma_3, \sigma_4} \frac{\Delta_1}{h} u^n(x) + f(x) \Big|_{\alpha_2}^{\sigma_3, \sigma_4} \frac{\Delta_1}{h} u^n(x-h)\right\} +$$

$$+ \frac{1}{h} p^{\sigma_5}(x - \frac{h}{2}) = 0, \qquad (4)$$

$$\rho^{\sigma_6}(x+\frac{h}{2})\frac{\Delta_0}{\tau} \varepsilon^{n-\frac{1}{2}}(x+\frac{h}{2}) + p^{\sigma_{10}}(x+\frac{h}{2})\frac{\Delta_1}{h} u^{\sigma_{11}}(x) +$$

$$+ \alpha_3 f(x+h) \Big|_{\alpha_4}^{\sigma_7, \sigma_8} \frac{\Delta_1}{h} \varepsilon^{\sigma_9}(x+\frac{h}{2}) + (1-\alpha_3) \Big|_{\alpha_5}^{\sigma_7, \sigma_8} \frac{\Delta_1}{h} \varepsilon^{\sigma_9}(x-\frac{h}{2}) = 0$$

approximating with the first order of accuracy the system of equations of gas dynamics

$$B\frac{\partial W}{\partial t} + A\frac{\partial W}{\partial x} = 0,$$

$$W = \begin{pmatrix} \rho \\ u \\ \varepsilon \end{pmatrix}, B = \begin{pmatrix} 1 & 0 & 0 \\ 0 & \rho & 0 \\ 0 & 0 & \rho \end{pmatrix}, A = \begin{pmatrix} u & 0 & 0 \\ p_\rho & f & p_\varepsilon \\ 0 & p & f \end{pmatrix}$$

$$f = \rho u \quad p = p(\varepsilon, \rho),$$

here x is the Eulerian variable, $\Delta_0 = T_0 - E$,

$$f(x)\Big|_{\alpha_1}^{\delta_1,\delta_2} = (\alpha_1 \rho^{\delta_1}(x + \tfrac{h}{2}) + (1-\alpha_1)\rho^{\delta_1}(x - \tfrac{h}{2}))u^{\delta_2}(x),$$

$$\rho^{\delta_1}(x + \tfrac{h}{2}) = \delta_1 \rho^{n+\tfrac{1}{2}}(x + \tfrac{h}{2}) + (1-\delta_1)\rho^{n-\tfrac{1}{2}}(x + \tfrac{h}{2}),$$

$$u^{\delta_2}(x) = \delta_2 u^{n+1}(x) + (1-\delta_2) u^n(x)$$

the rest of notations were introduced analogously or are conventional; $\delta_1, \ldots, \delta_{11}, \alpha_1, \ldots, \alpha_5$ are the parameters, $0 \le \delta_i \le 1$, $0 \le \alpha_j \le 1$, $i = 1,\ldots,11$, $j = 1,\ldots,5$. All thermodynamical values are taken at the half-integer points ($n+\tfrac{1}{2}$, $i+\tfrac{1}{2}$), and velocity at the integer points (n, i).

In accordance with the definition of complete conservatism, there were considered transformations of difference scheme (4), assigned by the following matrices:

$$\begin{pmatrix} 1 & 0 & 0 \\ u^n(x) & 1 & 0 \\ 0 & 0 & 1 \end{pmatrix}, \begin{pmatrix} 1 & 0 & 0 \\ 0 & 1 & 0 \\ \varepsilon^{\delta_1}(x+\tfrac{h}{2}) & 0 & 1 \end{pmatrix},$$

$$\begin{pmatrix} 1 & 0 & 0 \\ \tfrac{1}{2}(u^n(x))^2 & u^n(x) & 0 \\ 0 & 0 & 1 \end{pmatrix}, \begin{pmatrix} 1 & 0 & 0 \\ 0 & 1 & 0 \\ -\dfrac{p^{\delta_2}(x+\tfrac{h}{2})}{\rho^{\delta_3}(x+\tfrac{h}{2})} & 0 & 1 \end{pmatrix},$$

(5)

$$\begin{pmatrix} 1 & 0 & 0 \\ u^n(x) & 1 & 0 \\ \tfrac{1}{2}(u^n(x))^2 + \varepsilon^{\delta_1}(x+\tfrac{h}{2}) & u^n(x) & 1 \end{pmatrix}.$$

It is shown that at the parameters

$$\alpha_1 = \alpha_2 = \alpha_3, \quad \alpha_4 = 1, \quad \alpha_5 = 0,$$
$$1 - \alpha_6 = \delta_3 = \delta_7 = \delta_9 = \delta_1,$$
$$\delta_2 = \delta_4 = \delta_8 = \delta_{11} = 0, \quad \delta_5 = \delta_{10},$$
$$\delta_1 = 0 \text{ or } 1$$

difference scheme (1) is completely conservative. There were obtained constraints on the parameters of the pransformations $\chi_i (i=1,2,3)$:

$$\gamma_1 = \gamma_3 = \sigma_1, \quad \gamma_2 = \sigma_3.$$

On applying the transformation, assigned by matrix (5), to difference scheme (4) it changes into a scheme approximating gas dynamics equations in a divergent form.
Write the difference equations obtained with the above transformations:

1) Difference equation of impulse:

$$\frac{\Delta_0}{2\tau}\left[\rho^{n-\frac{1}{2}}(x + \tfrac{h}{2})(u^n(x) + u^{n-1}(x))\right] +$$

$$+ \frac{\Delta_1}{2h}\left[f(x)\Big|_{\alpha_1}^{\sigma_1,0}(u^n(x) + u^n(x - h))\right] + \frac{\Delta_1}{h}p^{\sigma_5}(x - \tfrac{h}{2}) = 0.$$

2) Difference equation of internal energy balance:

$$\frac{\Delta_0}{\tau}\left[\rho^{n-\frac{1}{2}}(x + \tfrac{h}{2})\varepsilon^{n-\frac{1}{2}}(x + \tfrac{h}{2}) + p^{\sigma_5}(x + \tfrac{h}{2})\frac{\Delta_1}{h}u^n(x) +\right.$$

$$\left.+ \frac{\Delta_1}{h}\left\{[\alpha_1\rho^{\sigma_1}(x + \tfrac{h}{2})\varepsilon^{\sigma_1}(x + \tfrac{h}{2}) + (1-\alpha_1)\rho^{\sigma_1}(x-\tfrac{h}{2})\varepsilon^{\sigma_1}(x-\tfrac{h}{2})]u^n(x)\right\} = 0.\right.$$

3) Difference equation of kinetic energy balance:

$$\frac{\Delta_0}{\tau}\left[\rho^{n-\frac{1}{2}}(x + \tfrac{h}{2})u^{n-1}(x)\right] +$$

$$+ \frac{\Delta_1}{2h}\left[f(x)\Big|_{\alpha_1}^{\sigma_1,0} u^n(x)u^n(x - h)\right] + u^n(x)\frac{\Delta_1}{h}p^{\sigma_5}(x - \tfrac{h}{2}) = 0.$$

4) Entropy equation

$$\rho^{1-\sigma_1}(x + \tfrac{h}{2})\left[\frac{\Delta_0}{\tau}\varepsilon^{n-\frac{1}{2}}(x + \tfrac{h}{2}) + p^{\sigma_5}(x+\tfrac{h}{2})\frac{\Delta_0}{\tau}\left(\frac{1}{\rho^{n-\frac{1}{2}}(x\,\tfrac{h}{2})}\right)\right] +$$

$$+ \alpha_1 u^n(x+h)\rho^{\sigma_1}(x+\tfrac{3h}{2})\left[\frac{\Delta_1}{h}\varepsilon^{\sigma_1}(x+\tfrac{h}{2}) + p^{\sigma_5}(x+\tfrac{h}{2})\frac{\Delta_1}{h}\left(\frac{1}{\rho^{\sigma_1}(x+\tfrac{h}{2})}\right)\right] +$$

$$+ (1 - \alpha_1)u^n(x)\rho^{\sigma_1}(x - \tfrac{h}{2})\left[\frac{\Delta_1}{h}\varepsilon^{\sigma_1}(x-\tfrac{h}{2}) + p^{\sigma_5}(x-\tfrac{h}{2})\frac{\Delta_1}{h}\left(\frac{1}{\rho^{\sigma_1}(x-\tfrac{h}{2})}\right)\right] = 0$$

Having added the second and the third equations obtain a difference scheme approximating an equation of full energy. The proposed family of complete conservative schemes possess the same undesirable property as the scheme from work [3], namely for approximating kinetic energy $\frac{1}{2}\rho u^2$ the product of velocities at different layers is taken.

The present method of constructing conservative schemes can be applied to other multi-parametrical families of difference schemes.

LITERATURE

1. Samarsky A.A., Popov Yu.P. Raznostnye metody reshenia zadach gazovoi dinamiki.- M.: Nauka, 1980, 352 s.
2. Ivanov F.V., Fedotova Z.I., Shokin Yu.I. On complete conservatism of difference schemes.- In: Numerical methods in fluid dynamics / Ed.by N.N.Yanenko and Yu.I.Shokin.-Moscow: Mir, Publishers, 1984, p.225-244.
3. Kuzmin A.V., Makarov V.L., Meladze G.V. Ob odnoi polnostyu konservativnoi skheme dlya uravnenii gasovoi dinamiki v peremennykh Ailera.- Zhurn. vychisl. matem.fiz., 1980, t.20, N 1, s.171-181.

A NUMERICAL ANALYSIS OF A NONLINEAR EIGENVALUE PROBLEM OCCURRING IN VISCOUS OSCILLATIONS OF A SUPPORTED DROP

Massimo STRANI, Filippo SABETTA
Dipartimento di Meccanica e Aeronautica
Università "La Sapienza", ROMA

Under the assumption of zero gravity and small surface deformation, the problem of axisymmetric viscous vibrations of a liquid drop, immersed in an outer fluid and in partial contact with a spherical bowl may be described by the set of linearized Navier Stokes equations

$$\nabla \cdot \underline{v} = 0$$

$$\rho \gamma \underline{\omega} = \mu \nabla \times \nabla \times \underline{\omega}$$

$$\underline{\omega} = \nabla \times \underline{v}$$

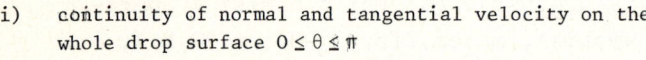

associated with the following set of boundary conditions

i) continuity of normal and tangential velocity on the whole drop surface $0 \leq \theta \leq \pi$

$$v_r^i(R,\theta) = v_r^o(R,\theta) = -\gamma z(\theta)$$

$$v_\theta^i(R,\theta) = v_\theta^o(R,\theta)$$

ii) continuity of tangential stresses and balance of the normal momentum on the free portion of the drop surface $0 < \theta < \theta_0$

$$s_{r\theta}^i(R,\theta) = s_{r\theta}^o(R,\theta)$$

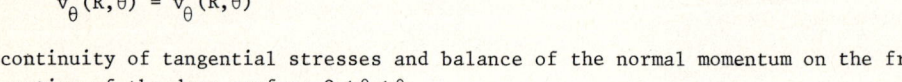

where

$$s_{r\theta} = \mu \left[\frac{v_{r,\theta}}{r} + r \left(\frac{v_\theta}{r}\right)_{,r} \right]$$

$$s_{rr} = -p + 2\mu v_{r,r}$$

iii) no slip condition and no radial deformation on the supported portion of the drop surface $\theta_0 \leq \theta \leq \pi$

$$v_\theta(R,\theta) = 0$$

$$z(\theta) = 0$$

Moreover the following integral condition expresses the conservation of the drop volume

$$\int_0^\pi z \sin\theta \, d\theta = 0$$

The expansion of the scalar and vector potential in series of Legendre polynomials and spherical Bessel functions and the inversion of the differential equation for the free surface shape by the use of the Green function method, reduce the problem

[Strani & Sabetta 1986] to the following nonlinear eigenvalue problem

$$\frac{1}{\lambda^2} z(x) = \int_{-1}^{1} K(x,\xi,\sigma) z(\xi) d\xi$$

where $\lambda = b + i\omega$ is the eigenvalue, $z(x)$ is the free surface shape and the analytical expression of the kernel $K(x,\xi,\sigma)$ is known. A projection of the above equation on the Legendre polynomials in $[-1,1]$ leads to

$$\frac{1}{\lambda^2} z_i = \sum_{j=1}^{\infty} K_{ij}(\sigma) z_j \qquad (1)$$

which allows, by truncating the sums to N terms, for a numerical study of the eigenvalue problem to be carried on.

The matrix K_{ij} depends on the support parameter $a = \cos\theta_0$, on the nondimensional densities and viscosities ρ^i/ρ^*, ρ^o/ρ^*, μ^i/μ^*, μ^o/μ^*, with μ^* a reference viscosity, on a viscosity parameter

$$\varepsilon = \mu^*/(\rho^* R\sigma)^{1/2}$$

and on the nondimensional value of the eigenvalue itself

$$\lambda = (\frac{\rho^* \gamma^2 R^3}{\sigma})^{1/2}$$

Numerical solution of the eigenvalue problem

The eigenvalue problem (1) may be numerically solved, as is usually done, by truncating the expansions to N terms (typical values used in the calculation $N = 10$, 20, 30), and finding the roots of the characteristic equation

$$\det\left[\frac{1}{\lambda^2} U_N + K_N(\lambda,\varepsilon,a)\right] = 0$$

where U_N is the identity $N \times N$ matrix, K_N is the $N \times N$ truncation of the infinite matrix K in (1), and the non dimensional values of densities and viscosities are assumed to be given. For each value of the support angle ψ_0 and $\varepsilon = 0$ (inviscid case) the eigenvalues are known [Strani & Sabetta 1984]. To evaluate the function $\lambda = \lambda(\varepsilon)$ at $\varepsilon = n\Delta\varepsilon$ ($n = 1, 2, \ldots$) we used a Newton-Ralphson procedure which starts with the approximate value $\lambda(n\Delta\varepsilon) \simeq \lambda((n-1)\Delta\varepsilon)$. For the higher values of ε where $|d\lambda/d\varepsilon|$ is greater, we could avoid the use of very small values of $\Delta\varepsilon$ by giving to $\lambda(n\Delta\varepsilon)$ an initial approximate value obtained with a linear extrapolation from the values at $(n-1)\Delta\varepsilon$ and $(n-2)\Delta\varepsilon$.

We present here a numerical study of the eigenvalue problem, for the case of a drop in a vacuum ($\rho^o = \mu^o = 0$). In fact, for small values of the support angle ψ_0 and large values of ε, the behaviour of the real and imaginary part of the eigenvalues is rather cumbersome as shown in Fig. 1.

The most interesting features of the plot are the wiggles appearing for large ε values on the branches of the solution with $\omega \neq 0$ and the peak of ω that appears before the bifurcation points a, b, c in Fig. 1. Moreover it may be observed that along each of the branches with $\omega = 0$ and b decreasing with ε, the mode number varies in the discontinuous way indicated in Fig. 1. All these features could suggest the occurrence of spurious solutions due to the finite truncation. To ascertain if this was the cause, an analysis of the influence of the truncation number on the second eigenvalue has been performed.

When N is increased from 10 to 20 the wiggle amplitude is strenghtened instead of smoothened (Fig. 2.a) and for N = 30 the wiggles seems to converge to a shape indepen

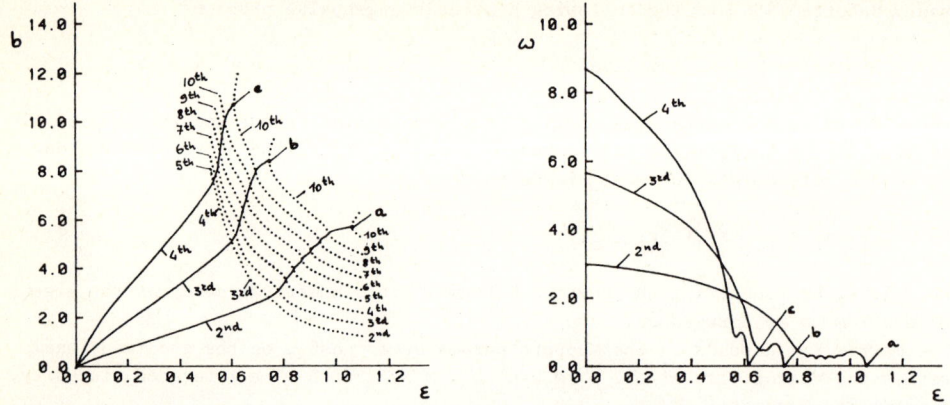

Fig. 1 - Roots $\lambda = b + i\omega$ of the characteristic equation for $\psi_0 = .1$ and truncation N = 10. ——— $\omega \neq 0$, ... $\omega = 0$.
The ordinal number marked near each branch indicates the type of the corresponding eigenmode.

dent on N (Fig. 2.b). The consequent induction that the wiggles are not due to the finite truncation should be supported by more numerical results than we could do, due to the high costs of the calculations when N > 20. We postpone to the end a more complete discussion on the origin of the wiggles.

On the contrary a great influence of the truncation number N on the last bifurcation points (like points a, b, c of Fig. 1) is observed, as shown in Fig. 2 for the second eigenvalue. Numerical evidence (see also Fig. 2.a) shows that these bifurcation points occur at values of ε that increase in a regular way with N and should disappear as $N \to \infty$. In fact from these points a branch bifurcates with $\omega = 0$, $db/d\varepsilon < 0$ and the mode number equal to the truncation number N. When N is increased, points like a, b, c etc. occur at larger values of ε, and a more complete and accurate plot

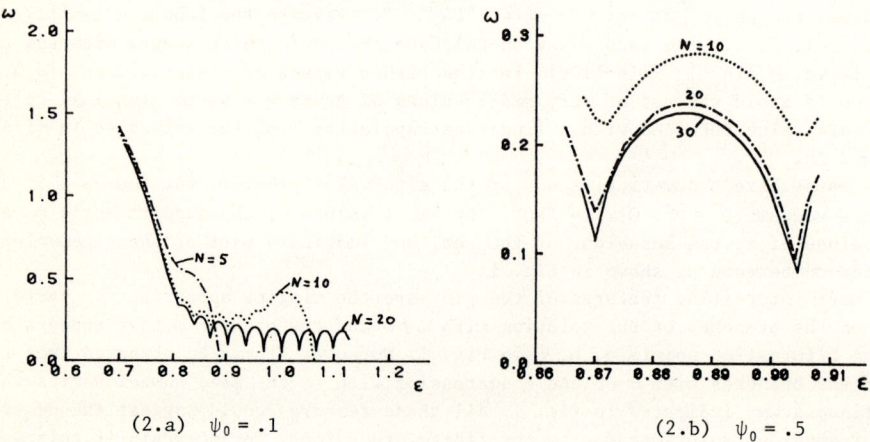

(2.a) $\psi_0 = .1$ (2.b) $\psi_0 = .5$

Fig. 2 - Influence of the truncation number N on the eigenvalue corresponding to a second mode for the drop. The real parts of the eigenvalues are not plotted.

of the solutions of the eigenvalue problem is obtained, especially for the purely real, b-decreasing branches. Numerical results suggest that, in the limit $N \to \infty$, the plot of the solutions is made of an infinite set od branches with $\omega \neq 0$ and $db/d\varepsilon > 0$ along which the mode number is substantially independent on ε (solid lines in Fig.1): and an infinite set of purely real, b-decreasing branches, each one extending in the whole range $0 < \varepsilon < \infty$, along which an iterated transition from an n-th mode to an (n-1)-th mode is observed (dotted lines in Fig. 1).

In order to have a better insight into the phenomenon of the above-mentioned wiggles we also made a comparison between the cases $\psi_0 = 0$ and $\psi_0 = 10^{-6}$ (Fig. 3).

For $\psi_0 = 0$ and, say, the second mode a couple of branches of complex conjugate solution bufurcates, at a certain $\bar{\varepsilon}$ (points a in Fig. 3) into two purely real branches along which b approaches 0 and ∞, respectively, as $\varepsilon \to \infty$. Moreover the intersection of these last two branches with the decreasing purely real branches corresponding to higher modes gives rise to other bifurcation points (points b, c, d in Fig. 3).

The corresponding plot for $\psi_0 = 10^{-6}$, even if very close to the one for $\psi_0 = 0$, appears to be much more complicated. Some bifurcation point has disappeared, someone else has only slightly changed its position, some new one has suddenly appeared.

A careful inspection of Fig. 3 indicates that the central point is to understand the evolution of the bifurcation points such as a, b, c, d when ψ_0 is increased, and the way they all can disappear for large values of ψ_0. This mechanism is qualitatively shown in the three-dimensional plot of Fig. 4.

Bifurcation points such as b) disappear as soon as $\psi_0 \neq 0$ in the simple way indicated in Fig. 4.a and 4.b.

At the same time each of the bifurcation points like c, d, e (say c) gives rise to a couple of points (say c' and c") which are connected by two branches of complex conjugate solutions where the imaginary part first increases from zero in c' to a maximum to decrease again to zero in c". The points d' and d" corresponding to d, are clearly seen in Fig. 3.

At larger values of ψ_0 points c' and a coalesce into the bifurcation point c' (also indicated in fig. 3), which disappears when ψ_0 is still increased. In an analogous way we observe coalescence and then disappearance of points c"-d', d"-e' and so on for all the other bifurcation points.

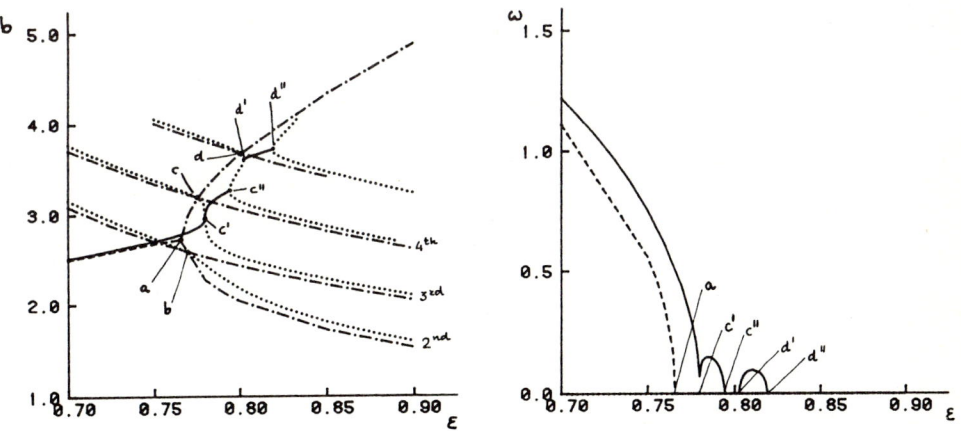

Fig. 3 - Comparison between solution curves of the eigenvalue problem for $\psi_0 = 0$ [--------- $\omega \neq 0$, - - - - - - $\omega = \bar{0}$] and for $\psi_0 = 10^{-6}$ [——— $\omega \neq 0$, ·········· $\omega = 0$].

The wiggles in the branches of the solution are thus seen to be intimately connected with the fact that all bifurcation points in the plot for $\psi_0 = 0$ must disappear when ψ_0 is increased. Since each branch of the solution obtained for $\psi_0 = 0$ is not affected by the truncation number N we are led to the conclusion that the wiggles observed at large values of ε and relatively small values of ψ_0 do not depend on errors introduced by the finite truncation but indicate a singular behaviour of the eigenvalue problem when the perturbation $(a+1)$ tends to zero.

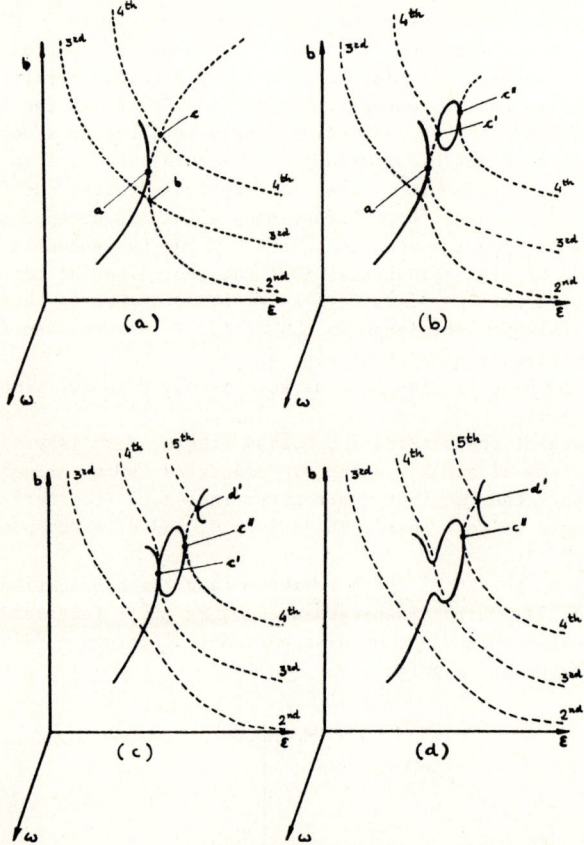

Fig. 4 - Qualitative plot of the dependence on ε of the eigenvalues $\lambda = b + i\omega$. The plot indicates the way the bifurcation points a, b, c... in fig. 4.a evolve and disappear when increasing ψ_0. [——— $\omega \neq 0$, —·—·— $\omega = 0$].

References

1. Strani M., Sabetta F., 1984, Journ. Fl. Mech., **141**, 233.
2. Strani M., Sabetta F., 1986, Dipartimento di Meccanica e Aeronautica, Roma, Internal Report.

Algebraic Model of Large Eddy Simulation

M.D.Su

Qinghua University
Beijing, China

1. Introduction

Large Eddy Simulation (LES) is a recently developed and hopeful method to study the turbulence. The new idea of LES was first advanced by Smagorinsky in 1963. In 1970 it was first used to calculate the turbulence in straight channel and planetary boundary layer by Deardorff. Then the method of LES is improved and used by Moin, Schumann, Antonopoulos-Dannis et al. Some advances were achieved in LES of turbulence. Modelling of the influence of wall and its curvature on structure of turbulence is still a difficult problem of LES. The purpose of this paper is that the subgrid Reynolds's stress in LES is calculated by means of algebraic model and the method of LES is used to simulate the turbulence in a straight channel. The results will be compared with Schumann's result. We hope this method can be expanded in the LES of turbulence in a curvature channel.

2. Principle Equations for the large scale values

According to the basic ideal of LES, the physical values as u,v,w,p, are divided into two parts: a large scale value and a small scale value. The large scale value is filtered value of the original value. The small scale value is difference between the original value of the physical value and its filtered value. In this paper the following filter function is used:

$$G(x|x') = \begin{cases} \dfrac{1}{\Delta_1 \Delta_2 \Delta_3} & |x_i - x_i'| \leq \Delta_i/2 \quad i=1,2,3 \\ 0 & |x_i - x_i'| > \Delta_i/2 \quad i=1 \text{ or } 2 \text{ or } 3 \end{cases} \quad (1)$$

The filtered value is written as $\bar{f}$ and

$$\bar{f} = \int_{\Delta V} G \cdot f \, d(\Delta V) \qquad f' = f - \bar{f} \qquad (2)$$

ΔV is volume of grid and assumed to be so small that h_i in it can be considered as a constant. Therefore

$$\overline{g(q_i) \cdot f} = \int_{\Delta V} G g(q_i) \cdot f \, d(\Delta V) \qquad (3)$$

where $g(q_i)$ is a function of q_i, and g is a geometric value (for example, h_i).

Filtering the Navier-Stokes equations with filter function (1), we can get the princple equation for the large scale values u,v,w, p:

$$\frac{1}{h_k}\frac{\partial \bar{v}_k}{\partial g_k} + \frac{\bar{v}_k}{h_e h_k}\frac{\partial h_e}{\partial g_k} - \frac{\bar{v}_k}{h_k^2}\frac{\partial h_k}{\partial g_k} = 0 \qquad (4)$$

$$\frac{\partial \bar{v}_i}{\partial t} + \frac{\bar{v}_k}{h_k}\frac{\partial \bar{v}_i}{\partial g_k} + \frac{\bar{v}_i \bar{v}_k}{h_i h_k}\frac{\partial h_i}{\partial g_k} - \frac{\bar{v}_k \bar{v}_k}{h_i h_k}\frac{\partial h_k}{\partial g_i}$$

$$= -\frac{1}{h_i}\frac{\partial}{\partial g_i}\left(\frac{\bar{p}}{\rho}\right) + \left\{\frac{1}{h_k}\frac{\partial \tau_{ik}}{\partial g_k} + \frac{\tau_{ik}}{h_e h_k}\frac{\partial h_e}{\partial g_e} - \frac{\tau_{ik}}{h_k^2}\frac{\partial h_k}{\partial g_k} - \frac{\tau_{kk}}{h_i h_k}\frac{\partial h_k}{\partial g_i} + \frac{\tau_{ik}}{h_i h_k}\frac{\partial h_i}{\partial g_k}\right\} \qquad (5)$$

where $\tau_{ik} = \bar{\tau}_{ik} + \tau_{ik}^* = 2\nu \bar{S}_{ik} + (-\overline{v_i' v_k'}) \qquad v_i' = v_i - \bar{v}_i$

$$\frac{\partial \bar{E'}}{\partial t} + \frac{\bar{v}_k}{h_k}\frac{\partial \bar{E'}}{\partial g_k} + \frac{\overline{v_k' v_i'}}{h_i}\frac{\partial \bar{v}_i}{\partial g_k} + \frac{\overline{v_k' \partial E'}}{h_k \partial g_k} - v_k \left(\frac{\overline{v_k' v_k'}}{h_e h_k}\frac{\partial h_e}{\partial g_k} - \frac{\overline{v_k' v_k'}}{h_e h_k}\frac{\partial h_k}{\partial g_e}\right)$$

$$-v_\ell \left(\frac{\overline{v_\ell' v_k'}}{h_\ell h_k} \frac{\partial h_\ell}{\partial g_k} - \frac{\overline{v_k' v_k'}}{h_\ell h_k} \frac{\partial h_k}{\partial g_\ell} \right) = -\frac{1}{h_k} \frac{\partial}{\partial g_k} \left(\overline{v_k' \frac{p'}{\rho}} \right) + \frac{\overline{p'}}{\rho h_k} \frac{\partial v_k}{\partial g_k} - 2\nu \frac{1}{\hbar} \frac{\partial}{\partial g_k} \left(\overline{v_\ell' S_{k\ell}'} \frac{\hbar}{h_k} \right) - \varepsilon \quad (6)$$

where $\hbar = h_1 h_2 h_3$; $\varepsilon = 2\nu \overline{S_{k\ell}' S_{k\ell}'}$; $\overline{E'} = \frac{1}{2} \overline{v_i' v_i'}$

$$\frac{\partial \overline{v_i' v_j'}}{\partial t} + \frac{\overline{v_k}}{h_k} \frac{\partial \overline{v_i' v_j'}}{\partial g_k} + \overline{v_i' v_k'} \frac{1}{h_k} \frac{\partial \overline{v_j}}{\partial g_k} + \overline{v_j' v_k'} \frac{1}{h_k} \frac{\partial \overline{v_i}}{\partial g_k} + \overline{v_i' v_j'} \left(\frac{\overline{v_k}}{h_i h_k} \frac{\partial h_i}{\partial g_k} + \frac{\overline{v_k}}{h_j h_k} \frac{\partial h_j}{\partial g_k} \right) + \overline{v_j' v_k'} \left(\frac{\overline{v_i}}{h_i h_k} \frac{\partial h_i}{\partial g_k} - \frac{2\overline{v_k}}{h_i h_k} \frac{\partial h_k}{\partial g_i} \right)$$

$$+ \overline{v_i' v_k'} \left(\frac{\overline{v_j}}{h_j h_k} \frac{\partial h_j}{\partial g_k} - \frac{2\overline{v_k}}{h_j h_k} \frac{\partial h_k}{\partial g_j} \right) + \frac{1}{h_k} \frac{\partial}{\partial g_k} \overline{v_i' v_j' v_k'} + \overline{v_i' v_j' v_k'} \left(\frac{1}{h_k h_\ell} \frac{\partial h_\ell}{\partial g_k} - \frac{1}{h_k^2} \frac{\partial h_k}{\partial g_k} \right)$$

$$-\left(\frac{\overline{v_k' v_k' v_j'}}{h_i h_k} \frac{\partial h_k}{\partial g_i} - \frac{\overline{v_k' v_k' v_i'}}{h_j h_k} \frac{\partial h_k}{\partial g_j} \right) = -\left\{ \frac{1}{h_i} \frac{\partial}{\partial g_i} \left(\overline{v_j' \frac{p'}{\rho}} \right) + \frac{1}{h_j} \frac{\partial}{\partial g_j} \left(\overline{v_i' \frac{p'}{\rho}} \right) \right\} + \frac{\overline{p'}}{\rho} \left(\frac{1}{h_i} \frac{\partial v_j'}{\partial g_i} + \frac{1}{h_j} \frac{\partial v_i'}{\partial g_j} \right)$$

$$+ \nu \left\{ \Delta \overline{v_i' v_j'} + \nabla \left[\left(v_i' \frac{h_k}{h_j} \frac{\partial}{\partial g_i} \left(\frac{v_j'}{h_k} \right) + v_j' \frac{h_k}{h_i} \frac{\partial}{\partial g_j} \left(\frac{v_i'}{h_k} \right) - \frac{\overline{v_i' v_j'}}{h_i h_j h_k} \frac{\partial h_i h_j}{\partial g_k} \right) \vec{e}_k \right] + \frac{2}{\hbar} \frac{\partial}{\partial g_j} \left(\frac{\hbar}{h_j} \cdot \frac{\overline{v_i' v_\ell'}}{h_\ell} \frac{1}{h_j} \frac{\partial h_j}{\partial g_\ell} \right) \right.$$

$$\left. + \frac{2}{\hbar} \frac{\partial}{\partial g_i} \left(\frac{\hbar}{h_i} \cdot \frac{\overline{v_j' v_\ell'}}{h_\ell} \frac{1}{h_i} \frac{\partial h_i}{\partial g_\ell} \right) \right\} - 2\nu \left\{ \frac{\overline{S_{j k}'}}{h_k} \frac{\partial}{\partial g_k} \left(\frac{v_i'}{h_j} \right) + \frac{\overline{S_{i k}'}}{h_k} \frac{\partial}{\partial g_k} \left(\frac{v_j'}{h_i} \right) + \overline{S_{k k}' \left(\frac{v_i'}{h_j h_k} \frac{\partial h_k}{\partial g_j} + \frac{v_j'}{h_i h_k} \frac{\partial h_k}{\partial g_i} \right)} \right\} \quad (7)$$

It should be noted that in all above equations the summation convention is applied only for indices k and 1, not for indices i,j.

3. Algebraic model

To close the equation system (4)-(7), the another modelling relaxation for small scale values must be added.

The simplest model of subgrid Reynolds stress $\overline{v_i' v_j'}$ is

$$-(\overline{v_i' v_j'})_{iso} = \nu^* \overline{S}_{ij} + \frac{2}{3} \delta_{ij} \overline{E'} \quad (8)$$

where

$$\nu^* = f(c\Delta)^2 S ; \quad S = \left\{ \frac{1}{2} \overline{S}_{k\ell} \overline{S}_{k\ell} \right\}^{1/2} ; \quad \Delta = \left\{ \hbar \Delta g_1 \Delta g_2 \Delta g_3 \right\}^{1/3} \quad (9)$$

c is a selected constant, f is a parameter depending on the field of large scale values. $\overline{E'}$ and f will be determined according to the following equation of algebraic model:

After the Rotta's and Lauder suggestion, it is assumed

$$\frac{\overline{p'}}{\rho} \left(\frac{1}{h_i} \frac{\partial v_j'}{\partial g_i} + \frac{1}{h_j} \frac{\partial v_i'}{\partial g_j} \right) = -C_1 \frac{\varepsilon}{E'} \left(\overline{v_i' v_j'} - \frac{2}{3} \delta_{ij} \overline{E'} \right)$$

$$- \frac{C_2 + 8}{11} \left(P_{ij} - \frac{2}{3} P \delta_{ij} \right) - \frac{30 C_2 - 2}{55} \cdot 2 \overline{E'} S_{ij} - \frac{8 C_2 - 2}{11} \left(D_{ij} - \frac{2}{3} P \delta_{ij} \right)$$

$$+ \left\{ \frac{1}{8} \frac{\varepsilon}{E'} \left(\overline{v_i' v_j'} - \frac{2}{3} \delta_{ij} \overline{E'} \right) + \frac{3}{200} \left(P_{ij} - D_{ij} \right) \right\} \cdot \frac{E'^{3/2}}{\varepsilon} \cdot f(x_2) \quad (10)$$

where x_2 is a normal distance to wall. we assume $f(x_2) = 1/x_2$, u_τ is shearing stress velocity, ν is kinetic viscous, and

$$-P_{ij} = \frac{\overline{v_i' v_k'}}{h_k} \frac{\partial \overline{v_j}}{\partial g_k} + \frac{\overline{v_j' v_k'}}{h_k} \frac{\partial \overline{v_i}}{\partial g_k} - C_p \overline{v_i} \left(\frac{\overline{v_j' v_k'}}{h_j h_k} \frac{\partial h_j}{\partial g_k} - \frac{\overline{v_k' v_k'}}{h_j h_k} \frac{\partial h_k}{\partial g_j} \right) - C_p \overline{v_j} \left(\frac{\overline{v_i' v_k'}}{h_i h_k} \frac{\partial h_i}{\partial g_k} - \frac{\overline{v_k' v_k'}}{h_i h_k} \frac{\partial h_k}{\partial g_i} \right)$$

$$-D_{ij} = \frac{\overline{v_i' v_k'}}{h_j} \frac{\partial \overline{v_k}}{\partial g_j} + \frac{\overline{v_j' v_k'}}{h_i} \frac{\partial \overline{v_k}}{\partial g_i} - C_p \overline{v_i} \left(\frac{\overline{v_j' v_k'}}{h_j h_k} \frac{\partial h_j}{\partial g_k} - \frac{\overline{v_k' v_k'}}{h_j h_k} \frac{\partial h_k}{\partial g_j} \right) - C_p \overline{v_j} \left(\frac{\overline{v_i' v_k'}}{h_i h_k} \frac{\partial h_i}{\partial g_k} - \frac{\overline{v_k' v_k'}}{h_i h_k} \frac{\partial h_k}{\partial g_i} \right)$$

$$P = \frac{1}{2} P_{\ell\ell}$$

(11)

Because the discussed flow is flow with large Re number and steady turbulence (in the statistic mean), so

$$\frac{\partial \overline{v_i' v_j'}}{\partial t} \sim 0 \qquad \frac{\partial \overline{v_i' v_j'}}{\partial g_k} \sim 0 \qquad i \neq j \neq k$$

The third order relaxation of velocity and the relaxation between the fluctuating velocity and fluctuating pressure are ignored. In addition, it is assumed

$$\varepsilon_{ij} = 2\nu \left\{ \frac{\overline{S_{jk}'}}{h_k} \frac{\partial}{\partial g_k} \left(\frac{v_i'}{h_j}\right) + \frac{\overline{S_{ik}'}}{h_k} \frac{\partial}{\partial g_k} \left(\frac{v_j'}{h_i}\right) + \overline{S_{kk}'} \left(\frac{v_i'}{h_j h_k} \frac{\partial h_k}{\partial g_j} + \frac{v_j'}{h_i h_k} \frac{\partial h_k}{\partial g_i}\right) \right\} = \frac{\overline{v_i' v_j'}}{E'} \varepsilon \delta_{ij} \tag{12}$$

$$\varepsilon = C_E \frac{\overline{E'}^{3/2}}{\Delta} \tag{13}$$

Substituting (8)–(13) into (6) (7), we get equations

$$\overline{v_i' v_j'} \left(\frac{\overline{v_k}}{h_i h_k} \frac{\partial h_i}{\partial g_k} + \frac{\overline{v_k}}{h_j h_k} \frac{\partial h_j}{\partial g_k} + (C_1 + \delta_{ij}) C_E \frac{\overline{E'}^{1/2}}{\Delta} \right)$$

$$+ \overline{v_i' v_k'} \left(\frac{1}{h_k} \frac{\partial \overline{v_j}}{\partial g_k} + \frac{\overline{v_j}}{h_j h_k} \frac{\partial h_j}{\partial g_k} - 2 \frac{\overline{v_k}}{h_j h_k} \frac{\partial h_k}{\partial g_j} + \frac{C_2+8}{11} \left[\frac{1}{h_k} \frac{\partial \overline{v_j}}{\partial g_k} - \frac{C_P \overline{v_j}}{h_k h_i} \frac{\partial h_i}{\partial g_k} \right] + \frac{8C_2-2}{11} \left[\frac{1}{h_j} \frac{\partial \overline{v_k}}{\partial g_j} - \frac{C_P \overline{v_j}}{h_i h_k} \frac{\partial h_i}{\partial g_k} \right] \right)$$

$$+ \overline{v_j' v_k'} \left(\frac{1}{h_k} \frac{\partial \overline{v_i}}{\partial g_k} + \frac{\overline{v_i}}{h_i h_k} \frac{\partial h_i}{\partial g_k} - 2 \frac{\overline{v_k}}{h_i h_k} \frac{\partial h_k}{\partial g_i} + \frac{C_2+8}{11} \left[\frac{1}{h_k} \frac{\partial \overline{v_i}}{\partial g_k} - \frac{C_P \overline{v_i}}{h_k h_j} \frac{\partial h_j}{\partial g_k} \right] + \frac{8C_2-2}{11} \left[\frac{1}{h_i} \frac{\partial \overline{v_k}}{\partial g_i} - \frac{C_P \overline{v_i}}{h_j h_k} \frac{\partial h_j}{\partial g_k} \right] \right)$$

$$+ \overline{v_k' v_k'} \frac{9C_2+6}{11} \left(\frac{C_P}{h_k} \left\{ \frac{\overline{v_i}}{h_j} \frac{\partial h_k}{\partial g_j} + \frac{\overline{v_j}}{h_i} \frac{\partial h_k}{\partial g_i} \right\} - \frac{2}{3} \frac{C_P}{h_k h_\ell} \frac{\overline{v_k}}{\partial g_\ell} \delta_{ij} \right) - \overline{v_\ell' v_k'} \frac{6C_2+4}{11} \frac{1}{h_k} \left(\frac{\partial \overline{v_\ell}}{\partial g_k} - \frac{C_P \overline{v_\ell}}{h_\ell} \frac{\partial h_\ell}{\partial g_k} \right) \delta_{ij}$$

$$\frac{2}{3} C_1 C_E \frac{\overline{E'}^{3/2}}{\Delta} \delta_{ij} - \frac{60 C_2 - 4}{55} \overline{E'} \overline{S}_{ij}$$

$$+ \left\{ \frac{C_E}{8} \frac{\overline{E'}^{3/2}}{\Delta} (\overline{v_i' v_j'} - \frac{2}{3} \delta_{ij} \overline{E'}) + \frac{3}{200} (P_{ij} - D_{ij}) \right\} \frac{\Delta}{C_E} f(x_2) \tag{14}$$

and

$$\overline{v_\ell' v_k'} \left(\frac{1}{h_k} \frac{\partial \overline{v_\ell}}{\partial g_k} - \frac{\overline{v_\ell}}{h_\ell h_k} \frac{\partial h_\ell}{\partial g_k} \right) + \overline{v_k' v_k'} \frac{\overline{v_\ell}}{h_\ell h_k} \frac{\partial h_k}{\partial g_\ell} + C_E \frac{\overline{E'}^{3/2}}{\Delta} = 0 \tag{15}$$

The subgrid Reynolds stress $\overline{v_i' v_j'}$ can be divided into two parts:

$$\overline{v_i' v_j'} = (\overline{v_i' v_j'})_{iso} + (\overline{v_i' v_j'})_{in} \; ; \qquad -(\overline{v_i' v_j'})_{in} = \nu^{(ij)} \langle \overline{S}_{ij} \rangle \tag{16}$$

where $(\ )_{iso}$ is isotropic part and $(\ )_{in}$ is inhomogeneous part.
The inhomogeneous part satisfies following equation

$$\overline{v_i' v_k'} \left(\frac{1}{h_k} \frac{\partial \overline{v_j}}{\partial g_k} + \frac{\overline{v_j}}{h_j h_k} \frac{\partial h_j}{\partial g_k} - \frac{2 \overline{v_k}}{h_j h_k} \frac{\partial h_k}{\partial g_j} \right) + \overline{v_j' v_k'} \left(\frac{1}{h_k} \frac{\partial \overline{v_i}}{\partial g_k} + \frac{\overline{v_i}}{h_i h_k} \frac{\partial h_i}{\partial g_k} - \frac{2 \overline{v_k}}{h_i h_k} \frac{\partial h_k}{\partial g_i} \right)$$

$$\left\{ \frac{C_E}{8} \frac{\overline{E'}^{1/2}}{\Delta} (\overline{v_i' v_j'} - \frac{2}{3} \delta_{ij} \overline{E'}) + \frac{3}{200} (P_{ij} - D_{ij}) \right\} \frac{\Delta}{C_E} f(x_2) \tag{17}$$

Deleting the term with $\{\ \}$ on the right side of equation (14), we get the equation of $(\ )_{iso}$. Solving this equation, we have

$$(\overline{v_i' v_j'})_{iso} = F_{ij} (\overline{v_1}, \overline{v_2}, \overline{v_3}, \overline{E'}^{1/2}/\Delta) \tag{18}$$

Substituting it into equation (15), the equation for $\overline{E'}^{1/2}/\Delta$ is gotton

$$G(\overline{v_1}, \overline{v_2}, \overline{v_3}, \overline{E'}^{1/2}/\Delta) = 0 \tag{19}$$

The minimum non-negative root of equation (18) is the value of $\overline{E'}^{1/2}$
Then from

$$f = \frac{C}{S} \frac{\overline{E'}^{1/2}}{\Delta} \qquad (20)$$

we can calculate the value of f. Substituting this value of f into eq. (8), $(\overline{v_i' v_j'})_{iso}$ can be calculated. And substituting the value of E' into eq. (15), the value of $(\overline{v_i' v_j'})_{in}$ can be solved. According to (8),(16), ν^* and ν^{ij} can be determined, where < > notes statistic average value in time.

4. Numerical simulation of turbulence in a straight channel

First we use this model in the numerical simulation of turbulence in a straight channel. The geometric scale and coordinate system are shown in sketch. We assume Re→∞, in other side the pressure is divided into two parts:

$$P = P_0 + P_1 \qquad (21)$$

where dP_0/dx=const.
Substituting (21) into (9) and using the above described algebraic mode, the equation can be arranged in the following form:

$$\frac{\partial \overline{u}}{\partial x} + \frac{\partial \overline{v}}{\partial y} + \frac{\partial \overline{w}}{\partial z} = 0 \qquad (22)$$

$$\left. \begin{array}{l} \frac{\partial \overline{u}}{\partial t} + [\overline{v}]\frac{\partial \overline{u}}{\partial n} - \frac{\partial}{\partial n}\left([\nu^*]\frac{\partial \overline{u}}{\partial n}\right) + \frac{\partial \overline{P_1}}{\partial x} = H_x \\ \frac{\partial \overline{v}}{\partial t} + [\overline{v}]\frac{\partial \overline{v}}{\partial n} - \frac{\partial}{\partial n}\left([\nu^*]\frac{\partial \overline{v}}{\partial n}\right) + \frac{\partial \overline{P_1}}{\partial y} = H_y \\ \frac{\partial \overline{w}}{\partial t} + [\overline{v}]\frac{\partial \overline{w}}{\partial n} - \frac{\partial}{\partial n}\left([\nu^*]\frac{\partial \overline{w}}{\partial n}\right) + \frac{\partial \overline{P_1}}{\partial z} = H_z \end{array} \right\} \qquad (23)$$

where [ ] notes the average value at plane y=const. H_x, H_y, H_z note the summations of the right terms. The boundary conditions for eqs. (22) (23) are:

$$\overline{u} = \overline{v} = \overline{w} = 0 \; ; \qquad \partial \overline{P_1}/\partial y = 0 \qquad (24)$$

and periodic conditions in x- and z-directions.

In the calculation the variable scale of grid in y-direction is used and in x- and z-directions the scale of grid is constant. The values of u, v, w, p are defined in the centre of grid.

The periods in x- and z-directions are respectively L and W, where L=8T and W=4T. Using the Fourier fransformation method and implicit (in n-direction) and explicit (in x-and z-direction) scheme, the equations (22) and (23) are solved.

To provide the stability of calculation, the time step should satisfy the following inequality

$$\Delta t = \frac{f_k}{\frac{|u|_{max}}{\Delta x} + \frac{|w|_{max}}{\Delta z} + 2\nu^*_{max}\left(\frac{1}{\Delta x^2} + \frac{1}{\Delta y^2}\right)} \qquad (25)$$

where we take f_k=0.5-0.2
The values of other parameters in above eqs. are shown in Table.

parameters	C_E	C_P	C_1	C_2	C
numbers of eq.	(13)	(11)	(10)	(10)	(9)
Values	1.10	0.3	2.5	0.75	1.5

To reduce the computer time, the value $\overline{E'}^{1/2}/\Delta$ is calculated every ten time steps. The calculation has been completed on M-150 with a grid of 32X16X16 volumes. 500 time steps were performed, consuming 50 hours of CPU time. The results of computation are shown in Figs. 1-6.

5. Conclusion

For the turbulence with large Reynolds number the LES with algebraic model is useful. The comparison between our results and Schumann's shonws that the present model is correct. We hope this model can be expanded into the LES of the turbulence in the curvature channel.

References

1. Smagorinsky, J.Mon. Wea. Rev. 91, 99(1963).
2. Deardorff, J.W.,Fluid Mech., 41, pp450-480.
3. Ferziger, J.H.TF-16 Dept. of Mech.Eng., Stanford University.
4. Schumann, U.,J.Comp. Phys.,18, 376(1975).
5. Antonopoulor-Donnis, M.J.Fluid Mech.,104, 55(1981).
6. Su Mingde, Advance in Mech. 3 (1984) (in chinese).
7. Rotta, J.C.Z.,fur Phys.,129, 547(1951).
8. Launder, B.E., Reece G.J., Rodi, W.J.,Fluid Mech.,68, part 3,pp537-566(1975).

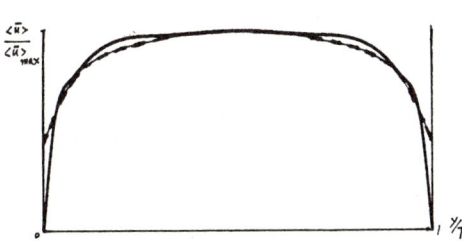

Fig.1 $\langle \tilde{u} \rangle$ - y

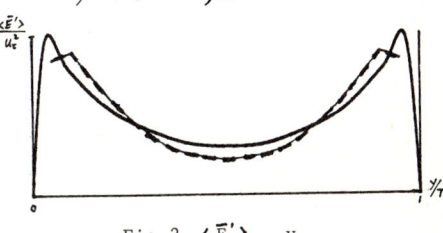

Fig.2 $\langle \bar{E}' \rangle$ - y

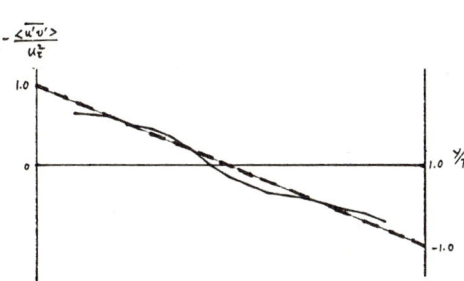

Fig.3 $\langle u', v' \rangle$ - y

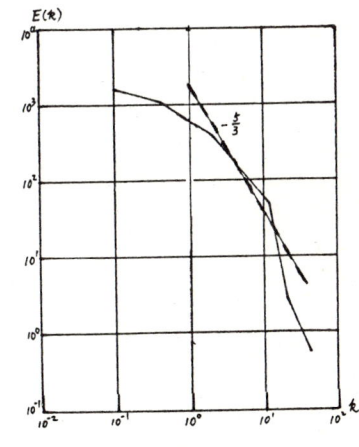

Fig.4 E(k)

E'-isoline

Fig.5

p'-isoline

Fig.6

A THREE-DIMENSIONAL INCOMPRESSIBLE FLOW SOLVER

Yukimasa Takemoto
Institute of Plasma Physics
Nagoya University, Nagoya, Japan
and
Yoshiaki Nakamura
Department of Aeronautical Engineering
Nagoya University, Nagoya, Japan

Abstract

Recently, accurate and stable numerical solutions for time-dependent convection-dominated fluid flows are urgently required. A calculation procedure to solve the unsteady incompressible Navier-Stokes equations at relatively high Reynolds numbers in arbitrary shapes is presented. The Adams-Bashforth scheme is used for the time integration and a third-order upwind differencing for the convective terms which have the advantage of suppressing the instabilities. The third-order method for the space derivatives is based on the Quadratic Upstream Interpolation for Convective Kinematics(QUICK) algorithm. The previous two-dimensional algorithm was extended to the three-dimension. As a test problem, the solution for a time-marching cubic cavity flow at the Reynolds number of 10^4 was obtained. The Taylor-Gortler type secondary vortices and the corner vortices were well simulated. The agreement with experiment was qualitatively good.

I. Introduction

The main difficulties in handling incompressible high Reynolds number fluid flow in arbitrary shapes are the modelling procedure for the first-derivative convective terms and the treatment of the boundary conditions on irregular boundaries.
For the computation of high Reynolds number flows, second-order central differencing for the first-derivative convective terms leads to unphysical oscillations or numerical instabilities. The classical remedy to overcome this instability has been to use first-order upwind differencing. This, however, has introduced artificial numerical diffusion effect which obscures the accurate modelling of physical processes at high Reynolds numbers. The second-order upwind scheme might be considered to be better in this sense[1]. On the other hand, Kawamura and Kuwahara developed a third-order upwind scheme and have successfully applied it to various examples[2].
In this paper, the solution procedure uses spatially third-order accurate upwind finite difference technique which was developed by Leonard[3,4] and has been successfully extended to generalized coordinates by authors. It is well known that the transformed equations take a compact form[5,6]. The present paper presents the validity of the above calculation procedure to solve the convection-dominated incompressible flow problems in arbitrary shapes. In order to examine its validity, a three-dimensional

lid-driven cubic cavity flow at Reynolds number Re=10^4 has been calculated.

Section II addresses the transformed governing equations, and Section III devotes to the QUICK algorithm extended to the generalized coordinates. In section IV the solution procedure is described, and the results are shown in the section V.

II. Governing Equations

The time-dependent incompressible Navier-Stokes equations in the conservative form can be written in the non-dimensional form[5]. It is convenient to define the following velocities:

$$U=\xi_x u+\xi_y v+\xi_z w, \quad V=\eta_x u+\eta_y v+\eta_z w, \quad W=\zeta_x u+\zeta_y v+\zeta_z w. \tag{1}$$

The contravariant velocity components U, V and W along the ξ, η and ζ coordinates are calculated in the above equation. Using these velocities, the governing equations are written in the compact form as

$$\hat{q}_t+\hat{E}_\xi+\hat{F}_\eta+\hat{G}_\zeta=Re^{-1}(\hat{R}_\xi+\hat{S}_\eta+\hat{T}_\zeta), \tag{2}$$

where

$$\hat{q}=J^{-1}\begin{bmatrix}0\\u\\v\\w\end{bmatrix}, \quad \hat{E}=J^{-1}\begin{bmatrix}U\\uU+\xi_x p\\vU+\xi_y p\\wU+\xi_z p\end{bmatrix}, \quad \hat{F}=J^{-1}\begin{bmatrix}V\\uV+\eta_x p\\vV+\eta_y p\\wV+\eta_z p\end{bmatrix}, \quad \hat{G}=J^{-1}\begin{bmatrix}W\\uW+\zeta_x p\\vW+\zeta_y p\\wW+\zeta_z p\end{bmatrix},$$

$$\hat{R}=J^{-1}\begin{bmatrix}0\\g_1 u_\xi+g_2 u_\eta+g_3 u_\zeta\\g_1 v_\xi+g_2 v_\eta+g_3 v_\zeta\\g_1 w_\xi+g_2 w_\eta+g_3 w_\zeta\end{bmatrix}, \quad \hat{S}=J^{-1}\begin{bmatrix}0\\g_2 u_\xi+g_4 u_\eta+g_5 u_\zeta\\g_2 v_\xi+g_4 v_\eta+g_5 v_\zeta\\g_2 w_\xi+g_4 w_\eta+g_5 w_\zeta\end{bmatrix}, \quad \hat{T}=J^{-1}\begin{bmatrix}0\\g_3 u_\xi+g_5 u_\eta+g_6 u_\zeta\\g_3 v_\xi+g_5 v_\eta+g_6 v_\zeta\\g_3 w_\xi+g_5 w_\eta+g_6 w_\zeta\end{bmatrix},$$

$$g_1=\xi_x^2+\xi_y^2+\xi_z^2, \quad g_2=\xi_x\eta_x+\xi_y\eta_y+\xi_z\eta_z, \quad g_3=\xi_x\zeta_x+\xi_y\zeta_y+\xi_z\zeta_z,$$

$$g_4=\eta_x^2+\eta_y^2+\eta_z^2, \quad g_5=\eta_x\zeta_x+\eta_y\zeta_y+\eta_z\zeta_z, \quad g_6=\zeta_x^2+\zeta_y^2+\zeta_z^2.$$

III. Application of QUICK Algorithm to Generaized Coordinates

In the present paper, the solution procedure uses spatially third-order accurate upwind finite difference technique based on the QUICK algorithm that has been extended to generalized coordinates[7]. The QUICK method was developed by Leonard[3,4], and this scheme has a third-order accuracy for the convective terms. The error terms include the fourth-order derivative, and these terms have a function to suppress the high frequency disturbances. The fourth-order terms have been positively added as smoothing in the compressible flow solver[6].

In the present code, we use a regular grid where the velocities and pressure are located at the same grid points, altough the staggered grid might be better.

We apply the three-dimensional QUICK algorithm to the

convective terms of Eq.(2). For example, the estimated value of $(J^{-1}uU)_{i+1/2,j,k}$ is given as follows:

$$(J^{-1}uU)_{i+\frac{1}{2},j,k} = (J^{-1}U)_{i+\frac{1}{2},j,k} \{ \frac{1}{2}(u_{i+1,j,k}+u_{i,j,k}) - \frac{\Delta\xi^2}{8}CURVu^{\xi}_{i+\frac{1}{2},j,k}$$

$$+\frac{\Delta\eta^2}{24}CURVu^{\eta}_{i+\frac{1}{2},j,k}+\frac{\Delta\zeta^2}{24}CURVu^{\zeta}_{i+\frac{1}{2},j,k} \} ,\quad (3)$$

where the curvature terms (CURV) in the quadratic upwind scheme depend on a direction of the contravariant velocity U:

$$CURVu^{\xi}_{i+\frac{1}{2},j,k}$$

$$=\begin{cases}(u_{i+1,j,k}-2u_{i,j,k}+u_{i-1,j,k})/\Delta\xi^2 , & if \quad J^{-1}U_{i+\frac{1}{2},j,k}>0 \\ (u_{i+2,j,k}-2u_{i+1,j,k}+u_{i,j,k})/\Delta\xi^2 , & if \quad J^{-1}U_{i+\frac{1}{2},j,k}<0.\end{cases} \quad (4)$$

Using these relations, the third-order upwind finite difference expressions for the generalized QUICK method are obtained.

IV. <u>Solution Procedure</u>

The governing equations under the appropriate boundary and initial conditions are numerically solved by means of a time-dependent finite difference method which will be decribed briefly.
A time-implicit numerical algorithm is used to solve the governing equations, because it is desirable to take a larger time-step, Δt, than that by a explicit scheme. We used the second-order Adams-Bashforth scheme for the convective terms at the first step and the implicit Crank-Nichlson scheme on the fractional step.

The first step gives

$$(\tilde{u}/J-u^n/J)/\Delta t = \frac{3}{2}H^n - \frac{1}{2}H^{n-1} + \frac{1}{2}Re^{-1}Q(u^n) , \quad (5)$$

where

$$Q=\hat{R}_\xi + \hat{S}_\eta + \hat{T}_\zeta , \quad H = -(J^{-1}uU)_\xi - (J^{-1}uV)_\eta - (J^{-1}uW)_\zeta$$

The second step is as follows:

$$(u^{n+1}/J-\tilde{u}/J)/\Delta t = -[(J^{-1}\xi_x p)_\xi + (J^{-1}\eta_x p)_\eta + (J^{-1}\zeta_x p)_\zeta]^{n+1/2} + \frac{1}{2}Re^{-1}Q(u^{n+1}) \quad (6)$$

The pressure field is calculated by the Poisson equation.

$$\nabla^2 p^{n+1/2} = \frac{\tilde{D}}{\Delta t}, \quad (7)$$

where $\tilde{D}=\xi_x\tilde{u}_\xi+\eta_y\tilde{v}_\eta+\zeta_y\tilde{w}_\zeta$.
In the above equations (6) and (7), the pressure and velocity are

adjusted iterartively to get converged solutions. As a consequence, the divergence at t=n+1 is satisfied: $D^{n+1} = 0$.

V. Results and Discussions

We applied the present scheme to a three-dimensional lid-driven cavity flow. It was computed to test the combination of numerical algorithm, grid mapping, initial conditions and boundary conditions for the transformed equations. In the present case, 35×35×35 non-uniform grid points were used in the x, y, and z directions. The dimensions of the simulated cavity are $0 \leq x,y,z \leq 1$. The minimum grid size is 0.002 near the wall. A uniform flow(u=1) in the x-direction was imposed on the top surface(y=1). For Re=10^4 chosen, the computation was performed up to t=60 with Δt = 0.001.

The velocity vector distributions in several planes are shown in Figs. 1 to 3. Figure 1 is the velocity vector(u,v) distribution on the planes at z=0.5, 0.81, and 0.965. Generally, the flow is recirculating in each plane except near the side-wall plane(z=1). At z=0.5 and 0.81 planes, the downward flow impinges upon the bottom surface(y=0) at right bottom. This suggests that the flow is three-dimensional and deflects in other directions. The corner vortices are also shown at bottom right and left.

Further interesting thing is that at the bottom right the source-like flow is occurring near the end plate: z=0.965. The interface line is clearly seen to be horizontal and straight where the downward and upward flows collide and move toward the negative x.

Figure 2 shows the velocity vector(w,v) distributions at planes with x=const. At x=0.995, we can see the relatively high velocity fluid is coming downward in the upper region. Another interesting pattern is seen at plane x=0.85, where the Taylor-Gortler type vortices are clearly shown near the bottom surface.

Figure 3 shows the velocity vectors(w,u) inside the planes with y=const. It is interesting to see that the firework patterns occur near the bottom surface: y=0.02, where the flow strongly impinges upon the bottom surface, and deflects in different directions.

We, then, show the velocity distribution along two lines. Figure 4a shows the y-direction velocity, v, along the line with y=0.5 and z=0.5. The comparisons with Freitas et al's simulation[8] and Koseff et al's experiment[9,10] are shown. The present result is similar to those results. Freitas et al's simulation shows too higher velocity around x=0.2 than others. The peak value and its location are fairly deviated from our result and experiment. Their flow condition is at Reynolds number=3200, while our result is at Reynolds number=10000. Furthermore, in their flow, the ratio of the span of the cavity with its chord is three, while our result is unity.

Figure 4b shows the u-velocity distribution along the y-coordinate at z=0.5 and x=0.5. Although three results have similar characteristic, Freitas's result is higher than the experiment, and our result is lower than the experiment.

Finally, we show the vortcity contours at a plane with x=0.85 in Fig. 5, and the pressure contours at a plane with y=0.02 in Fig. 6.

VI. Concluding Remarks

A new upwind Adams-Bashforth QUICK algorithm in three

dimensions that was extended to generalized coordinates has been developed. A three-dimensional cubic lid-driven cavity flow was calculated at Re = 10^4 with non-uniform 35×35×35 grid. The agreement between the present prediction, other simulation, and the experiment was qualitatively good.

The present code could be a useful tool for other time-dependent incompressible flow problems at high Reynolds numbers in the generalized coordinates.

References

1) W. Shy: Journal of Computational Physics, Vol. 57(1985), 415.
2) T.Kawamura and K.Kuwahara : AIAA paper 84-0340 (1984).
3) B.P.Leonard : Proceedings of the 2nd National Symp. on Numerical Properties and Methodologies in Heat Transfer. (Univ. of Maryland,1983) 211.
4) B.P.Leonard : Proceedings of the 1983 Int. Conf. on Computational Techniques and Applications (Univ. of Sydney, Australia,1984) 106 .
5) R.Peyret and H.Viviand : AGARD-AG-212 (1975) .
6) J.L.Steger : AIAA Journal Vol. 16 (1978) 679.
7) Y.Nakamura and Y.Takemoto : Proceedings of Int. Symp. Compu. Fluid Dynam. Tokyo, (1985) .
8) C.J.Freitas, R.L. Street, A.N. Findikakis and J.R. Koseff: Inter. Journal for Numer. Methods in Fluids, 5(1985), 561.
9) J.R. Koseff and R.L. Street: Journal of Fluids Engineering, Transactions of the ASME, Vol. 106(1984) 385.
10) J.R. Koseff and R.L. Street: Journal of Fluids Engineering, Transactions of the ASME, Vol. 106(1984) 390.

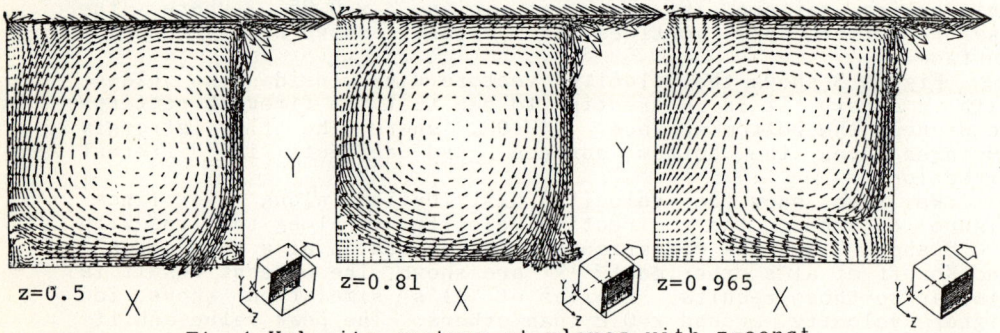

Fig.1 Velocity vectors at planes with z=const.

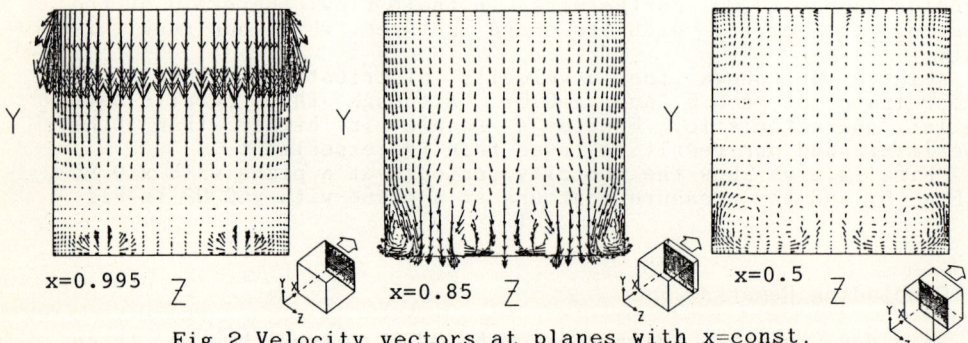

Fig.2 Velocity vectors at planes with x=const.

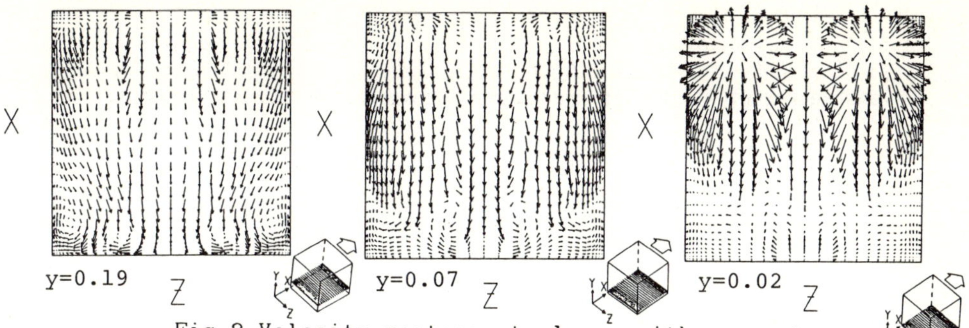

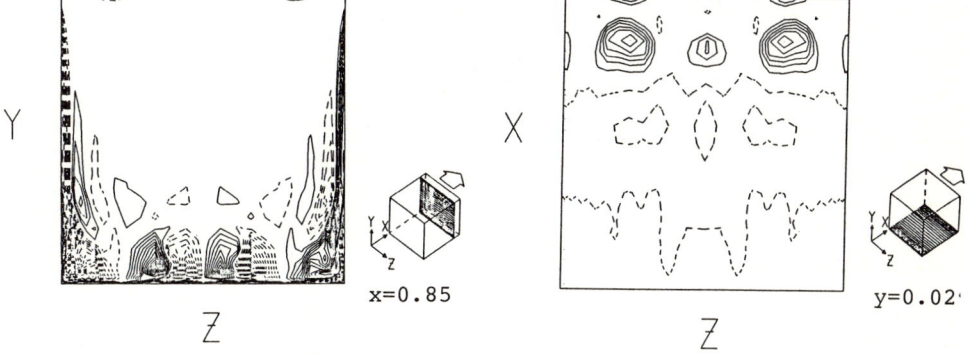

Fig.3 Velocity vectors at planes with y=const.

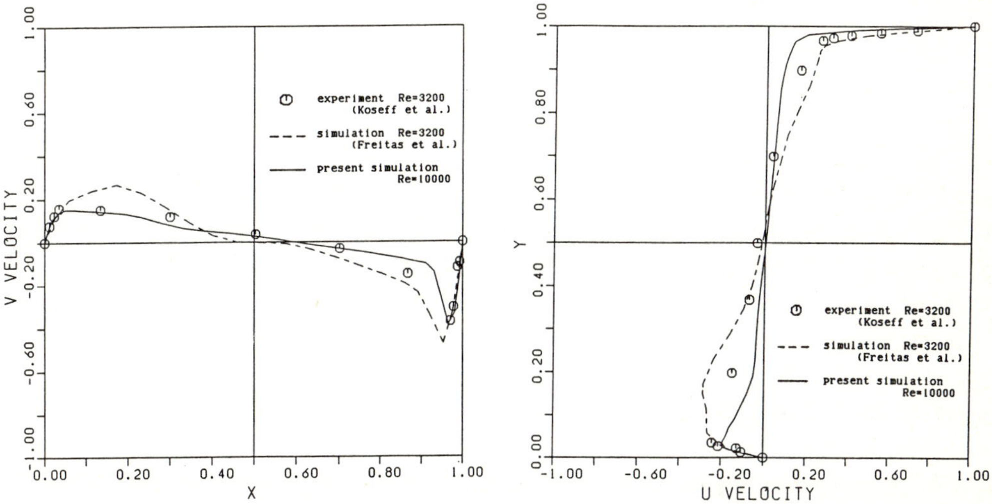

Fig. 4 Velocity distribution along lines.
(a) V velocity distribution along a line with y=0.5 and z=0.5.
(b) U velocity distribution along a line with x=0.5 and z=0.5.

Fig.5　Vorticity contours at plane with x=0.85.

Fig. 6　Pressure contours at plane with y=0.02.

VARIABLE-ELLIPTIC-VORTEX METHOD FOR INCOMPRESSIBLE FLOW SIMULATION

Zhen-huan Teng
Department of Mathematics, Peking University
Beijing, P.R. China

I. Introduction

Vortex methods have provided an effective approach for the numerical simulation of incompressible fluid flow at high Reynolds number. Point vortex method was first introduced by Rosenhead[7] and subsequently vortex blob methods were developed by Chorin[2], and Kuwahara and Takami[5]. There have been a large number of successful flow simulation by vortex blob methods[6].

It is however noticed that up to now all of the numerical vortex blobs are assumed to retain fixed shape for all time while the actual flow can undergo substantial distortion. Even though the "unphysical behavior" of vortex blobs does not interfere with the convergence of vortex methods ([4],[1]), it may reduce the accuracy of the vortex methods. Thus there is considerable interest in finding an appropriate approach to form a method with variable vortex blobs, which can follow the distortion of actual vortex blobs.

In this paper we present a variable-elliptic-vortex method in meeting this need, which is a genelization of elliptic-vortex method proposed by the author in [8]. The most attractive feature of the new model is that the numerical vortex blobs can be translated, rotated and deformed in elliptic shape according to the decomposition theorem of velocity in a small neighborhood. The main merits of the new model are as follows: (1) it provides a more flexible and more reasonable approach to mimic physical flow; (2) it has higher order accuracy in space than the fixed shape vortex method.

II. Approximate Motion of a Small Elliptic Blob

We are mainly concerned with incompressible inviscid flow in two dimension. Let Ω_0 be an elliptic blob at t=0 centred at $\underline{\alpha}_0 = (\alpha_{10}, \alpha_{20})$ defined by $\Omega_0 = \{\underline{\alpha} | (\underline{\alpha} - \underline{\alpha}_0) A (\underline{\alpha} - \underline{\alpha}_0)^T \leq 1, \underline{\alpha} \in R^2\}$, where $A = (a_{ij})$ is a 2×2 positive definite matrix. A particle starting at t=0 from the position $\underline{\alpha} \in \Omega_0$ follows a trajectory $\underline{r}(t;\underline{\alpha})$ determined by

$$\begin{cases} \dfrac{d\underline{r}}{dt} = \underline{u}(\underline{r},t) \\ \underline{r}(0;\underline{\alpha}) = \underline{\alpha} \in \Omega_0 \end{cases} \tag{1}$$

We expand the velocity vector $\underline{u}(\underline{r},t)$ at $\underline{r}_0(t)=\underline{r}(t;\underline{\alpha}_0)$, the trajectory of the center $\underline{\alpha}_0$ of Ω_0, by Taylor's theorem

$$\underline{u}(\underline{r},t) = \underline{u}(\underline{r}_0,t) + (\underline{r}-\underline{r}_0)\cdot\nabla\underline{u}(\underline{r}_0,t)^T + O(|\underline{r}-\underline{r}_0|^2).$$

Substituting the expression into (1) and neglecting $O(|\underline{r}-\underline{r}_0|^2)$, we get an approximate system of (1) with second order accuracy in space

$$\begin{cases} \dfrac{d(\underline{r}-\underline{r}_0)}{dt} = (\underline{r}-\underline{r}_0)\cdot\nabla\underline{u}(\underline{r}_0,t)^T \\ \underline{r}(0;\underline{\alpha}) = \underline{\alpha} \in \Omega_0 \end{cases} \tag{2}$$

If assume $\underline{r}_0(t)$ is a known trajectory, then (2) is a linear system of o.d.e. We denote the 2×2 fundamental matrix of

$$\begin{cases} \dfrac{dZ}{dt} = Z\cdot\nabla\underline{u}(\underline{r}_0,t)^T \\ Z(0) = E \end{cases} \tag{3}$$

by $Z(t)$, E being a unitary matrix. In view of Liouville theorem and incompressibility div $\underline{u}=0$, we get det $Z(t)=1$. A calculation shows that the approximate motion of Ω_0 can be expressed as

$$\Omega(t) = \{\underline{r}\mid (\underline{r}-\underline{r}_0(t))Z^{-1}(t)AZ^{-1}(t)^T(\underline{r}-\underline{r}_0(t))^T \leq 1,\ \underline{r}\in R^2\}, \tag{4}$$

where $A(t)\equiv Z^{-1}(t)AZ^{-1}(t)^T$ is a positive definite matrix and its determinant is constant i.e. $\det A(t)=\det A$. This implies that $\Omega(t)$ is deformed in the elliptic shape and its area is invariant in time. So we arrive at the following conclusion:

<u>Proposition:</u> If the motion of a small elliptic blob Ω_0 is approximated by the second order accuracy system (2), then the approximate motion of Ω_0 is the sum (4) of a (rigid) translation following its center trajectory $\underline{r}_0(t)$ and a distortion in elliptic shape with conserved area.

III. Variable Elliptic Vortex Model

Suppose the vorticity field ξ is now represented by a sum of elliptic vortex blobs

$$\xi(\underline{r},t) = \sum_{j=1}^{N} \Gamma_j \gamma(\underline{r};\Omega_j(t)) \tag{5}$$

where Γ_j are their respective circulation, γ is a uniform vorticity distribution over an ellipse $\Omega_j(t)$, which is given by

$$\gamma(\underline{r};\Omega_j(t)) = \begin{cases} \dfrac{1}{\sigma_j}, & \underline{r} \in \Omega_j(t) \\ 0, & \underline{r} \bar\in \Omega_j(t) . \end{cases}$$

Here σ_j is the area of $\Omega_j(t)$ and the ellipse $\Omega_j(t)$ is defined by

$$\Omega_j(t) = \{ \underline{r} \mid (\underline{r}-\underline{r}_j(t))A_j(t)(\underline{r}-\underline{r}_j(t))^T \le 1, \underline{r} \in R^2 \},$$

where $A_j(t)$ is a 2×2 positive definite matrix and $\underline{r}_j(t) = (x_j(t), y_j(t))$ is the center position of $\Omega_j(t)$.

It follows from [8] that the induced velocity field by vorticity field (5) is expressed as

$$\underline{u}(\underline{r},t) = \sum_{j=1}^{N} \Gamma_j \left(\frac{\partial}{\partial y}, - \frac{\partial}{\partial x} \right) \phi(x,y; \Omega_j(t)), \qquad (6)$$

where the explicit expression of ϕ is given in [8]. The formula shows that once the motions of elliptic blobs $\Omega_j(t)$ $\{j=1,2,\ldots,N\}$ are known the velocity field can be explicitly solved by (6). We will use system (2) to approximate $\Omega_j(t)$. To this end replacing $\underline{r}_0(t)$ in (3) with $\underline{r}_j(t)$ yields the equations of fundamental matrix Z_j

$$\begin{cases} \dfrac{dZ_j}{dt} = Z_j \cdot \nabla \underline{u}(\underline{r}_j(t),t)^T \\ Z_j \big|_{t=0} = E, \end{cases} \qquad (j=1,2,\ldots,N) \qquad (7)$$

where the center trajectories $\underline{r}_j(t)$ of $\Omega_j(t)$ are described by

$$\begin{cases} \dfrac{d\underline{r}_j}{dt} = \underline{u}(\underline{r}_j(t),t) \\ \underline{r}_j \big|_{t=0} = \underline{r}_j(0). \end{cases} \qquad (j=1,2,\ldots,N) \qquad (8)$$

$\underline{r}_j(0)$ being the initial center position. Here $\underline{u}(\underline{r}_j(t),t)$ and $\nabla\underline{u}(\underline{r}_j(t),t)$ are defined by (6). As showed in section II the approximate motion of $\Omega_j(t)$ is expressed as

$$\Omega_j(t) = \{\underline{r} \mid (\underline{r}-\underline{r}_j(t))A_j(t)(\underline{r}-\underline{r}_j(t))^T \le 1, \underline{r} \in R^2\} \qquad (9)$$

where $A_j(t) = Z_j^{-1}(t) A_j(0) Z_j^{-1}(t)^T$ and $A_j(0)$ is the initial coefficient matrix of ellipse Ω_j. In view of the proposition in section II each of $\Omega_j(t)$ is preserved in elliptic shape and its area is conserved but its axis ratio and orientation may be changed in time. Systems (7) and (8) with (6) and (9) yield a time continuous semidiscrete simulation to Euler equation with second order accuracy in space. There are 6N equations (7),(8) with 6N unknowns $\underline{r}_j(t)$, $Z_j(t)$ ($j=1,2,\ldots,N$).

A full discrete version of variable-elliptic-vortex method may be derived from approximating (8) by Euler's method

$$\underline{r}_j^{n+1} = \underline{r}_j^n + \underline{u}(\underline{r}_j^n, nk)k \ , \ \underline{r}_j^0 = \underline{r}_j(0), \tag{10}$$

where k is a time step, $\underline{r}_j^n = \underline{r}_j(nk)$ is the center of $\Omega_j^n = \Omega_j(nk)$. For approximating (7) we freeze $\nabla \underline{u}(\underline{r}_j(t),t) = \nabla \underline{u}(\underline{r}_j^n, nk)$ for $t \in [nk,(n+1)k]$, and then get a constant coefficient system

$$\begin{cases} \dfrac{d\tilde{Z}_j}{dt} = \tilde{Z}_j \cdot \nabla \underline{u}(\underline{r}_j^n, nk)^T \ , \ t \in [nk,(n+1)k] \\ \tilde{Z}_j \big|_{t=nk} = E \ . \end{cases} \tag{11}$$

From o.d.e. we know that $\tilde{Z}_j$ can be integrated out in close form. In terms of $Z_j^{n+1} = \tilde{Z}_j((n+1)k)$ one gets

$$\Omega_j^{n+1} = \{\underline{r} \mid (\underline{r}-\underline{r}_j^{n+1}) A_j^{n+1} (\underline{r}-\underline{r}_j^{n+1})^T \leq 1, \ \underline{r} \in R^2\} \ , \tag{12a}$$

where

$$A_j^{n+1} = (Z_j^{n+1})^{-1} A_j^n ((Z_j^{n+1})^{-1})^T \ , \ A_j^0 = A_j(0) \ . \tag{12b}$$

In summary the procedure to mimic incompressible inviscid flows is as follows. The elliptic vortex blobs Ω_j^n are moved by the law (10) and deformed by the map (12), and then the new velocity field (6) at $t=(n+1)k$ is determined by the vortex blobs at their new positions $\underline{r}_j^{n+1}$ and their new elliptic shapes Ω_j^{n+1}. Repeating the procedure (10) and (11) one gets an approximate solution of Euler equation.

IV. Numerical Results

We present some numerical experiments to demostrate the performance of variable-elliptic-vortex method.

1. Dynamic motion of two Rankine vortices: We use two numerical vortex blobs to simulate their motion. Numerical results are displayed in Fig.1. From this example we can see how the elliptic vortex blobs are translated and deformed in time.

2. Rotation of a vortex sheet: The sheet is defined as the limit of an infinitely thin elliptic vortex with a uniform vorticity distribution. The sheet is replaced by a row of N=30 equidistant elliptic vortex blobs at t=0. The results are

displayed in Fig.2. By symmetry only half of blobs are shown. The solid lines indicate the exact solution of the rotational sheet. The periods and the small circles represent the center positions of vortex blobs. We can see the numerical solutions are in very close agreement with exact solution at t=1 except the end bolb. At t=2 the pattern becomes chaotic. This occurrence may be caused by the unstable rotation of sheet.

3. Rolling-Up of vortex sheet: The initial vorticity distribution is given by

$$\xi(x,y,0) = -\frac{d}{dx}(1-x^2)^{\frac{1}{2}}, \quad -1 \leq x \leq 1, \quad y=0.$$

The sheet is approximated by a row of N=60 elliptic blobs. Results showed in Fig.3 are reasonable in view of what is known from the numerical experiments in [3] and [5].

V. Conclusion

We have presented a new variable-elliptic-vortex method for mimicing incompressible inviscid flows. This method can be easily used to approximate high Reynolds number flows by incorporating a random walk algorithm to mimic the viscosity effect and a vortex generation algorithm to maintain the no-slip boundary condition ([2],[8]). The deformable behavior of elliptic-vortex blobs may provide a versatile approach for flow simulation. We hope that this method will be applied to other kinds of inviscid flow problems as well as high Reynolds flow problems.

References

[1] J.T. Beal and A. Majda: "Vortex methods I,II," Math. Comp., 39(1982), pp. 1-52.

[2] A.J. Chorin: "Numerical study of slightly viscous flow," J.Fluid Mech., 57(1973), pp. 785-796.

[3] A.J. Chorin and P. Bernard: "Discretization of vortex sheet with an example of roll-up," J. Comput. Phys., 13(1973), pp. 423-428.

[4] O. Hald: "The convergence of vortex methods II," SIAM J. Numer. Anal., 16(1979), pp. 726-755.

[5] K. Kuwahara and H. Takami: "Numerical studies of two-dimensional vortex motion by a system of point vortices," J. Phys. Soc. Japan, 34(1973), pp. 247-253.

[6] A. Leonard: "Vortex methods for flow simulations," J. Comput. Phys., 37(1980), pp. 289-335.

[7] L. Rosenhead: "The point vortex approximation of a vortex sheet," Proc. Roy. Soc. London Ser. A. 134(1932), pp. 170-192.

[8] Z.H. Teng: "Elliptic-vortex method for incompressible flow at high Reynolds number," J. Comput. Phys. 46(1982), pp. 54-68.

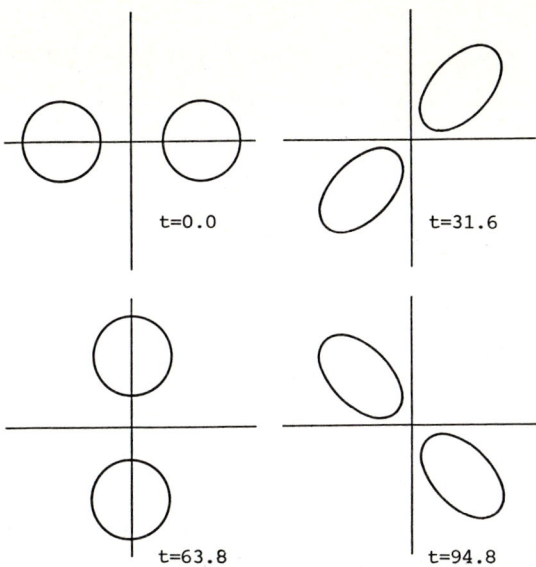

Fig.1. Two Vortices processing around each other

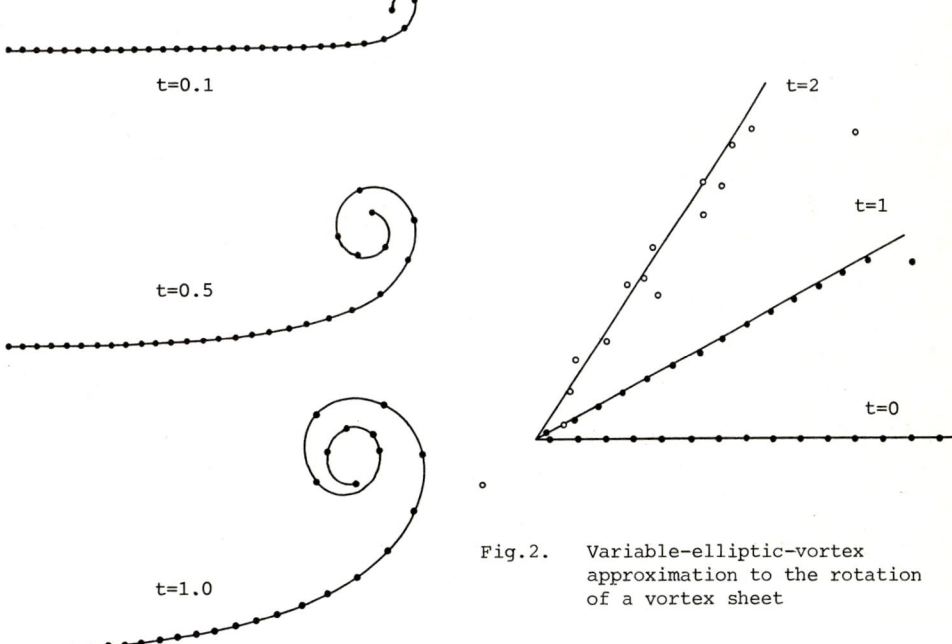

Fig.2. Variable-elliptic-vortex approximation to the rotation of a vortex sheet

Fig.3. Variable-elliptic-vortex approximation to the rolling-up of a vortex sheet

APPLICATION OF THE FAST ADAPTIVE COMPOSITE GRID METHOD TO COMPUTATIONAL FLUID DYNAMICS

J. W. Thomas
Roland Schweitzer
Mike Heroux
Department of Mathematics
Colorado State University

Steve McCormick
Department of Mathematics
University of Colorado at Denver

Ann M. Thomas
Department of Mathematics
University of Northern Colorado

1. Introduction

The need for local resolution occurs often in the area of computational fluid dynamics. Special local features of the forcing function, boundary and boundary conditions can demand resolution in restricted regions of the domain that is much finer than the required global resolution. It is important that the discretization and solution processes account for this locally; that is, that the local phenomena do not precipitate a dramatic increase in the overall computation times and storage needs. Unfortunately, the objective of efficiently adapting to local features is often in conflict with the solution process: equation solvers can degrade or even fail in the presence of varying discretization scales; data structures that account for irregular grids can be cumbersome; the computer architecture may not readily account for irregularity (e.g. in terms of "vectorizability" or "parallelizability"), etc.

There are several responses to this problem ranging from composite grid methods and multigrid techniques to a large number of ad hoc approaches which are usually problem specific. The fast adaptive composite grid method (FAC) [1, 2] is a discretization and solution method designed to achieve efficient local resolution by systematically constructing the discretization based on various regular grids and using them as a basis for fast solution. It shares all of the attributes of the composite grid and multigrid methods, especially in terms of the local resolvability and speed of the latter, but it is systematic and can be fully automated for problem solution. The objective of this paper is to demonstrate the utility of FAC for solving aerodynamics problems. We treat a model boundary layer problem and the thin disturbance transonic flow problem.

2. **The FAC Algorithm**

The FAC algorithm was originally developed to solve linear problems. The scheme presented in [1] is a correction scheme that relies heavily upon this fact. The algorithm presented here is actually an extension of the linear algorithm that applies both to linear and nonlinear problems. The scheme used to pass information from the fine to the coarse grid is analogous to that used in the full approximation scheme in multigrid methods [3].

To be specific, suppose the problem to be solved is the thin disturbance transonic flow problem described in Figure 1. Suppose the composite grid, denoted $\mathcal{G}$, on which this problem is to be solved is that pictured in Figure 2. Suppose further that the finite difference equations on $\mathcal{G}$ approximating the problem described in Figure 1 are given by

$$(1) \qquad \mathcal{L}(u) = \ell.$$

Here the composite grid operator $\mathcal{L}$ consists of the usual type differencing used by Murman and Cole in [4] and an appropriate nonuniform differencing at the interface of the two grids which we shall describe later. The composite grid can be partitioned so that $\mathcal{G} = \mathcal{G}_C \cup \mathcal{G}_I \cup \mathcal{G}_F$ where $\mathcal{G}_I$ consists of the points along the internal boundary of the fine grid region, and $\mathcal{G}_C$ and $\mathcal{G}_F$ consists of the points outside and inside the internal boundary $\mathcal{G}_I$, respectively.

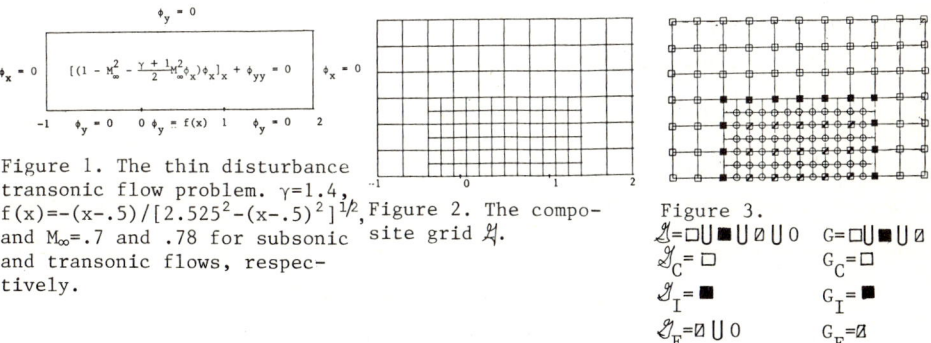

Figure 1. The thin disturbance transonic flow problem. $\gamma=1.4$, $f(x)=-(x-.5)/[2.525^2-(x-.5)^2]^{1/2}$, and $M_\infty=.7$ and $.78$ for subsonic and transonic flows, respectively.

Figure 2. The composite grid $\mathcal{G}$.

Figure 3.
$\mathcal{G}=\square\cup\blacksquare\cup\boxtimes\cup 0 \qquad G=\square\cup\blacksquare\cup\boxtimes$
$\mathcal{G}_C=\square \qquad\qquad\qquad G_C=\square$
$\mathcal{G}_I=\blacksquare \qquad\qquad\qquad G_I=\blacksquare$
$\mathcal{G}_F=\boxtimes\cup 0 \qquad\qquad G_F=\boxtimes$

Since the fine grid patch was chosen to be aligned with the coarse grid lines, there is an underlying coarse grid covering the entire region. The coarse grid is denoted by G and is decomposed into $G = G_C \cup G_I \cup G_F$ in analogy to the decomposition of $\mathcal{G}$. These grids are illustrated in Figure 3.

$\mathcal{G}$ and G are the principal grids used in the a gorithm. To be able to pass information between grids, assume that an interpolation operator I (linear interpolation) and a restriction operator I^T (injection) has been defined so that

$$I: G \to \mathcal{G}$$

and

$$I^T: \mathcal{G} \to G.$$

The coarse grid operator L is then defined as

$$L(u) = I^T \ell(Iu).$$

Based on the partitioning of the grids $\mathcal{G}$ and G, $\mathscr{u}$, u, ℓ and L can be partitioned as $\mathscr{u} = (\mathscr{u}_C, \mathscr{u}_I, \mathscr{u}_F)$, $u = (u_C, u_I, u_F)$,

$$\ell(\mathscr{u}) = \begin{bmatrix} \ell_{CC} & \ell_{CI} & H \\ \ell_{IC} & \ell_{II} & \ell_{IF} \\ H & \ell_{FI} & \ell_{FF} \end{bmatrix}(\mathscr{u})$$

and

$$L(u) = \begin{bmatrix} L_{CC} & L_{CI} & H \\ L_{IC} & L_{II} & L_{IF} \\ H & L_{FI} & L_{FF} \end{bmatrix}(u)$$

where each ℓ_{KL} and L_{KL} are functions of $\mathscr{u}$ and u, respectively. Note that $\ell_{CC} = L_{CC}$, $\ell_{IC} = L_{IC}$, $\ell_{CI} = L_{CI}$ and $\ell_{II} = L_{II}$. ℓ_{FF} is similar to L_{FF} except that there are many more grid points in $\mathscr{u}_F$. The significant difference between ℓ and L is in the blocks ℓ_{IF} and L_{IF}.

With this machinery in place and an initial guess u^0, an FAC cycle can be written as:

<u>Step 1:</u> $u^{k+1} = L^{-1}[L(I^T \mathscr{u}^k) + I^T(\ell - \ell(\mathscr{u}^k))]$

<u>Step 2:</u> $\mathscr{u}^{k+1} = Iu^{k+1}$

<u>Step 3:</u> $\mathscr{u}_F^{k+1} = \ell_{FF}^{-1}(\ell_F - \ell_{FI}(\mathscr{u}^{k+1})\mathscr{u}_I^{k+1}).$

Actually, the inverse in Step 1 is taken loosely for our prototype problem since neither L^{-1} nor ℓ^{-1} exist. In the solution scheme where L^{-1} is calculated iteratively by line SOR, the matrix inverse used should be replaced by the Moore-Penrose generalized inverse, which produces the solution with minimum norm.

Note that the solution procedures necessary in the FAC scheme are applied to rectangular regions. Thus, given the correct data structure, the presence of the fine grid patches will in no way inhibit vectorization or parallelization of the solution scheme. Also, if more than one patch is present, the patches can be solved in parallel. Finally, note that no particular solver is prescribed for Steps 1 or 3. For the transonic problem presented in the next section, line SOR was used because that is the scheme that has been successfully used for the usual global grid applications. (It is not really necessary to use the same solver in both places. In fact, the FAC software developed presently allows for the choice of six different solvers.)

3. Applications

3.1 <u>Boundary Layer</u>. The first result involves a rather simple model problem used to illustrate the capability of FAC to resolve a boundary layer

$$\epsilon \phi_{xx} + \phi_{yy} = 0 \quad (x, y) \in (0, 1) \times (0, 1)$$
$$\phi(1, y) = 0, \phi(0, y) = 1, \quad y \in [0, 1]$$
$$\phi(x, 1) = \phi(x, 0) = (1 - x)e^{-x/\epsilon}, \quad x \in [0, 1].$$

The problem was solved using FAC with two levels of patches. One patch covers $0 \leq x \leq 2/12$ and the finer patch $0 \leq x \leq 1/12$. The scheme used to cycle between the patches is $\{C, F, F^2, F, C, F, F^2, F, F^2\}$ where C, F and F^2 represent a solution on the coarse, fine and finest grids, respectively. The solution for $\epsilon = .01$ with $\Delta x = \Delta y = 1/12$ is given in Figure 4. Figures 5 and 6 are graphs of the solution on the two patches so Figures 4, 5 and 6 collectively give the solution to the composite grid problem. (The solution on regions that are covered by patches should be taken from the finest patch covering the region.)

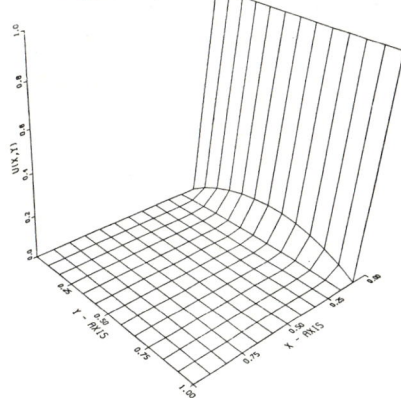

Figure 4. The coarse grid solution (after convergence).

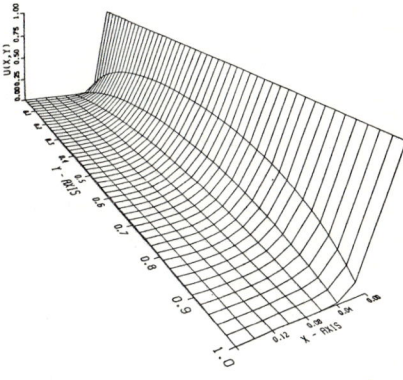

Figure 5. The solution on the patch $0 \leq x \leq 2/12$.

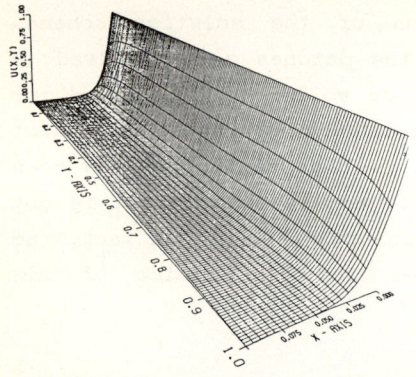

Figure 6. The solution on the patch $0 \leq x \leq 1/12$. The composite grid solution is Figure 4, $2/12 \leq x \leq 1$; Figure 5, $1/12 \leq x \leq 2/12$; and Figure 6, $0 \leq x \leq 1/12$.

3.2 Transonic Small Disturbance Equation.

The small disturbance potential flow problem was chosen as an example because while it is relatively easy to implement, the problem is nonlinear, is mixed elliptic-hyperbolic and contains a shock. The problem was solved by FAC using two patch sizes: one large enough to contain the whole hyperbolic region and one containing a strip along the top of the airfoil. Vertical line SOR was used on both the coarse grid and the patch. The difference operator at interior boundary was defined as a scaled fine grid operator using interpolated points from the coarse grid where necessary. The cycling scheme used was C, F, C, F.

The problem was solved for $M_\infty = .7$ (subsonic) and $M_\infty = .78$ (transonic) and the results in each case were compared to those with a coarse grid and an extensive fine grid. Figures 7 and 8 contain the plots of the pressure coefficients for the three grid configurations. As can be seen, the solutions of the subsonic problem are all very good. For the supersonic problem that is not the case. The pressure coefficient for the FAC solutions are not as large as that for the extensive fine grid solution and do not resolve the shock as well. However, it should also be noted that the FAC solutions perform much better than the coarse grid solution. What we feel is the most important point is that the FAC solution with a very thin layer is much better than the coarse grid solution and nearly as good as the FAC solutions with larger patches.

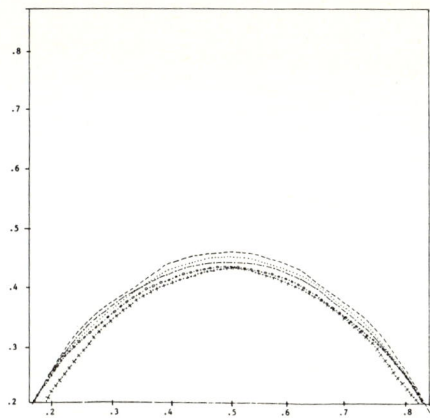
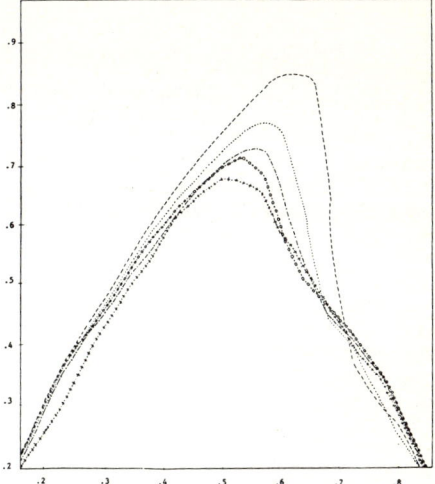

Figure 7. Plots of the pressure coefficient along the top of the airfoil for $M_\infty = .7$.
--- = FAC solution with a patch height of eight coarse grid points,
-·- = FAC solution with a patch height of five coarse grid points,
°·° = FAC solution with a patch height of three coarse grid points, and
+·+ = the extensive coarse grid solution ($\Delta x = .075$).

Figure 8. Plots of the pressure coefficient along the top of the airfoil for $M_\infty = .78$ using the same notations and patch sizes as in Figure 7.

References

1. S. McCormick and J. W. Thomas, The Fast Adaptive Composite Grid (FAC) Method for Elliptic Equations, Math. of Comp. Vol. 49, No. 174, April, 1986.

2. L. Hart, S. McCormick, A. O'Gallagher and J. W. Thomas, The Fast Adaptive Composite Grid Method (FAC): Algorithms for Advanced Computers, to appear.

3. Achi Brandt, Guide to Multigrid Development, <u>Multigrid Methods</u>, W. Hackbusch and U. Trottenberg, eds., Springer-Verlag, 1982.

4. Earl M. Murman and Julian D. Cole, Calculation of Plane Steady Transonic Flows, AIAA Journal, Vol. 9, No. 1, Jan., 1971.

Merging of Vortices with Decaying Cores and
Numerical Solutions of Navier-Stokes Equations

L. Ting
Courant Institute of Mathematical Sciences
New York University, New York, NY 10012

G.C. Liu
NASA Langley Research Center
Hampton, VA 23665

From recent studies of the merging of vortices with decaying cores, a new numerical method for solving the two dimensional incompressible Navier-Stokes equations is formulated. The solution is assumed to be composed of Lamb vortices with cores which may overlap each other. The N-S equations are satisfied approximately under a minimum principle. The latter in turn yields the equations for the velocities of the vortex centers. The solution of the flow field is

thereby obtained by numerical integration of the equations of motion of the vortex centers. The numerical method is employed to study the merging of vortices and the roll up of a trailing vortex sheet. Comparisons of the results with the corresponding finite difference solutions of the N-S equations are presented.

Introduction

Matched asymptotic solutions of N-S equations were constructed by Ting and Tung [1] to study the motion and decay of two dimensional incompressible vortices. Thus, the singularities of the inviscid solution are removed and the velocity of the vortex center is defined, with the leading term in agreement with the inviscid theory. It is also shown that the leading solution of the core structure can be approximated by that of a single vortex in ambient fluid, an optimum Lamb vortex defined by the rule of merging.

The validity of the asymptotic solution will be in question when the vortices are not too far apart and the vortical cores of adjacent vortices overlap or merge with each other. The merging of vortices is a canonical viscous flow problem for which numerical solutions of N-S were constructed [2,3]. They show that the merging of vortices takes place in two stages. In the first stage, the contour lines of constant vorticity are centered around points of local maximum. These points spiral gradually towards each other. When they coincide, the second stage begins and the identities of individual vortices disappear. The contour lines approach concentric circles of The "optimum" Lamb vortex predicted directly from the initial data by the Poincare relationships [4]. They are: the conservation the total strength Γ and the first moments of the vorticity distribution and the increment of its polar moment at the constant rate $4\nu\Gamma$ where ν is the kinematic viscosity. Consequently, we arrive at the rule of merging [5]:

"When several vortices merge to a single one, the latter is an optimum Lamb vortex of the same total strength and centered at their center of gravity. The initial size of the single vortex is defined by the matching of its polar moment with that of those vortices at the beginning of merging."

When the asymptotic solution is extended to the merging stages, it was found that [5]: i) the asymptotic solution deviates from the numerical solution only in the second stage of merging and ii) when it is extended beyond the second stage of merging, i.e. well beyond its region of validity, it is again in good agreement with the numerical solution. This happens because the "extended" asymptotic solution converges to the same optimum Lamb vortex as time t increases.

The "extended" asymptotic solution was employed to study the roll up of trailing vortex sheet [5]. The contour lines agree quite well with the corresponding numerical solution [6]. These facts prompt us to recast the asymptotic solution as an approximate solution to the N-S equations, investigate its error and formulate a method of improvement.

Approximate Solutions of N-S Equations

We introduce the ansatz that the flow field is represented by the superposition of N Lamb vortices. The vortical cores can overlap each other. We then show that the solution satisfies approximately the N-S equations under a minimum principle which in turn defines the velocities of the vortex centers.

Let $\zeta^*(x,y,\delta(t))$ denote the vorticity distribution of a single Lamb vortex of unit strength centered at the origin, i.e.,

$$\zeta^*(x,y,\delta(t)) = \frac{1}{\pi\delta^2(t)} e^{-r^2/\delta^2(t)} \tag{1}$$

with $\delta^2(t) = 4\nu t$ and $r^2 = x^2 + y^2$. The Lamb vortex is created at $t = 0$ since the core size δ vanishes at $t = 0$. The corresponding stream function and induced velocity are denoted by $\psi^*(x,y,\delta)$ and $\vec{V}^*(x,y,\delta)$ respectively. The ansatz on the vorticity distribution ζ of a viscous flow induced by an initial distribution $\zeta_0(x,y)$ is:

$$\zeta(x,y,t) = \sum_{k=1}^{N} \Gamma_k \zeta^*_k \quad \text{and} \quad \psi(x,y,t) = \sum_{k=1}^{N} \Gamma_k \psi^*_k \tag{2}$$

where $\quad \zeta^*_k = \zeta^*(x - X_k(t), y - Y_k(t), \delta_k(t + t^*_k))$

and $\quad \psi^*_k = \psi^*(x - X_k(t), y - Y_k(t), \delta_k(t + t^*_k))$. $\tag{3}$

Since solution (1) for a Lamb vortex fulfills the N-S equations and the far field conditions, solution (2) fulfills all the linear conditions except the nonlinear vorticity diffusion equation which reduces to

$$F(x,y,t) = \sum_k \Gamma_k \{ (\sum_{\ell \neq k} \Gamma_\ell \vec{V}^*_\ell) - \dot{X}_k(t)\hat{i} - \dot{Y}_k(t)\hat{j} \} \cdot \nabla \zeta^*_k = 0 \tag{4}$$

To fulfill Eq. (4) approximately, we seek $\dot{X}_k(t), \dot{Y}_k(t), k = 1 \cdots N$, such that the function

$$H(\dot{X}_1 \cdots \dot{X}_N, \dot{Y}_1 \cdots \dot{Y}_N, t) = \langle F^2(x,y,t) \rangle = \min . \tag{5}$$

for $t \geq 0$ where $\langle \ \rangle$ denotes the area integral over the xy plane. In this sense, we say that ζ and ψ of (2) is the "best" approximate solution of the N-S equations. At each instant t, $H(\dot{X}_1 \cdots, t)$ is minimized by $\dot{X}_1 \cdots \dot{Y}_N$, if and only if the following 2N linear equations for $\dot{X}_m$ and $\dot{Y}_m$ are fulfilled:

$$\sum_m (a_{km}\dot{X}_m + e_{km}\dot{Y}_m) = C_k \quad \text{and} \quad \sum_m (e_{mk}\dot{X}_m + b_{km}\dot{Y}_m) = D_k \tag{6}$$

for $k = 1 \cdots N$ with $a_{km} = \Gamma_k \Gamma_m \langle \partial_x \zeta^*_k \partial_x \zeta^*_m \rangle$, $e_{km} = \Gamma_k \Gamma_m \langle \partial_x \zeta^*_k \partial_y \zeta^*_m \rangle$,

$C_k = \sum_m c_{km}$, $D_k = -\sum_m d_{km}$, $c_{km} = \Gamma_k \Gamma_m \langle \partial_x \zeta^*_k (\partial_x \zeta^*_m \sum_{\ell \neq m} \partial_y \psi^*_\ell - \partial_y \zeta^*_m \sum_{\ell \neq m} \partial_x \psi^*_\ell) \rangle$

and b_{km} and d_{km} are the same as a_{km} and c_{km} respectively with ∂_x interchanged with ∂_y. These coefficients are elementary functions of $X_k - X_m$, $Y_k - Y_m$, δ_m and δ_k (see Appendix B of Ref. 8). In particular, we have $a_{kk} = b_{kk} = 2\Gamma_k^2/(\pi\delta_k^4)$ and $e_{kk} = 0$. Since ζ^* decays exponentially in length scale δ, all the non-diagonal elements, $a_{km} \cdots e_{km}$, contain a factor $\exp(-\Lambda_{km}^2)$ where $\Lambda_{km}^2 = R_{km}^2/(\delta_k^2 + \delta_m^2)$ denotes the ratio of the square of the distance between the k-th and m-th center and the sum of the squares of their core sizes. The N pairs of equations (6) can be rearranged as

$$G Z = H \qquad (7)$$

where Z and H denote the column matrices of $\{\dot{X}_1, \dot{Y}_1, \dot{X}_2, \ldots, \dot{Y}_N\}$ and $\{C_1, D_1, C_2 \cdots D_N\}$ respectively. G is the corresponding 2N×2N matrix. For example, its first two rows are $\{a_{11}, 0, a_{12}, e_{12}, \cdots, e_{1N}\}$ and $\{0, b_{11}, \cdots, b_{1N}\}$. Due to the minimum principle, G is symmetric and positive definite. Since the nondiagonal elements vanishes as $\exp(-\Lambda_{km}^2)$, G is dominated by its diagonal elements and is sparse. Equation (7) can be solved readily for $\dot{X}_k$ and $\dot{Y}_k$.

The ratio Λ_{km}^2, in the exponential factor, can be used to characterize the k-th and m-th vortices. When $\Lambda_{km}^2 \gg 1$ these two vortices are isolated from each other. When $\Lambda_{km}^2 \ll 1$, they have merged to a single one. For a k-th vortex, which is far part from the others ($\Lambda_{km}^2 \gg 1$, $m \neq k$), both the (2k - 1)-th and 2k-th equation of (7) are decoupled from the rest of (7) and yield the asymptotic or the classical result for the velocity of the k-th vortex center. The nondiagonal elements of G and those of $\{c\}$ and $\{d\}$ in H account for the interaction effects of adjacent vortices.

The above observations explain why the agreement between the extended asymptotic solution and the finite difference solution for the simultaneous merging of three vortices is better than that for two vortices [2,3]. The agreement is even better for four vortices as the number of pivot points in minimizing $\langle F^2 \rangle$ increases. They also explain why in the study of the roll up of vortex sheet, the extended asymptotic solution agrees quite well with the finite difference solution even at the late stage of roll up [5]. The reason is as follows. There are 100 vortices used to model the sheet. The core of a vortex will overlap at most one or two adjacent vortices while the velocity of a vortex center is the resultant of the velocities induced by all the vortices. In the next section we present numerical examples to illustrate those points.

Numerical Examples

Figs. 1 and 2 show the contour lines of constant vorticity, ζ/ζ_{max}, in the roll up of a vortex sheet behind a wing with elliptical load [6,7]. The thickness of the sheet is 2.5% of the semispan S. The Reynolds number is 200,000 based on the circulation Γ_0 around the root section. Fig. 1 shows the contour lines at the station 11 S downstream where more than half of the initial vortices, accounting for 86% of Γ_0 are rolled up around an "eye" of the "tip vortex." The contour lines in the extended asymptotic solution using 100 vortices equally spaced initially differ from those in the finite difference solution [7] in details because of the waviness in the former. To improve the solution, we introduce 160 vortices at the initial station with $\Lambda_{km} = 1/32$. We apply the rule of merging when $\Lambda_{km} < 1/32.1$. At the station 11S, there are 94 vortices. Fig. 2 shows the contour lines which are in better agreement with the finite difference solution [7]. Also shown in the insert are the contour lines in the inner core of the eye enlarged five times. Thus, we demonstrate that the rule of merging enables us to allow the vortices to merge in the eye while those in the outer fringes will not. Consequently, the waviness of the contour lines will be reduced.

Figures 3 and 4 compare the results of different solutions for the merging of two identical vortices, N = 2, with strength $\Gamma = 25$, Re = 100, core size $\delta = 1$ and centered at (2,0) and (-2,0) when t = 1. We study the merging of two vortices extensively because the probability of the merging of more than two vortices

(N > 2) simultaneously is very small. Another reason is that among all $N \geq 2$, the extended asymptotic solution for N = 2 is the least satisfactory and the effect of merging will be the most pronounced.

Fig. 3 compares the trajectories of the points of maximum vorticity during the first stage of merging. The curves in heavy line show the trajectory given by the approximate solution based on the minimum principle. These curves are much closer to those of finite difference solution than those of the extended asymptotic solution. Figure 4a shows the contour lines of $\zeta/\zeta_{max} = 1/\sqrt{e}$, $1/e$ and $1/\sqrt{e^3}$ of the finite difference solution. The heavy dotted lines show the lines based on the optimum single vortex. Hence, the merging to a single vortex is nearly completed by t = 25. Fig. 4b shows the contour lines of the approximate solution based upon the minimum principle. They agree with those in 4a in size and in overall orientation (of principal axes) while differing in details. Also shown in dotted lines are the corresponding axes based on the extended asymptotic solution. We see that the contour lines of the latter will be completely out of phase with the finite difference solution. Thus the approximate solution does improve the extended asymptotic solution substantially by correcting the velocity of the vortex centers based on the minimum principle.

Further improvements will be implemented, for example by reinitializing the vorticity distribution using more vortices, see Ref. 8, and by extending the ansatz to allow the strength of each vortex $\Gamma_k(t)$ to be time dependent and governed likewise by the minimum principle.

References

1. Ting, L. and Tung, C. Phys. Fluids, Vol. 8, 1965, 1039-1051.
2. Lo, R. and Ting, L. Phys. Fluids, Vol. 19, 1976, 912-913,.
3. Liu, G.C. and Hsu, C.H. Proc. 4th Intern. Conf. Appl. Numer. Modeling, Chengkung Univ., Taiwan, Dec. 1984, 656-665.
4. See Truesdell, C. "The Kinematics of Vorticity" 1954, Indiana Univ. Press.
5. Ting, L. Proc. Intern. Symp. Comput. Fluid Dynamics, Tokyo, 1985, Preprints Vol. II, 582-595.
6. Weston, R.P. and Liu, C.H. AIAA Paper No. 82-0951, June, 1982.
7. Weston, R.P., Ting, L. and Liu, C.H. AIAA Paper No. 86-0558, Jan., 1986.
8. Liu, G.C. and Ting, L. AIAA Paper No. 86-1072, May, 1986.

Acknowledgement. L.T. was supported by NASA Grant NCCI-58.

VORTEX ROLL UP FOR ELLIPTICAL SPANLOAD AT 5.5 SPAN DOWNSTREAM

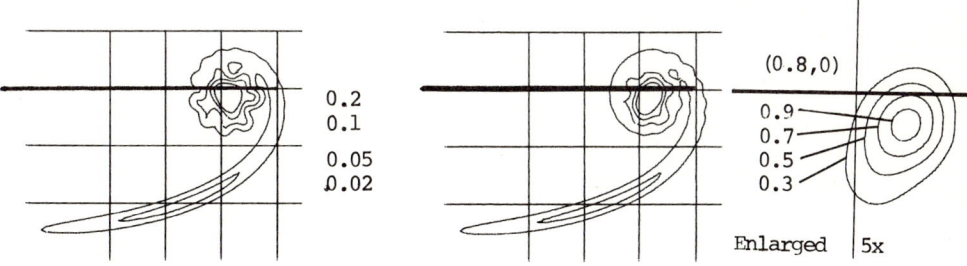

Fig. 1, 100 vortices   Fig.2, 200 vortices initially merged to 149 vortices

MERGING OF TWO VORTICES

2.5
1.5
.5
Y
0
-.5
-1.5
-2.5
X

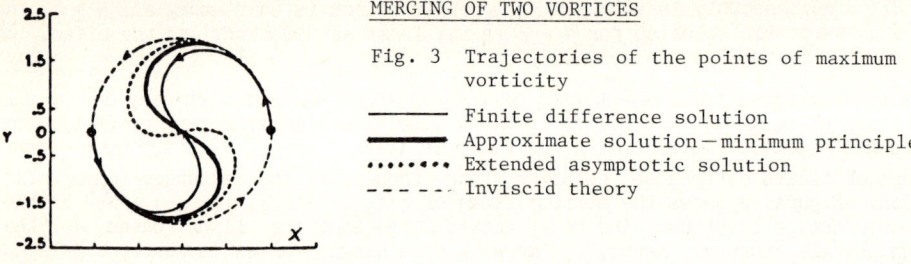

Fig. 3 Trajectories of the points of maximum vorticity

—— Finite difference solution
━━ Approximate solution — minimum principle
······ Extended asymptotic solution
- - - - Inviscid theory

Fig. 4 Contour lines of constant vorticity
a) Finite difference solution

$e^{-1/2}$
e^{-1}
$e^{-3/2}$
(a) t=1.0
(b) t=1.97
(c) t=2.93
(d) t=5.34
(e) t=8.72
(f) t=14.51
(g) t=20.30
(h) t=25.13

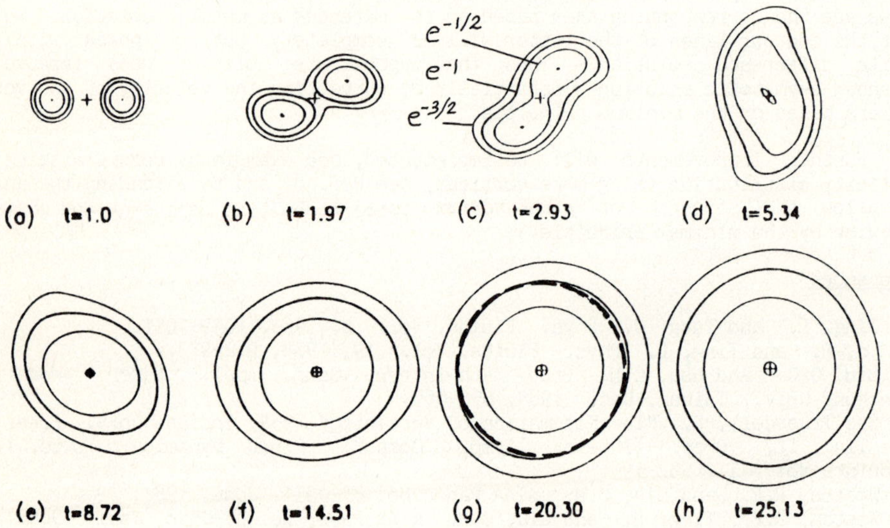

b) Approximate solution — minimum principle

$e^{-1/2}$
e^{-1}
$e^{-3/2}$
t=1.0
t=1.97
t=2.93
t=5.34
t=8.72
t=14.51
t=20.30
t=25.13

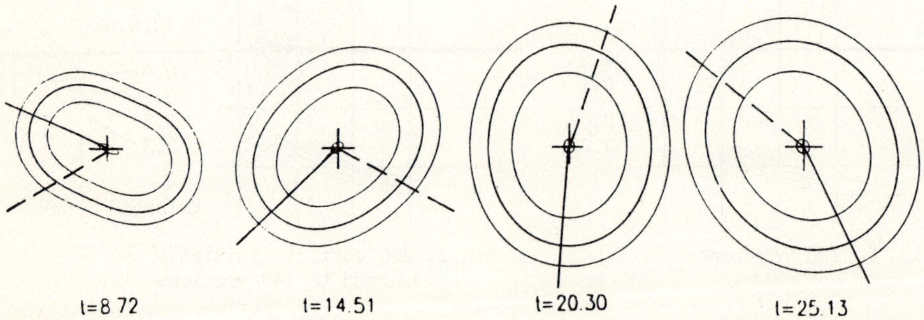

DIRECT SIMULATION OF SHEAR FLOW TURBULENCE IN A PLANE CHANNEL
BY SIXTH ORDER ACCURATE METHOD OF LINES WITH NEW
SIXTH ORDER ACCURATE MULTI-GRID POISSON SOLVER

H. Tokunaga, N. Satofuka and H. Miyagawa[†]
Department of Mechanical Engineering, Kyoto Institute of Technology
Matsugasaki, Sakyo-ku, Kyoto 606, JAPAN

I. INTRODUCTION

It is shown that the higher order method of lines is an efficient method for the numerical simulation of homogeneous and isotropic turbulence [1]. In the present paper we extend the method of lines, develop a new efficient higher order accurate Poisson solver and carry out direct simulation of shear flow turbulence in a plane channel which is at first dealt with by making use of the pseudo-spectral and Chebyshev polynomial expansion method [2]. The new method consists of the sixth order accurate modified differential quadrature (MDQ) method to the spatial discretization, which is already successfully applied to direct simulation on instability of laminar flows in the fourth order accuracy [3,4], with the new sixth order accurate multi-grid Poisson solver and the explicit Runge-Kutta-Gill time integration scheme.

II. GOVERNING EQUATIONS

The motion of 3-D viscous and incompressible flow in a plane channel is governed by the Navier-Stokes equation with respect to the velocity field $\vec{u} = (u,v,w)$ in the Cartesian coordinate as

$$\frac{\partial \vec{u}}{\partial t} + (\vec{u} \cdot \nabla)\vec{u} = -\nabla p + \frac{1}{R} \Delta \vec{u}, \tag{1}$$

and the Poisson equation for the pressure p

$$\Delta p = f(\frac{\partial u}{\partial x}, \frac{\partial u}{\partial y}, \ldots, \frac{\partial w}{\partial z}), \tag{2}$$

where R denotes the Reynolds number defined as $R = Uh/\nu$, by means of the maximum fluid velocity in the channel, h the half width of the channel and ν the kinematic viscosity. In Fig. 1 the computational domain is shown. Since there exist solid walls in the y-direction, we adopt an unequally spaced grid as

$$y_j = \tanh(\eta_j \tanh^{-1} a)/a, \quad \eta_j = -1+2(j-1)/(L-1), \quad j = 1,2,\ldots,L, \tag{3}$$

where $0<a<1$ and L denotes the number of grid points in the y-direction.

III. NUMERICAL PROCEDURE

The first and the second derivative of u with respect to y, for example, at a grid point $y = y_j$ are approximated by making use of the sixth order accurate MDQ method [1,3,4] as

$$\frac{\partial u}{\partial y}\bigg|_{y=y_j} = \sum_{m=\gamma}^{\gamma+M-1} a_{jm} u(y_{j+m}) \equiv D_M(u_j),$$

$$\frac{\partial^2 u}{\partial y^2}\bigg|_{y=y_j} = \sum_{m=\gamma}^{\gamma+M-1} \sum_{l=\gamma}^{\gamma+M-1} a_{jl} a_{lm} u(y_{j+m}) \equiv D_M^2(u_j). \tag{4}$$

The weighting coefficients a_{jm} are expressed as the Lagrange interpolation as

$$a_{jm} = \Pi'(y_j)/[(y_j-y_m)\Pi'(y_m)], \text{ for } j \neq m, \text{ and } a_{jj} = \Pi''(y_j)/[2\Pi'(y_j)], \tag{5}$$

$$\Pi(y) = (y-y_{j+\gamma})(y-y_{j+\gamma+1})\cdots(y-y_{j+\gamma+M-1}).$$

[†] Present Address: Ishikawajima-Harima Heavy Industries Co., Ltd., Mizuho-machi, Nishitama-gun, Tokyo 190-12, JAPAN

We adopt $\gamma = -(M-1)/2$ at inner grid points and on the solid wall boundary $\gamma = 0$. The index $M = 7$ corresponds to the sixth order accuracy. Substituting approximate relations to spatial derivatives shown in (4) into the 3-D Navier-Stokes equation (1), we obtain a set of ordinary differential equations(ODEs),

$$\frac{d\vec{U}}{dt} = \vec{F}(U_i, D_M(U_i), D_M^2(U_i), D_M(P_i)) \equiv \vec{F}(\vec{U}, \vec{P}), \tag{6}$$

where $\vec{U}$ includes 3-D velocity components at 3-D grid points N, and $\vec{P}$ represents pressure values at 3-D grid points expressed as

$$\vec{U} = (u_1, u_2, \ldots, u_N, v_1, v_2, \ldots, v_N, w_1, w_2, \ldots, w_N)^T, \quad \vec{P} = (p_1, p_2, \ldots, p_N)^T. \tag{7}$$

The set of ODEs (6) is integrated by using the fourth order accurate explicit Runge-Kutta-Gill time stepping scheme.

Sixth Order Accurate Multi-Grid Poisson Solver

In the discretization of the Poisson equation (2), same approximate relations of (4) are used newly in the present paper. The relaxation of the discretized Poisson equation is dealt with the multi-grid technique in order to accelerate its convergence. In this method we use g levels computational grids Ω^k, $k = h, 2h, \ldots, 2^{g-1}h$, in which the grid spacing increases from h, the finest grid spacing, by the factor 2 successively. In the finest Ω^h the Poisson equation is discretized as

$$\Delta^h p^h = f^h, \tag{8}$$

where Δ^h denotes the difference operator with the sixth order accuracy. Several relaxation sweeps are carried out in Ω^h and we obtain an approximate value p^h for p. Next on Ω^{2h} we solve

$$\Delta^{2h} p^{2h} = I_h^{2h}(f^h - \Delta^h p^h) + \Delta^{2h} p^{2h}, \tag{9}$$

where I_h^{2h} denotes the contraction operator from Ω^h to Ω^{2h}. Thirdly we modify the approximate value from p^h to

$$p^h \leftarrow p^h + I_{2h}^h (p^{2h} - I_h^{2h} p^h), \tag{10}$$

where I_{2h}^h represents the prolongation operator from Ω^{2h} to Ω^h. The same procedure is applied between different grids, and this cycle is continued untill the convergent solution is obtained.

The model 3-D Poisson equation is solved by the present multi-grid solver. The L_2 error is shown in Fig. 2 where computational grids used are 64x64x64, 32x32x32, 16x16x16 and 8x8x8, respectively. Even in 8x8x8 grid, L_2 error equals to 1.20×10^{-4}, which is smaller than that of the SOR method in 32x32x32 grid, and from Fig. 2 the present Poisson solver is shown to have the sixth order accuracy. In Fig. 3 we show the convergence history of the Poisson solver in 32x32x32 grid. The abscissa denotes the work which corresponds to the number of sweeps in the finest grid. Coaser grids 16x16x16, 8x8x8 and 4x4x4 are used in order to apply the multi-grid technique. In the three levels grid we obtain the fast convergence in comparison with the SOR method. The CPU time of the present Poisson solver equals to 42,746 mill-sec in FACOM VP-200 using the four levels grid which is approximately a third CPU time of the SOR method.

IV. DIRECT SIMULATION OF SHEAR FLOW TURBULENCE

Initial and Boundary Condition

In order to deal with the subcritical transition of plane Poiseuille flow to turbulence, we adopt the Benney-Lin type initial condition, which is composed of solutions for 2-D and 3-D Orr-Sommerfeld equation expressed as

$$u(x,y,z) = (1-y^2, 0, 0) + \text{Real}[u_{2D}(y)\exp(i\alpha x)]$$
$$+ \text{Real}[u_{+3D}(y)\exp(i(\alpha x + \beta z))] + \text{Real}[u_{-3D}(y)\exp(i(\alpha x - \beta z))], \tag{11}$$

The solutions $u_{2D}(y)$, $u_{+3D}(y)$ and $u_{-3D}(y)$ are calculated by the new direct simulation method of the laminar flow instability in [3,4].

Although the present method make possible a variety of boundary conditions from its nature, we adopt at first the periodic boundary condition to the x- and z-direction since results obtained by the present method should be compared with other results [2,6]. With respect to the y-direction the non-slip boundary condition for the velocity components and the Neumann boundary condition of the pressure are imposed.

Results

The present computations are carried out in the 32x32x32 grid. At first we show the result on direct simulation of 3-D shear flow in a plane channel at R = 750. Fig. 4 depicts the temporal development of the maximum amplitude of the x-directional velocity for 2-D primary, 3-D primary and 2-D harmonic mode. It is shown that the 2-D primary mode decays rapidly untill t = 20 and gradually after this period. Both the 3-D primary and 2-D harmonic mode do not grow at all, which means that the transition of plane Poiseuille flow to turbulence is not seen at R = 750. The present computation is carried out by FACOM VP-200, and its computing time equals to 36 minutes untill t = 79.

In Fig. 5 the temporal development of the maximum amplitude is depicted at R = 1250. In the present case the 2-D primary mode only gradually decreases, while the 2-D harmonic and the 3-D harmonic mode are excited. From this fact we conclude that the transition of plane Poiseuille flow to turbulence occurs at R = 1250 though it is not observed at R = 750. This result shows an excellent agreement with Orszag and Kells' result [2], which confirms the accuracy and the effectivity of the present computational method.

We show the equi-shear lines at t = 0, 17 and 22 for the direct simulation at R = 1500 in Fig. 6. The present computation clearly shows the formation of a horseshoe vortex which is found in the experiment by Nishioka et al. [5] and in the computation by Biringen [6] using the pseudo-spectral method in the x- and z-direction and the central difference method in the y-direction. In Fig. 7 the contours of vorticities in the x- and z-direction are shown in the position y = -1+ Δy_{min} and t = 17. The horseshoe vortex is found in the x-z and x-y plane. The comparison of the present result with that of Biringen [6] shows that with respect to the accuracy in the y-direction the present method is superior to Biringen' method but the situation is reversed in the x- and z-direction owing to the use of the pseudo-spectral method. However, the present method is easily extended to the arbitrary order accuracy.

V. CONCLUSIONS

It is shown that the newly developed multi-grid Poisson solver has the sixth order accuracy and is very efficient. The new computational method, which is composed of the sixth order accurate MDQ method to the spatial discretization, the present Poisson solver and the RKG time stepping scheme, is applied to the direct simulation of shear flow turbulence in a plane channel. The result obtained shows an excellent agreement with the methods based on the pseudo-spectral method, and the accuracy and the efficiency of the present method are confirmed. From now on the extension of the present method to LES and other shear flow turbulence is desired.

REFERENCES

[1] N. Satofuka, H. Nakamura and H. Nishida, Proc. Ninth Int. Conf. Num. Meth. Fluid Dyn.. Saclay, France, (1984).
[2] S. A. Orszag and L. C. Kells, J. Fluid Mech., 96(1980), 159.
[3] H. Tokunaga, N. Satofuka, Y. Tanimura, Proc. Int. Symp. Comp. Fluid Dyn.-Tokyo, Japan, (1985).
[4] H. Tokunaga, N. Satofuka and H. Miyagawa, Memoirs of Faculty of Engineering and Design, Kyoto Institute of Technology, 34(1985), 72.
[5] M. Nishioka, M. Asai and S. Iida, Transition of Turbulence, ed. by R. E. Meyer, Academic Press, (1981), 113.
[6] S. Biringen, J. Fluid Mech., 148(1984), 413.

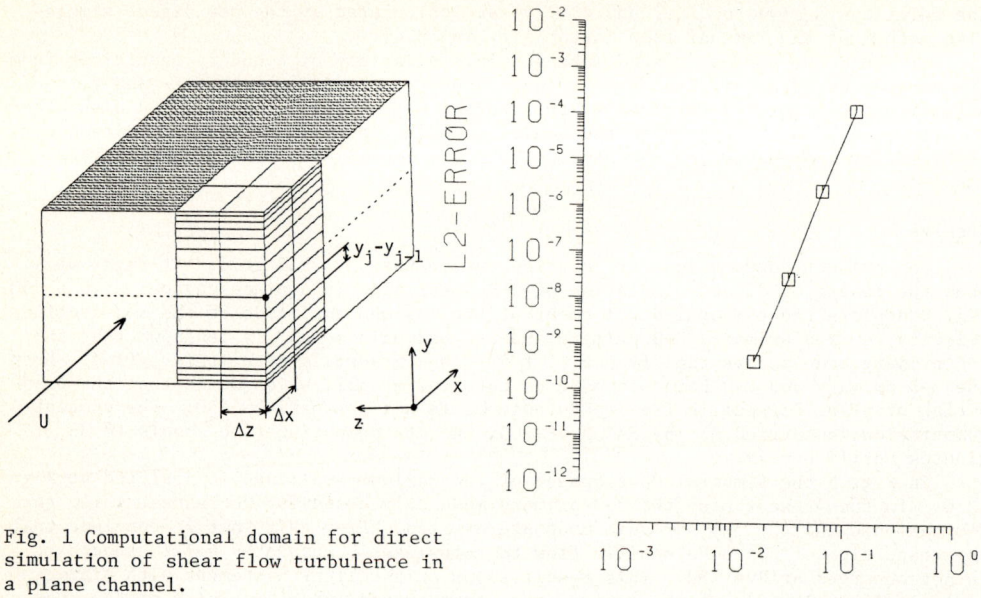

Fig. 1 Computational domain for direct simulation of shear flow turbulence in a plane channel.

Fig. 2 L_2 error of new higher order accurate multi-grid Poisson solver in 64x64x64, 32x32x32, 16x16x16 and 8x8x8 grid.

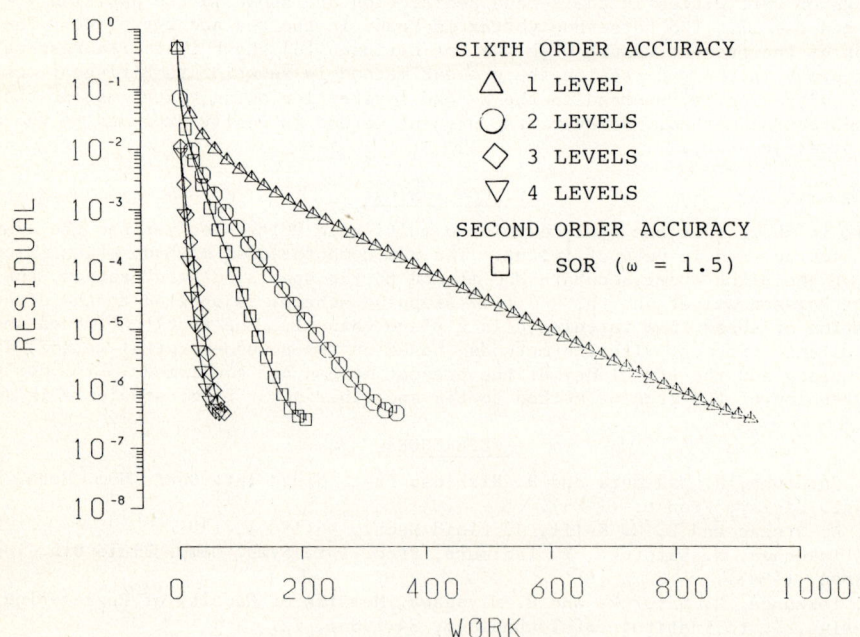

Fig. 3 Convergence history of sixth order accurate multi-grid Poisson solver in 32x32x32 grid.

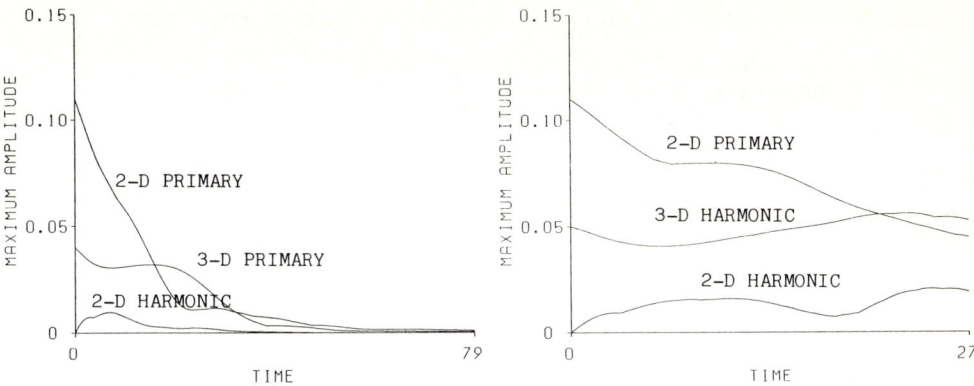

Fig. 4 Temporal development of maximum amplitude at R = 750.

Fig. 5 Temporal development of maximum amplitude at R = 1250.

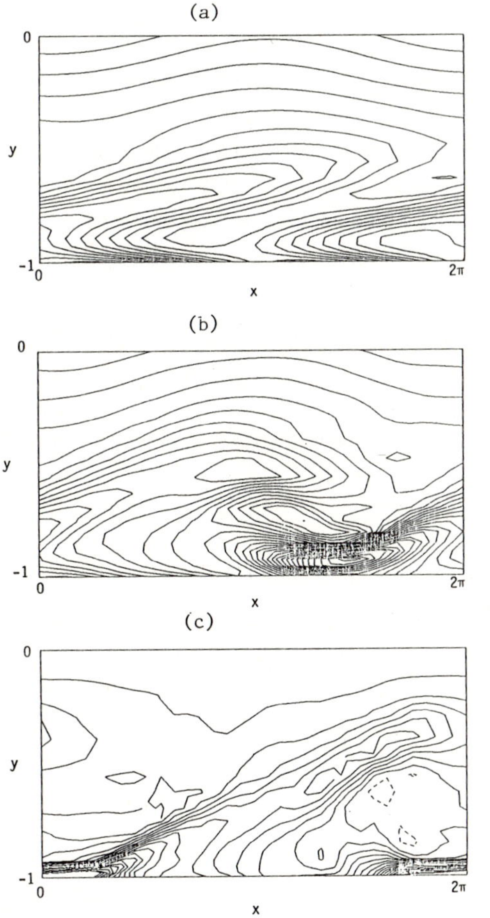

Fig. 6 Equi-shear lines for u_y at R = 1500 and at (a) t = 0, (b) t = 17 and (c) t = 22.

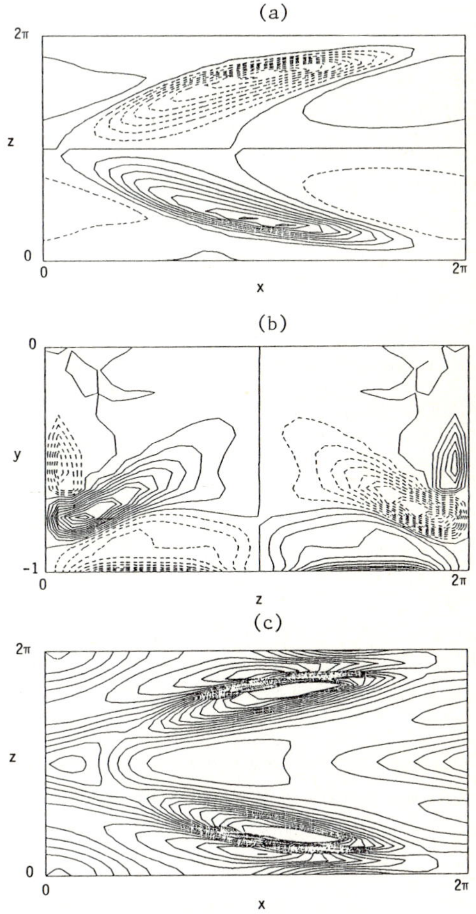

Fig. 7 Vorticity contours at R = 1500 and t = 17. (a) and (b) vorticity in x-direction, (c) in z-direction.

SOLUTIONS OF THE NAVIER STOKES EQUATIONS USING AN EFFICIENT SPECTRAL METHOD

R. Verstappen, J. ten Thije, R.W. de Vries, P.J. Zandbergen.
(Twente Univ. of Technology, Enschede, 7500AE, The Netherlands)

Introduction.

In this paper we will concentrate on the solution of two dimensional Navier-Stokes problems using spectral methods. A discussion of these methods can be found in refs. 1 and 2. Use will be made of Chebysev expansions and the boundary conditions will be taken into account by using the so called τ approximation. In order to find a solution of the Navier-Stokes equations the so called pressure correction method will be used together with a time integration of second order. (Ref. 3). The pressure will be solved by a method due to Kleiser (ref. 4). Using spectral methods together with Kleiser's method in general will result in very expensive calculations. This especially is true for general boundary conditions. It seems therefore worthwhile to take account of any symmetry that the method poses and to do the utmost to economize upon the computational work. Since we use a separation of variables technique to solve the problem, the accurate and efficient calculation of eigenvalues and eigenvectors is crucial in order to maintain the convergence properties of the spectral method. This poses strong demands for increasing numbers of coefficients. The ultimate purpose of this investigation was to obtain an insight in the complexity of the method, the advantages which can be gained by using a vector-processor and to present some results to illustrate the findings.

Description of the method.

The Navier-Stokes equations for an incompressible fluid inside a domain B are written as

$$\frac{\partial \underline{u}}{\partial t} = \underline{u} \times \text{curl } \underline{u} - \text{grad } s + \nu \Delta \underline{u} \tag{1}$$

$$\text{div } \underline{u} = 0 \tag{2}$$

$$\text{where } s = p + \frac{1}{2} (\underline{u}.\underline{u}) \tag{3}$$

It will be assumed that along the edges ∂B of B the normal and tangential components of $\underline{u}$ are given. By taking the divergence of eq. (1) and applying eq. (2) we readily find

$$\Delta s = \text{div } (\underline{u} \times \text{curl } \underline{u}) = \text{div } \underline{M} \tag{4}$$

As Kleiser (ref.4) has shown the system (1) - (3) is equivalent with (1) - (2) if we require

$$\text{div } \underline{u} \ (t=0) = 0 \tag{5}$$

$$\text{div } \underline{u} = 0 \quad \text{on } \partial B \tag{6}$$

In order to obtain a second order accurate discretization in time we use a Crank-Nicholson like technique. Eq.(1) is written for $t = (n+1) \Delta t$ as

$$\underline{u}^{n+1} - \underline{u}^n = \Delta t \ \{\tfrac{3}{2} \underline{M}^n - \tfrac{1}{2} \underline{M}^{n-1} - \text{grad } s^{n+\frac{1}{2}} + \tfrac{\nu}{2} (\Delta \underline{u}^{n+1} + \Delta \underline{u}^n)\} \tag{7}$$

where $s^{n+\frac{1}{2}}$ is determined from

$$\Delta s^{n+\frac{1}{2}} = \text{div } (\tfrac{3}{2} \underline{M}^n - \tfrac{1}{2} \underline{M}^{n-1}) \tag{8}$$

The final system to be solved is then a system of 3 Helmholtz equations and a Poisson equation given by

$$\Delta \underline{u}^{n+1} - \lambda \underline{u}^{n+1} = -\underline{r}^n + \text{grad } q^{n+\frac{1}{2}} \tag{9}$$

$$\Delta q^{n+\frac{1}{2}} = \text{div } \underline{r}^n \tag{10}$$

with $\lambda = \frac{2}{\nu \Delta t}$; $\underline{r}^n = \frac{2}{\nu}[\tfrac{3}{2} \underline{M}^n - \tfrac{1}{2} \underline{M}^{n-1} - \tfrac{1}{\Delta t} \underline{u}^n] + \Delta \underline{u}^n$; $q^{n+\frac{1}{2}} = \tfrac{2}{\nu} s^{n+\frac{1}{2}}$.

Up to now we are missing boundary conditions to solve eq.(10) for q. This boundary condition is in fact eq.(6). In order to solve this problem we split it into a homogeneous and a particular part

$$\underline{u} = \underline{u}_p + \underline{u}_h \tag{11}$$

$$q = q_p + q_h \tag{12}$$

The particular solution $\underline{u}_p$ we determine from eq.(9) with the prescribed boundary conditions for $\underline{u}$, whereas q_p is determined from eq.(10) and the boundary condition $q_p = 0$. We can now calculate $D = \text{div } \underline{u}_p$.
The homogeneous solution is determined as follows. We write

$$q_h = \Sigma \ \alpha_i \ q_i, \quad \underline{u}_h = \Sigma \ \alpha_i \ \underline{u}_i \tag{13a, 13b}$$

and solve

$$\Delta q_i = 0 \qquad \text{with } q_i = 1 \text{ in edge point } i \qquad (14)$$
$$q_i = 0 \text{ in edge point } i \ne j$$

$$\Delta \underline{u}_i - \lambda \underline{u}_i = \text{grad } q_i \quad \text{with } \underline{u}_i = 0 \text{ on the edge} \qquad (15)$$

From eq. (6) it follows that α_i can be solved from

$$\sum_i \alpha_i (\text{div } \underline{u}_i)_j = -D_j \qquad (16)$$

It will be clear that the coefficients of the left hand side can be determined once, and that the right hand changes every time step. It is hence advantageous to decompose the left hand side once. A further remark which can be made is that by introducing $\hat{\underline{u}} = \underline{u}_i + \frac{1}{\lambda} \text{grad } q_i$ we obtain a homogenious equation for $\hat{\underline{u}}_i$.

<u>Solution for a Helmholtz problem.</u>

We will now solve $\Delta u - \lambda u = f$ on a square $-1 \le x,y \le 1$ by posing for instance

$$u = \sum_{n=0}^{N} \sum_{m=0}^{M} x_{nm} T_n(x) T_n(y) \qquad (17)$$

If we substitute these expressions into eq.(8) and take care of the boundary conditions the 2N quantities $u_{N-i,m}$ $u_{n,N-i}$; $i,j=0,1$ can be eliminated. This leads to

$$AU + U^{0,2} - \lambda U = F - R^* \qquad (18)$$

Here A is a (N-1) x (N-1) matrix representing the $\frac{\partial^2}{\partial x^2}$ operator with Diriclet boundary conditions. $U^{0,2}$ is the $\frac{\partial^2}{\partial y^2}$ operator, F is the source term and R^* is the result of the elimination procedure. If we now diagonalize A and introduce F and G by

$$A = C V C^{-1}, \quad U = C \cdot Z, \quad G = C^{-1}(F-R^*) \qquad (19)$$

we find

$$VZ + Z^{0,2} - \lambda Z = G \qquad (20)$$

It will be clear that the efficient solution of (20) is crucial for the complexity of the method. We first observe that it is possible to split A in two matrices A_1 and A_2, and that a transformation matrix T can be found having only nonzero elements in the last two columns and on the diagonal such that $B = TAT^{-1}$ has a nice structure. If the components of B are rearranged we find that we can write

$$Mw = \mu w \qquad (21)$$

with $M = \begin{bmatrix} M_1 & 0 \\ 0 & M_2 \end{bmatrix}$ and where both M_1 and M_2 have the following structure

$$M_1 = \begin{bmatrix} 0 & * & * & & * \\ & 0 & * & & * \\ 0 & 0 & 0 & 0 & * \\ * & * & * & * & * \end{bmatrix}$$

The eigenvalues μ are found by applying the QR algorithm on both M_1 and M_2. Due to the nice structure of M_1 and M_2 we can also easely determine the eigenvektors of M_1 and M_2 and hence those of A.

In order to find Z we can write eq.(20) as two quasi tridiagonal systems for the even and the odd components of Z where the structure of the matrix is as follows

$$\begin{bmatrix} * & * & * & * & * & * \\ * & * & * & & & \\ & * & * & * & & \\ & & * & * & * & \\ & & & * & * & * \\ & & & & * & * \end{bmatrix}$$

These systems can be solved quite easily in $O(N^2)$ operations. So the real time consuming work is the determiniation of G and U according to (19) since this takes $O(N^3)$ operations. It should be observed that due to the special structure of R^* in (19) the multiplication $C^{-1} R^*$ takes only $O(N^2)$ operations and if we have no source term F the calculation of G and U is determined by $U = C \dot{Z}$ and hence twice as fast as usually. It will be understood that the transformations from the spectral space to the fysical space and vice versa are performed by using an F.F.T. method.

The results so far obtained were tested for a model problem. It is very difficult to obtain a bound on the accuracy of the eigenvalues, but an indication can be obtained by calculating the diagonalisation in half, single and double precision which accounts to 7, 14 and 28 decimal places, whereas the rest of the calculations is performed in single precision. Consider the following problem

$$\frac{\partial^2 q}{\partial x^2} + \frac{\partial^2 q}{\partial y^2} = -2\pi^2 \sin \pi \, 2 \sin \pi \, y \qquad (22)a$$

$$q = 0 \qquad \text{along the edge} \qquad (22)b$$

The solution is q = sin Π x sin Π y
The maximum absolute error as a function of N is given by

N	half	single	double
4	5,76 10^{-1}	5,76 10^{-1}	5,76 10^{-1}
8	9,12 10^{-2}	8,53 10^{-2}	8,31 10^{-2}
16	6,31 10^{-6}	3,29 10^{-6}	1,1 10^{-6}
32	4,83 10^{-5}	8,25 10^{-13}	2,31 10^{-13}
64	7,73 10^{-2}	6,53 10^{-11}	1,03 10^{-13}

The conclusion is that for N > 64 the diagonalisation even in double precision cannot be performed accurately enough.

The influence matrix of Kleiser

We know that we have to solve eq.(16) in order to get the influence coefficients α_i. By using symmetry along the lines x=0, y=0 and x=±y the problem can be reduced considerably since there is a 8-fold symmetry. Moreover we can use the fact that we can avoid a source term for the Helmholtz equations for the homogenious problem. It should be observed that the influence matrix as written down in eq.(16) is five-fold singular. The homogenious problem voor q and $\underline{u}$ is solved by

$$q_h = \sum_{i=1}^{5} a_i f_i(x,y) \qquad u_h = v_h = 0 \qquad (23)$$

with

$$f_1 = T_0(x) T_0(y) = 1 \quad f_2 = T_{N-1}(x) T_{N-1}(y), \quad f_3 = T_N(x) T_N(y)$$
$$f_4 = T_N(x) T_{N-1}(y), \quad f_5 = T_{N-1}(x) T_N(y)$$

In order to make the system non-singular we have choisen $q_{0,0}=0$, $q_{n,m}=0$ (n,m)= (N-1,N)
The resulting matrix is solved by a Gauss-elimination which reduces the matrix to an upper triangular matrix. The total amount of work can be estimated as follows.
For the determination of (q_i, $\underline{u}_i$) about $\frac{3}{2}$ N^4 operations.
For the Gauss-elimination   about 21 N^3 operations.
The determination of $\underline{\alpha}$ about 16 N^2 operations.

Some results

The above method has been used to calculate an exact solution of the Navier-Stokes equations given by ref.(5).

$$u(x,y,t) = -\cos(\lambda x) \sin(\lambda y) \, e^{-2\lambda^2 t/Re} \qquad (24)a$$

$$v(x,y,t) = \sin(\lambda x) \cos(\lambda x) \, e^{-2\lambda^2 t/Re} \qquad (24)b$$

$$p(x,y,t) = -\frac{1}{4}\{\cos(2\lambda x) + \cos(2\lambda y)\} \, e^{-4\lambda^2 t/Re} \qquad (24)c$$

We start with $\lambda = \frac{\Pi}{2}$ and $Re = 50$

The solution is determined for $t=1$

The results both for $\underline{u}$ and p agree in 8 digits for N=32

The total amount of operations is estimated as

$$(\tfrac{3}{2} N + 23) N^3 + 9 k N^3$$

There is only a small advantage in using a super computer since only a small part can be vectorized and also since for N=32 not a real gain is available.

References

1. Gottlieb, D. & Orzag, S.A., Numerical analysis of spectral methods: Theory and Applications, Philadelphia 1977.

2. Peyretrand, R. & Taylor, T.D. Computational methods for fluid flow, Springer Verlag, New York, 1983.

3. Chorin, A.J., Numerical solution of the Navier-Stokes equations, Math. Comp. 22, 1968.

4. Kleiser, L., Numerische Simulationen zum laminar - turbulenten Umschlags prozess der ebenen Poiseuille-Strömung, dissertation Universität Karlsruhe, 1982.

AN IMPLICIT FLUX-SPLIT ALGORITHM
FOR THE COMPRESSIBLE EULER AND NAVIER-STOKES EQUATIONS

Robert W. Walters
Virginia Polytechnic Institute and State University
Blacksburg, VA 24061 USA

James L. Thomas
NASA Langley Research Center
Hampton, VA USA

Bram Van Leer
Delft University of Technology
Delft, The Netherlands

INTRODUCTION

The purpose of the present paper is to present recent progress in the development of implicit upwind schemes for the Euler and Navier-Stokes equations. The upwind-differencing scheme presented here is based on the concept of flux-vector splitting and uses the splitting developed by Van Leer [1], not only because it is continuously differentiable through sign changes of eigenvalues but also because it leads to normal-shock profiles with at most two transition zones, and generally one. A general class of upwind spatial-differencing schemes, ranging from standard first- or second-order fully-upwind to third-order upwind-biased differencing, is considered for the convective and pressure terms, the viscous terms are represented with second-order accurate central differences. The method is implemented in generalized coordinates according to a finite-volume discretization, following the developments in [2]. The two basic implicit schemes considered are both approximations to the unfactored backward-time integration scheme in delta form and correspond to a spatially-split approximate-factorization (AF) method [3] and several types of relaxation methods [4-8], respectively.

GOVERNING EQUATIONS

The time-dependent compressible Navier-Stokes equations express the conservation of mass, momentum, and energy for an ideal gas in the absence of external forces. The non-dimensional form of the equations in conservation law form can be written in generalized coordinates as

$$\frac{\partial Q}{\partial t} + \frac{\partial F}{\partial \xi} + \frac{\partial G}{\partial \eta} + \frac{\partial H}{\partial \zeta} = \frac{1}{Re}\left(\frac{\partial R}{\partial \xi} + \frac{\partial S}{\partial \eta} + \frac{\partial V}{\partial \zeta}\right)$$

where $Q = [\rho, \rho u, \rho v, \rho w, e]^T/J$ represents the standard form of the conserved variables and J is the Jacobian of the transformation. The elements of the flux-vectors F, G, and H contain the inviscid and pressure terms and the vectors R, S, and V contain the shear-stress and heat-flux terms. The equation set is closed by using the Sutherland law to evaluate the molecular viscosity, Stokes hypothesis for the bulk viscosity, and finally an ideal gas law to evaluate the pressure. Though the governing equations are written in generalized coordinates, a finite volume formulation is employed so that the discrete equations are exactly satisfied for freestream flow on arbitrary grids.

SPATIAL DIFFERENCING

The upwind differencing of the convective and pressure terms is based on the flux splitting technique developed by Van Leer [1] for Cartesian coordinates. The multi-dimensional extension to generalized coordinates is developed in [2] and [4]. The shear stress and heat flux terms are central differenced.

As an example of the upwind differencing, the gradient of the flux, F, is written as a flux balance across a cell, i.e.

$$\left(\frac{\partial F}{\partial \xi}\right)_i = \frac{[F^+(Q^-) + F^-(Q^+)]_{i+\frac{1}{2}} - [F^+(Q^-) + F^-(Q^+)]_{i-\frac{1}{2}}}{\Delta \xi_i}$$

where the notation $F^{\pm}(\overline{Q^+})$ indicates that the value of the split fluxes are obtained by upwind-biased interpolations of the state variables, Q, to the cell interfaces. In the finite volume formulation, Q is a cell centered quantity, i.e. Q_{ijk}^n denotes the average value of the dependent variables at time n in the center of the cell denoted (ijk). In order to obtain Q at a cell face, the following interpolation formula is used:

$$\overline{Q}_{i+\frac{1}{2}}^{\pm} = Q_i \pm \frac{s}{4}[(1 \mp \kappa s)\nabla_\xi + (1 \pm \kappa s)\Delta_\xi]Q_i$$

The operators Δ_ξ and ∇_ξ are the standard forward and backward difference operators respectively. The parameter κ ranges from $[-1,1]$ and determines the spatial accuracy of the difference approximation. For fully supersonic inviscid flow, the choice $\kappa = -1$ is commonly used because it corresponds to fully upwind differencing, thus allowing one to recover a space marching method [5]. For any other flow regime, $\kappa = \frac{1}{3}$ is the most common choice of the parameter because it yields the smallest truncation error [6]. In order to avoid oscillations in large gradient regions, such as in the vicinity of a shock wave, the interpolation formula is modified by the parameter s, which limits the higher-order terms in the interpolation. The value of s is determined locally from

$$s = \frac{2\Delta_\xi Q_i \nabla_\xi Q_i + \varepsilon}{(\Delta_\xi Q_i)^2 + (\nabla_\xi Q_i)^2 + \varepsilon}$$

where ε is a small number (10^{-6}) which is used in order to prevent division by zero in regions of zero gradients. In the η and ζ directions, analogous formulas hold.

The vectors R, S, and V are treated conservatively as differences across the cell interfaces of first derivative terms. In the finite volume formulation, this corresponds to second-order accurate central differencing of the cell-centered quantities. Additional details of this approach can be found in [6].

TEMPORAL INTEGRATION

The governing equations are integrated in time by first applying the backward Euler time integration scheme. Linearizing the equations about the previous time level and rearranging into the 'delta' form [3] results in

$$\left[\frac{I}{J\Delta t} + \delta_\xi \frac{\partial \hat{F}}{\partial Q} + \delta_\eta \frac{\partial \hat{G}}{\partial Q} + \delta_\zeta \frac{\partial \hat{H}}{\partial Q}\right] \Delta Q = \bar{R}^n$$

where, for simplicity, $\hat{F} = (F - \frac{1}{Re} R)^n$, etc., and $\bar{R}$ is the steady-state residual, i.e.

$$\bar{R} = -(\delta_\xi \hat{F} + \delta_\eta \hat{G} + \delta_\zeta \hat{H}).$$

The difference operator δ implies upwind differencing for the convective and pressure terms and central differencing for the viscous terms.

In two dimensions (η, ζ), approximate factorization (AF) of the above results in:

$$\left[\frac{I}{J\Delta t} + \delta_\eta \frac{\partial \hat{G}}{\partial Q}\right] \left[\frac{I}{J\Delta T}\right]^{-1} \left[\frac{I}{J\Delta t} + \delta_\zeta \frac{\partial \hat{H}}{\partial Q}\right] \Delta Q = \bar{R}^n$$

The AF method has the advantage that it is a time-accurate algorithm, applicable to a wide class of problems, and essentially independent of the type of spatial differencing. Moreover, the algorithm is completely vectorizable on modern supercomputers such as the CYBER-205 or the CRAY XMP. The disadvantage of the algorithm is that the time step is limited due to the presence of the splitting error associated with the method.

Relaxation strategies based on the properties of the coefficient matrix with upwind differencing [6] have been considered for the linearized equations. The convergence and stability properties of the relaxation scheme at large time steps is improved compared to the approximate factorization method [4]. Another advantage of the relaxation approach is that the algorithm recovers Newton's method for supersonic, inviscid flows in two-dimensions at large time steps with fully upwind differencing and thus quadratic convergence can be obtained and space marching employed [5]. The chief disadvantage of line Gauss-Seidel relaxation is that it is not a completely vectorizable algorithm. The LU decomposition can be vectorized over the entire field with the vector length being the number of lines; however, the back substitution step is not vectorizable because the right hand side depends on information from the previous line. In AF, both the LU decomposition and back-substitution vectorize because the lines are completely uncoupled. Thus there exists a trade-off between the two algorithms. A more complete discussion concerning these key issues can be found in [4].

In three dimensions, relaxation is applied in one coordinate direction (taken to be the ξ-direction here). This results in a system of equations to be solved in the cross-flow plane of the form

$$[M + \delta_\eta \frac{\partial \hat{G}}{\partial \eta} + \delta_\zeta \frac{\partial \hat{H}}{\partial \zeta}] \Delta Q = \bar{R}^n(Q^{n+1}, Q^n)$$

where

$$M = \frac{I}{J \Delta t} + \frac{\partial \hat{F}^+}{\partial Q} - \frac{\partial \hat{F}^-}{\partial Q}$$

and the notation $\bar{R}(Q^{n+1}, Q^n)$ indicates that the most recent value of Q is used to evaluate $\bar{R}$. The relaxation is effected by sweeping in the ξ-direction through the mesh, alternating between forward and backward passes when using higher-order differencing in order to maintain stability.

The cross-flow plane equation can be solved with either AF or line-relaxation, but is solved here with the spatially-split AF algorithm. The resulting equation is:

$$[M + \delta_\eta \frac{\partial \hat{G}}{\partial \eta}][M]^{-1}[M + \delta_\zeta \frac{\partial \hat{H}}{\partial \zeta}] \Delta Q = \bar{R}^n$$

For completely supersonic, inviscid flow, the cross-flow plane equation can be solved iteratively at each streamwise location resulting in a space-marching method. For subsonic flows, several iterations could also be performed efficiently in the cross-flow plane before advancing to the next plane but the effect of this on the total efficiency of the algorithm has not yet been investigated.

RESULTS

As a demonstration of quadratic convergence to the steady-state for supersonic inviscid flow, Mach 3 flow over a 5° wedge is considered. Pressure contours from the second-order accurate fully upwind solution are shown in figure 1a and the convergence of the scheme, measured by the L_2-norm of the steady-state residual is shown in figure 1b. Free-stream conditions were used as initial conditions for the calculation and as figure 1b indicates, as the intermediate solution becomes sufficiently close to the final solution, the residual decays quadratically.

For viscous flows, the method produces accurate results even on highly-stretched non-orthogonal meshes and converges rapidly to the steady-state. In figure 2a, Mach number contours obtained from the solution of a Re = 1600 laminar flow in a double throat nozzle are shown. The grid contained 246 mesh points in the streamwise direction and 81 points highly clustered near the wall, in the normal direction. The geometry of the nozzle is described in [9] and the computational boundary conditions and other Reynolds number solutions presented in [10]. The streamwise distribution of the surface heat transfer coefficient, C_H, at Re = 1600 is compared to a Re = 400 case in figure 2b. The convergence rate of the method for the Re = 1600 case is shown in figure 2c.

The three-dimensional algorithm produces solutions in reasonable computing times. The transonic flow over a swept, tapered ONERA M6 wing, using a 161 x 41 x 33 mesh was obtained in approximately $\frac{1}{2}$ hour on the VPS32 (CYBER 205) supercomputer at NASA Langley. Several chordwise pressure distributions and Mach number contours on the surface of the wing are shown in figure 3. Additionally, supersonic, inviscid, three-dimensional flow can be space-marched by the algorithm. Figure 4a shows Mach number contours over an analytically defined forebody at a free-stream Mach number of 1.7 and zero angle-of-attack. Top and bottom surface pressure distributions on the plane of symmetry are compared with the experimental values[11] in figure 4b. This computation utilized the space-marching feature of the algorithm in order to obtain a computationally efficient solution.

CONCLUDING REMARKS

The upwind scheme presented here has proven to be a versatile method for aerodynamic research. The implicit formulation results in rapid convergence to the steady-state even for high Reynolds number viscous flow on highly stretched meshes. Additional work aimed at further efficiency improvements in three-dimensional simulation is underway.

REFERENCES

1. Van Leer, B., "Flux-Vector Splitting for the Euler Equations," ICASE Report No. 82-30, September 1982; also: Lecture Notes in Physics, Vol. 170, 1982, pp. 507-512.
2. Anderson, W.K., Thomas, J.L. and Van Leer, B., "A Comparison of Finite Volume Flux Vector Splittings for the Euler Equations," AIAA Paper No. 85-0122, January 1985.
3. Beam, R.M. and Warming, R.F., "An Implicit Factored Scheme for the Compressible Navier-Stokes Equations," AIAA Journal, Vol. 16, No. 4, 1978, pp. 393-402.
4. Thomas, J.L., Van Leer, B. and Walters, R.W., "Implicit Flux-Split Schemes for the Euler Equations," AIAA 85-1680, July 1985.
5. Walters, R.W. and Dwoyer, D.L., "An Efficient Iteration Strategy Based on Upwind/Relaxation Schemes for the Euler Equations," AIAA 85-1529, July 1985.

6. Thomas, J.L. and Walters, R.W., "Upwind Relaxation Algorithms for the Navier-Stokes Equations," AIAA 85-1501, July 1985.
7. Van Leer, B. and Mulder, W.A., "Relaxation Methods for Hyperbolic Equations," Report 84-20, Delft University of Technology, 1984.
8. Chakravarthy, S.R., "Relaxation Methods for Unfactored Implicit Schemes," AIAA 84-0165, January 1984.
9. Announcement of the 1985 GAMM Workshop on Numerical Simulation of Compressible Navier-Stokes Equations.
10. Thomas, J.L., Walters, R.W., Van Leer, B. and Rumsey, C.L., "An Implicit Flux-Split Algorithm for the Compressible Navier-Stokes Equations," Proceedings of the 1985 GAMM Workshop on Numerical Simulation of Compressible Navier-Stokes Flows.
11. Townsend, J.C., Howell, D.T., Collins, I.K. and Hayes, C., "Surface Pressure Data on a Series of Analytic Forebodies at Mach Numbers from 1.70 to 4.50 and Combined Angles of Attack and Sideslip," NASA TM 80062, June 1979.

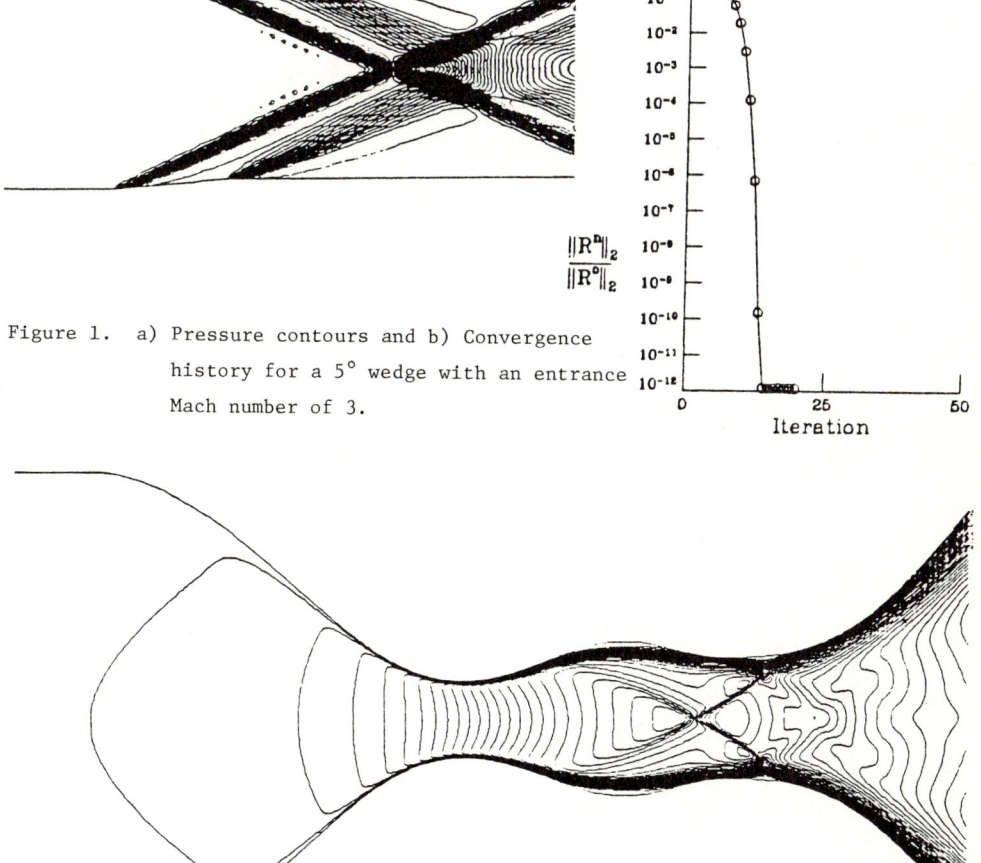

Figure 1. a) Pressure contours and b) Convergence history for a 5° wedge with an entrance Mach number of 3.

Figure 2. a) Mach number contours for a dual throat nozzle; Re = 1600.

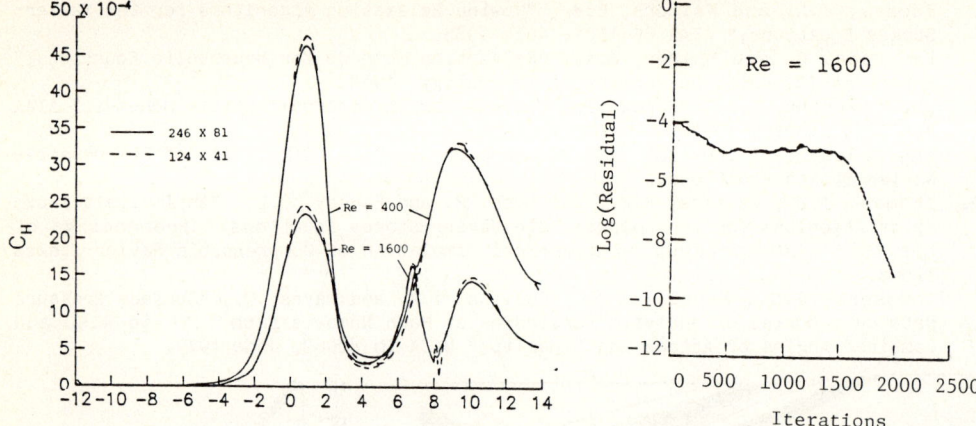

Figure 2. b) Streamwise distribution of the surface heat transfer coefficient and c) Convergence history starting from a freestream initialization on the fine mesh (246 x 81).

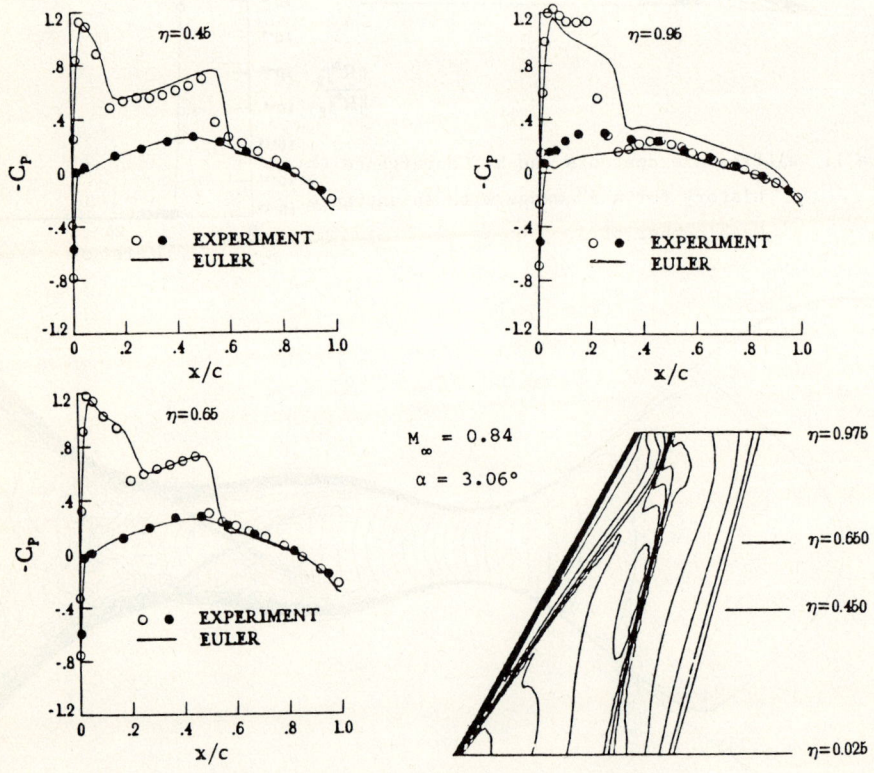

Figure 3. Chordwise pressure distributions and Mach contours for the ONERA M6 wing.

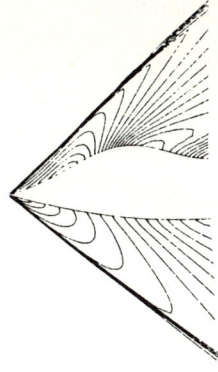 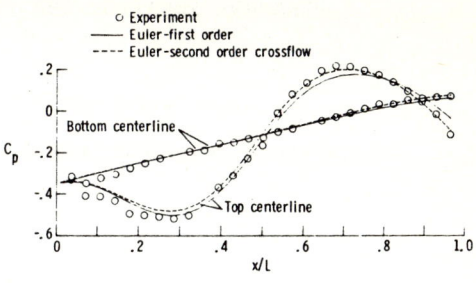

Figure 4. a) Mach contours over analytically defined forebody and b) pressure distributions in the plane of symmetry; $M_\infty = 1.7$, $\alpha = 0°$.

NUMERICAL SOLUTION OF TRANSONIC SMALL DISTURBANCE PRESSURE EQUATION AND ITS APPLICATIONS

L.X. Wang and S.J. Luo

Northwestern Polytechnical University, Xi'an, China

I. Introduction

In the past decade, several new approaches [1-3] to the wind tunnel wall interference problem have been introduced. A common feature of them requires static pressure distributions measured on or near the walls to be used as boundary conditions imposed on flow computations. Meanwhile, for many transonic design problems we need to calculate airfoil from the given pressure distributions. For such problems, only the pressure distributions on the model and near the walls as well as in the flow fields are dealt with. If potential equation is still used as the governing equation, the boundary conditions and the computing process are relatively complicated. In order to convert boundary conditions into Dirichlet form, a transonic small disturbance pressure equation is proposed for computing transonic flow fields in wind tunnel or free streams.

Transonic small disturbance pressure equation is a partial differential equation like potential equation in transonic regime. Two equations are similar in form and have some mathematical properties in common, but the former has stronger nonlinearity than the latter, which is introduced by the term u_x^2. How to solve the pressure equation has not been studied theoretically or numerically. In the present paper, the mixed difference method and relaxation techniques are used to solve the pressure equation numerically. The basic procedure follows the work developed by Murman and Cole [4], while some improvements have been studied to fit the new equation.

Comparisons with transonic small disturbance potential equation indicate both the accuracy of the procedure as well as its ease of implementation. Applications to assessment of wall interference and design for airfoil from the given pressure distributions are illustrated. The results are generally satisfactary.

II. Transonic Small Disturbance Pressure Equation

Two dimensional transonic small disturbance potential equation is
$$(1-M_\infty^2 - \frac{\gamma+1}{V_\infty}M_\infty^2 \phi_x)\phi_{xx} + \phi_{yy} = 0 \tag{1}$$
where M_∞ and V_∞ are free stream Mach number and velocity, ϕ is small disturbance potential, γ is ratio of specific heats, x and y are Cartesian coordinates.

Differentiating eq.(1) with respect to x and using $u = \phi_x$, where u is disturbance velocity in x direction, we get
$$(1-M_\infty^2 - \frac{\gamma+1}{V_\infty}M_\infty^2 u)u_{xx} + u_{yy} = \frac{\gamma+1}{V_\infty}M_\infty^2 u_x^2 \tag{2}$$
Considering the relation $Cp = -2u/V_\infty$, the pressure coefficient Cp can be deduced as soon as u has been known. For convenience, here, we define eq.(2) as transonic small disturbance pressure equation.

III. Boundary Conditions Of Dirichlet Form

Two types of Dirichlet boundary value problems are easily formulated for a subsonic free stream Mach number.

The outer boundary conditions to be applied along the up and down control surfaces in wind tunnel are determined from static pressure measurements. The actual condition is given by $u/V_\infty = -Cp/2$.

On the upstream and downstream boundaries, u is defined as a fourth degree polynomial in y, whose coefficients are computed from the values of u_y and u_{yy} at four corners.

For free air calculation, the disappearing velocity disturbance on outer boundary are approximately employed.

The airfoil boundary condition for computing flow fields in wind tunnel or free stream is also imposed in Dirichlet form, which is given by measured or design pressure distributions. Because there is no jump in u at the trailing edge, Kutta condition is unnecessary.

IV. Numerical Formulation

Let $F_{i,j}$, $P_{i,j}$ and $Q_{i,j}$ be centered difference approximations to u_{yy}, u_{xx} and u_x, respectively:

$$F_{i,j} = \frac{u_{i,j+1} - 2u_{i,j} + u_{i,j-1}}{\Delta y^2} \qquad (3)$$

$$P_{i,j} = A_{i,j} \frac{u_{i+1,j} - 2u_{i,j} + u_{i-1,j}}{\Delta x^2} \qquad (4)$$

$$Q_{i,j} = \frac{\gamma+1}{V_\infty} M_\infty^2 \left(\frac{u_{i+1,j} - u_{i-1,j}}{2\Delta x} \right)^2 \qquad (5)$$

where $A_{i,j} = 1 - M_\infty^2 - \frac{\gamma+1}{V_\infty} M_\infty^2 u_{i,j}$

Define a switching function μ with the value unity at supersonic points and zero at subsonic points

$$\mu_{i,j} = \begin{cases} 0, & \text{if } A_{i,j} > 0 \\ 1, & \text{if } A_{i,j} < 0 \end{cases} \qquad (6)$$

Then, the finite difference approximation to eq.(2) can be written as a unified form.

$$(1-\mu_{i,j})P_{i,j} + \mu_{i,j}P_{i-1,j} + F_{i,j} = (1-\mu_{i,j})Q_{i,j} + \mu_{i,j}Q_{i-1,j} \qquad (7)$$

In actual calculation a variable spacing mesh is used. At each mesh point, the differential operators are replaced by either centered or upwind difference operators, which are simply realized by a **switch** depending on that the local velocity at a mesh point is subsonic or supersonic. Then, Dirichlet boundary conditions are easily incorporated in. The resulting large set of simultaneous algebraic equations are solved in an iterative fashion using a line relaxation algorithm. Values of u are calculated along a vertical line. Each calculation is started with an initial guess, usually the zero disturbance velocity. The iteration terminates when the solutions converge to the final accuracy.

In addition to the familiar procedure mentioned above, several extensions are indicated specifically by numerical experiments.

1. Shock-capturing technique is still used here. Embedded supersonic

zone and shock develope natually in iterative process. The shock jump lies between two mesh points $(i-1,j)$ and (i,j), when $A_{i-1,j} < 0$ and $A_{i,j} > 0$. Let point (i,j) be the shock point. **From eq. (6) we can see that the difference operator at subsonic points is employed at the shock. Theoretical analysis shows** that the difference scheme of eq.(7) does not satisfy shock jump relation at the shock point. But fortunately the shock position and supersonic zone calculated by eq.(7) agree well with those calculated by potential equation.

There is another type of difference scheme.

$$(1-\mu_{i,j})P_{i,j} + \mu_{i,j}P_{i-1,j} + F_{i,j} = (1-\mu_{i,j})Q_{i,j} + \mu_{i-1,j} Q_{i-1,j} \qquad (8)$$

which is exactly the transonic small disturbance shock relation at shock point. The converged solutions satisfy the proper jump condition.

2. In iterative line relaxation algorithm, **for u_x^2 we need use the value calculated from** a previous stage to ensure convergence of both sub- and super-critical flow.

3. Many numerical experiments for computing eq.(2) show that the line relaxation may be over-relaxed for sub-critical flow and must be under-relaxed for super-critical flow to get a converged solution.

The reliability and the accuracy of the above method used in sub- and super-critical flows have been proved by a large number of numerical examples. Fig.1 illustrates the pressure distribution along a line in the flow fields calculated by eq.(1) and eq.(2), for a NACA0012 airfoil at $M_\infty = 0.9$, $\alpha = 0$, and $H = 105$ mm, where α is the angle of attack, H is the distance between the line and the airfoil. This typical example shows good agreement of two equations.

V. Applications

Two applications for the present study are given below.

1. Assessing transonic wind tunnel wall interference

The procedure for assessing wall interference are simply stated.

a). Experimentally measure the pressure distributions on the model and on the control surfaces near the walls.

b). Calculate the pressure fields in free stream by eq.(2) from the measured airfoil pressure distributions.

c). Compare the measured and computed pressure distributions on the control surfaces. If the two are equal, there is interference free in wind tunnel. Otherwise some wall interference exists.

Fig.2 shows the comparisons between experimental and computing pressure distributions on control surface for a NACA0012 airfoil at $\alpha = 0$, $\sigma = 2\%$ and $M_\infty = 0.6, 0.75$ and 0.9, where σ is the ventilation ratio.

2. Designing airfoil from the given pressure distributions

The basic idea of designing airfoil can be conveniently illustrated for a symmetric airfoil of no lift.

a). Calculate u distributions by eq.(2) from the designing pressure distributions on airfoil. Thus, u_{yy}, u_{xx} and u_x at any point can be immediately calculated.

b). On the airfoil (y=0)
$$u_{yi,1} = \frac{u_{i,2} - u_{i,1}}{\Delta y_1} - \frac{\Delta y_1}{2} u_{yyi,1} \tag{9}$$

c). From irrotational condition $v_x = u_y$, the disturbance velocity in y direction $v(x)$ can be computed by integrating u_y along the chord.

d). Finally, we get
$$Y(x) = \int_0^x \frac{v(\xi)}{V_\infty} d\xi \tag{10}$$
where $Y(x)$ is the shape curve of airfoil section.

Fig.3 illustrates the airfoil designed by present method. The given pressure distributions are the data obtained by Knechtel [5]. The precise circular-arc airfoil is also sketched for comparison.

VI. References

1. Kemp, W.B. Jr., NASA TM 81819, 1980.
2. Mokry, M., Peake,D.J. and Bowker,A.J.,NAE Aeronautical Rept. LR-575,1974.
3. Stahara, S.S., AIAA J. 18, 63-71, 1980.
4. Murman, E.M. and Cole, J.D. , AIAA J. 9, 114-121, 1971.
5. Knechtel, E.D., NASA TND-15, 1959.

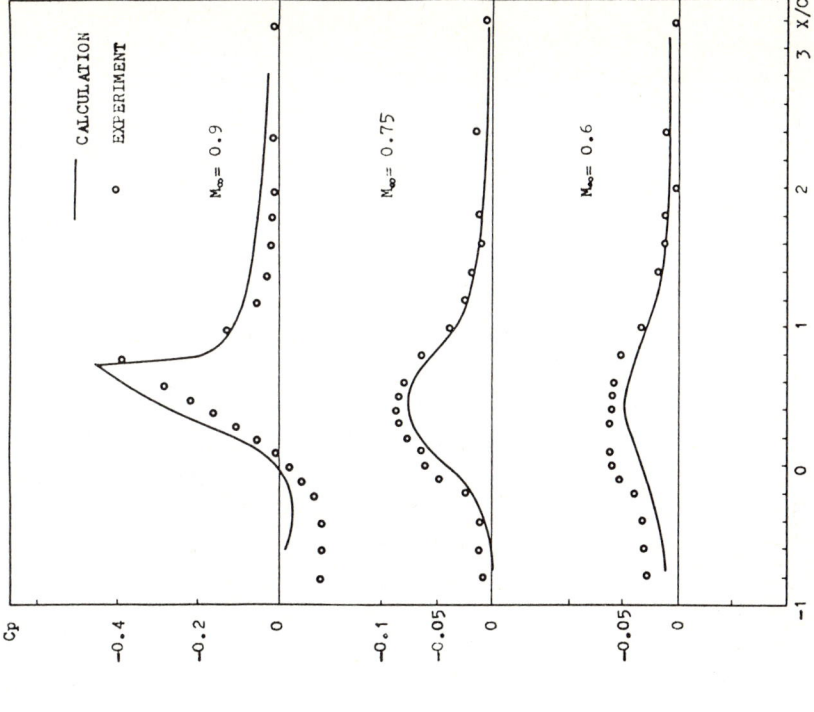

Fig.2 Comparison of calculated and experimental pressure distribution on control surface, $\sigma=2\%, \alpha=0$

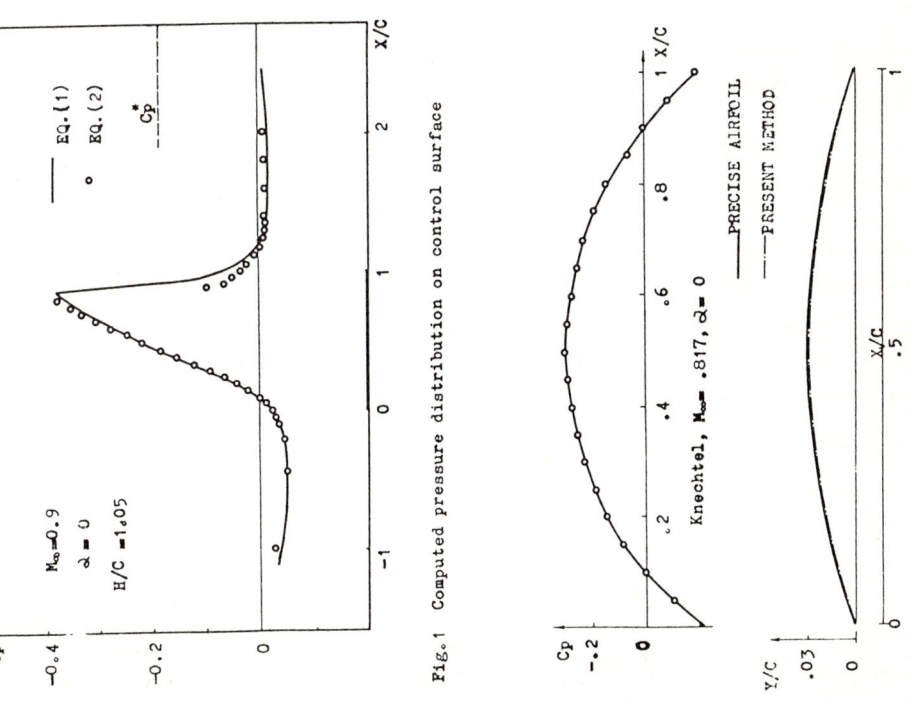

Fig.1 Computed pressure distribution on control surface

Fig.3 Circular-arc airfoil

A NEW SWITCH-SCHEME FOR CONVECTION-DIFFUSION EQUATIONS*

R.Q.Wang and Y.G.Han   Computing Center of Academia Sinica, Beijing
B.M.Zhou   Chinese Academy of Space Technology, Beijing
J.A.Sun   Northwestern Normal College, Lanchow, China

INTRODUCTION

At present the switch-scheme, combining first-order upwind difference (UDS) and second-order central difference (CDS) has been widely developed and applied with success to solve the convection-diffusion equations and the Navier-Stokes equations [1-2]. However, the choice of a switching factor is a key problem for using the switch-scheme, because it has a significant effect on the numerical results. In this paper, the authors point out the fact that the numerical solution, obtained by using the switching factors based on the solution itself as used in [2], is strongly dependent on the switching factors, and then give a new way to construct the switching factors based on the cell Reynolds number. As a result, our new switch-scheme possesses some advantages over previous ones, i.e., it has good stable and monotonic properties and frees of the restriction on the cell Reynolds number. The famous Samarskii's scheme[3] is a special case of our schemes. In some sense, our schemes are a fair approximation to the Allen-Southwell exponential type scheme[3] (AS).

MOTIVATION

Let us consider a simple model problem

$$\frac{\partial u}{\partial t} + a \frac{\partial u}{\partial x} = \varepsilon \frac{\partial^2 u}{\partial x^2} \qquad 0 < x < 1, \quad t > 0 \tag{1}$$

$$u(0,x) = \varphi(x), \quad u(t,0) = 0 \quad \text{and} \quad u(t,1) = 1 \tag{2}$$

where a and $\varepsilon (\varepsilon > 0)$ are constants, $\varphi(x)$ is a known function, and discretize Eq.(1) using the following switching scheme

$$\frac{u_i^{n+1} - u_i^n}{\Delta t} + \theta(a_+ D_- u_i^{n+1} + a_- D_+ u_i^{n+1}) + (1-\theta) a D_0 u_i^{n+1} = \varepsilon D_+ D_- u_i^{n+1} \quad (i=1,2,\cdots N) \tag{3}$$

where $\theta (0 \leq \theta \leq 1)$ may be a variable factor called the switching factor to be determined. $D_\pm$, D_0, $a_\pm$ are defined below

$$D_+ u_i = \frac{u_{i+1} - u_i}{h}, \quad D_- u_i = \frac{u_i - u_{i-1}}{h}, \quad D_0 u_i = \frac{u_{i+1} - u_{i-1}}{2h}, \quad a_\pm = \frac{a \pm |a|}{2}$$

The scheme (3) can be rewritten as follows

$$\frac{u_i^{n+1} - u_i^n}{\Delta t} + a D_0 u_i^{n+1} = \varepsilon \sigma D_+ D_- u_i^{n+1} \tag{4}$$

---

* This work is supported by science fund of Academia Sinica, contract No.R850337.

where $\sigma = (1+ \frac{|R|}{2} \theta)$ is a fitting factor related to θ ($R = \frac{a\Delta x}{\varepsilon}$ — the cell Reynolds number). When $\sigma = \frac{R}{2} \coth \frac{R}{2}$, (4) is the Il'in's scheme (IS). For a nonlinear Burgers' equation, Kellogg et al.[4] have suggested a modified scheme

$$\frac{u_i^{n+1} - u_i^n}{\Delta t} + D_0(\frac{u^2}{2})_i^{n+1} = \varepsilon D_+ D_- (\sigma u)_i^{n+1} \qquad (5)$$

An improved switching factor with parameter k has been given in [2] as the following (WY)

$$\theta = (\frac{|u_{i+1} - 2u_i + u_{i-1}| + 10^{-30}}{|u_{i+1} - u_i| + |u_i - u_{i-1}| + 10^{-10}})^k \qquad k=1,2,3,\cdots \qquad (6)$$

In fact, only k=3 has been taken in the calculations in [2]. Our numerical experiments with (6) show that the numerical results are strongly dependent on the value of k, and that spurious oscillations occur for any k, when the cell Reynolds number and the time increment Δt are big enough. As a result, it is difficult to reach a steady solution (see Table). Such a phenomenon is also observed for nonlinear equations. So, in order to obtain a stable and monotonic numerical solution for any R, we would give a new way to construct the switching factor.

NEW SWITCH-SCHEME

To overcome the above mentioned difficulties, we propose another form of the switching factor θ, based on the cell Reynolds number instead of the unknown solution itself as (6), the formulation of θ is

$$\theta = (\frac{|R|}{c+|R|})^\ell \qquad (7)$$

where c and ℓ are two positive parameters to be determined according to the monotonicity requirement for the switch-scheme. This can be easily done by examining the roots of the characteristic equation associated with the difference equation (3) without the time difference term. Fixed c, ℓ is only a function of R. Particularly, taking c=2 and ℓ=1, we obtain

$$\frac{u_i^{n+1} - u_i^n}{\Delta t} + aD_0 u_i^{n+1} = \varepsilon(1+ \frac{|R|}{2} \frac{|R|}{2+|R|}) D_+ D_- u_i^{n+1} \qquad (8)$$

which is the Samarskii's scheme (SMS)[3].

In this paper we have investigated two monotonic schemes with

$$\theta = (\frac{|R|}{1+|R|})^2 \qquad (9)$$

and

$$\theta = \begin{cases} (\frac{|R|}{1+|R|})^3 & \text{for } |R| \leq 4 \\ 1 - \frac{2}{|R|} & \text{for } |R| > 4 \end{cases} \qquad (10)$$

The first scheme (WC1) is more accurate than the Samarskii's scheme and the second (WC2) can be considered as a modified Spalding's hybrid scheme (SPS)[5] with the switching factor

$$\theta = \begin{cases} 0 & \text{for } |R| \leq 2 \\ 1 - \dfrac{2}{|R|} & \text{for } |R| > 2 \end{cases} \quad (11)$$

Fig.1 compares different fitting factors for different switch-schemes. For application to time-dependent multidimensional problems, semi-implicit switch-schemes of Gauss-Seidel type are used. They keep the favourable monotonic feature and have a rapid convergence rate for any R.

NUMERICAL TESTS

Detailed comparisons of accuracy, stability and convergence rate of different switch-schemes are made for one and two-dimensional Burgers' equations and for the stream function-vorticity Navier-Stokes equations. The same cases used in [6] for nonlinear Burgers' equations are also tested by different switch-schemes, numerical phenomena are similar to the linear case demonstrated in Sec.2. For space is limited, numerical results are not given here in detail. In order to check and compare different switch-schemes for solving the incompressible Navier-Stokes equations, steady flow in a driven square cavity (see Fig.2) and flow over a backward-facing step in a two-dimensional channel flow measured and computed by Denham et al. in [8] (see Fig.7) have been computed for different Reynolds number with the time-dependent technique. Some results are shown in Figs. 3-7. Fig.3 shows the profile of streamwise velocity of the cavity flow on the line x=0.5 at R_e=400 with a 41×41 mesh for different difference schemes. It should be pointed out that the switch-scheme with the switching factor (6) with k = 3 requires a small time increment Δt and will break down, if $\Delta t > 0.2$ at R_e=400. However, our schemes still work well with any large time step and from Fig. 3 it can be found that our results are in good agreement with those predicted by Burggraf[7]. The flow over a step is computed with 21×21 grid points at R_e=50, 100, 200 and 229 with parabolic inlet flow. A comparison of the position L_R/h (h-step height) of reattachment point is shown in Fig.6. It is clear that our results with scheme WC2 are close to experimental data[8]. Fig.7 shows the backward step flow patterns obtained by WC2 and the Il'in scheme at R_e=100. They are quite coincident.

In conclusion, the new switch-scheme with switching factors based on the cell Reynolds number is stable, monotonic, accurate and fast convergent.

References

[1] C.W.Hirt, in Numerical Methods for P.D.E. (ed. by S.V. Pater), 1980.
[2] H.M.Wu et al., Lecture Notes in Phys., No. 141, 1981.
[3] E.P. Doolan et al., Uniform Numerical Methods for Problems with Initial and Boundary Layers, Boole press, Dublin, 1980.

[4] R.B.Kellogg et al., SIAM J.Numer. Anal. 17, No.6, 1980.
[5] S.V. Patankar, Numerical Heat Transfer and Fluid Flow, McGraw-Hill, 1980.
[6] R.L. Roach, AIAA paper No. 82-102, 1982.
[7] O. Burggraf, J.Fluid Mech., 24 (1966) 133-151.
[8] M.K. Denham et al., Tran. Instit. Chem. Engrs., 52 (1974) 129-144.

TABLE	Example of computing a steady solution to Eqs. (1) and (2) $(a=1, \varepsilon=0.01, \Delta x=0.025, \max\lvert(u^{n+1}-u^n)/u^n\Delta t\rvert < 10^{-4})$						
scheme x	0.025	0.05	0.075	0.925	0.950	0.975	N
Eq.(6) $k = 1$ $\Delta t=0.025$	-.409-34	-.327-33	.299-32	.714-2	.371-1	.193	> 500
$\Delta t=1.0$							31
$k = 2$ $\Delta t=0.025$	-.432-37	.346-36	-.315-35	.329-2	.221-1	.149	> 500
$\Delta t=1.0$							78
$k = 3$ $\Delta t=0.025$	.756-37	-.605-36	.552-35	.183-2	.149-1	.122	> 500
$\Delta t=1.0$	-.244-38	.197-37	-.182-36	.508-3	.196-1	-.207-2	> 500
Eq.(9) $\Delta t=0.025$	.850-34	.718-33	.543-32	.242-2	.180-1	.134	219
$\Delta t=1.0$							35
Eq.(10) $\Delta t=0.025$	.205-43	.290-42	.383-41	.439-3	.557-2	.760-1	236
$\Delta t=1.0$							30
Exact solution	.416-42	.548-41	.672-40	.553-3	.674-2	.821-1	

N - number of time steps

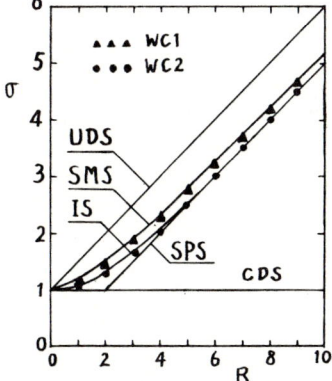

Fig.1. Comparison of the fitting factors for different schemes.

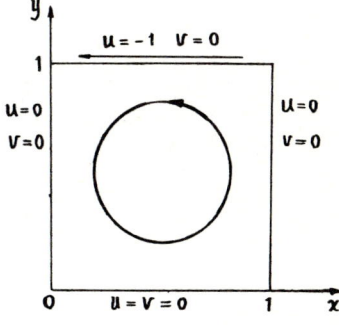

Fig.2. Geometry of the driven cavity flow

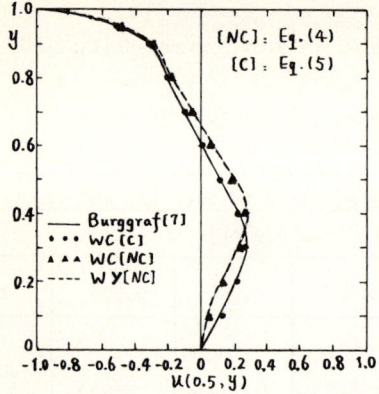

Fig.3. Profile of u-velocity on the line x = 0.5 of the cavity

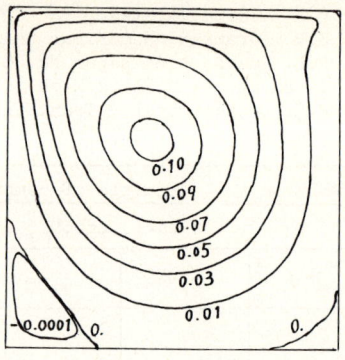

Fig.4. Streamlines for Reynolds number 400.

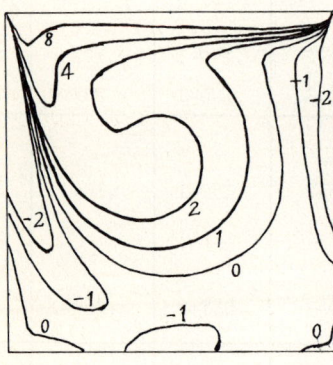

Fig.5. Vorticities for Reynolds number 400.

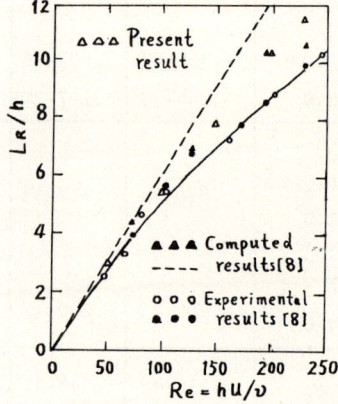

Fig.6. Variation of recirculation zone length with Reynolds number.

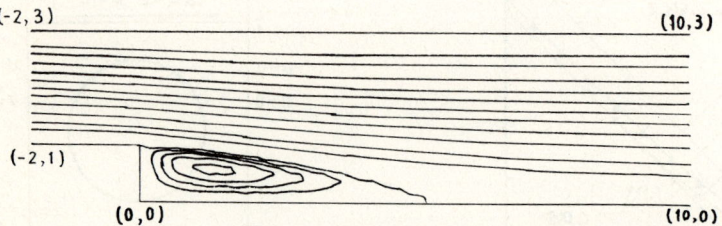

Fig.7. Flow streamlines over a backward step at Re=100

STABILITY OF SEMIDISCRETE APPROXIMATIONS FOR HYPERBOLIC INITIAL-BOUNDARY-VALUE PROBLEMS: AN EIGENVALUE ANALYSIS

ROBERT F. WARMING AND RICHARD M. BEAM

NASA Ames Research Center
Moffett Field, CA 94035, USA

1. Introduction

An evolutionary (time-dependent) partial differential equation (PDE) can be reduced to a system of ordinary differential equations (ODEs) by replacing the spatial derivatives with finite-difference approximations. The resulting approximation is called *semidiscrete* since the time variable is left continuous. The procedure of reducing a PDE to an ODE system is often called the *method of lines* since a solution of the ODE system gives an approximation to the PDE solution along x equals constant lines in (x,t) space.

A semidiscrete approximation for a hyperbolic initial-boundary-value problem (IBVP) leads to a complication since, in general, more boundary conditions are required for the semidiscrete approximation than are specified for the PDE. The additional boundary conditions are often called *numerical* boundary conditions. Any numerical procedure used to provide a numerical boundary condition is called a *numerical boundary scheme* (NBS).

An essential requirement for a semidiscrete approximation is the convergence (as the spatial mesh is refined) of the approximate solution to the solution of the PDE. If an approximation is *consistent*, then by the Lax equivalence theorem *stability* is a necessary and sufficient condition for *convergence*. It is generally easy to check the consistency of the approximation; however, the stability analysis can be a formidable problem.

Improper treatment of the NBS can lead to instability of the semidiscrete approximation even though one starts with a stable approximation for the pure initial-value problem (IVP) or Cauchy problem. For the purposes of this paper, we will assume that the approximation is consistent and stable (i.e. convergent) for the IVP and will consider only the effect of the NBS on the stability of the semidiscrete IBVP.

For a linear IVP or an IBVP with homogeneous boundary conditions, a semidiscrete approximation (on J spatial mesh intervals) results in a system of ODEs of the form $d\mathbf{u}(t)/dt = A\mathbf{u}(t)$ where $\mathbf{u}$ is a J-component vector, and A is a $J \times J$ matrix. The solution of the homogeneous ODE system can be written as $\mathbf{u}(t) = e^{At}\mathbf{u}(0)$ where e^{At} is an exponential matrix. The classical Lax-Richtmyer stability definition is equivalent to the requirement that the matrix norm of e^{At} be uniformly bounded for $0 \leq t \leq T$ independent of the spatial mesh size. The practical problem is that there is no known simple algebraic test for the uniform boundedness of the matrix norm for hyperbolic IBVPs. Necessary (and sufficient in special cases, e.g., IVPs) conditions can be expressed in terms of the eigenvalues of the matrix A; however, for the IBVP the eigenvalue analysis is, in general, intractable and the resulting conditions are not sufficient for stability.

In the 1960s and early 1970s, a stability theory for fully discrete IBVPs was developed by Godunov and Ryabenkii [1], Kreiss [3], Osher [4], and Gustafsson, Kreiss, and Sundström [2]. For the purposes of this paper, we refer to this theory as the GKS theory. Trefethen [6] showed that the main result of the theory has a physical interpretation in terms of group velocity, and Strikwerda [5] extended the theory to semidiscrete approximations.

The GKS theory reduces the stability analysis of a semidiscrete IBVP on a finite domain to the study of three auxiliary problems: the Cauchy problem and the related left- and right-quarter-plane problems. The stability of the left- and right-quarter-plane problems is checked by the normal mode analysis. An advantage of the GKS theory is that it provides an algebraic test which is necessary and sufficient for stability but does so at the expense of using a more complicated norm (in the stability definition).

Since the Lax-Richtmyer and GKS stability definitions differ, the connection between the eigenvalue analysis for the finite domain problem (which gives necessary Lax-Richtmyer stability conditions) and the eigenvalue or normal mode analysis for the quarter plane problems (which gives necessary and sufficient GKS stability conditions) is rather obscure. In this paper, we consider a direct algebraic comparison between the stability polynomial of the finite domain problem and the polynomials associated with the quarter-plane problems of the GKS theory. We show under what asymptotic conditions the finite-domain analysis leads to a stability polynomial which is consistent with the quarter plane analysis. The asymptotic eigenvalue analysis (finite-domain, $J \to \infty$) establishes the connection between the algebraic tests of the GKS stability theory and the Lax-Richtmyer stability definition. In addition, it also leads to a conjecture which gives necessary and sufficient conditions for Lax-Richtmyer stability in terms of the algebraic tests of the GKS theory.

This short paper gives only a brief treatment of our analysis. A detailed exposition (including plots of eigenvalue distributions for various NBSs) is given by the authors in [8]. This paper is restricted to semidiscrete approximations. For fully discrete difference approximations we have recently stated a conjecture [7] which relates the Lax-Richtmyer and GKS stability theories by a generalization of a theorem from linear algebra.

2. Initial-Boundary-Value Problem for a Model Hyperbolic Equation

For simplicity we restrict our attention to the stability of semidiscrete approximations to the IBVP for the model hyperbolic equation

$$\frac{\partial u}{\partial t} = c\frac{\partial u}{\partial x}, \quad 0 \leq x \leq L, \quad t \geq 0 \tag{2.1}$$

where c is a real constant. Initial data $u(x,0) = f(x)$, $0 \leq x \leq L$ are given at $t = 0$ and for $c > 0$ the problem is well-posed if an *analytical* boundary condition is prescribed at $x = L$

$$u(L,t) = g(t) \quad \text{for} \quad c > 0. \tag{2.2}$$

3. A Prototype Semidiscrete Approximation for the Model IBVP

Let the spatial interval L be divided into J subintervals of length Δx, i.e., $J\Delta x = L$, $x = x_j = j\Delta x$. On the interior mesh points the spatial derivative u_x is replaced by a second order central difference quotient and we obtain the system of ODEs

$$\frac{du_j}{dt} = \frac{c}{2\Delta x}(u_{j+1} - u_{j-1}), \quad j = 1, 2, \cdots, J-1 \tag{3.1}$$

where $u_j(t) = u_j$ denotes the semidiscrete approximation to $u(x,t)$. The right boundary ($x = L$) is advanced by using the analytical boundary condition (2.2). We assume that the boundary condition is homogeneous, i.e., $g(t) = 0$ and for the semidiscrete problem we write

$$u_J = 0. \tag{3.2}$$

The spatial computational stencil of (3.1) uses the 3 points $j, j \pm 1$. If we apply (3.1) at the left boundary ($j = 0$), then the stencil protrudes one point to the left of the boundary. It is clear that an additional *numerical boundary scheme* (NBS) is required to determine the semidiscrete solution. At this boundary ($j = 0$) we change from a centered approximation to a one-sided spatial differencing approximation of u_x:

$$\left.\frac{\partial u}{\partial x}\right|_0 = \frac{1}{\Delta x}[-\alpha u_2 + (1+2\alpha)u_1 - (1+\alpha)u_0] \tag{3.3}$$

where α is a parameter. The approximation (3.3) is first-order accurate for any α except $\alpha = 1/2$ in which case it is second-order accurate. If we insert (3.3) into the PDE (2.1) evaluated at $j = 0$, there follows the NBS

$$\frac{du_0}{dt} = \frac{c}{\Delta x}[-\alpha u_2 + (1 + 2\alpha)u_1 - (1 + \alpha)u_0]. \tag{3.4}$$

The system of ODEs (3.1) together with the analytical boundary condition (3.2) and the NBS (3.4) can be written in vector-matrix form as

$$\frac{d\mathbf{u}(t)}{dt} = A\mathbf{u}(t) \tag{3.5}$$

where $\mathbf{u}$ is a J-component vector and A is a $J \times J$ matrix.

4. Lax-Richtmyer Stability of a Semidiscrete Approximation

The essential element in the stability of a semidiscrete approximation represented by (3.5) is the behavior of the solution as the spatial mesh is refined ($\Delta x \to 0$, or $J \to \infty$). Consequently, one must consider an infinite sequence of ODE systems. The J-th member of the sequence is the ODE system (3.5) of dimension J. In order to define stability, we need some measure of the magnitude of the solution vector and we use a conventional vector norm $\|\cdot\|$.

Lax-Richtmyer stability for a semidiscrete approximation is defined as follows:

Definition 4.1. *A semidiscrete approximation represented by the sequence of ODEs (3.5) is said to be Lax-Richtmyer stable if there exists a constant $K > 0$ such that for any initial condition $\mathbf{u}(0)$*

$$\|\mathbf{u}(t)\| \leq K\|\mathbf{u}(0)\| \tag{4.1}$$

for all J, $J\Delta x = L$ with L fixed and for all t, $0 \leq t \leq T$ with T fixed.

The eigenvalues s_ℓ of A are, in general, complex and we write $s_\ell = \Re(s_\ell) + i\Im(s_\ell)$. Any semidiscrete approximation for (2.1) will have the factor $c/\Delta x$ on the right-hand side (see e.g. (3.1)). In considering the eigenvalues of the matrix A it is convenient to define a matrix $\hat{A}$ such that $A = (|c|/\Delta x)\hat{A}$. Consequently, the eigenvalues $\hat{s}$ of $\hat{A}$ are related to the eigenvalues s of A by $\hat{s} = s\Delta x/|c|$.

A necessary condition for Lax-Richtmyer stability can be stated as follows:

Lemma 4.1. *A necessary condition for a semidiscrete approximation to be Lax-Richtmyer stable is that for all J, $J\Delta x = L$ and all t, $0 < t \leq T$ there exists a nonnegative constant w such that*

$$\max_\ell \Re(\hat{s}_\ell) \leq \frac{w}{J}. \tag{4.2}$$

If the matrix A is a normal matrix, i.e., $AA^* = A^*A$, then (4.2) is both necessary and sufficient for stability in the L_2 norm. It is important to realize that a stable semidiscrete approximation can have eigenvalues $\hat{s}_\ell$ with positive real parts, but the real parts must approach zero at least as fast as $1/J$.

5. Normal Mode Analysis (Quarter-Plane Problems)

The GKS theory (see introduction) reduces the stability analysis of an IBVP on a finite domain to the study of the three auxiliary problems: the Cauchy problem and the related left- and right-quarter-plane problems. For example, one can obtain the related right-quarter-plane problem from the finite domain semidiscrete problem by fixing Δx and the left boundary at $x = 0$ and letting $J \to \infty$. Note that now L is not fixed but $L \to \infty$ as $J \to \infty$, and the resulting spatial domain is $[0 \leq x < \infty)$.

The algebraic tests of the GKS theory are carried out by means of the *normal mode analysis* which is based on the resolvent equations. The resolvent equations for a quarter-plane problem are obtained by substituting

$$u_j(t) = e^{st}\phi_j, \tag{5.1}$$

where s is a complex constant, into the semidiscrete approximation (i.e., the interior scheme) and the boundary conditions. The resolvent equations consist of a difference equation and boundary conditions for the eigenfunction ϕ_j. The general solution of the resolvent equations which is in L_2 has the form

$$\phi_j = \phi_0 \kappa^j, \qquad |\kappa| < 1. \tag{5.2}$$

The semidiscrete approximation is GKS stable if there are no nontrivial solutions of the form

$$\Re(s) > 0, \qquad |\kappa| < 1 \tag{5.3}$$

and

$$\Re(s) = 0, \quad |\kappa| = 1 \quad \text{such that if} \quad |\kappa^*| \to 1^- \quad \text{then} \quad \Re(s^*) \to 0^+ \tag{5.4}$$

where κ^* indicates a perturbation off the unit circle and s^* a perturbation off the imaginary axis (complex s-plane). We refer to an eigenvalue of the form (5.3) or (5.4) as a GKS eigenvalue or a GKS generalized eigenvalue, respectively.

To illustrate the application of the quarter-plane normal mode analysis we compute the right quarter-plane eigensolutions for the semidiscrete approximation described in section 3. If we substitute (5.1) into the interior scheme (3.1) and into the NBS (3.4) we obtain

$$\frac{2s\Delta x}{c}\phi_j = \phi_{j+1} - \phi_{j-1}, \qquad \frac{s\Delta x}{c}\phi_0 = -\alpha\phi_2 + (1+2\alpha)\phi_1 - (1+\alpha)\phi_0. \tag{5.5a,b}$$

Since (5.5a) is a difference equation for ϕ, we look for a solution of (5.5) of the form $\phi_j = \kappa^j$ and obtain

$$\frac{2s\Delta x}{c} = \kappa - \frac{1}{\kappa}, \qquad \frac{s\Delta x}{c} = (\kappa - 1)[-\alpha\kappa + (1+\alpha)]. \tag{5.6a,b}$$

By eliminating $s\Delta x/c$ between (5.6a,b), one obtains the following cubic equation for κ:

$$q(\kappa) = (\kappa - 1)^2(2\alpha\kappa - 1) = 0. \tag{5.7}$$

The roots of (5.7), i.e., the zeros of $q(\kappa)$, are

$$\kappa = 1, 1, \frac{1}{2\alpha} \qquad (\alpha \neq 0). \tag{5.8}$$

We then check to see if there is a GKS eigenvalue or a GKS generalized eigenvalue. We omit the details of the analysis (see [8]) and summarize the GKS stability results as follows:

$$\alpha < -1/2 \quad \text{unstable (GKS eigenvalue)} \tag{5.9a}$$
$$\alpha = -1/2 \quad \text{unstable (GKS generalized eigenvalue)} \tag{5.9b}$$
$$\alpha > -1/2 \quad \text{stable (no GKS eigenvalue or generalized eigenvalue).} \tag{5.9c}$$

6. Normal Mode Analysis (Finite-Domain Problem)

The normal mode analysis for the finite domain proceeds in the same way as the right-quarter-plane normal mode analysis except that there are two boundaries and, consequently, the analytical boundary condition (3.2) is imposed at $x = L$. The results of the analysis (see [8]) are summarized as follows. An eigenfunction is

$$\phi_j = \kappa^j + (-1)^{J+1}\kappa^{2J}(-1/\kappa)^j \tag{6.1}$$

and the corresponding eigenvalue is

$$2\hat{s} = \kappa - \frac{1}{\kappa}, \quad \hat{s} = \frac{\Delta x}{|c|}s \tag{6.2}$$

where κ is a zero of the polynomial $f(\kappa)$ defined by

$$f(\kappa) = (\kappa - 1)^2(2\alpha\kappa - 1) + \kappa^{2J}[-(-1)^J\kappa^{-1}(\kappa+1)^2(\kappa+2\alpha)]. \tag{6.3}$$

A complete solution of the eigenvalue problem requires that we find the zeros of the polynomial (6.3). Unfortunately, one cannot obtain the zeros of $f(\kappa)$ analytically and so the eigenvalue problem for the finite domain is intractable. On the other hand, a knowledge of the precise values of the zeros is more information than we need since the algebraic tests for stability require only asymptotic $(J \to \infty)$ values of the eigenvalues.

7. Finite Domain vs. Quarter Plane: the Eigenvalue Connection

The κ-polynomial (6.3) for the finite domain problem can be written as

$$f(\kappa) = q(\kappa) + \kappa^{2J}h(\kappa) \tag{6.4}$$

where

$$q(\kappa) = (\kappa - 1)^2(2\alpha\kappa - 1), \qquad h(\kappa) = [-(-1)^J\kappa^{-1}(\kappa+1)^2(\kappa+2\alpha)]. \tag{6.5a,b}$$

We have intentionally split the polynomial $f(\kappa)$ into two parts where the polynomial $q(\kappa)$ is precisely the κ-polynomial (5.7) associated with the right-quarter-plane problem of the GKS theory.

The notion of stability for the finite domain problem is intimately associated with the solution behavior as the spatial mesh is refined, i.e., $J \to \infty$. Hence we are primarily interested in the zeros of $f(\kappa)$ for large J. In particular, we are interested in the conditions under which the polynomial $f(\kappa)$ reduces to the quarter-plane polynomial $q(\kappa)$ in the limit $J \to \infty$. The reduction obviously occurs when $|\kappa|^{2J} \to 0$ as $J \to \infty$. The resulting conditions (i.e., the asymptotic behavior of κ) determine the connection between the normal mode analysis of the finite domain problem and the normal mode analysis of the quarter plane problem.

If a zero of the polynomial $f(\kappa)$ can be estimated asymptotically (for large J), then the corresponding eigenvalue $\hat{s}$ is given by (6.2). One can restate the necessary condition (4.2) for Lax-Richtmyer stability by requiring that every eigenvalue $\hat{s}$ of the matrix $\widehat{A}$ satisfy

$$\Re(\hat{s}) \leq \frac{w}{J}. \tag{6.6}$$

Lax-Richtmyer instability occurs if inequality (6.6) is not satisfied. In the asymptotic analysis to follow, we relate the presence of a GKS eigenvalue or GKS generalized eigenvalue to Lax-Richtmyer instability.

What values of κ do we actually need to consider? One can show that there is no loss in generality in assuming that $|\kappa| \leq 1$. In general, the zeros of $f(\kappa)$ which are crucial to the stability (actually instability) of a semidiscrete approximation depend on J and we write $\kappa = \kappa(J)$. Since we are assuming $|\kappa| \leq 1$, we let

$$|\kappa| = |\kappa(J)| = 1 - \epsilon, \quad 0 \leq \epsilon(J) < 1. \tag{6.7}$$

There are three possible cases to consider:

$$\text{case (1):} \quad \epsilon(J) \geq \delta > 0 \quad \text{as} \quad J \to \infty \tag{6.8a}$$
$$\text{case (2):} \quad \epsilon(J) \to 0 \quad \text{as} \quad J \to \infty \tag{6.8b}$$
$$\text{case (3):} \quad \epsilon(J) = 0 \quad \text{for all } J. \tag{6.8c}$$

From (6.2) it easy to show that the real part of $\hat{s}$ can be written as

$$\Re(2\hat{s}) = \Re(\kappa - 1/\kappa) = \frac{a(|\kappa|^2 - 1)}{|\kappa|^2} \tag{6.9}$$

where $\kappa = a+ib$, $|\kappa|^2 = a^2+b^2$. In case (1) $\Re(\hat{s}) \neq 0$ as $J \to \infty$, in case (2) $\Re(\hat{s}) \to 0$ as $J \to \infty$, and in case (3) $\Re(\hat{s}) = 0$ for all J.

case (1)

In the first case we assume that ϵ is strictly bounded away from zero as $J \to \infty$, i.e. $|\kappa| < 1$ as $J \to \infty$ and consequently

$$\lim_{J \to \infty} |\kappa|^{2J} = 0. \tag{6.10}$$

Then in the limit $J \to \infty$, the polynomial $f(\kappa)$ reduces to the right-quarter-plane polynomial

$$q(\kappa) = (\kappa - 1)^2(2\alpha\kappa - 1). \tag{6.11}$$

We next check to see if the above cubic polynomial has one or more zeros with $|\kappa| < 1$. If there is no zero $|\kappa| < 1$, then the assumption which led to (6.11) is invalid and we drop consideration of this case. A necessary condition for Lax-Richtmyer stability is that all the eigenvalues must satisfy inequality (6.6) and hence $\Re(\hat{s}) \leq 0$ in the limit $J \to \infty$. But in obtaining (6.11) we have already taken the limit $J \to \infty$ and the semidiscrete approximation will clearly be unstable if $\Re(\hat{s}) > 0$. Consequently, if $q(\kappa) = 0$ and (6.2), i.e., the resolvent equations, have a nontrivial solution of the form

$$\Re(\hat{s}) > 0, \quad |\kappa| < 1, \tag{6.12}$$

then the semidiscrete approximation is Lax-Richtmyer unstable. It is clear that (6.12) is identical to (5.3) of the GKS theory. Consequently, case (1) with $\Re(\hat{s}) > 0$ corresponds to a GKS eigensolution.

case (2)

In case (1), $|\kappa|$ was assumed to be strictly less than unity as $J \to \infty$. The second case of interest is $|\kappa| < 1$ for any finite J but $|\kappa| \to 1^-$ as $J \to \infty$. From (6.7) and (6.8b) we have

$$|\kappa| = 1 - \epsilon(J), \quad \epsilon(J) \to 0 \quad \text{as} \quad J \to \infty, \tag{6.13}$$

and consequently

$$|\kappa|^{2J} = [1 - \epsilon(J)]^{2J} = e^{2J \ln[1-\epsilon(J)]} \approx e^{-2J\epsilon(J)}, \quad J \to \infty. \tag{6.14}$$

Considering the product $J \cdot \epsilon(J)$ in the limit $J \to \infty$ there are only two possibilities: either

$$J \cdot \epsilon(J) \leq K, \quad J \to \infty, \quad \text{and} \quad \lim_{J \to \infty} |\kappa|^{2J} = \text{constant} > 0 \tag{6.15a}$$

where K is a positive constant independent of J, or

$$J \cdot \epsilon(J) \to \infty, \quad J \to \infty, \quad \text{and} \quad \lim_{J \to \infty} |\kappa|^{2J} = 0. \tag{6.15b}$$

In general, $\Re(\hat{s})$ is a function of κ, e.g., (6.9), and using (6.13) we obtain

$$\Re(\hat{s}) \approx W\epsilon, \quad J \to \infty \tag{6.16}$$

where W is a constant.

First we consider the possibility given by (6.15a). The κ-polynomial $f(\kappa)$ does not reduce to the quarter-plane polynomial $q(\kappa)$, however

$$\Re(\hat{s}) \leq \frac{|W|K}{J}$$

which satisfies the *necessary* condition (6.6) for Lax-Richtmyer stability.

For the second possibility given by (6.15b) the κ-polynomial $f(\kappa)$ does, in fact, reduce to the quarter-plane polynomial $q(\kappa)$. But from (6.15b) and (6.16) we have for any finite value of K, and J sufficiently large,

$$\Re(\hat{s}) > \frac{WK}{J} \qquad \text{if} \quad W > 0 \qquad (6.17)$$

which violates the *necessary* Lax-Richtmyer stability condition (6.6).

In summary, if the reduced κ polynomial $q(\kappa)$ has a zero $|\kappa| = 1$ and a perturbation of the κ inside the unit circle ($|\kappa| = 1 - \epsilon$, $\epsilon > 0$) leads to a positive $\Re(\hat{s})$ ($W > 0$ in (6.16)) the scheme is Lax-Richtmyer unstable. This algebraic test sequence is precisely the GKS test for a generalized eigenvalue (see (5.4))!

case (3)

This final case has the distinct feature that $|\kappa| = 1$ is a solution of the κ polynomial for all J and consequently $\hat{s} = 0$ for all J. The κ polynomial does not reduce to the quarter-plane polynomial. The GKS analysis does not distinguish this case from case (2) and both are considered in the GKS generalized eigenvalue test. From the point of view of an eigenvalue analysis, the two cases must be treated separately since any instability in the present case derives not from an eigenvalue with a positive real part but from the algebraic growth (as $J \to \infty$) of the norm of the solution due to the normal mode with eigenvalue $\hat{s} = 0$. Details of the analysis for this case will be presented elsewhere [8].

8. Concluding Remarks

We have presented a brief outline of an analysis that correlates Lax-Richtmyer stability and the GKS normal mode stability analysis. A more complete analysis leads us to conjecture that, except for certain identifiable *borderline* cases

$$\text{GKS stability} \iff \text{Lax-Richtmyer stability},$$

i.e., the GKS algebraic tests can be used to check Lax-Richtmyer stability.

References

[1] S. K. Godunov and V. S. Ryabenkii, "Special stability criteria of boundary value problems for non-selfadjoint difference equations," Russ. Math. Surv. 18, 1-12 (1963).

[2] B. Gustafsson, H.-O. Kreiss and A. Sundström, "Stability theory of difference approximations for mixed initial boundary value problems. II," Mathematics of Computation 26, 649-686 (1972).

[3] H.-O. Kreiss, "Difference approximations for the initial-boundary value problem for hyperbolic differential equations," "Numerical Solutions of Nonlinear Differential Equations, Proceedings Adv. Symposium," Madison, Wis., John Wiley and Sons, New York, 141-166 (1966).

[4] S. Osher, "Stability of difference approximations of dissipative type for mixed initial-boundary value problems. I," Mathematics of Computations 23, 335-340 (1969).

[5] J.C. Strikwerda, " Initial boundary value problems for the method of lines," J. Comput. Physics 34, 94-107 (1980).

[6] L. N. Trefethen, "Group velocity interpretation of the stability theory of Gustafsson, Kreiss, and Sundström," J. Comput. Physics 49, 199-217 (1983).

[7] R. F. Warming and R. M. Beam: "Proceeding of the Workshop on Oscillation Theory, Computation, and Methods of Compensated Compactness," IMA, Minneapolis, Minn., April 1985, Springer-Verlag, New York, (in press).

[8] R. F. Warming and R. M. Beam: "Stability of semidiscrete approximations for hyperbolic initial boundary value problems I: an eigenvalue analysis", NASA TM, 1986.

3-D and 2-D Solutions of the Quasi-Conservative Euler Equations

C. Weiland and M. Pfitzner

Messerschmitt-Boelkow-Blohm GmbH
Ottobrunn, FRG

SUMMARY

We present 3-D and 2-D calculations of inviscid transonic and supersonic flows around bodies and through nozzles. The method is based on the integration of the unsteady Euler equations formulated quasi-conservatively. A finite difference scheme with split-matrix upwind biased discretization is employed, which is third order accurate in space. Explicit first order time marching with local time steps is used. Flows about bodies at Mach numbers above 2 are calculated with a front shock fitting version of our code. It is shown that the scheme is capable of capturing shocks and vortex sheets.

INTRODUCTION

The numerical calculation of general inviscid unsteady flows around missiles, through nozzles and about space- and aircrafts requires an accurate and efficient algorithm capable of capturing nonlinear phenomena like vortex sheets and imbedded shocks. To ensure the correct jump relations at shocks the conservative or the quasi-conservative formulation of the Euler equations have to be used. Since the governing equations are of hyperbolic character, it is advantageous to utilize the direction of signal propagation for the discretization of the partial derivatives [1,2]. This can be done by using a split matrix upwind discretization scheme yielding a robust and stable algorithm [3,4]. Although a pure upwind scheme has superior stability property, there are restrictions in its ability to describe an upstream moving shock wave. A new upwind biased scheme, which is third order accurate in space supplemented by a nonlinear artificial viscosity model acting only near strong gradients allows the shock to move to its correct position. In the hypersonic regime with very strong front shocks a procedure where this shock is fitted while the (weaker) imbedded shocks are captured is preferred.

GOVERNING EQUATIONS

The quasi-conservative Euler equations in generalized coordinates read

$$Q_\tau + AQ_\xi + BQ_\eta + CQ_\zeta = 0 \qquad (1)$$

where

$$\xi = \xi(x, y, z, t), \quad \eta = \eta(x, y, z, t),$$
$$\zeta = \zeta(x, y, z, t), \quad \tau = t \qquad (2)$$
$$Q = (\rho, \rho u, \rho v, \rho w, e)^T$$

and

$$K = a\,k_x + b\,k_y + c\,k_z + k_t$$
$$K = (A, B, C) \quad \text{for} \quad k = (\xi, \eta, \zeta) \qquad (3)$$

with

$$a = \partial f/\partial Q, \quad f = (\rho u, \rho u^2 + p, \rho uv, \rho uw, (e + p) u)^T$$
$$b = \partial g/\partial Q, \quad g = (\rho v, \rho vu, \rho v^2 + p, \rho vw, (e + p) v)^T \quad (4)$$
$$c = \partial h/\partial Q, \quad h = (\rho w, \rho wu, \rho wv, \rho w^2 + p, (e + p) w)^T$$

and

$$p = (\gamma - 1)(e - \rho(u^2 + v^2 + w^2)/2) \quad (5)$$

if ideal gas is assumed.
The dependent velocity variables u,v,w are taken to be the cartesian components $u=v_x$, $v=v_y$, $w=v_z$. The matrices K may be diagonalized analytically [1-3]:

$$K = T_k \Lambda_k T_k \quad (6)$$

with

$$\Lambda_k = \text{diag}(U_k, U_k, U_k, U_k + c_s|\nabla k|, U_k - c_s|\nabla k|) \quad (7)$$

$$U_k = k_x u + k_y v + k_z w + k_t$$
$$|\nabla k| = (k_x^2 + k_y^2 + k_z^2)^{1/2} \quad (8)$$
$$c_s = (\gamma p/\rho)^{1/2}, \quad k = (\xi, \eta, \zeta)$$

U_k is the contravariant velocity component corresponding to the coordinate direction k and c_s is the local speed of sound. The diagonalizing matrices of eigenvectors T_k are not unique because of the three identical eigenvalues in Λ_k. One explicit form of the T_k may be found in ref.[3].

The matrices K can be split according to the sign of the eigenvalues in Λ_k:

$$K = T_k (\Lambda_k^+ + \Lambda_k^-) T_k^{-1} \equiv K^+ + K^- \quad (9)$$

where

$$\Lambda_k^\pm = (\Lambda_k \pm |\Lambda_k|)/2$$

Eq. (1) becomes

$$Q_\tau + A^+ Q_\xi^+ + A^- Q_\xi^- + B^+ Q_\eta^+ + B^- Q_\eta^- + C^+ Q_\zeta^+ + C^- Q_\zeta^- = 0 \quad (10)$$

UPWIND BIASED DISCRETIZATION

The derivatives Q^+_k and Q^-_k, $k=(\xi,\eta,\zeta)$ are approximated by the upwind biased formula

$$Q_\xi^\pm \Big|_m = \pm \frac{1}{6\Delta\xi}(Q_{m\mp 2} - 6Q_{m\mp 1} + 3Q_m + 2Q_{m\pm 1}) \quad (11)$$

and analogous for $Q^\pm_\eta$ and $Q^\pm_\zeta$.
To preserve the monotonicity near captured shocks a nonlinear artificial diffusion term d is added to the left hand side of eq.(1) [5]:

$$d = d_\xi + d_\eta + d_\zeta$$

$$d_k = \alpha_k \, \Delta k^3 \max_{i=4,5} |\lambda_k^i| \frac{\partial}{\partial k} \left(\frac{|\partial^2 p / \partial k^2|}{4\bar{p}} \frac{\partial Q}{\partial k} \right) \qquad (12)$$

$\alpha_k \cong 0.7, \qquad k = (\xi, \eta, \zeta)$

This term is small in regions where the flow variables are smooth and acts only at imbedded discontinuities. The time integration is done by the explicit one step formula

$$Q^{n+1} = Q^n + \Delta\tau \, (Q_\tau^n) \qquad (13)$$

with local timesteps

$$\Delta\tau = \left[\sum_{k=\xi,\eta,\zeta} (\max_{i=4,5} |\lambda_k^i|/\Delta k) \right]^{-1} \qquad (14)$$

Global timesteps are used for truely unsteady calculations or when strong transients are anticipated, which could produce very unphysical intermediate disturbances containing negative densities or pressures if local timesteps are used.

BOUNDARY CONDITIONS

At subsonic and supersonic inflow and outflow boundaries characteristic compatibility relations based on the one-dimensional characteristic equations in quasi-conservative form are used. The shock fitting version of our code employs the algorithm described in [6], where also the compatibility equations are used to derive expressions for the velocity of the shock and for the flow variables at this boundary.

At impermeable walls the compatibility equations are evaluated a half cell width off the body to circumvent the difficulties connected with the vanishing of three eigenvalues at points where the corresponding contravariant velocity goes to zero [4].

Cylindrical symmetry axes are treated by shifting the coordinate lines a half cell width off the symmetry line and simply applying continuation conditions.

RESULTS

In fig.1 a comparison of pressure isolines of flow about a sphere calculated with the shock-fitting and -capturing versions are shown. The number of grid points between the shock and the body is slightly higher in the second case. The shock is captured within 3 grid lines and is smooth even in regions where it crosses the coordinate lines skewly. The shock fitting result is more accurate in spite of the slightly coarser mesh. Unclustered grids of (17*33) and (33*32) points were used, respectively, in these calculations. In addition a high accuracy shock fitting result (sphere flow at M_∞ =1.5 using a 33*65 grid) is displayed on fig.1. The lines of constant total pressure are plotted, which have to be identical with the streamlines in the exact solution.

In fig. 2 the pressure and total pressure distributions on the body contour of flow about a NACA0012 airfoil at M_∞ =0.85, α =1° using a coarse unclustered (31*145) C-type mesh is shown. The shocks are captured crisply within two grid points and their locations are correct [13]. The jump in the total pressure is correct for the lower shock and somewhat to high for the upper shock due to the coarseness of the mesh.

Fig.3 displays Mach number contours of flow through a nozzle [8] calculated with different values of the ratio of specific heats γ =1.4 and γ = 1.17, representing the range of experimentally measured data. One can see that a smaller γ weakens the compression wave near the wall and moves it toward the center of the nozzle. Since the actual value of γ in the expansion region can be expected to have risen significantly, a more accurate desciption of the experimental configuration can be expected if real gas effects are included in the code.

In fig.4 the isolines of the Mach number and the characteristics of flow about a hemisphere cylinder at M_∞ =1.4 and α=10° is displayed. The embedded recompression shock on the leeside, which was also found in experiments, can be recognized clearly. Also shown is a comparison of the pressure distribution on the body contour with experimental data [7] which yields good agreement.

Fig.5 shows isolines of Mach number and total pressure and the pressure distribution on the surface of flow about a delta wing (Dillner) at M_∞ =0.7, α = 15° (x/c=0.488). Also shown is the coarse mesh (19*17*22) used in this calculation. It was generated by analytically defining the coordinate surfaces, which intersect the wing contour perpendicularly and applying a 2-D elliptic solver [9] on these surfaces. The leeside vortices found in calculations with finite volume [10] and with finite difference space marching schemes [12] can also bee seen here. Due to the coarseness of the grid, however, the pressure drop on the leeside and the vortex structure are less pronounced than in result from finer grids [11].

CONCLUSION

A variety of transonic and supersonic flow configurations have been calculated demonstrating the capabilities of our quasi-conservative finite difference upwind biased scheme. To extend the range of the code to realistic calculations in the hypersonic regime real gas effects have to be included in the scheme and this is planned in the near future.

Acknowledgement:
We would like to thank Mr. S. Riedelbauch for doing the nozzle flow calculations using the presented scheme as part of his diploma thesis.

REFERENCES

[1]: Chakravarthy S.R.: The Split-Coefficient Method for Hyperbolic Systems of Gas Dynamic Equations, AIAA paper 80-0268 (1980)
[2]: Steger J.L. and Warming R.F.: Flux Vector Splitting of the Inviscid Gasdynamic Equations with Application to Finite Difference Methods, J. Comp. Phys. **40**, p.263-293 (1981)
[3]: Whitfield D.L. and Janus J.M.: Three-Dimensional Unsteady Euler Equations Solution Using Flux Vector Splitting, AIAA paper 84-1552 (1984)
[4]: Weiland C.: A Split Matrix Method for the Integration of the Quasi-Conservative Euler Equations, Notes on Numerical Fluid Mechanics, Vol.13, Vieweg 1986
[5]: Pulliam T.H.: Artificial Dissipation Models for the Euler Equations, AIAA paper 85-0438 (1985)
[6]: Weiland C.: Numerical Integration of the Governing Equations for the Domain of Pure Supersonic Flow, ESA TT-380 (1977)
[7]: Hieh T.: Low Supersonic Three-Dimensional Flow About a Hemisphere-Cylinder, AIAA paper 75-836 (1975)
[8]: Riedelbauch S.: Berechnung von Duesenstroemungen, MBB-UR-834/860 (1986)
[9]: Halter E.: Programmdokumentation FCBFC Benutzerhandbuch, unpublished report of the KFK Karlsruhe (1985)
[10]: Smith J.H.B.: Numerical Solutions for Three-Dimensional Cases-Delta Wings, AGARD-AR-211 paper no.8 (1985)
[11]: Rizzi A.: Euler Solutions of Transonic Vortex Flow Around the Dillner Wing-Compared and Analyzed, AIAA paper 84-2142 (1984)
[12]: Weiland C.: Vortex Flow Simulations Past Wings Using Euler Equations, AGARD-CP-342, paper no.19 (1983)
[13]: Pulliam T.H. and Barton J.T.: Euler Computations of Agard Working Group 07 Airfoil Test Cases, AIAA paper 85-0018 (1985)

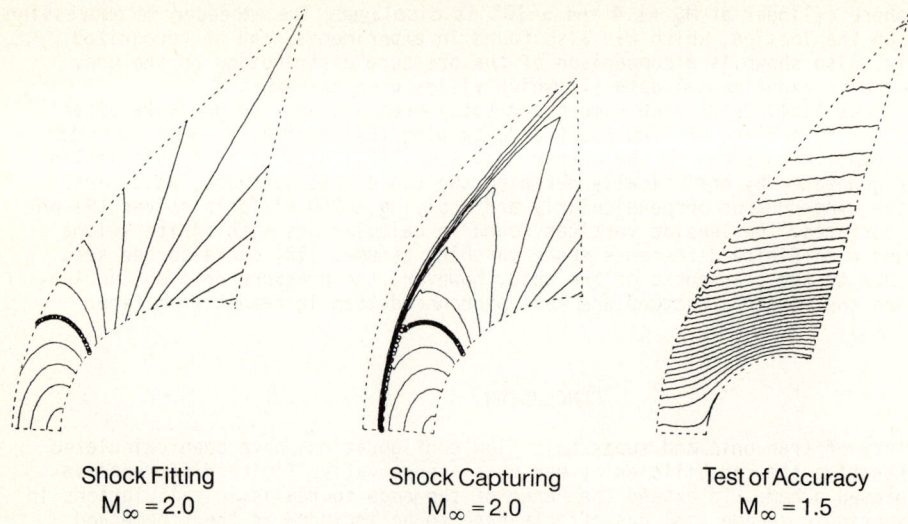

Shock Fitting
$M_\infty = 2.0$

Shock Capturing
$M_\infty = 2.0$

Test of Accuracy
$M_\infty = 1.5$

Fig. 1: Sphere. Comparison of shock fitting and shock capturing results

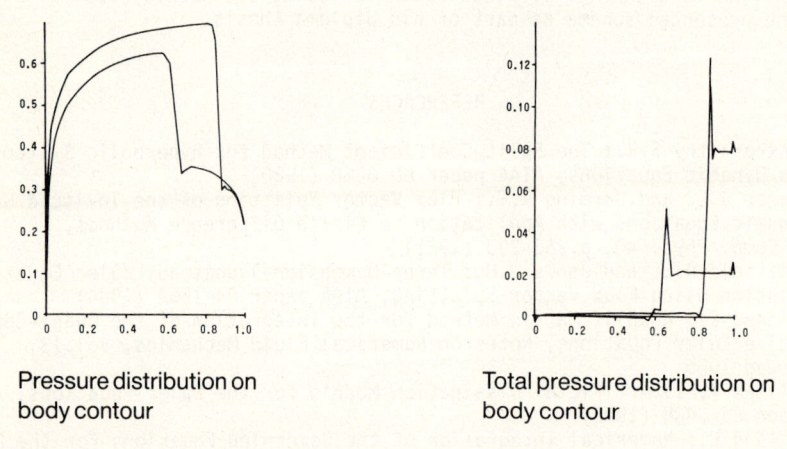

Pressure distribution on
body contour

Total pressure distribution on
body contour

Fig. 2: NACA 0012 airfoil, $M_\infty = 0.85$, $\alpha = 1°$

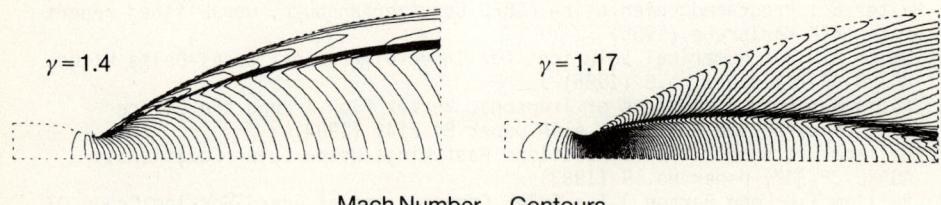

$\gamma = 1.4$

$\gamma = 1.17$

Mach Number   Contours

Fig. 3: Nozzle. Isentropic Coefficients $\gamma = 1.4$, $\gamma = 1.17$

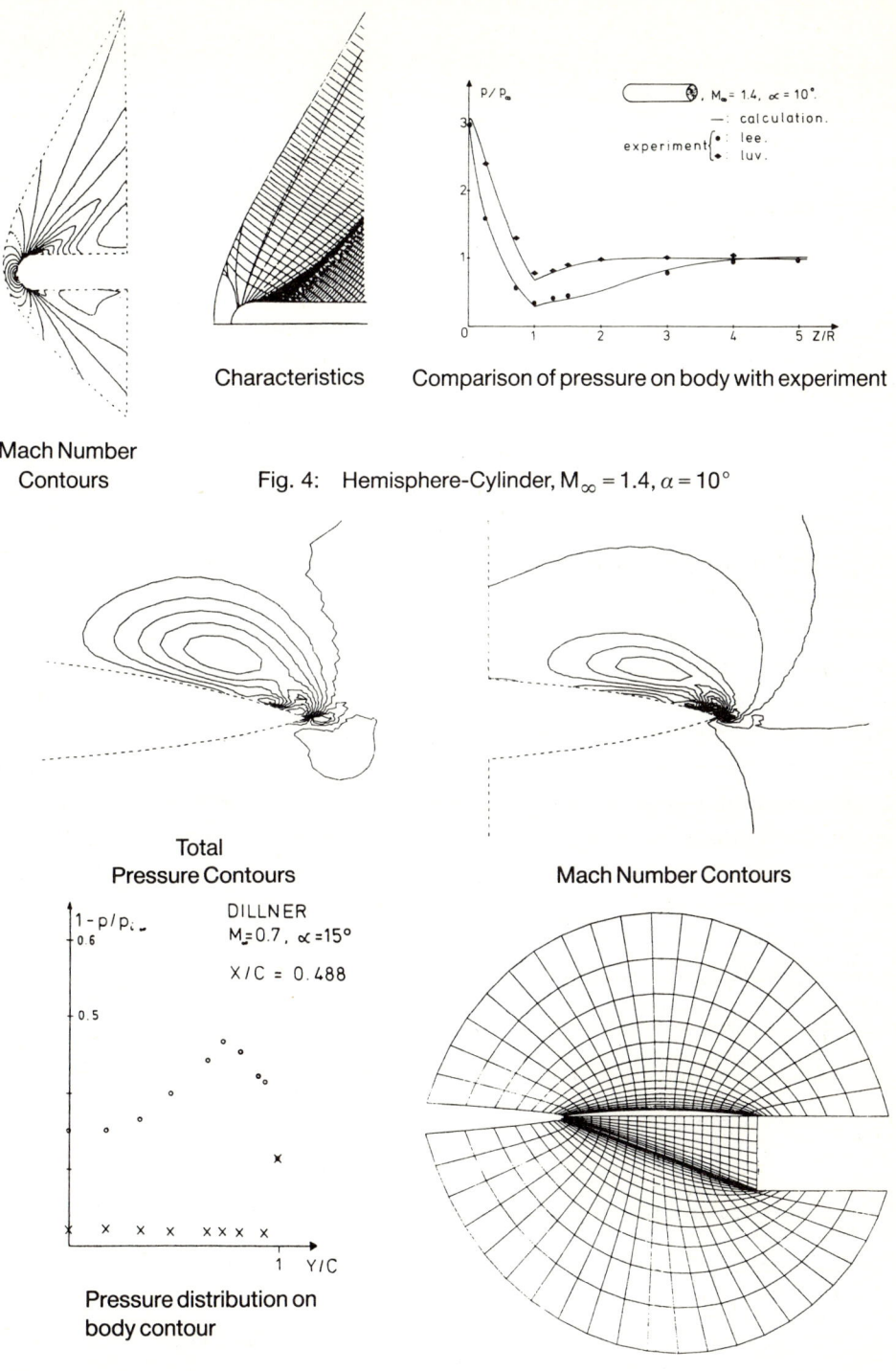

Mach Number Contours

Characteristics

Comparison of pressure on body with experiment

Fig. 4: Hemisphere-Cylinder, $M_\infty = 1.4$, $\alpha = 10°$

Total Pressure Contours

Mach Number Contours

Pressure distribution on body contour

Grid

Fig. 5: Dillner wing, $M_\infty = 0.7$, $\alpha = 15°$

An Unconditionally L_∞ - Stable Method of Fractional Steps for
Numerical Solution of Convective Diffusion Problems

J. H. Wu

Department of Mechanics, Peking University, Beijing, P.R. China

Developing a stable and efficient numerical method for solving the convective diffusion problems continues to be an active and challenging research area because of its importance in prediction of heat and mass transfer and fluid flow. Tests of the diverse methods of this area in the literature reveal that two types of problems appear quite frequently at high Péclet numbers: (i) Parasitic oscillation and/or artificial numerical attenuation; (ii) Negative values of concentrations or excess temperature.

The author's paper [1] presented a hybrid method of fractional steps with L_∞-stability to overcome these difficulties. But it requires $\Delta t \leq d^2/3K$ and is not economical for sufficiently fine grids which we have to adopt in many practical problems. The purpose of this paper is to propose another unconditionally L_∞-stable method of fractional steps, in which a modified characteristics scheme is used for convection calculation on triangular grids as in [1,2] and a finite-difference method of implicit scheme for diffusion calculation. The difference equations were derived as well for nonuniform triangular grids. If all the angles of triangles are not greater than $\frac{\pi}{2}$, we are able to obtain the discrete extremum principle for unsteady problems, which plays an important role to prevent numerical solutions from parasitic oscillations and negative values of concentrations or excess temperature.

Applying the technique of fractional steps conceived by N.N. Yanenko [3] to the convective diffusion problem:

$$\frac{\partial C}{\partial t} + \vec{V} \cdot \nabla C = \nabla \cdot (K \nabla C) + f \quad \text{in} \quad \Omega$$
$$C = \bar{C} \quad \text{on} \quad \Sigma_1 = \Gamma_1 \times (0, T)$$
$$(K \nabla C) \cdot \vec{n} = 0 \quad \text{on} \quad \Sigma_2 = \Gamma_2 \times (0, T) \qquad (1)$$
$$C = C^0 \quad \text{in} \quad \bar{\Omega} \quad \text{at} \quad t = 0$$

one can reduce the problem to a successive solution of the following equations:

$$\frac{1}{2} \frac{\partial C^{(1)}}{\partial t} + \vec{V} \cdot \nabla C^{(1)} = 0 \quad \text{if} \quad n \Delta t < t \leq (n+\tfrac{1}{2})\Delta t \qquad (2)$$

$$\frac{1}{2} \frac{\partial C^{(2)}}{\partial t} = \nabla \cdot K \nabla C^{(2)} + f \quad \text{if} \quad (n+\tfrac{1}{2})\Delta t < t \leq (n+1)\Delta t \qquad (3)$$

with initial data

$$c^{(1)}(X, n\Delta t) = C(X, n\Delta t) \qquad n = 0, 1, 2, \ldots, N_T$$

$$c^{(2)}(X, (n+\tfrac{1}{2})\Delta t) = c^{(1)}(X, (n+\tfrac{1}{2})\Delta t)$$

Here $C(X, n\Delta t) = c^{(2)}(X, n\Delta t)$ (for $n = 1, 2, \ldots, N_T$) is the final solution in the whole step $t = n\Delta t$.

The equation (2) means that the function $c^{(1)}$ remains constant along the characteristic curves defined by

$$\begin{aligned}\frac{1}{2}\frac{dx}{dt} &= u \\ \frac{1}{2}\frac{dy}{dt} &= v\end{aligned}, \qquad n\Delta t < t \leq (n+\frac{1}{2})\Delta t \qquad (4)$$

If $\vec{V} = (u, v)$ is assumed to be constant along characteristics for $n\Delta t < t \leq (n+\tfrac{1}{2})\Delta t$ (It will take place, if momentum Eqs. are solved also by the technique of fractional steps.) and $P_i(x_i, y_i)$ is a node of the triangular grid at $t = (n+\tfrac{1}{2})\Delta t$, the foot $\tilde{P}_i(\tilde{x}_i, \tilde{y}_i)$ of the characteristic curve AB (Fig.1) passing through $P_i(x_i, y_i)$ is reduced to be defined by

$$\begin{aligned}\tilde{x}_i &= x_i - u^n(\tilde{x}_i, \tilde{y}_i)\Delta t \\ \tilde{y}_i &= y_i - v^n(\tilde{x}_i, \tilde{y}_i)\Delta t\end{aligned} \qquad (5)$$

and assumed in the element (e_i) with three nodes $(x_\alpha^{(e_i)}, y_\alpha^{(e_i)})$, $\alpha = 1, 2, 3$.

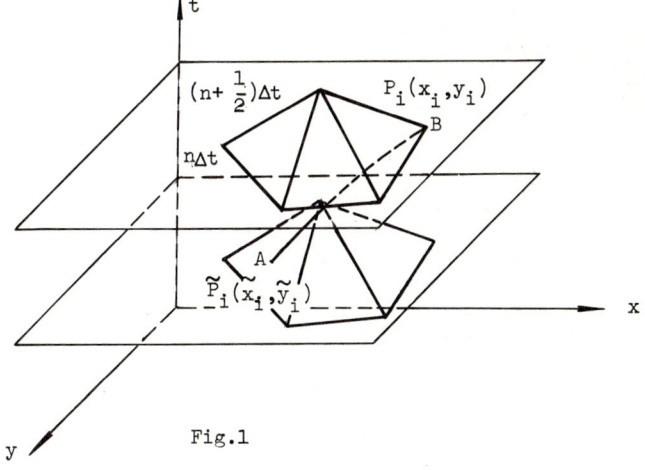

Fig.1

The original Cartesian coordinates of the point $P_i(x_i, y_i)$ in the element (e_i) are linearly related to the area coordinates as follows:

$$\tilde{x}_i = \sum_{\alpha=1}^{3} \tilde{L}_\alpha^{(e_i)} x_\alpha^{(e_i)}$$

$$\tilde{y}_i = \sum_{\alpha=1}^{3} \tilde{L}_{\alpha}^{(e_i)} y_{\alpha}^{(e_i)} \tag{6}$$

$$1 = \sum_{\alpha=1}^{3} \tilde{L}_{\alpha}^{(e_i)}$$

Linear interpolation of $\vec{V} = (u,v)$ and $C^{(1)}$ on the grid at time $t = n\Delta t$ gives:

$$\tilde{L}_1^{(e_i)} = [(a_{22} x_i - a_{12} y_i) - f] / d$$

$$\tilde{L}_2^{(e_i)} = [(a_{11} y_i - a_{21} x_i) - g] / d \tag{7}$$

$$\tilde{L}_3^{(e_i)} = 1 - \tilde{L}_1^{(e_i)} - \tilde{L}_2^{(e_i)}$$

$$C_i^{n+\frac{1}{2}} = C^{(1)}(P_i, (n+\tfrac{1}{2})\Delta t) = C(\tilde{P}_i, n\Delta t) = \sum_{\alpha=1}^{3} \tilde{L}_{\alpha}^{(e_i)} C_{\alpha}^{(e_i)}(n\Delta t) \tag{8}$$

where

$$a_{11} = [x_1^{(e_i)} - x_3^{(e_i)}] + [u_1^{(e_i)}(n\Delta t) - u_3^{(e_i)}(n\Delta t)] \Delta t ,$$

$$a_{12} = [x_2^{(e_i)} - x_3^{(e_i)}] + [u_2^{(e_i)}(n\Delta t) - u_3^{(e_i)}(n\Delta t)] \Delta t ,$$

$$a_{21} = [y_1^{(e_i)} - y_3^{(e_i)}] + [v_1^{(e_i)}(n\Delta t) - v_3^{(e_i)}(n\Delta t)] \Delta t ,$$

$$a_{22} = [y_2^{(e_i)} - y_3^{(e_i)}] + [v_2^{(e_i)}(n\Delta t) - v_3^{(e_i)}(n\Delta t)] \Delta t ,$$

$$b_1 = x_3^{(e_i)} + u_3^{(e_i)}(n\Delta t) \Delta t ,$$

$$b_2 = y_3^{(e_i)} + v_3^{(e_i)}(n\Delta t) \Delta t ,$$

$$f = a_{22} b_1 - a_{12} b_2 ,$$

$$g = a_{11} b_2 - a_{21} b_1 ,$$

$$d = a_{11} a_{22} - a_{12} a_{21} .$$

From (7) (8) and $L_{\alpha}^{(e_i)} \geq 0$, $\alpha = 1, 2, 3$ we have

$$\min_j C_j^n \leq C_i^{n+\frac{1}{2}} \leq \max_j C_j^n \tag{9}$$

The finite-difference equations for solving Eq.(3) on the nonuniform triangular grids are based on the integral derivation with the assumptions: (i) C is assumed to vary linearly over each triangle; and (ii) K is assumed to be constant over each triangle. For every interior nodal point in a triangular grid we define a secondary grid of polygonal figures whose vertices are alternately the centroids of its neighbouring triangles and midpoints of its neighbouring sides, as shown in Fig.2. Consider the k-th neighbouring triangle (e_{i_k}) of the point P_i defined by the two side vectors $\vec{s}_{i_k}$, $\vec{s}_{i_{k+1}}$ with values C_i, C_{i_k} and $C_{i_{k+1}}$ at the respective vertices as shown in Fig.3. Since C is assumed to be

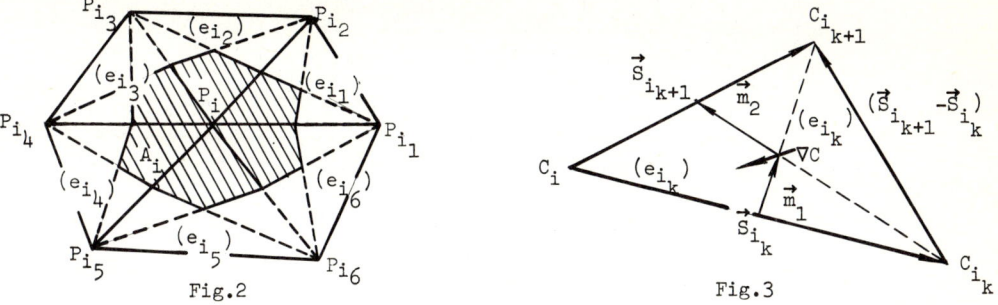

Fig.2   Fig.3

a linear function of position, each such triangle has a vector $\nabla C^{(e_{i_k})}$ associated with it which satisfies the equation

$$C_j = C_i + \vec{S}_j \cdot \nabla C^{(e_{i_k})} \qquad j = i_k, \quad i_{k+1} \tag{10}$$

and is given by

$$\nabla C^{(e_{i_k})} = \frac{(C_i - C_{i_k})\vec{S}^{\perp}_{i_{k+1}} - (C_i - C_{i_{k+1}})\vec{S}^{\perp}_{i_k}}{\vec{S}^{\perp}_{i_k} \cdot \vec{S}_{i_{k+1}}} \tag{11}$$

where the vector $\vec{S}^{\perp}$ represents the vector $\vec{S}$ rotated counterclockwise by an angle $\pi/2$. Within each triangle the flux diffusing quantity is given by

$$\vec{F}^{(e_{i_k})} = -K^{(e_{i_k})} \nabla C^{(e_{i_k})} \tag{12}$$

The flux contribution $R^{(e_{i_k})}$ from the triangle (e_{i_k}) shown in Fig.3 to the rate of change $\frac{\partial C}{\partial t}$ in the secondary grid element is

$$R^{(e_{i_k})} = \vec{F}^{(e_{i_k})} \cdot (\vec{m}_1^{\perp} + \vec{m}_2^{\perp}) = \frac{1}{2} \vec{F}^{(e_{i_k})} \cdot (\vec{S}^{\perp}_{i_{k+1}} - \vec{S}^{\perp}_{i_k}) \tag{13}$$

Summing around the vertex and using the central vertex value C_i and f_i as the average value over the polygon we obtain

$$A_i \frac{C_i^{n+1} - C_i^{n+\frac{1}{2}}}{\Delta t} = \frac{1}{2} \sum_{k=1}^{E_i} \vec{F}^{(e_{i_k})} \cdot (\vec{S}^{\perp}_{i_{k+1}} - \vec{S}^{\perp}_{i_k}) + A_i f_i^{n+1} \tag{14}$$

where

$$A_i = \sum_{k=1}^{E_i} \Delta^{(e_{i_k})} / 3 \text{ is the area of the polygon,}$$

$$\Delta^{(e_{i_k})} = \frac{1}{2} \vec{S}^{\perp}_{i_k} \cdot \vec{S}_{i_{k+1}} \text{ is the area of the triangle } (e_{i_k}).$$

The Eq. (14) can be reduced to

$$(C_i^{n+1} - C_i^{n+\frac{1}{2}}) = -b_{ii} C_i^{n+1} \Delta t + \sum_{k=1}^{N_i} b_{ii_k} C_{i_k}^{n+1} \Delta t + f_i^{n+1} \Delta t \tag{15}$$

due to (11) and (12). In Eq. (15)

$$b_{ii} = \frac{1}{2A_i} \sum_{k=1}^{E_i} \frac{K^{(e_{i_k})}(\vec{S}_{i_{k+1}}^\perp - \vec{S}_{i_k}^\perp)\cdot(\vec{S}_{i_{k+1}}^\perp - \vec{S}_{i_k}^\perp)}{\vec{S}_{i_k}^\perp \cdot \vec{S}_{i_{k+1}}^\perp}$$

$$b_{ii_k} = \frac{1}{2A_i}[\frac{K^{(e_{i_k})}\vec{S}_{i_{k+1}}^\perp \cdot (\vec{S}_{i_{k+1}}^\perp - \vec{S}_{i_k}^\perp)}{\vec{S}_{i_k}^\perp \cdot \vec{S}_{i_{k+1}}^\perp} - \frac{K^{(e_{i_{k-1}})}\vec{S}_{i_{k-1}}^\perp \cdot (\vec{S}_{i_k}^\perp - \vec{S}_{i_{k-1}}^\perp)}{\vec{S}_{i_{k-1}}^\perp \cdot \vec{S}_{i_k}^\perp}]$$

If all the angles of triangles are taken less than or equal to $\pi/2$, we have

$$b_{ii} \geq 0, \quad b_{ii_k} \geq 0, \quad k = 1,2,\cdots,N_i.$$

It is obvious that

$$-b_{ii} + \sum_{k=1}^{N_i} b_{ii_k} = 0$$

(15) is rewritten as:

$$C_i^{n+1} = \{C_i^{n+\frac{1}{2}} + \sum_{k=1}^{N_i} b_{ii_k} \Delta t \, C_{i_k}^{n+1}\} / (1+b_{ii}\Delta t) + f_i^{n+1}\Delta t/(1+b_{ii}\Delta t) \quad (16)$$

Noticing that all the coefficients of $C_i^{n+\frac{1}{2}}$ and $C_{i_k}^{n+1}$ in the right side of (16) are nonnegative and their sum is equal to identity, we obtain

$$\min \{C_i^{n+\frac{1}{2}}, \min_j C_j^{n+1}\} + \frac{f_i^{n+1}}{(1+b_{ii}\Delta t)} \leq C_i^{n+1}$$

$$\leq \max \{C_i^{n+\frac{1}{2}}, \max_j C_j^{n+1}\} + \frac{f_i^{n+1}\Delta t}{(1+b_{ii}\Delta t)} \quad (17)$$

From (9) and (17) follows the extremum property:

$$\min \{C^0, \bar{C}\} \leq C_i^n \leq \max \{C^0, \bar{C}\} \quad \text{for } f \equiv 0,$$
$$C_i^n \geq \min \{C^0, \bar{C}\} \quad \text{for } f \geq 0,$$
$$C_i^n \leq \max \{C^0, \bar{C}\} \quad \text{for } f \leq 0.$$

If $f \geq 0$ (existing only source of C), then the nonegativity of concentrations will not appear for any C_i^n.

The error equations of (8) and (16) are

$$\varepsilon_i^{n+\frac{1}{2}} = \sum_{\alpha=1}^{3} \tilde{L}_\alpha^{(e_i)} \varepsilon_\alpha^{(e_i)}(n\Delta t)$$

$$\varepsilon_i^{n+1} = \{\varepsilon_i^{n+\frac{1}{2}} + \sum_{k=1}^{N_i} b_{ii_k} \Delta t \, \varepsilon_{i_k}^{n+1}\} / (1+b_{ii}\Delta t) \quad (18)$$

From this follows:

$$\min_j \varepsilon_j^0 \leq \varepsilon_i^n \leq \max \varepsilon_j^0 \quad (19)$$

This implies L_∞ - stable.

To test numerically the convergence and accuracy of this method for nonlinear problem the method was run for a one-dimensional nonlinear convection diffusion with known analytic solution. The test problem is

$$u \frac{\partial u}{\partial x} = \nu \frac{\partial^2 u}{\partial x^2} \qquad \nu = \text{const} > 0$$

$$u(0) = 1, \qquad u(1) = 0.$$

In this case, the exact solution is

$$u(x) = C_1 \tanh(C_2 - \frac{C_1}{2\nu} x)$$

with $C_1 = 1$ and $C_2 = 12.5$ if $\nu = \frac{1}{25}$. Although the test problem is limited to one-dimensional steady state, calculations have been performed using the two-dimensional time-dependent computer program. Computational grid and comparison of numerical results with analytic solution and upwind difference and central difference solutions are presented in Fig.4 (a) and Fig.4 (b), respectively. It is seen that the parasitic oscillation is eliminated and the numerical damping effect is insignificant.

Unconditional L_∞-stability, simplicity for code and flexibility in handling complex geometrics make this method applicable to many practical problems. The numerical examples with applications will be given in subsequent papers.

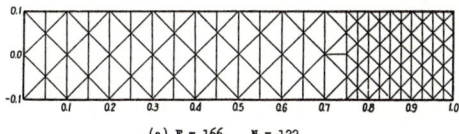

(a) E = 166, N = 122

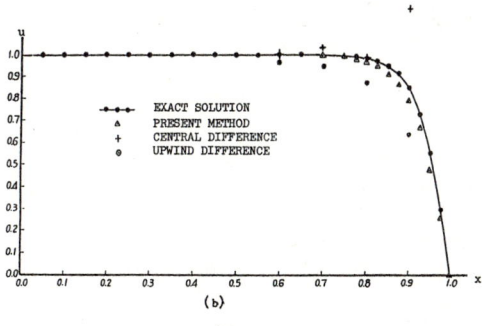

(b)

Fig. 4

References

[1] Wu Jianghang, "A Hybrid Method of Fractional Steps with L_∞-Stability for Numerical Modelling of Aquatic Environments", Chinese Journal of Computational Physics, Vol. 2, No.2, 1985.
[2] Wu Jianghang and Chen Kaiqi, "A Hybrid Method of Fractional Steps with L_∞-Stability for Numerical Modelling of Harbours and Bays", Papers presented at the International Conference on NUMERICAL AND HYDRAULIC MODELLING OF PORTS AND HARBOURS, Birmingham, England, 23-25 April 1985.
[3] N.N. Yanenko, The Method of Fractional Steps, English Translation Edited by M. Holt, Springer-Verlag Berlin Heidelberg NewYork 1971.

A HYBRID UPWIND SCHEME FOR THE COMPUTATION OF SHOCK-ON-SHOCK INTERACTION AROUND BLUNT BODIES

J. Y. Yang
PEDA Corporation
Palo Alto, CA USA

Introduction

This paper presents a hybrid upwind scheme suitable for the computation of nonstationary shock-on-shock interaction around blunt bodies in gasdynamics. The hybrid scheme combines the flux vector splitting methods [5,6], the modified flux method [1], and the characteristic flux difference splitting method [7]. The thus combined scheme has several desirable features for the computation of shock-on shock interaction which involves many discontinuities such as shocks and sliplines (in 2-D). First, it is a two time-level scheme and is second-order accurate in time and space almost everywhere except for some critical points. Second, the hybrid scheme is entropy-satisfying, thus it excludes expansion shocks and leads only to physically relevant solutions. The later property is particularly important for shock-on-shock type problems because for every interaction of two type of discontinuities, there are additional discontiniuties being generated and may become rather complicated.

With entropy-satisfying and shock-capturing capabilities and with suitable fine grid system, high resolution of flowfields for complex shoch-on-shock interaction problems can be obtained. It is the purpose of this paper to report some of our recent computations of nonstationary shock-on-shock interaction problems around blunt bodies such as circular cylinders, ellipses and reentry vehicle geometries using the present proposed hybrid upwind scheme.

Theoretical Considerations

The governing equations of the two-dimensional unsteady gas dynamics, neglecting the effects of viscosity and heat transfer, in general curvilinear coordinate systems (ξ, η) are written in conservation form as

$$\partial_\tau Q + \partial_\xi F + \partial_\eta G = 0 \tag{1}$$

where $Q = \hat{Q}/J$ and $F = (\xi_t \hat{Q} + \xi_x \hat{F} + \xi_y \hat{G})/J$, $G = (\eta_t \hat{Q} + \eta_x \hat{F} + \eta_y \hat{G})/J$, and $J = \xi_x \eta_y - \xi_y \eta_x$, the metric Jacobian. $\hat{Q} = (\rho, \rho u, \rho v, e)^T$ is the conservative variables, $\hat{F} = (\rho u, \rho u^2 + p, \rho uv, u(e+p))^T$ and $\hat{G} = (\rho v, \rho vu, \rho v^2 + p, v(e+p))^T$ are the flux vectors in Cartesian coordinates.

Here ρ is the gas density, u, v are the gas velocity components, e the internal energy and p the gas pressure. The pressure is related to other variables by the equation of state, for a perfect gas, $p = (\gamma - 1)[e - 0.5\rho(u^2 + v^2)]$, where γ is the specific heats ratio.

Due to the hyperbolicity of system (1), the Jacobian coefficient matrix $A_\xi = \partial F / \partial Q$ of the transformed equations has real eigenvalues $a_1 = U$, $a_2 = U + c_\xi$, $a_3 = U$, and $a_4 = U - c_\xi$ with $U = \xi_t + \xi_x u + \xi_y v$ and $c_\xi = c\sqrt{\xi_x^2 + \xi_y^2}$, where $c = \sqrt{\gamma p/\rho}$, is the speed of sound.

(Similarly, the eigenvalues of $B_\eta = \partial G/\partial Q$ are $b_1 = V, b_2 = V, b_3 = V + c_\eta$, and $b_4 = V - c_\eta$, with $V = \eta_t + \eta_x u + \eta_y v$ and $c_\eta = c\sqrt{\eta_x^2 + \eta_y^2}$.)

One can also find similarity transformation matrices T_ξ and T_η such that $T_\xi^{-1} A_\xi T_\xi = \Lambda_\xi = \text{diag}\{a_l\}$ and $T_\eta^{-1} B_\eta T_\eta = \Lambda_\eta = \text{diag}\{b_l\}$.

It is well known that Eq. (1) is a hyperbolic conservation laws hence both features of hyperbolicity and conservation-law can be ultilized in constructing numerical methods for solving them. A simple and natural way to unify these two aspects is to put Eq.(1) in the following form:

$$\partial_\tau Q + (\hat{A}_\xi^+ + \hat{A}_\xi^-)\partial_\xi F + (\hat{B}_\eta^+ + \hat{B}_\eta^-)\partial_\eta G = 0 \tag{2}$$

where $\hat{A}_\xi^\pm$ and $\hat{B}_\eta^\pm$ are the split normalized Jacobian coefficient matrices and can be symbolically viewed as " $A_\xi^\pm/A_\xi$ " and "$B_\eta^\pm/B_\eta$" where $A_\xi^\pm$ and $B_\eta^\pm$ are the same as those defined in the flux vector splitting method of Steger and Warming [5].

A first-order upwind scheme for Eq.(2) results when standard backward and forward spatial difference operators are used. Both explicit and implicit schemes for Eq.(2) can be fond in [8]. Relation between several upwind methods including Roe scheme [4], Huang scheme [2], the conservative supra-characterisitcs method [3] and the characteristic flux difference splitting method [7] was discussed in [3]. Several ways of constructing higher-order schemes based on (2) have been investigated and applied to practical unsteady aerodynamic problems with strong shocks [9-11].

In this study, we further elaborate a second-order scheme considered in [11] using a hybrid approach which combines the flux vector splitting [5,6], the characteristic flux difference splitting [7] and a modified flux approach similar to Harten's [1].

To achieve second-order accuracy, one can further approximate Eq.(2) by

$$\partial_\tau Q + (\hat{A}_\xi^+ + \hat{A}_\xi^-)\partial_\xi F^M + (\hat{B}_\eta^+ + \hat{B}_\eta^-)\partial_\eta G^M = 0 \tag{3}$$

where $F^M = F + E$ and $G^M = G + H$ are modified fluxes to be defined later.

Numerical Algorithms

Let us define a uniform computational mesh system (ξ_j, η_k) with mesh sizes $\Delta\xi$, and $\Delta\eta$ and let $Q_{j,k}^n$ denote the value of Q at time level $n\Delta\tau$ and at position $(j\Delta\xi, k\Delta\eta)$.

A conservative scheme for Eq.(2) can be expressed in terms of numerical fluxes F^N and G^N as follows:

$$\frac{Q_{j,k}^{n+1} - Q_{j,k}^n}{\Delta\tau} + \frac{F_{j+\frac{1}{2},k}^N - F_{j-\frac{1}{2},k}^N}{\Delta\xi} + \frac{G_{j,k+\frac{1}{2}}^N - G_{j,k-\frac{1}{2}}^N}{\Delta\eta} = 0 \tag{4}$$

Using dimensional splitting, the solution procedure becomes locally one-dimensional and can be represented as

$$Q_{j,k}^{n+2} = L_\xi(\Delta\tau)L_\eta(\Delta\tau)L_\eta(\Delta\tau)L_\xi(\Delta\tau)Q_{j,k}^n \tag{5}$$

For the L_ξ operator in the ξ-direction, we have

$$L_\xi Q_{j,k}^n = Q_{j,k}^* = Q_{j,k}^n - \lambda_\xi(F_{j+\frac{1}{2},k}^N - F_{j-\frac{1}{2},k}^N), \quad \lambda_\xi = \Delta\tau/\Delta\xi \tag{6}$$

The numerical flux $F_{j+\frac{1}{2},k}^N$ is given by

$$F_{j+\frac{1}{2},k}^N = F_{j+1,k}^n - (\Delta F)_{j+\frac{1}{2},k}^+ + (E_{j+1,k} - \hat{A}_{\xi_{j+\frac{1}{2},k}}^+ \Delta_{j+\frac{1}{2},k} E) \tag{7a}$$

$$= F_{j,k}^n + (\Delta F)_{j+\frac{1}{2},k}^- + (E_{j,k} + \hat{A}_{\xi_{j+\frac{1}{2},k}}^- \Delta_{j+\frac{1}{2},k} E) \tag{7b}$$

with $\Delta_{j+\frac{1}{2}}() = ()_{j+1} - ()_j$. Similar expression can be given for the L_η operator.

In Eq.(7), the split normalized Jacobian coefficient matrix $\hat{A}_\xi^\pm$ are given by

$$T_\xi^{-1}\hat{A}_\xi^\pm T_\xi = \hat{\Lambda}_\xi^\pm = \text{diag}\{\hat{a}_l^\pm\}, \quad \hat{a}_l^\pm = \frac{1 \pm \text{sgn}(a_l)}{2} \tag{8}$$

The value of E at nodal point j, k is $E_{j,k} = (e_1, e_2, ..., e_4)^T_{j,k}$ and its l components is given by

$$e_{l_{j,k}} = \text{sgn}(\tilde{e}_{l_{j+\frac{1}{2},k}}) \min(|\tilde{e}_{l_{j+\frac{1}{2},k}}|, |\tilde{e}_{l_{j-\frac{1}{2},k}}|), \quad \text{if} \quad \tilde{e}_{l_{j+\frac{1}{2},k}} \tilde{e}_{l_{j-\frac{1}{2},k}} \geq 0,$$

$$= 0, \quad \text{if} \quad \tilde{e}_{l_{j+\frac{1}{2},k}} \tilde{e}_{l_{j-\frac{1}{2},k}} \leq 0, \tag{9}$$

where $\tilde{e}_{l_{j+\frac{1}{2},k}}$ ($l = 1, ..., 4$) are components of a colume vector of either one of the following:

$$\tilde{E}_{j+\frac{1}{2},k} = \text{sgn} A_{\xi_{j+\frac{1}{2},k}} (I - \lambda_\xi |A_{\xi_{j+\frac{1}{2},k}}|) \Delta_{j+\frac{1}{2},k} F/2 \tag{10a}$$

$$\tilde{E}_{j+\frac{1}{2},k} = |A_{\xi_{j+\frac{1}{2},k}}| (I - \lambda_\xi |A_{\xi_{j+\frac{1}{2},k}}|) \Delta_{j+\frac{1}{2},k} Q/2 \tag{10b}$$

In Eq.(10), $|A_\xi| = T_\xi \text{diag}\{|a_l|\} T_\xi^{-1}$ and $\text{sgn} A_\xi = T_\xi \text{diag}\{\text{sgn} a_l\} T_\xi^{-1}$
The quantities $(\Delta F)^\pm_{j+\frac{1}{2}}$ can be defined by one of the following:

$$(\Delta F)^\pm_{j+\frac{1}{2}} = \hat{A}^\pm_{\xi_{j+\frac{1}{2}}} (F_{j+1} - F_j) \tag{11a}$$

$$(\Delta F)^\pm_{j+\frac{1}{2}} = (A^\pm_\xi Q)_{j+1} - (A^\pm_\xi Q)_j \tag{11b}$$

$$(\Delta F)^\pm_{j+\frac{1}{2}} = F(M_\xi)^\pm_{j+1} - F(M_\xi)^\pm_j \tag{11c}$$

In Eq.(11c), the flux is split according to the contravariant Mach number in the ξ-direction, defined as $M_\xi = U_\xi/c$.
For $|M_\xi| < 1$,

$$F^\pm = f^\pm_\rho (1, \frac{\tilde{\xi}_x(-U_\xi \pm 2c)}{\gamma} + u, \frac{\tilde{\xi}_y(-U_\xi \pm 2c)}{\gamma} + v, \frac{[(\gamma-1)U_\xi \pm 2c]^2}{2(\gamma^2 - 1)} + \frac{V_\xi^2}{2})^T \tag{12a}$$

with $f^\pm_\rho = \frac{\rho c}{4} (M_\xi \pm 1)^2 |\nabla \xi|/J$, $U_\xi = U/|\nabla \xi|$, and $V_\xi = (v\xi_x - u\xi_y)/|\nabla \xi|$, and $\tilde{\xi}_x = \xi_x/|\nabla \xi|$.
For $|M_\xi| \geq 1$,

$$F^+ = F, \quad F^- = 0, \quad M_\xi \geq +1, \tag{12b}$$

$$F^+ = 0, \quad F^- = F, \quad M_\xi \leq -1. \tag{12c}$$

Results and Discussion

A series of calculations with a wide range of shock Mach number were carried out to study the transient shock wave diffraction phenomena produced by a blast wave impinging on blunt bodies.

First we validate the computer code by comparing results with experiments.

Fig. 1 shows the comparison of present computation (1a) and experimental result using holographic interferometry (1b) (Courtesy of K. Takayama, Tohoku University, Japan) for a shock wave passing a circular cylinder with shock Mach number $M_s = 1.7$. Excellent agreement is fond in every aspect except for some viscous phenomena which are not accounted for by the Euler equations.

The second example as shown in Fig. 2 is the density contours of shock diffraction by an elliptical cylinder with shock Mach number $M_s = 2.81$ and $30°$ angle of attack at a sequence of time. Particular attention is paid to the aft body region where the two Mach shocks collide. Complex shock structures such as Mach shock, sliplines and vortex and their interactions are well presented.

For references, a non-entropy **satisfy**ing weak solution, e.g., stable expansion shock solution of the Euler equations is depicted in Fig. 3 for a circular cylinder.

Finally, an aeroassisted orbital transfer vehicle (AOTV) geometry is considered. Here the incident shock Mach number M_s is 30. The density, pressure and Mach number contours are shown in Fig. 4 for the case of zero degrees of angle of attack with $\gamma = 1.4$. Symmetrical flow patterns were observed. Strong expansion over the shoulder accelerates the flow to become supersonic. In Fig. 5 density and Mach number contours are shown for the case of 15° angle of attack with $\gamma = 1.1$ which can be considered as an approximation for a chemically frozen and vibration excited gas. The peculiar flow structures at the lee side is rather different from the symmetrical case and may have significant effect on the aerobrake mechanism.

The above calculation examples demonstrate some capability of the present scheme proposed for solving the unsteady gasdynamic problems and the quality of results are quite satisfactory.

References

[1] Harten, A., J. Comp. Phys., 49, p357, 1983.
[2] Huang, L. C., J. Comp. Phys., 42, p195, 1981.
[3] Lombard, C.K., Oliger, J. and Yang, J.Y. Proc 8th Int. Conf. on Numerical Methods in Fluid Dynamics, Aachen, 1982, (Ed. E. Krause), Lect. Notes in Physics 170.
[4] Roe, P.L., Proc 7th Int. Conf. on Numerical Methods in Fluid Dynamics, Stanford, 1980, (Ed. R. C. Reynolds and R. W. MacCormack), Lect. Notes in Physics 141.
[5] Steger, J.L. and Warming, R.F., J. Comp. Phys., 40, p263, 1981.
[6] Van Leer, B. Proc 8th Int. Conf. on Numerical Methods in Fluid Dynamics, Aachen, 1982, (Ed. E. Krause), Lect. Notes in Physics 170.
[7] Yang, J.Y.,"A Characteristic Flux Difference Splitting Method for the Hyperbolic Conservation Laws," Ph. D. Thesis, Stanford University, April, 1982.
[8] Yang, J.Y., Lombard, C.K., and Bershader, D. AIAA Paper 83-0040, AIAA 21st Aerospace Sciences Meeting, Reno, 1983.
[9] Yang, J. Y.,"Second- and Third-Order Upwind Flux Difference Splitting Schemes for the Euler Equations," NASA TM-85959, July, 1984.
[10] Yang, J. Y. AIAA Paper 85-0292, AIAA 23rd Aerospace Sciences Meeting, Reno, 1985.
[11] Yang, J.Y. Proc 6th GAMM Conf. on Numerical Methods in Fluid Mechanics, Göttingen, September 1985.

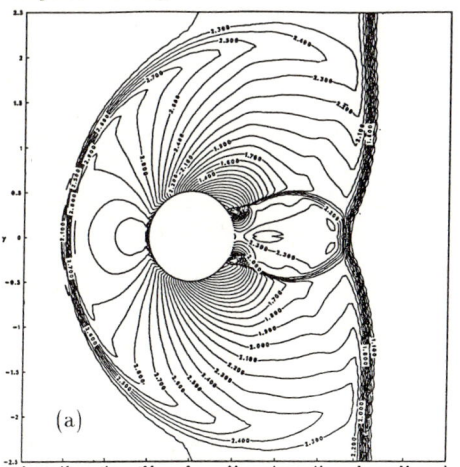
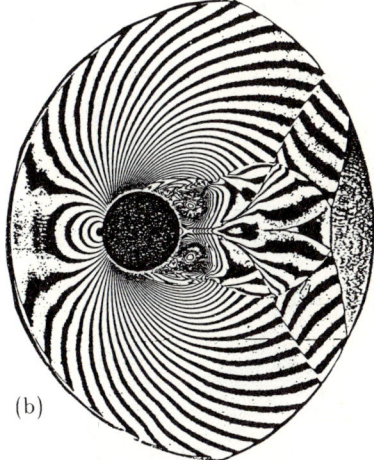

Fig. 1 Comparison of computed and experimental isopycnics for shock reflection by a circular cylinder. $M_s = 1.7$, $\gamma = 1.4$ (a) Computation (b) Experiment

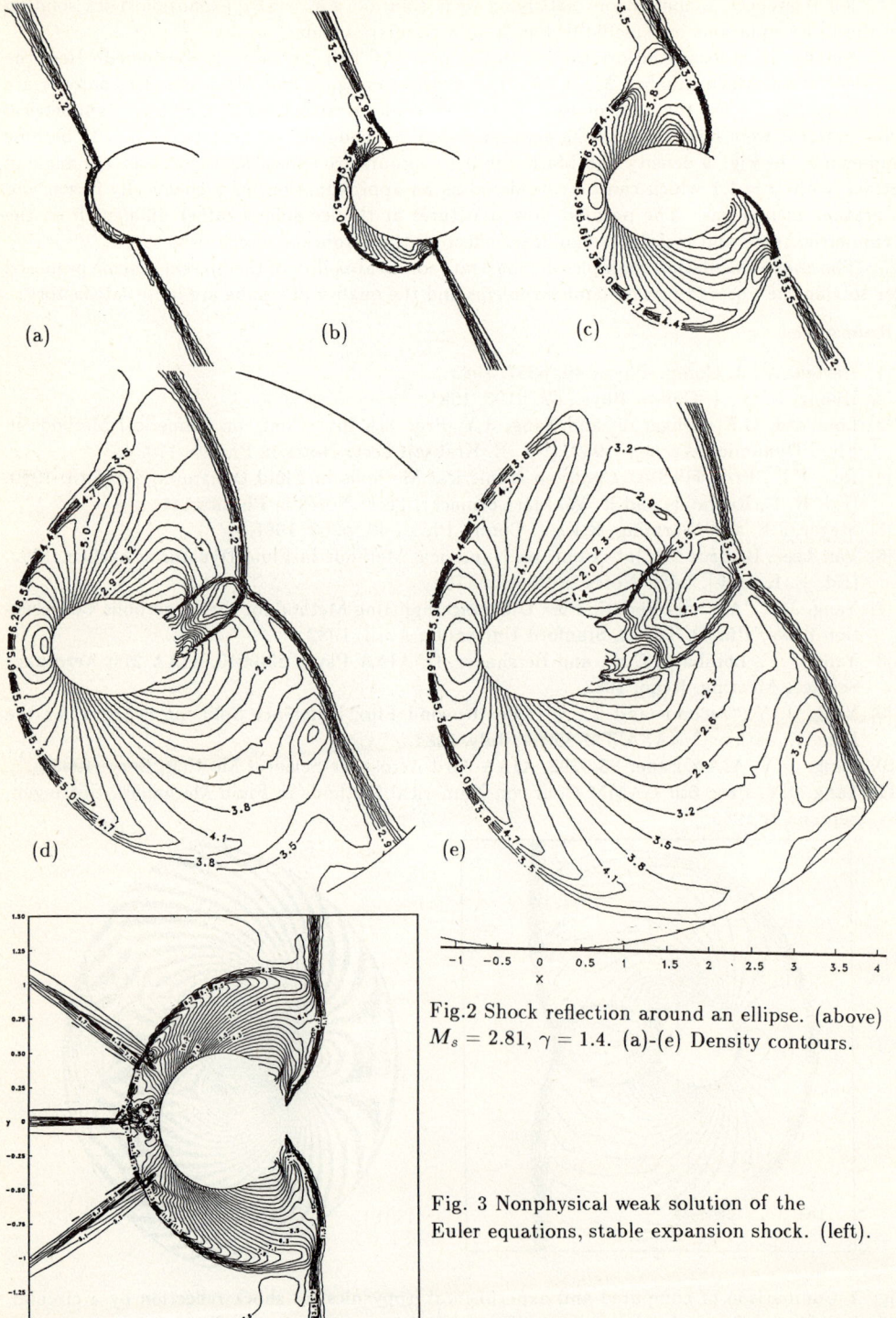

Fig.2 Shock reflection around an ellipse. (above) $M_s = 2.81$, $\gamma = 1.4$. (a)-(e) Density contours.

Fig. 3 Nonphysical weak solution of the Euler equations, stable expansion shock. (left).

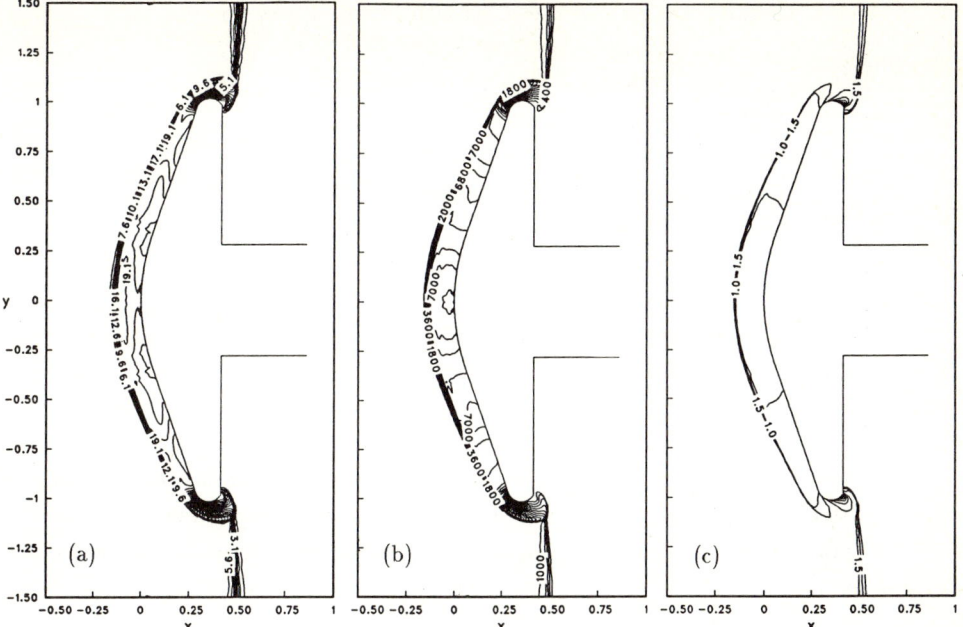

Fig. 4 Shock reflection around AOTV; $M_s = 30$, $\gamma = 1.4$, and $0°$ angle of attack.
(a) Density contours (b) Pressure contours (c) Mach nubmer contours.

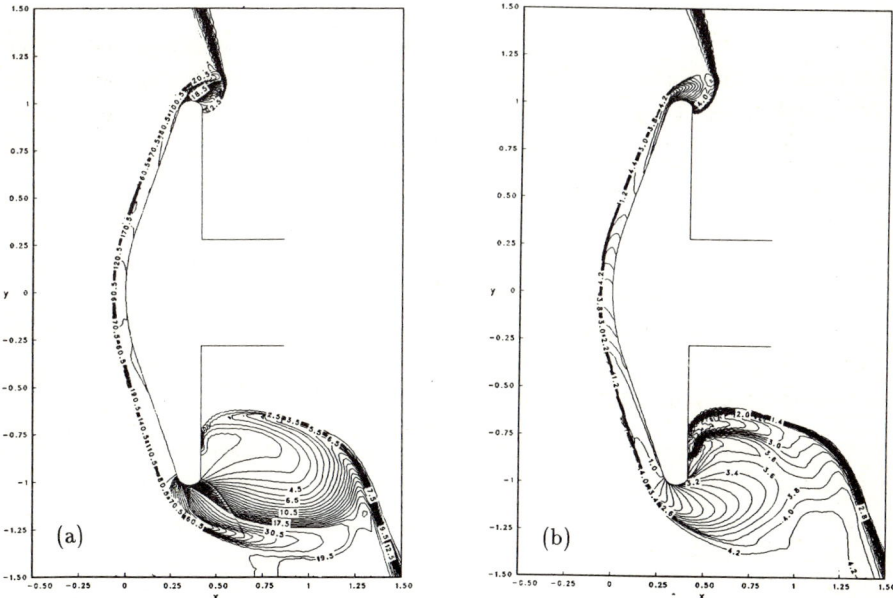

Fig. 5 Shock reflection around AOTV; $M_s = 30$, $\gamma = 1.1$ and $15°$ angle of attack.
(a) Density contours (b) Mach number contours.

Multiple Laminar Flows Through Curved Pipes

Z.H. Yang & H.B. Keller
Dept. of Math.   Appl. Math.
Shanghai Univ. of Science   Caltech
and Technology   Pasadena, CA 91125, U.S.A
Shanghai, China

1. Introduction

Following the early work of Dean (1927,1928) there have been several numerical studies of the steady, laminar, viscous flow of an incompressible fluid through a slightly curved pipe of circular cross section. In particular, Dennis (1980) with Collins (1975) and with Ng (1982) have computed such flows when the coiling ratio a/L is small. Here a is the pipe radius and L is the radius of curvature of the axis of the pipe. Also Van Dyke has applied the Stokes series and Dombes-Sykes technique (1978) to this problem. In all of this work the crucial parameter is the Dean number D, defined as

$$D \equiv Ga^3 (\frac{2a}{L})^{1/2} / \mu \nu \qquad (1.1)$$

where G is the constant pressure gradient driving the flow, μ is the viscosity and ν is the coefficient of kinematic viscosity. For small D and a/L << 1 all of the results agree.

In particular for a straight pipe, a/L = 0, the flow is the classical Poiseuille flow. However a slight curvature of the pipe axis induces a centrifugal force on the fluid which then forms a secondary flow, moving outward along the symmetry axis and returning along the upper and lower curved surfaces. Thus a pair of symmetric vortices is superposed on the Poiseuille flow. These qualitative features are observed in all of the previously cited references for small D and a/L <<1. What happens as D and a/L increase? Few of the previous studies consider a/L = O(1). Further, Van Dyke's expansions disagree with the finite difference calculations for larger values of D. And in Dennis & Ng(1982) dual solutions are found for the range 957.5 < D < 5000; that is, a four vortex solution is computed in addition to the standard two vortex flow described above.

In this paper we attempt to clarify the situation by determining the structure of the families of solutions that exist as D varies. In addition we show how this structure changes as a/L increases (to 0.3). For this purpose we must retain the full Navier-Stokes equations and do not make the simplification of a/L << 1.

2 General Formulation

We employ the notation used in Collins & Dennis (1975) and Dennis & Ng (1982) as indicated in Figure 1. The circular cross section of the tube in (x,y) plane has radius a with center at L on the x-axis. The tube is coiled along a circle of radius L in the (x,z)-plane. With no pitch in the coil the tube thus forms a torus. Our equations are exact for this case. Dimensionless velocity components of the fluid are (u,v,w) at a point P with dimensionless polar coordinates (r α). Here u is the radial and v is the angular component of velocity in the pipe cross section, w is the axial velocity normal to the cross section and $r \equiv r'/a$, where r' is the dimensional radius.

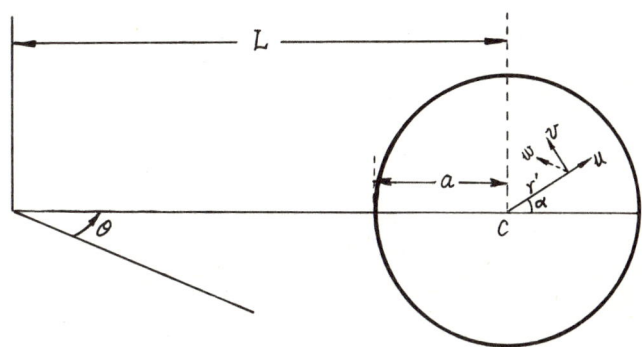

Fig. 1

We seek flows independent of θ, the angular deviation from the (x,y)-plane. A stream function $\phi(r, \alpha)$ is introduced in terms of which the transverse velocity components are given by:

$$u(r, \alpha) = \frac{1}{r(1+ \varepsilon r \cos \alpha)} \frac{\partial \phi}{\partial \alpha} ,$$
$$v(r, \alpha) = \frac{-1}{(1+ \varepsilon r \cos \alpha)} \frac{\partial \phi}{\partial r} .$$
(2.1)

Here $\varepsilon \equiv a/L$ is the "coiling ratio" and the continuity equation is thus satisfied. Using these velocity components in the Navier-Stokes equations we introduce the modified Laplacian

$$\tilde{\nabla}^2 \equiv \frac{1+\varepsilon r \cos \alpha}{r} [\frac{\partial}{\partial r}(\frac{r}{1+ \varepsilon r \cos \alpha} \frac{\partial}{\partial r}) + \frac{\partial}{\partial \alpha}(\frac{\varepsilon \sin \alpha}{1+ \varepsilon r \cos \alpha} \frac{\partial}{\partial \alpha})] \quad (2.2)$$

and the vorticity

$$\Omega = -\tilde{\nabla}^2 \phi \quad (2.3)$$

to get for the w-momentum equation

$$\tilde{\nabla}^2 w + \frac{1}{r(1+ \varepsilon r \cos \alpha)} (\frac{\partial \phi}{\partial r} \frac{\partial w}{\partial \alpha} - \frac{\partial \phi}{\partial \alpha} \frac{\partial w}{\partial r}) = -D \quad (2.4)$$

and on elimination of the pressure from the other momentum equations:

$$\tilde{\nabla}^2 \Omega + \frac{1}{r(1+\varepsilon r \cos \alpha)} (\frac{\partial \phi}{\partial r} \frac{\partial \Omega}{\partial \alpha} - \frac{\partial \phi}{\partial \alpha} \frac{\partial \Omega}{\partial r})$$
$$+ \frac{2 \varepsilon \Omega}{(1+\varepsilon r \cos \alpha)^2} (\sin \alpha \frac{\partial \phi}{\partial r} + \frac{\cos \alpha}{r} \frac{\partial \phi}{\partial \alpha}) \quad (2.5)$$

$$= \frac{w}{(1+\varepsilon r\cos\alpha)^2}(\sin\alpha\frac{\partial w}{\partial r} + \frac{\cos\alpha}{r}\frac{\partial w}{\partial \alpha}) \ .$$

The equations used in Dennis (1980) are obtained by setting $\varepsilon = 0$ in (2.1)-(2.5)(i.e. they use the small coiling ratio approximation but we do not).

Boundary conditions on the wall of the tube, $r=1$, yield:

$$w(1,\alpha) = \phi(1,\alpha) = \frac{\partial \phi}{\partial \alpha}(1,\alpha) = 0, \ 0 \leq \alpha \leq \pi \tag{2.6}$$

We study here only flows symmetric about the x-axis for which:

$$w(r,\alpha) = w(r,-\alpha), \quad \phi(r,\alpha) = -\phi(r,-\alpha), \quad \Omega(r,\alpha) = -\Omega(r,-\alpha). \tag{2.7}$$

Thus on the symmetry axis we have:

$$\frac{\partial w}{\partial \alpha}(r,0) = \frac{\partial w}{\partial \alpha}(r,\pi) = 0,$$

$$\phi(r,0) = \phi(r,\pi) = 0, \tag{2.8}$$

$$\Omega(r,0) = \Omega(r,\pi) = 0.$$

3. Fourier Series Expansions

To solve the boundary value problem posed in (2.2)-(2.8) we seek Fourier expansions of the stream function, axial velocity and vorticity in the forms:

$$\begin{aligned}
\text{a)} \quad & \phi(r,\alpha) = \sum_{k=1}^{\infty} f_k(r)\sin k\alpha \ ; \\
\text{b)} \quad & w(r,\alpha) = \sum_{k=0}^{\infty} w_k(r)\cos k\alpha \ ; \\
\text{c)} \quad & \Omega(r,\alpha) = \sum_{k=1}^{\infty} g_k(r)\sin k\alpha \ .
\end{aligned} \tag{3.1}$$

With these forms the symmetry conditions (2.7) and the implied boundary conditions (2.8) are satisfied.

Using the expansions (3.1) in the differential equations (2.3)-(2.5) and applying the orthogonality properties and other identities for the trigonometric functions yield an infinite system of coupled nonlinear, second order ordinary differential equations for the coefficient functions $\{f_k(r), w_k(r), g_k(r)\}$.

4. Numerical Procedures

To solve, or rather, to approximate the solution of the problem formulated in section 3 we first truncate the Fourier expansions, we then use difference approximations on the resulting system of O.D.E.s and finally we solve the nonlinear difference equations by means of Newton's method and continuation procedures.

5. Results of Calculations

In addition to the stream function and axial flow velocity we have computed Re, the Reynolds number based on the mean axial velocity:

$$Re = 2\sqrt{2} \int_0^1 w_0(r)r\,dr;$$

and the friction ratio (ratio of curved, γ_c, to straight, γ_s, wall friction):

$$\frac{\gamma_c}{\gamma_s} = 4\sqrt{2}\ \frac{Re}{D}$$

We have computed solution paths with D varying for the following sets of values of Fourier truncation, K, mesh spacing, h, and coiling ratio, ε:

I. $K = 10$, $h = \frac{1}{40}$; $\varepsilon = 0$;

II. $K = 10$, $h = \frac{1}{60}$; $\varepsilon = 0$, $\varepsilon = 0.1$;

III. $K = 20$, $h = \frac{1}{60}$; $\varepsilon = 0$, $\varepsilon = 0.1$, $\varepsilon = 0.2$.

Starting from the trivial state with $u = v = w = 0$ for $\varepsilon = 0$ and $D = 0$ we used continuation with increasing D as described in Section 4. In each of the three cases a simple fold was found and arclength continuation was used to accurately locate the fold and to traverse it. The solution branches were then continued with decreasing D and, in each case, another fold was found. Again these folds were located accurately and traversed to obtain a third branch in each of the three cases, now with increasing D. For cases I and II, extensions of these third branches continued well beyond where we could trust the numerical results. However for case III a third and fourth fold were found, leading to five branches of solutions. In Table 1 we list the critical value of the Dean number, Dm, at the m-th fold. For cases II and III the fold solutions found for $\varepsilon = 0$ were continued in ε up to 0.1 and for case III the continuation went up to $\varepsilon = 0.2$. These results are also given in Table 1.

We call the family of solutions varying continuously with D in $D_{m-1} < D < D_m$ the "m-th branch" ($D_0 \equiv 0$). Our calculations seem to suggest that the analytic problem has infinitely many branches although we have computed only five of them. On the first branch, that emanating from $D = 0$, the solutions are of the classic form described by Dean - we call these "two-vortex" flows. These two-vortex flows persist on the entire first branch and over most of the second branch down (in D values) to about $D \approx 5000$ where four-vortex solutions gradually appear. These four-vortex flows are formed in the calculations by the development, as D decreases on the second branch, of a small weak pair of vortices about the axis of symmetry near the outer edge of the tube. The four-vortex flows remain on the entire third branch and onto the fourth branch down to $D \approx 14,000$ where six-vortex flows appear. We believe that, as this process continues, 2n-vortex flows can form for all n=1,2,... . Indeed on the fifth branch we have computed 8-vortex solutions at $D \approx 25,000$.

In Table 3 we compare our computed values of γ_c/γ_s on the first branch with various values reported in the literature (for two-vortex flows). The agreement is quite good. Dennis and Ng(1982) have also obtained four-vortex solutions over 957.5 < D < 5000. We claim that these solutions are on the third branch. They were obtained accidentally in Dennis and Ng(1982) as a result of convergence difficulties with increasing D values near 5000. Then as D was decreased the solution "jumped" back onto the first branch. This is typical of the behavior to be expected near folds if no special technique for traversing them is used. Thus the intermediate second branch was not obtained in Dennis and Ng (1982). In Table 2 we compare the values of the four-vortex solutions obtained in Dennis and Ng (1982) with our values on the third branch. The agreement leaves no doubt as to the identity of the two results. The somewhat larger discrepancies at D=5000 is due, we believe, to inaccuracies in Dennis and Ng where convergence difficulties occurred. Graphs of the stream function and axial velocity contour lines on the third branch also agree well with those in Dennis and Ng.

During the course of this work we have benefitted from conversations with Prof. A.Acrivos. We also wish to thank Prof. S.C.R.Dennis who first brought the matter of multiple solution to our attention and suggested that we work on it.

Table 1

K	h	ε	D_1	D_2	D_3	D_4
10	1/40	0	12120	951	---	---
10	1/60	0	12752	950	---	---
20	1/60	0	25146	955	15642	7725
20	1/60	0.1	27508	1138	18179	10576
20	1/60	0.2	30071	1358	20440	14807

Table 2

	γ_c/γ_s		$W_0(0)$		Re	
D	Dennis& Ng(1982)	This Work	Dennis& Ng(1982)	This Work	Dennis& Ng(1982)	This Work
2000	1.8329	1.8338	1.0795	1.0803	192.9	192.8
3000	2.0463	2.0472	1.0514	1.0522	259.2	259.1
4000	2.2177	2.2172	1.0390	1.0389	318.8	318.9
5000	2.3662	2.3527	1.0332	1.0368	373.5	375.7

Table 3
Comparisons of γ_c/γ_s

D	Collins & Dennis[1]	Dennis & Ng[5]	Dennis[4]	This Work
1000	1.550	1.548	1.546	1.548
2000	1.852	1.847	---	1.848
3000	---	2.064	2.063	2.065
4000	---	2.237	2.237	2.238
5000	2.392	2.377	2.383	2.383

References

[1] W.M.Collins & S.C.R.Dennis, The Steady Motion of a Viscous Fluid in a Curved Tube, Q.Jl.Mech. Appl. Math. 28 (1975), pp.133-156.
[2] W.R.Dean, Note on the Motion of Fluid in a Curved Pipe, Phil. Mag. 4 (1927), pp. 208-223.
[3] W.R.Dean, The Stream-line Motion of Fluid in a Curved Pipe, Phil. Mag. 5 (1928), pp.673-695.
[4] S.C.R.Dennis, Calculation of Steady Flow through a Curved Tube Using a New Finite-difference Method, J. Fluid Mech. 99 (1980), pp.449-467.
[5] S.C.R.Dennis & M.Ng, Dual Solutions for Steady Laminar Flow through a Curved Tube, Q.Jl. Mech. Appl. Math. 35 (1982), pp.305-324.
[6] H.B.Keller, Numerical Solutions of Bifurcation and Nonlinear Eigenvalue Problems, in Applications of BifurcationTheory(ed. by Rabinowitz), Academic Press,New York (1977), pp.359-384.
[7] M.D.Van Dyke, Extended Stokes Series: Laminar Flow through a Loosely Coiled Pipe , J. Fluid Mech. 86 (1978), pp129-145.

NUMERICAL EXPERIMENTS WITH A SYMMETRIC HIGH-RESOLUTION SHOCK-CAPTURING SCHEME

H.C. YEE[1]

MS 202A-1, NASA Ames Research Center
Moffett Field, CA. 94035 USA

I. Introduction

Characteristic-based explicit and implicit total variation diminishing (TVD) schemes for the two-dimensional compressible Euler equations have recently been developed [1,2]. This is a generalization of recent work of Roe and Davis [3,4] to a wider class of symmetric (non-upwind) TVD schemes other than Lax-Wendroff. Roe and Davis's schemes can be viewed as a subset of the class of explicit methods. The main properties of the present class of schemes are that they can be implicit, and, when steady-state calculations are sought, the numerical solution is independent of the time step. In reference [2], a comparison of a linearized form of the present implicit symmetric TVD scheme with an implicit upwind TVD scheme originally developed by Harten [5,6] and modified by Yee [7] was given. The results favored the symmetric method. It was found that the symmetric method is just as accurate as the upwind method while requiring less computational effort. It is emphasized that the generic use of the notion upwind and symmetric TVD schemes here pertains to the schemes without the limiter present. With a limiter present, an upwind TVD scheme no longer has its traditional upwinding meaning. The same situation also applies to symmetric TVD schemes. Another way of distinguishing an upwind from a symmetric TVD scheme is that the numerical dissipation term corresponding to an upwind TVD scheme is upwind weighted [5-10], as opposed to the numerical dissipation term corresponding to a symmetric TVD scheme which is centered [1-4].

Currently, more numerical experiments are being conducted on time-accurate calculations and on the effect of grid topology, numerical boundary condition procedures, and different flow conditions on the behavior of the method for steady-state applications. The purpose of this paper is to report our experiences with this type of scheme and give some basic guidelines for using the scheme. A description of upwind and symmetric TVD schemes, including formulation and extension to the two-dimensional Euler equations of gas dynamics in curvilinear coordinates, can be found in references [2,7,11]. This paper contains a brief description of the TVD algorithm and a summary of the numerical computations.

II. Description of Algorithm in One Dimension

Consider a one-dimensional system of hyperbolic conservation laws

$$\frac{\partial U}{\partial t} + \frac{\partial F(U)}{\partial x} = 0. \tag{1}$$

Here U and $F(U)$ are column vectors of m components and $A = \partial F/\partial U$. Let the eigenvalues of A be $(a^1, a^2, ..., a^m)$. Denote R as the matrix whose columns are eigenvectors of A, and R^{-1} as the inverse of R. Let the grid spacing be denoted by Δx such that $x = j\Delta x$, and let $U_{j+\frac{1}{2}}$ denotes some symmetric average of U_j and U_{j+1} (for example, $U_{j+\frac{1}{2}} = \frac{1}{2}(U_{j+1} + U_j)$). Let $a^l_{j+\frac{1}{2}}$, $R_{j+\frac{1}{2}}$, $R^{-1}_{j+\frac{1}{2}}$ denote the quantities a^l, R, R^{-1} related to A evaluated at $U_{j+\frac{1}{2}}$. Define

$$\alpha_{j+\frac{1}{2}} = R^{-1}_{j+\frac{1}{2}}(U_{j+1} - U_j) \tag{2}$$

as the difference of the local characteristic variables.

A one-parameter family of conservative explicit and implicit second-order TVD schemes can be written as

$$U^{n+1}_j + \lambda\theta\left[\tilde{H}^{n+1}_{j+\frac{1}{2}} - \tilde{H}^{n+1}_{j-\frac{1}{2}}\right] = U^n_j - \lambda(1-\theta)\left[\tilde{H}^n_{j+\frac{1}{2}} - \tilde{H}^n_{j-\frac{1}{2}}\right], \tag{3a}$$

---

[1]Research Scientist, Computational Fluid Dynamics Branch.

where $\lambda = \frac{\Delta t}{\Delta x}$, Δt is the time step and θ is a parameter. When $\theta \neq 0$, the scheme is implicit. The numerical flux function $\tilde{H}_{j+\frac{1}{2}}$ can be expressed as

$$\tilde{H}_{j+\frac{1}{2}} = \frac{1}{2}\left[F_j + F_{j+1} + R_{j+\frac{1}{2}}\Phi_{j+\frac{1}{2}}\right]. \tag{3b}$$

The elements of the $\Phi_{j+\frac{1}{2}}$ denoted by $(\phi^l_{j+\frac{1}{2}})^S$ for a general second-order symmetric TVD scheme [1] are

$$(\phi^l_{j+\frac{1}{2}})^S = -\lambda\beta(a^l_{j+\frac{1}{2}})^2 \hat{Q}^l_{j+\frac{1}{2}} - \psi(a^l_{j+\frac{1}{2}})\left[\alpha^l_{j+\frac{1}{2}} - \hat{Q}^l_{j+\frac{1}{2}}\right], \tag{4a}$$

where $\alpha^l_{j+\frac{1}{2}}$ are elements of (2). When $\beta = 1$, the method is best suited for time-accurate calculations, and when $\beta = 0$, it is mainly for steady-state applications. The function ψ is

$$\psi(z) = \begin{cases} |z| & |z| \geq \epsilon \\ (z^2 + \epsilon^2)/2\epsilon & |z| < \epsilon \end{cases}. \tag{4b}$$

Here $\psi(z)$ in (4b) is an entropy correction to $|z|$ where ϵ is a small positive parameter (see reference [12] for a formula for ϵ). Examples of the 'limiter' function $\hat{Q}^l_{j+\frac{1}{2}}$ can be expressed as

$$\hat{Q}^l_{j+\frac{1}{2}} = \text{minmod}(\alpha^l_{j-\frac{1}{2}}, \alpha^l_{j+\frac{1}{2}}) + \text{minmod}(\alpha^l_{j+\frac{1}{2}}, \alpha^l_{j+\frac{3}{2}}) - \alpha^l_{j+\frac{1}{2}} \tag{4c}$$

$$\hat{Q}^l_{j+\frac{1}{2}} = \text{minmod}(\alpha^l_{j-\frac{1}{2}}, \alpha^l_{j+\frac{1}{2}}, \alpha^l_{j+\frac{3}{2}}) \tag{4d}$$

$$\hat{Q}^l_{j+\frac{1}{2}} = \text{minmod}\left[2\alpha^l_{j-\frac{1}{2}}, 2\alpha^l_{j+\frac{1}{2}}, 2\alpha^l_{j+\frac{3}{2}}, \frac{1}{2}(\alpha^l_{j-\frac{1}{2}} + \alpha^l_{j+\frac{3}{2}})\right]. \tag{4e}$$

The minmod function of a list of arguments is equal to the smallest number in absolute value if the list of arguments is of the same sign, or is equal to zero if any arguments are of opposite sign.

The elements of the $\Phi_{j+\frac{1}{2}}$ denoted by $(\phi^l_{j+\frac{1}{2}})^U$ for a second-order upwind TVD scheme, originally developed by Harten, and later modified and generalized by the author [1,13], are

$$(\phi^l_{j+\frac{1}{2}})^U = \sigma(a^l_{j+\frac{1}{2}})(g^l_{j+1} + g^l_j) - \psi(a^l_{j+\frac{1}{2}} + \gamma^l_{j+\frac{1}{2}})\alpha^l_{j+\frac{1}{2}}. \tag{5a}$$

The function $\sigma(z) = \frac{1}{2}\psi(z) + (\theta - \frac{1}{2})\lambda\beta z^2$ and

$$\gamma^l_{j+\frac{1}{2}} = \sigma(a^l_{j+\frac{1}{2}})\begin{cases} (g^l_{j+1} - g^l_j)/\alpha^l_{j+\frac{1}{2}} & \alpha^l_{j+\frac{1}{2}} \neq 0 \\ 0 & \alpha^l_{j+\frac{1}{2}} = 0 \end{cases}. \tag{5b}$$

Examples of the 'limiter' function g^l_j can be expressed as

$$g^l_j = \text{minmod}(\alpha^l_{j-\frac{1}{2}}, \alpha^l_{j+\frac{1}{2}}) \tag{5c}$$

$$g^l_j = \left(\alpha^l_{j+\frac{1}{2}}\alpha^l_{j-\frac{1}{2}} + |\alpha^l_{j+\frac{1}{2}}\alpha^l_{j-\frac{1}{2}}|\right)/\left(\alpha^l_{j+\frac{1}{2}} + \alpha^l_{j-\frac{1}{2}}\right) \tag{5d}$$

$$g^l_j = S \cdot \max\left\{0, \min(2|\alpha^l_{j+\frac{1}{2}}|, S \cdot \alpha^l_{j-\frac{1}{2}}), \min(|\alpha^l_{j+\frac{1}{2}}|, 2S \cdot \alpha^l_{j-\frac{1}{2}})\right\}; \quad S = \text{sgn}(\alpha^l_{j+\frac{1}{2}}). \tag{5e}$$

The method of extending scalar TVD schemes to nonlinear systems of hyperbolic conservation laws (1) in equation (3) is sometimes referred as the local characteristic approach and is a variant of Roe's linear Riemann solver [14]. The advantages of this approach as opposed to Davis's simplified approach [4] to systems are that: (a) this approach in effect uses scalar schemes on each characteristic field; (b) the limiter used need not be the same for each field; e.g., one can use a more compressive limiter for the linear fields and use a less compressive limiter for the nonlinear fields; (c) one can even use different schemes for different fields; (d) it is more efficient than the exact or approximate Riemann solvers; (e) it provides a natural way to linearize the implicit TVD schemes. For the one-dimensional Euler equation of gas dynamics, the characteristic fields consist of two nonlinear fields $u \pm c$ and a linear field u.

For two-dimensional time-accurate calculations, the explicit scheme $\beta = 1, \theta = 0$ is implemented by the Strang type of time splitting method [15]. For steady-state numerical study, the implicit scheme $\beta = 0, \theta = 1$ considered here is implemented in a conservative noniterative ADI form [2]. The numerical solution is independent of the time step. The implicit operator has a regular block tridiagonal structure and the resulting block tridiagonal matrix is diagonally dominant. One can modify a classical ADI central difference code by simply changing the linear numerical dissipation term into the one designed for the TVD scheme; i.e., the third term of equation (3b). See references [7,11] for more details.

Let the grid indexing at the left boundary in the x-direction be j=0. For the explicit operator, this is a five-point scheme in each coordinate direction and one needs values of g_0^l as well as U_0. For simplicity and illustration purposes, the values of U_0 are assumed to be updated by the procedure described in reference [11]. The following three ways of obtaining g_j^l at the boundaries are discussed.

$$g_0^l = 0; \quad g_0^l = g_1^l; \quad g_0^l = \alpha_{\frac{3}{2}}^l. \qquad (6a,b,c)$$

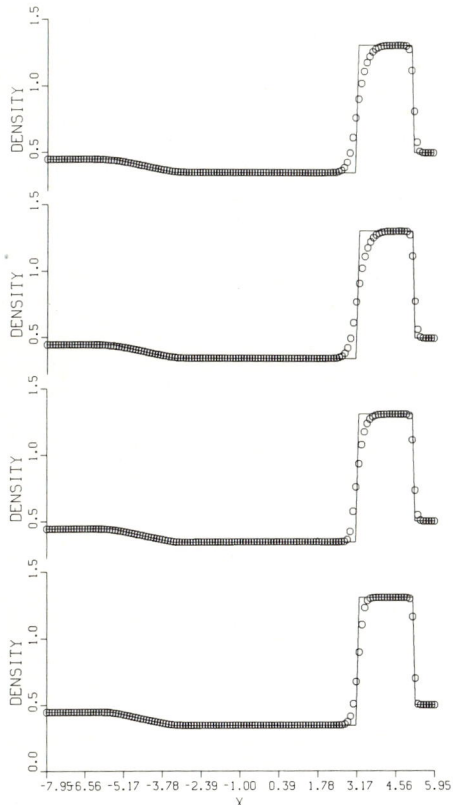

One can interpret (6a) as a first-order TVD scheme at the boundary, whereas (6b) is obtained by zeroth-order extrapolation. Boundary scheme (6c) is the least dissipative among the three. It is obtained by simply removing the logic in the limiter and using the next availiable α^l.

III. Numerical Results

3.1 Time Accurate Calculations: The numerical experiments were mainly performed on the one-dimensional shock tube problem and the NACA0012 and NACA0018 airfoils. Limiter (4e) produces slightly sharper shocks and contact discontinuities than (4c) or (4d).

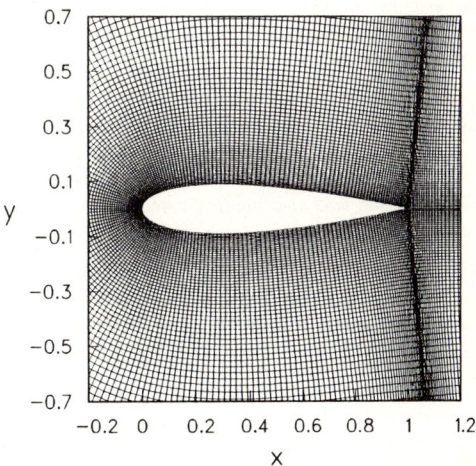

Fig. 1 The Density Distribution (– exact, o computed) of the shock tube problem using limiter (4c-4e) for the symmetric scheme and using (5d) for the upwind scheme (top to bottom)

Fig. 2 The 299 × 79 C grid for the NACA0018 airfoil.

One Dimension: Figure 1 shows the density distribution for the same shock tube problem as used in reference [5]. The solid lines are the exact solutions and the circles are the results computed by using limiters (4c-4e) for the symmetric TVD scheme and (5d) for the upwind scheme respectively (from top to bottom). With the use of the same limiter, the study shows that in general the symmetric scheme is slightly more diffusive than the upwind scheme. However, the resolution (not shown) of the symmetric scheme with limiter (4e) is better than the upwind scheme with limiter (5c). Note that not all of the limiters that are designed for upwind schemes are applicable for symmetric schemes.

Two Dimensions: In the experiment, a time dependent curved shock wave collides with the NACA0018 airfoil at an angle of attack $\alpha = 30°$. The Mach number M_S is approximately 1.5 at the moment of collision. The experiment was performed by Mandella and Bershader [16] of Stanford University. In the numerical simulations, a constant planar incident shock of $M_S = 1.5$ and $\alpha = 30°$ is used as an initial condition just before the leading edge of the airfoil. This work was done jointly with Young Moon of U.C. Berkeley. Figure 2 shows the 299×79 C grid for the NACA0018 airfoil. Figure 3 shows the numerical results in the form of density contour plots using (4e) with the symmetric scheme, and (5d) for the nonlinear fields and (5e) for the linear fields with the upwind scheme at (approximately) the four sequential instances of the interferograms. The incident and reflected shocks, Mach stems, and slip surface on the lower surface are captured within 3 grid points. Also the vortices at the upper nose and the trailing edge of the airfoil are well captured in the simulation. On the upper surface the experimental results show a weak slip surface which it is not captured in the simulation. By increasing the grid resolution around that region, this slip surface is also expected to be captured. Even though a planar shock was used to model the curved shock, the shape and location of the discontinuities compare favorably with the experiments. The shock resolution of the symmetric TVD scheme is similar to the upwind scheme except that it is slightly more diffusive. This leads us to believe that even closer agreement between the numerical simulation and the experiments is possible if a time dependent curved shock is used. A detail description of the physical problem and a more closely simulated result with the exact experiment can be found in a paper by Moon and Yee [17]. Aside from the difference in CPU time (at least 35 – 40% less), one advantage of symmetric TVD schemes over upwind schemes is that the symmetric TVD schemes are less sensitive to numerical boundary condition treatments for higher Mach number flows. Figure 4 shows the computations of the symmetric TVD scheme (4e) at a CFL number of 0.98 for a planar incident shock of $M_S = 20$ and $\alpha = 30°$. With the same numerical boundary conditions, the upwind scheme diverges; i.e., a characteristic type of boundary condition is needed for the upwind scheme.

3.2 Steady-State Calculations: Some steady-state computations have been carried out to illustrate the sensitivity of the implicit symmetric (and upwind) scheme on the various numerical boundary condition procedures, grid topologies and flow conditions. A typical solution computed by the symmetric scheme using (4e) is shown in figure 5. Figure 6 shows the 163×49 C grid with the outer boundary 24 chord lengths away from the body. Some of the numerical study using (4c) and (4d) with $\beta = 0$ and $\theta = 1$ using the same grid is summarized below.

Accuracy: To demonstrate how numerical boundary conditions affect the accuracy of the scheme, the computations of the NACA0012 airfoil at a freestream Mach number $M_\infty = 0.8$ and $\alpha = 1.25°$ with (4c) are shown in figure 7. This figure compares the Mach contours of boundary schemes (6a) with (6b). One can see a drastic difference in resolution near the solid body. Away from the body, the resolution is indistinguishable. A comparison was also made between boundary schemes (6b) and (6c). No visible difference in accuracy was found.

Stability: Computation of the NACA0012 airfoil with freestream Mach numbers ranging from 0.8 to 1.8 with and without lift show that boundary schemes (6a) and (6b) have a similar stability and convergence rate. However boundary scheme (6c) has a lower stability bound than (6a) and (6b). This is due to the fact that boundary scheme (6c) is the least dissipative among the three procedures.

Grid Topology: The grid shown in figure 1 was generated by a hyperbolic grid method. This grid is fairly regular and for the most part is nearly orthogonal. The present method appears to have no difficulty in obtaining a fairly accurate solution. If one were to use a highly irregular and nonorthogonal grid, the present scheme would require a long time to reach steady state or, in an extreme case, fail to converge.

Convergence Rate: A numerical study was also conducted on the effect of a freestream Mach number on the convergence rate of the scheme. The following is based on $0.7 \leq M_\infty \leq 1.8$, $0° \leq \alpha \leq 7°$, and grid sizes ranging from 163×33 to 249×49 with either a C or O grid. For $0.7 \leq M_\infty \leq 0.8$, a L_2-norm residual of

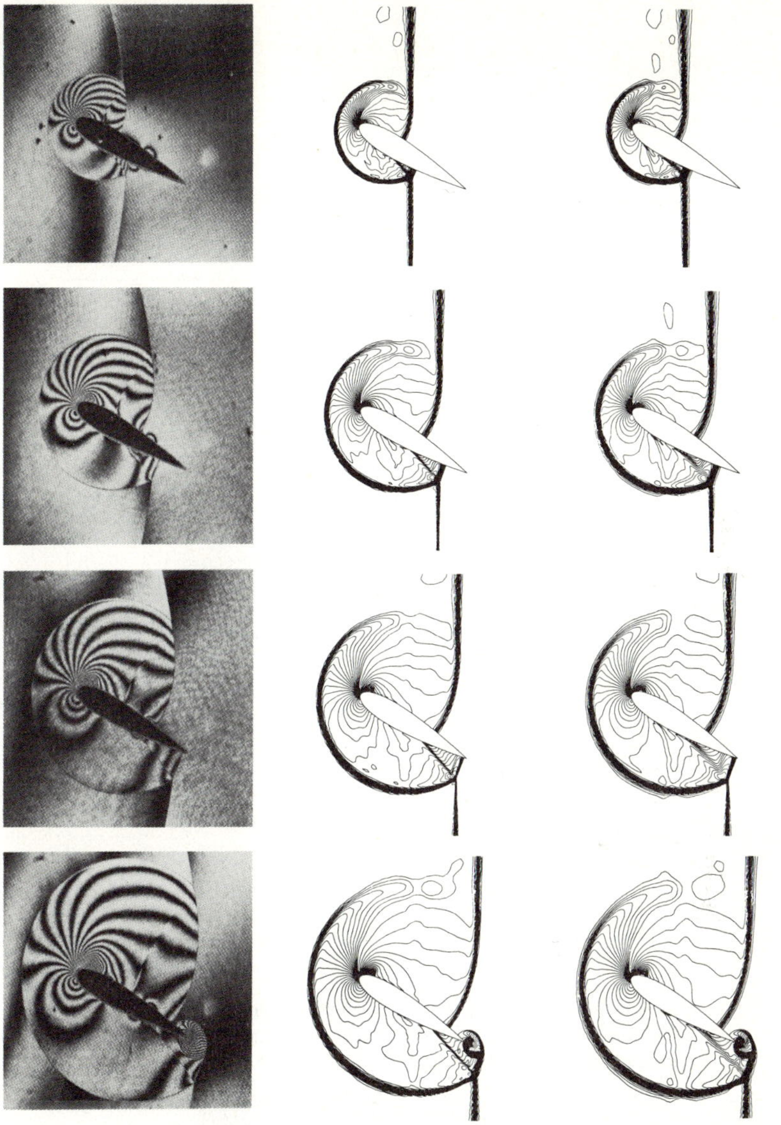

Fig. 3 The Density contours by the upwind scheme (5a,d,e) (middle) and by the symmetric scheme (4a,e) (right) for the NACA0018 airfoil with $M_S = 1.5$, $\alpha = 30°$ compared with the interferograms at approximately the same sequential instances.

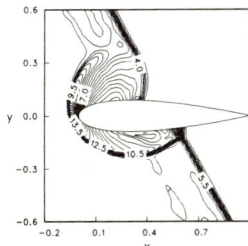
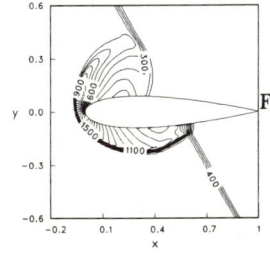

Fig. 4 The density contours (left) and Mach contours (right) by the symmetric scheme (4a,e) for the NACA0018 airfoil with $M_S = 20$, $\alpha = 30°$.

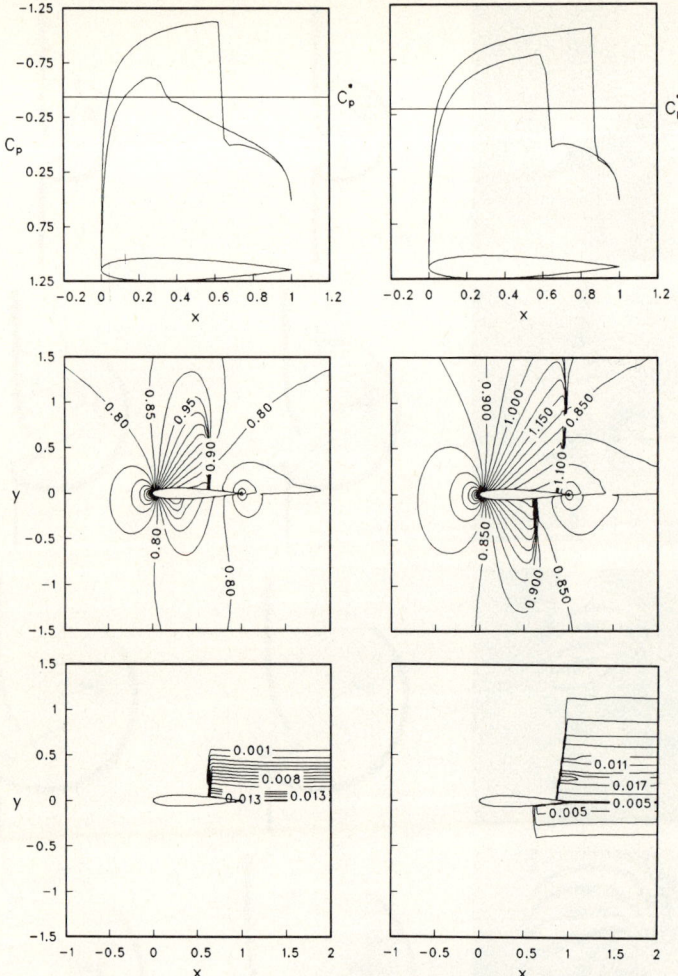

Fig. 5 The pressure coefficients, Mach contours and entropy contours (top to bottom) computed by the symmetric scheme (4a,e) for the NACA0012 airfoil with $M_\infty = 0.8$, $\alpha = 1.25°$ (left) and $M_\infty = 0.85$, $\alpha = 1°$ (right).

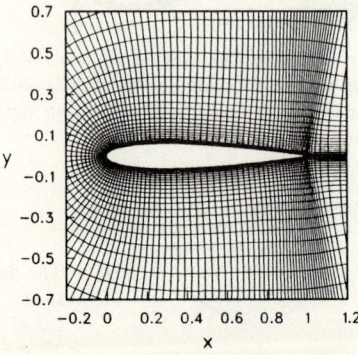

Fig. 6 The 163×49 C grid for the NACA0012 airfoil.

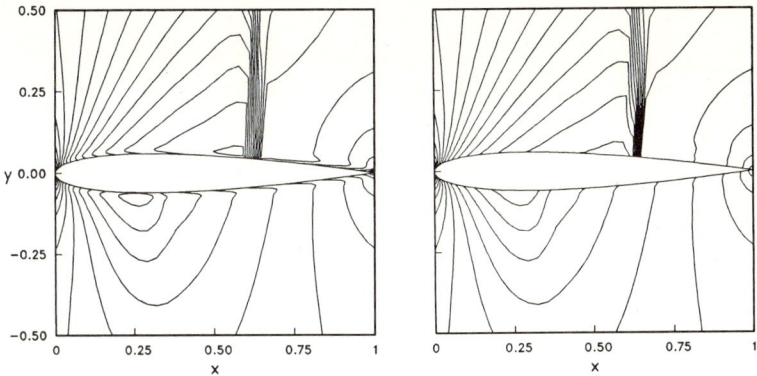

Fig. 7 The Mach contours computed by boundary schemes (6a) (left) and (6b) (right) for the NACA0012 airfoil with $M_\infty = 0.8$, $\alpha = 1.25°$.

10^{-7} can be reached in 600-1200 steps depending on the grid size, time step and α. For $0.85 \leq M_\infty \leq 0.98$, a residual of 10^{-7} can be reached in 1500-3500 steps. For $1.2 \leq M_\infty \leq 1.8$, a residual of 10^{-7} can be reached in 200-500 steps.

IV. Concluding Remarks

For problems containing shocks only, numerical experiments for steady-state calculations show that the implicit symmetric TVD schemes, while requiring less computational effort, are just as accurate as the implicit upwind TVD scheme originally developed by Harten [5] and modified by Yee [7]. However, for problems containing both shocks and contact discontinuities, numerical experiments show that the explicit symmetric TVD schemes are slightly more diffusive than the explicit upwind TVD scheme especially at the contact surfaces. The current study on the various steady-state airfoil calculations also indicates the degree of sensitivity of the method to grid topology, freestream Mach number, and angle of attack. Moreover, proper numerical boundary condition procedures are essential for numerical simulation of fluid flow. Improper treatment of boundary conditions can lead to instability and inaccuracy, even if one starts with a stable high-resolution scheme that catered to problems with shocks. Furthermore, for certain flow regimes, different linearization or relaxation procedures other than ADI should be investigated to improve the convergence rate of steady-state calculations.

References

1. H.C. Yee, NASA-TM 86775, July 1985, also to appear, J. Comput. Phys., 1986.
2. H.C. Yee, Proc. 6th GAMM Conf. Numer. Meth. in Fluid Mechanics, W. Germany, Sept. 1985.
3. P.L. Roe, ICASE Report No. 84-53, October 1984.
4. S.F. Davis, ICASE Report No. 84-20, June 1984.
5. A. Harten, NYU Report, Oct., 1982; SIAM J. Num. Anal, Vol. 21, 1984, pp. 1-23.
6. A. Harten, J. Comput. Phys., Vol. 49, 1983, pp 357-393.
7. H.C. Yee, Int. J. Comput. Math. Appl., in press.
8. P.L. Roe, AMS publications, Lectures in Appl. Math., Vol. 22, 1985.
9. P.K. Sweby, SIAM J. Num. Analy., Vol. 21, 1984.
10. S. Osher and S. Chakravarthy, SIAM J. Num. Analy., Vol. 21, 1984.
11. H.C. Yee and A. Harten, AIAA Paper No. 85-1513, July 1985, to appear AIAA J.
12. A. Harten and J.M. Hyman, J. Comp. Phys., Vol. 50, 1983.
13. H.C. Yee, "A comparative Study of Flux Limiters for Two-Dimensional Time-Accurate Calculations," in preparation.
14. P.L. Roe, J. Comp. Phys. Vol. 43, 1981.
15. H.C. Yee and P. Kutler, NASA TM-85845, August, 1983.
16. M. Mandella and D. Bershader, "Generation and Aerodynamic Interaction of Compressible Vortices," in preparation.
17. Y.J. Moon and H.C. Yee, "Numerical Simulation by TVD Schemes of Complex Shock Reflections from Airfoils at High Angle of Attack," in preparation.

A DESIGN AND TEST OF A NUMERICAL COUPLED LAND-ATMOSPHERE-OCEAN MODEL

Q.C. Zeng,   X.H. Zhang
C.G. Yuan and X.Z. Liang
(Institute of Atmospheric Physics,
Academia Sinica, Beijing, China)

I. Introduction

In order to simulate and predict long-term variability of the atmospheric and oceanic motions one has to adopt a coupled land-atmosphere-ocean model. Such a model should be able to describe dynamical and physical processes in each model components and interactions between different components, and the computational scheme should be stable for a very long time numerical integration and have satisfactory accuracy. Therefore, we make an emphasis on the formulation of finite-difference scheme free from false source of energy and computational modes and keeping the consistency of boundary conditions at the interfaces between the different mediums. We have already carried out a series of basic numerical experiments to test the model's performance and obtained satisfactory results.

II. Description of Model

The atmosphere and ocean both are baroclinic fluids laying on a rotating sphere with gravity. For describing their large scale motions quasi-static approximation in the vertical is accurate enough. After some transformations of coordinates and variables and some other suitable approximations the governing equations can be rewritten as follows.

The A-equation set:

$$\frac{\partial \vec{V}}{\partial t} + \mathcal{L}(\vec{V}) = -P\nabla_2\phi' - \frac{\Pi}{P^2}\nabla_2 p_s - (2\omega\cos\theta + \frac{V\lambda}{a}\text{ctg}\theta)\vec{k}\times\vec{V}$$

$$+ \mu\Delta_2\vec{V} + \frac{g^2}{P^2}\frac{\partial}{\partial\zeta}(\frac{1-\zeta}{RT}k_{a0}\frac{\partial\vec{V}}{\partial\zeta}),$$

$$\frac{\partial\Pi}{\partial t} + \mathcal{L}(\Pi) = \frac{c_e^2}{(1-\zeta)P}[-PW + (1-\zeta)(\frac{\partial p_s'}{\partial t} + \frac{1}{P}\vec{V}\cdot\nabla_2 p_s)]$$

$$\mu_1\Delta\Pi + \frac{g^2}{P^2}\frac{\partial}{\partial\zeta}(\frac{1-\zeta}{RT}k_{a1}\frac{\partial\Pi}{\partial\zeta}) + \frac{R}{c_p}Pq_a, \qquad (2.1)$$

$$\frac{\partial\rho_a M}{\partial t} + \mathcal{L}(\rho_a M) = \mu_2\Delta_2\rho_a M + \frac{g^2}{P^2}\frac{\partial}{\partial\zeta}(\frac{1-\zeta}{RT}k_{a2}\frac{\partial\rho_a M}{\partial\zeta}) + Pq_v,$$

$$\frac{p_s'}{\partial t} = -\nabla\cdot P\vec{V}_3,$$

$$\Pi = P(1-\zeta) \frac{\partial \phi'}{\partial \zeta}, \quad (\Pi \equiv PRT');$$
$$(0 \leq \zeta \leq 1)$$

the O-equation set:

$$\frac{\partial \vec{V}}{\partial t} + \mathcal{L}(\vec{V}) = -\frac{\phi}{\rho_0} \nabla_2 p' - \frac{\phi}{\rho_0} \zeta g \rho' \nabla_2 H - (2\omega\cos\theta + \frac{v_\lambda}{a} \text{ctg}\theta)\vec{K} \times \vec{V}$$
$$+ \nu \Delta_2 \vec{V} + \frac{g^2}{\rho_0 \phi^4} \frac{\partial}{\partial \zeta}(k_{00} \frac{\partial \vec{V}}{\partial \zeta}),$$

$$\frac{\partial \Xi}{\partial t} + \mathcal{L}(\Xi) = -\nu_1[W + \phi(1+\zeta)\frac{\partial Z_0}{\partial t} + \zeta \vec{V}\cdot\nabla_2 H]$$
$$+ \nu_1 \Delta_2 \Xi + \frac{g^2}{\rho \phi^4}\frac{\partial}{\partial \zeta}(k_{01}\frac{\partial \Xi}{\partial \zeta}) + q_0,$$

$$\frac{\partial S}{\partial t} + \mathcal{L}(S) = -\nu_2[W + \phi(1+\zeta)\frac{\partial Z_0}{\partial t} + \zeta \vec{V}\cdot\nabla_2 H]$$
$$+ \nu_2 \Delta_2 S + \frac{g^2}{\rho_0 \phi^4}\frac{\partial}{\partial \zeta}(k_{02}\frac{\partial S}{\partial \zeta}), \quad (2.2)$$

$$\frac{\partial Z_0}{\partial t} + \nabla\cdot\phi\vec{V}_3 = 0,$$

$$\frac{\partial p'}{\partial \zeta} = -\rho' g H,$$

$$\rho' = c_1 T' + c_2 s',$$

$$\Xi \equiv \phi T',$$

$$S \equiv \phi s';$$

$(-1 \leq \zeta \leq 0;$ for the ocean)

and the L-equation set:

$$\vec{v} = 0,$$

$$\frac{\partial \Xi}{\partial t} = \frac{\partial}{\partial \zeta}(\frac{k_{ss}}{\rho_{ss}}\frac{\partial \Xi}{\partial \zeta})\frac{g^2}{\phi^4},$$

$$\frac{\partial Q}{\partial t} = \frac{\partial}{\partial \zeta}(\frac{k_{ss}}{\rho_{ss}}\frac{\partial Q}{\partial \zeta})\frac{g}{\phi^4} + q_w, \quad (2.3)$$

$$\Xi \equiv \phi T',$$

$$Q \equiv \phi q',$$

$(-1 \leq \zeta \leq 0;$ for the land)

In these sets of equations $0 \leq \zeta \leq 1$ corresponds to the atmosphere, and $\zeta \equiv 1 - p/p_s$; $-1 \leq \zeta \leq 0$ to the ocean, and $\zeta \equiv (Z - Z_0)/(H + Z_0)$, or to the land, and $\zeta \equiv Z/H_s$, where p_s is the atmospheric pressure p at the bottom surface, Z_0 and $-H$ are the elevations of the top and the bottom of the ocean respectively, and H_s is the thickness of the active layer of the soil. $P \equiv \sqrt{p_s}$, $\phi \equiv \sqrt{gH}$ for the ocean or $\phi \equiv \sqrt{g H_s}$ for the land; $M = P_m$, m is the mixing ratio of water vapour in the atmosphere, T', q', p' and s' are the departures of temperature, moisture, oceanic water density and salinity from their standards respectively; c_e, v_1 and v_2 are the functions of stratifications; q_a, q_v, q_0, q_w are the source functions; $\vec{V}_3 = \vec{V} + \vec{K}W$, $\vec{V} = P\vec{v}$ (for the atmosphere) or $\Phi\vec{v}$ (for the ocean), $\vec{v} = \vec{\theta}^0 v_\theta + \vec{\lambda}^0 v_\lambda$, and $W = P\dot{\zeta}$ (for the atmosphere) or $\Phi\dot{\zeta}$ (for the ocean), $\nabla \equiv \nabla_2 + \vec{K}\partial/\partial\zeta$, ∇_2 and Δ_2 are the gradient operator and Laplacian on spherical surface respectively, and the operator $\mathcal{L}$ takes the following form

$$\mathcal{L}(F) \equiv \frac{\alpha_1}{2a\sin\theta}\left(\frac{\partial}{\partial\theta}v_\theta F \sin\theta + v_\theta \sin\theta \frac{\partial F}{\partial\theta}\right)$$
$$+ \frac{\alpha_2}{2a\sin\theta}\left(\frac{\partial}{\partial\lambda}v_\lambda F + v_\lambda \frac{\partial F}{\partial\lambda}\right) + \frac{\alpha_3}{2}\left(\frac{\partial}{\partial\zeta}\dot{\zeta}F + \dot{\zeta}\frac{\partial F}{\partial\zeta}\right) \quad (2.4)$$

where αS are some flexible coefficients for controlling the computational processes and the computational errors. Note that even the vector $(v_\theta, v_\lambda, \dot{\zeta})$ in $\mathcal{L}$ can also be flexibly chosen.

At the interfaces between the atmosphere and the ocean or the land a set of boundary conditions is determined by requiring the continuity of pressure and the fluxes of momentum, moisture, heat and salinity. Besides, the shore, which is the interface between the ocean and the land, is considered as a rigid wall. Since the formulas of boundary conditions are too complicated, we will not present them here.

III. Computational scheme

The governing equations, including all the boundary conditions, are then approximated by a set of finite-difference equations written at a three-dimensional staggered grid. The finite-difference scheme is designed carefully to conserve the total available energy of the numerical solutions under certain constrains. This means that there is no false computational source and dissipation of energy, and the scheme is suitable for the study of long-term variations.

In the design of such energy conservative (or quadraticly conservative in a general sense) scheme the main attention should be paid to constructing finite difference analogue of the operator $\mathcal{L}$ which possesses following very distinctive integrated property

$$\int_0^1 \int_S F \mathcal{L}(F) \, dS = 0, \quad (3.1)$$

where the integration domain S is extended over the whole spherical surface for the atmosphere or over that part of spherical surface occupied by the world ocean.

Figs. 1.a and 1.b show the arrangement of variables in the grids on spherical surface and along the vertical respectively. In our model finite difference analogue of the operator $\mathcal{L}$ is written as follows

$$\mathcal{L}(V_\lambda)_{i,j,k+\frac{1}{2}} = \frac{\alpha_1}{2a\sin\theta_j\delta\theta}\left(v_{\theta j+\frac{1}{2}}\sin\theta_{j+\frac{1}{2}}V_{\lambda j+1} - v_{\theta j-\frac{1}{2}}\sin\theta_{j-\frac{1}{2}}V_{\lambda j-1}\right)_{i,k+\frac{1}{2}}$$
$$+ \frac{\alpha_2}{2a\sin\theta_j\delta\lambda}\left(v_{\lambda i+\frac{1}{2}}V_{\lambda i+1} - v_{\lambda i-\frac{1}{2}}V_{\lambda i-1}\right)_{j,k+\frac{1}{2}}$$
$$+ \frac{\alpha_3}{\delta\zeta}\left(\dot{\zeta}_{k+1}V_{\lambda k+\frac{3}{2}} - \dot{\zeta}_k V_{\lambda k-\frac{1}{2}}\right)_{i,j}, \quad (3.2)$$

where $\alpha_1, \alpha_2, \alpha_3$ are some flexible coefficients for controlling the truncational errors. It is not difficult to prove that

$$\sum_i \sum_j \sum_k (V_\lambda)_{ijk+\frac{1}{2}} \mathcal{L}(V_\lambda)_{ijk+\frac{1}{2}} a^2 \sin\theta_j \delta\theta\delta\lambda\delta\zeta = 0. \qquad (3.3)$$

This is the finite difference analogue of formula (3.1).

Note that the mass conservation must be also satisfied during the numerical integration. This is expressed by the following formula

$$\sum_i \sum_j \sum_k [\ \{\ (PV_\lambda)_{i+1,j,k+\frac{1}{2}} - (PV_\lambda)_{i-1,j,k+\frac{1}{2}}\ \}$$
$$+ \{(PV_\theta \sin\theta)_{i+\frac{1}{2},j+\frac{1}{2},k+\frac{1}{2}} - (PV_\theta \sin\theta)_{i+\frac{1}{2},j-\frac{1}{2},k+\frac{1}{2}}\ \}]\ \delta\theta\delta\lambda\delta\zeta = 0, \quad (3.4)$$

for the atmosphere. The formula for the ocean is the same as (3.4) but the summation is extended only for the area occupied by the ocean.

IV. R-H Wave Experiments

In order to test the computational stability and accuracy of the finite-difference scheme we have performed a series of numerical experiments using the atmospheric model with linear and nonlinear Rossby-Haurwitz waves as initial conditions. Fig. 2 shows the result of an 100-day's numerical integration (19200 time steps) with a nonlinear baroclinic Haurwitz wave, which is a travelling wave solution to the primitive equations. During the long-term time integration the wave keeps its shape very well. The computed phase-velocity averaged over the 100 days is $7.38°$ per day, which is just $0.32°$ per day less than the theoretical value of the phase speed.

V. Simulation of the Climate

Another test of the model capability is a climate simulation by taking all the physical processes in the atmosphere and the land into account but with a prescribed sea surface temperature. Fig.3 shows the simulated 500 mb monthly mean geopotential height for January. Fig.4 shows the zonally averaged monthly mean surface temperature simulated and observed. The simulated fields are realistic.

IV. Simulation of Oceanographic Circulation

The third test of the model capability is a simulation of oceanographic circulation by taking the wind stress observed or simulated by using a comprehensive atmospheric model as the forcing on the ocean surface. The ocean initially is motionless, then a circulation is driven by the wind stress. Fig. 5 shows the 45th day oceanographic circulation driven by a simulated atmospheric motion in a barotropic ocean. The large scale circulations are very realistic and quasi-stationary.

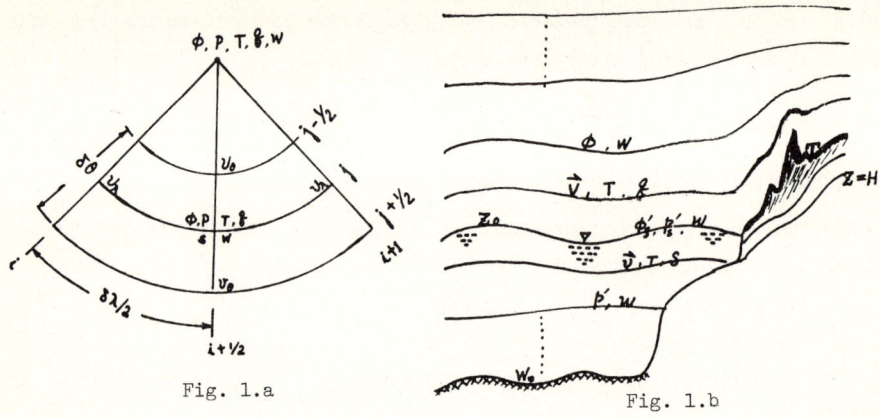

Fig. 1.a

Fig. 1.b

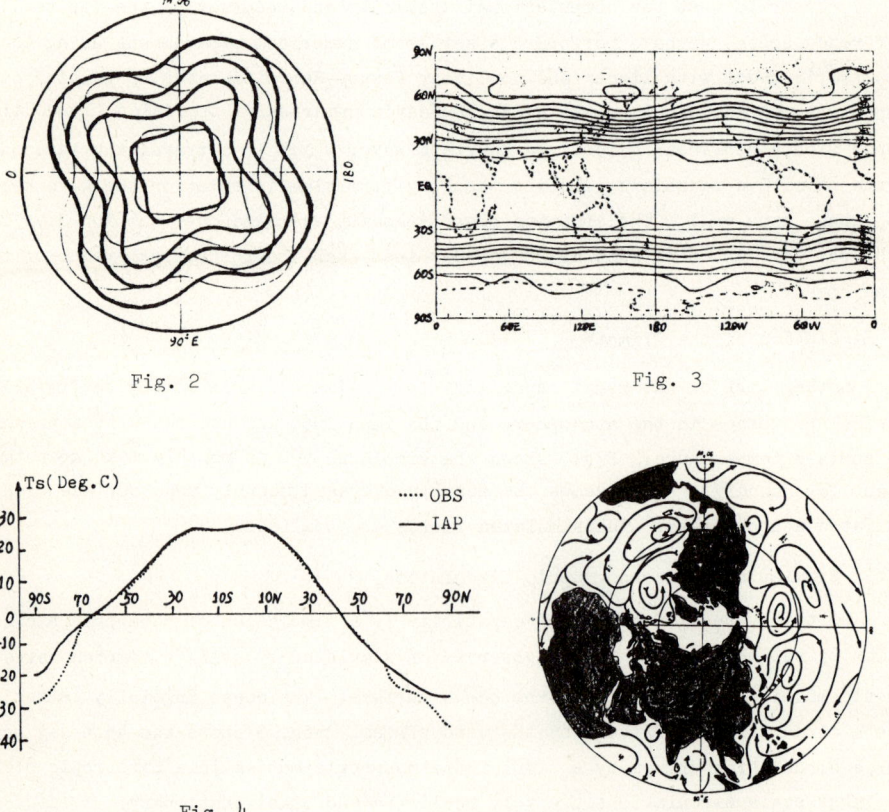

Fig. 2

Fig. 3

Fig. 4

Fig. 5

A MIXED ANTIDISSIPATIVE METHOD

SOLVING THREE DIMENSIONAL SEPARATED FLOW

H.X. Zhang and M. Zheng
China Aerodynamics Research and
Development Center
Mianyang, Sichuan, China

Introduction

When viscous flow fields are computed using the difference method, the following three demands must be satisfied: The first, in order to calculate exactly the viscous effects, the numerical viscous term (or additional artificial viscous term) must be much less than the real physical viscous term. Because the real physical viscous term is very small comparing with the numerical viscous term of the first order scheme when the Reynolds number is very large, the second order scheme or high order scheme must be used in the viscous region. In this paper, the second order scheme is used for simplicity. The second, the difference scheme should be able to capture the shock wave automatically if there is a shock wave in the flow field. Because the oscillations occur near the shock wave when the second order and third order schemes are used, the first order or fourth order scheme, which can capture the shock wave smoothly or have very small oscillations near the shock wave, can be used. For simplicity, the first order scheme is used in this paper. The third, the stable time step must be large for saving computational time, particularly when the mesh size is small. Therefore, it is suitable to use implicit scheme which is unconditional stable.

According to the above demands, the satisfying difference scheme L is

$$L = (1 - \theta)L_2 + \theta L_1 \qquad (1.1)$$

where L_1 and L_2 express first order and second order difference schemes respectively; θ is an automatically adjustable switch function, $\theta = 1$ near the shock wave, and $\theta = 0$ in the viscous region.

Obviously, equation (1.1) is a mixed scheme. In this scheme, the second order scheme L_2 can be given using antidissipative method[1], so the method given in this paper can be known as mixed antidissipative implicit method.

Using this method, the three dimensional separated flow over a blunt fin mounted on a flat plate is solved. The results are satisfactory.

Computational Method

1. One dimensional flow

For simplicity, let us study one dimensional flow at first. The Navier-Stokes equations are:

$$\frac{\partial U}{\partial t} + \frac{\partial F}{\partial x} + \frac{\partial Fv}{\partial x} = 0 \qquad (2.1)$$

where U is vector function, and $F=F(U)$, $Fv=Fv(U_x)$. In order to capture the shock wave smoothly, we can use the first order scheme with positive dissipative term

L_1:
$$U_i^{n+1} = U_i^n - \frac{\Delta t}{4\Delta x}[(F_{i+1}^n - F_{i-1}^n) + (F_{i+1}^{n+1} - F_{i-1}^{n+1})]$$
$$- \frac{\Delta t}{2\Delta x}[(Fv_{i+\frac{1}{2}}^n - Fv_{i-\frac{1}{2}}^n) + (Fv_{i+\frac{1}{2}}^{n+1} - Fv_{i-\frac{1}{2}}^{n+1})] \qquad (2.2)$$
$$+ \frac{1}{2}Q_x(U_{i+1}^n - 2U_i^n + U_{i-1}^n)$$

where Q_x is the coefficient of second order dissipation. Using the antidissipative method, i.e. subtracting dissipative term from the first order scheme, we can obtain the second order scheme L_2:

$$
\begin{aligned}
U_i^{n+1} = U_i^n &- \frac{\Delta t}{4\Delta x}[(F_{i+1}^n - F_{i-1}^n) + (F_{i+1}^{n+1} - F_{i-1}^{n+1})] \\
&- \frac{\Delta t}{2\Delta x}[(Fv_{i+\frac{1}{2}}^n - Fv_{i-\frac{1}{2}}^n) + (Fv_{i+\frac{1}{2}}^{n+1} - Fv_{i-\frac{1}{2}}^{n+1})] \\
&+ \frac{1}{2}Q_x(U_{i+1}^{n+1} - 2U_i^{n+1} + U_{i-1}^{n+1}) - \frac{1}{2}Q_x(U_{i+1}^n - 2U_i^n + U_{i-1}^n)
\end{aligned} \quad (2.3)
$$

Substituting equations (2.2) and (2.3) into equation (1.1), the mixed implicit scheme is

$$
\begin{aligned}
\delta U_i^{n+1} &+ \frac{\Delta t}{4\Delta x}(\delta F_{i+1}^{n+1} - \delta F_{i-1}^{n+1}) + \frac{\Delta t}{2\Delta x}(\delta Fv_{i+\frac{1}{2}}^{n+1} - \delta Fv_{i-\frac{1}{2}}^{n+1}) \\
&- \frac{1}{2}Q_x(1-\theta)(\delta U_{i+1}^{n+1} - 2\delta U_i^{n+1} + \delta U_{i-1}^{n+1}) \\
&= -\frac{\Delta t}{2\Delta x}(F_{i+1}^n - F_{i-1}^n) - \frac{\Delta t}{\Delta x}(Fv_{i+\frac{1}{2}}^n - Fv_{i-\frac{1}{2}}^n) \\
&+ \frac{1}{2}Q_x\theta(U_{i+1}^n - 2U_i^n + U_{i-1}^n)
\end{aligned} \quad (2.4)
$$

Using the relations between F and U, Fv and U_x, we obtain

$$
\begin{aligned}
\delta F^{n+1} &= A^n \delta U^{n+1} \\
\delta Fv^{n+1} &= -D^n \delta U_x^{n+1}
\end{aligned} \quad (2.5)
$$

where A and D are Jacobian matrices for $F(U)$ and $Fv(U_x)$ respectively, and

$$\delta U^{n+1} = U^{n+1} - U^n \quad (2.6)$$

Substituting equations (2.5) and (2.6) into equation (2.4), we obtain

$$
\begin{aligned}
\delta U_i^{n+1} &+ \frac{\Delta t}{4\Delta x}(A_{i+1}^n \delta U_{i+1}^{n+1} - A_{i-1}^n \delta U_{i-1}^{n+1}) - \frac{\Delta t}{2\Delta x^2}[D_{i+\frac{1}{2}}^n \delta U_{i+1}^{n+1} \\
&- (D_{i+\frac{1}{2}}^n + D_{i-\frac{1}{2}}^n)\delta U_i^{n+1} + D_{i-\frac{1}{2}}^n \delta U_{i-1}^{n+1}] - \mathcal{E}_{x2}(\delta U_{i+1}^{n+1} - 2\delta U_i^{n+1} + \delta U_{i-1}^{n+1}) \\
&= -\frac{\Delta t}{2\Delta x}(F_{i+1}^n - F_{i-1}^n) - \frac{\Delta t}{\Delta x}(Fv_{i+\frac{1}{2}}^n - Fv_{i-\frac{1}{2}}^n) \\
&+ \mathcal{E}_{x1}(U_{i+1}^n - 2U_i^n + U_{i-1}^n)
\end{aligned} \quad (2.7)
$$

where $\mathcal{E}_{x1} = \frac{1}{2}Q_x\theta$, $\mathcal{E}_{x2} = \frac{1}{2}Q_x(1-\theta)$, and from ref.[2] the expression of θ is

$$\theta = \frac{|P_{i+1}^n - 2P_i^n + P_{i-1}^n|}{|P_{i+1}^n + 2P_i^n + P_{i-1}^n|}$$

or

$$\theta = \frac{||P_{i+1}^n - P_i^n| - |P_i^n - P_{i-1}^n||}{|P_{i+1}^n - P_i^n| + |P_i^n - P_{i-1}^n|} \quad (2.8)$$

We have proved that the above difference scheme (2.7) is unconditional stable if equation (2.1) is linear.

Similarly, setting out from other first order explicit schemes with positive dissipative terms, one can also establish the corresponding second order explicit schemes by means of the antidissipative method.

If the implicit scheme is used in the fine mesh region near the wall and the explicit scheme is used in the coarse mesh region on the outside, the larger time step can be adopted.

2. Three dimensional flow

It can be shown that the time-split method, ADI method and factorization method are equivalent to each other in second order accuracy, we use the time-split method to solve three dimensional NS equations. The three dimensional NS equations are

$$\frac{\partial U}{\partial t} + \frac{\partial F}{\partial \xi} + \frac{\partial G}{\partial \eta} + \frac{\partial H}{\partial \zeta} + \frac{\partial Fv}{\partial \xi} + \frac{\partial Gv}{\partial \eta} + \frac{\partial Hv}{\partial \zeta} = 0 \qquad (2.9)$$

where $Fv=Fv(U_\xi)$, $Gv=Gv(U_\xi, U_\eta)$, $Hv=Hv(U_\xi, U_\eta, U_\zeta)$.

According to the time-split method, solving equation (2.9) is equivalent to solving following three equations:

$$\frac{\partial U}{\partial t} + \frac{\partial F}{\partial \xi} + \frac{\partial Fv}{\partial \xi} = 0 \qquad (2.10)$$

$$\frac{\partial U}{\partial t} + \frac{\partial G}{\partial \eta} + \frac{\partial Gv}{\partial \eta} = 0 \qquad (2.11)$$

$$\frac{\partial U}{\partial t} + \frac{\partial H}{\partial \zeta} + \frac{\partial Hv}{\partial \zeta} = 0 \qquad (2.12)$$

Using the above difference scheme (2.7) for one dimensional flow, the difference scheme solving equation (2.10) can be given, then U_ξ in Gv can be calculated, and the difference scheme solving equation (2.11) can be given. After that, U_η, U_ξ in Hv can be calculated, and the difference scheme solving equation (2.12) can be given.

Simulation of Blunt Fin Induced Shock Wave and Turbulent Boundary Layer Interaction

Let us study three dimensional separated flow over the blunt fin mounted on a flat plate (fig. 1). When supersonic or hypersonic flow passes over th blunt fin, the fin bow shock causes the boundary layer to separate from the surface ahead of the fin, resulting in a separated flow region composed of horseshoe vortices near the surface and a lambda-type shock pattern ahead of the fin. The flow phenomena of this problem are quite complex.

1. Mesh system

For simplicity, the flow is assumed to be symmetrical with respect to the center plane of the fin, hence only half of the flow is calculated. Fig. 2 shows a mesh system (30x29x29), where the I-direction corresopnds to the coordinate ξ along the fin. the J-direction corresponds to the coordinate η which is outward from the fin, the K-direction corresponds to the coordinate ζ which is normal to the flat plate. A fine mesh near the wall is required for an adequate resolution of the viscous effects.

2. Boundary conditions and initial condition

The fin is assumed infinite in height and length, so zero gradient boundary condition are imposed at the outer boundaries in corresponding directions. On the plane of symmetry, the symmetrical conditions are imposed. On the wall, the no-slip condition is applied. The wall is assumed to be adiabatic.

The outer boundary of J=Jmax is set far away enough from the fin to avoid any influence on the interaction. Here, the profile of U can be given using the boundary layer profile on the flat plate along the outer boundary. In addition, as for the initial condition, we can take it as uniform field.

3. Turbulence model

The flow is assumed to be turbulent. The above system of NS equations is valid for turbulent flow, as well as laminar flow, replacing the laminar viscosity μ_L

and μ_L/Pr_L with turbulent viscosity ($\mu_L+\mu_t$) and ($\mu_L/Pr_L+\mu_t/Pr_t$) respectively. Sutherland's formula is used to evaluate the laminar viscosity μ_L. A two layer turbulent model developed by Baldwin and Lomax is applied to evaluate the turbulent viscosity μ_t. The laminar Prandtl number is assumed to be 0.72 for air, and Pr_t=0.90.

4. Results

The flow to be simulated is at free stream conditions M_∞=2.95, T_∞=98.33°K and Re_D= 5x10^5 based on the diameter D of the blunt fin. The incoming boundary layer thickness =0.18D.

The results show that the fin bow shock causes the boundary layer to separate from the surface, resulting in a separated flow region composed of primary horseshoe vortex, secondary vortex and lambda-type shock ahead of the fin. Fig.3 shows the streamlines on the plane of symmetry. Fig.4 is a magnified streamlines on the plane of symmetry near the corner. It is clear to exist a primary horseshoe vortex and a small secondary vortex. All calculated flow features —— such as surface pressure, horseshoe vortices, " oil flow " on the surface and so on —— are satisfactorily in agreement with the experiment [3].

Brief Conclusion

From the above study, the following conclusions can be given:
1. When viscous flowfields with shock wave are computed, the mixed antidissipative implicit method given in this paper is suitable. The antidissipative method for establishing the second order scheme is universal.

2. The computational results of three dimensional separated flow over the blunt fin mounted on a flat plate show that all main features of the flow can be well simulated.

References

1. Zhang H.X., et.al, Applied Mathematics and Mechanics, 4, 1, 1983.
2. Zhang H.X., et.al, Mechanica sinica (China), 4, 1981.
3. Dolling D.S., et.al, AIAA J. 20, 12, 1982.

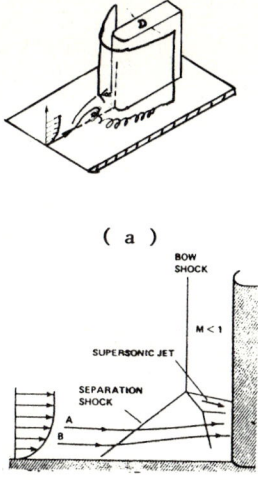

Fig. 1 Blunt fin on a flat plate and a simple sketch on the plane of symmetry.

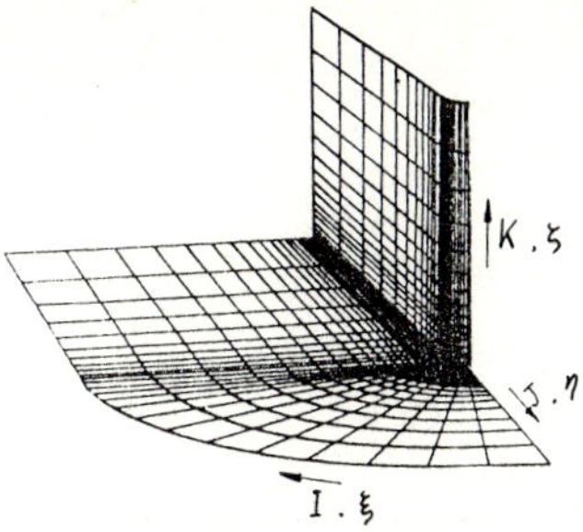

Fig. 2 Mesh distribution on the blunt fin and the flat plate

Fig. 3 The streamlines on the plane of symmetry

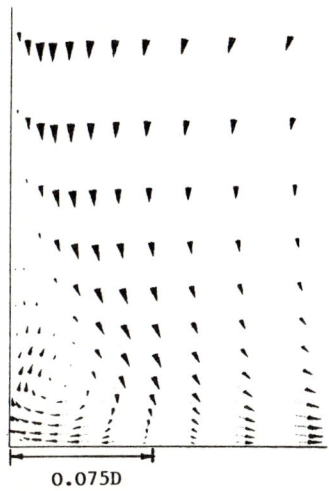

Fig. 4 The magnified streamlines on the plane of symmetry near the corner.

Pointwise Finite Element Method
and Its Applications to Compressible Flows

J. Zhang

Department of Mathematics
Xian Jiaotong University, Xian, China

It is known to all that in traditional finite element method (TFEM) the form of the coefficient matrix of approximate algebra formulae entirely depends on the form of the differential equations approximated. I call this phenomenon the "Form Dependence". Form Dependence might cause great trouble in programming software package for the coefficient matrix types corresponding to different differential equations included might be too many.

The main purpose of pointwise finite element method (PFEM) is to avoid the Form Dependence of TFEM. The basic difference between TFEM and PFEM is that in TFEM, what is approximated is directly the differential equation as a whole—its equivalent integral form, that is why Form Dependence exists, while in PFEM, the finite element interpolants are used to construct approximate differential operators and then establish the approximate formulas simply by substituting the operators into the differential equation. In other words, PFEM is designed first to be a method of approximating derivatives. Such an idea makes PFEM at least formally most generalized, just like finite difference method, because in form, any differential equation is a combination of the solution function and its derivatives of different order.

Another feature of PFEM is that it changes the average element matrix assembly of TFEM into a weighted one. If there were no such a concept, PFEM's generality would only be theoretical for it would have no means to make the approximate equation suit the type of continuous problem.

The basis of PFEM are the following two theorems, whose detailed proof can be found in [1].

<u>Theorem 1</u>. If the finite element interpolation basis $\{\Phi_\beta^{(\sigma)}, |\sigma| \leq r\}$ is of degree $k-1$ and uniformity to order q, and is orthogonal to another function basis $\{\Psi_{(\gamma)}^\alpha, |\gamma| \leq r\}$, then the matrix $[B_{\beta\gamma\theta}^{\alpha\sigma}]$ can be an approximate differential operator for smooth function $u(x)$ with the following estimation

$$|B_{\beta\gamma\theta}^{\alpha\sigma} u_\sigma^\beta - [D_\theta u(x)]_\gamma^\alpha| \leq C \cdot h^{k-|\theta|} \max_{\substack{x \in \Omega \\ |\theta_0|=k}} |D_{\theta_0} u(x)| \qquad (1)$$

where $D_\theta = \dfrac{\partial^{\theta_1+\theta_2+\cdots+\theta_n}}{\partial x_1^{\theta_1} \partial x_2^{\theta_2} \cdots \partial x_n^{\theta_n}}$, $h = \max\limits_{e} \text{diam}(\Omega_e)$, $u_\sigma^\beta = D_\sigma u(x)\big|_{x=x^\beta}$,

and $B_{\beta\gamma\theta}^{\alpha\sigma} = \int_\Omega \Psi_{(\gamma)}^\alpha D_\theta \Phi_\beta^{(\sigma)} d\Omega / \int_\Omega \Psi_{(\gamma)}^\alpha \Phi_\alpha^{(\gamma)} d\Omega$ (with no sum on α and γ) (2)

$$|\theta| = \theta_1 + \theta_2 + \cdots + \theta_n \leq q \qquad (3)$$

$\{\Psi_{(\gamma)}^\alpha, |\gamma| \leq r\}$ can be constructed locally on element with the element-orthogonal condition

$$\int_{\Omega_e} \psi_{(\gamma),e}^\ell \varphi_{m,e}^\sigma d\Omega = C \cdot \delta_m^\ell \cdot \delta_{\gamma_1}^{\sigma_1} \cdot \delta_{\gamma_2}^{\sigma_2} \cdots \delta_{\gamma_n}^{\sigma_n} \qquad (4)$$

and the weighted assembly

$$\Psi(x)_{(\gamma)}^\alpha = \sum_{i=1}^{M(x^\alpha)} W_{e_i, m_e} \psi(x)_{(\gamma), e_i}^{m_e} / M(x), \quad \delta \alpha \longleftrightarrow m_e \qquad (5)$$

where $\{\psi_{(\gamma)}^\ell, |\gamma| \leq r\}$, $\{\varphi_m^{(\sigma)}, |\sigma| \leq r\}$, $M(x)$ and W_{e_i, m_e} are respectively the shape functions of $\{\Psi_{(\gamma)}^\alpha, |\gamma| \leq r\}$ and $\{\Phi_\beta^{(\sigma)}, |\sigma| \leq r\}$, the coincidence degree at x of different elements, the weight coefficient of element Ω_e to its node m_e. We call $\{W_{e_i, m_e}\}$ the scheme parameters for the fact that their values can be adjusted to controll the numerical domains of the approximate scheme. It should be emphasized that it is unnecessary for PFEM to make the choice of schemes a priori as the finite difference method does. PFEM software will automatically run in accordance with the input values of scheme parameters.

Having this theorem, we have already been able to realize the idea of PFEM, but this means we must calculate a series of approximate differential operators corresponding to the derivatives of different orders appeared in the differential equation to be solved. And also the restriction (3) on element type is too strict. Hence the following one is needed.

<u>Theorem 2</u>. If the finite element interpolation basis $\{\phi_\beta^{(\sigma)}, |\sigma| \leq r\}$ is of degree k-1 and uniformity to order one or more, which the function basis $\{\psi_{(\gamma)}^\alpha, |\gamma| \leq r\}$ is orthogonal to, then for smooth function u(x), first order approximate differential operators $[BX1_\beta^\alpha], [BX2_\beta^\alpha], \cdots [BXN_\beta^\alpha]$ can be constructed with the accuracy

$$\left| BXI_\beta^\alpha u^\beta - \frac{\partial}{\partial x_i} u(x) \right|_{x=x^\alpha} \leq C \cdot h^{k-1} \max_{\substack{x \in \Omega \\ |\theta_0|=k}} \left| D_{\theta_0} u(x) \right|$$

where $BXI_\beta^\alpha = \int_\Omega \Psi_{(0)}^\alpha \dfrac{\partial}{\partial x_i} \Phi_\beta^{(0)} d\Omega / \int_\Omega \Psi_{(0)}^\alpha \Phi_\alpha^{(0)} d\Omega$ (with no sum on α)

Thus, the derivative of any order can be approximated by the sequent action on u(x) of $[BX1_\beta^\alpha], [BX2_\beta^\alpha], \cdots [BXN_\beta^\alpha]$ without any decrease of accuracy.

Theorem 2 indicates that PFEM can solve any differential equation with same finite element matrix structure. Although the form of the coefficient matrix of the final approximate formulas still depends on that of the differential equation to be solved, its process of finite element production has been standardized. Another advantage of PFEM implied in the theorem 2 is that PFEM has almost complete freedom in choosing element type regardless of how many the differential equation's order is, because from Theorem 1,first order approximate differential operator can be constructed with any non-piecewise-constant finite element basis.

What the above theorems have solved is the consistency. In practical numerical calculation, the crux is the stability. Although the principle of finding stable scheme of PFEM,which is to adjust the scheme parameters, has been included in these theorems, how should we use such a principle? I suggest a procedure: search a stable finite difference scheme first, whose stability is usually easier to analyse than that of finite element mode, and then, choose the scheme parameters of PFEM to imitate it. As a great number of excellent finite difference schemes have been developed, working in such a procedure will make us transform in passing this resource into a more applicable finite element mode.

The partial differential equations solved here are the Navier-Stokes written in Conservation Law form for a Cartesian coordinate system. The viscosity coefficients and the thermal conductivity are all assumed temperature-dependent laminar values for the viscous flows treated here. For inviscid transonic flows with shock waves, these terms are still used, but with artificial viscosity coefficients, to facilitate the capture of shocks.

The PFEM software has run with scheme parameters imitating many popular finite difference schemes, such as the Lax-Wendroff, the flux-upstream with stress-centre, the flux-upstream with pressure-downstream, etc.. Calculations using all these schemes show similar stability and convergence speed to the ones described by other writers using corresponding finite difference schemes. But here in this paper, only some results using the following finite element basis and the scheme parameters imitating MacCormack finite difference scheme ([4]) are reported.

$$\{\varphi_i\} = \begin{Bmatrix} (1-\eta_1)(1-\eta_2) \\ \eta_1(1-\eta_2) \\ \eta_1 \eta_2 \\ (1-\eta_1)\eta_2 \end{Bmatrix} \quad \text{and} \quad \{\psi_i \times |\Delta J|\} = \begin{Bmatrix} (2/3 - \eta_1)(2/3 - \eta_2) \\ (1/3 - \eta_1)(2/3 - \eta_2) \\ (1/3 - \eta_1)(1/3 - \eta_2) \\ (2/3 - \eta_1)(1/3 - \eta_2) \end{Bmatrix}$$

where ΔJ is the Jacobian matrix of the element transformation, and η_1, η_2 represent the local coordinates on the master element.

Figs. 1-2 show the inviscid numerical flow through a transonic steam turbine cascade of Von Karmann Institute. In Figure 2, it is evident that the shock captured by PFEM is steeper than that by Denton finite volume scheme. Perhaps this has something to do with the fact that the numerical diffusion of PFEM used in this calculation, just the laminar values, is weaker than that included in Denton method under the grid used. The blade-surface M distribution of either PFEM's or Denton's coincides poorly with experimental data. This might be caused by the fact that the Mach number is so high (≥ 1.42) before the shock that a sudden increase of boundary layer thickness, even separation, must have taken place.

Figs. 3-4 show the viscous numerical flow through a plate cascade. The Reynolds number is 3.8973×10^6. The grid shown in Figure 3 is too coarse to analyze the detailed viscous structure under such a high Reynolds number, but fine enough to calculate those bigger vortices which are what I want. The initial velocity field has a constant-norm distribution and piece-constant directions: parallel to the inlet flow and the blade surface respectively in the blade-upwind area and the rest. The calculated vortex developing process and final Karmann vortex shedding, which can be seen clearly after $t \geq 1600\Delta t$, are all very reasonable.

I believe that I have reached the following objectives.
- obtaining a most generalized numerical procedure by basing the approximation of differential equation on its most essential factor: derivative.
- making such a procedure practical by establishing the automatic scheme flexibility, the freedom in choosing element type and the standardization of producing finite element matrix.

Working together, these two aspects make PFEM especially suitable for programming large software package including widely different, e.g., in dimension, in type and in the physical phenomenon described, differential equations.

The scheme control principle of PFEM has also been developed into weighted residual finite element method and shows similar success in calculating flow problems.

REFERENCES

[1] J.Zhang, "Pointwise Finite Element Method", to appear as the thesis for Ph.D degree, Xi'an Jiaotong University.

[2] Spradley, L.W., J.F.Stalnaker and A.W.Ratliff, "Computation of Three-Dimensional Viscous Flows with Navier-Stokes Equations", AIAA paper 80-1348, 1980.

[3] Kawahara, M., H. Hirano and K. Tsubuta, "Selective Lumping Finite Element Method for Shallow Water flow", Int. J. Num. Meth. Engng. Vol.2, 89-112, 1982.

[4] MacCormack, R.W., "The Effect of Viscosity in Hypervelocity Impact Cratering", AIAA paper 69-354, 1969.

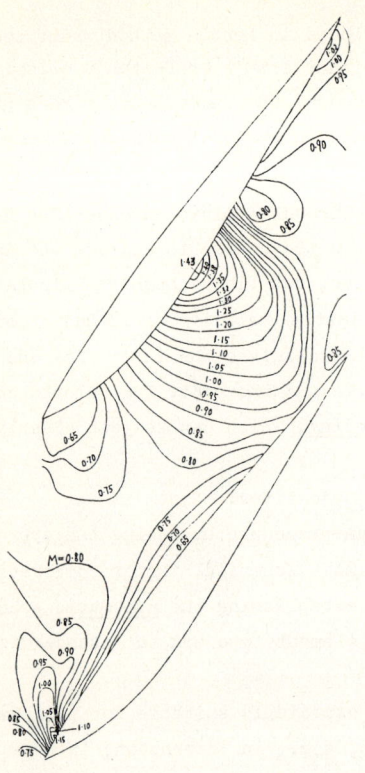
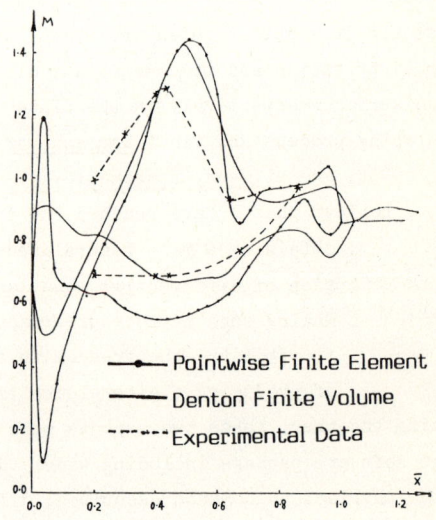

Figure 1. 0.65-1.5 isomact lines in VKI cascade

Figure 2. M distribution on VKI blade surface

Figure 3. Grid for a plate cascade attached by a 45° inflow

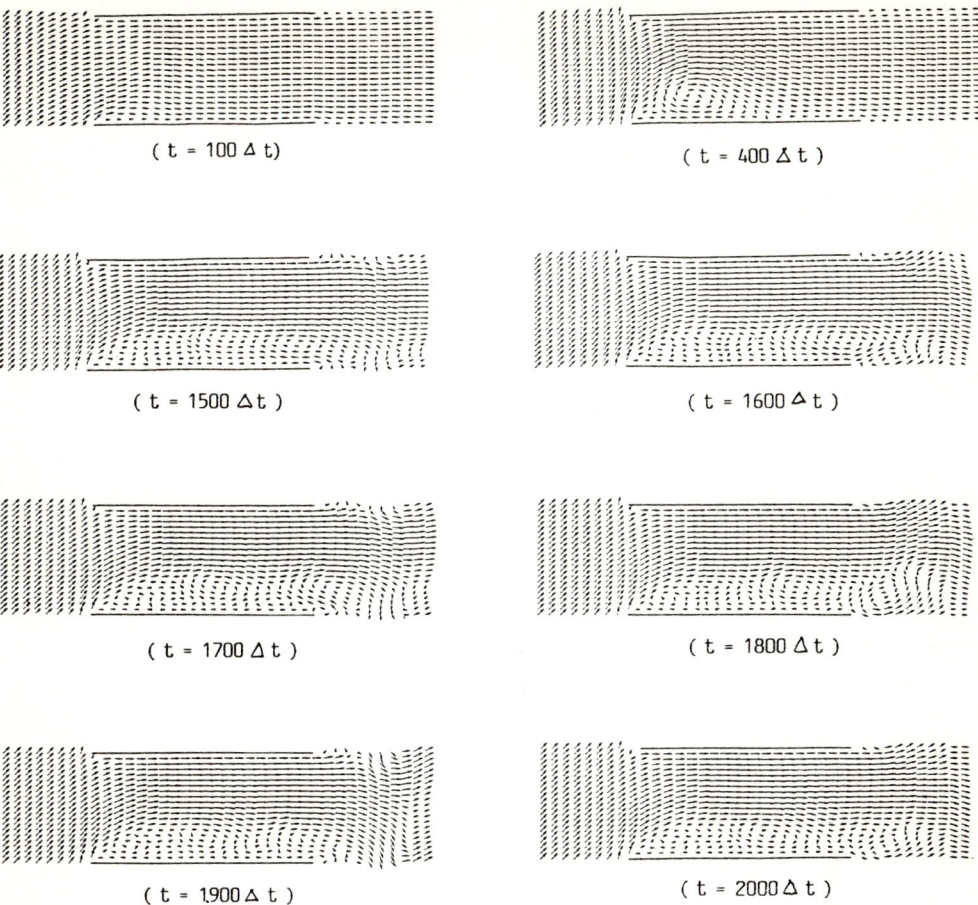

Figure 4. The velocity field development of the problem shown in Fig.3

UNSTEADY TRANSONIC FLOWS AROUND OSCILLATING WINGS

J.B. Zhang

(China Aerodynamic Research and Development Center, Mianyang, Sichuan, China)

INTRODUCTION

As well known, main difficulties in calculating unsteady transonic flows are: (a) the non-linearity of the governing equation of the transonic flow; (b) the unsteady shock wave in the flow field. Since 1970s, remarkable progresses have been making on computational methods of unsteady transonic flows (Refs.1-5). We attach importance to two factors in selecting or developing computational methods of unsteady transonic flows, i.e. (a) nonlinearities, including that caused by the shock wave movement, (b) as high as possible computational efficiency. In our recent works two finite difference schemes for both 2-D and 3-D cases are used in the calculation. One is a conventional ADI scheme, and the other is our new variant of factorization scheme. The fundamental equation and numerical scheme will be introduced in the following sections and some numerical results to show the satisfactory performance of our schemes for test problems will also be presented.

FUNDAMENTAL EQUATION AND BOUNDARY CONDITIONS

Based on the potential assumption, the conservation form of the low frequency transonic small disturbance equation can be written as:

$$-\frac{\partial f_0}{\partial t} + \frac{\partial f_1}{\partial x} + \frac{\partial f_2}{\partial y} + \frac{\partial f_3}{\partial z} = 0, \quad (1)$$

where

$$f_0 = B\varphi_x \quad (2a)$$
$$f_1 = E\varphi_x + F\varphi_x^2 + G\varphi_y \quad (2b)$$
$$f_2 = \varphi_y + H\varphi_x\varphi_y \quad (2c)$$
$$f_3 = \varphi_z \quad (2d)$$

φ being the small disturbance potential and

$$B = 2kM^2 \quad (3a)$$
$$E = 1 - M^2. \quad (3b)$$

Spatial coordinates are non-dimensionized by chord length c; time t is non-dimensionized by $1/\omega$, where ω is angular frequency; $k = \omega c/U$ is reduced frequency, where U is the freestream velocity. If putting $F = -\frac{1}{2}(\gamma+1)M$, $G = H = 0$, the equation becomes its classical form. In order to capture shock wave correctly, a modified small disturbance theory is used, selecting the following NLR coeffcients:

$$F = -\tfrac{1}{2}(3-(2-\gamma)M^2)M^2 \qquad (4a)$$
$$G = -\tfrac{1}{2}M^2 \qquad (4b)$$
$$H = -M^2 \qquad (4c)$$

For rectangular wings one can assume G=H=0. For two-dimensional aerofoils, $\partial f_2/\partial y$ vanishes.

The corresponding pressure coefficient is

$$Cp = -2(\varphi_x + k\,\varphi_t) \qquad (5)$$

Boundary conditions for flow field are:

Far Upstream: $\quad \varphi = 0 \qquad\qquad (6a)$

Far Downstream: $\quad \varphi_x = 0 \qquad\qquad (6b)$

Far Spanwise: $\quad \varphi_y = 0 \qquad\qquad (6c)$

Far Above and Below: $\quad \varphi_z = 0 \qquad\qquad (6d)$

Wing Root: $\quad \varphi_y = 0 \qquad\qquad (6e)$

On the wing surface, we use the linearized unsteady boundary condition:

$$\varphi^{\pm} = f_x^{\pm} + k\,\varphi_t^{\pm}, \quad \text{on } z = 0 \text{ for } x_{LE} \leq x \leq x_{TE} \qquad (7)$$

For $z = 0$ and $x > x_{TE}$, the condition for the wake is

$$\Delta \varphi_x + \Delta(k\,\varphi_t) = 0 \qquad (8)$$

We choose a steady flow solution as an initial condition, i.e.

$$\varphi(x,y,z,0) = g(x,y,z) \qquad (9)$$

The above mentioned equation and initial boundary conditions form a well-posed problem.

NUMERICAL ALGORITHM

For clarity we present only numerical scheme for the 2-D flow problem. Scheme for 3-D case is straightforward. Equation (1) will be

$$\frac{\partial}{\partial t}(B\varphi_x) = \frac{\partial F}{\partial x} + \frac{\partial}{\partial z}(\varphi_z) \qquad (10)$$

where

$$F = -\frac{(\gamma^*+1)M^2}{2}\left[\frac{M^2-1}{(\gamma^*+1)M^2} + \varphi_x\right]^2 \qquad (11a)$$

$$\gamma^* = 2 - (2-\gamma)M^2 \qquad (11b)$$

Using trapezoidal formula for the time integral, Eq. (10) can be written as:

$$\frac{B\delta_x}{\Delta t}(\varphi^{n+1} - \varphi^n) = \tfrac{1}{2}\delta_x(F^{n+1} + F^n) + \tfrac{1}{2}\delta_{zz}(\varphi^{n+1} + \varphi^n) \qquad (12)$$

Two difference schemes are used in the numerical solution.

(1) ADI Scheme

x-sweep

$$\frac{B\overleftarrow{\delta}_x}{\Delta t}(\varphi^{n+\frac{1}{2}} - \varphi^n) = \frac{1}{2} D_x(F^{n+\frac{1}{2}} + F^n) + \delta_{zz}\varphi^n \tag{13a}$$

z-sweep

$$\frac{B\overleftarrow{\delta}_x}{\Delta t}(\varphi^{n+1} - \varphi^{n+\frac{1}{2}}) = \frac{1}{2} \delta_{zz}(\varphi^{n+1} - \varphi^n) \tag{13b}$$

where the upper subscript $n + \frac{1}{2}$ means intermediate values; D_x is a type-dependent mixed difference operator introduced by Murman and Cole, the central difference being used in subsonic regions and the backward difference being used in supersonic regions. $\overleftarrow{\delta}_x$ is a backward difference operator, this will be useful for the stability of the solution process.

(2) A New Factorization Scheme

If operator $B\overleftarrow{\delta}_x$ is applied to both sides of Eq.(12) an implicit factorization scheme may be written as:

$$(\frac{B\overleftarrow{\delta}_x}{\Delta t} - \frac{1}{2}\delta_x\delta_x S_1)(B\overleftarrow{\delta}_x - \frac{\Delta t}{2}\delta_{zz})(\varphi^{n+1}-\varphi^n)=B\overleftarrow{\delta}_x[\frac{1}{2}\delta_x(S_1+S_2)\delta_x + \delta_{zz}]\varphi^n \tag{14}$$

where

$$S_1 = 1-M^2 -(\gamma^* + 1)M^2\varphi_x; \quad S = 1-M^2$$

Its two step scheme can be easily written as:

x-sweep
$$(\frac{B\overleftarrow{\delta}_x}{\Delta t} - \frac{1}{2} D_x\delta_x)(\varphi^{n+\frac{1}{2}} - \varphi^n) = \frac{B\overleftarrow{\delta}_x}{2}[D_x(S_1+S_2)\delta_x + \delta_{zz}]\varphi^n \tag{15a}$$

z-sweep
$$B\overleftarrow{\delta}_x(\varphi^{n+1} - \varphi^n) - \frac{\Delta t}{2}\delta_{zz}(\varphi^{n+1}-\varphi^n) = \varphi^{n+\frac{1}{2}} - \varphi^n \tag{15b}$$

This scheme needs a specification of φ_x, φ_{xx}, φ_z, at $x = -\infty$. In the present calculation, on the basis of physical reason they are set to zeroes.

These schemes are non-iterative. The solution is advanced by the two-step procedure from time n to time level n+1. During x-sweep these would result in a quadri-diagonal matrix. In the z-sweep, these would result in a tri-diagonal matrix. Thomas algorithm is employed for solving these algebraic equations.

The extension of this finite difference scheme to flow aroung a 3-D wing gives also satisfactory results.

NUMERICAL TESTS

The calculation for NACA 64A 006 airfoil with a quarter-chord oscillating flap is carried out. The steady solution used as initial values for this test case is also calculated by the present method. Fig.1 shows the numerical results by our new scheme and its comparison with those by other methods. Our solution simulates the intermittent movement (type B) of the shock, which was observed from Tijdeman's experiment (Ref.6), and is in good agreement with other results.

Calculations were made for 3-D unsteady transonic flow around a rectangular wing.

The wing has circular-arc airfoil sections of thickness-to-chord ratio of 5%, and it was subjected to oscillatory motion in the first bending mode. In Fig.2 magnitudes and phase angles of the unsteady pressure jump obtained by present method, LTRAN3, experiment, and kenel-function method are plotted at 50% semispan station for M=0.9 and k=0.26. In general, the three sets of curves obtained by the nonlinear calculations are fairly consistent with the experimental results, and the peaks in pressure jumps occur at almost same locations for both nonlinear calculations and experimental results.

CONCLUSIONS

This report is a brief review of a part of our work on unsteady 2-D and 3-D transonic flow calculation in connection with flutter analysis: (1) An alternating direction implicit scheme is applied to the solution for the low frequency transonic small disturbance equation. (2) A new implicit factorization scheme is developed to solve the problem. From results of quite a few test cases and their comparisons with others, these methods are shown to be encouraging.

REFERENCES

[1] Magnus, R.J., and Yoshihara, H., Unsteady Transonic Flows Over An Airfoil, AIAA Journal Vol.13, Dec. 1975, pp. 1622-1628.
[2] Ballhaus, W.F., and Goorjian, P.M., Implicit Finite-Difference Computation of Unsteady Transonic Flow about Airfoil, AIAA Paper 77-205, 1977.
[3] Houwenk, R., and Van der Vooren, J., Results of an Improved Verson of LTRAN2 for Computing Unsteady Airloads on Airfoil Oscillating in Transonic Flow, AIAA Paper 79-1553, 1979.
[4] Borland, C., Rizzetta, D., and Yoshihara, H., Numerical Solution of Three-Dimensional Unsteady Transonic Flow Over Swept Wings, AIAA Paper 81-0329, 1981.
[5] Guruswamy, P., and Goorjian, P.M., Comparison Between Computations and Experimental Data in Unsteady Three-Dimensional Transonic Aerodynamics, Including Aeroelastic Applications, J. AIRCRAFT Vol. 21, No.5, 1984.
[6] Tijdeman, H.C., On the Motion of Shock Waves on an Airfoil with Oscillating Flap, Symposium Transsonicum II, Springer-Verlag, 1975 pp. 49-56.
[7] Lessing, H.C., et al., NASA TN D-344, 1960.

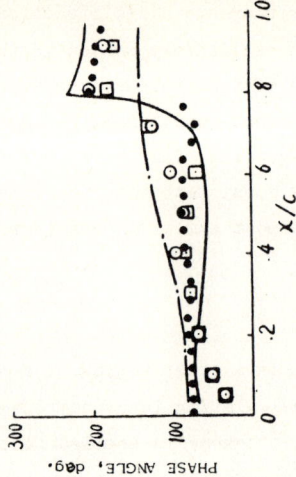
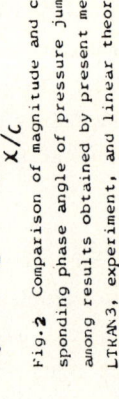

Fig.2 Comparison of magnitude and corresponding phase angle of pressure jumps among results obtained by present method, LTRAN3, experiment, and linear theory

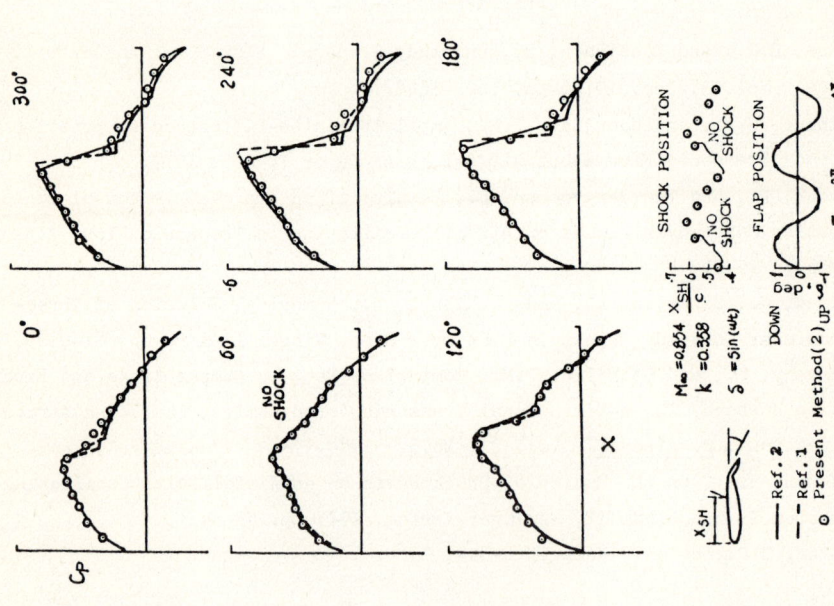

Fig.1 Unsteady Upper Surface Pressure Coefficient for An NACA 64A006 Airfoil with Oscillating Trailing-Edge Flap, Type B.

A LAGRANGIAN-EULERIAN PARTICLE MODEL FOR TURBULENT TWO-PHASE FLOWS WITH REACTING PARTICLES

L.X.Zhou and J. Zhang

(Department of Engineering Mechanics,
Tsinghua University, Beijing, China)

For gas-particle reacting two-phase flows numerical models with Lagrangian treatment of particle phases [1] [2] (PSIC method) can track the particle history and reduce the memory storage of computers. However, predicted particle velocities and concentration are difficult to be verified by experiments in which only Eulerian measurements can be made. On the other hand, models with unified Eulerian treatment of particle and gas phases (two-fluid model) [3] [4] [5] can easily account for the turbulent transport of particle mass and momentum, but can not identify the particle history effects, and therefore are difficult to be used in such problems as pulverized-coal combustion where particles may have different behaviors during different periods. The basic idea of the Lagrangian-Eulerian particle model proposed here is to consider the turbulent transport of particle mass and momentum by Eulerian conservation laws and to track the particle mass and temperature history along their paths (in Lagrangian coordinates) simultaneously.

The time-averaged conservation equations based on the concept of a multi-fluid model for particles of size group k are given as follows [6]

$$\frac{\partial \rho_k}{\partial t} + \frac{\partial}{\partial x_j}(\rho_k v_{kj}) = \frac{\partial}{\partial x_j}(\frac{\nu_k}{\sigma_k} m_k \frac{\partial n_k}{\partial x_j}) + S_k \tag{1}$$

$$\frac{\partial}{\partial t}(\rho_k v_{ki}) - \frac{\partial}{\partial t}(\frac{\nu_k}{\sigma_k} m_k \frac{\partial n_k}{\partial x_i}) + \frac{\partial}{\partial x_j}(\rho_k v_{kj} v_{ki}) = \frac{\rho_k}{\tau_{rk}}(v_i - v_{ki})$$

$$+ v_i S_k + \rho_k g_i + \frac{\partial}{\partial x_j}[\nu_k \rho_k (\frac{\partial v_{kj}}{\partial x_i} + \frac{\partial v_{ki}}{\partial x_j})]$$

$$+ \frac{\partial}{\partial x_j}[\frac{\nu_k}{\sigma_k} m_k (v_{ki} \frac{\partial n_k}{\partial x_j} + v_{kj} \frac{\partial n_k}{\partial x_i})] \tag{2}$$

where $\rho_k = n_k m_k$ -- apparent density of particle group k, n_k -- number density, m_k -- particle mass, $S_k = -n_k \dot{m}_k = -n_k \frac{dm_k}{dt}$ -- mass source due to particle mass change, v_k -- particle velocity, v -- gas velocity, g -- gravitational acceleration, ν_k -- turbulent viscosity of particle group k, $\sigma_k = \nu_k/D_k$, D_k -- particle diffusivity, $\nu_k = \nu_g(1 + \tau'_{rk}/\tau_T)^{-1}$, $\tau'_{rk} = \rho_{pk} d_k^2/(18\mu)$, $\tau_{rk} = \tau'_{rk}\frac{\exp(B_k)-1}{-B_k}(1+Re_k^{2/3}/6)^{-1}$, $Re_k = |\vec{v}-\vec{v}_k|d_k/\nu$, $\tau_T = \sqrt{\frac{3}{2}} c_\mu^{3/4} k/\varepsilon$, The particle mass and temperature history, as an example, in pulverized-coal combustion are described by the following equations [6]

$$\dot{m}_k = \dot{m}_w + \dot{m}_v + \dot{m}_h \tag{3}$$

$$\dot{m}_w = \pi d_k Nu D \rho \ln[1 + (Y_{w,p} - Y_{w,g})/(1 - Y_{w,p})] \tag{4}$$

$$Y_{w,p} = B_w \exp(-E_w/RT_k)$$

$$\dot{m}_v = m_c[\alpha_1 B_{v1} \exp(-E_{v1}/RT_k) + \alpha_2 B_{v2} \exp(-E_{v2}/RT_k)]$$

$$\frac{dm_c}{dt} = -m_c[B_{v1}\exp(-E_{v1}/RT_k) + B_{v2}\exp(-E_{v2}/RT_k)] \tag{5}$$

$$\dot{m}_k = \pi d_k Nu D \rho \ln[(\dot{m}_s/\dot{m}_k - Y_{s,g})/(\dot{m}_s/\dot{m}_k - Y_{s,p})]$$

$$\dot{m}_s = \pi d_k^2 \rho Y_{s,p} \sum_\ell [B_\ell \exp(-E_\ell/RT_k)]$$

$$\dot{m}_h = \sum \alpha_s \dot{m}_s \tag{6}$$

$$m_k c_k \frac{dT_k}{dt} = \pi d_k^2 \epsilon \sigma(T^4 - T_k^4) + m_k c_p(T-T_k)[\exp(\dot{m}_k c_p/(\pi d_k Nu \lambda)-1]^{-1}$$

$$- \dot{m}_w L_w - \dot{m}_{v1} \Delta h_1 - \dot{m}_{v2} \Delta h_2 + \sum_\ell \dot{m}_{h,\ell} Q_\ell \tag{7}$$

where $\dot{m}_k$, $\dot{m}_w$, $\dot{m}_v$, $\dot{m}_h$, $\dot{m}_s$, --- mass changing rate, moisture evaporation rate, char combustion rate and s - species consuming rate of coal particles respectively, m_c -- mass of raw coal (dry and ash free), Y_s -- mass fraction of s - species, α_1, α_2, -- devolatilization coefficients, L_w, Δh_1, Δh_2, Q_ℓ -- heating effects, B_w, E_w, B_{v1}, B_{v2}, E_{v1}, E_{v2}, B_1, B_2, B_3, E_1, E_2, E_3, -- kinetic constants, d_k --particle diameter, T_k -- particle temperature, T--gas temperature, ϵ --emissivity, σ -- Stefan-Boltzmann constant, C_p--gas specific heat, C_k--particle specific heat, subscripts p, g--particle surface and gas respectively. The conservation equations of gas phase based on the k - ϵ turbulence model and EBU-Arrhenius reaction model can be expressed as

$$\frac{\partial}{\partial t}(\rho \varphi) + \frac{\partial}{\partial x_j}(\rho v_j \varphi) = \frac{\partial}{\partial x_j}(\Gamma_\varphi \frac{\partial \varphi}{\partial x_j}) + S_\varphi + S_{p\varphi} \tag{8}$$

where φ denotes 1, v_i, k, ϵ, $\bar{f}$, g, Ys, h, $\bar{f}$-- mixture fraction, $g = \overline{f'^2}$, h--gas enthalpy, Γ_φ--transport coefficient of φ, S_φ-- source term of gas phase itself, being the same as that for single-phase flows, $S_{p\varphi} = Spm, (\Sigma n_k \dot{m}_k)$, v_i $Spm, - \Sigma \rho_k/\tau_{rk}(v_i - v_{ki})$, $\alpha_s Spm$, $\Sigma n_k Q_k + c_p T Spm$ for gas continuity, momentum, species and energy equations respectively, and $S_{p\varphi} = 0$ for other equations. The boundary conditions for gas phase are the same as for single-phase flows. Uniform inlet distributions, zero gradients and fully-developed flows are taken as the inlet, wall and outlet conditions for the particle phases. The basic features of the solution procedure (so-called LEAGAP Algorithm) are: Solving particle velocities and concentration from eq. (1) and (2); solving particle mass and temperature from eq.(3) to (7); Solving gas field from eq. (8) by using source terms $S_{p\varphi}$ both from Eulerian and Lagrangian predictions of particle phases. The SIMPLE algorithm [7] is used inside both the gas phase and particle phases. Besides, the iterations are made between particle phases and the gas phase. As an example, predictions are made for a dump combustor shown in Fig.1. Predicted axial velocity profiles and the size of recirculation zone for cold flow are in good agreement with experiments from [8] (Fig.2,3). There is an obvious velocity slip between the gas phase and particle phases (Fig.4). And the particle mass flux reaches its maximum value at the axis as in the simple jet, but with a stronger mixing rate (Fig.5). In case of gas-phase combustion the size of the wall recirculation zone is substantially reduced compared with that in cold flow (Fig.6). Predicted temperature field has the same feature as the experimental data from [9], but the actual turbulent mixing is stronger than predicted (Fig.7). Figs. 8-9 are predicted particle paths for a dump combustor with an annular jet. Figs.10 and 11 give the gas velocity, particle velocity and concentration distributions in case of coal combustion. All this results are plausible.

References

[1] A.S.Abaas, S.S.Kouss, F.C.Lockwood, 18th Symp. on Comb. pp1427, 1981.
[2] L.D.Smoot, P.J.Smith, Coal Combustion and Gasification, Pergamon, 1985.
[3] T.R.Blake et al, DOE/FE-1770-32, 1977.
[4] D.B.Spalding, ICDME, HTS/81/2, 1981.
[5] A.A.Mostafa, S.E.Elghobashi, Fall Meeting of the West. St. Sec. Comb. Inst., 1983.
[6] L.Zhou, A Lagrangian-Eulerian particle model for pulverized-coal combustion, Research Rep., Dep. Eng. Mech., Tsinghua Univ. (In Chinese), 1984.
[7] S.V.Patanker, Numerical Heat Transfer and Fluid Flow, Hemisphere, 1980.
[8] L.E.Moon, G.Rudinger, G.R.Salter, AFOSR-TR-75-1648, 1975.
[9] El Banhawy et al, Combustion and Flame, V.50, pp153-165, 1983.

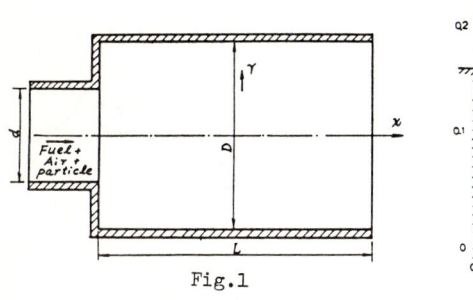

Fig.1

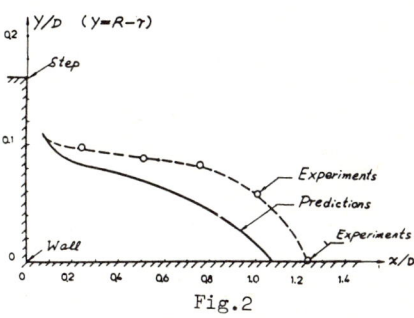

Fig.2

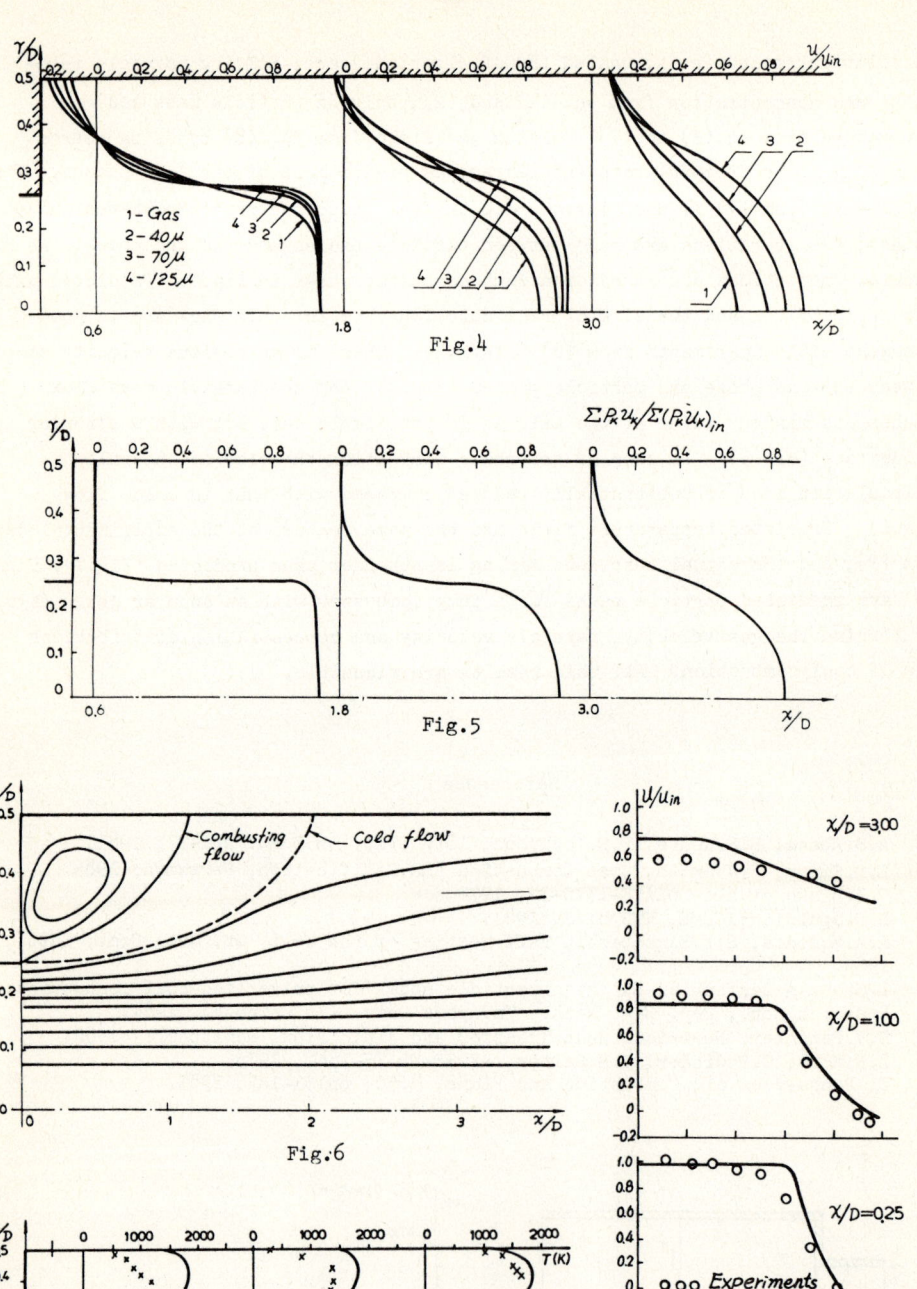

Fig.4

Fig.5

Fig.6

Fig.7

Fig.3

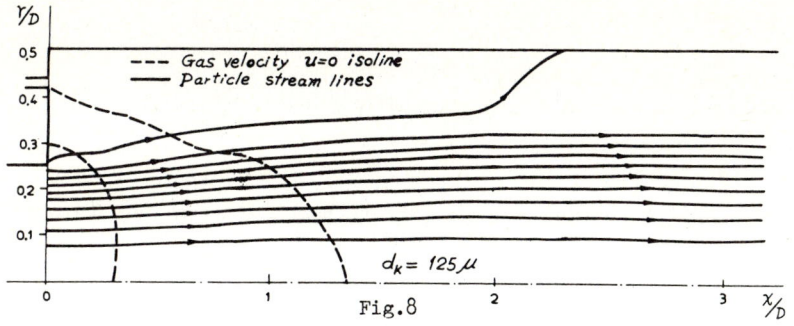

Fig. 8

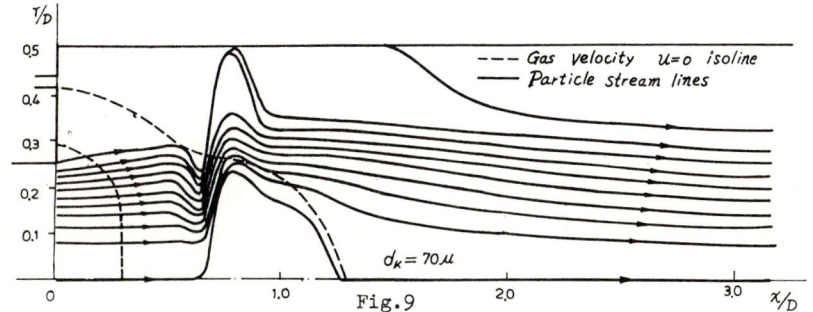

Fig. 9

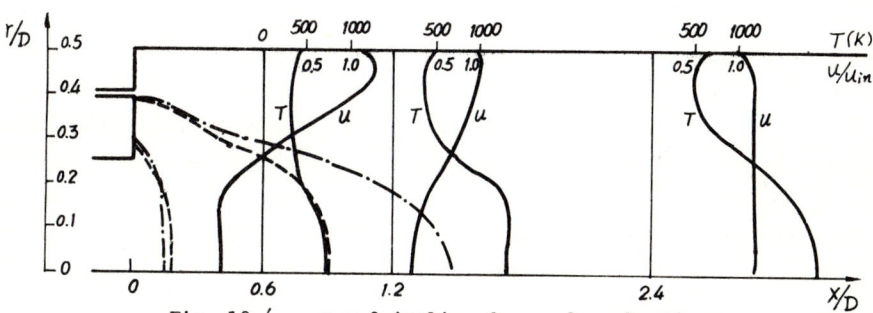

Fig. 10 (--- $u = 0$ isoline for coal combustion,
—·— $u = 0$ isoline for cold gas flow)

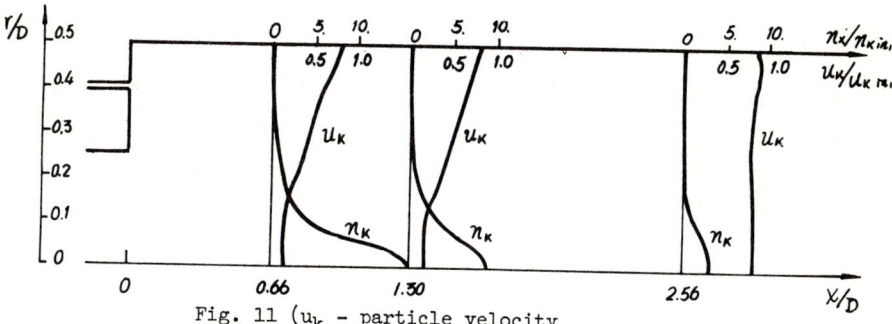

Fig. 11 (u_k – particle velocity,
n_k – particle number density)

Analysis of Transonic Wings including Viscous Interaction

Z. Q. Zhu
BIAA, Beijing, China

H. Sobieczky
DFVLR, 3400 Göttingen West Germany

Transonic flow computations past aerodynamic configurations are complicated by interactions between the outer inviscid flow -- usually containing shock waves -- and effects created by the viscous flow near the surface and the wake. Although the solutions of Navier-Stokes equation to account for these effects are conceptually possible, the cost is still too high to allow their use in practice. On the other hand, the boundary layer method, which is reasonably cheap, has been shown to produce good results when coupled to outer inviscid flow solutions. C.L.Street [1] and T.Cebeci et al [2] have made such calculations. In [1] the FLO30 code and a displacement thickness concept were used. In [2] the strip theory for the boundary layer essentially was used, although in an inverse mode, which allows to compute the separation region.

Recently more results of this coupling scheme have been published [3,4,5]. In our efforts to improve computational tools, for the first step, some potential flow methods [6,7,8,9] have been coupled with a 3D integral boundary layer method [10] for wings. This coupling models a "weak interaction" between viscous and inviscid flow (Fig.1). The nonconservative and robust FLO22 code for isolated wings is a popular tool of applied aerodynamics. Examples of conservative codes are FLO27 and FLO30 for wing and wing-fuselage configurations respectively. An improved version of FLO27, termed here "E92", for wing-fuselage combinations with the capacity to observe details at the wing root is used (Fig.2). Each one of these codes has been coupled with Stock's integral boundary layer code. Both displacement thickness and transpiration velocity concepts were used to construct the coupling scheme, and the latter was found to be more economical and useful not only for the analysis but also for the design extension. During the iterative process the displacement thickness must be underrelaxed and smoothened to minimize oscillations.

As calculated examples we have selected three configurations -- ONERA M6 wing, DFVLR F4 wing-body combination and a new test wing DFVLR F5. They are shown in Fig.3. The experimental data of the well known M6 wing and F4 wing-body configuration are available, so the comparison between calculated results and experimental data has been made. As an example, some results by different codes on selected spanwise positions are shown in Figs. 4-5 for comparison. From these figures, it seems that:

1. Agreement of calculated results by these codes with experimental data is reasonably good for M6 and is only partially reasonably good for F4(FLO30), except

near the trailing edge.

2. Viscous effects make the shock wave move forward in both cases.

3. For the F4 configuration shock position and strength predicted by FLO22 are better and in good agreement with experimental data. FLO30M overestimates shock position and strength on the upper surface. The conservative scheme is one major reason, but besides this, FLO30M obviously does not account for obtaining a body lift-equivalent to establishing a Kutta condition at a wing edge.

4. Trailing edge pressure is too high, which indicates a breakdown of the concept of wake-trailing-edge interaction. Fig. 6 shows that a better overall pressure distribution can be obtained by adjusting the displacement thickness near the trailing edge according to typical results from 2D analysis with a strong (trailing edge) viscous interaction mode [11]. This illustrates the importance of introducing 3D strong interaction near the trailing edge.

More detailed comparison was given in [12]. DFVLR F5 wing was generated analytically by software (E88M) used also in our analysis programs. The aim of this test case is mainly development of new viscous flow analysis algorithms. Here we use the configuration which includes a symmetrical, nearly shock-free design condition and has a slender uncambered trailing edge, to develop one of the analysis codes capable of treating the wing root flow quality.

This latter configuration is also used to develop a new design version of a potential flow code, based on the "fictitious gas" concept, but simpler and more practice-oriented. The modifications of boundary conditions necessary to simulate boundary layer effects -- a transpiration velocity concept -- can also be used for modelling surface geometry alterations resulting in a favourable pressure distribution with reduced shock waves. [13].

Our present conclusions are drawn from comparing results obtained by the different codes for the different configurations. This approach allows to isolate persistent problems of poor phenomena modelling: with the three viscous-inviscid computer codes available we try to make improvements currently and in the near future in the following fields:

1. Computational grid quality:

    Fine grids are essential especially near strong pressure gradients and thus influence the extent of local supersonic domains. With the constraint of keeping the number of grid points constant, a change from CH - to CO - type grids is economical because it avoids a waste of grid points at the wing tip extension.

2. Shock wave models:

    Non-conservative shock models seem to represent better shock strength than a fully conservative calculation, but both models should be improved by an entropy correction if potential flow methods are to be kept competitive with Euler codes.

3. Fuselage lift:

    Body lift needs to be accounted for in wing-body codes. Body boundary layer results and open separation modelling seem essential for these improvements.

4. Interaction models:

   Viscous interaction has been found to be essential but still insufficient if the "strong" interaction model at the trailing edge is lacking. The same is true for interactions between boundary layer and strong shock waves. Inverse boundary layer computation is a promising approach for the near future until practically useful results from solving the Navier-Stokes equations will be available. Local solutions to the Navier-Stokes equations, embedded into inviscid potential (or Euler) solutions and coupled to upstream boundary layer method results, will certainly be used in future coupling strategies if computer speed, memory and/or CPU costs will still not permit a global solution of the Navier-Stokes equations.

References:
1. Street, C.L.: AIAA 81-1266, 1981
2. Cebeci, T. et al: 3. Symp. on "Num. and Phy. aspects of Aerodynamic flow" Long Beach, 1985
3. Wai, J.C. et al : 3. Sump. on "Num. and Phy. aspects of Aerodynamic flow" Long Beach, 1985
4. Wigton, L. et al : 3. Symp. on "Num. and Phy. aspects of Aerodynamic flow" Long Beach, 1985
5. Samat, S.S. et al : AIAA 83-1806, 1983
6. Jameson, A.; Caughey, D.A. : NASA CR-153297, 1977
7. Jameson, A.; Caughey, D.A. : AIAA 77-635, 1977
8. Caughey, D.A.; Jameson, A.: AIAA 77-677, 1977
9. Caughey, D.A.: AIAA 83-0374, 1983
10. Stock, H.W. : NASA TM 75320, 1978
11. Melnik, R.E. et al: AIAA 77-680, 1977
12. Zhu,Z.Q.;Sobieczky, H.: DFVLR IB 85A23, 1985
13. Zhu,Z.Q.;Sobieczky, H.: (in print)

	Geometry Grid	INVISCID potential flow			VISCOUS 3D Boundary layer
Source code	E88M Sobieczky	FLO22 Jomeson, Coughey	FLO27M Coughey, Jomeson	FLO30M Caughey, Jomeson	3D INTEGRAL METHOD Stock
Given extension	—	shock-free design exten. E83	external grid version E92	—	—
New extension viscous-inviscid coupling		←——→	←——→	←——→	←——→

Fig. 1: Table of source codes and program development

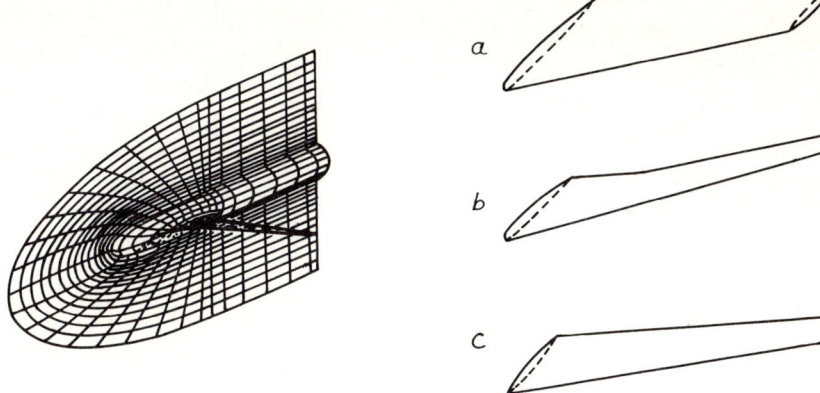

Fig. 2: Wing - body computational grid (CH type)

Fig. 3: Test wing configurations:
a) ONERA M6
b) DFVLR F4
c) DFVLR F5

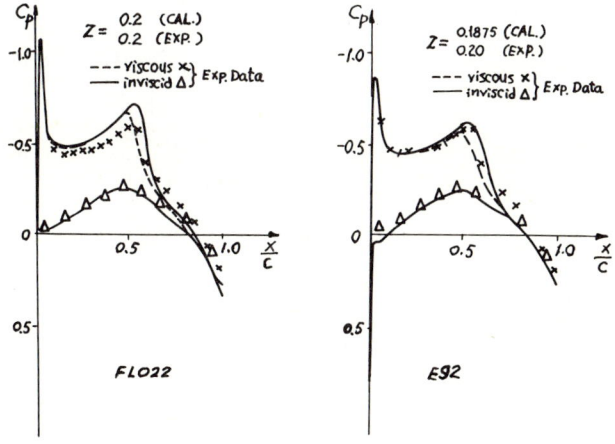

Fig. 4: ONERA M6 wing: results of two viscous analysis codes.
Mach = 0.84, Ang. att. = 3.06, Re = 2.6 Mill.

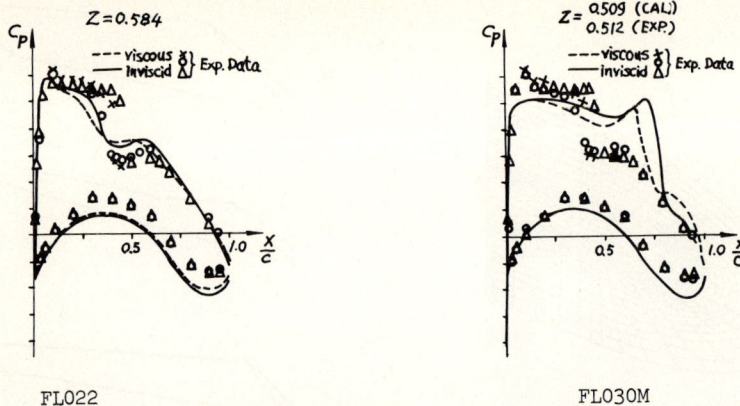

FL022   FL030M

Fig. 5: DFVLR F4 wing: results of two viscous analysis codes.
Mach = 0.77, Ang. att. = 0.01, Re = 3.0 Mill.

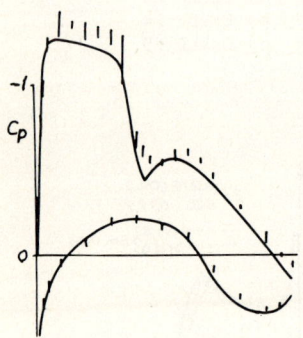

Fig. 6: Improvements of pressure distribution and shock location by a viscous displacement modification at the trailing edge. (FL030M)
DFVLR F 4 wing, Mach = 0.75, Re = 3.0 Mill., Z = 0.584

LIST OF PARTICIPANTS

Dr. Takayuki Aki
National Aerospace Laboratory
7-4-41, Jindaiji-Higashi-Machi
Chofu-Shi, Tokyo 182, Japan

Prof. Claude Bardos
Ecole Normale Superieure
45 Rue d'Ulm
75230 Paris Cedex 05, France

Mr. Michael Barton
Sverdrup Technology, Inc.
P.O. Box 30650
Midpark Branch
Middleburg Heights, OH 44130, U.S.A.

Dr. Francesco Bassi
Dipartimento di Energetica
Politecnico di Milano
Piazza Leonardo da Vinci, 32
20133 Milano, Italy

Prof. Michel Bercovier
E.N.S.
5 Rue dr Roux
75015 Paris, France

Prof. Y.G. Bian
Institute of Mechanics
Academia Sinica
Beijing, China

Dr. Frederick Blottner
Division 1636
Sandia National Laboratories
Albuquerque, NM 87185, U.S.A.

Dr. Jean-Paul Boujot
CISI
BP 24
91190 Gif sur Yvette, France

Dr. John E. Bowcock
Department of Mathematics
University of Birmingham
P.O. Box 363
Birmingham, B15 2TT, England, U.K.

Dr. Stuart Bramley
Department of Mathematics
University of Strathclyde
Livingstone Tower, 26 Richmono Street
Glasgow, G1 1XH, Scotland, U.K.

Dr. Gerald Browning
NCAR
P.O. Box 3000
Boulder, CO 80307, U.S.A.

Dr. Charles-Henri Bruneau
Laboratoric D'Analyse Numerique
Mathematique Batiment 425
Universite de Paris Sud
91405 Orsay Cedex, France

Prof. Henri Cabannes
Universite Pierre et Marie Curie
Mecanique Theorique
Tour 66-4, Place Jussieu
75230 Paris Cedex 05, France

Prof. H.S. Cao
Changsha Institute of Technology
Changsha, Hunan, China

Dr. J.T. Cao
Aeronautical Engineering Institute
Jinxi, Liaoning, China

Dr. John Carroll
School of Mathematical Sciences
National Institute for Higher Education
Dublin 9, Ireland

Dr. Thierry Cartage
Unite de Mecanique Appliquee (UCL)
Place du Levant 2
1348 Louvain-la-Neuve, Belgium

Dr. James Carter
United Technologies Research Center
MS 16
Silver Lane
East Hartford, CT 06108, U.S.A.

Dr. C.L. Chang
Dept. of Math.
Cleveland State University
Euclid Avenue at East 24th Street
Cleveland, OH 44115, U.S.A.

Dr. James L.C. Chang
Rocketdyne, Rockwell International
6633 Candga Avenue
Canoga Park, CA 91303, U.S.A.

Dr. Q.S. Chang
Institute of Applied Mathematics
Academia Sinica
Beijing, China

Dr. B.M. Chen
Computing Center, Academia Sinica
P.O. Box 2719
Beijing, China

Mr. F.S. Chen
Department of Mathematics
Shanghai Institute of Electric Power
Shanghai, China

Mr. K.M. Chen
Shanghai Mechanical Engineering College
Shanghai, China

Mr. T.H. Chen
Dept. of Mathematics
Fudan University
Shanghai, China

Dr. Z.B. Chen
China Aerodynamics Research
and Development Center
P.O. Box 211
Mianyang, Sichuan, China

Dr. Z.L. Chen
Beijing Institute of Aerodynamics
P.O. Box 7215
Beijing, China

Prof. G. Chernyi
Institute of Mechanics
Moscow University
Michurinskii PR. 1
Moscow, 119899, U.S.S.R.

Ms. Shenaz Choudhury
Department of Mathematics
Carnegie Mellon University
Schenley Park
Pittsburgh, PA 15213, U.S.A.

Prof. C.K. Chu
Department of Applied Physics
and Nuclear Engineering
Columbia University
New York, NY 10027, U.S.A.

Dr. Robert Clark
Los Alamos National Laboratory
MS B257
Los Alamos, NM 87544, U.S.A.

Dr. Thomas J. Coakley
NASA Ames Research Center
Moffett Field, CA 94035, U.S.A.

Prof. Richard Collins
Technical University of Nova Scotia
P.O. Box 1000
Halifax, Nova Scotia, B3J 2X4, Canada

Dr. Bertrand Costes
ONERA, BP 72
92322 Chatillo Cedex, France

Dr. Cornelis Cuvelier
Delft University of Technology
Mathematics
P.O. Box 356
2600 AZ Delft, The Netherlands

Prof. Andrea Dadone
Istituto di Macchine
Universita' di Bari
Via Re David, 200
I-70125, Bari, Italy

Prof. Hisaaki Daiguji
Department of Mechanical Engineering
Tohoku University
AZA Aoba, Aramaki, Sendai 980, Japan

Dr. Tran Khoa Dang
ONERA, 29 AV. de la Division Leclerc
BP72
92322 Chatillon Cedex, France

Dr. Herman Deconinck
Vrije Universiteit Brussel
Dept. of Fluid Mechanics
Pleinlaan 2
1050 Brussels, Belgium

Prof. Stanley C.R. Dennis
Department of Applied Mathematics
University of Western Ontario
London, Ontario, N6A 5B9, Canada

Dr. S.S. Dong
Institute of Applied Physics
and Computational Mathematics
P.O. Box 8009-13
Beijing, China

Prof. John Dorning
University of Virginia
Reactor Facility
Charlottesville, VA 22901, U.S.A.

Dr. Philip Drummond
NASA Langley Research Center
Mail Stop 156
Hampton, VA 23665, U.S.A.

Prof. D.R. Du
Institute of Aerodynamics
Northwestern Polytechnical University
Xian, Shanxi, China

Mr. J.G. Du
China Aerodynamics Research
and Development Center
P.O. Box 211
Mianyang, Sichuan, China

Dr. Douglas L. Dwoyer
NASA Langley Research Center
Computational Methods Branch
Hampton, VA 23665, U.S.A.

Prof. Harry Dwyer
Dept. of Mechanical Engineering
University of California
Davis, CA 95616, U.S.A.

Dr. V.P. Dymnikov
Head of the Laboratory
Department of Numerical Mathematics
USSR Academy of Sciences
Gorky St., 11
Moscow, 103009, U.S.S.R.

Dr. Knut Stale Eckhoff
University of Bergen
Dept. of Mathematics
Allegt 55
N-5000 Bergen, Norway

Dr. Y. Eguchi
Dept. of Nuclear Engineering
Faculty of Engineering
The University of Tokyo
7-3-1 Hongo, Bunkyo-ku, Tokyo 113, Japan

Dr. G. Erlebacher
NASA Langley Research Center
Mail Stop 156
Hampton, VA 23665, U.S.A.

Dr. Vance Faber
Los Alamos National Laboratory
C-3, MS B265
Los Alamos, NM 87545, U.S.A.

Dr. Paul A. Farrell
Dept. of Mathematical Science
Kent State University
Kent, OH 44242, U.S.A.

Dr. Bernardo Favini
Dipartimento di Meccanica e Aeronautica
Via Eudossiana, 18
00184 Roma, Italy

Prof. K. Feng
Computing Center, Academia Sinica
P.O. Box 2719
Beijing, China

Dr. Clive Fletcher
University of Sydney
Dept. of Mechanical Engineering
New South Wales, 2006, Australia

Prof. Karl Förster
Institut fur Aerodynamik und Gasdynamik
Pfaffenwaldring 21
D-7000 Stuttgart 80, West Germany

Dr. Jacob E. Fromm
IBM Almaden Research Center K34/802
650 Harry Road
San Jose, CA 95120-6099, U.S.A.

Dr. D.X. Fu
Beijing Institute of Aerodynamics
P.O. Box 7215
Beijing, China

Dr. R. F. Fu
Department of Engineering Physics
Tsinghua University
Beijing, China

Prof. Laszlo Fuchs
Dept. of Gasdynamics
The Royal Institute of Technology
S-100 44 Stockholm, Sweden

Dr. J.C. Gan
The Scientific Research
and Development of the Air Force
PLA
Beijing, China

Dr. Y.K. Gao
Aeronautical Engineering Institute
Jinxi, Liaoning, China

Dr. Francesco Grasso
Institute of Gasdynamics
80 P. le Tecchio
80125 Naples, Italy

Dr. S.F. Han
Institute of Mathematical Sciences
Chengdu Branch Academia Sinica
Chengdu, Sichuan, China

Mr. Y.Q. Han
Computing Center, Academia Sinica
P.O. Box 2719
Beijing, China

Dr. Dieter Hanel
Aerodynamisches Institut of RWTH Aachen
5-7, Wullnerstraße
D 5100 Aachen, West Germany

Mr. Iwao Harada
Energy Research Laboratory
Hitachi, Ltd
1168 Moriyama-Cho
Hitachi, Ibaraki 316, Japan

Dr. C.S. He
Chinese Aerodynamics Research Society
P.O. Box 2425
Beijing, China

Dr. Pieter W. Hemker
Centrum Voor Wiskunde en Informatica
P.O. Box 4079
1009 AB Amsterdam, The Netherlands

Prof. Maurice Holt
Mechanical Engineering
University of California
Berkeley, CA 94720, U.S.A.

Prof. T.X. Hou
Beijing Meteorological Institute
Beijing, China

Prof. D. Huang
The Institute of Mathematics Research
Peking University
Beijing, China

Mr. D.T. Huang
Department of Engineering Mechanics
Tsinghua University
Beijing, China

Dr. M.K. Huang
Dept. of Aerodynamics
Nanjing Aeronautical Institute
Nanjing, China

Dr. Mohammed Y. Hussaini
ICASE M/S 132C
NASA Langley Research Center
Hampton, VA 23665, U.S.A.

Prof. A. Jameson
Dept. of Mechanical
and Aerospace Engineering
Princeton University
Princeton, NJ 08544, U.S.A.

Prof. Adrien Jami
ENSTA-CNRS
Groupe Hydrodynamique Navale
Chemin de la Huniere
91120 Palaiseau, France

Dr. L.Q. Jiao
Computing Center, Academia Sinica
P.O. Box 2719
Beijing, China

Dr. Gary M. Johnson
Institute for Computational Studies
P.O. Box 1852
Fort Collins, CO 80522, U.S.A.

Dr. Upender K. Kaul
NASA Ames Research Center
MS 202A-14
Moffett Field, CA 94035, U.S.A.

Dr. David Kerlick
Mail Stop 202A-14
NASA Ames Research Center
Moffett Field, CA 94086, U.S.A.

Prof. P.K. Khosla
Department of Aerospace Engineering
and Engineering Mechanics
University of Cincinnati
Cincinnati, OH 45221-0070, U.S.A.

Dr. Paul Kutler
NASA Ames Research Center
M.S. 229-2
Moffett Field, CA 94035, U.S.A.

Dr. Dochan Kwak
NASA Ames Research Center
202 A-14
Moffett Field, CA 94035, U.S.A.

Ms. Sylvia Lee
Q6, MS K557
Los Alamos National Laboratory
Los Alamos, NM 87544, U.S.A.

Dr. W.H. Lee
Los Alamos National Laboratory
Los Alamos, NM 87544, U.S.A.

Dr. C.-P. Li
NASA-Johnson Space Center
Houston, TX 77058, U.S.A.

Mr. C.W. Li
Computing Center, Academia Sinica
P.O. Box 2719, Beijing, China

Dr. F.W. Li
Institute of Aerodynamics
Northwestern Polytechnical University
Xian, Shanxi, China

Mr. J. Li
Computing Center, Academia Sinica
P.O. Box 2719
Beijing, China

Mr. J.X. Li
Computing Center, Academia Sinica
P.O. Box 2719
Beijing, China

Mr. K.N. Li
Dept. of Mathematics, Physics
and Mechanics
Nanjing Aeronautical Institute
Nanjing, China

Mr. W.L. Li
Department of Engineering Mechanics
Tsinghua University
Beijing, China

Mr. X.Y. Li
Computing Center, Academia Sinica
P.O. Box 2719
Beijing, China

Prof. Y.F. Li
Computing Center, Academia Sinica
P.O. Box 2719
Beijing, China

Prof. Q. Lin
Institute of Systems Science
Academia Sinica
Beijing, China

Dr. B.Y. Ling
Nanchang Aircraft Company
Nanchang, China

Dr. C.H. Liu
NASA Langley Research Center
Mail Stop 128
Analytical Methods Branch
Hampton, VA 23665-5225, U.S.A.

Dr. G.C. Liu
NASA Langley Research Center
Mail Stop 128
Hampton, VA 23665-5225, U.S.A.

Mr. J.G. Liu
Dept. of Math.
Fudan University
Shanghai, China

Dr. X.Z. Liu
Computing Center, Academia Sinica
P.O. Box 2719
Beijing, China

Dr. Charles Lombard
PEDA Corporation
4151 Middlefield Road, Suite 7
Palo Alto, CA 94303, U.S.A.

Dr. Raymond Luh
PEDA Corporation
4151 Middlefield Road, Suite 7
Palo Alto, CA 94303, U.S.A.

Dr. Y.W. Ma
Beijing Institute of Aerodynamics
P.O. Box 7215
Beijing, China

Dr. Mujeeb Malik
High Technology Corporation
P.O. Box 7262
Hampton, VA 23666, U.S.A.

Dr. Daniela Mansutti
Istituto per le Applicazioni del Calcolo
Via del Policlinico, 137
00161, Roma, Italy

Dr. Thomas Manteuffel
Los Alamos National Laboratory
C-3, MS B265
Los Alamos, NM 87545, U.S.A.

Dr. R.E. Melnik
Grumman Corporate Research Center
M/S A08-35
Bethpage, NY 11714, U.S.A.

Prof. Charles L. Merkle
The Pennsylvania State University
Dept. of Mechanical Engineering
University Park, PA 16802, U.S.A.

Prof. John Miller
Numerical Analysis Group
University of Dublin, Trinity College
Dublin 2, Ireland

Dr. Nimai-Kumar Mitra
Institut fur Thermo-und Fluiddynamik
RUHR-Universitat Bochum
Postfach 102148
4630 Bochum, West Germany

Dr. J.P. Monnet
Institute of Systems Science
Academia Sinica
Beijing, China

Prof. Masatake Mori
Institute of Information Sciences
University of Tsukuba
Sakura, Niihari, Ibaraki 305, Japan

Prof. Keith William Morton
Oxford University
Computing Laboratory
8-11 Keble Road
Oxford, OX13 5LF, England, U.K.

Dr. Kazuhiro Nakahashi
National Aerospace Laboratory
7-44, Jindaiji-Higashi
Chofu, Tokyo 182, Japan

Dr. Yoshiaki Nakamura
Department of Aeronautical Engineering
Nagoya University
Chikusa-ku, Nagoya 464, Japan

Prof. M Napolitano
Universita Degli Studi di Bari
Facolta di Ingegneria
Istituto di Macchine ed Energetica
Via Re David, 200
I-70125 Bari, Italy

Prof. R.A. Nicolaides
Department of Mathematics
Carnegie-Mellon University
Schenley Park
Pittsburgh, PA 15213, U.S.A.

Prof. Helmer L. Nielsen
Dept. of Mechanical Engineering
San Jose State University
1 Washington Square
San Jose, CA 95192-0086, U.S.A.

Prof. Nobuhide Nishikawa
Faculty of Engineering
Chiba University
1-33 Yayoi
Chiba 260, Japan

Mr. Jau Nordstrom
The Aeronautical Res. Inst. of Sweden
Box 11021
S-16111 Bromma, Sweden

Dr. Maurice O'Reilly
Regional Technical College
Dublin Road
Dundalk, Ireland

Dr. Eugene O'Riordan
Regional Technical College
Dublin Road
Dundalk, Ireland

Prof. Koichi Oshima
Institute of Space
and Astronautical Science
6-1 Komaba 4-Chome
Meguro-Ku, Tokyo 153, Japan

Dr. Gary A. Osswald
University of Cincinnati
Aerospace Engineering
M.L. #70
Cincinnati, OH 45221, U.S.A.

Mr. G.W. Ou
Department of Engineering Mechanics
Tsinghua University
Beijing, China

Dr. J. Periaux
AMD/BA
HP300
78 Quai Carnot
92214 Saint-Cloud, France

Dr. Michael Pfitzner
MBB-ERNO
Postfach 801169
8000 Munchen
West Germany

Dr. Daniel Pryor
Institute for Computational Studies
P.O. Box 1852
Fort Collins, CO 80522, U.S.A.

Mr. E.P. Qian
Computing Center, Academia Sinica
P.O. Box 2719
Beijing, China

Mr. N. Qin
Room 421, James Watt Building
Glasgow University
Glasgow, G12 8QQ, Scotland, U.K.

Mr. Luigi Quartapelle
Istituto di Fisica
Politecnico di Milano
Piazza Leonardo da Vinci, 32
20133 Milano, Italy

Mr. P. Que
China Aerodynamics Research
and Development Center
P.O. Box 211
Mianyang, Sichuan, China

Prof. Karl, G. Roesner
Technische Hochschule Darmstadt
Institut fur Mechanik
Hochschulstrasse 1
D-6100 Darmstadt, West Germany

Prof. V. Ruas
Pontificia Universidade Catolica
Departamento de Informatica
Rio de Janeiro, 22453, Brazil

Prof. Stanley G. Rubin
Department of Aerospace Engineering
and Engineering Mechanics
University of Cincinnati
Cincinnati, OH 45221-0070, U.S.A.

Prof. Viktor V. Rusanov
Keldysh Inst. of Applied Math.
USSR Academy of Sciences
Miusskaya Pl. 4
125047 Moscow A-47, U.S.S.R.

Prof. Robert Russell
Department of Mathematics & Statistics
Simon Fraser University
Burnaby, British Columbia, V5A 1S6, Canada

Prof. Filippo Sabetta
Dipartimento di Meccanica e Aeronautica
Via Eudossiana 18
00184 Roma, Italy

Prof. Nobuyuki Satofuka
Kyoto Institute of Technology
Matsugasaki, Sakyo-Ku, Kyoto 606, Japan

Mr. G.S. Shen
The Scientific Research
and Development of Air Force
PLA
Beijing, China

Prof. M.Y. Shen
Department of Engineering Mechanics
Tsinghua University
Beijing, China

Prof. Y.D. Shevelev
U.S.S.R. Academy of Sciences
Institute of Mechanics
Moscow, U.S.S.R.

Prof. Z.C. Shi
Dept. of Mathematics
Chinese University of Science
and Technology
Hefei, China

Dr. Z.C. Shi
Institute for Engineering Structure Studies
Tongji University
Shanghai, China

Prof. Yurii Shokin
Akademgorodok, Computing Center
USSR Academy of Sciences, Siberian Division
660036 Krasnoyarsk 36, U.S.S.R.

Mr. C. Shu
Dept. of Aerodynamics
Nanjing Aeronautical Institute
Nanjing, China

Mr. L.D. Shu
Hunan Computing Center
Changsha, Hunan, China

Dr. Soubbaramayer
19 Parc D'Ardenag
91120 Palaiseau, France

Prof. Massimo Strani
Universita' di Roma la Sapienza'
Dipartimento di Meccanica e Aeronautica
Via Eudossiana 18
00184 Roma, Italy

Dr. M.D. Su
Department of Engineering Mechanics
Tsinghua University
Beijing, China

Dr. H.S. Sun
Beijing Institute of Special
and Electrical Devices
P.O. Box 9213
Beijing, China

Dr. S.F. Sun
Institute of Mechanics
Academia Sinica
Beijing, China

Dr. Y.J. Sun
Beijing Institute of Special Mechanical
and Electrical Devices
P.O. Box 9213
Beijing, China

Ms. Julie Swisshelm
Institute for Computational Studies
P.O. Box 1852
Fort Collins, CO 80522, U.S.A.

Mr. Yukimasa Takemoto
Institute of Plasma Physics
Nagoya University
Chikusa-ku, Nagoya 464, Japan

Mr. B. Tan
Beijing Institute of Electronic
System Engineering
Beijing, China

Mr. G.F. Tang
Hunan University
Changsha, Hunan, China

Mr. J.Z. Tang
Commission for Science and Technology
Qingdao, Shandong, China

Prof. Thomas Taylor
The Johns Hopkins University
Applied Physics Laboratory
Johns Hopkins Road
Laurel, MD 20707, U.S.A.

Prof. Z.H. Teng
Department of Mathematics
Peking University
Beijing, China

Prof. L. Ting
Courant Institute of Mathematical Sciences
New York University
251 Mercer Street
New York, NY 10012, U.S.A.

Dr. Hiroshi Tokunaga
Department of Mechanical Engineering
Kyoto Institute of Technology
Matsugasaki, Sakyo-Ku, Kyoto 606, Japan

Prof. Giovanni Torella
Accademia Aeronautica
Dipartimento di Scienze Applicate al Volo
Direzione Studi
80078 Pozzudli Napoli, Italy

Prof. Leonid Turchak
U.S.S.R. Academy of Sciences
Comput. Centre
Vavilov Str. 40
Moscow, 117333, U.S.S.R.

Dr. R.W. Walters
Aerospace & Ocean Engineering
Virginia Polytechnic Institute
and State University
Blacksburg, VA 24061, U.S.A.

Dr. D.Q. Wang
Institute of Aerodynamics
Northwestern Polytechnical University
Xian, Shanxi, China

Dr. G.X. Wang
Beijing Institute of Special Mechanical
and Electrical Devices
P.O. Box 9213
Beijing, China

Prof. J.H. Wang
Institute of System Science
Academia Sinica
Beijing, China

Ms. L.X. Wang
Institute of Aerodynamics
Northwestern Polytechnical University
Xian, Shanxi, China

Ms. L.Y. Wang
National Defence Industrial Publishing House
Beijing, China

Prof. P.S. Wang
Northwestern Polytechnical University
Xian, Shanxi, China

Mr. Q. Wang
Institute of Applied Physics and
Computational Mathematics
P.O. Box 8009-13
Beijing, China

Prof. R.Q. Wang
Computing Center, Academia Sinica
P.O. Box 2719
Beijing, China

Mr. Z.Q. Wang
P.O. Box 725-202
Shenyang, China

Dr. Z.X. Wang
Institute of Applied Physics
and Computational Mathematics
P.O. Box 8009-13
Beijing, China

Dr. Robert Warming
NASA Ames Research Center
Mail Stop 202 A-1
Moffett Field, CA 94305, U.S.A.

Prof. Pieter Wesseling
Dept. of Math. & Inf.
University of Technology
P.O. Box 356
2600 AJ Delet
The Netherlands

Prof. J.H. Wu
Department of Mechanics
Peking University
Beijing, China

Dr. X. H. Wu
Department of Mathematics
Tongji University
Shanghai, China

Mr. Y.X. Wu
Institute of Mechanics
Academia Sinica
Beijing, China

Dr. D.Y. Xian
China Aerodynamics Research and
Development Center
P.O. Box 211
Mianyang, Sichuan, China

Mr. L.W. Xiang
7th Dept.
Shanghai Jiaotong University
Shanghai, China

Mr. L. Zhao
Institute of Mechanics
Academia Sinica
Beijing, China

Mr. M. Zheng
China Aerodynamics Research
and Development Center
P.O. Box 211
Mianyang, Sichuan, China

Dr. B. M. Zhou
Computational Station of
Fifth Research Institute
Ministry of Astronautics
Beijing, China

Prof. Y.L. Zhu
Computing Center, Academia Sinica
P.O. Box 2719
Beijing, China

Prof. Z.Q. Zhu
Beijing Institute of Aeronautics
and Astronautics
Beijing, China

Prof. F.G. Zhuang
Chinese Aerodynamics Research Society
P.O. Box 2425
Beijing, China

Dr. Y.S. Xiang
China Aerodynamics Research
and Development Center
P.O. Box 211
Mianyang, Sichuan, China

Mr. D.G. Xie
Dept. of Mechanics
Zhejiang University
Hangzhou, Zhejiang, China

Dr. G.R. Xu
Institute of Applied Physics
and Computational Mathematics
P.O. Box 8009-13
Beijing, China

Mr. R.J. Xu
Aeronautical Engineering Institute
Jinxi, Liaoning, China

Prof. S.R. Xu
Department of Computer Science
Zhongshan University
Guangzhou, China

Mr. Satoru Yamamoto
Graduate School, Tohoku University
AZA Aoba, Aramaki, Sendai 980, Japan

Dr. J.Y. Yang
PEDA Corporation
4151 Middlefield Road, Suite 7
Palo Alto, CA 94303, U.S.A.

Mr. T. Yang
Institute of Mechanics
Academia Sinica
Beijing, China

Prof. Z.H. Yang
Dept. of Mathematics
Shanghai University of Science
and Technology
Jiading, Shanghai, China

Dr. H.C. Yee
NASA Ames Research Center
Mail Stop 203A-1
Moffett Field, CA 94035, U.S.A.

Dr. Woon-Shing Yeung
Department of Mechanical Engineering
University of Lowell
Lowell, MA 01854, U.S.A.

Mr. W.A. Yong
Computing Center, Academia Sinica
P.O. Box 2719
Beijing, China

Mr. X. Yu
Institute of Mechanics, Academia Sinica
Beijing, China

Mr. X. Yuan
Dept. of Thermal Energy
Tsinghua University
Beijing, China

Prof. Pieter Zandbergen
Twente University of Technology
Dept. of Applied Mathematics
P.O. Box 217
Enschede 7500AE, The Netherlands

Mr. F.Z. Zeng
Computing Center, Academia Sinica
P.O. Box 2719
Beijing, China

Prof. Q.C. Zeng
Institute of Atmospheric Physics
Academia Sinica
Beijing, China

Mr. Y.N. Zeng
Department of Computer Science
Zhongshan University
Guangzhou, China

Dr. H.L. Zhang
Dept. of Aerodynamics
Nanjing Aeronautical Institute,
Nanjing, China

Mr. J. Zhang
Dept. of Engineering Mechanics
Tsinghua University
Beijing, China

Mr. J. Zhang
Dept. of Energy and Power Engineering
Xian Jiaotong University
Xian, China

Dr. J.B. Zhang
China Aerodynamics Ressarch
and Development Center
P.O. Box 211
Mianyang, Sichuan, China

Mr. X. Zhang
Automatical Control Department
Architectural Engineering College
Shenyang, China

Mr. J.B. Zhao
Dept. of Physics
Northwestern Teacher's College
Lanzhou, China

RAYMOND H. FOGLER LIBRARY
DATE DUE

BOOKS ARE SUBJECT TO
RECALL AFTER TWO WEEKS

JUL 1 5 1987